MEDICINAL CHEMISTRY - FUSION OF TRADITIONAL AND WESTERN MEDICINE

Third Edition

Authored By

Robert E. Smith

Adjunct Assistant Professor, Park University and Consultant
Science Advisor, United
States Food and Drug Administration (FDA)
U.S.A

Bentham Science Publishers Ltd.
Executive Suite Y - 2
PO Box 7917, Saif Zone
Sharjah, U.A.E.
Email: subscriptions@benthamscience.org

BENTHAM SCIENCE

CONTENTS

FOREWORD i

PREFACE ii
CONFLICT OF INTEREST ii
ACKNOWLEDGEMENTS ii

DEDICATION iv

CHAPTER 1 PERSONALIZED MEDICINAL CHEMISTRY 3
INTRODUCTION 3
METHODS 19
PCR 19
TOF-MS 21
Next-Generation Sequencing (NGS) 22
Pharmacogenomics 27
Effects of Drugs on Metabolism 28
APPLICATIONS 36
KEY POINTS 57
ABBREVIATIONS 57
REFERENCES 61

CHAPTER 2 BIOLOGY OF INFLAMMATION 69
KEY POINTS 83
ABBREVIATIONS 84
REFERENCES 85

CHAPTER 3 METABOLIC SYNDROME, DIABETES, HEART DISEASE AND STROKE 88
GLUCOSE AND ENERGY PRODUCTION 89
METABOLIC SYNDROME 92
DIABETES AND SMOLDERING INFLAMMATION 98
TREATMENTS AND POSSIBLE CURES FOR DIABETES 110
PEROXISOME PROLIFERATOR-ACTIVATED RECEPTORS, PPARS 111
RXR PROTEINS 113
LXR PROTEINS 114
FXR PROTEIN 115
CONSTITUTIVE ANDOSTANE RECEPTOR (CAR) 115
RETINOIC ACID-RELATED ORPHAN RECEPTOR (ROR) 116
TAK1 RECEPTOR 116
TREATMENTS FOR OBESITY 117
STROKE 119
MYOCARDIAL INFARCTION OR HEART ATTACK 120
PREVENTING AND TREATMENTING HEART DISEASE 125
KEY POINTS 131
ABBREVIATIONS 132
REFERENCES 134

CHAPTER 4 BIOLOGY OF CANCER: GENETICS, BIOMARKERS AND CLINICAL APPROACHES ...
138

INTRODUCTION 138
ANGIOGENESIS 144
ROLE OF THE IMMUNE SYSTEM 147
DIAGNOSIS AND PROGNOSIS 149
CLASSIFICATION FOR CANCER DEVELOPMENT 150
CAUSES OF CANCER 151
THE CELL CYCLE AND RTKS 152
ROLE OF VIRUSES IN CAUSING CANCER 154
THE ROLE OF THE P53 ONCOPROTEIN 155
ONCOGENES AND ONCOPROTEINS 157
ADVANCES IN TREATING BREAST CANCER 160
AROMATASE 163
VEGF 163
PI3K/AKT/MTOR PATHWAY 165
IGF-1 167
INFLAMMATORY BREAST CANCER 168
EARLY DETECTION 169
LUNG CANCER 172
SKIN CANCER 173
MICRORNAS 174
CANCER STEM CELLS 174
TELOMERASE 176
MUTATIONS IN SOMATIC AND GERMLINE CELLS 176
ROLE OF CYP450 ENZYMES 179
CANCER GENOMICS 179
CANCER EPIGENETICS 180
MUTATIONS 182
MEDICINAL CHEMISTRY POINT OF VIEW 183
KEY POINTS 188
ABBREVIATIONS: 189
REFERENCES 191

CHAPTER 5 MEDICINAL CHEMISTRY AND THE ENDOCRINE SYSTEM 198
HYPOTHALAMUS 199
PINEAL BODY (EPIPHYSIS) 203
PITUITARY GLAND (HYPOPHYSIS) 204
THYROID 206
ALIMENTARY SYTEM 208
KIDNEYS 210
ADRENAL GLANDS 213
REPRODUCTIVE SYSTEM 214
THE HEART AND SKELETAL MUSCLES 217
BONES AND SKELETAL SYSTEM 219
CALCIUM REGULATION 222
TARGETS 223
DIFFUSE NEURO-ENDOCRINE SYSTEM 224
MAJOR ENDOCRINE SYSTEMS 224
DISEASES 225
KEY POINTS 226
ABBREVIATIONS 228
REFERENCES 230

CHAPTER 6 REPRODUCTIVE MEDICINE, OSTEOPOROSIS 234
KEY POINTS 251
ABBREVIATIONS 252
REFERENCES 253

CHAPTER 7 THE NERVOUS SYSTEM 256
INTRODUCTION 256
Components of the Nervous System 259
Neurotransmitters 269
Neuropeptides 274
Nitric Oxide and Carbon Monoxide 277
Glial Cells 279
Pain Control 282
Drugs to Help People Stop Smoking [1] 287
Drugs That Treat Neurodegenerative Diseases 287
Inflammatory Cytokines 289
Parkinson's Disease 290
Alzheimer's Disease 290
Epilepsy 296
Anxiety Disorders 297
Depression 297
Antipsychotics 299
Autism 299
Amyotrophic Lateral Sclerosis (ALS) [1] 300
Damaged Spinal Cords 301
Medicinal Chemistry Point of View 301
KEY POINTS 310
ABBREVIATIONS 313
REFERENCES 315

CHAPTER 8 THE IMMUNE SYSTEM AND IMMUNE NETWORK 323
IMMUNE NETWORK 327
COMPONENTS OF THE IMMUNE SYSTEM 330
TYPES OF IMMUNE CELLS 334
CHEMICAL MEDIATORS OF THE IMMUNE SYSTEM 349
KEY POINTS 386
ABBREVIATIONS 387
REFERENCES 389

CHAPTER 9 INFECTIOUS DISEASES 395
KEY POINTS 408
ABBREVIATIONS 409
REFERENCES 410

CHAPTER 10 VACCINES 412
INTRODUCTION 412
VIRAL VECTORS 414
CANCER VACCINES 415
VACCINES FROM DEAD OR INACTIVE ORGANISMS OR VIRUSES 416
CHILDHOOD IMMUNIZATIONS 417

FASTER WAYS TO MAKE VACCINES 422
CLASSIFYING VIRUSES 423
FLU (INFLUENZA) VACCINES 423
MMR VACCINE THE MMR 424
VARICELLA VACCINE 425
HEPATITIS A VACCINES 425
DTAP VACCINE 425
MCV4 VACCINE 426
MENINGITIS VACCINES 426
HUMAN PAPILLOMA VIRUS VACCINE 426
RISKS TO PHARMACEUTICAL COMPANIES 427
EFFORT TO DEVELOP NEW VACCINES 428
DISCOVERING ANTIGENS AND EVALUATING CELL-MEDIATED IMMUNE RESPONSES 429
ADVANCES IN ANALYZING IMMUNE RESPONSES 430
KEY POINTS 431
ABBREVIATIONS 431
REFERENCES 432

CHAPTER 11 PREVENTING DISEASES BY PROPER NUTRITION 435
NUTRITION 435
LIMITS TO USA RECOMMENDATIONS 439
RISKS IN EATING MASS-PRODUCED MEAT 439
FISH AND SEAFOOD 442
OTHER SOURCES OF OMEGA-3 AND POLYUNSATURATED FATS 444
SOYBEANS AND SOY PRODUCTS 444
TURMERIC 445
GINGER ROOT 446
ANTIOXIDANTS IN FRUITS AND VEGETABLES 446
FIGHTING STARVATION AND MALNUTRITION 447
GREEN REVOLUTION 448
POLLUTION, SOIL CONTAMINATION, ECOLOGICAL RISKS AND HUMAN HEALTH 449
CHEMICAL COMPOSITION OF SOIL AND THE CHANGES INVOLVED IN THE PRODUCTION OF CROPS AND LIVESTOCK 450
GENETICALLY-MODIFIED ORGANISMS (GMOS) 450
NUTRIENTS IN FOODS 452
DIETARY SUPPLEMENTS 456
BREAST FEEDING 457
VITAMINS 458
TOXICITY OF POORLY CHELATED IRON 467
HAZARDOUS DIETARY SUPPLEMENTS 468
SOME OF THE MOST IMPORTANT AND POPULAR DIETARY SUPPLEMENTS 469
KEY POINTS 482
ABBREVIATIONS 484
REFERENCES 485

CHAPTER 12 NEW PROBLEMS AND SOLUTIONS 494
EMERGENT PROBLEMS 494
GLOBAL CLIMATE CHANGE 496
BIOTECHNOLOGY-BASED SOLUTIONS 497
NANOTECHNOLOGY 501
P4 MEDICINE 504

STEM CELL TECHNOLOGY 505
SHORTAGE OF MEDICINES 508
TESTING FOR PROSTATE SPECIFIC ANTIGEN 509
OTHER IMPORTANT RECENT DISCOVERIES 510
RECENTLY APPROVED DRUGS 511
KEY POINTS 526
ABBREVIATIONS 528
REFERENCES 530

CHAPTER 13 COMMUNICATION AND SIGNALING IN MEDICINAL CHEMISTRY 538
INTRODUCTION 538
G PROTEIN-COUPLED RECEPTORS 543
PROTEIN KINASES 546
ABBREVIATIONS 546
REFERENCES 550

CHAPTER 14 SYSTEMS THINKING IN MEDICINAL CHEMISTRY 552
SYSTEMS BIOLOGY 552
EMERGENT PROPERTIES 556
SYSTEMS THINKING 561
SINLE CELL TECHNOLOGIES 562
AUTOPOIESIS, ALLOSTASIS AND ATTRACTORS 566
GAIA 574
NETWORKS 576
NETWORKS IN AGING AS WELL AS NEURODEGENERATIVE AND PSYCHIATRIC DISEASES 585
NETWORK ANALYSIS OF MOTHER'S MILK 587
CONCLUSION 588
KEY POINTS 588
ABBREVIATIONS 590
REFERENCES 590

APPENDIX 594

SUBJECT INDEX 663

FOREWORD

As a practicing physician, I found this third edition to be very up-to-date, interesting and useful. It would be an excellent text book for the second semester in the standard two-semester course taught at most pharmacy schools. I was especially impressed with the first chapter on predictive, preventive, personalized and participatory (P4) medicine. It epitomizes the fusion of traditional and western medicine. That is, mathematics, the foundation of reductionist thinking, is used to quantify huge datasets from patients, while physicists, chemists, biologists and engineers develop the analytical tools needed to generate the data. All of this can be linked through the internet and used in mobile healthcare applications. Also, the book concepts and metabolic processes for which there was limited knowledge or was completely unknown when I was a medical student. The information presented provides undergraduate, medical and pharmacy students with useful information about the indications for and applications of modern medicinal chemistry. It will also prepare currently licensed physicians prepare for Board certification and recertification. Many of the old questions may have new answers. I also found the book to be very helpful in describing medicines that I currently prescribe to patients and new drugs that are being developed. On frequent occasions, a patient will ask, "Why aren't doctors doing more to find cures for common diseases". The information in this book provides useful answers. I am also frequently asked questions about nutrition, dietary supplements and environmental toxins. The information about these subjects was written in clear, simple language that most people can understand. For example, I am frequently asked if it is beneficial to take the popular supplement, açaí, to help lose weight. Not being from Brazil, I didn't know anything about it. This book talks about how açaí is 50% fat, as triglycerides. It is an excellent source of calories and antioxidants for undernourished natives and highly competitive athletes who need more calories. Taking açaí will actually make you gain weight. So, the book is very readable and has a multi-disciplinary approach. It teaches the kind of things that I would like to see medical and pharmacy students learn and could even be useful to lay people.

Mark D. Pilley

M.D. P.O. Box 241723 Omaha, NE 68124-3135

U.S.A

PREFACE

This book was written as a textbook for the second semester in a two-semester a course in medicinal chemistry that is taught in most pharmacy schools. It is preceded by the first edition, which was written for a one semester course and the second edition, which is for the first semester in a two semester course. They can be used in standard courses for pharmacy students and for students who are trying to get into medical, dental, pharmacy or graduate school. Moreover, people working in the pharmaceutical industry and doctors preparing for Medical Board Exams will also find it useful. The second edition discussed the fusion of traditional and western medicine and between systems thinking and reductionist thinking. This third edition contains chapters on personalized medicine and targeted drug discovery as well as the endocrine system that were not in the first edition. Unlike chapters in older books on medicinal chemistry, the chapter on the endocrine system talks about the alimentary tract and adipose tissues, which are now known to be secrete hormones. The importance of environmental toxins in autoimmune diseases is discussed. Practical advice is presented, especially when it helps illustrate an interdisciplinary approach and unexpected connections. The role of environmental toxins, such as bisphenol A (BPA) and perfluorooctanoic acid (PFOAA) in autoimmune diseases is discussed. Also, there are many updates based on research that was published since the first edition. New drugs have been approved and personalized medicine has benefited from next generation DNA sequencing methods. Also, three-dimensional printing was used to make a bioresorbable tracheal splint for an infant who was critically ill. So, advances in genomics, medical imaging, 3-D printing and regenerative medicine, along with increased computational power and the advent of mobile and wireless capabilities are allowing patients to be treated and monitored in ways that better meet their individual needs. There are also descriptions of the basic science behind cancer, heart disease, metabolic syndrome, infectious diseases, inflammation, reproductive medicine, the biology of information flow, the nervous system, immune network, vaccines, autopoiesis, systems biology and network theory. Finally, this and the topics discussed in it should not be taken as reflecting FDA policy or regulations.

CONFLICT OF INTEREST

The author confirms that this ebook contents have no conflict of interest.

ACKNOWLEDGEMENTS

Declared none.

Robert E. Smith
Park University,
Parkville,
USA,
E-mail: Robert.Smith05@park.edu

DEDICATION

This book is dedicated to my wife, Dee.

I also want to thank the staff of Bentham Science for their tremendous help and support.

CHAPTER 1

Personalized Medicinal Chemistry

Abstract: Perhaps nothing epitomizes the fusion of traditional and western medicine more than predictive, preventive, personalized and participatory (P4) medicine. It takes not just a holistic but also a quantitative and mathematical approach to practicing medicine. Personalized medicine is designed for the specific genetic, epigenetic and environmental properties of patients and their diseased cells. Diagnoses, treatments and cures are improving for diseases caused by a single gene (Mendelian). As the costs of genotyping with microarrays and complete DNA sequencing continue to drop, new collaborative projects become possible. Biomarkers are being discovered through advanced genomic, proteomic, metabolomic and imaging technologies. This has a very high priority because they can improve the diagnosis of a disease, define subsets of patients and use appropriate therapies for them. Clinical trials are being modernized by automation and improved data management. Instead of just making the medicine specific for the DNA that a person is born with, it can be made specific for the mutated DNA that is in a type of cancer or other disease. This is being done by developing monoclonal antibodies, which will bind to receptors that are specific for a particular type of cancer. Some of them are even parts of FDA-approved medications. Most can't kill cells by themselves, but they can still bind to cancer-specific antigens and deliver drugs that are covalently attached to the monoclonal antibody. Even treatments for diseases that are caused by many factors (genetic and environmental) are benefiting from P4 medicine.

Keywords: Avastin®, Biomarkers, Campath®, Erbitux®, Gene chips, Herceptin®, Metabolome, Next generation sequencing, Personalized medicine, Rituxan®, Vectibix®.

INTRODUCTION

Perhaps nothing epitomizes the fusion of traditional and western medicine more than predictive, preventive, personalized and participatory (P4) medicine [1]. "It takes not just a holistic but also a quantitative and mathematical approach to practicing medicine. At the same time, systems medicine emphasizes prevention and individual participation in one's own health care. It recognizes the important

Robert E. Smith

human need for patients and care givers to be actively involved in preventing and curing their diseases. At the same time, mathematics, the foundation of reductionist thinking, is used to quantify huge datasets from patients, while physicists, chemists, biologists and engineers develop the analytical tools needed to generate the data. All of this can be linked through the internet and used in mobile healthcare applications" [1]. So, systems medicine is "the application of systems biology to the study of human disease" [1]. It can use data about biomarkers from many people and many different tissues in an individual to help everyone work with their physicians to make their own medical decisions.

Personalized medicine has been defined as, "the ability to customize medicine using molecular information to more accurately understand disease patterns and diagnose disease, as well as to tailor preventive and therapeutic intervention more effectively with fewer side effects" [2]. "It includes not only prescribing medicines, but also maintaining the mental and physical well-being of the patient and care givers. Pre-emptive genome-based testing of adults and children in personalized healthcare is becoming very helpful, especially when studying diseases with Mendelian inheritance. Diagnostic tests are now available for over 2000 Mendelian conditions. These tests are changing the paradigms for screening and diagnosing rare conditions. Personalized medicine can help identify patients who are more susceptible to certain diseases or disease-related symptoms or are pre-symptomatic. It will identify patients who will respond to preventive treatments differently or whose diseases or symptoms may progress differently when compared with others in the general population. Just as important, personalized medicine engages patients and helps them prevent diseases, decide treatments and monitor recovery. As we continue to personalize healthcare, the public is expressing their desire to participate actively in healthcare decision-making that is based on analyzing their genomes" [2].

So, personalized medicine "tailors medical treatment to the individual characteristics, needs and preferences of each patient" [3]. Actually, it has been used for over 100 years to analyze blood types, to ensure that transfusions don't cause hemolytic reactions. Also, over 50 years ago, the genetic basis for the selective toxicities of fava beans and an antimalarial drug (primaquine) was discovered. It is a deficiency in the enzyme glucose-6-phosphate dehydrogenase

(G6PD) that is important in metabolism. Then, in 1977, different isozymes of cytochrome P450 2D6 (CYP450 2D6) were found to cause the effects of the anti-hypertensive drug debrisoquine to be exaggerated and last longer than in others. So, genetic differences can cause different pharmacokinetic parameters, such as area under the curve, or AUC. Pharmacogenomics is the "study of how variations of DNA and RNA characteristics affect responses to drugs. It has been a crucial part of personalized medicine for decades" [3].

The goal is to prescribe different drugs and medical devices for people with different types of nutrition, environment, genes, mRNA, miRNA, epigenetics and/or proteins. Such treatments are designed for the patient's specific anatomy (size), physiology and environment (home, hospital, ICU). Diagnostic devices can monitor vital signs, blood glucose, oxygen or other small molecules. They can perform electroencephalography (EEG) or electrocardiography (ECG or EKG) and do diagnostic imaging. Some can even determine part or all of the genome, epigenome, transcriptome, proteome and metabolome of the patient and/or his or her diseased cells. Also, the patient's blood or tissues can be analyzed for different types of enzymes (isozymes, like CYPs) that catalyze reactions that can metabolize drugs differently and affect their bioavailability, or ability to bind to different receptors. Such an approach led to the development and rapid approval of tratuzumab, or Herceptin®, for treating and curing patients who have the HER-2 gene that is involved in many cancer signaling pathways. More recently, it led to four anticancer drugs being approved by the FDA "for use in patients who have specific genetic characteristics that can be identified by a companion diagnostic test" [3]. "Individualized medical devices are being made, too. Three-dimensional printing was used to make a bioresorbable tracheal splint for an infant who was critically ill. Furthermore, research on induced pluripotent stem cells (iPSCs) may lead to people being able to use their own cells to biosynthesize their own organs when they need a transplant. So, advances in genomics, medical imaging, 3-D printing and regenerative medicine, along with increased computational power and the advent of mobile and wireless capabilities, are allowing patients to be treated and monitored in ways that better meet their individual needs" [3].

There are many other examples of personalized medical devices [3]. "A customized tinnitus masker tailors audio signals to suit the patient's hearing requirements. Pedicle screw spinal systems can be assembled by physicians to fit a patient's unique anatomy based on MRI and/or CT images. There is a software-based EEG analyzer that predicts a patient's response to various psychotropic drugs. The device estimates the probable "response to various medications to help guide clinical decisions. Also, there is a Zenith Fenestrated AAA Endovascular Graft for patients with abdominal aortic or aortoiliac aneurysms having morphology suitable for endovascular repair. Each device is tailored to the patient's individual aortic anatomy. Finally, there is an artificial pancreas device system under clinical investigation that automatically monitors patient glucose levels and delivers the proper dose of insulin to diabetics" [3].

The FDA and other governments' agencies evaluate applications for new medical devices and drugs. So, the following goals were described in a recent report from the FDA [3]. "Personalized medicine seeks to reduce the burden of disease by targeting prevention and treatment more effectively. With the help of personalized medicine, the health care management paradigm will focus on prevention, moving from illness to wellness, and from treating disease to maintaining health. By improving our ability to predict and account for individual differences in disease diagnosis, experience, and therapy response, personalized medicine offers hope for diminishing the duration and severity of illness, shortening product development timelines, and improving success rates. At the same time, it may reduce healthcare costs by improving our ability to quickly and reliably select the effective therapy for each patient while minimizing the costs associated with ineffective treatment and avoidable adverse events" [3].

To support this effort, the US FDA and other countries' regulatory agencies have worked with industry and academia to build an elaborate infrastructure to support personalized medicine. This has been described as the five pillars of systems medicine: cutting-edge technologies, digital infrastructure, personalized data clouds, new analytical tools and systems biology models [1]. "Cutting-edge technologies and algorithms can gather data, set up user-friendly databases and do analyses. For example, in 2003 the US National Institutes of Health (NIH) started a project called ENCODE to identify and define the functional DNA elements that are required for normal genome function. In 2012, about 40 articles were published describing the results of what had become an international effort. In the near future, the internet and mobile telecommunication will establish a data cloud – hopefully for everybody. Quantitative metrics can be established for what it means to be healthy, pre-disposed to a disease, or in various stages of the disease. The data will also help find biomarkers that can help decide on the best therapies for each individual, but also monitor the treatment process, making necessary changes when new conditions emerge" [1].

Moreover, systems medicine is making disease care more cost effective in human and financial terms. Treatments and potential cures for previously uncurable diseases are emerging. This continues to be done by separating people and diseases into distinct subgroups [1]. Genomic and other analyses can stratify people's disease risks, reactions to drugs and other clinically relevant factors into sub-groups. For example, diseases such as breast cancer, which were once classified as single diseases, are being stratified into clinically relevant subgroups based on interactions in genetic, molecular and cellular networks [4]. Prostate cancer [5] and Crohn's disease [6] stratifications are providing increasingly more accurate diagnoses and cost-effective interventions based on the underlying causes of disease [1]. Surgery and other aspects of traditional medicine will be informed by disease stratification and individual needs. By focusing on the causes rather than the symptoms of a disease, physicians and patients will be able to prevent diseases from occurring in the first place, or stop them before they can cause serious damage. Moreover, as we identify and understand the biological networks that are perturbed in diseases, systems medicine will continue to provide a stream of new drug targets for the pharmaceutical industry [1].

The drugs will be more effective and have fewer costly side effects, since they will be more personalized [1]. "They will target specific strata of the populations of people and types of diseases. It will be cheaper for pharmaceutical companies to do clinical trials if they are done on the correct patient populations. Interventions (including, but not limited to, pharmaceutical interventions) will start at an earlier stage in the disease process, often pre-symptomatically, where they are far more cost effective. The impacts of interventions will be more accurately monitored, allowing for adjustments to improve outcomes and reduce costs" [1].

As part of this system, everybody will have a personal data cloud, which will act as a medical record, with all of the health data for each individual - including the genome, blood chemistry, lifestyle data (activity levels, diet and stress), transcriptome and gut microbiome [1]. "The data will be collected and analyzed to produce a stream of highly personalized information about each person's health and disease. Furthermore, actionable information can be supplied back to individuals based on the analysis of data accumulated in their personal data cloud. This will be given not just to physicians and other professionals, but also to individuals and those with whom they confide. Finally, systems biology and medicine work together to create a cycle of innovation. As new biological insight inspires the development of new analytical tools, new tools and technologies produce new data, as new data drives the creation of more analytic tools that advance biological insight" [1].

This has been apparent since the decision was made in 1990 to sequence the human genome in 15 years. Advances in DNA sequencing produced produce data faster than expected. Bioinformatic tools were produced to sort the billions of fragmented sequences into the complete genome (shotgun sequencing). This helped the International Genome Sequencing Consortium and others to announce the complete DNA sequence in just 13 years, in 2003. Many government agencies, including the US FDA established collaborations such the Center of Excellence for Bioinformatics, Functional Genomics, and Structural Genomics, Office of *In Vitro* Diagnostic Device Evaluation and Safety and Voluntary Genomic Data Submission (VGDS) Program were established in 2012 and were quite helpful in solving the human genome. Then, in 2004, the FDA created the Genomics and Targeted Therapy Group. Numerous other groups were formed, including the Personalized Medicine Team in the Center for Biologics Evaluation and Research (CBER) in 2011 and the Division of Systems Biology at the National Center for Toxicological Research (NCTR). Also, the FDA has issued at least 21 guidances that relate to personalized medicines. This includes guidances on pharmacogenomics data submissions, tests, definitions, considerations, applying human factors and current Good Manufacturing requirements for combination products [3].

The FDA is also collaborating with other governments, academia and industry to develop regulatory standards, reference libraries, research methods, and tools that can be used to integrate biomarker identification into drug and device development and help make clinical decisions [3]. "The biomarker qualification program aims to provide a framework for scientific development and regulatory acceptance of biomarkers for use in drug development, facilitate integration of qualified biomarkers in the regulatory review process, and encourage the identification of new and emerging biomarkers. There is also a project on microarray and sequencing quality control, a genomic reference library for whole genome sequencing (WGS) platforms and a virtual physiological patient. They are also building a high performance integrated virtual environment (HIVE) for next generation sequencing analysis. Also, high resolution human leukocyte antigen (HLA) typing systems are being developed, as well as molecular tools to facilitate blood group typing. The FDA and other governments' health care agencies are working with others to design and conduct clinical trials better. They are refining the design of clinical trial and the statistical methods of analysis to address issues such as missing data, multiple endpoints, patient enrichment, and adaptive designs that often arise when developing targeted therapeutics. They are also looking closely at clinical trials anticancer drugs. This is complicated because many cancers are heterogeneous, each with their own specific genetic makeup. This heterogeneity is one reason why different people with cancer in the same organ respond differently to the same therapies. The I-SPY 2 trial is a collaborative initiative developed under a unique public-private partnership. It includes more than 20 cancer centers. They are trying to better understand heterogeneity and complexity to provide targeted therapies" [3].

It is also essential to have adequate and robust statistical methods to analyze data. So, scientists at Booz Allen Hamilton, the FDA supercomputer center, the Genomics Evaluations Team for Safety (GETS) and the Office of Vaccines Research and Review (OVRR) in CBER are comparing different methods to analyze genomic data to predict patient outcomes and/or prognosis. CDER and CDRH are also developing new device diagnostics to improve drug safety. They are assessing new device-based algorithms and biomarkers that can distinguish between benign and malignant drug-induced QT prolongation in an electrocardiogram [3].

The National Center for Toxicological Research (NCTR) is also doing research on the biology of cancer [3]. "One research project found that many tumors carry subpopulations of KRAS mutant cells, which can contribute to acquired resistance to some cancer therapeutics. "Experimental approaches are being developed to identify effective treatments to prevent drug resistance in tumors with defined genetic profiles. Researchers at CBER are identifying pharmacogenetic determinants of immunogenicity in patients with Hemophilia A. They may eventually be able to predict a patient's risk of immunological response to a given protein therapy before it is used in treatment. Other researchers at CBER are trying to understand better the effects of DNA modifications on the quality of protein products. Using proteins that are involved in blood clotting as models, they demonstrated that while "synonymous" or "silent" mutations do not affect the protein sequence, they may affect protein levels as well as protein folding and function. They want to know which mutations are deleterious and which may be safely employed in the design of therapeutic protein products. Their goal is to develop tools and methods to evaluate protein properties from the gene sequence. This could have many diverse implications for developing and evaluating safe and effective protein therapeutics, including biosimilar products" [3].

CBER's Office of Vaccines Research and Review (OVRR), together with the Genomics Evaluations Team for Safety (GETS) are collaborating with others to identify genetic risk factors that may be associated with adverse reactions to vaccines [3]. "Another study in collaboration with the Centers for Disease Control (CDC) and Northern California Kaiser looks at genetic risk factors of febrile seizures after MMR vaccination. Another study, in collaboration with the Innovation Center for Biomedical Informatics (ICBI) at George-town University, seeks to identify genes associated with vaccines, vaccine components, and several autoimmune diseases of interest. The goal is to help assess the plausibility that autoimmune diseases might occur as adverse reactions to vaccines. "Pathway models derived from this data may help predict autoimmune reactions to vaccines and other medical products in the future" [3].

Scientists at NCTR, in collaboration with scientists at the University of Liverpool (UK) and the Huashan Hospital (China), are performing whole genome sequencing and genetic analysis to identify susceptibilities to carbamazepine-induced hypersensitivity reactions [3]. "They are also collaborating with the University of Maryland to identify genetic factors that interact with common lifestyle factors in the Amish community that may contribute to heart disease. The metabolic responses of volunteers were examined after consuming certain diets and drugs that are associated with cardiovascular risk. This included blood triglyceride levels after a high fat meal, blood pressure after consuming much NaCl with a meal, and platelet aggregation response after taking aspirin or clopidogrel. The DNA from subjects who showed abnormal responses was sequenced using next-generation sequencing technology. Genetic association studies were also done. This work is ongoing, and as candidate genetic markers are discovered, they are being validated in another cohort. Identifying genetic factors that interact with drugs or certain diets to increase risks of cardiovascular disease or the efficacy of treatment will allow patients and their physicians to use personalized medicine to improve health" [3].

"The National Cancer Institute (NCI), the National Institute of General Medical Sciences (NIGMS) and the University of Maryland and FDA are trying to see if increasing the dose of clopidogrel increases antiplatelet responses and active metabolite exposure in individuals with genetically reduced CYP2C19 metabolism compared to those with normal metabolism [3]. "Researchers in the Office of Science and Engineering Laboratories at CDRH are using new methods to analyze electrocardiograms to predict which patients will benefit from cardiovascular therapies such as cardiac resynchronization therapy. They can diagnose electrical conduction problems and to quantify scar tissue in the heart, with different criteria for women and men, since women benefit significantly more than men from cardiac resynchronization therapy. The Office of Science and Engineering Laboratories at CDRH is collaborating with George Washington University to develop a microfluidic, high-throughput microchip to test the interaction of tears with contact lenses, care products, and microbes. They aim to provide individual testing results that can guide the prescription of lens materials and hygiene products for patients. So, from the FDA's perspective, the era of personalized medicine has arrived" [3].

The FDA has a website with information on personalized nutrition and medicine by the Division of Personalized Nutrition and Medicine (DPNM) [7]. "The

Division has two areas - Biometry and Biology. The main function of Biometry is to develop biometrical methods for all aspects of the FDA's mission, goals, and objectives. A subgroup within Biometry analyzes all data from the NTP National Toxicology Program (NTP). The Biology area is focusing on the broad areas of pharmacogenomics and nutrigenomics - how individuals respond to drugs and nutrients in foods. The overall goals of the DPNM are to develop and implement research strategies that will be able to account for genetic, environmental, and cultural factors that influence the expression of genetic makeups and produce knowledge for improving personal and public health. These overarching goals will be met with three parallel efforts that develop:

- Integration of omics methodologies to assess an individual's health status and, as importantly, susceptibility to specific chronic conditions influenced by environmental factors including diet
- Means to capture and assess an individual's nutritional, environmental, and activity exposures
- Classification algorithms that integrate the data from omics and environmental assessments that will result in evidence-based and validated biomedical decision making [7].

Genomics can be used to predict whether a person is more susceptible to a disease. For example, some people have a variation of a gene called BRCA1 have a higher risk of developing breast, ovarian, and possibly prostate, and colon cancers. BRCA was the first gene found to be correlated with breast cancer. Alterations in the second gene found to be correlated with breast cancer, BRCA2, have been associated with breast, pancreatic, gallbladder, and stomach cancers. The second major aspect of pharmacogenomics, also called tumor genomics, is targeted therapy. Tumors have different genomic variations, and tests based on genomics are helping doctors to identify cancers that are likely to respond to a particular treatment. The third aspect of pharmacogenomics includes testing for drug resistance. For example, the HIV virus genome is always changing, and resistance testing can help doctors choose the drug that will best match the virus and suppress it" [7].

As mentioned in a recent review, "In 2004, the FDA introduced the Critical Path Initiative, with the intent of modernizing drug development by incorporating recent scientific advances, such as genomics and advanced imaging techniques, into the process" [8]. "Important parts of this initiative are the identification and quantification of new biomarkers, clinical trial modernization, bioinformatics, drug manufacturing and collaborations between government, research institutes, academia and industry. The initiative began because society expects more and better drugs to be developed, but this is not happening. From 1994 to 2004, global spending on research and development of NMEs, or new molecular entities (drugs) increased from about $40 billion to $60 billion, while the total number of NMEs worldwide decreased from 40 per year to a 20-year low of just over 20 NMEs" [8].

The development of new biomarkers through advanced genomic, proteomic, metabolomic and imaging technologies has a very high priority [8]. "New biomarkers can improve diagnosis, define disease subsets that may differ in response, define individual variability in the drug's molecular target, and provide an early readout of response to therapy. The Biomarker Consortium (http://www.biomarkersconsortium.org) has been formed at the Foundation for NIH, or FNIH, and is funding biomarker trials for PET scanning in non-Hodgkin's lymphoma" [8].

The next part of the critical initiative is the modernization of clinical trials by establishing better standards for fully automating the trials and managing the data [8]. "An important part of this is the science of bioinformatics, or the computer analysis of biological data. The FDA holds the largest set of data on animal testing of new molecular entities (NMEs), but it is not in a user-friendly form. One improvement on this is developing digitized electrocardiograms that may help scientists evaluate NMEs for adverse cardiac events. Another aspect of bioinformatics will be to establish quantitative models of disease processes. This will use data on biomarkers, clinical outcomes and the effects of various interventions. The last part of the critical initiative is to improve drug manufacturing by incorporating new science and technology, especially the use of modern process control technologies. Process control is used widely in some industries, such as petroleum refining, production of hydroelectric energy and the manufacturing of electronics. Automated analytical instruments, such as GC, GC-MS, ion chromatography, ICP-AES, ICP-MS and LC-MS can continuously monitor parameters such as the composition of petroleum distillates, the chloride content of water and the metal content of electroplating baths. If critical parameters fall outside of control limits, the process is automatically corrected, or stopped, until it can be corrected. Similar process control analyses would be useful in drug manufacturing" [8].

It is no surprise that there is a big difference between brain cancer and cancer in other tissues. However, even for a given tissue or cell type, such as leukocyte cancer (leukemia), there are important differences. So, the National Cancer Institute (NCI) started a $100 million research program to determine the genomes of different cancers [9]. They are doing this in collaboration with the National Human Genome Research Institute to determine the genomes of brain, lung and ovarian cancers [10]. Also, the M.D. Anderson Cancer Center at the University of Texas had a research project called the Biomarker-Integrated Approaches for Targeted Therapy for Lung Cancer Elimination (BATTLE) [9]. "They collected tissue samples (biopsies) of lung cancer cells at different stages of the disease. Tissue and cancer biopsies identified DNA biomarkers. They established the proof-of-principle that molecular-based, individualized, targeted therapy can work for lung cancer patients" [9].

The NCI Office of Cancer Genomics' Cancer Genome Characterization Initiative, CGCI supports cutting-edge genomics research on adult and pediatric cancers. Researchers develop and apply advanced sequencing and other genome-based methods to identify novel genetic abnormalities in tumors. The extensive genetic profiles generated by CGCI may inform better cancer diagnosis and treatment [10]. "They completed a pilot lung cancer project. They used next generation transcriptome sequencing, as well as gene expression and epigenome profiling to study early-stage lung carcinogenesis. They discovered some changes in early-stage lung epithelial tumor cells and tested them to discover their roles in cancer development. They also compared candidate alterations between the various pathologically-determined epithelial phenotypes (normal, dysplastic, neoplastic, and malignant) to identify alterations that associate with those phenotypes. Alterations that were found to correlate with early-stage phenotypes (dysplastic and neoplastic) "are likely to play a role in the initiation of lung cancer. "Completion of this comparison" analysis could identify a new set of regulatory pathways in lung carcinogenesis, which may be pursued in future prospective biomarker studies for risk assessment, early diagnosis, and targeted therapies" [10].

The CGCI also has a Cancer Genome Anatomy Project (CGAP) that generated a wide range of genomics data on cancerous cells that are accessible through the CGAP website (ocg@mail.nih.gov) [10]. "Part of this is the SNP500 database whose goal is to resequence 102 reference samples to find known or newly discovered single nucleotide polymorphisms (SNPs) which are of immediate importance to molecular epidemiology studies in cancer. Together with the Initiative for Chemical Genetics (ICG), the CGCI established Chembank, an interactive database for small molecules. It contains data from hundreds of biomedically relevant small molecule screens of hundreds-of-thousands of compounds. Chembank also provides analytical tools to facilitate data mining. They also have a Childhood Cancer center that has "comprehensive information on childhood cancers, including current treatments, clinical trials, prevention, genetics, testing, and more. There is also a Cancer Genetic Markers of Susceptibility (CGEMS) initiative. It is identifying common inherited genetic variations associated with a number of cancers, including breast and prostate. Data from these genome-wide association studies (GWAS) are available through the Division of Cancer Epidemiology Genetics website. (http:// dceg.cancer.gov/ research/how-we-study/genomic-studies/cgems-summary). The CGCI also provides the scientific community with validated, full open reading frame (ORF) cDNA clones for all of the currently defined human genes. There is also a mammalian cDNA Library from the NIH Mammalian Gene Collection (MGC). It is providing full-length clones for most of the defined human and mouse genes, along with selected clones of cow and rat genes" [10].

Finally, the Office of Cancer Genomics has an SOP manual [10]. "It is a set of guidelines for investigators participating in OCG projects to characterize tumors on a molecular level. "The SOP provides templates and protocols that apply to all projects, as well as some that apply specifically to individual projects. Investigators must follow the protocols in the SOP when contributing samples and data. The sample and data acquisition process is explained in comprehensive detail to ensure that all materials contributed will be of sufficient quality to be utilized in the projects" [10].

In 2009, researchers began examining the genomes of medulloblastoma, a brain tumor that occurs primarily in children, and B-cell non-Hodgkin lymphoma (NHL), a white blood cell cancer [10]. "Both projects revealed many genetic alterations not detected in previous studies that are potentially important in cancer onset and/or progression. Additionally, CGCI partners with the OCG TARGET initiative to help solve the genetic mysteries of some pediatric cancers that do not respond well to standard therapies. These hard-to-treat cancers include "refractory to treatment" acute myeloid leukemia and two rare kidney tumors. Findings from some or all of these projects may translate into improved "survival and quality-of -life for patients afflicted with these diseases. They have begun projects to better characterize cancers from HIV-positive patients, early-stage lung cancer, and Burkitt lymphoma. Data is publicly available through "National Institutes of Health (NIH) and National Cancer Institute (NCI) databases" [10].

In December 2012, British Prime Minister David Cameron announced a project to sequence the genomes of 100,000 Britons (100,000 Genome Project or 100kGP) affected with cancer or rare diseases [11]. "In July 2013, the Department of Health announced that this initiative will be coordinated by the newly-established, government-owned company Genomics England (London, UK). They will try to incorporate whole-genome and whole exome sequencing into clinical studies. They also plan to develop strategies for the effective and safe use of whole-genome sequencing. This will require excellent patient communication, given the breadth and value of information it can generate" [11].

For example, it is important to remember that genetics has its limits. Most people already know that almost all of us have unique genomes, or genetics. Of the three billion base pairs, about six million are unique to the individual. Only the so-called identical twins (or triplets, quadruplets, *etc*) have the same sequence of three billion base pairs. They are more properly called homozygotic twins. Even though they begin life with the same DNA, they are not identical people. They can have very different susceptibilities to diseases, including diseases with a known genetic component. So, even if current genetic testing techniques show that a person is susceptible to a disease, it does not mean they will get the disease. At the same time, just because genetic (DNA) testing indicates that a person does not have a genetic defect that makes them susceptible to a disease, that person may still get the disease. For example, the entire genome of the scientist, J. Craig Venter, has been reported. He did not have a genetic defect making him susceptible to skin cancer, but he still got it. He was exposed to direct sunlight (and UV radiation) when he was a young surfer in California. So, environment is very important. It can affect the microbiome, virome, epigenetics, post-translational modifications and mobile genetic elements, which can all affect the disease phenotype. Moreover, diseased cells have different genetics and epigenetics. Cancer and heart disease are really many different diseases. So, medicine needs to be personalized not just for individuals, but also for their particular types of cancers.

The Sheik Khalifa Bin Zayed Al Nahyan Institute for Personalized Cancer Therapy at the MD Anderson Center supports preclinical research and clinical trials in which a patient's tumor biopsy is assayed for abnormal genes and gene products to select therapy with agents targeting the product of those particular abnormal genes [12]. "This program integrates research into clinical trials. Its goal is to implement personalized cancer therapy and improve patient outcomes. The seminal BATTLE trial in lung cancer demonstrated the practicality of this approach. They did a series of Phase I/II trials. Experimental drugs were assigned based on biomarkers that were detected in the patients' cancers. Several clinical trials based on genetic and molecular biomarkers in patients' cancers are now underway. They significantly improved response rates in combined targeted therapy/ chemotherapy Phase I clinical trials in multiple disease sites. For example, patients treated with a drug targeting the PI3K pathway had a "response rate of 35% in patients with mutations compared to a 4 to11% response rate generally observed in phase I trials" [12].

Stanford University and a company called Nodality started looking at leukemia [9]. "They prepared fluorescent-labeled antibodies that bind to specific phosphoproteins, which are important in signaling pathways in cancer cells. They used a cell flow technology in which cells flowed past a fluorescence detector. Those that had phosphoproteins fluoresced and were detected. This established their amount, which controlled the "extent to which the signaling pathways are activated in cancer cells" [9]. Phosphoproteins will be discussed further in Chapter 3. Nodality recently described a Single Cell Network Profiling (SCNP) technology that characterized signaling and drug sensitivity and resistance profiles in mast cell leukemia (MCL) bone marrow mast cells [13]. "They also showed quantitative measurements of EGFR pathway signaling and modulation from "non-small cell lung cancer (NSCLC) patients. The epidermal growth factor (EGF) binds to its receptor (EGFR), which is also a tyrosine kinase. It catalyzes the phosphorylation of some of its own tyrosine residues, which activates downstream signaling cascades" [13].

The Harvard Partners Center for Genetics and Genomics was founded in 2001 with the specific goal of accelerating the realization of personalized medicine. Its name changed in 2008 to the Center for Personalized Genetic Medicine [14]. "The change reflects an emphasis on translational medicine. That is, they are trying to translate laboratory results into clear medical treatments and cures. The Personal Genome Project was announced by George Church in 2006. It will publish full genome sequences and medical records of volunteers to enable research into personalized medicine. The Laboratory for Personalized Molecular Medicine was founded in 2007 to identify specific mutations in genes linked to clinical outcomes in patients with leukemia and lymphoma. Identifying the presence or absence of these mutations is becoming a standard of care for patients with acute myeloid leukemia. LabPMM also develops patient-specific molecular tests from patient tumor DNA samples. The ultra-sensitive tests are used by leading cancer treatment centers world-wide to monitor residual disease and treatment" [14].

The Mayo Clinic has also a Center for Individualized Medicine [15]. "About 100,000 people in the USA die each year from adverse reactions to medications and another two million are hospitalized. Hopefully, research at the intersection of pharmacology and genetics (pharmacogenomics) will make it possible to predict who is likely to have an adverse reaction to a drug before it is given to them. It may also be able to predict whether a patient will respond well to a medicine" [15].

There are also Centers for Personalized Medicine at the Roswell Park Cancer Institute (RPCI) [16], Duke [17], Stanford [18], the University of Pennsylvania

(Penn) [19], the Sheikh Khalifa Bin Zayed Al Nahyan Institute for Personalized Cancer Therapy [12] and many others. Next, some of the analytical methods that are being used in these organizations will be discussed.

METHODS

PCR

Many of the gene deletions, mutations and duplications were first detected by Southern blot and multiplex polymerase chain reaction (PCR). Because Southern blotting was a labor intensive method, it was replaced by the multiplex PCR and DNA microarrays. The PCR can amplify, or make many copies of a select gene through a chain reaction initiated by the proper primer and catalyzed by DNA polymerase. Multiplex PCR is a modification, in which several sets of primers are used to produce amplicons of different DNA sequences. By targeting multiple genes at once, more data can be acquired from a single assay. Multiplex PCR can rapidly detect deletions or duplications in genes. It can also look for microsatellites and single nucleotide polymorphisms (SNPs). It was first used in 1988 to detect deletions in the dystrophin gene that causes muscular dystrophy, and is still quite useful [20]. "The most commonly used multiplex PCR protocols look for 18 exons at deletion hot spots and detect 90 98% of them. However, it is not as good at identifying gene duplications in the Duchenne muscular dystrophy *DMD* in males or heterozygous deletions and duplications in females. Instead, multiplex ligation dependent probe amplification (MLPA) analysis has become the method of choice for detecting deletions and duplications" [20].

Multiplex PCR can be combined with "microfluidic mixers, valves, pumps, channels, chambers, heaters, and DNA microarray sensors in a single biochip" [21]. "It can do sample preparation, including magnetic bead-based cell capture, cell preconcentration and purification, and cell lysis. It can also do PCR and detect DNA hybridization. So, a DNA microchip is a collection of microscopic DNA spots attached to a solid surface. Each spot contains picomoles (10^{-12} moles) of a specific DNA sequence, often called a probe, reporter or oligo. It can be a short section of a gene or other piece of DNA. It is used to hybridize a cDNA or cRNA (also called anti-sense RNA) sample (called the target). Probe-target hybridization is usually quantified by detection of fluorescence-, silver-, or chemilluminescence-labeled targets. This provides the relative abundance of nucleic acid sequences in the target. Since an array can contain tens of thousands of probes, many genetic tests can be run in parallel" [21].

The probes are attached covalently to a solid surface that can be glass or a silicon chip [21]. "Photolithography is used to make Affymetrix quartz chips. Other microarray platforms, such as Illumina, use microscopic beads, instead of a larger solid support. Alternatively, microarrays can be made by directly synthesizing DNA probes onto solid surfaces. DNA arrays either measure DNA or use DNA as part of its detection system. DNA microarrays can be used to measure changes levels of expression, to detect SNPs, or for genotyping or targeted resequencing of DNA. In addition to the traditional microarray on a biochip, there is a bead array. It is a collection of microscopic polystyrene beads, each with a specific probe and two or more dyes, which do not interfere with the fluorescent dyes used on the target sequence. That way, the ratio of fluorescence intensities can help in the quantitation" [21].

Small microarrays can be used to look for microorganisms that can contaminate food, mycoplasms in cell cultures or other pathogens. A chromatin immuno-precipitation (ChIP) can detect DNA sequences that are bound to a particular protein [21]. "It can look for histone modifications or Polycomb-group proteins when studying epigenetics or RNA polymerase, when studying the transcriptome. There is also a chip for identifying methylation of adenine in DNA. It is called the DamID (DNA adenine methyltransferase identification). It can identify binding sites by expressing the proposed DNA-binding protein as a fusion protein with a bacterial DNA methyltransferas. When the protein of interest binds to DNA, it places the methyltransferase near the binding site. Adenosine methylation does not occur naturally in eukaryotes, so adenine methylation in any region must have been caused by the fusion protein, implying the region is located near a binding site. Unlike ChIP, DamID does not require antibodies, but makes use of adenine methylation near the protein's binding sites to selectively amplify those regions that are bound by the protein of interest. There are also arrays for detecting alternate splicing sites of predicted exons, fusion gene microarrays (to detect fusion transcripts in cancer cells) and tiling arrays that have overlapping probes designed to study a region of interest, sometimes as large as an entire human chromosome. It is used to detect the expression of transcripts or alternate splice forms which may not have been previously known or predicted" [21].

MLPA is a variation of the multiplex PCR, in which target sequences can be amplified using just one primer pair [21]. "One oligonucleotide probe contains the base sequence that is recognized by the forward primer, the other the sequence recognized by the reverse primer. Only when both probe oligonucleotides are hybridized to their respective targets, can they be ligated into a complete probe. Only the ligated oligonucleotides, but not the unbound probe oligonucleotides, are amplified if the probe is split into two parts. If they were not split, the primer sequences at either end would cause the probes to be amplified whether they were hybridized to the template DNA or not. The amplification product would not depend on the number of target sites present in the DNA sample. So, in the MLPA method, the resulting amplicons are separated by capillary electrophoresis. This provides better resolution than multiplex PCR. Since the forward primer used for probe amplification is fluorescent, each amplicon produces a peak which can be detected by the fluorescence detector on the capillary sequencer. After calibration with reference samples, the relative quantity of each amplicon can be determined" [21].

TOF-MS

Another method uses "time-of-flight mass spectrometry (TOF-MS) to separate and detect amplicons. Matrix assisted laser desorption and ionization (MALDI) produces volatile ions that the TOF-MS can separate due to the differences in mass (molecular weight) of analytes, such as DNA, RNA, proteins and other biopolymers. It is especially useful when looking for SNPs, the most frequent genetic polymorphism. There is a database of SNPs, dbSNP, which accumulates information on single-nucleotide genome variation. It describes about 38 million validated SNPs in the human genome (build 137) [22]. Thus, in the human genome, on average, one nucleotide in 100 is variable. In current research and laboratory practice, SNPs are most commonly genotyped by real-time PCR using TaqMan probes with a fluorescent label. Large-scale studies have used high-density biochips for genotyping [23]. While the first approach can genotype individual markers in hundreds or thousands of specimens in a single experiment, high-density biochips provide data on dozens or hundreds of thousands of SNPs in dozens of specimens [23]. However, massARRAY™ analysis with MALDI-TOF MS is better for intermediate-scale genotyping of dozens or hundreds of markers in hundreds of specimens, which is most commonly required for the majority of purposes [24].

Next-Generation Sequencing (NGS)

Multiplex PCR, MLPA and massARRAY™ all look for specific genes and/or SNPs. However, there are newer methods that can sequence one to 43 billion short reads (50-400 bases each) per analysis [25]. They are called next-generation sequencing (NGS) and massively parallel sequencing methods. They use spatially separated, clonally amplified DNA templates or a single molecule of DNA in a flow cell. They are quite different from the very first generation that used Sanger sequencing based on chain-terminating dideoxynucleotides that were radio-actively or fluorescence-labeled. The pieces of DNA that were synthesized were separated by capillary electrophoresis and detected by UV light or auto-radiography [25].

Instead, sequencing can be done by two different ways. One uses PCR and multiple primer pairs "that are combined with the genomic DNA of interest [25]. "This multiplex approach preserves precious small (ng) amounts of DNA. Using multiplex primer pairs leverages the advantage of the high throughput of NGS platforms with the fact that each sequence read represents a single DNA product. The analysis is done in three steps. First, sequencing libraries are generated by PCR clonal amplification. Second, the base sequence is determined by DNA synthesis - by adding nucleotides to the complementary strand rather than chain termination. Third, the spatially segregated, amplicons are sequenced simultaneously in a massively parallel fashion without needing a physical separation step. This generates hundreds of megabases to gigabases of nucleotide sequences in a single instrument run and increases the amount of data produced and fundamentally changed genome sequencing. Newly emerging NGS technologies and instruments decreased the cost of sequencing to where it approaches the goal of only $1000 per genome, which is making it quite popular" [25].

The second approach is hybrid capture [25]. "It hybridizes DNA fragments from a whole-genome library to complementary sequences that are synthesized and combined into a mixture of probes specifically designed for the matching regions in the genome. These probes typically have covalently linked biotin moieties. This enables a secondary capture by mixing the complexes between the probe and library with streptavidin-coated magnetic beads. Hence, the targeted regions of the genome can be selectively captured from solution by applying a magnetic field, whereas most of the rest of the genome is washed away in the supernatant. Subsequent denaturation releases the captured library fragments from the beads into solution. Then, they are ready for postcapture amplification, quantitation, and sequencing. When the probes are designed to capture essentially all of the known coding exons in a genome, it is called exome sequencing. More probes can be designed, synthesized, and added to an exome reagent. This is called exome plus. When only a subset of the exome or of the genome outside of the exome is targeted, it is called a targeted panel" [25].

Three major platforms for NGS are Illumina's (CA, USA) sequencing by DNA synthesis, Roche's (CT, USA) pyrosequencing system and Life Technologies' (NY, USA) semiconductor-based sequencing [26]. "The Illumina Genome Analyzer IIX™, MiSeq™ and HiSeq™, as well as Helicos Biosciences Heliscope platforms use reversible dye terminator chemistry. It is a cyclic method that includes incorporating nucleotides, fluorescence imaging and cleavage. A fluorescence-labeled terminator is imaged as each dNTP is added and then cleaved so that the next base can be added. These nucleotides are chemically blocked so that each addition is a unique event. An imaging step follows each step of base incorporation. Then, the blocked group is removed chemically to prepare each strand for the next addition reaction catalyzed by DNA polymerase. This series of steps continues for the number of cycles that are specified by the user" [26].

Sequencing by reversible terminator chemistry can use four colors in the Illumina platform, or a one-color cycle in the Helicos BioSciences [26] instrument. "Helicos BioSciences used virtual Terminators, which are unblocked terminators with a second nucleoside analogue that acts as an inhibitor. These terminators have the appropriate modifications for terminating or inhibiting groups so that DNA synthesis is terminated after adding a single base" [26]. So, this is one of the sequencing methods that are based on DNA synthesis.

The DNA template is prepared by emulsion PCR, also known as clonal-emPCR. A DNA library is first generated by fragmenting genomic DNA into random pieces [27]. "Single-stranded DNA fragments (templates) are attached to the surface of micrometer-sized beads with adaptors or linkers. One bead is attached to a single DNA fragment from the DNA library. The surface of the beads contains oligonucleotide probes with sequences that are complementary to the adaptors that bind to the DNA fragments. The beads are then mixed into water-oil emulsion droplets. In the aqueous water-oil emulsion, each of the droplets capturing one bead is a PCR microreactor that produces amplified copies of the single DNA template. They can be recovered, arrayed and analyzed" [27].

Sequencing can also be done by ligation, catalyzed by DNA ligases [27]. "In its simplest form, a fluorescence labelled probe hybridizes to its complementary sequence that is adjacent to the primed template. Then, DNA ligase is added, followed by the dye-labelled probe. Non-ligated probes are washed away, followed by fluorescence imaging to determine the identity of the ligated probe. The cycle can be repeated by using either cleavable probes to remove the fluorescent dye and regenerate a 5′-phosphate group for subsequent ligation cycles (chained ligation, Life Technologies SOLiD4™) or by removing and hybridizing a new primer to the template (unchained ligation, Complete Genomics™)" [27].

Pyrosequencing is another approach. It is a non-electrophoretic, bioluminescence method that measures the release of inorganic pyrophosphate by proportionally converting it into visible light using a series of enzymatic reactions [28]. "Unlike other sequencing approaches that use modified nucleotides to terminate DNA synthesis, pyrosequencing manipulates DNA polymerase by adding a limiting amount of deoxynucleotide triphosphate (dNTP). Upon incorporation of the complementary dNTP, a DNA polymerase catalyzes the extension of the primer and pauses. DNA synthesis is reinitiated after adding the next complementary dNTP in the dispensing cycle. The order and intensity of the light peaks reveal the

DNA sequence" [28].

Pyrosequencing can be done using a Roche/454, which "uses DNA templates prepared by emPCR, with 1–2 million beads deposited into titanium-coated pico titer plate (PTP) wells" [28]. "They increase DNA read length and improve data quality by reducing crosstalk between adjacent wells that contain single clonally amplified beads. Smaller beads which have sulfurylase and luciferase attached to them so they can emit light are loaded into wells that surround the template beads. Individual dNTPs are then streamed across the wells and dispensed in a predetermined order. The bioluminescence is imaged with a charge-coupled device (CCD) camera" [28].

DNA can also be sequenced rapidly using "an integrated circuit to perform non-optical detections" [29]. "Sequence data are obtained by directly sensing the H^+ ions produced by template-directed synthesis catalyzed by DNA polymerase and using all-natural nucleotides on a massively parallel semiconductor-sensing device or ion chip. The ion chip contains ion-sensitive, sensors based on field-effect transistors with 1.2 million wells. They provide confinement and allow parallel, simultaneous detection of independent sequencing reactions. The complementary metal-oxide semiconductor (CMOS) production process enables low-cost, large-scale production and scaling to higher densities and larger array sizes" [29].

This technology was used to analyze cytology specimens with an Ion Personal Genome Machine (PGM) [30]. "An amplicon library was generated from 10 ng of DNA from each sample using the Ion Ampliseq Cancer Panel (Life Technologies). They looked for 46 hotspot mutations in genes. Library quantification was performed using the Bioanalyzer High Sensitivity DNA Chip (Agilent Technologies, Santa Clara, CA). It should be noted that some of the components in the kit contain dimethylsulfoxide (DMSO)" [30]. Even though pure DMSO is relatively safe, it does penetrate through the skin and rapidly enters the bloodstream, carrying any solutes present with it. So, it should be handled with care and workers should always wear gloves. Emulsion PCR was performed manually using the Ion Xpress Template Kit (Life Technologies), followed by manual breaking the emulsion to isolate the ion spheres [31]. "Sequence alignment and base calling was performed by Torrent Suite software V2.0.1 (Life Technologies) using Human Genome Build 19 (Hg19) as the reference. Torrent Variant Caller software V1.0 (Life Technologies) was used for the detection of variants, whereas the Integrative Genomics Viewer (IGV) was used to visualize variants" [31].

Recently, it was shown that NGS methods can be used in a clinical setting [32]. "They analyzed somatic mutations within exons of 19 genes in patients. Blood, tumor biopsy and archived tumor samples were collected from 50 patients, who were recruited from four cancer centers. Samples were analyzed using targeted exon sequencing, multiplex somatic mutation genotyping using Sequenom MassARRAY and Sanger sequencing. Nineteen actionable mutations were identified in 16 (32%) patients. Results using the three technologies agreed with each other in 100% of the biopsy specimens and 95% of the archived specimens. Profiling results from paired archival/biopsy specimens were concordant in 30/34 (88%) patients. It was shown that next generation sequencing can be used for real-time genomic profiling in advanced cancer patients is feasible. Moreover, actionable mutations identified in this study were relatively stable between archival and biopsy samples. This implies that cancer mutations that are good predictors of drug response may remain constant across clinical stages" [32].

Recent advances in PCR technology have also been shown to be clinically relevant [33]. "It is called digital PCR because it can count DNA molecules simply and directly. The sample being analyzed is diluted very much and divided into many aliquots, so that some receive at least one molecule of the target (positive aliquots), while others do not. The number of positive aliquots reflects the abundance of the target DNA locus in the sample. In those aliquots that contain a rare variant, it can be detection even in the presence of the more common variant, as it would be if bulk DNA were amplified" [33]. A droplet digital PCR system (ddPCR) was used with 20,000 partitioned reactions (microdroplets) to obtain a linear response with respect to DNA concentrations in the interval of 0.16–99.6% [34]. Similar results were obtained in a study using a microfluidic based approach [35].

Pharmacogenomics

It is also important to remember that pharmacogenomics is important. The concentrations of therapeutic drugs should be monitored to see if there are important differences in different populations of patients. This can be especially important in pediatrics [36]. "Absorption, distribution, metabolism, and excretion are variable and undergo rapid changes as neonates mature. Neonates demonstrate wide inter-individual variability in maturational factors and pharmacokinetics, which is often unexplained. Neonatal intensive care interventions, including therapeutic hypothermia and extracorporeal membrane oxygenation, may increase pharmacokinetic variability further. Therapeutic drug monitoring (TDM) has proven to be especially useful with agents such as aminoglycosides, anti-convulsants, caffeine, digoxin, and vancomycin. Moreover, TDM was used to evaluate everolimus for treating pediatric patients who had "subependymal giant cell astrocytoma associated with tuberous sclerosis. Everolimus was recently approved by the US FDA" [36]. This application of TDM was adopted from its use in transplant recipients to learn how to reduce variability in everolimus exposure. Several factors contributed to the high inter-individual variability in everolimus disposition in pediatric patients.

First, wide ranges of ages and body sizes were included in the clinical study. Also, minimum plasma concentrations (C_{min}) of everolimus were lower in the younger patients, after being normalized to the surface area of the body. The adjusted doss may be due to age-related differences in the functional activity of CYP3A4. Finally, patients with tuberous sclerosis–associated subependymal giant cell astrocytoma often require coadministration of antiepileptic drugs that can induce CYP enzymes. This can decrease everolimus blood concentrations. A target range of everolimus exposure was supported by an exposure–response (E–R) analysis for efficacy and safety, which supported the adaptation of TDM for pediatric patients. A logistic regression model demonstrated that subependymal giant cell astrocytoma tumor response reached a plateau when Cmin > 5 ng/ml, whereas the E–R profile for safety indicated that there was no safety concern when the C_{min} was <15 ng/ml. A standardized validated assay for everolimus has been available since its approval for transplant patients. It facilitated the development of TDM for this pediatric indication [36].

A recent example of a TDM program that was developed after FDA approval was the use of TDM to manage infliximab therapy in pediatric inflammatory bowel disease patients [36]. The TDM program with infliximab was expected to be effective, given the validated E–R relationship that was established during its development in both adults and pediatric patients [36].

Still, one of the most important, but often overlooked issues in analytical biochemistry is sample preparation. That is, an analysis is only as good as the sample preparation method. So, it is essential that tissue samples be taken properly. So, tissue microarrays (TMAs) have been developed for robust assay development and quality control in the clinical laboratory setting" [37]. "They can analyze very small amounts of tissues and dramatically increase the efficiency of tissue-based studies since they can examine tens to hundreds of different patient samples on the same slide. By linking data on patient samples to pathological and clinical data, including disease follow-up and treatment response, TMAs have expanded translational research, enabling large-scale biomarker and disease progression studies, leading to more informed and improved clinical hypotheses. TMAs are coupled with high-throughput and objective analysis of biomarker expression. This is done with robust imaging technologies and has led to further improvement in biomarker data. This will ensure the continued use of TMAs, especially in translational medicine" [37].

Effects of Drugs on Metabolism

It is also important to determine the effects of drugs on patient's metabolism. It is essential to monitor the concentration of a therapeutic drug and its metabolites before a drug can be approved. Moreover, as many biochemicals and metabolites as possible should be detected and quantified by LC-MS, GC-MS and/or NMR. These methods complement each other. LC-MS and GC-MS are much more sensitive than NMR and are quite useful in monitoring the concentrations of therapeutic drugs, hormones, neurotransmitters and other biochemical that are present at relatively low levels. A mass spectrometer can be tuned so that it can detect a selected compound or set of compounds with optimal sensitivity, while avoiding interferences from the rest of the sample, or sample matrix. That is, MS is excellent when searching for a needle in a haystack. However, to see the haystack, one needs NMR. Unlike MS, ^{1}H-NMR can detect all ^{1}H nuclei with equal sensitivity. So, unlike MS, analysts can determine the relative concentrations of compounds that contain hydrogen without needing a standard compound to calibrate the instrument. So, LC-MS and GC-MS can be excellent tools in pharmacagenomic studies, but NMR is essential when analyzing urine for analytes that are present at a wide range of concentrations.

Their concentrations can vary, depending on the patient's genetics, health, age, exposure to toxins, gut microbiome, medical care, dietary supplements, nutrition and even level of stress. For example, the metabolite phenylpyruvate (also known as phenylketone) accumulates in the urine of patients who have phenylketonuria. It is caused by a mutated gene coding for phenylalanine hydroxylase. When the activity of this enzyme is too low, phenylalanine can accumulate, so the treatment is to restrict dietary phenylalanine. Metabolomics can evaluate the interactions between genotype and environmental factors to help choose which drug (or drugs) will be the most effective and have the least toxicity and/or adverse effects [38]. "It may also predict a patient's likely response to medical treatments and monitor his or her progress during therapy. Discovering important new metabolites can lead to discovering biomarkers that can be used to predict the low incidence of idiosyncratic toxicity. Then, metabolomics can be combined with genetic data. One can look for biomarkers of Mendelian disease traits. Metabolomics can use quantitative trait locus (QTL) mapping of individual metabolites to identify associations with genetic polymorphisms. It can also do metabolic "quantitative trait locus (mQTL) mapping and metabolomic genome-wide associated studies (mGWAS) that examines associations between multiple metabolites and many genes. QTL and mQTL are similar approaches, but mQTL provides a larger set of associations than found with QTL mapping, so it is more similar to GWAS. The associations are often between SNPs and metabolite concentrations or ratios. Metabolomics is especially useful in personalized medicine because the metabolome at any given time reflects the meta-bolic phenotype that is downstream of genetic and environmental effects. So, it is another important tool for personalized medicine" [38].

NMR is one of the most widely-used methods to do metabolomics studies. NMR measures the absorbance of radio frequency (rf) electromagnetic radiation by organic compounds that are placed in a strong magnetic field. So, an NMR contains a strong magnet, along with rf emitters and receivers that can be set to the frequencies at which magnetically susceptible nuclei (like ^{1}H and ^{13}C) absorb. The exact frequency that each nucleus in an analyte absorbs depends on its chemical environment and chemical properties. However, unlike UV-Visible (UV-Vis) and infrared (IR) spectroscopy, one does not plot the absorption *vs* the frequency or wavelength of the electromagnetic radiation. That is, the peak widths and range of frequencies (or wavelengths) absorbed are much smaller in NMR than UV-Vis (200 – 800 nm) or IR (2.5 – 20 μm). That is, in an NMR that has a magnet with a field strength of 21.1 tesla, ^{1}H nuclei will absorb at or near 900 MHz. The compound tetramethylsilane (TMS) is used as a standard to set the rf scale on the x-axis (horizontal axis) of the spectrum. The hydrogens (^{1}H isotope) in methyls ($-CH_3$) attached to long chain alkyl groups will absorb slightly higher frequencies. They are shielded slightly from the magnetic field, so the frequencies that they absorb are shifted due to their chemical and magnetic properties. So, if TMS absorbs at 90.0000 MHz, the ^{1}H nuclei in CH_3 will absorb at or near 900.0009 MHz, which is 1 ppm away from where TMS absorbs. Phenyl protons (^{1}H) will absorb at or near 900.0063 MHz. It would be very inconvenient to label the horizontal axis from 900.0000 to 900.00063 MHz, so it is labeled as ppm. Also, instead of using the term peaks, like one does in UV-Vis and IR, the term chemical shift is used in NMR. Unlike its usual usage as a unit of concentration (such as μg/g), here ppm is a unit of the rf that is absorbed. Another important difference is that quantitation by NMR is not done using peak heights (or absorbances at a specific wavelength), but peak areas. This is done by integration, so one often uses the chemical shift and the integrals to interpret NMR spectra. If the experiment is done properly, all the ^{1}H nuclei will have the same sensitivity to exposure rf energy. On the other hand, each compound and functional group will have a unique ability to absorb UV-Vis or IR electromagnetic radiation. This is called absorptivity. It is the same for all UV-Vis and IR spectrophotometers, so molar absorptivities are published and can be used for quantitation. In MS, the mass spectrometer is tuned so that it has optimal sensitivity for the analytes of interest. Sensitivities will vary from one MS to another and will be different on the same MS if the tuning parameters are changed. So, standards are required for quantitation by MS, UV-Vis and IR, but not for NMR, especially if one is only interested in the relative concentrations of the metabolites.

There are other important differences between NMR and UV-Vis or IR spectroscopies. First, in UV-Vis and IR, electrons absorb electromagnetic energy and go from the ground state to an excited state, which are separated by enough energy at room temperature so that almost all of the electrons are excited. However, the difference in "energy between the ground state and excited state of nuclei is much smaller, so only a few parts per million of the nuclei can be excited in NMR. This makes NMR a much less sensitive technique than UV-Vis, IR and especially MS. Also, it can take a few seconds for ^{1}H nuclei to return to the ground state after being excited, so a delay in rf pulses may be required for proper quantitation. Another way to think of the NMR phenomenon is that a nucleus is tipped away from the magnetic field when an rf pulse is applied. To get the maximum amount of rf energy absorbed, the nucleus should be exposed to an rf pulse that lasts long enough to tip the nucleus by 90°. This is called the pulse width and is measured in μsec. However, it can take several seconds for a nucleus to return to the ground state (tip angle of 0°), so a relatively long pulse delay would be needed for quantitation. By using repeated pulses and Fourier transforms, many acquisitions can be acquired. The signal to noise ratio (S/N) is proportional to the square root of the number of acquisitions. So, the S/N doubles after four acquisitions, increases 10-fold after 100 acquisitions and 100 fold after 10^4 acquisitions. If there is a 6 sec pulse delay, 60 000 seconds are needed for 10^4 acquisitions. An alternative is to tip the nucleus by a smaller angle, such as 30°, so it takes much less time for a nucleus to return to the ground state. So, many NMR spectroscopists will use a pulse width 30° pulse width (about 3 μsec) and a 0.5 sec pulse delay when acquiring ^{1}H-NMR spectra.

There is another important difference between NMR and MS, UV-Vis and IR. The locations (chemical shifts) of the peaks in an NMR spectrum depend on the concentration, pH and ionic strength of the sample. So, the chemical shifts are different in dilute *vs* concentrated urine and in blood from different people who have different concentrations of proteins and electrolytes. So, NMR spectroscopists look for the chemical shifts that are due to hydrogens in the most abundant analyte, such as creatinine in urine. In contrast, when MS is coupled to a separation method (GC or LC), the analytes are separated from each other.

Proteins, triglycerides and lipids can be removed by solid phase extraction or other methods. So, an analyte will reach the MS detector at the same time, regardless of the concentration or ionic strength of the sample. That is, when doing research under good laboratory practices (GLP), methods based on GC-MS and/or LC-MS should specify the expected retention times of the analytes of interest, as well as the expected mass spectra. On the other hand, a GLP method for analyzing an aqueous, physiological sample, an NMR method should not specify that all the peaks due to analytes have the same chemical shifts in every sample. Instead, it should specify that the chemical shifts of the peaks due to the most abundant analytes (like creatinine) should be set to a standardized value. Alternatively, a concentrated buffer, such as 1M phosphate, can be added to keep the pH constant for each urine sample. So, when 640 μL of urine was mixed with 80 μL of 1M phosphate buffer, pH 7.0, the $-CH_3$ and $-CH_2-$ groups in creatinine produced chemical shifts of 3.05 and 4.07 ppm, respectively in the ^1H spectrum [11]. Since TMS is not soluble in water, zero ppm was set using a similar water-soluble compound, "3-(trimethylsilyl) propionic-(2,2,3,3-d_4) acid sodium salt (TSP)" [39].

Mass spectrometry (MS) is another very popular tool in metabolomics. It can be used without a column, as in flow injection analysis-MS (FIA-MS) [40] or in direct infusion MS (DIMS) [41]. "In FIA-MS, the sample is pumped directly into the MS through an atmospheric pressure ionization (API) source. Its advantages are that very little sample is required and the analysis time is very short. However, matrix effects can cause problems with quantitation when using many MS instruments. However, high resolution MS (HRMS) overcomes this problem, due to its superb ability to its ability to distinguish between ions with m/z that are within a ±0.5 Daltons of each other and its larger dynamic range (ability to detect analytes whose concentrations differ by several orders of magnitude). When combined with post-acquisition data filtering, FIA-HRMS can identify metabolites in biological fluids" [40]. On the other hand, DIMS uses a chip-based direct infusion nanoelectrospray ion source [41]. "Recently, the reproducibility of a multi-batch DIMS metabolomics study was established and a dataset created. It may serve as a benchmark for data processing and quality control for rapid, large-scale metabolomics studies" [41].

Of course, MS can also be coupled to CE, GC and/or LC. In CE-MS, uncharged metabolites are separated from those that are charged. The charged metabolites are separated from each other based on differences in their charges and size, which affects their migration time through a fused silica capillary [42]. CE-MS has been used to determine the identities and concentrations of metabolites in the saliva of patients with oral, breast and pancreatic cancers [43]. On the other hand, volatile uncharged metabolites and metabolites that can be derivatized to make them volatile can be analyzed by GC-MS [44]. "It is non-selective and can detect every steroid that is excreted, but it has been difficult in the past to present the data in a way that it can be interpreted easily. So, a simplified graphical representation of quantitative urinary steroid hormone profiles and diagnostic ratios was developed" [44].

It is also possible to do direct infusion MS (DIMS) from a silicon chip [45]. Aliquots (10 μL) of the supernatant can be placed on the back plane of the chip. An electrospray can be applied to the front of the chip, located near an atmospheric pressure ionization (API) interface [46]. The miniaturization reduces waste disposal and the consumption of solvents while eliminating sample carry over [46]. DIMS has been used to analyze drugs and free fatty acids in blood plasma [45, 46], as well as metabolites in muscles [47]. A simple, rapid high-throughput DIMS method was used to identify free fatty acid biomarkers in the serum of patients with pancreatic cancer [46]. It has also been used to screen biomarkers for lung cancer [48], colorectal cancer [49] and diabetes [50].

Samples can also be introduced into an MS by direct infusion, which is done without using a column to separate the analytes [51]. FIA-MS is especially useful for rapid analysis of many samples for targeted metabolites [51]. When looking for chlorine-containing xenobiotics and their metabolites, a data filtering method was used that was based on the accurate masses of the ^{35}Cl and ^{37}Cl isotopes [51]. Although many earlier studies used older mass spectrometers, high resolution spectrometers based on Fourier transform ion cyclotron resonance (FTICR) and/or Orbitrap are being used more these days [52 - 54]. They can separate many compounds and have large dynamic ranges [51]. One group reported being able to screen 10 000 samples per week [55]. Still, the lack of a column to separate the analytes is a limitation.

GC-MS is a very popular method for doing metabolomics studies, because GC has excellent separating power for analytes and electron impact (EI) ionization makes it an almost universal, nonselective detector with similar sensitivities for all volatile compounds that are stable at the high temperatures used [56]. A recent study of steroid hormone analysis used GC-MS and a new graphical method of data presentation of urinary steroid hormones and the diagnostic ratios of their concentrations [57].

Perhaps the most popular method for doing metabolomics studies is LC-MS [58, 59]. Even though LC does not produce peaks that are as sharp and well separated as GC, it is preferred when analyzing samples for ionic and polar metabolites, as well as compounds that are thermally labile. When using ion exchange columns to separate analytes, LC is often called ion chromatography, or IC [60]. It was used to analyze head and neck cancer cells for anionic metabolites with about 100-times the sensitivity of reversed-phase LC or hydrophilic interaction chromatography (HILIC) [60]. It found significant differences in glycolysis and tricarboxylic acid energy metabolism [60]. Reversed-phase LC columns packed with octadecylsilica (ODS or C_{18}) have been used more than others [60]. When used with a UV-Vis detector it is called HPLC, but when a MS detector is used, it is called LC-MS. When the particle size of the C_{18} (or any other column packing material) is below 2 μm, it is called ultra-performance liquid chromatography (UPLC) or UPLC-MS [59]. "The smaller particle sizes improve the resolution of the peaks in the chromatogram. There are also HILIC UPLC columns, as well as LC and UPLC columns that are packed with other materials. Two different columns can be used in two-dimensional LC-MS, in which part or all of the effluent off the first column is transferred to a second column. It works best when the two columns separate analytes based on very different mechanisms. These are called orthogonal columns" [59].

After doing the spectral analyses and finding putative biomarkers, it is important to do proper quality control [61]. "This includes making receiver operator characteristic (ROC) curves. That is, very few biomarkers can unequivocally diagnose a disease. There are almost always some false positives and false negatives" [61]. Moreover, "there is very little consistency and relatively little rigor in how researchers select, assess or report their candidate biomarkers. In particular, few studies report any measure of sensitivity, specificity, or provide receiver operator characteristic (ROC) curves with associated confidence intervals. Even fewer studies explicitly describe or release the biomarker model used to generate their ROC curves" [61].

The simplest, but most naïve way to determine the quality of biomarkers is to indicate the percentage of individuals who are correctly classified as having a disease or not [61]. "Unfortunately, this forces the analyst to predetermine the optimal criteria for the classification. There may be a difference between a mathematically and ethically optimal boundary. That is, for ebola virus disease, it might be essential to identify all people who have the disease, even though there might be some false positives. So, it is often better "to consider the frequency with which a test for biomarkers produces true positives (TP), true negatives (TN), false positives (FP) and false negatives (FN)" [61].

A handbook of metabolomics markers for many forms of cancer, neuro-degenerative diseases, cardiovascular diseases, neurological diseases, diseases of inborn errors of metabolism, osteoarthritis, celiac disease, Crohn's disease, ulcerative colitis, diabetes, interstitial cystitis and uremia was published recently [62]. Some of these markers are quite likely to be accepted as biomarkers for diagnosing diseases. This includes 3-hydroxybutyrate, N-acetyl-glycine, 3-hydroxy-2-methylbutanoic acid and nonanedioic acid for breast cancer; sarcosine for prostate cancer; desmosterol for Alzheimer's disease; and indoxyl sulphate, *meta*-hydroxyphenyl propionic acid and phenylacetyl glycine to differentiate between celiac and potencial celiac diseases [62].

Perhaps one of the most important uses of metabolomics is to help discover new drugs. A recent review article stated that, Metabolism represents the 'sharp end' of systems biology, because changes in metabolite concentrations are necessarily amplified relative to changes in the transcriptome, proteome and enzyme activities, which can be modulated by drugs [63]. Systems biology is an integral part of systems medicine [64, 65]. It is also an important part of preventive, personalized and participatory (P4) medicine [64, 65]. Models of human metabolic networks are needed to understand drug metabolism [63]. "One such model, Recon2, can predict the effects of inborn errors of metabolism, identify exometabolites and show how metabolism varies between tissues and cellular compartments. Small molecule transporters for drugs can also transport metabolites, so they are of special importance. Many pharmaceutical compounds enter or leave cells through such transporters. Also, some dietary compounds, such as naringen in grapefruit can be metabolized to produce compounds like naringenin that can alter cytochrome P450 metabolism of statins. So, it is anticipated that the emerging field of systems pharmacology and metabolomics may help to discover and develop new drugs" [63].

APPLICATIONS

As mentioned previously, personalized medicine has been used for over 100 years to ensure that people who need a blood transfusion get compatible blood types. It has also been used to match organ donors with recipients. Then in 1977, Cytochrome P450 2D6 (CYP450 2D6) was found to cause the effects of the anti-hypertensive drug debrisoquine to be exaggerated and last longer in some patients. People with different genotypes and phenotypes can metabolize drugs differently. Drug metabolizing enzymes can be divided into phase I and phase II [66]. "Phase I occurs before Phase II. The Phase I enzymes catalyze the transformation of functional groups by oxidation, reduction and hydrolysis. Phase I metabolism can either detoxify compounds, or make others more toxic. Phase I enzymes include several cytochrome P450s, alcohol dehydrogenase, aldehyde dehydrogenase, dihydropyrimidine dehydrogenase, NADPH:quinone oxidoreductase, esterases and epoxide hydrolases" [66]. Phase II enzymes detoxify compounds and make them more water-soluble, so they can be excreted. Phase II enzymes catalyze the addition of hydrophilic groups, such as glutathione, sulfate, histamine, catechol, and uridine triphosphate. Phase II enzymes include N-acetyl transferase, glutathione S-transferases, sulfotransferases, histamine methyltransferase, and catechol-O-methyl transferase.

We all have genes that code for CYPs, but there is much polymorphism in them [66]. "They catalyze the metabolism and biotransformation of drugs and other molecules. There are different CYPs, which catalyze different reactions that oxidize or reduce various substrates. For example, CYP2D6 catalyzes the metabolism of the cancer drug called tamoxifen. There are multiple forms, or isozymes, of CYP2D6 and other CYPs. Some people have a genetic variation that codes for an isozyme of CYP2D6 that can make them not respond properly to tamoxifen. Genetic tests are also done on patients suffering from other diseases. Tests for CYP genes and/or isozymes are done on people taking antidepressants, anticoagulants, and proton pump inhibitors. Some people metabolize these drugs faster than others, so doses need to accommodate this" [15].

For example, β-blockers, antipsychotics, codeine and many other drugs are metabolized at different rates by different CYP2D6 enzymes [66]. "The phase I metabolism of many other drugs are affected by different forms of CYP1A1, CYP1A2, CYP1B1, CYP2A6, CYP2B6, CYP2C9, CYP2C19, CYP2E1, CYP3A4, aldehyde dehydrogenase, alcohol dehydrogenase, dihydropyrimidine dehydrogenase, NADPH:quinone oxidoreductase, and plasma pseudocholinesterase. The phase II metabolism of many drugs are affected by different forms of glutathione S-transferases, histamine methyl transferase, N-acetyl transferases, sulfotransferases, catechol-O-methyl transferase and UTP transferases" [66].

Another enzyme, called thiopurine methyltransferase breaks down thiopurines that are used to treat some types of cancer. Some people don't have this enzyme and need to have their doses lowered, so that toxic levels of thiopurine drugs don't build up [67]. "Another enzyme, called UGT1A1, or UDP-glucuronosyltransferase, is responsible for breaking down another anticancer drug, Camptosar, which is used on colorectal cancer. Some people have a genetic variation that causes Camptosar to be broken down slowly, causing doctors to prescribe lower doses" [67].

Another application that has been around for decades is the development and use of of Herceptin® and Gleevec® for the personalized treatment of breast cancer. Herceptin targets breast cancer patients who have the type 2 form of receptor for the human epidermal growth factor, or HER2 receptor [67]. "Gleevec® is used to treat chronic myelogenous leukemia, a cancer that is caused by a definite chromosomal abnormality, called the Philadelphia chromosome. There is a reciprocal translocation between chromosomes 9 and 22. This produces a *bcr-abl* fusion gene that codes for an abnormal tyrosine kinase that is not properly regulated. Myeloid cells are a feature of most tissues. Myeloid cells have a wide array of biological activities that include immune activation, arteriogenesis, and regulation of salt balance and blood pressure. Myeloid cells also regulate vascularity. Tumor associated macrophages influence the growth of blood vessels in part because they are a source of vascular endothelial growth factor, or VEGF. Myeloid cells also promote angiogenic branching and anastamosis. Depending on the context, myeloid cells can be either anti-angiogenic or pro-angiogenic. A myeloid cell is a large cell of the bone marrow that is a precursor of the mature granulocytes of the blood" [67]. Gleevec binds to the active site and inhibits tyrosine kinases. It is indicated for the treatment of not just leukemia, but also gastrointestinal tumors and other types of cancer.

Still, there have been many advances in personalized medicine for treating and even curing breast cancer. There were about 232,340 new cases and 39,620 deaths in 2013 in the USA as well as 1,300,000 worldwide, with 450,000 deaths each year [68 - 69]. Despite unacceptable failure rates and adverse side effects, chemotherapy is still widely used [70]. "Chemotherapeutic drugs (doxorubicin, cisplatin, gemcitabine, bevacizumab and trastuzumab) are used to help manage breast cancer. However, there are important inter-patient differences in responses to these and other drugs. Even the blockbuster drugs (> $1billion dollar annually) are effective in only about half of the patients, while the other half get no benefits. This can be due to acquired multidrug resistance (MDR), DNA polymorphisms and inter-tumor heterogeneous subpopulations that respond differently to radiation, chemo- and targeted therapies in different patients with the same type of cancer" [70].

In 2014, breast cancer is said to consist of many subsets of genetically definable and/or molecularly distinct syndromes [70]. "They have different natural histories that require different treatments. There are estrogen receptor positive (ER+), human epidermal growth factor receptor2 (HER2; also called ERBB2 and HER2+) and triple-negative (ER^-, progesterone receptor negative, PR^-, and $HER2^-$). There are also six independent intrinsic molecular subtypes: normal-like, HER2-enriched (HER2E), luminal (A and B), basal A/basal-like and basal B/claudin-low. However, new methods may discover other subtypes. A statistical analysis called pathway-assisted clustering was used to analyze LC-MS data on plasma proteome samples from breast cancers. It found that cancers can change from one subtype to another. This can be seen by looking for changes in biological and signaling pathways. Similar subtypes share distinct yet similar pathway activation networks, while dissimilar subtypes use different networks. Moreover, the distance or similarity of cancer subtypes based on pathway analysis might elaborate the relationships between breast cancer subtypes" [70].

So, instead of just making the medicine specific for the DNA that a person is born with, it can be made specific for a type of cancer. This is being done by developing monoclonal antibodies, which will bind to receptors that are specific for a particular type of cancer. Antibodies are proteins that are made by immune cells, and will be discussed further in Chapter 7. Antibodies bind to receptors on the surface of cells that the body recognizes as being foreign and harmful. When a person is infected by a harmful virus or microorganism, memory cells will make an antibody against the invader.

Moreover, some B cells can be used to make therapeutic monoclonal antibodies. Most B lymphocytes produce a mixture of antibody proteins in response to a single toxic antigen. These are polyclonal antibodies. However, a nearly immortal cell line of myeloma B cells can be fused with antibody-producing spleen cells from a mouse. These hybridomas can produce large amounts of a single monoclonal antibody (mAb) because they can survive and grow in cell cultures. These mAbs can be directed against a specific site (epitope) on antigens that are produced by infectious viruses and bacteria. Some of them are even approved medications such as Herceptin®, and T-DM1® for breast cancer; Rituxan® for B-cell lymphomas; Campath® for leukemia and three others: Erbitux®, Vectibix® and Avastin® for colorectal cancer [15].

Avastin® targets the vascular endothelial growth factor (VEGF), which helps stimulate blood vessel formation in the growing tumor. Ranibizumab (Lucentis®) also targets VEGF. It was approved for treating macular degeneration, which occurs frequently in diabetic patients. Herceptin® targets the human epidermal growth factor receptor 2 (HER2), which is expressed at high levels in some breast tumors [67]. "These mAb are unusual in that they can target and sometimes even kill cells. Most can't kill cells by themselves, but they can still bind to cancer-specific antigens and deliver drugs that are covalently attached to the mAb. Thus, there is much research to bind drugs to mAbs, so that the drug only reaches the target organ. The success of this approach depends on not only the mAb, but also the type of drug that is bound to it, and the molecule that is used to link or connect them. One mAb drug delivery system in development has the class of drugs called maytanisoids bound to a mAb. Maytanisoids are natural products, derived from an Ethiopian shrub. Some of them contain disulfide (S-S) bonds that enable them to be covalently bound to mAbs. The first drug of this class to be isolated and tested was called Maytansine. It had excellent activity *in vitro*, but not *in vivo*, when it was administered without a mAb. So, chemists synthesized derivatives of maytansine and attached them to lysine groups on mAbs. One of the derivatives, called maytanisoid DM1, was coupled to a humanized mAb (hyC242), which binds to antigens on human colorectal, pancreatic, gastric and some lung cell cancers. The target cancer cells internalized the drug- mAb complex. The mAb was broken off (hydrolyzed) inside the cell, releasing the cytotoxic drug. The free drug was able to move back out of the cell and into neighboring cancer cells, killing them. Conjugates of DM1 are in various stages of clinical evaluation. The first one, called Cantuzumab mertansine or huC242-DM1, has been through Phase I evaluation in cancer patients. More clinical trials of other conjugates are on-going" [67].

Other drug-antibody conjugates are being developed [15]. One of them is a conjugate with a derivative of a natural bacterial product, called CC-1065. The derivative has a disulfide (S-S) bond and a phosphate group. The derivative is a prodrug called DC4. It is converted to the active drug in cells by an enzyme (phosphatase) catalyzed reaction. The active drug binds in a sequence-selective manner to the minor groove of DNA and alkylates adenine bases.

Another drug that has been bound to a mAb is a derivative of paclitaxel. Paclitaxel is a drug that stabilizes microtubules. It does this by stabilizing GDP-bound tubulin in the microtubule. This stops mitosis. Paclitaxel is not cytotoxic enough to be used without mAbs, but the derivative has shown promising results [67].

The antitumor antibiotic called calicheamicin has also been bound to mAbs [67].

A reactive group called a hydrazide was added to calicheamicin, so it could be bound covalently to a mAb. The drug- mAb conjugate called anti-CD3- -calicheamicin has been approved for use in treating acute myeloid leukemia [67].

So, recent progress in cancer immunotherapy has produced complete responses in patients who did not respond to standard therapies. However, due to tumor heterogeneity and inter-individual variations in anti-tumor immunity, only subsets of patients experience clinical benefit [44]. Each patient's immune network is genetically unique and subject to changes caused by epigenetics, nutrition, lifestyle, and exposure to environmental toxins. Part of its job is to detect tumor-associated antigens (TAMs), which can be either tumor-specific or self-antigens that are expressed on healthy and cancerous cells. Tumor antigens have been divided into five classes: tissue-differentiation antigens (Melan-A/MART-1, TRP-2), overexpressed antigens (survivin, telomerase), antigens derived from epigenetic changes (or cancer–testis antigens, MAGE-A3, NY-ESO-1), antigens derived from mutated genes (p53, RAS) and viral antigens (derived from human papillomaviruses or Epstein-Barr viruses). The melanoma associated antigen MAGE-A1 was identified by a genetic approach, which paved the way for the discovery of many antigens on a broad range of tumors. Most are self-antigens that are usually expressed on a wide variety of tumors of different histological origins [44].

The cancer–testis antigens (CTAs) may be ideal targets for immunotherapy because they are expressed in so many different types of cancer, while their expression on normal cells is restricted to germ line cells in the testis that are protected against immune responses [71]. "They are "foreign-like" antigens to which T cell tolerance is limited. They "elicit spontaneous humoral and cell-mediated immune responses in cancer patients. Also, CTA expression is switched on by demethylation of the promotor region of the gene that codes for CTA. This is important because there are some drugs that modulate the tumor epigenome. This includes azacitidine and decitabine, two DNA methyl transferase inhibitors, as well as the immunomodulatory drug lenalidomide. These drugs enhance the expression of CTA thus causing a more immunogenic tumor cell phenotype. Differentiation and overexpressed are also self-antigens. Differentiation antigens are expressed on both tumor cells and normal cells from the same tissue.

Examples include Melan-A/MART-1 and gp100, which are expressed in malignant melanoma cells as well as healthy melanocytes in the skin, eye and ear. However, immunotherapy towards differentiation antigens can cause toxicity due to inappropriate immune responses directed against normal cells" [71].

Most tumor-specific antigens are ideal targets for immunotherapy since they come from somatic mutations in the tumor cells and are patient-specific. Mutations that are specific to cancer cells can be identified by comparing the whole genome of cancer cells with those of the original cells [72]. "On the other hand, whole-exome sequencing (WES) is more affordable for larger sample sizes, since only 1% of the genome needs to be sequenced. WES found several mutations in cancer of the breast cancer, ovaries, kidneys, head and neck. Recently, mutated antigens that are recognized by anti-tumor T cells were identified by mining whole exome sequence data to identify the "mutated proteins that were expressed in patients' tumors. The mutated T cell epitopes that were identified using an MHC binding algorithm were synthesized and tested to see if they could be recognized by tumor infiltrating T-lymphocytes (TIL). Mutated antigens expressed on autologous tumor cells were identified as targets of three TIL that were associated with tumor regressions following adoptive transfer of the lymphocytes. This simplified approach, which avoids the need to generate and laboriously screen cDNA libraries from tumors, may become quite useful in identifying mutated T cell antigens that are expressed in many types of tumors" [72].

Viral antigens are also tumor-specific and it is relatively easy to prime T cells to be active against them [71]. For example, the human papilloma virus (HPV) produces two oncoproteins (E6 and E7) that are possible targets for cancer immunotherapy. There are two prophylactic HPV vaccines that are available that protect against cervical cancer and other diseases that are associated with HPV.

Once tumor antigens are identified, active and passive immunotherapies can be devised. Active immunotherapies induce a tumor-directed immune response by vaccinating patients with tumor antigens [71]. "Passive immunotherapies use cytokines, mAbs and/or adoptive T cells to stimulate the immune system. For example, the cytokine IL-2 is a growth factor for T-cells. It was approved by the FDA because it caused complete "cancer regression in 8% of the patients with metastatic renal cancer or melanoma. The mAbs tratuzumab and cetuximab have been discussed. Another type of mAb targets the immunosuppressive molecules CTLA-4 and PD-1. They block immunoinhibitory pathways, which may enhance normal endogenous anti-tumor responses. CTLA-4 down-regulates T cell activation. Ipilimumab is a human mAb that blocks CTLA-4, which enhances the proliferation and effector activity of T cells. The FDA approved it in 2011 for the treatment of patients with metastatic melanoma. Two other promising targets for antibody therapy are PD-1 and its ligand PD-L1. When PD-L1 binds to its PD-1 receptor on T cells, it inhibits the activation of T cells and induces apoptosis in them. PD-L1 is expressed on APCs and several solid tumors such as renal cell, colon, breast, ovarian and hematological cancers. Nivolumab, anti-PD-1 mAb, was evaluated in a phase I clinical trial on patients with metastatic renal cell carcinoma, melanoma and non-small cell lung carcinoma" [71].

Passive immunotherapy can also use adoptive T cell therapy (ACT), in which patients are treated with tumor-reactive T cells that are selected and expanded *ex vivo* [71]. "Both effector and memory T cells can induce complete remission in patients that are incurable by other anticancer treatments. Two major sources of T lymphocytes for ACT are the tumor itself and the peripheral blood of the patient. When a tumor is infiltrated by anti-tumor T cells, these cells can be isolated from the resected tumor mass. For example, TILs were isolated from 93 metastatic melanoma patients who had failed chemotherapy and other immunotherapies such as IL-2 and anti-CTLA-4 treatments. The T cells isolated from the resected tumor were first selected for anti-tumor reactivity and then expanded *in vitro*. Twenty patients (22%) achieved a complete regression with ongoing complete responses in 19 patients after 3 years. In spite of this success, a major limitation of TIL therapy is the absence of sufficient numbers of TILs in other types of cancer. To extend this type of therapy to other types of cancer, genes coding for the antigen-specific T cell receptor (TCR) can be transferred into lymphocytes that are isolated from the patient's peripheral blood. The T cell transduction is done with retro- or lentiviral vectors to make T cells that recognize tumor antigens and eradicate tumor cells that express them in many patients. "There were objective clinical responses in patients with metastatic synovial cell sarcoma and melanoma treated with TCRs towards NY-ESO-1 antigens. Two melanoma patients out of 11 were in complete regression that persisted for more than 1 year" [71].

Also, T cells can be modified by chimeric antigen receptors (CAR)s [73]. "In contrast to TCRs that are specific for a certain HLA-peptide complex, CARs can recognize antigens without depending on them. This may make CARs useful for patients with different HLA haplotypes, and increase the number of potential targets, including carbohydrates and glycolipids. Although one of the first successful CAR therapies with complete remissions was reported on a solid neuroblastoma, they may also be effective against hematological malignancies. The most investigated CAR target is CD19, which is expressed on normal B cells as well as on most B cell leukemia's and lymphomas. Several clinical studies on patients with B cell malignancies such as chronic lymphocytic leukemia caused partial and complete responses in a subset of patients. Also, two children with relapsed and refractory acute lymphoblastic leukemia were successfully treated with CARs that targeted CD19. One child was reported with an ongoing complete regression while the other one relapsed due to the growth of blast cells that no longer expressed CD19. Still, CD19 CARs were able to kill aggressive leukemia cells. So, using T cells with CARs against other targets in combination with CD19 CARs may be useful on tumors that downregulate the CD19 CAR target. Based on the success of CAR therapies in hematological malignancies, the expansion of this therapy to solid tumors may be possible. However, it will be more challenging due to the immunosuppressive tumor microenvironment that can hinder cell-based immunotherapy" [73].

The tumor microenvironment builds up barriers such as impeding T cell infiltration and inhibiting T cell stimulation [71]. "It can up-regulate suppressive pathways in immunosuppressive cells, such as myeloid-derived suppressor cells (MDSCs) and T_{regs}. MDSCs are a heterogeneous population of immune cells that promote tumor growth by their immunosuppressive roles, including suppressing the activation of anti-tumor T cells. Also, T_{regs} inhibit T cell anti-tumor functions. So, prior to adoptive T cell transfer of either TIL or TCR/CAR engineered T cells, they are eliminated by lymphodepleting regimens. Despite some clinical benefit there can be toxic side effects. One way to overcome this toxicity is to use genetic engineering on isolated anti-tumor T cells so that they can produce their own pro-inflammatory or pro-proliferative cytokines. Another option is the chemical conjugation of interleukin-loaded nanoparticles onto the cell membrane of adoptively transferred T cells. The sustained release of the interleukins from the stably attached particles provides the cells directly with these immunostimulatory agents" [71].

Although immunostimulating cytokines expand and activate anti-tumor T cells and lymphodepletion eliminates T_{regs}, several other immunosuppressive pathways may be able to impede cancer immunotherapy [71]. "Blocking these pathways with immunomodulating antibodies may be effective. These inhibitory pathways, also known as immune checkpoints, maintain self-tolerance while regulating the amplitude and duration of immune responses. Since several immunoinhibitory pathways are involved in tumor development, a combination of antibodies may enhance and prolong the anti-tumor effects. For example, a combination of CTLA-4 and PD-1 pathway inhibition caused a synergistic effect that further enhanced the activation of anti-tumor T cells and improved long-term survival rates compared to the inhibition of both pathways alone. Moreover, the tumor rejection was further improved by including a cancer vaccine" [71].

However, patients can still have different immune responses against cancer, based on mutations or polymorphisms in immune regulatory genes [71] or exposure to environmental toxins. So, researchers are looking for both prognostic and predictive biomarkers. Predictive biomarkers can be used to help make therapeutic decisions. For example, indoleamine 2,3-dioxygenase (IDO) and FoxP3 are two suggested biomarkers for patients being treated with ipilimumab. IDO is an immunosuppressive enzyme that is expressed by immune cells and cancer cells, while FoxP3 is a marker of T_{Regs}. It is a transcription factor that enables regulatory T cells to limit immune responses and favor self-tolerance, including in cancer.

The increased expression of both of both of them was associated with the positive clinical response to ipilimumab. In another study, the pre-existence of antibody responses and the presence of tumor antigen-specific T cells isolated from the peripheral blood of a patient were suggested as predictive biomarkers for treatment with ipilimumab because the presence of these biomarkers was highly correlated with the clinical benefit the anti-CTLA-4 treatment [71]. "Furthermore, the PD-L1 ligand was suggested as a potential predictive biomarker for the antibody therapy against the PD-1 pathway. In general, a T cell inflamed microenvironment can be a predictive biomarker for all cancer immunotherapies because patients that have T cells in the tumor site are more likely to benefit from additional immunotherapies such as blocking the PD-1 pathway. Therefore, the presence of TILs combined with the "expression of PD-L1 on the tumor cells is a better predictive biomarker for patients receiving PD-1 immune checkpoint blockers than PD-L1 expression alone. This indicates that a combination of predictive biomarkers is better for making clinical decisions" [71].

Another way to look for biomarkers is to study the cancer metabolome, as was done with colorectal cancer (CRC) [74]. "Most cases of CRC are sporadic and affected by lifestyle. Unfortunately, less than 40% of CRC patients are diagnosed at the localized stage, when the five-year survival rate is still very good. So, efforts are being made to use metabolomics, which reflects a patient's functional status and phenotype. Perturbations in biological pathways and changes in concentrations of metabolites will occur in cancer. Studying altered metabolism in the colon and rectum could improve diagnostic tests and help to discover ways to treat or even prevent CRC" [74].

So, one study analyzed metabolites in blood serum from 64 CRC patients and 65 healthy controls [74]. "A total of 33 differential metabolites were identified, and oleamide was the most depleted. Oleamide enhances the activity of certain types of serotonin receptors. Also, glycolysis, arginine, proline metabolism and fatty acid metabolisms were perturbed. A subsequent study also found pertubations in the tricarboxylic acid (TCA) cycle, urea cycle, fatty acids and gut flora metabolism. Moreover, the alteration in lipid (including fatty acid) metabolism was seen. The levels of several long-chain fatty acids (linoleic acid, elaidic acid, tetracosanoic acid, pentacosanoic acid, hexacosanoic acid and octacosanoic acid) were higher in CRC patients. It should be noted "that the proliferation of cancer cells is promoted by high fat diets. The increased levels of fatty acids might be related to the increased degradation of lysophosphatidylcholines (LysoPCs). The reduction of polyunsaturated ultra-long-chain fatty acids and the increase of long-chain fatty acids suggest the alteration of enzymes, such as fatty acid desaturases, which may be involved in the pathogenesis of CRC. The significance of one of the polyunsaturated ultra-long-chain fatty acids, GI tract acid-446, has been implicated in the diagnosis of CRC and in protecting the body against CRC. Gastrointestinal (GI) tract acid-446 (GTA-446) is one of several circulating long-chain fatty acid metabolites referred to as GTAs. They were first discovered by metabolomic profiling of serum from patients with CRC and asymptomatic controls. GTAs, and GTA-446 in particular, have been detected at reduced levels in CRC patient serum compared to disease-free subjects. The GTA family has over 20 members containing between 28 and 36 carbons and ranging in size between 446 and 596 atomic mass units (amus). Concentrations of GTA (including GTA-446) do not go back up in CRC patients after surgery, chemo or radiation therapy, and are not the result of tumor burden. So, the initial reduction is thought to precede tumor formation" [74].

The concentrations of tryptophan, tyrosine and phenylalanine metabolism, and of several aromatic metabolites, which are related to gut microbial host metabolism, fluctuated in the serum of CRC patients [74]. "One of them, 5 hydroxytryptamine, was the most significantly depleted metabolite. Another study identified elevated concentrations of 3 hydroxybutyrate, formate, acetate, glycerol, phenylalanine, proline, lipids and glycoproteins, and depleted concentrations of several potential precursors of glucose in gluconeogenesis, including lactate, pyruvate, alanine, citrate, creatine, tyrosine, valine, glutamine, and other gluconeogenic amino acids in the serum of CRC patients. The decreased levels of lactate, pyruvate, alanine, and glutamine suggest an increased uptake of glucose by the liver and possibly an increase in hepatic gluconeogenesis" [74].

Others used both a metabolomic and proteomic approach to define oncofetal biomarkers for CRC by analyzing serum samples from CRC patients, healthy control adults and fetuses [74]. "Valine, glycine, threonine, 1 deoxyglucose and ribitol decreased while 3 hydroxybutyric acid increased in the serum of CRC patients. Quantitative linkages were established between the five metabolites, 3 hydroxybutyric acid, valine, threonine, 1 deoxyglucose and glycine, and the five proteins, MACF1, APOH, A2M, IGL and VDB. This integrated approach may promote the use of biomarkers with clinical potential in clinical practice" [74].

Others have analyzed urine for metabolites [74]. "These studies observed a number of dysregulated metabolic pathways, such as the TCA cycle, glycolysis, urea cycle, histidine metabolism, pyrimidine metabolism, tryptophan metabolism and polyamine metabolism. Moreover, the levels of several metabolites derived from gut microbial metabolism were altered. The concentrations of 5 hydroxyindoleacetate, 2 hydroxyhippurate, phenylacetate, phenylacetylglutamine and p-hydroxyphenylacetate increased, while phenol, hippurate, indoleacetate, indole, tyrosine, 4 aminohippurate and trimethylamine N oxide decreased. This is consistent with the altered community of gut microbiota seen in a previous metagenomic study on CRC patients" [74].

Others also study fecal water, a cheap, reproducible and effective method for detecting CRC markers [74]. "A reduced concentration of short-chain fatty acids was the most significant marker in it. These compounds (mainly acetate, propionate and butyrate) are associated with the fermentation of complex carbohydrates by the gut bacteria. Of these, butyrate is the most important in the development, treatment and prevention of CRC. That is, butyrate influences gene expression, reduces cell proliferation and induces both differentiation and apoptosis. Butyrate and acetate may also appear inhibit oxidative damage to DNA. The human fecal water metabolomic study supports the finding that short-chain fatty acids may help to keep CRC from developing" [74].

Metabolomics can also be done on tumors. Amino acids, inorganic acids, aldehydes, fatty acids, fatty acid esters and polyols have been found in tumors [74]. "Most of these metabolites are associated with factors influencing tumorigenesis, such as increased tissue hypoxia, altered glycolysis, lipid/steroid metabolism, and inflammation. Levels of the metabolites involved in the pentose phosphate pathway, as well as the TCA and urea cycles were also altered in CRC tumors. Among these metabolites, the concentrations of lactate, taurine and glycine increased, while those of glucose and myoinositol decreased. Lactate and glycine concentrations were higher in CRC patients, while glucose and most other monosaccharides were lower. This observation was consistent with the Warburg effect. That is, the rate of glucose uptake and its conversion to lactate increased, along with glycolysis within the CRC, when compared to normal tissue. This shows that tumor hypoxia is caused, not by the excessive oxygen consumption of cancer cells, but by an inadequate blood supply caused by angiogenesis" [74].

Hopefully, tumor markers with higher specificity and sensitivity for early diagnosis will significantly improve the clinical outcome of CRC patients [74]. "Seven metabolites, citrate, hippurate, p-cresol, 2-aminobutyrate, myristate, putrescine and kynurenate, may be good biomarkers for diagnosing CRC early. Also, another study claimed to have measured the serum levels of fatty acids in patients with CRC. As happens quite often in papers published by food chemists, the authors did not really measure the fatty acid content, but the fatty acyl content of triglycerides. They claimed that the concentrations of nine fatty acids fluctuated (increased levels of C14:0, C15:0, C18:3 n 6 and C18:0, and decreased levels of C24:0, C25:0, C26:0, C28:0 and C30:0) in CRC patients at a very early clinical stage. They claimed that this showed that profiling serum fatty acids can be a useful way to diagnose CRC early. However, they used GC-MS to analyze fatty acid methyl esters (FAMEs) that were produced after hydrolyzing fatty acyls off of triglycerides. If there actually were fatty acids in the blood serum, they would have also been converted to FAMEs" [74]. It is possible that there were some fatty acids in the blood serum, but GC-MS analysis can't tell if there are.

So, if another group were to try to duplicate their results using NMR or LC-MS, they would find important differences. NMR can easily distinguish between fatty acids and fatty acyls that are part of mono-, di- and/or triglycerides. LC-MS can also separate and detect fatty acids, but mono-, di- and triglycerides will bind tightly to columns used in LC-MS. They would have to be removed by solid phase extraction before the blood sample could be analyzed. So, unfortunately, the effort to find biomarkers for CRC or any other disease will be hindered by the poor choice of nomenclature that appears so often in peer reviewed articles. Fatty acids and fatty acyls that are part of triglycerides have very different physical, chemical and biological properties.

Cancer is not the only disease with genetic components. Another such case is Parkinson's disease (PD), a neurodegenerative disease. Sporadic idiopathic PD (OMIM entry #168600) is a complex multifactorial disease that can be caused by a combination of genetic and environmental factors, including diet [75, 76]. PD is characterized by several motor (bradykinesia, rigidity, tremor, postural instability) and non-motor symptoms (hyposmia, autonomic dysfunction, REM sleep behavior disorder, depression, cognitive decline and others) [76]. "Dopaminergic neurons in the substantia nigra (SN) that conatin neuromelanin degenerate. Aggregates called Lewy bodies (LB) that contain proteins and lipids form in the cytoplasmic. However, the genetic contributions to PD are no longer seen only in a mono-causal or Mendelian way. The monogenic forms of PD make up only a small fraction of the cases. Instead, low-frequency variants often have some effect" [76]. Moreover, dietary toxins such as poorly liganded iron and neurotoxic acetogenins [77, 78] can be important causes of PD. So, a systems biology approach is important when studying PD [76].

Still, mutations in either of two genes cause autosomal dominant forms of PD [75]. "They are the α-synuclein (*SNCA*) and leucine-rich repeat kinase 2 (*LRRK2*) genes. The neurotoxic LBs contain α-synuclein. It is a cytoplasmic lipid binding protein that is involved in storing and regulating neurotransmitters, as well as in recycling synaptic vesicles. The *LRRK2* gene has 51 exons that code for a very large protein, called lrrk2 (or dardarin). It contains two domains with enzyme activities (GTPase and kinase) and several protein-protein interaction domains" [75]. Many variants have been identified in this gene in PD patients, but only six of these (Asn1437His, Arg1441Cys, Arg1441Gly, Tyr1699Cys, Gly2019Ser, and Ile2020Thr) have been shown to cause PD [76]. "Moreover, homozygous or compound heterozygous mutations in each of the following three genes: *PRKN* (parkin, *PARK2*), *PINK1* (*PARK6*), and *DJ-1* (*PARK7*) cause autosomal recessive forms of PD, usually with no atypical clinical features. One the other hand, mutations in any of these three genes: *ATP13A2* (*PARK9*), *PLA2G6* (*PARK14*), *FBXO7* (*PARK15*), can cause rarer forms of recessive parkinsonism, usually with very early-onset (<30 years) and atypical clinical features (pyramidal, dystonic, ocular movement, and cognitive disturbances). Mutations in parkin are the most frequent. They account for about half of the familial PD cases due to recessive inheritance and disease onset before the age of 45 years, and also ~15% of the sporadic cases with onset before 45. Mutations in PINK1 and PARK7 are less frequent, accounting for about 1–8%, and 1–2%, respectively, of the sporadic cases with early-onset" [75].

Also, several genome-wide studies have been done [75]. "There are data on more than 12,000 cases and 20,000 controls that provided evidence for 16 independent risk loci. Each individual allele presents only a small risk or protective factors, conferring 1.1 to 1.4 fold increases and 0.95 to 0.7 fold decreases in risk respectively" [75]. This data was based on voluntary participation of patients, who provided saliva or blood samples for genetic analysis.

One important question is whether a person will really want to know if they are susceptible to a serious disease, such as cancer. When a person finds out that they are susceptible to a disease, it can increase the chance that he or she will get the disease. This is a real problem with DNA testing for diseases with a known genetic component. It is very helpful to research if relatives of patients with genetic diseases allow themselves to be tested for the presence of a defective gene. The scientific information will help researchers develop a cure, but it can be devastating (physically and psychologically) to find out that you have a genetic defect that will probably cause a disease in yourself and may be passed on to one's children.

Fortunately, the popular media can be of much assistance. In 1995, scientists were on the Larry King show, asking for volunteers who had a history of prostate cancer to allow their genes to be analyzed. The response was wonderful. About 250 families were recruited from around the world, and their genes were mapped [79].

Cystic fibrosis is another disease that might be able to be cured by personalized medicine [3]. "In 2012, the FDA approved a new therapy for cystic fibrosis (CF), a serious inherited disease that impairs the lungs and digestive system. The drug, Kalydeco™ (known generically as ivacaftor), was approved for patients with a specific genetic mutation (glycine to aspartic acid on amino acid 551, or G551D). It is in a gene that codes for the cystic fibrosis transmembrane conductance regulator (CFTR) protein. It is important for regulating the transport of NaCl and water in the body. There are hundreds of known mutations that can lead to CF. The G551D mutation causes only 4% of cases in the USA (about 1200 people). In these patients, Kalydeco helps to restore the function of the CFTR protein that is coded for by the mutated gene. It binds to this CFTR chloride channel to induce an unconventional form of gating, which keeps the channel open. It allows a proper flow of salt and water on the surface of the lungs and helps prevent the buildup of sticky mucus that can lead to life-threatening lung infections and digestive problems in patients with CF" [3].

Kalydeco® was the first drug to address the underlying cause – rather than the symptoms – of CF [3]. "The skillful application of genomic science helped researchers understand at a molecular level the reasons why the CFTR protein fails to function. This helped them develop a medicine to improve its function, and to use the results of a genetic test to select the right patients for the drug. While it is too soon to say whether Kalydeco® will be a complete cure for those eligible to receive it, patients are experiencing significantly improved lung function and weight gain. It is also important to note that the path of development that ultimately led to the approval of Kalydeco® driven by patients. The drug emerged from a decade-long collaboration between the Cystic Fibrosis Foundation (CFF) and the drug's manufacturer, Vertex Pharmaceuticals. The Foundation had been at work several decades previously, organizing and advocating on behalf of the patient community. They funded research that led to the discovery of the *CFTR* gene in 1989. They built an extensive patient registry and clinical trial network for studying the genetics of the disease. Starting in the late 1990s, CFF funded a major drug screening effort that led to the discovery of the compound and invested a total of $75 million toward the development of the drug. Finally, the FDA approved Kalydeco® in a relatively short time. Elegant science and a well-designed program allowed the agency to apply several mechanisms to streamline and expedite the review process. First, the drug application was granted "priority review," a designation that is given to candidate drugs that offer major advances in treatment or provide a treatment where no adequate therapy exists. The goal for completing a priority review is six months, but the well-prepared submission, strong evidence, and a commitment by all of the parties involved enabled the review to be completed, and the drug approved, in half that time" [3].

Personalized medicine is also important in pediatrics [80]. "Genetic tests are available before conception that can detect mutations that are associated with over 150 genetic disorders [80]. "There are *in vitro* fertilization techniques to help couples have a baby. If the mother has a mitochondrial disease, the nucleus from her fertilized egg can be transplanted into the enucleated embryo of another woman. The embryo has functioning mitochondria and can be transplanted back into the mother. After conception, prenatal tests of the mother's blood can diagnose many genetic diseases. Still, many childhood diseases have multiple causes, but still need personalized medicine. For example, children with type-1 diabetes must monitor their consumption of carbohydrates and monitor their blood glucose concentrations. Unlike adult-onset type-2 diabetes, which can be caused by decades of over-consumption of sugars, fats and calories, type-1 diabetes is not. Still, both types of diabetes require P4 medicine. It must be predictive, preventive, personalized and participatory. Unfortunately, type-1 diabetes is not so predictable or preventable, but type-2 can be prevented in many cases by adjusting one's diet and lifestyle. Both are personalized and participatory as each patient or their caregiver must provide medicine, control their lifestyle and diet, and monitor their blood glucose. At the same time, research continues to try to find ways to prevent and even cure both types" [80].

For example, it was discovered recently that alterations in the metabolism of branched-chain amino acids is linked to impaired fasting glucose levels (IFG) [81]. That is, an unbiased metabolomics study identified not only glucose, but also the branched chain ketoacid 3-methyl-2-oxovalerate was associated with IFG, which is important for the estimated one out of three adults in the USA who are pre-diabetic [81]. Also, it was found that different medications for diabetes have different effects on the metabolism of branched-chain amino acids (BCAAs) [55]. That is, the concentrations of BCAAs and aromatic amino acids (AAA) are correlated with insulin resistance and progression toward type-2 diabetes. Moreover, after treatment with metformin, the BCAAs and AAAs that are formed changed in a way that depended on the insulin sensitivity of the patent [82].

An important part of this research is to do genetic screens and DNA sequencing of as many people as possible, but with special emphasis on newborns. A universal screening panel of tests was recommended by the Secretary's Advisory Committee for Heritable Diseases of Newborns and Children includes over 50 conditions, of which most are heritable [80]. So far, newborn screening assays determine the concentrations of metabolites and proteins or assay for enzyme activities. However, it has been proposed that next-generation sequencing should be used. The goal is to increase the number of conditions that can be detected before symptoms begin. Still, some remarkable progress has already been made

[80, 83]. "For example, a patient was diagnosed in 2010 with a progressive, inflammatory bowel disease–like condition that required repeated abdominal surgery and severe, progressive disability. After sequencing his and his parents' genomes, a defect in the gene coding for the protein called XIAP (X-linked inhibitor of apoptosis protein) was found. After this discovery, a bone marrow stem cell transplant resulted in cure. It may also be important to look at the epigenome. For example, inappropriate DNA methylation can cause inappropriate gene expression that leads to Prader-Willi syndrome. That is, the pattern of methylation of cytosine at a single genomic locus determines which parental allele is expressed, which determines whether the disease occurs. Moreover, changes in DNA methylation can play an important role in many types of cancer, so ways to effectively drug the epigenome are being developed [83].

Massive parallel sequencing (MPS) is also quite useful. It was used to study glycogen storage diseases (GSDs) [25]. A MPS gene panel for known GSDs was developed by capture-based enrichment followed by MPS. It was used to facilitate the molecular diagnosis of patients with suspected GSDs. For a cohort of 17 patients who were suspected of GSDs, but without clear molecular diagnosis, definite diagnoses were made for 65% of them. Notably, GSD type Ib can usually be distinguished from GSD type Ia. So, patients who exhibit hepatic dysfunction can be distinguished from those who primarily have exercise intolerance, muscle weakness and cramps. Due to overlapping symptoms between these two subtypes, undiagnosed or misdiagnosed GSD type Ib patients were found to be rather common [25].

MPS can also be used to diagnose mitochondrial disorders, which have many different phenotypes [25]. Good outcomes for patients with mitochondrial disorders and their family members depend on prompt, accurate and early diagnosis. So, a one-step approach can be used clinically. It uses one long-range PCR to enrich the entire mitochondrial DNA (mtDNA), followed by MPS. This approach can avoid the pitfalls of the Sanger sequencing method. For example, the contamination from nuclear homologs of mtDNA sequences can lead to misdiagnosis of a patient. The comprehensive nature of this assay has led it to becoming the first-line test for the analysis of disorders related to mtDNA. Still, defects in the nuclear genome are responsible for most suspected mitochondrial diseases. Naturally, this can complicate the molecular diagnosis even more. To overcome these issues, WES and WGS are becoming valuable tools when trying to discover new genes [25].

In a completely different approach, mitochondrial diseases may be avoidable using a new *in vitro* fertilization technique, in which the nucleus of a fertilized embryo is transplanted into a donor embryo from another woman after having the nucleus removed. Mitochondria are inherited from the mother, not the father. The embryo will contain nuclear DNA from both the biological mother and father. It will grow in the mother, who will give birth to her own child, who will have healthy mitochondria. This can prevent mitochondrial diseases, such as muscular dystrophy. The procedure has been approved in Great Britain and is being considered in the USA.

There is also much heterogeneity in nonsyndromic intellectual disabilities [25]. More than 100 genes on the X chromosome have been found to be responsible for X-linked intellectual disabilities [25]. A multiplex PCR amplicon-based approach was tested in a pilot project for a set of 86 genes [84]. It had a detection rate of 33% in a cohort of patients being tested. It was robust and cost effective for clinical use [84].

Preventive, personalized medicine can also play a role in treating other neurodegenerative diseases. Recent research has discovered a key protein, activity dependent neuroprotective protein (ADNP) [85]. “It is essential for brain formation and regulating >400 important genes, including the one that codes for apolipoprotein E. When mutated, it is a major risk factor for Alzheimer’s disease. ADNP is a part of complex that is essential for proper gene expression. In men, the ADNP gene is mutated in autism, its expression is dysregulated in schizophrenia and significantly reduced in the blood of patients with “neurodegenerative diseases such as Alzheimer’s disease and multiple sclerosis. New results indicate that monitoring the concentrations of ADNP can be helpful in monitoring the progression of the disease, and may even provide ADNP replacement therapies. So, it is noteworthy that the ADNP peptide derivative NAP (davunetide) can protect against axonal transport deficits in several animal models of neurodegeneration as well as in a recent clinical trial. So, it is important to further investigate the use of ADNP as a predictive marker and ADNP replacement as a personalized therapy to prevent and protect against neuro-degenerative diseases" [85].

So, science continues to advance. An interesting example of such advancement is that tobacco plants can be genetically engineered to produce patient-specific antibodies against follicle B cell lymphoma, a subtype of non-Hodgkin's lymphoma [86]. "Vaccines were administered by subcutaneous injection of a fragment of the variable chain of tumor-specific immunoglobulin antigen that was obtained from biopsies of patients' tumors. The antigen was incorporated into a plant virus vector that took over the tobacco plant's metabolism and induced it to make many copies of the antigen in a relatively short time" [86].

One measure of the advancement in P4 medicine is that about 2% of the population of the USA uses or have used genetic tests [87]. "There are about 1000-1300 tests available for about 2500 conditions and new tests are emerging rapidly. A survey of healthcare claims data for 32 million people in the USA from 2008–2011 was done. They found that thecost grew by 14% per year between 2008 and 2011, primarily resulting from increased utilization. They predicted that genetic testing and molecular diagnostics use will continue to grow in the next 5 years. Moreover, by strengthening the ability to collect and analyze data in P4 medicine we will promote positive changes that benefit patients [87].

In 2008, it was hoped that medicine would be P4; that is it will be predictive, personalized, preventive and participatory [79]. By 2014, a convergence of three megatrends is driving P4 medicine [1]. "First, systems biology and systems medicine can better understand the complexities of human diseases, Second, the digital revolution has greatly improved our ability to collect, integrate, store, analyze and communicate data and information. This includes medical histories, results of clinical tests and mobile applications to inform patients and help personalize their care. Third, consumers have an amazing access to information that stimulates their interest in managing their own health care. Patients and their care givers are driving this transformation of health care" [1].

Within the next ten years, individual genomes will probably be available and the data will be able to predict the individual's risk factors and future health [9]. "Scientists are working to develop hand-held devices that will prick the finger and quantify about 2500 organ-specific proteins from all the approximately 50 human organs. The data will be sent by wireless communication to a file server, which will analyze the data and email a report to the physician and patient. Hopefully, this will enable a rapid evaluation of current health. "Also, our genomes and metabolomes will enable a predictive and personalized medicine. Physicians will learn how to use drugs to re-engineer metabolic networks that have been perturbed by the beginnings of a disease, before symptoms can emerge. Moreover, education will put patients in the position to take more responsibility for charting their own future health choices" [9].

Eventually, we hope to have technologies that will enable people to take some saliva or a drop of blood from their fingertips, analyze it at home and send it to a central computer. The data will be analyzed and compared to the individual's genome, epigenome, transcriptome, proteome and metabolome. By using systems thinking, biomarkers for diseases will be identified before the diseases are manifested in the phenotype. Instead of reacting to diseases after symptoms appear, we will be proactive in preventing them.

KEY POINTS

1. **Perhaps nothing epitomizes the fusion of traditional and western medicine more than predictive, preventive, personalized and participatory (P4) medicine. Personalized medicine is when different medicines are prescribed for people with different types of nutrition, environment, genes, mRNA, epigenetics and/or proteins. Such medicine is designed for the specific genetic, epigenetic and environmental properties of the individual and their diseased cells.**

2. **Different drugs and medical devices can be prescribed for people with different types of nutrition, environment, genes, mRNA, miRNA, epigenetics and/or proteins.**

ABBREVIATIONS

DNA = Deoxyribonucleic Acid;

cDNA = Complementary DNA;

RNA	=	Ribonucleic Acid;
P4	=	Medicine Predictive, Preventive, Personalized and Participatory Medicine;
FDA	=	Food and Drug Administration;
G6PD	=	Glucose-6-Phosphate Dehydrogenase;
VYP450	=	Cytochrome P450;
AUC	=	Area Under the Curve;
ICU	=	Intensive Care Unit;
EEG	=	Electroencephalography;
EKG	=	Electrocardiography;
HER-2	=	Human Epidermal Growth Factor Receptor-2;
iPSCs	=	Induced Pluripotent Stem Cells;
MRI	=	Magnetic Resonance Imaging;
CT	=	Computed Tomography;
ENCODE	=	Encyclopedia of DNA Elements;
VGDS	=	Voluntary Genomic Data Submission;
CBER	=	Center for Biologics Evaluation and Research;
WGS	=	Whole Genome Sequencing;
HIVE	=	High Performance Integrated Virtual Environment;
HLA	=	Human Leukocyte Antigen;
GETS	=	Genomics Evaluations Team for Safety;
OVRR	=	Office of Vaccines Research and Review;
NCTR	=	National Center for Toxicological Research;
KRAS	=	K Rat Sarcoma Oncogene;
CDC	=	Centers for Disease Control;
MMR	=	Measles, Mumps and Rubella;
ICBI	=	Innovation Center for Biomedical Informatics;
NCI	=	National Cancer Institute;
the NIGMS	=	National Institute of General Medical Sciences;
CDRH	=	Center for Devices and Radiological Health;
DPNM	=	Division of Personalized Nutrition and Medicine;
NTP	=	National Toxicology Program;
BRCA1 and BRCA 2	=	Breast Cancer Genes 1 and 2;
NME	=	New Molecular Entity, HIV Human Immunodeficiency Virus;
NIH	=	National Institutes of Health;

PET	=	Positron Emission Tomography;
GC	=	Gas Chromatography, MS Mass Spectrometry;
ICP-AES	=	Inductively Coupled Plasma Atomic Emission;
NMR	=	Nuclear Magnetic Resonance;
CGCI	=	Cancer Genome Characterization Initiative;
CGAP	=	Cancer Genome Anatomy Project;
ICG	=	Initiative for Chemical Genetics;
CGMS	=	Cancer Genetic Markers of Susceptibility;
GWAS	=	Genome-Wide Association Sudy;
ORF	=	Open Reading Frame;
MGC	=	Mammalian Gene Collection;
SOP	=	Standard Operating Procedure;
SCNP	=	Single Cell Network Profiling;
MCL	=	Mast Cell Leukemia;
EGFR	=	Epidermal Growth Factor Receptor;
NSCLC	=	Non-Small Cell Lung Cancer;
RPCI	=	Roswell Park Cancer Institute;
PCR	=	Polymerase Chain Reaction;
SNPs	=	Single Nucleotide Polymorphisms;
DMD	=	Duchenne Muscular Dystrophy;
MLPA	=	Multiplex Ligation Dependent Probe Amplification;
ChIP	=	Chromatin Immunoprecipitation;
DamID DNA a	=	Denine Methyltransferase Identification;
TOF-MS	=	Time-of-Flight Mass Spectrometry;
MALDI	=	Matrix Assisted Laser Desorption and Ionization;
NGS	=	Next-Generation Sequencing;
dNTP	=	Deoxynucleotide Triphosphate;
PTP	=	Pico Titer Plate;
CMOS	=	Complementary Metal-Oxide Semiconductor;
PGM	=	Personal Genome Machine;
DMSO	=	Dimethylsulfoxide;
ddPCR	=	Droplet Digital PCR;
TDM	=	Therapeutic Drug Monitoring;
TMA	=	Tissue Microarray;
NMR	=	Nuclear Magnetic Resonance;

MS	=	Mass Spectrometry;
high	=	Resolution MS;
ultra-performance	=	Liquid Chromatography UPLC;
LC-MS	=	Liquid Chromatography–MS;
GC-MS	=	Gas Chromatography-MS;
QTL	=	Quantitative Trait Locus;
mQTL	=	Metabolic Quantitative Trait Locus;
mGWAS	=	Metabolomic Genome-Wide Associated Studies;
UV-Vis	=	UV-Visible;
IR	=	Infrared; MHz MegaHerz;
GLP	=	Good Laboratory Practices;
FIA-MS	=	Flow Injection Analysis-MS;
DIMS	=	Direct Infusion MS;
API	=	Atmospheric Pressure Ionization;
m/z	=	Mass to Charge Ratio;
CE	=	Capillary Electrophoresis;
FTICR	=	Fourier Transform Ion Cyclotron Resonance;
ROC	=	Curves Receiver Operator Characteristic Curves;
TP	=	True Positives;
TN	=	True Negatives;
FP	=	False Positives;
FN	=	False Negatives;
NADPH	=	Nicotinamide Adenine Dinucleotide Phosphate Reduces;
UTP	=	Uridine Triphosphate;
MDR	=	Multidrug Resistance;
ER+	=	Estrogen Receptor Positive;
mAb	=	Monoclonal Antibody;
VEGF	=	Vascular Endothelial Growth Factor;
GDP	=	Guanosine Diphosphate;
CTA	=	Cancer–Testis Antigens;
WES	=	Whole-Exome Sequencing;
TIL	=	Tumor Infiltrating T-Lymphocytes;
MHC	=	Multiple Histocompatibility Complex;
HPV	=	Human Papilloma Virus;
ACT	=	Adoptive T Cell Therapy;

CTLA4	=	Cytotoxic T-Lymphocyte-Associated Protein 4;
TCR	=	T Cell Receptor;
CAR	=	Chimeric Antigen Receptor;
MDSCs	=	Myeloid-Derived Suppressor Cells;
Tregs	=	Regulatory T Cells;
IDO	=	Indoleamine 2,3-Dioxygenase;
CRC	=	Colorectal Cancer;
TCA	=	Cycle Tricarboxylic Acid Cycle;
LysoPC	=	Lysophosphatidylcholine;
GI	=	Tract Gastrointestinal Tract;
PD	=	Parkinson's Disease;
REM	=	Rapid Eye Movement;
GTPase	=	GTP Phosphatase;
CFTR	=	Cystic Fibrosis Transmembrane Conductance Regulator;
CFF	=	Cystic Fibrosis Foundation;
IGF	=	Impaired Fasting Glucose;
BCAAs	=	Branched-Chain Amino Acids;
AAA	=	Aromatic Amino Acid;
XIAP	=	X-Linked Inhibitor of Apoptosis Protein;
MPS	=	Massive Parallel Sequencing;
GSD	=	Glycogen Storage Disease;
ADNP	=	Activity Fependent Neuroprotective Protein

REFERENCES

[1] Flores M, Glusman G, Brogaard K, Price ND, Hood L. P4 medicine: how systems medicine will transform the healthcare sector and society. Per Med 2013; 10(6): 565-76. [http://dx.doi.org/10.2217/pme.13.57] [PMID: 25342952]

[2] Daak-Hirsch S, Campbell CA. The role of patient engagement in personalized healthcare. Per Med 2014; 11: 1-4. [http://dx.doi.org/10.2217/pme.13.102]

[3] Hamburg MA. Paving the Way for Personalized Medicine: FDA's Role in a New Era of Medical Product Development. Bethesda, MD: US FDA 2013.

[4] Perou CM. Molecular stratification of triple-negative breast cancers. Oncologist 2010; 15 (Suppl. 5): 39-48. [http://dx.doi.org/10.1634/theoncologist.2010-S5-39] [PMID: 21138954]

[5] Wright JL, Lin DW. Genetics: Prostate cancer risk stratification by genotype and PSA. Nat Rev Urol

2009; 6(12): 641-2.
[http://dx.doi.org/10.1038/nrurol.2009.229] [PMID: 19956193]

[6] Tamboli CP, Doman DB, Patel A. Current and future role of biomarkers in Crohn's disease risk assessment and treatment. Clin Exp Gastroenterol 2011; 4: 127-40.
[http://dx.doi.org/10.2147/CEG.S18187] [PMID: 21753895]

[7] US FDA. Personalized Nutrition and Medicine http://www.fda.gov/. , [Accessed Feb 20, 2014];

[8] Woodcock J, Woosley R. The FDA critical path initiative and its influence on new drug development. Annu Rev Med 2008; 59: 1-12.
[http://dx.doi.org/10.1146/annurev.med.59.090506.155819] [PMID: 18186700]

[9] Carlson B. What the devil is personalized medicine? Biotechnol Healthc 2008; 5(1): 17-9.
[PMID: 22478696]

[10] National Cancer Institute. Office of Cancer Genomics http://ocg.cancer.gov/. , [Accessed Feb 22, 2014];

[11] Haga SB. 100k Genome Project: Sequencing and much more. Per Med 2013; 10: 761-4.
[http://dx.doi.org/10.2217/pme.13.80]

[12] Farhangfar C. The Sheik Khalifa Bin Zayed Al Nahyan Institute for Personalized Cancer Therapy http://www.mdanderson.org/education-and-research/research-at-md-anderson/personalized-advanced-therapy/ institute-for-personalized-cancer-therapy/ipct-2pager-industry-final.pdf. , [Accessed Feb 22, 2014];

[13] Annual Meeting AA. www.businesswire.com/news/home/20130411005344/en/Nodality-Announces-Data-Presentations-AACR-2013-Annual#.UwuW0CLnbIU. 2013.

[14] Care PH. Center for Personalized Genetic Medicine. http://pcpgm.partners.org/.

[15] Clinic M. Personalized Medicine http://mayoresearch.mayo.edu/center-for-individualized-medicine/. 2014.

[16] Roswell Park Cancer Institute. http://www.roswellpark.edu/research/center-personalized-medicine/about. 2014.

[17] Duke Center for Personalized and Precision Medicine. http://dukepersonalizedmedicine.org/.

[18] Stanford Center for Genomics and Personalized Medicine. http://scgpm.stanford.edu/.

[19] Penn Medicine's Center for Personal Diagnosis. http://www.pennmedicine.org/personalized-diagnostics/.

[20] Nouri N, Fazel-Najafabadi E, Salehi M, *et al.* Evaluation of multiplex ligation-dependent probe amplification analysis *versus* multiplex polymerase chain reaction assays in the detection of dystrophin gene rearrangements in an Iranian population subset. Adv Biomed Res 2014; 3: 72.
[http://dx.doi.org/10.4103/2277-9175.125862] [PMID: 24627880]

[21] Liu RH, Yang J, Lenigk R, Bonanno J, Grodzinski P. Self-contained, fully integrated biochip for sample preparation, polymerase chain reaction amplification, and DNA microarray detection. Anal Chem 2004; 76(7): 1824-31.
[http://dx.doi.org/10.1021/ac0353029] [PMID: 15053639]

[22] US National Institutes of Health. http://www.ncbi.nlm.nih.gov/projects/SNP/snp_sum_mary.cgi.

[23] Steemers FJ, Gunderson KL. Whole genome genotyping technologies on the BeadArray platform. Biotechnol J 2007; 2(1): 41-9.
[http://dx.doi.org/10.1002/biot.200600213] [PMID: 17225249]

[24] Stepanov VA, Trifonova EA. Multiplex SNP genotyping by MALDI-TOF mass spectrometry: Frequencies of 56 immune response gene SNPs in human populations. Mol Biol 2013; 47: 852-62.
[http://dx.doi.org/10.1134/S0026893313060149]

[25] Koboldt DC, Steinberg KM, Larson DE, Wilson RK, Mardis ER. The next-generation sequencing revolution and its impact on genomics. Cell 2013; 155(1): 27-38.
[http://dx.doi.org/10.1016/j.cell.2013.09.006] [PMID: 24074859]

[26] Zhang VW. Massively parallel sequencing for diagnosing clinically and genetically heterogeneous disorders. Per Med 2013; 10: 613-9.
[http://dx.doi.org/10.2217/pme.13.40]

[27] Shendure J, Ji H. Next-generation DNA sequencing. Nat Biotechnol 2008; 26(10): 1135-45.
[http://dx.doi.org/10.1038/nbt1486] [PMID: 18846087]

[28] Metzker ML. Sequencing technologies - the next generation. Nat Rev Genet 2010; 11(1): 31-46.
[http://dx.doi.org/10.1038/nrg2626] [PMID: 19997069]

[29] Rothberg JM, Hinz W, Rearick TM, *et al.* An integrated semiconductor device enabling non-optical genome sequencing. Nature 2011; 475(7356): 348-52.
[http://dx.doi.org/10.1038/nature10242] [PMID: 21776081]

[30] Kanagal-Shamanna R, *et al.* Next-generation sequencing-based multi-gene mutation profiling of solid tumors using fine needle aspiration samples: promises and challenges for routine clinical diagnostics. Mod Pathol 2013; 1-14.
[PMID: 23907151]

[31] Agilent. High Sensitivity DNA Kit. www.genomics.agilent.com/files/Manual/G2938-90321_ HighSensitivity DNA_KG.pdf.

[32] Tran B, *et al.* Feasibility of real time next generation sequencing of cancer genes linked to drug response Results from a clinical trial 2013.

[33] Day E, Dear PH, McCaughan F. Digital PCR strategies in the development and analysis of molecular biomarkers for personalized medicine. Methods 2013; 59(1): 101-7.
[http://dx.doi.org/10.1016/j.ymeth.2012.08.001] [PMID: 22926236]

[34] Pinheiro LB, Coleman VA, Hindson CM, *et al.* Evaluation of a droplet digital polymerase chain reaction format for DNA copy number quantification. Anal Chem 2012; 84(2): 1003-11.
[http://dx.doi.org/10.1021/ac202578x] [PMID: 22122760]

[35] Bhat S, Herrmann J, Armishaw P, Corbisier P, Emslie KR. Single molecule detection in nanofluidic digital array enables accurate measurement of DNA copy number. Anal Bioanal Chem 2009; 394(2): 457-67.
[http://dx.doi.org/10.1007/s00216-009-2729-5] [PMID: 19288230]

[36] Momper JD, Wagner JA. Therapeutic drug monitoring as a component of personalized medicine:

applications in pediatric drug development. Clin Pharmacol Ther 2014; 95(2): 138-40.
[http://dx.doi.org/10.1038/clpt.2013.227] [PMID: 24253660]

[37] Gustafson MD, Rimm DL, Dolled-Filhart M. Tissue microarrays: leaping the gap between research and clinical adoption. Per Med 2013; 10: 441-51.
[http://dx.doi.org/10.2217/pme.13.42]

[38] Sun J, Beger RD, Schnackenberg LK. Metabolomics as a tool for personalizing medicine: 2012 update. Per Med 2013; 10: 149-61.
[http://dx.doi.org/10.2217/pme.13.8]

[39] Gu H, Pan Z, Xi B, *et al.* 1H NMR metabolomics study of age profiling in children. NMR Biomed 2009; 22(8): 826-33.
[http://dx.doi.org/10.1002/nbm.1395] [PMID: 19441074]

[40] Rathahao-Paris E, Paris A, Bursztyka J, Jaeg J-P, Cravedi J-P, Debrauwer L. Identification of xenobiotic metabolites from biological fluids using flow injection analysis high-resolution mass spectrometry and post-acquisition data filtering. Rapid Commun Mass Spectrom 2014; 28(24): 2713-22.
[http://dx.doi.org/10.1002/rcm.7066] [PMID: 25380493]

[41] Kirwan JA, Weber RJM, Broadhurst DI, Viant MR. Direct infusion mass spectrometry metabolomics dataset: a benchmark for data processing and quality control. Sci Data 2014 Jun 10; 1: 140012.
[http://dx.doi.org/10.1038/sdata.2014.12]

[42] Hirayama A, Wakayama M, Soga T. Metabolome analysis based on capillary electrophoresis-mass spectrometry. Trends Analyt Chem 2014; 61: 215-22.
[http://dx.doi.org/10.1016/j.trac.2014.05.005]

[43] Sugimoto M, Wong DT, Hirayama A, Soga T, Tomita M. Capillary electrophoresis mass spectrometry-based saliva metabolomics identified oral, breast and pancreatic cancer-specific profiles. Metabolomics 2010; 6(1): 78-95.
[http://dx.doi.org/10.1007/s11306-009-0178-y] [PMID: 20300169]

[44] Krone N, Hughes BA, Lavery GG, Stewart PM, Arlt W, Shackleton CH. Gas chromatography/mass spectrometry (GC/MS) remains a pre-eminent discovery tool in clinical steroid investigations even in the era of fast liquid chromatography tandem mass spectrometry (LC/MS/MS). J Steroid Biochem Mol Biol 2010; 121(3-5): 496-504.
[http://dx.doi.org/10.1016/j.jsbmb.2010.04.010] [PMID: 20417277]

[45] Dethy J-M, Ackermann BL, Delatour C, Henion JD, Schultz GA. Demonstration of direct bioanalysis of drugs in plasma using nanoelectrospray infusion from a silicon chip coupled with tandem mass spectrometry. Anal Chem 2003; 75(4): 805-11.
[http://dx.doi.org/10.1021/ac0260692] [PMID: 12622370]

[46] Zhang Y, Qiu L, Wang Y, Qin X, Li Z. High-throughput and high-sensitivity quantitative analysis of serum unsaturated fatty acids by chip-based nanoelectrospray ionization-Fourier transform ion cyclotron resonance mass spectrometry: early stage diagnostic biomarkers of pancreatic cancer. Analyst (Lond) 2014; 139(7): 1697-706.
[http://dx.doi.org/10.1039/c3an02130k] [PMID: 24551873]

[47] Kirwan JA, Weber RJ, Broadhurst DI, Viant MR. Direct infusion mass spectrometry metabolomics

dataset: a benchmark for data processing and quality control. Scientific Data 2014; 1: 140012. [http://dx.doi.org/10.1038/sdata.2014.12]

[48] Guo Y, Wang X, Qiu L, *et al.* Probing gender-specific lipid metabolites and diagnostic biomarkers for lung cancer using Fourier transform ion cyclotron resonance mass spectrometry. Clin Chim Acta 2012; 414: 135-41.
[http://dx.doi.org/10.1016/j.cca.2012.08.010] [PMID: 22906735]

[49] Li F, Qin X, Chen H, *et al.* Lipid profiling for early diagnosis and progression of colorectal cancer using direct-infusion electrospray ionization Fourier transform ion cyclotron resonance mass spectrometry. Rapid Commun Mass Spectrom 2013; 27(1): 24-34.
[http://dx.doi.org/10.1002/rcm.6420] [PMID: 23239314]

[50] Zhang Y, Wang Y, Guo S, Guo Y, Liu H, Li Z. Ammonia-treated N-(1-naphthyl) ethylenediamine dihydrochloride as a novel matrix for rapid quantitative and qualitative determination of serum free fatty acids by matrix-assisted laser desorption/ionization-Fourier transform ion cyclotron resonance mass spectrometry. Anal Chim Acta 2013; 794: 82-9.
[http://dx.doi.org/10.1016/j.aca.2013.07.060] [PMID: 23972979]

[51] Rathahao-Paris E, Paris A, Bursztyka J, Jaeg J-P, Cravedi J-P, Debrauwer L. Identification of xenobiotic metabolites from biological fluids using flow injection analysis high-resolution mass spectrometry and post-acquisition data filtering. Rapid Commun Mass Spectrom 2014; 28(24): 2713-22.
[http://dx.doi.org/10.1002/rcm.7066] [PMID: 25380493]

[52] Madalinski G, Godat E, Alves S, *et al.* Direct introduction of biological samples into a LTQ-Orbitrap hybrid mass spectrometer as a tool for fast metabolome analysis. Anal Chem 2008; 80(9): 3291-303.
[http://dx.doi.org/10.1021/ac7024915] [PMID: 18351782]

[53] Lucio M, Fekete A, Weigert C, *et al.* Insulin sensitivity is reflected by characteristic metabolic fingerprints--a Fourier transform mass spectrometric non-targeted metabolomics approach. PLoS One 2010; 5(10): e13317.
[http://dx.doi.org/10.1371/journal.pone.0013317] [PMID: 20976215]

[54] Erve JC, Demaio W, Talaat RE. Rapid metabolite identification with sub parts-per-million mass accuracy from biological matrices by direct infusion nanoelectrospray ionization after clean-up on a ZipTip and LTQ/Orbitrap mass spectrometry. Rapid Commun Mass Spectrom 2008; 22(19): 3015-26.
[http://dx.doi.org/10.1002/rcm.3702] [PMID: 18763271]

[55] Fuhrer T, Zamboni N. High-throughput discovery metabolomics. Curr Opin Biotechnol 2015; 31: 73-8.
[http://dx.doi.org/10.1016/j.copbio.2014.08.006] [PMID: 25197792]

[56] Halket JM, Waterman D, Przyborowska AM, Patel RK, Fraser PD, Bramley PM. Chemical derivatization and mass spectral libraries in metabolic profiling by GC/MS and LC/MS/MS. J Exp Bot 2005; 56(410): 219-43.
[http://dx.doi.org/10.1093/jxb/eri069] [PMID: 15618298]

[57] Krone N, Hughes BA, Lavery GG, Stewart PM, Arlt W, Shackleton CH. Gas chromatography/mass spectrometry (GC/MS) remains a pre-eminent discovery tool in clinical steroid investigations even in the era of fast liquid chromatography tandem mass spectrometry (LC/MS/MS). J Steroid Biochem Mol

Biol 2010; 121(3-5): 496-504.
[http://dx.doi.org/10.1016/j.jsbmb.2010.04.010] [PMID: 20417277]

[58] Gika HG, Theodoridis GA, Plumb RS, Wilson ID. Current practice of liquid chromatography-mass spectrometry in metabolomics and metabonomics. J Pharm Biomed Anal 2014; 87: 12-25.
[http://dx.doi.org/10.1016/j.jpba.2013.06.032] [PMID: 23916607]

[59] Yin P, Xu G. Current state-of-the-art of nontargeted metabolomics based on liquid chromatography–mass spectrometry with special emphasis in clinical applications. J Chromatogr Part A (in press)

[60] Wang J, Christison TT, Misuno K, *et al.* Metabolomic profiling of anionic metabolites in head and neck cancer cells by capillary ion chromatography with Orbitrap mass spectrometry. Anal Chem 2014; 86(10): 5116-24.
[http://dx.doi.org/10.1021/ac500951v] [PMID: 24766394]

[61] Xia J, Broadhurst DI, Wilson M, Wishart DS. Translational biomarker discovery in clinical metabolomics: an introductory tutorial. Metabolomics 2013; 9(2): 280-99.
[http://dx.doi.org/10.1007/s11306-012-0482-9] [PMID: 23543913]

[62] Medina S, Domínguez-Perles R, Gil JI, Ferreres F, Gil-Izquierdo A. Metabolomics and the diagnosis of human diseases--a guide to the markers and pathophysiological pathways affected. Curr Med Chem 2014; 21(7): 823-48.
[http://dx.doi.org/10.2174/09298673206661311119124056] [PMID: 24251573]

[63] Kell DB, Goodacre R. Metabolomics and systems pharmacology: why and how to model the human metabolic network for drug discovery. Drug Discov Today 2014; 19(2): 171-82.
[http://dx.doi.org/10.1016/j.drudis.2013.07.014] [PMID: 23892182]

[64] Hood L, Heath JR, Phelps ME, Lin B. Systems biology and new technologies enable predictive and preventative medicine. Science 2004; 306(5696): 640-3.
[http://dx.doi.org/10.1126/science.1104635] [PMID: 15499008]

[65] Hood L, Balling R, Auffray C. Revolutionizing medicine in the 21st century through systems approaches. Biotechnol J 2012; 7(8): 992-1001.
[http://dx.doi.org/10.1002/biot.201100306] [PMID: 22815171]

[66] Dingermann T, Steinhilber D, Folkers G. Molecular Biology in Medicinal Chemistry. New York: Wiley-VCH 2004; pp. 330-8.

[67] Chari RV. Targeted cancer therapy: conferring specificity to cytotoxic drugs. Acc Chem Res 2008; 41(1): 98-107.
[http://dx.doi.org/10.1021/ar700108g] [PMID: 17705444]

[68] Siegel R, Naishadham D, Jemal A. Cancer statistics, 2013. CA Cancer J Clin 2013; 63(1): 11-30.
[http://dx.doi.org/10.3322/caac.21166] [PMID: 23335087]

[69] Koboldt DC, *et al.* Comprehensive molecular portraits of human breast tumours. Nature 2012; 490(7418): 61-70.
[http://dx.doi.org/10.1038/nature11412] [PMID: 23000897]

[70] Nandy A, Gangopadhyay S, Mukhopadhyay A. Individualizing breast cancer treatment-The dawn of personalized medicine. Exp Cell Res 2014; 320(1): 1-11.

[http://dx.doi.org/10.1016/j.yexcr.2013.09.002] [PMID: 24051330]

[71] Wayteck L, Breckpot K, Demeester J, de Smedt SC, Raemdon K. A personalized view on cancer immunotherapy. Cancer Lett 2013. (in press)
[PMID: 24051308]

[72] Robbins PF, Lu YC, El-Gamil M, *et al.* Mining exomic sequencing data to identify mutated antigens recognized by adoptively transferred tumor-reactive T cells. Nat Med 2013; 19(6): 747-52.
[http://dx.doi.org/10.1038/nm.3161] [PMID: 23644516]

[73] Rosenberg SA, Yang JC, Sherry RM, *et al.* Durable complete responses in heavily pretreated patients with metastatic melanoma using T-cell transfer immunotherapy. Clin Cancer Res 2011; 17(13): 4550-7.
[http://dx.doi.org/10.1158/1078-0432.CCR-11-0116] [PMID: 21498393]

[74] Zheng X, Xie G, Jia W. Metabolomic profiling in colorectal cancer: Opportunities for personalized medicine. Per Med 2013; 10: 741-55.
[http://dx.doi.org/10.2217/pme.13.73]

[75] Singleton AB, Farrer MJ, Bonifati V. The genetics of Parkinson's disease: progress and therapeutic implications. Mov Disord 2013; 28(1): 14-23.
[http://dx.doi.org/10.1002/mds.25249] [PMID: 23389780]

[76] Funke C, Schneider SA, Berg D, Kell DB. Genetics and iron in the systems biology of Parkinson's disease and some related disorders. Neurochem Int 2013; 62(5): 637-52.
[http://dx.doi.org/10.1016/j.neuint.2012.11.015] [PMID: 23220386]

[77] Potts LF, Luzzio FA, Smith SC, Hetman M, Champy P, Litvan I. Annonacin in Asimina triloba fruit: implication for neurotoxicity. Neurotoxicology 2012; 33(1): 53-8.
[http://dx.doi.org/10.1016/j.neuro.2011.10.009] [PMID: 22130466]

[78] Smith RE, Tran K, Richards KM, Champy P. Bioactive annonaceous acetogenins. Stud Nat Prod Chem 2014; 41: 93-117.

[79] Hood L. A personal journey of discovery: developing technology and changing biology. Annu Rev Anal Chem (Palo Alto, Calif) 2008; 1: 1-43.
[http://dx.doi.org/10.1146/annurev.anchem.1.031207.113113] [PMID: 20636073]

[80] Feero WG, Guttmacher AE. Genomics, personalized medicine, and pediatrics. Acad Pediatr 2014; 14(1): 14-22.
[http://dx.doi.org/10.1016/j.acap.2013.06.008] [PMID: 24369865]

[81] Menni C, Fauman E, Erte I, *et al.* Biomarkers for type 2 diabetes and impaired fasting glucose using a nontargeted metabolomics approach. Diabetes 2013; 62(12): 4270-6.
[http://dx.doi.org/10.2337/db13-0570] [PMID: 23884885]

[82] Walford GA, Davis J, Warner AS, *et al.* Branched chain and aromatic amino acids change acutely following two medical therapies for type 2 diabetes mellitus. Metabolism 2013; 62(12): 1772-8.
[http://dx.doi.org/10.1016/j.metabol.2013.07.003] [PMID: 23953891]

[83] Worthey EA, Mayer AN, Syverson GD, *et al.* Making a definitive diagnosis: successful clinical application of whole exome sequencing in a child with intractable inflammatory bowel disease. Genet Med 2011; 13(3): 255-62.

[http://dx.doi.org/10.1097/GIM.0b013e3182088158] [PMID: 21173700]

[84] Hu H, Wrogemann K, Kalscheuer V, *et al.* Mutation screening in 86 known X-linked mental retardation genes by droplet-based multiplex PCR and massive parallel sequencing. HUGO J 2009; 3(1-4): 41-9.
[http://dx.doi.org/10.1007/s11568-010-9137-y] [PMID: 21836662]

[85] Gozes I. Preventive, personalized medicine at the level of key regulatory proteins: Updates on activity dependent neuroprotective protein (ADNP) as a case study. EPMA J 2014; 5 (Suppl. 1): A97.
[http://dx.doi.org/10.1186/1878-5085-5-S1-A97]

[86] Arntzen CJ. Plant science. Using tobacco to treat cancer. Science 2008; 321(5892): 1052-3.
[http://dx.doi.org/10.1126/science.1163420] [PMID: 18719274]

[87] de Sa J, *et al.* Growth of molecular diagnostics and genetic testing in the USA, 2008–2011: Analysis and implications. Per Med 2013; 10: 785-92.
[http://dx.doi.org/10.2217/pme.13.84]

CHAPTER 2

Biology of Inflammation

Abstract: Inflammation can be a cause or a symptom of many diseases. When controlled, it is also an important part of maintaining good health. Reactive oxygen substances (ROS) are produced as part of normal, healthy aerobic metabolism, such as electron transport in the mitochondria of cells. Many foods, spices, herbs and dietary supplements contain antioxidants that can destroy ROS and help to prevent diseases. Some important dietary antioxidants include vitamins A and E, oleic acid, poly-unsaturated fats, omega-3 fats, resveratrol and polyphenols. Inflammation is a significant factor in many diseases, including arthritis, cancer, diabetes, heart disease, stroke, Alzheimer´s disease, hormonal diseases, osteoporosis, inflammatory bowel disease, pelvic inflammatory disease, and many others. Smoldering inflammation is a relatively low level of inflammation that occurs in obesity, type-2 diabetes, asthma, and atherosclerosis. In many cases, these diseases of inflammation can be prevented by avoiding obesity, trans fats and saturated fats that are in the typical fast food diet that many people in the USA consume. Instead, unsaturated fats are much better. Omega-3 fats can be taken as dietary supplements such as fish oil and flaxseed oil. They are also present in fatty fish, such as salmon. Omega-3 fats are also very important in the brain, where they play an important role in cognitive function and behavior. Inflammation plays an important role in all stages of atherosclerosis and cardiovascular disease. Inflammation is also an important factor in stroke

Keywords: Arthritis, Alzheimer´s disease, Cancer, Heart disease, Inflammation, Omega-3 fats, ROS, Saturated fats, Stroke, Trans fats.

Inflammation is also a part of traditional and western medicine. Infusions of willow bark were used by Native Americans to treat headaches, fever and overall inflammation. Unfortunately, the acidity of salicylic acid caused severe stomach problems in many people. While working in the lab of Adolf von Baeyer, Felix Hoffmann acetylated the carboxylic acid to make acetylsalicylic acid, or aspirin, thus reducing this harmful side effect.

Inflammation is also an area that has been misunderstood. For many years, the U.S. Department of Agriculture (USDA) maintained a website that listed the

Robert E. Smith

antioxidant potential of many foods and spices. This information was used in advertisements to help sell many products. One fruit, açaí, is still advertised as being the food with the highest antioxidant capacity. However, a few years ago, the USDA removed the data from the website because there is no evidence that consuming large amounts of antioxidants have little or no preventive or therapeutic value. Still, smoldering inflammation is known to be a cause of many diseases.

So, let us look at the underlying cause of many diseases – inflammation, along with some of the enzymes, oxidizing substances and antioxidants that affect it. When controlled properly, inflammation is an important part of maintaining good health. To get energy, human cells contain mitochondria, which make proinflammatory free radicals and reactive oxygen substances (ROS) as by-products of the TCA cycle and oxidative phosphorylation. Other subcellular organelles, called peroxisomes, oxidize fatty acids that have more than eight carbons and produce H_2O_2, which is broken down by the enzyme called catalase. Before it is broken down, H_2O_2 can oxidize compounds such as phenols, aldehydes and alcohols. Also, it can act as a second messenger that is produced in response to extracellular stimuli, and can regulate various biological processes.

Also, when disease-causing (pathogenic) microorganisms invade the body, immune cells kill the microbes by causing oxidative damage and inflammation. Although there is some collateral damage to the surrounding healthy cells and tissues, this does get repaired in healthy people. However, when the immune system is overactive, it can incorrectly identify environmental chemicals, foods and even one's own cells as foreign and mount a potentially fatal autoimmune response or allergic reaction. So, inflammation, like so much else, must be carefully controlled. Although pathogenic bacteria can cause inflammation and disease, remember that there are non-pathogenic bacteria that are essential for healthy human life. This includes *Bifidobacterium* and *Bacteriodes*, which help the immune system protect against the development of inflammatory diseases. They do this by helping to digest fermentable dietary fiber.

One definition of inflammation is that it occurs when parts of the body become red, warm, swollen, and damaged. This can happen when the immune system responds to foreign materials, irritation, bone or nerve damage, infection by microorganisms, and ischemia (lack of blood flow), followed by reperfusion. Inflammation is caused by the production of reactive oxygen species (or substances) and other chemicals, such as histamine, pro-inflammatory proteins (cytokines) and eicosanoids [1]. Reactive oxygen substances (ROS) are produced as part of normal, healthy aerobic metabolism, such as electron transport in the mitochondria of cells. Please note that much of the chemical literature uses the term reactive oxygen species (ROS), instead of reactive oxygen substances, but many biologists would consider this to be a misuse of the word species. When oxygen reacts with nutrients and metabolites, it produces ROS such as hydrogen peroxide, the superoxide anion, the hydroxyl radical and nitric oxide. Hydrogen peroxide (H_2O_2) is made in mitochondria as a byproduct of biochemical reactions that produce energy for cells. It can react with unchelated iron (Fe^{2+}) and copper (Cu^{2+}) to produce the superoxide anion (O_2^-), which can also be produced by the reaction catalyzed by NADPH oxidases in immune cells.

$$2\ O_2 + \text{NADPH} \rightarrow 2\ O_2^- + \text{NADP}^+ + H^+$$

The superoxide anion can then react with other molecules to produce other ROS. When a pathogenic organism enters the body, NADPH oxidase in phagocytes and T lymphocytes that originate in lymph nodes produce ROS which kill the invading organism.

NADPH oxidase also exists in other tissues and its activation can be harmful. For example, NADPH oxidase in the lungs can cause oxidative damage after cardiopulmonary bypass operations [2]. A chemical carcinogen, diethylnitrosamine, can activate a signaling pathway in liver Kupffer cells (a type of macrophage) that causes inflammation and tissue damage [3]. In the brain, there is an NADPH oxidase in glial cells that is responsible for oxidative damage after suffering from a stroke. This can be prevented in gerbils by giving them apocynin, an NADPH oxidase inhibitor from *Picrorhiza kurroa*, a native plant grown in the mountains of India, Nepal, Tibet, and Pakistan [4]. Inflammation is a hallmark of acute stroke. Proinflammatory cytokines are released into the blood and brain. The immune system is also adversely affected, so stroke patients are susceptible to infections. In order to prevent this, regulatory T-helper cells release the anti-inflammatory cytokine IL-10, which helps to limit the amount of brain damage [5, 6].

In Alzheimer's disease, there is an oxidative damage caused by a defective amyloid-β peptide (Aβ). The oxidative damage is caused by activating NADPH oxidase in neurons in the cerebral cortex [7]. Several antioxidants, including apocynin and its metabolite, diapocynin inhibit NADPH oxidase *in vitro* [8]. Apocynin derivatives have also been shown to interrupt intracellular signaling, which caused decreased migration in breast cancer cells [9]. Apocynin can also protect kidney damage by inhibiting renal NADPH oxidase [10]. When activated by a ligand, the dopamine receptor, D_5, also inhibits NADPH oxidase to control high blood pressure, along with kidney and the adrenal gland function [11].

Our cells contain natural antioxidants, such as the D_5 receptor, superoxide dismutase (SOD) and the glutathione. SOD and glutathione help destroy ROS, such as the superoxide anion, O_2^-. The following reaction is catalyzed by SOD:

$$2\,O_2^- + 2\,H^+ \rightarrow O_2 + H_2O_2$$

Fortunately, many foods, spices, herbs and dietary supplements contain antioxidants that can destroy ROS and help to prevent diseases. Some important dietary antioxidants include vitamins A and E, oleic acid, polyunsaturated fats (often called fatty acids), omega-3 fats, resveratrol, polyphenols, and many other molecules that contain one or more C=C double bonds. Inflammation is a significant factor in many diseases, including arthritis, cancer, diabetes, heart disease, stroke, Alzheimer's disease, hormonal diseases, osteoporosis, inflammatory bowel disease, pelvic inflammatory disease, and many others.

Hydrogen peroxide does not cause much direct oxidative damage, but it can react with free iron in the +2 oxidation state (Fe^{2+}) to produce the hydroxyl radical ($OH^\cdot$) in the Fenton reaction, eq (**1**):

$$Fe^{2+} + H_2O_2 \rightarrow Fe^{3+} + OH^- + OH^\cdot \quad (\mathbf{1})$$

The hydroxyl radical is highly reactive and damaging. It can oxidize lipids and lead to atherosclerosis and heart disease, as stated in the iron hypothesis [12]. That is, differences in the amounts of stored iron can explain why men are more prone to heart disease than women. Iron stores in men rise after adolescence, but remain low in women. It only begins to rise after the age of about 45. The maximum sex difference in heart disease is also at age 45. Moreover, medicines like aspirin that cause gastrointestinal blood loss may protect against heart disease by decreasing iron stores. This may also partly explain the protective effect of NSAIDs on Alzheimer's disease. That is, a systems biology analysis showed that poorly liganded iron and copper (free iron and copper) can help cause not just Alzheimer's, but also Parkinson's and Huntington's disease. Free Fe^{2+} and Cu^{2+} can also lead to diseases caused by prions, bacteriocides and chemical toxicity [13]. That is, there is a continuous autocatalytic production of hydroxyl radicals involving poorly liganded iron, leading to cell death *via* apoptosis. That is, once Fe^{3+} is produced in the Fenton reaction, eq (**1**), it can react with superoxide anions, $O_2^{\cdot-}$, (also produced by mitochondria) to regenerate Fe^{2+} eq (**2**)

$$Fe^{3+} + O_2^{\cdot-} \rightarrow Fe^{2+} + O_2 \quad \textbf{(2)}$$

Vitamin C (ascorbic acid or ascorbate at physiological pH) can also reduce Fe^{3+} to Fe^{2+}. The hydroxyl radical is quite reactive and can damage many cellular components – not just lipids. It can also release Fe^{2+} from Fe-S catalytic centers in proteins like ferritin, thus producing more free Fe^{2+} that can react with more H_2O_2 in eq (**1**) to produce more hydroxyl radicals in an autocatalytic reaction. In a related reaction, the superoxide anion can react with nitric oxide (NO) to make the highly reactive peroxynitrite anion, $ONOO^-$, which can react with tyrosine or cysteine in proteins (to make nitrosylated proteins), with DNA (to make 8-hydroxy-20-deoxyguanosine) and with fatty acyls to make nitrosylated fats. So, to prevent the formation of excess hydroxyl radicals and superoxide anions, it is important to keep Fe^{2+} fully liganded.

That is, iron has up to six individual chelation sites, arranged octahedrally, and many ligands will bind to only some of them. For example, many dietary phenolic compounds can chelate iron and copper. However, partial chelation by ascorbate will transform vitamin C into a pro-oxidant (promoting the production of hydroxyl radicals) and not an antioxidant, which is the healthy form of it. So, the binding of incompletely liganded iron (usually bivalent) to inappropriate cellular structures can cause catalytic activities that kill cells, leading to not just atherosclerosis, but also stroke, age-related macular degeneration, prion diseases, sepsis, septic shock, viral infections and neurodegenerative diseases. It may also contribute to microbial, plant, animal and chemical toxicities [13].

Prostaglandins are important pro-inflammatory biochemicals. The prostaglandins, eicosanoids, and leukotrienes are made from arachidonic acid. Arachidonic acid is present in the brain and is often attached to carbon number two of phospholipids. The enzyme called phospholipase A_2 catalyzes the hydrolysis of fatty acyl groups on this carbon. Thus, phospholipase A_2 is involved in the oxidative damage that is caused by the Aβ peptide in Alzheimer's disease [7].

Cyclooxygenase, or COX, catalyzes the conversion of arachidonic acid to prostaglandins. COX-1 produces prostaglandins that stimulate the production of stomach acid, but COX-2 produces different prostaglandins that cause pain and arthritis. We learned that non-steroid anti-inflammatory drugs (NSAIDs) treat pain caused by inflammation by inhibiting COX-2.

Prostaglandins were first discovered in the prostate gland. They all have 20 carbons, so they are eicosanoids, as are leukotrienes. Prostaglandins have a five-carbon ring and are derived from arachidonic acid. COX catalyzes their synthesis. They are found in almost all tissues and organs. There are nine known prostaglandin receptors. They are can be ligated (or bound) to G-protein coupled receptors. One type of prostaglandin receptor regulates inflammation. Prostaglandins have short half-lives, so they exert their effects locally, as either paracrine (locally active) or autocrine (active on the same cell that makes them). They can produce fever and dilate blood vessels.

Leukotrienes are also 20-carbon derivatives of arachidonic acid, but their synthesis is catalyzed by 5-lipoxygenase (5-LOX). As the name implies, leukotrienes have three C=C double bonds and are made by leukocytes (white blood cells). They help regulate blood flow. One of them, leukotriene 4, or LBT-4, attracts white blood cells to the site of an infection, and it increases the permeability of blood vessels [1].

Inflammation can be caused (or induced) by infection, tissue injury or by tissue stress and malfunction [14]. Inducers interact with sensors, such as toll-like receptor 4 (TLR4), immunoglobulin E (IgE), a subcellular particle called the NALP3 inflammasome or a protein called the Hageman factor. The sensors then initiate the production of several inflammatory mediators. These include tumor necrosis factor alpha (TNF-α), IL-6 (interleukin-6), prostaglandin E_2 (PGE_2), vasoactive amines, IL-1β and bradykinin. Tocilizumab (Acterma®) is a monoclonal antibody that binds to the IL-6 receptor and is approved for treating rheumatoid arthritis. So, inflammatory mediators affect the cells and tissues that produce an effect, and are also called effectors. Target cells include endothelial cells, hepatocytes, leukocytes, the hypothalamus and smooth muscle cells. So, bacteria can produce lipopolysaccharides, which interact with TLR4, which initiates the production of TNF-α, IL-6 and PGE_2. This affects endothelial cells, hepatocytes, leukocytes, the hypothalamus and other cells. Bacterial inducers include pathogen-associated molecular patterns (PAMPs) and virulence factors. Host cells have receptors for PAMPs, which trigger inflammation [15].

Microbes stimulate the innate immune system and its TLR and nucleotide-binding oligomerization domain (also known as NOD-like receptors). The cytoplasmic receptor NOD2 senses peptidoglycan fragments and activates nuclear factor κB (NF-κB) signaling, which lead to inflammatory immune responses. Dysregulation of NOD2 signaling is associated with inflammatory diseases, such as Crohn's disease and Blau syndrome. Macrophages and mast cells that reside in the affected tissue produce inflammatory modulators. These include cytokines, chemokines, vasoactive amines (histamine and serotonin), eicosanoids and products of proteolytic cascades. So, there is an interconnected network of genes in NF-kB signaling [16]. At the same time, allergens are sensed by IgE, which produces vasoactive amines (such as histamine), which activate endothelial cells and smooth muscle cells. Another class of inducers, crystals of monosodium urate and calcium pyrophosphate dehydrate are sensed by NALP3, which produces IL-1β, which activates endothelial cells, hepatocytes, leukocytes, the hypothalamus and other cells. These crystals are formed in the diseases called gout and pseudogout. The fourth type of inducer, the protein called collagen, is sensed by the Hageman factor, which is mediated by bradykinin, which activates endothelial cells and smooth muscle cells. The Hageman factor (also known as factor XII) becomes activated by contact with collagen and other components of the extracellular matrix that are produced when the vascular endothelium is damaged in stressed or malfunctioning tissues [15].

There are also some lipids, called lipoxins, resolvins and protectins that are anti-inflammatory [1]. A low dose of aspirin (81 mg) stimulates the production of lipoxins. Higher doses have no effect on lipoxin synthesis. Anti-inflammatory lipids are released in the later stages of inflammation and restore the normal, healthy condition. The successful initial inflammatory response eliminates infectious agents. This is followed by a resolution and repair phase, which is mediated by macrophages that are already in the tissue and macrophages that are drawn to the damaged site [15]. They switch the lipid mediators from pro-inflammatory prostaglandins to anti-inflammatory lipoxins, which inhibit the recruitment of neutrophils and promote the recruitment of monocytes. They work together with resolvins, protectins, transforming growth factor β and other growth factors to remove dead cells and start remodeling tissues [15].

There are also some endogenous inducers of inflammation that come from stressed, damaged, or malfunctioning tissues [15]. There is something called para-inflammation, or smoldering inflammation. This is a relatively low level of inflammation that occurs in diseases such as obesity, type-2 diabetes, asthma, and atherosclerosis [15]. In many cases, these diseases of inflammation can be prevented by avoiding obesity, *trans* fats and saturated fats that are in the typical fast food diet that many people in the USA consume. Instead, unsaturated fats are much better.

Lipoxins are made from arachidonic acid, but the resolvins and protectins are made from omega-3 fats, such as alpha linolenic acid (ALA), eicosapentaenoic acid (EPA) and docosahexaenoic acid (DHA). Omega-3 fats can be taken as dietary supplements such as fish oil and flaxseed oil. They are also present in fatty fish, such as salmon. Eskimos eat much fatty fish and they have a very low incidence of arthritis and heart disease. Inflammation is important in all stages of heart disease, from its beginning, development and final, tragic stages [14]. Proper diet and cardiovascular exercise can limit inflammation and help prevent heart disease.

Lipoxin A4 is important in asthma, a prevalent disease of chronic inflammation [17]. It regulates the activation of natural killer (NK) cells and type 2 innate lymphoid cells (ILC2s). ILCs express high levels of cytokines that are important in the pathogenesis of the inflammatory response. There are also type 1 and type 3 ILCs (ILC1 and ILC3). ILC1s include natural killer (NK) cells and interferon-γ (IFN-γ)–producing ILC1s. Group 3 ILCs produce IL-17A and/or IL-22. All ILCs express the IL-2Rα (CD25) and IL-7Rα (CD127) subunits, as well as the common γ chain (CD132), which is the other subunit for the receptors for IL-2 and IL-7. ILC2s also express the receptors for IL-33 (ST2) and IL-25 (IL-17RB). ILC2s cells secrete high levels of interleukin 5 (IL-5) and IL-13 in response to the epithelial-derived cytokines IL-25, IL-33, and thymic stromal lymphopoietin (TSLP). IL-5 and IL-13 are crucial for causing the allergic response. Both have been targets of monoclonal antibodies used in human trials for treating asthma [18]. IL-5 is the most potent growth, differentiation, and survival factor for eosinophils and is important in eosinophil chemotaxis. IL-13 has a role as a central mediator in asthma pathogenesis. ILC2s secrete IL-13 in response to prostaglandin D_2 (PGD_2) by activating a GPCR. They also express the receptor for lipoxin A_4. NK cells are highly activated in severe asthma. They down-modulate the airway inflammatory response by inducing apoptosis in eosinophils. On the other hand, ILC2s promote airway inflammation by secreting interleukin-13 (IL-13). Despite their opposite functions, both NK cells and ILC2s express the receptor for lipoxin A4. In fact, lipoxin A4 increased the ability of NK cells to induce eosinophil apoptosis and decreased IL-13 production by ILC2 cells. Lipoxins are anti-inflammatory for neutrophils and eosinophils and may help clear inflamed tissue. Also, lipoxins promote the restoration of an injured airway epithelium by indirectly blocking the release of the proinflammatory cytokines IL-6 and IL-8 and by inhibiting neutrophil transmigration. So, lipoxin D_4 receptor agonists and/or synthetic lipoxin A_4 analogues might be good therapeutic agents for asthma [17, 18].

Inflammation plays an important role in all stages of atherosclerosis and cardiovascular disease [14]. Atherosclerosis is the main process underlying macrovascular disease. It starts when endothelial cells that line the intima are activated. This can be induced by saturated free fatty acids or cholesterol, but it leads to the expression of leukocyte adhesion molecules. So, leucokytes bind, then migrate through the endothelium to the intima where they can attract monocytes which ultimately transform into lipid-laden foam cells. This process may be enhanced in people who have type-2 diabetes. After immune cells and inflammatory mediators interact, the process can continue to atheroma and make rupture-prone atherosclerotic plaques. Inflammatory pathways are also involved in thrombosis, the late stage of atherosclerosis. It is responsible for most of the clinical complications of macrovascular disease. So, macrovascular disease is can be a major consequence of obesity-induced inflammation and type-2 diabetes [19]. Smoldering inflammation can also lead to diabetes, stroke and cancer, along with neurodegenerative and autoimmune diseases, which will discussed further in Chapters 2, 3, 6 and 7. Inflammation may be more important than cholesterol levels in causing heart disease. Fifty percent of all heart attacks and strokes occur in people with normal or even low levels of cholesterol [20]. Normal, healthy endothelial cells (ECs) on the innermost surface of arterial walls resist adhesion by leukocytes [14]. "When a person smokes, consumes many saturated fats, is obese, has high blood pressure, is hyperglycemic or resistant to insulin, adhesion molecules (such as vascular cell adhesion molecule-1, or VCAM-1) are expressed by ECs. This allows leukocytes to attach to the arterial wall. VCAM-1 binds to monocytes and T lymphocytes, which are found in early atherosclerotic plaques. Oxidized lipids, nuclear factor-κB (NF-κB) and TNF-α can also stimulate VCAM-1 expression. These things can be prevented by laminar blood flow, which causes some anti-atherosclerotic mechanisms, such as expression of the natural anti-oxidant, superoxide dismutase and an increase in nitric oxide (NO) synthase. There is an increase in the concentration of NO, which causes vasodilation and limits VCAM-1 gene expression" [14].

However, when blood flow is disturbed, lesions can form and monocytes can adhere to the arterial endothelium [14]. "Monocytes can then enter the vessel walls, causing a fatty streak to start developing. This is the first stage of atherosclerosis. Monocytes mature into macrophages, which express scavenger receptors and engulf modified lipoproteins. Cholesteryl esters accumulate in the cytoplasm and macrophages become foam cells, which have an increased concentration of lipids. Next, T-lymphocytes of the adaptive immune system enter the arterial cell walls and participate in further inflammation. They are attracted to the arterial cells by chemokines (γ-IP-10, MIG and I-TAC) that are induced by interferon-γ" [14].

The next stage is progression to the formation of plaque, which can begin early in life, when a person is still in their early twenties [14]. "A mixture of cytokines: interleukins (IL-1β, IL-1α, IL-6, IL-18) and tumor necrosis factors (TNF-α, TNF-β) are involved in this process, as are the macrophage colony stimulating factor (MCSF) and the monocyte chemoattractant protein-1, MCP-1. Next, the plaques can rupture, leading to thrombosis. In healthy arteries, there is a fibrous cap that contains collagen for strength. The strong cap protects blood from contacting the lipid core. When the cap ruptures, a thrombus, or blood clot, can form. Inflammation interferes with the integrity of the collagen matrix by blocking the creation of new collagen fibers and by stimulating the destruction of existing collagen. T-lymphocytes participate in this pro-inflammatory process. They produce CD40, a cell surface protein and IL-1. These biomolecules promote the production of enzymes in macrophages that catalyze the destruction of collagen. The CD40 ligand also stimulates macrophage production of tissue factor VII, which initiates the coagulation cascade" [14].

To prevent all these things from happening, one should try to lower the risk factors by not smoking, engaging in regular strenuous cardiovascular exercise and by consuming a healthy diet, with fewer calories, and more fresh fruits and vegetables and other sources of antioxidants [14]. An important biomarker for generalized inflammation is the high-sensitivity C-reactive protein, or hsCRP, which binds to phosphatidyl choline that is present on the exterior surface of dead or dying cells and activates the complement system of the innate immune system [14]. In healthy postmenopausal women, it is a more accurate predictor of risk for a heart attack than total cholesterol or LDL [20].

Obesity is also linked to chronic, low-grade inflammation in adipose and liver tissues [21]. "This can lead to insulin resistance and type-2 diabetes through the pro-inflammatory nuclear factor-κB (NF-κB), which is activated by the phosphorylation of the regulatory protein IκB by IκB kinases (IKKs). One of these kinases, IKKε, is elevated in adipose tissue when on a high fat diet. In an attempt to identify a small molecule inhibitor of IKKε, a library of 150,000 compounds was screened and one of them (amlexanox) bound with high affinity to the IKKε receptor. This drug is currently used to treat asthma allergic rhinitis and aphthous ulcers. When given to mice on a high fat diet, it prevented weight gain, improved insulin sensitivity, attenuated hepatic steatosis, reduced adipose tissue inflammation and promoted energy expenditure in adipose tissue through increased thermogenesis" [21].

Another group studied the receptor activator of NF-κB ligand (RANKL), a potent activator of NF-κB that is involved in bone homeostasis [22]. They showed that binding of RANKL to its receptor RANK in the liver activates NF-κB and promotes pro-inflammatory cytokine expression in the liver. On the other hand, blocking RANKL signaling improved insulin resistance in the liver and prevented diabetes [22]. Moreover, RANKL antagonists such as denosumab have been proven to be safe in patients and are currently used to treat osteoporosis. So, they may be effective in breaking the link between inflammation and diabetes [21].

Inflammation is also important in neurodegenerative diseases. Omega-3 fats are very important anti-inflammatory nutrients that are important in the brain, where they play an important role in cognitive function and behavior [1]. The typical American (USA) fast food diet contains little or no omega-3 fats [23]. Instead it contains high levels of *trans* fats and saturated fats, which increase the risk of heart disease and may be partly responsible for the poorer academic performance of children in the USA, compared to other countries that don't eat so much unhealthy food.

Inflammation is also an important factor in Alzheimer's disease [4, 24, 25]. This was shown recently in a continuation of the Framingham study [24]. The goal of the Framingham study was (and still is) to monitor the health of thousands of people for many years, and for several generations. Even though the original goal was to see if there were certain lifestyles that made a person more or less likely to suffer from cardiovascular disease, it now looks at Alzheimer's disease and dementia as well. Scientists and doctors have monitored the health of thousands of people for many years, and for several generations. They found that people who had higher concentrations of pro-inflammatory cytokines in their blood were more likely to get Alzheimer's disease [24]. It has also been shown that people who take NSAIDs earlier in life have a lower incidence of Alzheimer's disease [25]. However, once a person gets Alzheimer's disease, NSAIDs are no longer effective.

Inflammation is also an important factor in stroke, which can be caused by the temporary lack of blood flow to the brain (ischemia), followed by reperfusion. Ischemia-reperfusion causes glia and astrocytes to become activated and produce ROS, which cause oxidative damage and inflammation. Antioxidants can prevent this from happening [26].

Reactive oxygen substances (ROS) and inflammation are probably involved in cancer, too [22, 4]. In the 19th century, doctors observed that tumors often arose at sites of chronic inflammation [4]. In the 20th and 21st centuries, epidemiological studies showed that chronic inflammation is an important risk factor in many types of cancer [4]. At the same time NSAIDs have been shown to help prevent cancer [4]. Damage to DNA by ROS is a widely accepted cause of cancer [27]. Most mutations caused by ROS convert deoxyguanosine into deoxythymidine (G $\rightarrow$ T). Another mutation is the accumulation of 8-hydroxydeoxyguanosine, or 8-OHdG, which can be caused by nitrosamines and polynuclear aromatic hydrocarbons produced by smoking tobacco. Some viruses, such as hepatitis B and C, can cause chronic inflammation and liver cancer. Also, G $\rightarrow$ T transitions can be caused by aflatoxins, which are known as human carcinogens. Some types of cancer cells also produce ROS of their own. ROS production can be induced by oncogenes, such as *Ras* [27]. Also, pro-inflammatory cytokines, such as IL-1β, IL-6, IL-23 and TNF-α are produced after genes are activated by transcription factors, such as transcription factor nuclear factor NF-κB and STAT3, or signal transducer and activator of transcription activator 3 [4]. Phase I and II trials of antagonists of IL-6 and other chemokines are being conducted [27]. NF-κB is a coordinator of innate immunity and inflammation and is a tumor promoter. It activates the transcription of genes that code for pro-inflammatory cytokines, adhesion molecules, enzymes (such as COX and inducible nitric oxide synthase, iNOS) and angiogenic factors [4].

There are tumor-associated macrophages, or TAMs, which infiltrate most, if not all, tumors [4]. "They usually have an M2 phenotype, which promotes tumor growth, remodels tissues, promotes angiogenesis and suppresses adaptive immunity. Angiogenesis is required to supply blood to tumors after they grow to a certain size. Vascular endothelial growth factor (VEGF) not only stimulates angiogenesis, but also attracts monocytes into the primary tumor and the metastatic tumor. Then the monocytes secrete more pro-angiogenic factors, further stimulating angiogenesis. Chemokine receptors and their ligands also direct the movement of tumor cells during inflammation. When normal cells are transformed into cancer cells, they start to express chemokine receptors that help cause cell migration and survival at their metastatic site" [4].

Also, microRNAs (miRNAs) are involved in endothelial cell migration and angiogenesis [28]. "They can be found in the human bloodstream. When they are altered, it can lead to cancer, so they may be good biomarkers to detect early signs of it. Moreover, some of the miRNAs in tumor cells are packaged into microvesicles and delivered to endothelial cells. One of them, miR-9, suppressor of cytokine signaling 5 (SOCS5), reduces the activity of the JAK-STAT pathway, which is one of the major oncogenic signaling pathways that is activated in a several types of cancer. STAT proteins play a crucial role in tumor cell proliferation, survival and invasion, and help form a unique tumor micro-environment. Also, in endothelial cells, JAK1 is a proangiogenic protein and a target of miR-17-92. Even though it is probably difficult to target miRNAs, this study suggests that targeting the JAK-STAT pathway could be fruitful. So, several inhibitors of this pathway are in later phases of clinical trials for treating other diseases and have shown acceptable toxicological profiles" [28].

At the same time, the adaptive immune system can survey the body and eliminate newly produced tumors, in a process called tumor editing [4]. "However, if this fails and the tumor gets large enough to be detected in a clinical setting, the adaptive immune system can be suppressed by signals such as IL-10. Such tumors are also infiltrated by regulatory T cells which suppress both the innate and adaptive immune systems" [4].

Another class of molecules that can stimulate tumors is the sex steroid hormones [4]. "For example, prostate cancer depends on androgens and IL-1β. The inflammatory cytokine, IL-1β, is produced by macrophages in the tumor microenvironment. It converts androgen receptors from being inhibitors to becoming stimulators of prostate cancer growth. Also, females are less susceptible to cancer at sites such as the liver that are not targets of sex hormones" [4].

So, inflammation can be good or bad. It is needed to kill invading bacteria, as part of a healthy immune response. It is bad when it is unresolved, in what is often called smoldering inflammation. This occurs during obesity and chronic infections and can be caused by periodontal disease. Omega-3 fats and polyunsaturated fats help protect against smoldering inflammation because they are converted into resolvins and protectins, which resolve the smoldering inflammation and protect against further inflammation. So, omega-3 fats and polyunsaturated fats can help prevent diseases of inflammation, including arthritis, cancer, diabetes, heart disease, stroke, Alzheimer's disease, hormonal diseases, osteoporosis, inflammatory bowel disease, pelvic inflammatory disease, and many other diseases.

KEY POINTS

1. **Inflammation can be a cause or a symptom of many diseases. When controlled, it is also an important part of maintaining good health. Reactive oxygen substances (ROS) are produced as part of normal, healthy aerobic metabolism, such as electron transport in the mitochondria of cells.**

2. **Many foods, spices, herbs and dietary supplements contain antioxidants that can destroy ROS and prevent diseases, including arthritis, cancer, heart disease, stroke and brain diseases. Some important dietary antioxidants include vitamins A and E, oleic acid, poly-unsaturated fats, omega-3 fats, resveratrol and polyphenols. Inflammation is a significant factor in many diseases, including arthritis, cancer, diabetes, heart disease, stroke, Alzheimer's disease, hormonal diseases, osteoporosis, inflammatory bowel disease, pelvic inflammatory disease, and many others.**

3. **There are also some endogenous inducers of inflammation that come from stressed, damaged, or malfunctioning tissues. Smoldering inflammation is a relatively low level of inflammation that occurs in diseases such as obesity, type-2 diabetes, asthma, and atherosclerosis. In many cases, these diseases of inflammation can be prevented by avoiding obesity, *trans* fats and saturated fats**

that are in the typical fast food diet that many people in the USA consume. Instead, unsaturated fats are much better.

4. **Omega-3 fats can be taken as dietary supplements such as fish oil and flaxseed oil. They are also present in fatty fish, such as salmon. Omega-3 fats are also very important in the brain, where they play an important role in cognitive function and behavior.**

5. **Inflammation plays an important role in all stages of atherosclerosis and cardiovascular disease. Inflammation is also an important factor in stroke.**

ABBREVIATIONS

ROS	=	Reactive Oxygen Substances
NADPH	=	Nicotinamide Adenine Dinucleotide Phosphate Reduced
Aβ	=	Amyloid-β Peptide
SOD	=	Superoxide Dismutase
NSAID	=	Non-Steroid Anti-Inflammatory Drug
COX	=	Cyclooxygenase
5-LOX	=	5-Lipoxygenase
TLR4	=	Toll-Like Receptor 4
IgE	=	Immunoglobulin E
NALP3	=	Nacht Domain-, Leucine-Rich Repeat-, and PYD-Containing Protein 3
TNF-α	=	Tumor Necrosis Factor Alpha
IL-6	=	Interleukin-6
PGE_2	=	Prostaglandin E2
PAMP	=	Pathogen-Associated Molecular Pattern
NOD	=	Nucleotide-Binding Oligomerization Domain
NF-κB	=	Nuclear Factor κB
ALA	=	Alpha Linolenic Acid
EPA	=	Eicosapentaenoic Acid
DHA	=	Docosahexaenoic Acid
NK	=	Cells Natural Killer Cells
ICL2	=	Type 2 Innate Lymphoid Cells
GPCR	=	G-Protein Coupled Receptor
EC	=	Endothelial Cells

VCAM-1	=	Vascular Cell Adhesion Molecule-1
NO	=	Nitric Oxide
MCSF	=	Macrophage Colony Stimulating Factor
MCP-1	=	Monocyte Chemoattractant Protein-1
hsCRP	=	High-Sensitivity C-Reactive Protein
IKK	=	IκB Kinase
LDL	=	Low Density Lipoprotein
RANKL	=	Receptor Activator of NF-κB Ligand
TAM	=	Tumor-Associated Macrophage
microRNA	=	Micro Ribonucleic Acid
miRNA	=	microRNA
SOCS5	=	Suppressor of Cytokine Signaling 5

REFERENCES

[1] Corey EJ, Czako B, Kürti L. Molecules and Medicine. New York: John Wiley & Sons 2007.

[2] Dodd-o JM, Welsh LE, Salazar JD, *et al.* Effect of NADPH oxidase inhibition on cardiopulmonary bypass-induced lung injury. Am J Physiol Heart Circ Physiol 2004; 287(2): H927-36. [http://dx.doi.org/10.1152/ajpheart.01138.2003] [PMID: 15277207]

[3] Mantovani A, Allavena P, Sica A, Balkwill F. Cancer-related inflammation. Nature 2008; 454(7203): 436-44. [http://dx.doi.org/10.1038/nature07205] [PMID: 18650914]

[4] Wang Q, Tompkins KD, Simonyi A, Korthuis RJ, Sun AY, Sun GY. Apocynin protects against global cerebral ischemia-reperfusion-induced oxidative stress and injury in the gerbil hippocampus. Brain Res 2006; 1090(1): 182-9. [http://dx.doi.org/10.1016/j.brainres.2006.03.060] [PMID: 16650838]

[5] Planas AM, Chamorro A. Regulatory T cells protect the brain after stroke. Nat Med 2009; 15(2): 138-9. [http://dx.doi.org/10.1038/nm0209-138] [PMID: 19197285]

[6] Liesz A, Suri-Payer E, Veltkamp C, *et al.* Regulatory T cells are key cerebroprotective immunomodulators in acute experimental stroke. Nat Med 2009; 15(2): 192-9. [http://dx.doi.org/10.1038/nm.1927] [PMID: 19169263]

[7] Shelat PB, Chalimoniuk M, Wang JH, *et al.* Amyloid beta peptide and NMDA induce ROS from NADPH oxidase and AA release from cytosolic phospholipase A2 in cortical neurons. J Neurochem 2008; 106(1): 45-55. [http://dx.doi.org/10.1111/j.1471-4159.2008.05347.x] [PMID: 18346200]

[8] Steffen Y, Gruber C, Schewe T, Sies H. Mono-O-methylated flavanols and other flavonoids as inhibitors of endothelial NADPH oxidase. Arch Biochem Biophys 2008; 469(2): 209-19. [http://dx.doi.org/10.1016/j.abb.2007.10.012] [PMID: 17996190]

[9] Klees RF, *et al.* Apocynin derivatives interrupt intracellular signaling resulting in decreased migration in breast cancer cells. J Biomed Biotechnol 2006; 2006(2): 87246.

[10] Attia DM, Verhagen AM, Stroes ES, *et al.* Vitamin E alleviates renal injury, but not hypertension, during chronic nitric oxide synthase inhibition in rats. J Am Soc Nephrol 2001; 12(12): 2585-93. [PMID: 11729226]

[11] Yang Z, Asico LD, Yu P, *et al.* D5 dopamine receptor regulation of reactive oxygen species production, NADPH oxidase, and blood pressure. Am J Physiol Regul Integr Comp Physiol 2006; 290(1): R96-R104. [http://dx.doi.org/10.1152/ajpregu.00434.2005] [PMID: 16352863]

[12] Sullivan JL. Iron and the sex difference in heart disease risk. Lancet 1981; 1(8233): 1293-4. [http://dx.doi.org/10.1016/S0140-6736(81)92463-6] [PMID: 6112609]

[13] Kell DB. Towards a unifying, systems biology understanding of large-scale cellular death and destruction caused by poorly liganded iron: Parkinson's, Huntington's, Alzheimer's, prions, bactericides, chemical toxicology and others as examples. Arch Toxicol 2010; 84(11): 825-89. [http://dx.doi.org/10.1007/s00204-010-0577-x] [PMID: 20967426]

[14] Libby P. Inflammation and cardiovascular disease mechanisms. Am J Clin Nutr 2006; 83(2) (Suppl.): 456S-60S. [PMID: 16470012]

[15] Medzhitov R. Origin and physiological roles of inflammation. Nature 2008; 454(7203): 428-35. [http://dx.doi.org/10.1038/nature07201] [PMID: 18650913]

[16] Warner N, Burberry A, Franchi L, *et al.* A genome-wide siRNA screen reveals positive and negative regulators of the NOD2 and NF-κB signaling pathways. Sci Signal 2013; 6(258): rs3. [http://dx.doi.org/10.1126/scisignal.2003305] [PMID: 23322906]

[17] Barnig C, *et al.* Lipoxin A4 regulates natural killer cell and type 2 innate lymphoid cell activation in asthma. Sci Transl Med 2013; 5(174) 174ra26 [http://dx.doi.org/10.1126/scitranslmed.3004812]

[18] O'Byrne PM, Naji N, Gauvreau GM. Severe asthma: future treatments. Clin Exp Allergy 2012; 42(5): 706-11. [http://dx.doi.org/10.1111/j.1365-2222.2012.03965.x] [PMID: 22515391]

[19] van Greevenbroek MM, Schalkwijk CG, Stehouwer CD. Obesity-associated low-grade inflammation in type 2 diabetes mellitus: causes and consequences. Neth J Med 2013; 71(4): 174-87. [PMID: 23723111]

[20] Gorman RM. Always On. Proto Dispatches from the Frontiers of Medicine 2008;; 26-31.

[21] Crunkhorn S. Metabolic disorders: Breaking the links between inflammation and diabetes. Nat Rev Drug Discov 2013; 12(4): 261. [http://dx.doi.org/10.1038/nrd3988] [PMID: 23493085]

[22] Kiechl S, Wittmann J, Giaccari A, *et al.* Blockade of receptor activator of nuclear factor-κB (RANKL) signaling improves hepatic insulin resistance and prevents development of diabetes mellitus. Nat Med 2013; 19(3): 358-63. [http://dx.doi.org/10.1038/nm.3084] [PMID: 23396210]

[23] Schlosser E. Fast Food Nation. New York: Houghton-Mifflin 2001.

[24] Tan ZS, Beiser AS, Vasan RS, *et al.* Inflammatory markers and the risk of Alzheimer disease: the Framingham Study. Neurology 2007; 68(22): 1902-8. [http://dx.doi.org/10.1212/01.wnl.0000263217.36439.da] [PMID: 17536046]

[25] McGeer EG, McGeer PL. Innate immunity in Alzheimer's disease: a model for local inflammatory reactions. Mol Interv 2001; 1(1): 22-9. [PMID: 14993335]

[26] Wang Y, Chang CF, Chou J, *et al.* Dietary supplementation with blueberries, spinach, or spirulina reduces ischemic brain damage. Exp Neurol 2005; 193(1): 75-84. [http://dx.doi.org/10.1016/j.expneurol.2004.12.014] [PMID: 15817266]

[27] Waris G, Ahsan H. Reactive oxygen species: role in the development of cancer and various chronic conditions. J Carcinog 2006; 5: 14. [http://dx.doi.org/10.1186/1477-3163-5-14] [PMID: 16689993]

[28] Zhuang G, *et al.* Tumour-secreted miR-9 promotes endothelial cell migration and angiogenesis by activating the JAK-STAT pathway. EMBO J 2012; 31: 3513-23.

CHAPTER 3

Metabolic Syndrome, Diabetes, Heart Disease and Stroke

Abstract: Heart disease is the biggest killer in the world. Its major cause is obesity or metabolic syndrome, which can also lead to diabetes and stroke - the third leading killer (after cancer) and the most frequent cause of disability in the world. All of these involve imbalances in energy metabolism. Early symptoms of metabolic syndrome include excessive weight, high blood pressure, elevated blood glucose and elevated levels of lipids, especially low density lipoprotein (LDL). People who have one or more of these symptoms are at a higher risk of developing heart disease, stroke and type-2 diabetes. Childhood obesity is such a big problem that it will probably mean that this generation will be the first to have a shorter life expectancy than their parents. Ischemic strokes need to be treated as soon as possible, so it is very important to get a stroke victim to a hospital as soon as possible, so the blockage can be removed and blood can begin to flow properly. Drugs such as aspirin, clopidrogel and dipyridamole can be given to prevent the formation of more clots by preventing blood platelets from aggregating. If the patient can get to a hospital soon enough, tPA, or tissue plasminogen activator, can be given. When blood flow to the heart is interrupted, it causes a myocardial infarction, or heart attack. The most common cause is a blockage in the coronary artery that is usually caused by the rupture of an atherosclerotic plaque.

Keywords: Diabetes, Heart disease, Metabolic syndrome, Stroke, Tissue plasminogen activator, tPA.

Next, let us look at the biggest killer in the world, heart disease, along with its major cause, obesity and metabolic syndrome, which can also lead to diabetes and stroke, which is the third leading killer (after cancer) and the most frequent cause of disability in the world. All of these involve imbalances in energy metabolism. However, let us also remember that somewhere in the world a child dies every 5 seconds from malnutrition or starvation, as described by the UN Secretary General, Ban Ki-moon in a three-day summit on food security in Rome in Nov. 2009. That is, hunger kills 17,000 children daily, while people in the USA and

Robert E. Smith

many other countries over-eat and suffer the effects of obesity.

GLUCOSE AND ENERGY PRODUCTION

For a person to be healthy, it is important that he/she is able to convert food into needed energy. Dietary carbohydrates are converted to glucose, which is the primary fuel that provides energy to our cells (especially in the brain), and as a source of intermediates for the biosynthesis of other carbohydrates, as well as nucleic acids, fats, amino acids and proteins. Glucose homeostasis is maintained by several hormones, with insulin and glucagon being the most important. When the concentration of glucose in the blood rises in a healthy person, the β-cells of the pancreas secrete insulin, which inhibits hepatic synthesis of glucose and/or increases glucose uptake into the muscles, liver and adipose tissue. This causes blood glucose levels to decrease. Glucagon is secreted by the α-cells of the pancreas when the concentration of glucose in the blood is low. Glucagon has the opposite effect of insulin. It stimulates the liver to make more glucose by breaking down some of its stored glycogen.

Glucose can be broken down further by glycolysis to produce energy, or it can be converted into glucose-3-phosphate and then into phospholipids and/or triglycerides, depending on the needs of the cell. Triglycerides can be stored for later use, or hydrolyzed into glycerol and free fatty acids, which can be used for fuel. When energy is no longer needed, fatty acids can react with glycerol to form triglycerides, which store energy.

Dietary proteins are broken down into amino acids, which can be used to make human proteins and polypeptides. Carbohydrates, nucleic acids, fats and proteins are interconvertible. So, people on a low fat, high carbohydrate diet can still biosynthesize the fats that they need. Similarly, people on a low carbohydrate, high protein and fat diet can make all the carbohydrates that they need. Since our bodies are continuously breaking down and re-making their internal components, we need food to supply the raw materials. Also, we need energy to live, so our diets must provide molecules like glucose, which can be broken down and converted to energy. The energy balance is closely regulated. More than 700,000 kcal are ingested each year by most well-fed adults, and "body weight is kept within a range of ± 1 kg, or an equivalent of 7000 kcal. That is, "the regulation of energy operates with an accuracy of about 99% [1].

So, glucose can be broken down by the metabolic pathway called glycolysis, which occurs in the cytosol. Under aerobic conditions, glycolysis makes pyruvic acid, or pyruvate, along with a coenzyme called NADH, and a high-energy molecule called adenosine triphosphate, or ATP. The overall reaction is:

$$C_6H_{12}O_6 + 2\ NAD^+ + 2\ ADP + 2\ H_2PO_4^- \rightarrow 2C_3H_4O_3 + 2\ NADH + 2H^+ + 2ATP + 2\ H_2O$$

where $C_6H_{12}O_6$ is glucose and $C_3H_4O_3$ is pyruvic acid, which exists as pyruvate at physiological pH of 7.3 – 7.4. The structure of pyruvic acid is shown in Fig. (**1**).

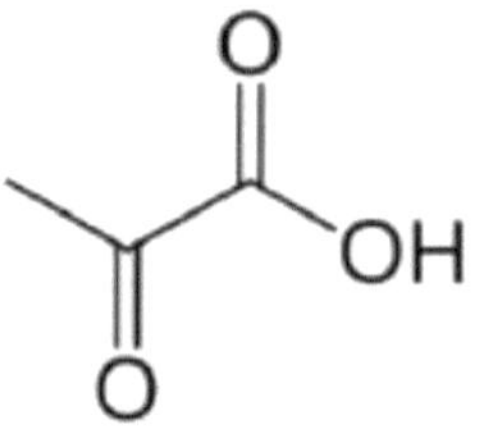

Fig. (1). Pyruvic acid, also written as $CH_3COCOOH$.

If there is not enough oxygen, as in over-worked muscle cells, pyruvate is converted to another carboxylic acid, lactic acid (lactate at pH 7.3). The lactate causes the burning sensation in over-worked muscles. Under anaerobic conditions, such as fermentation in yeast, the end product of glycolysis is ethanol. Also, remember that carboxylic acids, such as acetic acid, pyruvic acid and lactic acid are relatively weak acids and they lose one hydrogen ion (H^+) at pH 7.3, and become carboxylate anions, *i.e.*

$$CH_3COOH \rightleftharpoons CH_3COO^- + H^+ \quad pK_a = 4.75 \text{ (Acetic acid)}$$

$$CH_3COCOOH \rightleftharpoons CH_3COCOO^- + H^+ \quad pK_a = 2.39 \text{ (Pyruvic acid)}$$

$$CH_3CHOHCOOH \rightleftharpoons CH_3CHOHCOO^- + H^+ \quad pK_a = 3.86 \text{ (Lactic acid)}$$

Once the pyruvate is formed in the cytosol of healthy cells, it goes to the mitochondria, where it is converted to acetyl-CoA, catalyzed by an enzyme called pyruvate decarboxylase. Acetyl-CoA is central to the balance between carbohydrate, amino acid and fat metabolism. "Acetyl-CoA feeds into another metabolic process called the citric acid cycle (also known as the tricarboxylic acid cycle, TCA cycle, and the Krebs cycle). The citric acid cycle provides more energy and it is linked to amino acid and protein metabolism.

In fact, fatty acids can also be broken down to acetyl-CoA, and fed into the citric acid cycle. Acetyl-CoA is broken down into CO_2 and H_2O, while high energy molecules, GTP, $FADH_2$ and NADH are produced. This is the overall reaction of one turn of the cycle:

Acetyl-CoA + 3 NAD^+ + FAD + GDP + phosphate + 2 H_2O → CoA-SH + 3 NADH + 3 H^+ + $FADH_2$ + GTP + 2 CO_2

During the cycle, some of intermediate molecules can be transformed into amino acids, which are the building blocks of proteins. Likewise, proteins can be broken down into amino acids, which can feed into the TCA cycle.

The mitochondria are often called the power plants of the cell. That is, one of their major functions is to produce energy in the form of adenosine triphosphate, or ATP. They do this in an electron transport chain. NADH and $FADH_2$ are oxidized and ATP is produced, along with reactive oxygen substances (ROS). ATP activates many biomolecules and powers much of the biosynthesis, development, regulation and repair that are required to sustain life. As a result, inefficiency or poor regulation of energy metabolism can have a profound effect on human health. However, natural history and human history can work against us. We evolved with an ability to store energy at times when food was not plentiful. So, when modern humans go on a diet, our genes code for proteins which slow down our metabolism and limit our ability to lose weight.

Throughout most of human history, malnutrition and starvation have been serious problems. In the late 1800s, skinny people had little chance of social or political success. Obesity was thought to be a sign of good health. Teddy Roosevelt had to gain much weight before he could become a viable candidate for president. Only in the middle to late 1900s did Americans begin to change this thinking. In the last 30 years, metabolic syndrome has become more common, affecting more than 30 million people in the USA [1]. An estimated two-thirds of the US population and one billion people throughout the world are overweight and/or obese. Obesity kills more people than tobacco [1]. Much of this can be attributed to higher caloric intake, reduced physical activity and the aging of our population. Early symptoms of metabolic syndrome include excessive weight, high blood pressure, elevated blood glucose and elevated levels of lipids, especially low density lipoprotein (LDL). People who have two or more of these symptoms are at a higher risk of developing heart disease, stroke and type-2 diabetes [1]. Childhood obesity is such a big problem that it will probably mean that this generation will be the first to have a shorter life expectancy than their parents.

METABOLIC SYNDROME

Metabolic syndrome was first described by G.M. Reaven in 1988 [1]. In addition to the symptoms just mentioned, metabolic syndrome is characterized by many abnormalities, including inefficient use of glucose as an energy source, excessive conversion of glucose to fat, lowered sensitivity to insulin, loss of mitochondrial function and energy production, low levels of high density lipoprotein (HDL), excessive appetite and weight gain, impaired circulation because of clogged arteries, inflammation, impaired organ function and depressed immune response [1].

Modern American life tends to encourage metabolic disorders. Instead of burning calories to grow and produce food, we tend to get in a car and buy fast food that tastes good to many people, but is loaded with calories [2]. At the same time, some people develop eating disorders (anorexia and bulemia), which kill people much sooner than diseases of obesity, such as heart disease, stroke and type-2 diabetes. So, it is important to remember that good health requires a good balance. Not just our diets, but also our lifestyles need proper balance, with an emphasis on the wholeness of life [3]. Although our modern American society has eliminated the need for hard, physical activity to make a living, it has also discouraged us from getting proper exercise, but still wants us to look very skinny. Our rapid, high pressure lifestyles can cause stress and leave us little time for good eating habits. The typical fast food diet is high in calories and flavor, but low in nutrition

[3]. As a result, many of us have high levels of glucose in our blood, which can cause problems.

It is important to note that the American (USA, not South American) fast food life style is rejected by many of our European (and South American) allies. For example, when the first McDonald's restaurant opened in Rome, Italy, there was a large demonstration against it. The demonstrators started a Slow Food organization (website: http://www.slowfood.com/), dedicated to the very Italian process of spending much time preparing the food, enjoying lively discussions, slowly eating the food and singing songs after the meal was over. Many European tourists in the USA are unpleasantly surprised by how fast they are supposed to eat their meals in American restaurants. Similarly, our Mexican, Central and South American friends reject the fast foods of the USA. When instant soups were introduced into South American markets, they failed miserably, because cooks felt sure that the only way to prepare healthy meals was to do it slowly. Moreover, it is much better to eat slowly and chew, rather than gulp. It takes about a half hour for ingested food to affect our brains and make us feel satiated. When we eat a large meal in just a few minutes, we can eat much more than if we eat slowly. Another very common problem in the USA is sweetened soda pop, which has been shown to stimulate the appetite. In fact, most of us have been brain washed into thinking that the only way to enjoy a movie is to have a large soda and box of buttered popcorn. The soda (or any sweetened beverage) can cause a rapid increase in blood glucose levels, which can eventually lead to diabetes. Even though sweetened beverages contain fruit sugar (high fructose corn syrup), they are very unhealthy. The fructose is rapidly converted to glucose, which does much damage.

One way that excess blood glucose causes problems is that glucose and its metabolites (such as methyl glyoxal) can react with proteins and body tissues [1]. These reactions can form advanced glycation end products (AGEs) as well as reactive oxygen and nitrogen products that are toxic. These products bio-accumulate and lower our sensitivity to insulin. High levels of glucose also inhibit the immune system. In fact, sweetened soda pop has been called public health enemy number one in the USA, since it not only has too much sugar, but also drinking it tends to make us feel hungrier. Sweetened soda pop has many calories, has no nutrients and rapidly increases blood glucose levels to dan-gerously high levels. The elevated glucose is converted to fat and is stored in fat cells called adipose cells [1].

Adipose tissue is an active endocrine and immune organ. It secretes lipid hormones (adipokines). When caloric intake exceeds energy consumed, adipocytes undergo increased recruitment, proliferation and differentiation. If adipogenesis is sufficient, metabolic syndrome can be avoided. When adipogenesis is impaired, adipocytes can become dysfunctional and help cause diseases related to metabolic syndrome. This is often called adiposopathy, or sick fat. This fat is especially dangerous when it accumulates in the abdomen. Adiposity increases the macrophage content of adipocytes.

Adipose cells can be either white or brown [4]. White adipocytes store fats as triglycerides in a relatively large, single lipid droplet. When nutritional sources of energy are scarce, the triglycerides in the lipid droplet are hydrolyzed, generating fatty acids that are exported into the blood and circulate to other tissues that need energy. Brown adipocytes, though, store lipids in many smaller droplets. Brown adipocytes also have more mitochondria than white adipocytes. Brown adipocytes will catalyze the hydrolysis of the triglycerides to make energy for themselves [4]. "However, having an excess (>150 mg/dL) of triglycerides in the blood is a risk factor for heart disease and strokes. That is, hypertriglyceridemia is diagnosed by measuring the fasting level of triglycerides in the blood. Mild and moderate hypertriglyceridemia (150–999 mg/dL) increase the risk of cardiovascular diseases. Severe and very severe hypertriglyceridemia (triglycerides > 1000 mg/dL) is also considered to be a risk for pancreatitis" [4].

Adipose cells not only store energy, but also produce regulatory molecules, including a small protein hormone called leptin [1]. It is derived from adipose tissue. When leptin is secreted in a healthy person, it goes to the brain and it causes her (or him) to feel full and stop eating. Leptin has a crucial role in regulating energy intake and expenditure. Leptin affects the movement (or flux) of nutrients between organs, as well as the conversion and storage of nutrients [5, 6]. Leptin-mediated signals from the brain regulate lipid metabolism in adipocytes [6]. "The signaling mechanism operates through the phosphoinositide-3 kinase

(PI3K) system in the hypothalamus. The hypothalamus engages the sympathetic nervous system, inducing signaling in adipocytes through the endocannabinoid system" [6]. During starvation, the concentration of leptin falls, signaling a decline in energy stores to the brain [5]. "This causes changes in the expression of genes in the hypothalamus that affects thyroid hormone levels. Unfortunately for modern humans who have ready access to food, the opposite does not occur. When leptin concentrations increase (after eating), we develop a resistance to leptin and gain weight" [5].

In the brain, leptin targets many sites, including hypothalamic neurons in the paraventricular, dorsomedia, ventromedial and premammillary nuclei [5]. "Leptin also regulates neurons in the hindbrain and striatal areas. It also has opposing actions on two different populations of arcurate nuclei in the brain. One population synthesizes two neuropeptides (neuropeptide Y and agouti-regulated peptide) that stimulate appetite. Another population produces pre-proopiomelanocortin (POMC), which is a precursor of the appetite-suppressing peptide, called the α-melanocyte stimulating hormone" [5].

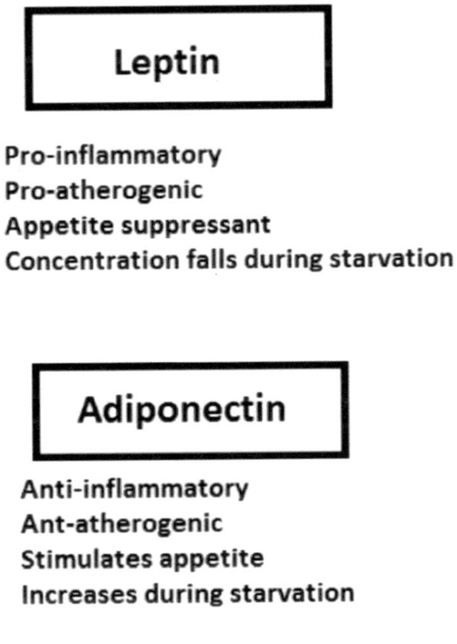

Fig. (2). Differences between leptin and adiponectin.

Another regulatory molecule produced by adipose cells is called adiponectin, which increases insulin sensitivity, decreases glucose production in the liver and decreases glucose uptake in the muscles [1]. "All of these are antidiabetic activities. Adiponectin is abundant in human blood (5–20 μg/ml), and its plasma concentration decreases as fat accumulates in the body. Plasma adiponectin concentrations are lower in patients with diabetes and ischemic heart disease. It activates PI3K and glucose uptake in muscles, enhances β-oxidation of fatty acids, and suppresses glucose production in the liver. Adiponectin receptors are found in skeletal muscle and the liver and they have important roles in the regulation of glucose and lipid metabolism, along with inflammation and oxidative stress. Adiponectin also decreases the concentration of lipids and cholesterol in the circulating blood and is anti-inflammatory" [1]. The different roles of leptin and adiponectin are highlighted in Fig. (**2**).

Excess body fat (especially abdominal fat) is associated with dysregulation of energy metabolism and diminished adiponectin secretion. Excess body fat also increases the production of a pro-inflammatory molecule called tumor necrosis factor alpha, or TNF-α [1]. Adipose cells (adipocytes) can become dysfunctional because of inflammation and improper regulation of crucial enzymes. So, adipocytes are not just storage cells for triglycerides, but complex cells that must be properly regulated [1]. Adipose tissue is highly vascularized [7]. "Several crucial angiogenic factors, including leptin, adiponectin, hepatocyte growth factor 1 (HGF-1), angiopoietin-2, and vascular endothelial growth factor A "(VEGF-A) are secreted by adipocytes, suggesting an autoregulatory function for angiogenesis in adipose tissue. VEGF-A accounts for most of the proangiogenic activity in adipose tissue. When VEGF-A is up-regulated in adipocytes, it improves vascularization and causes a "browning" of white adipose tissue, with an increase in energy expenditure and resistance to high fat diet-mediated metabolic insults. However, the effects of angiogenesis inhibitors depend on the biological context. Angiogenesis facilitates healthy fat pad expansion. Adipocytes have a smaller average size. There is an absence of hypoxia, minimal fibrosis, and essentially none of the hallmarks of inflammation that are characteristic of dysfunctional adipose tissue. VEGF-A also upregulates mitochondrial uncoupling protein 1 (UCP-1) and peroxisome proliferator-activated receptor-γ coactivators levels, prompting white adipose tissue to assume a phenotype resembling more closely brown adipose tissue" [7]. So, dietary angiogensis inhibitors in mushrooms and onions may have a negative health effect when gaining weight, but are probably helpful when one's weight is stable, even if obese. When an adult is gaining weight, inhibiting angiogenesis might decrease healthy fat pad expansion, but once one's weight is stabilized, inhibiting angiogenesis might inhibit this critical step in cancer metastasis.

Obesity is a risk factor in rheumatoid arthritis. TNF-α is found at relatively high concentrations in the joints of arthritis sufferers. The drug called etanercept (Enbrel®) blocks the TNF-α receptor. It has been used for years to treat rheumatoid arthritis and non-Hodgkin lymphoma. Enbrel® is called a biological because it is produced by living (biological) cells that have been cloned to produce it. It is a fusion protein that is made by translating a piece of DNA that links the human gene for the TNF receptor to the gene for the part of the immunoglobulin G1 (IgG1) protein that is crystallizable (Fc). More recently, a completely human protein, also a monoclonal antibody, called Adalimab (Humira®) has been approved for treating diseases by binding to TNF-α. It was found that the biologics, (etanercept, infliximab and adalimumab) have been more effective in treating rheumatoid arthritis when used in combination with methotrexate [8].

More recently (Jan 2010) a completely different drug, called tocilizumab (Acterma®), was approved by the FDA to treat patients with severe to moderate rheumatoid arthritis who had an inadequate response to TNF blockers. Tocilizumab is a monoclonal antibody that binds to the interleukin-6 receptor.

Another important class of excitatory peptide hormones that helps control appetite and energy levels is called the orexins, or hypocretins. They bind to orexin receptors (members of the GPCR class) and increase the craving for food and increases wakefulness. Hypocretin-producing cells can be inhibited by leptin, but are activated by hypoglycemia and ghrelin, another peptide hormone (secreted by the hypothalamus). Ghrelin increases food intake and fat mass and activates the reward network mesolimbic system of the brain.

In a recent study, an antagonist for both orexin 1 and 2 receptors was described [9]. It treated insomnia but did not disrupt cognition, as do current treatments for insomnia, zolpidem (Ambien®) and eszopiclone (Lunesta®). It had a wider therapeutic margin for sleep *versus* cognitive impairment [9]. In May, 2013 a US FDA committee voted in favor of approving an orexin receptor antagonist, suvorexant, for treating insomnia by turning off wakefulness. However, the FDA rejected the application, but left the door open for approving a lower dose.

One of the indicators of metabolic syndrome is an elevated level of LDL and cholesterol in blood, which are associated with plaque to build up. Plaque can develop in the arteries, which lowers blood circulation and proper oxygenation of tissues. This plaque can lead to hardening of the arteries and atherosclerosis. The heart uses fats as a major source of energy, so people with high LDL are susceptible to heart attacks. *Trans* fats and saturated fats increase the susceptibility to heart attacks and stroke. Unsaturated fats, polyunsaturated fats, as well as omega-3 and omega-6 fats lower our susceptibility to heart attack, stroke and other diseases of inflammation.

Another thing that happens in metabolic syndrome is a reduction in the amount of the enzyme called adenosine monophosphate-activated protein kinase, or AMP-APK [1]. AMP-APK is a critical regulator of fatty acid synthesis, energy production from fatty acids and glucose uptake. This enzyme can be activated when it is phosphorylated, or inactive when it is not phosphorylated. The activated form decreases the amount of fatty acid biosynthesis by phosphorylating key enzymes. The ratio of activated to inactivated AMP-APK increases with the amount of cardiovascular exercise that a person does [1]. Also, atypical antipsychotic drugs (olanzapine and clozapine) cause weight gain as a serious side effect by increasing the amount of activated AMP-APK in the brain [1].

DIABETES AND SMOLDERING INFLAMMATION

Metabolic syndrome and smoldering inflammation can also lead to type-2 diabetes, especially when a person has excess visceral, abdominal deep subcutaneous and ectopic fat [10]. The anti-inflammatory hormone insulin plays an important role in the development of type-2 diabetes [1]. When insulin binds to its receptor, it initiates a signaling cascade that must function properly to maintain good health. The insulin signaling cascade is shown in Figs. (**3 - 4**). When insulin binds to the α-subunit its membrane-bound receptor, it triggers the protein kinase activity of the β-subunit. Phosphorylated PI3K catalyzes the addition of a phosphate to PtdIns(4,5)P_2. This activates the PI3K cascade that regulates glycogen metabolism. The 3-phosphoinositide dependent protein kinase-1 (PDK-1) is subsequently activated [1]. "PDK-1 interacts with protein kinase B (PKB, also known as Akt) to phosphorylate glycogen synthase kinase-3 (GSK-3). In its active form, GSK-3 can catalyze the phosphorylation and inactivation of glycogen synthase. Also, the β-subunit catalyzes the phosphorylation of other adapter proteins, such as the insulin receptor substrate-1 (IRS-1) protein" [1].

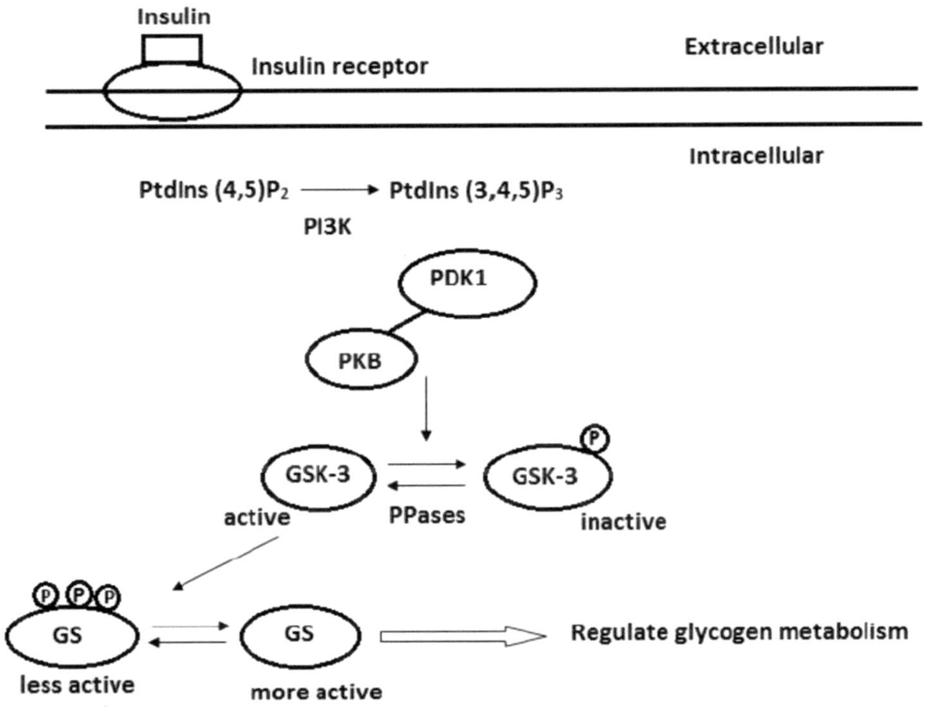

Fig. (3). Insulin signaling pathway and glycogen metabolism.

This then activates the Ras protein that activates the MAPK kinase cascade, activating genes and increasing enzyme synthesis. Also, PIP_2 binds to PKB, keeping it inactive. When phosphorylated, PKB is activated and stilutes glucose uptake by the GLUT4 glucose transporter. On the other hand, the α_{2A} adrenergic receptor mediates the suppression of insulin secretion, when it is appropriate [11]. When the signaling cascade is not regulated properly, it can contribute to metabolic syndrome, even before the onset of diabetes [1]. Insulin resistance and a decreased capacity for β cells to secrete insulin are central features of type-2 diabetes.

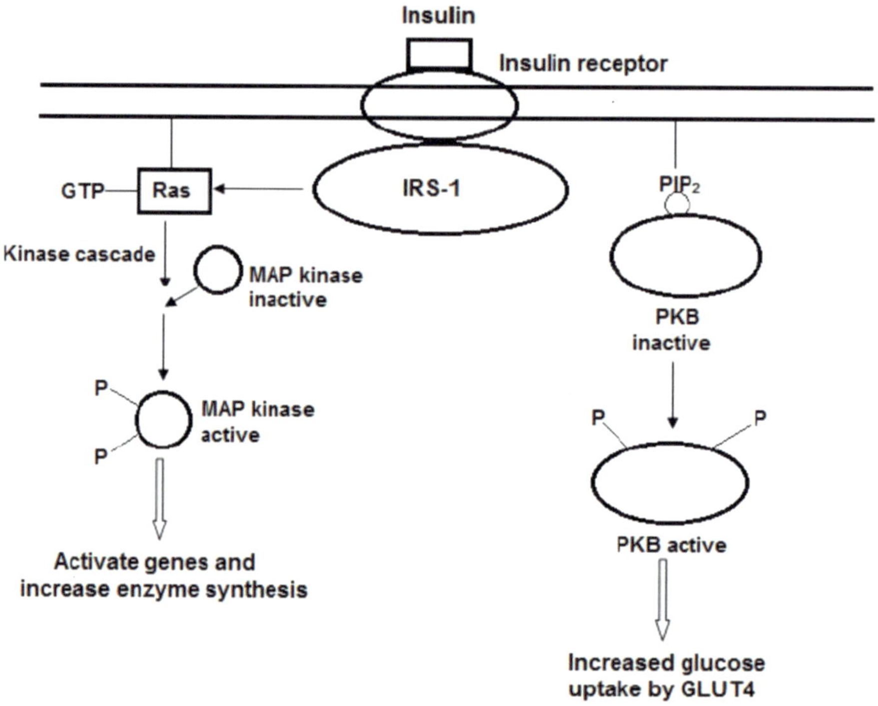

Fig. (4). Insulin signaling pathway and glucose metabolism.

It was estimated that 346 million people world-wide have diabetes [12]. It is characterized by hyperglycemia (fasting glucose >125 mg/dL) and an elevated percentage of glycated hemoglobin (≥6.5%), known as HbA_{1c}. That is, a chronic excess of glucose can cause tissue damage. Some of it is converted into fructose 6-phosphate, which feeds into the hexosamine pathway, which produces N-acetylglucosamine that can react with serine and tyrosine residues in proteins that are important in signal transduction. This is called OGlcNAcylation. Also, some of the excess glucose is converted to another sugar, sorbitol, as it enters the polyol pathway in a reaction catalyzed by aldol reductase, as shown in Fig. (**5**).

$$\text{Glucose} + \text{NADPH} + \text{H}^+ \rightarrow \text{Sorbitol} + \text{NADP}^+$$

Fig. (5). Reaction by which excess glucose enters the polyol pathway, catalyzed by aldol reductase.

The polyol pathway in which glucose is converted to sorbitol which is converted to fructose is shown in Fig. (**6**).

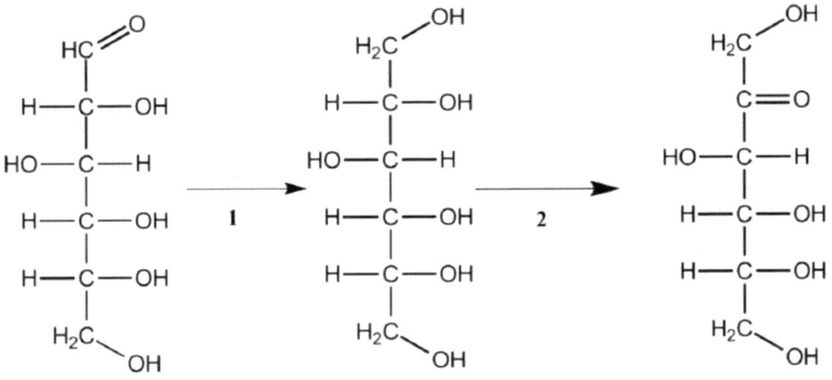

Fig. (6). Polyol pathway in which glucose is converted to sorbitol (1), which is then converted to fructose (2).

In the process, NADPH is converted to $NADP^+$, reducing the amount of NADPH that is needed to synthesize reduced glutathione, nitric oxide, *myo*-inositol and taurine. Nitric oxide (NO) is a vasodilator in blood vessels. In healthy endothelial cells, insulin activates endothelial nitric oxide synthase (eNOS) [13]. "In diabetes, endothelial insulin resistance impairs eNOS activation, reduces endothelium-dependent vasodilation, and promotes atherogenesis. Moreover, when the endothelium is inflamed due to activation by the nuclear factor kappa B (NFκB), insulin-mediated signaling pathways are perturbed, reducing NO bioactivity and increasing the release of cytokines" [13]. Also, glutathione is needed to destroy reactive oxygen and nitrogen substances (RONS, more commonly called reactive oxygen species, ROS) that are made by mitochondria as they produce energy. The elevated levels of RONS (or ROS) cause oxidative damage and smoldering inflammation. They can react with nitric oxide (NO) to make an especially reactive substance, peroxynitrate ($ONOO^-$), which can react with proteins, nitrosylating them. Glucose can be oxidized to methylglyoxal, which can cause proteins (like hemoglobin) to be glycated. These are called advanced glycation end products, or AGEs [14]. "Lipid oxidation can also form AGEs, including the most potent one, methylglyoxal. Another AGE, N-(Carboxymethyl)lysine (CML) accumulates in adipose tissue and fatty liver and provided evidence that this is a core mechanism leading to the dysregulation of cytokines production. CML is a major ligand for the receptor for AGE (RAGE). In addition to the effects in insulin resistance, AGEs have also been shown to induce β-cell dysfunction and apoptosis. AGEs can induce lipid peroxidization, damage mitochondria and cause apoptosis in cardiomyocytes. Sorbitol can also produce AGEs which can alter protein function and bind to specific receptors, changing of gene expression and causing toxicity. AGEs can also cause diabetic cardiomyopathy by activating the βII isoform of protein kinase C (PKC). This, in turn, alters cardiomyocyte contraction and the amount of Ca^{2+} that can be released from the endoplasmic reticulum" [14].

Furthermore, AGEs stimulate the formation of more RONS (ROS), such as the superoxide anion. Moreover, sorbitol cannot cross the cell membrane, so it builds up in the cell. This causes osmotic stresses as water enters the damaged cells. Sorbitol enters the sorbitol-aldolase reductase pathway, also known as the polyol pathway. When excess glucose is converted to sorbitol, less inositol is made. It is needed to make inositol phosphates (like IP_3) and phosphoinositides, which are especially important for properly functioning nerves and brain cells. Also, nitrosylated proteins and oxidized lipids are toxic and linked to neurodegenerative diseases and stroke.

Finally, excess glucose can lead to an increase in diacylglycerol (DAG), which activates PKC, which then stimulates the production of RONS. That is, glucose can enter glycolysis and produce an intermediate, dihydroxyacetone phosphate, DHAP. This can be converted to glycerol 3-phosphate, which can react with two fatty acids to form diacyl glycerol 3-phosphate. After the phosphate is hydrolyzed off, DAG is formed. In the vasculature, DAG and RONS (ROS) activate the β isoform of PKC. This leads to changes in cell signaling, the production of vasoconstrictors, and conversion of smooth muscle and endothelial cells to a proliferative phenotype in the retinal microcirculation and peripheral conduit vasculatures [15]. The activated DAG-PKC pathway can lead to vascular abnormalities in the retinal, renal, neural and cardiovascular tissues [16]. "This pathway regulates endothelial permeability, vasoconstriction, extracellular matrix (ECM) synthesis and turnover, cell growth, angiogenesis, cytokine activation and leucocyte adhesion in the heart. Increased DAG levels and PKC activity occur in diabetes. This is especially true for the α, β1/2 and δ isoforms of PKC in the retina, aorta, heart, renal glomeruli and circulating macrophages" [16].

Activated PKC has been associated with changes in blood flow, basement membrane thickening, extracellular matrix expansion, increases in vascular permeability, abnormal angiogenesis, excessive apoptosis and changes in enzymatic activity alterations such as Na^+/K^+–ATPase, cytosolic phospholipase 2 (cPLA2), PI3 Kinase and mitogen activated protein kinase (MAP kinase) [16]. Inhibition of PKC, especially the β-1/2 isoform, has been reported to prevent or normalize many vascular abnormalities [16]. Clinical studies have shown that ruboxistaurin, an orally available selective inhibitor of the PKCβ isoform, can normalize endothelial dysfunction, renal glomerular filtration rate and prevented loss of visual acuity in diabetic patients [17]. However, the FDA mandated another three-year clinical trial on ruboxistaurin, which the manufacturer may be reluctant to do, because of the cost. Still, they are looking into other indications for ruboxistaurin. Two phase III studies and one phase II study are ongoing in the USA. Two of these are looking at the effects of ruboxistaurin treatment on diabetic macular edema in patients with Type-1 and Type-2 diabetes [18]. The other is measuring the effects of ruboxistaurin treatment on early diabetic kidney in patients with Type-1 diabetes [18].

Eventually, diabetes can lead to heart disease, stroke, loss of vision, retinopathy, kidney failure, nervous system damage and even death [19]. "Type-1 diabetes, also known as juvenile onset diabetes and insulin dependent diabetes, is the most severe form. It is an autoimmune disease in which the body's immune system destroys the beta (β) cells in the pancreas. That is, the pancreas has endocrine and exocrine tissues. The exocrine cells secrete digestive enzymes which are delivered to the intestines. Clusters of endocrine cells (called islets of Langerhans) exist in the exocrine tissue. Within the islets, α-cells make glucagon; β-cells, insulin; δ-cells, somatostatin; and γ-cells, pancreatic polypeptide — all of which are delivered into the blood stream. Insulin and glucagon have opposite effects. Glucagon raises blood glucose and causes the liver to convert glycogen to glucose, which is released into the bloodstream. So, after eating a meal, when blood glucose increases, β-cells secrete insulin, which lowers the glucose. If levels fall too much, the α-cells secrete glucagon, elevating the glucose level. Obesity is correlated with insulin resistance, leading to type-2 diabetes. However, type-1 diabetes can affect infants and children long before they are old enough to become obese" [19].

Smoldering inflammation is characterized by high concentrations of circulating proinflammatory cytokines and free fatty acids [10]. "They can help to induce insulin resistance, and β-cell dysfunction. The sensitivity of target cells to insulin decreases. TNF-α and lipopolysaccharides can induce insulin resistance and the level of signal transduction through a physiological negative feedback mechanism of normal insulin signaling. When insulin binds to its receptor, tyrosine residues in the intracellular part are autophosphorylated. In the metabolic pathway of insulin signaling, the insulin receptor substrate (IRS) docks with the insulin receptor and is trans-phosphorylated in its tyrosine residues by the phosphorylated insulin receptor. "Subsequently, more members of the insulin signal transduction pathway, including phosphatidylinositol-3-kinase (PI3K) and Akt/protein kinase B (PKB), are recruited and activated, so they can induce downstream effects" [10].

The physiological negative feedback mechanism is induced when insulin activates mTOR and the ζ isoform of protein kinase C, PKCζ [10]. "These intracellular serine-threonine kinases can then either directly, or indirectly (*e.g.via* IkappaB kinase beta (IKKβ)), phosphorylate serine and threonine residues in the insulin receptor substrate, or IRS. Once it is phosphorylated, IRS interrupts, or at least reduces, insulin signal transduction, and induce dissociation of IRS proteins from the insulin, receptor, induce degradation of IRS proteins, remove IRS proteins from complexes that keep them in close proximity to the insulin receptor, and turn IRS proteins into inhibitors of insulin receptor kinases. The proinflammatory cytokines TNF-α, IL-670 and IL-1α, and saturated free fatty acids are all involved in obesity-associated low-grade inflammation. They can induce phosphorylation of the IRS proteins and disrupt insulin signaling, but with pathophysiological consequences. Obese, hypertrophic and/or insulin resistant adipocytes release saturated fatty acids that can activate the TLR-4/NF-kB pathways on macrophages in adipose tissue, which then release TNF-α, which in turn binds to TNF receptors on the adipocytes. This further stimulates fatty acid release and starts a vicious cycle of worsening inflammation and insulin resistance. However, unsaturated and omega-3 fats are anti-inflammatory" [10].

Insulin signal transduction by PI3 kinase mainly affects metabolic pathways such as translocating the glucose transporter (GLUT4 in the heart) and inhibiting hormone-sensitive lipase [10]. "The other main insulin signal transduction pathway of is the renin-angiotensin system/mitogen activated protein (RAS/MAP) kinase pathway. It primarily stimulates mitogenic rather than metabolic processes. The renin-angiotensin system (RAS) is a hormone system that regulates blood pressure and fluid (water) balance. When the blood volume is low, the kidneys activate prorenin and secrete renin into the blood plasma, where it catalyzes the conversion of angiotensinogen (released from the liver) to angiotensin I. It is then converted to angiotensin II in a reaction catalyzed by the angiotensin converting enzyme (ACE). It is vaso-active and causes blood vessels to constrict, raising the blood pressure. Moreover, angiotensin II stimulates the adrenal cortex to secret aldosterone and the RAS/MAP pathway in the vasculature."This causes the tubules of the kidneys to increase the reabsorption of Na^+ and water into the blood. This increases the volume of fluid in the body, and thus the blood pressure. If the RAS is too active, it can cause hypertension (high blood pressure). There are many anti-hypertensive drugs that target the RAS. They are also prescribed to treat heart failure, kidney failure and to the harmful effects of diabetes" [10]. Please note that RAS is quite distinct from the oncoprotein ras, or the gene that codes for it, RAS, will be discussed in more detail in Chapter 12.

Circulating proinflammatory cytokines and free fatty acids can also cause macrophages in adipose tissue to develop into pro-inflammatory macrophages, called M_1 macrophages [10]. "However, there are also anti-inflammatory M_2 macrophages that resolve or dampen inflammation caused by M_1 macrophages. In obesity, most of the macrophages in adipose tissue are the M_1 phenotype. Then, as triglycerides build up in the liver, steatosis (fat accumulation) can begin and lead to non-alcoholic fatty liver disease (NAFLD). This can develop further into steatohepatitis, fibrosis and eventually cirrhosis of the liver. Excess visceral fat exposes the liver to higher concentrations of proinflammatory cytokines and free fatty acids. They are transported through the portal vein and into the liver, where they contribute to the development of NAFLD. So, the occurrence and severity of visceral fat accumulation are highly correlated with NAFLD. Also, excess visceral fat may contribute directly to smoldering inflammation and increased systemic levels of free fatty acids. On the other hand, there is a protein hormone, adiponectin, which has anti-inflammatory effects in adipose tissue. It may exert these effects by regulating microRNAs that suppress intracellular proinflammatory pathways, such as toll-like receptor 4 signaling. So, microRNAs have become a potential therapeutic target or tool to treat insulin resistance and type-2 diabetes" [10].

Metabolic syndrome also important causes changes in cells and tissues [10]. "Healthy β-cells can maintain glucose homeostasis by increasing their size. However, when demand for insulin exceeds the ability to produce it, hyperglycemia can occur. It can start with β-cell failure due to acquired and/or genetic defects in existing β-cells or by reduced β-cell mass. Prolonged exposure of pancreatic β-cells to toxic levels of glucose and lipids (glucotoxicity and lipotoxicity) can cause oxidative stress. Inflammatory cytokines may also contribute to β-cell dysfunction and type-2 diabetes. They cause β-cells to produce IL-1β, which can cause the β-cells to deteriorate. Also, leptin, which circulates in considerably increased concentrations in obesity, was shown to increase the release of IL-1β by β-cells" [10].

At the same time, microvascular dysfunction is involved in a cycle of both causing type-2 diabetes and being caused by it [10]. "Obese insulin-resistant humans and rats have impaired capillary recruitment, which has been shown to be necessary for normal insulin-mediated glucose uptake by skeletal muscle. Elevated levels of free fatty acids and inflammatory cytokines, together with less adiponectin can lead to dysfunctional microvasculature. This can induce endothelial insulin resistance, reduce local NO production and lower insulin-mediated glucose uptake in muscle, as part of overall insulin resistance. Endothelial dysfunction: a shared factor underlying both micro- and macrovascular dysfunction. Macrovascular disease is a major consequence of obesity-induced inflammation and type-2 diabetes. Microvascular dysfunction may be further aggravated in the expanding adipose tissue, since it produces all the factors of the renin- angiotensin system (RAS) that are needed to produce angiotensin II. Moreover, RAS activity is enhanced in obesity. Perivascular fat around resistance arterioles of muscle may directly affect the function of these vessels. Microvascular dysfunction may also contribute to the vicious cycle of adipose tissue dysfunction and inflammation. Functional capillaries in the expanding adipose tissue are necessary to provide optimal blood flow as well as deliver nutrients and oxygen to adipocytes. Thus, insufficient adipose tissue angiogenesis and formation of capillaries may lead to hypoxia and induction of an inflammatory response. A relative reduction in the density of the capillary network, combined with microvascular dysfunction may aggravate the hypoxic and inflammatory processes in adipose tissue depots and lead to deterioration of insulin resistance and metabolic homeostasis. Microvascular dysfunction may also contribute to the development of type-2 diabetes due to its effects on β-cell function. For example, there are transient periods of (mild) hyperglycemia that coincide with insulin resistance and low-grade inflammation. This can lead to reduced perfusion and mild ischemia in the islets" [10].

Moreover, several processes induce the inflammatory response in adipose tissue [10]. "This includes adipocyte hypertrophy, apoptosis and macrophage infiltration, which most likely act simultaneously. When activated by the appropriate death ligands, receptors induce apoptosis that is mediated by caspase-8. If caspase-8 activity is blocked, however, then these receptors promote death by necroptosis (programmed necrosis). Necroptosis promotes inflammation, so inhibiting this pathway can limit extensive tissue damage and even lethality in inflammatory syndromes. On the subcellular level, inflammasomes, such as NOD-like receptor family pyrin domain containing 3 (NLRP3), may be central regulators of early adipose tissue inflammation. Also, inflammasomes are pattern-recognition receptors (PRRs) that assemble into larger structures that control the maturation and secretion of proinflammatory interleukins such as IL-1β. NLRP3 releases the cysteine protease caspase-1, which catalyzes the conversion of procytokines into their mature active forms. The expression of inflammasome NLRP3 components is increased in obesity. Several endogenous stress signals, including glucose, palmitate, cholesterol crystals, islet amyloid peptides and RONS (ROS), can induce inflammasome *in vivo*. Also cellular components of the innate and adaptive immune systems may contribute to adipose tissue dysfunction. The stromal vascular part of adipose tissue contains various types of immune cells, including macrophages. Still, human adipocytes and preadipocytes seem to be able to prime inflammation and attract T-cells independently of macrophages. Moreover, there are more macrophages, but also of pro-inflammatory T-cells in the subcutaneous adipose tissue of type-2 diabetes patients. T-cells that are infiltrated in adipose tissue may not only attract macrophages, but also skew their differentiation towards the M1 phenotyp. In contrast, induction of T-regulatory cells can be beneficial and reduce adipose tissue inflammation and insulin resistance. The anti-inflammatory master switch in adipocyte differentiation, PPAR-γ is helps to control visceral adipose-tissue-resident regulatory T-cells. The gut microbiome may also influence the inflammatory response that is seen in insulin resistance and type-2 diabetes. Gut-derived endotoxins can potentially trigger Toll-like receptors in adipose tissue or on pancreatic β-cells, thus contributing to both insulin resistance and β-cell failure. At the same time, in the liver, portal endotoxemia may contribute to inflammation in hepatic steatosis and be a relevant risk factor for nonalcoholic steatohepatitis (NASH)" [10].

There are two early activators of adipose tissue inflammation in obesity: the complement system and advanced glycation end products (AGEs) [10]. "The complement system is a complex protein network that is best known as being part of the innate immune system. It is produced in the liver, adipose tissue and endothelial cells. Complement activation is under tight control by many circulating and cell-bound inhibitors that prevent unrestrained progression of complement activation once any of three pathways is initiated. "They are the classical, the lectin and the alternative pathways. They all converge on complement C3, the central component of the complement system. The alternative pathway is an amplification loop that enhances complement activation once it is initiated by activation by any of the three pathways. All three pathways activate the terminal complement pathway. The complement system is an essential regulator of cell and tissue homeostasis, as well as increasing immunity. Adipose tissue expresses many complement components and regulators. The alternative complement pathway is activated in adipocytes. The expression of proximal components of the classical pathway is altered in the subcutaneous adipose cells of insulin resistant individuals. Much data suggest that complement activation in obese people may help recruit immune cells from the circulation to the adipose tissue. This may enhance the cellular immune response in adipose tissue and thereby contribute to adipose tissue dysfunction. In addition to the proposed effects of complement activation on adipose tissue dysfunction, complement activation may play a role in the progression of β-cell dysfunction in type-2 diabetes. Also, complement system activation leads to endothelial dysfunction. Since the microvasculature comprises approximately 98% of the total vascular surface area, complement activation will inevitably affect microvascular function. With respect to atherosclerosis, complement activation by the classical and perhaps also the lectin pathway may aid in removing apoptotic cells and cellular debris. This will have protective effects in the atherosclerotic plaque, while complement activation beyond C3 may be proatherogenic. Moreover, complement activation may be cause atherothrombosis to develop, since proteases of the coagulation and fibrinolysis systems may activate the complement system, and *vice versa*" [10].

Dyslipidemia (excess fatty acids and low ratio of LDL to HDL) is another feature of metabolic syndrome. This, together with accumulation of ectopic fat depot, can cause pancreatic lipotoxicity [10]. "Increased concentrations of saturated fatty acids can harm β-cells and produce other proinflammatory cytokines. Moreover, adipose tissue contains mature adipocytes, preadipocytes, endothelial cells, fibroblasts, macrophages and other immune cells. The number of adipose tissue macrophages increases in obesity. Thus, chemoattractants are produced locally and attract monocytes to adipose tissue depots, where they can contribute to inflammation and expansion of adipose tissue. This can lead to adipocyte hypertrophy, local hypoxia and adipocyte apoptosis. This generates signals to recruit macrophages. Hypertrophic adipocytes start to secrete low levels of TNF-α, which stimulate preadipocytes and endothelial cells to produce monocyte chemotactic protein. Inflammatory cytokines (TNF-α and IL-6) can cause insulin resistance" [10].

TREATMENTS AND POSSIBLE CURES FOR DIABETES

One important drug for diabetes, metformin (Glucophage®) acts by increasing the amount of one enzyme (adenosine monophosphate activated protein kinase, AMP-APK) and decreases the activity of another, protein kinase C [1]. "Glucophage® is the first line treatment for type-2 diabetes, also known as adult onset diabetes. It usually occurs in people who are over 40 and who are overweight. Glucophage® reduces the amount of glucose that is released from the liver and increases the transport of glucose into muscles, increasing insulin sensitivity and decreasing glucose absorption from the gastrointestinal tract. Metformin is also available as Janumet® in combination with the dipeptidyl peptidase (DPP-4) inhibitor, sitagliptin" [1].

Another important diabetes drug, glipizide (Glucotrol®) binds to K^+ ion channels in the β cells of the pancreas. This changes the membrane potential, which opens voltage-gated Ca^{2+} channels. This increases the concentration of intracellular Ca^{2+} which stimulates pancreatic β cells to increase the secretion of insulin [1]. "Glipizide can be used as a monotherapy or combination therapy for diabetes. A similar compound called Glimepiride was introduced in 1995. It is better for patients with impaired liver or kidney function. Glipizide is a so-called second generation sulfonyl urea, while Glimepiride is a third generation sulfonyl urea" [1].

Another important diabetes drug, pioglitazone (Actos®), is an agonist of the peroxisome proliferator-activated receptor-γ, or PPARγ [1]. "PPARs are found in skeletal muscle and the liver, which are affected by insulin. They are intracellular receptors that are activated by dietary fatty acids and eiconasoids. When PPARs are activated, they cause the transcription of several insulin-responsive genes that are involved in the control of glucose and lipid metabolism. So, when pioglitazone is given to a patient, it causes an increase in glucose metabolism and insulin sensitivity. It also preserves β cell function, protects cardiovascular and renal (kidney) function and reduces excess fat in the liver" [1].

One possible cure for diabetes would be to transplant islet or β-cells into patients, but protect them from the immune system [19]. "This can be done by encapsulating the cells in a protective semipermeable membrane, such as polytetrafluoroethylene (PTFE) with 400 nm pores laminated to another membrane that promotes neovascularization (angiogenesis). Angiogenesis can also be promoted by infusion of vascular endothelial growth factor (VEGF) transcutaneously. However, this has only been used in test animals. So, another approach is to encapsulate islet cells in a very thin (0.3 mm) sheet. Alginates are anionic polysaccharides usually extracted from seaweed. They are composed of homopolymeric regions of β-D-mannuronic acid ("M-blocks") and α- L-guluronic acid ("G-blocks") interspaced with regions of mixed sequence ("MG-blocks"). They have hydrogel-forming properties when mixed with with di- or trivalent cations (such as Ca^{2+}, Ba^{2+}, Fe^{3+}) that can be used as cross-linking agents. Alginates are biocompatibile and the encapsulation procedure can be performed under mild conditions not detrimental to cells. A clinical trial on non-immunosuppressed patients using alginate microcapsules (containing human islets) showed no sign of islet rejection 3 years after transplantation of the microcapsules. Blood glucose levels were stabilized for all patients and they didn't need as much exogenous insulin. However, these improvements dissipated over time and patients reverted to their original exogenous insulin therapy regimen at the end of the trial. Also, the company called Living Cells Technology in New Zealand has been using alginate microencapsulation of porcine islets over the last few years. They have completed a successful Phase I/IIa clinical trial in Russia and are currently conducting Phase IIb clinical trials in New Zealand and Argentina" [19].

PEROXISOME PROLIFERATOR-ACTIVATED RECEPTORS, PPARS

Metabolism is also regulated by peroxisome proliferator-activated receptors, or PPARs [20, 21]. "When PPARs are activated, they cause several insulin-responsive genes to be transcribed. These genes help control glucose and lipid metabolism. The PPAR family is composed of three proteins: PPAR-α, PPAR-δ (also known as PPAR-β) and PPAR-γ. They are nuclear receptors. Like other nuclear receptors, they interact with other proteins. The PPARs form heterodimers with the retionid X receptor, or RXR. Most PPAR/RXR heterodimers are constitutively bound to PPAR response elements (PPREs) that are in the promoter regions of target genes. When a ligand binds, it causes a conformational change that causes the degradation of co-repressor complexes, the recruitment of coactivator complexes, and the subsequent induction of target-gene expression. PPAR-α was the first isoform that was discovered. It is the target of the hypolipidemic fibrate drugs and other compounds that induce peroxisome proliferation. The endogenous ligands of PPARs are unsaturated fatty acids, eicosanoids, components of oxidized low density lipoproteins (LDLs) and very-low-density lipoproteins (VLDLs), and derivatives of linoleic acid" [21].

PPAR-α is predominantly expressed in the liver, brown fat and heart, but is also found in immune cells, muscles and kidneys [21 - 22]. It is important in dyslipidemia and diabetic cardiomyopathy [22]. "PPAR-γ is found in adipose, macrophages and muscle and is important in insulin resistance, obesity and metabolic syndrome. PPAR-δ is also found throughout the body. It helps control adipogenesis and lipid storage and is important in dyslipidemia and obesity. PPARδ enhances fatty acid catabolism and energy uncoupling in adipose tissue and muscle, and it suppresses macrophage-derived inflammation. It is a potential therapeutic target for the metabolic syndrome with the potential to control weight gain, enhance physical endurance, improve insulin sensitivity, and ameliorate atherosclerosis" [23].

PPARδ, increases the metabolic rate of skeletal muscle, improves insulin sensitivity, reduces the size of adipose cells and inhibits inflammation. It does this as part of a complex with another transcription factor protein called Krüppel-like factor 5 (KLF5). In a recent study, it was found that a protein called SUMO-1, small ubiquitin-related modifier type 1, binds to KLF5 and prevents it and PPAR(from activating transcription [24 - 25]. "SUMO-1 is similar to ubiquitin in that it binds to target proteins as part of a post-translational modification, but it doesn't cause proteins to be destroyed (or hydrolyzed). Instead, SUMO-1 is involved in transporting proteins to the cell nucleus, regulating transcription, apoptosis, and protein stability. In the case of PPARδ and KLF5, SUMO-1 prevents transcription from being activating. However, SUMO-1 can be removed from KLF5 by a protease, which catalyzes the hydrolysis of SUMO-1. When a ligand activates PPARδ, it recruits a protease called SENP1, which catalyzes the hydrolysis of SUMO-1" [24 - 25].

A class of drugs called thiazolodinediones is a group of highly selective and potent agonists of PPARγ. An example of a thiazolodinedione is pioglitazone (Actos®), which preserves the function of pancreatic β-cells and protects cardiovascular and renal (kidney) function in people with type-2 diabetes [1]. "Pioglitazone also reduces excess liver fat by over 50% in patients with nonalcoholic steatohepatitis, a type of liver disease that is characterized by inflammation and excess fat. This condition affects about 5% of the people in the USA and occurs primarily as a result of obesity, insulin resistance, diabetes and elevated cholesterol" [1].

Two distinct isoforms of PPAR-γ (PPAR-γ1 and PPAR-γ2) are derived from the same gene through alternative promoter usage or mRNA splicing [21]. "PPAR-γ has a crucial role in the control of adipocyte differentiation that cannot be compensated for by PPAR-α or PPAR-δ, and is essential for the formation of adipose tissue. Another repressive function of PPAR-γ is to inhibit inflammatory gene expression in a signal-specific manner. This process is called transrepression because inhibition does not depend on the binding of PPAR–RXR heterodimers to PPREs in target-gene promoters. PPAR-γ is also important in regulating the switch between M1 and M2 macrophages, where M1 macrophages produce many pro-inflammatory cytokines, while M2 macrophages generate anti-inflammatory products, mediate tissue repair and have antiparasite activity" [21].

RXR PROTEINS

The RXR protein interacts with other nuclear receptors to regulate target genes that mediate transport and catalyze the metabolism of dietary lipids [22]. "Approved drugs that target RXR heterodimers: include Bexarotene (for refractory cutaneous T-cell lymphoma), Alitretinoin (for Karposi's sarcoma, topical only), Tretinoin (for Acute promyelocytic leukemia and acne), Isotretinoin (for severe nodular acne), Calcitriol (for hypocalcemia due to chronic kidney failure and hypoparathyroidism), Ergocalciferol (for vitamin-D resistant rickets, hypoparathyroidism and familial hypoposphatemia), Levothyroxine (for hypothyroidism, euthyroid goiters and Hashimoto's thyroiditis) Pioglitazone and Rosiglitazone (for type-2 diabetes), Fenofibrate (for Types IIa, IIb, IV, and V hyperlipidemia) and Gemfibrozil for Types IIb, IV, and V hyperlipidemia. RXR heterodimers can have permissive binding partners (such as PPARs, liver X receptors, and farnesoid X receptor) that can be activated by agonists for both RXR and the partner receptor. For example, an RXR–PPAR heterodimer can be activated by both RXR and PPAR agonists independently or together to cause a synergistic activation. In contrast, RXR heterodimers that contain nonpermissive partners (such as vitamin D receptor and thyroid hormone receptor) can be activated only by the partner receptor's agonist, and not by an RXR agonist. Permissive partners serve as receptors for dietary lipids and may activate RXR to express appropriate levels of metabolic enzymes and transporters. In contrast, nonpermissive partners function primarily as hormone receptors and may inhibit RXR activation and keep target genes under tight hormonal control. So, a small change in hormone concentration substantially alters the level of target gene expression" [22].

There are three isoforms of RXR, RXRα, RXRβ and RXRγ [22]. Their natural ligands include 9-*cis*-retinoic acid and the omega-3 fat DHA. Ligands specific for RXRs have been developed as drugs to prevent breast cancer and they are called rexinoids [26]. They include bexarotene (targretin®) and panretin® [22, 26]. RXRs can also bind to two other nuclear receptors, the liver X receptors (LXRs) and farnesoid X receptor (FXR), which are permissive partners" [26]. "They are receptors for dietary lipids. They help RXRs establish steady-state expression levels for metabolic enzymes and transporters. On the other hand, non-permissive partners (vitamin D and thyroid receptors) may inhibit RXR activation and keep hormone levels under strict control" [26].

LXR PROTEINS

Liver X receptors (LXRs) are sterol sensors that bind oxidized derivatives of cholesterol, called oxysterols [26]. "There are two isoforms. LXRα is found primarily in the liver, adipose tissue, intestines, macrophages, and the kidneys, while LXRβ is ubiquitous. In response to an increased concentration of cellular oxysterols, LXRs activate genes that are involved in hepatic cholesterol metabolism and in transporting cholesterol from peripheral tissues to the liver. LXRs induce the expression of proteins that stimulate cholesterol efflux from macrophages, promote cholesterol transport in serum and uptake into liver, increase cholesterol catabolism into bile acids, increase biliary secretion of cholesterol, and inhibit the absorption of cholesterol in the intestine. LXRs also increase the synthesis of fatty acids and triglycerides by up-regulating sterol regulatory element-binding protein 1c, the master regulator of fatty-acid synthesis. Activation of LXR represses the induction of inflammatory mediators in macrophages, a mechanism with potential significance in atherosclerosis" [26].

FXR PROTEIN

The FXR is found in the enterohepatic system, kidney, and adrenals [27]. "It is a nuclear receptor for bile acids. FXR agonists can reduce levels of hepatic and serum triglycerides and may be useful in treating hypertriglyceridemia and cholesterol gallstone disease. FXR can be viewed as acting in a complementary and often reciprocal fashion to LXR in the control of lipid metabolism. FXR inhibits whereas LXR promotes bile acid production. Both receptors modulate glucose and lipid metabolism in a tissue-selective manner. Bile acids are toxic to cells, so FXR has a crucial role in controlling bile acid levels. At the same time, bile acids facilitate the efficient digestion and absorption of fats and cholesterol after a meal and help eliminate cholesterol from the body. So, FXR and LXRs help maintain an important balance. LXR and FXR influence numerous aspects of physiology. In addition to controlling sterol metabolism, LXR modulates fatty acid and carbohydrate metabolism in the liver, adipose and skeletal muscles. Similarly, FXR mediates effects on these pathways and regulates bile acid homeostasis" [27]. Moreover, FXR and LXR interact in several ways, including their effects on multidrug resistance, as well as sterol, fatty acid and glucose metabolism [28 - 29].

CONSTITUTIVE ANDOSTANE RECEPTOR (CAR)

Another anti-obesity hormone receptor that improves insulin sensitivity is the constitutive andostane receptor (CAR), which was first identified as a xenobiotic receptor [30]. It helps the body respond to environmental toxins. It also helps regulate energy metabolism by interacting with PPARα and its coactivator, PPARα coactivator 1α. It has unique anti-diabetic and anti-obesity effects, making it an attractive therapeutic target for drugs and a nutriceutical, Zhi Huang, which is a mixture of *Artemisia capillaris* and three other herbs [31].

RETINOIC ACID-RELATED ORPHAN RECEPTOR (ROR)

There is also a hormone receptor called the retinoic acid-related orphan receptor (ROR) [32]. "The RORγ1 isoform and PPARγ2 work together as key mediators of adipocytes differentiation. About 10-20% of the body's retinoic acid is stored as retinyl ester in adipocytes. Retinoids can modulate the structure and function of adipose tissue. For example, all-*trans* retinoic acid can inhibit adipogenesis by activating the RA nuclear hormone receptor, RAR. In contrast, activation of the retinoid X receptor (RXR) increases adipogenesis and insulin sensitivity. Also, retinoic acid can bind to the RORγ and inhibit transactivation of genes. The gene coding for RORγ can produce two isoforms, RORγ1 and RORγ2. RORγ1 is found in liver, adipose, and skeletal muscle, while RORγ2 was exclusively detected in a few distinct cell types of the immune system. RORγ1 levels were higher in visceral adipose tissue in morbidly obese people who were insulin resistant. In contrast, PPARγ2 levels were higher in the visceral adipose tissue in morbidly obese people who were not insulin resistant. So, it could be useful to measure the RORγ1 and PPARγ2 levels when evaluating new drugs for obesity-related type-2 diabetes" [32].

TAK1 RECEPTOR

Another nuclear receptor, TAK1 (TGFβ activated kinase 1), also plays a key role in the regulation of energy and lipid homeostasis, and promotes the development of metabolic syndrome [33]. "TAK1 and its closely related transcription factor TR2 form a subclass of the nuclear receptor superfamily. TAK1 is highly expressed in several tissues, including the testis, brain, kidneys, liver, and adipose tissue. Although TAK1 is still considered to be an orphan receptor, recent results indicate that some fatty acids and eicosanoids bind to it and enhance the transcriptional activity of TAK1. So, TAK1 might function as a lipid sensor and help cerebellar development and reproductive functions. It has also been shown to be important in lipid metabolism and gluconeogenesis and may promote metabolic syndrome. So, it may be a good therapeutic target for new drug development" [33].

TREATMENTS FOR OBESITY

Treatments for obesity have had limited success. Thyroid hormones were used as early as 1893, but excess use caused multiple toxicities [34]. Amphetamines became popular in the 1930s, because they suppress the appetite and increase energy expenditure, but they are addictive and toxic. They increase CNS dopamine, norepinephrine and serotonin activities. In fact, Japanese and German soldiers in World War II were given amphetamines, which kept them awake, made them aggressive and kept them from getting hungry. The Germans remembered their experience in World War I when a lack of food at the home front led to widespread protests, riots and eventual defeat. By the late 1930s, it was found that amphetamines could be made and transported much easier than food and they suppress the appetite. Also, Adolf Hitler was given intravenous doses of methamphetamine, which is sometimes called hillbilly heroin. It is one of the most deadly and addictive drugs known. However, drugs similar to amphetamines (phentermine and diethylpropion) have been approved for years for short term use [35]. They stimulate the release of dopamine and norepinephrine, but can have serious side effects.

Drugs other than amphetamines can also be problematic. A natural remedy, ephedrine, has caused deaths due to heart attacks. It was removed from the market for a while, but is now widely available. Sibutramine is an inhibitor of noradrenaline, serotonin and dopamine reuptake that was once available, but was withdrawn, due to cardiotoxicity and an increased incidence of stroke. Similarly, rimonabant blocks the cannabinoid-1 receptor, but has also been withdrawn due to excess toxicity.

Two effective and widely used drugs, fenfluramine and dexfenfluramine, were withdrawn from the market because of cardiotoxicity [30, 36]. They are agonists of the 5-HT_{2C} receptor, which helps mediate satiety. More recently, another

5-HT_{2C} receptor agonist, lorcaserin (Belviq®), has received FDA-approval. It was discovered after doing a structure-activity relationship (SAR) analysis of a set of compounds that contain a 3-benazepine structural motif [31]. This is bicyclic system that was based on the arylethylamine structure that is in serotonin (5-HT) and other agonists. The intramolecular Heck reaction was conducted on a series of readily available substituted phenethylamines.

More recently, bariatric surgery and the drug Orlistat have become popular. However, bariatic surgery to alter the digestive tract is currently the only effective treatment option with sustained beneficial effects on body weight and comorbidities such as type-2 diabetes [37]. "The rearranged gut has to handle an uncontrolled influx of undigested food. So, the gut wall adapts by increasing its size and mass. Because building new tissue requires energy, the rearranged gut takes up more glucose. Intestinal glucose metabolism is reprogrammed to meet increased bioenergetic demands; glucose transporter-1 is up-regulated, basolateral glucose uptake is enhanced, aerobic glycolysis is augmented, and glucose is directed toward metabolic pathways that support tissue growth. So, systemic glucose concentrations are reduced and the diabetic state is reversed" [37].

They decrease gastrointestinal absorption of energy. Orlistat (Xenical®) is the hydrogenated derivative of lipstatin, a lipase inhibitor of bacterial origin [37]. By inhibiting the hydrolysis of triglycerides, orlistat reduces the amount of free fatty acids taken up by the intestines [37]. The main side effects (diarrhea and flatulence) are due to the increased amount of undigested fat.

Another drug, Qnexa®, also known as Qsymia™, just received FDA approval. It is an extended-release capsule that contains a relatively low dose of phentermine and topiramate. It is approved for weight management in people with at least one comorbidity (such as hypertension) and with a body mass index (BMI) $\geq 27\ kg/m^2$ and weight loss for those with BMI $\geq 30\ kg/m^2$. As mentioned earlier, phentermine is a selective 5-HT_{2C} agonist. Topiramate is an anticonvulsant that was first synthesized in an attempt to discover good inhibitors of an important enzyme in gluconeogenesis, fructose-1,6-diphosphate [25]. It was not a good inhibitor, but its structure includes an O-sulfamate moiety that resembles the sulfonamide in other anticonvulsants that were already known. So, it was tested in an animal model and did so well that it eventually received FDA approval.

In many cases, it only takes a small (5-10%) decrease in body weight to cause a significant decrease in blood pressure and glycemic index [38]. This may make a doctor feel successful, but the patient may still feel that their physical appearance has not improved significantly. Young patients can be especially vulnerable to social stigma, making them susceptible to anorexia and bulimia, which are much worse than being obese.

STROKE

Remember that obesity is not a disease itself, but is a risk factor for other diseases, such as stroke and cardiovascular diseases. A stroke is the rapid loss of brain function caused by a lack of proper blood flow, needed to carry oxygen to the brain [1]. Strokes also cause inflammation. There are two main categories of stroke: ischemic and hemorrhagic. About 80% of all strokes are due to ischemia. Strokes can be caused by ischemia-reperfusion, thrombosis, embolism or hemorrhage. Ischemia is a lack of blood flow. It is followed be reperfusion, or renewed flow of the blood to the brain. Ischemia-reperfusion can cause extensive oxidative damage to the brain. The enzyme NADPH oxidase is activated in glial cells and reactive oxygen substances (ROS) are produced. Thrombosis is the formation of a blood clot, which can block the flow of blood at the site in the body in which it is formed. Embolisms, on the other hand, are due to blockage by an object called an embolus (plural emboli) that migrates to another part of the body, where it causes blockage. Hemorrhage is an internal bleeding, or loss of blood from the circulatory system.

High blood pressure is the most important modifiable risk factor that can cause stroke [39]. Often, blood pressure can be reduced by diet, exercise and medication. All these things can also be used to treat high cholesterol, another risk factor for stroke. Other risk factors, such as old age and diabetes, are not so easily controlled. Moreover, diabetes is the most common modifiable risk factor for chronic kidney disease, followed by systemic hypertension [39].

Ischemic strokes need to be treated as soon as possible, so it is very important to get a stroke victim to a hospital as soon as possible, so the blockage can be removed and blood can begin to flow properly. Drugs such as aspirin, clopidogrel and dipyridamole can be given to prevent more clots from forming by preventing blood platelets from aggregating. If the patient can get to a hospital soon enough, tPA, or tissue plasminogen activator, can be given. The drug called alteplase is a tPA that is produced by recombinant DNA technology. The complementary DNA (cDNA) from the gene that codes for tPA is inserted into Chinese Hamster Ovary cells (CHO cells), which secrete the tPA into the culture medium. Plasminogen is a serine protease that is made in the liver and released into the circulation. Plasminogen is converted to the enzyme called plasmin in a reaction that is catalyzed by tPA. Plasmin catalyzes the degradation of many proteins in blood plasma, especially those proteins that cause fibrin clots. So, tPA indirectly causes the breakdown of blood clots that can cause stroke. In a recent article, it was reported that the therapeutic window for alteplase (or tPA) can be extended from three hours to 4.5 hours [38]. On the other hand, strokes and heart attacks can be prevented by taking aspirin, which prevents the formation of blood clots. People with heart disease often take aspirin (100 mg), along with a beta-blocker (atenolol, 50 mg), a diuretic (a thiazide, 12.5 mg) [40], an ACE inhibitor (ramipril, 5 mg), and a statin (simvastatin, 20 mg). Thiazides include benzothiadiazine, chlorothiazide and hydrochlorothiazide. Recently, it was shown that these can be combined in one pill (called a polypill) and it was found to be safe and effective in reducing risk factors for stroke and heart disease [40].

MYOCARDIAL INFARCTION OR HEART ATTACK

When blood flow to the heart is interrupted, it causes a myocardial infarction, or heart attack. The most common cause is a blockage in the coronary artery that is usually caused by the rupture of an atherosclerotic plaque. Heart attacks are the leading cause of death in men and women world-wide [1]. Immediate treatment includes oxygen, aspirin and nitroglycerin. Heart attacks can be diagnosed by an electrocardiogram and by the analysis of blood for elevated levels of creatine kinase-MB and either troponin I or troponin T. Creatinine kinase has two subunits, one from the muscle (M) and one from the brain (B). So, there are three possible isozymes, MM, MB and BB. The heart muscle expresses both MM and MB isozymes. Troponin is a complex of three proteins and it is important in making muscles (including the heart) contract when the intracellular Ca^{+2}

concentration increases. Troponin, actin and myosin are the components of muscle fibers. Troponin is the protein to which Ca^{2+} binds. When a person shows symptoms of a heart attack, their blood is analyzed for troponin, which would appear if heart cells have died.

As discussed in Chapter 1 of this edition, smoldering inflammation can lead to heart attacks. It was once thought that the circulatory system was like the plumbing in our homes. If our arteries become clogged with plaques that contain cholesterol, it can cause a heart attack. However, we now know that inflammation comes first. First, adhesion molecules (such as vascular cell adhesion molecule-1, or VCAM-1) are expressed by endothelial cells [1]. This allows leukocytes to attach to the arterial wall. VCAM-1 binds to monocytes and T lymphocytes, which are found in early atherosclerotic plaques. Oxidized lipids, nuclear factor-κB (NF-κB) and TNF-α can also stimulate VCAM-1 expression. These things can be prevented by laminar blood flow, which causes some anti-atherosclerotic mechanisms, such as expression of the natural anti-oxidant, superoxide dismutase and an increase in nitric oxide (NO) Synthase. There is an increase in the concentration of NO, which causes vasodilation and limits VCAM-1 gene expression [41]. "However, when blood flow is disturbed, lesions can form and monocytes can adhere to the arterial endothelium. Monocytes can then enter the vessel walls, causing a fatty streak to start developing. This is the first stage of atherosclerosis. Monocytes mature into macrophages, which express scavenger receptors and engulf modified lipoproteins. Cholesteryl esters accumulate in the cytoplasm and macrophages become foam cells, which have an increased concentration of lipids. Next, T-lymphocytes of the adaptive immune system enter the arterial cell walls and participate in further inflammation. They are attracted to the arterial cells by chemokines (γ-IP-10, MIG and I-TAC) that are induced by interferon-γ" [41].

The next stage is progression to the formation of plaque, which can begin early in life, when a person is still in their early twenties [41]. "A mixture of cytokines: interleukins (IL-1β, IL-1α, IL-6, IL-18) and tumor necrosis factors (TNF-α, TNF-β) are involved in this process, as are the macrophage colony stimulating factor (MCSF) and the monocyte chemo attractant protein, MCP-1. Next, the plaques can rupture, leading to thrombosis. In healthy arteries, there is a fibrous cap that contains collagen for strength. The strong cap protects blood from contact with the lipid core. When the cap ruptures, a thrombus, or blood clot, can form. Inflammation interferes with the integrity of the collagen matrix by blocking the creation of new collagen fibers and by stimulating the destruction of existing collagen. T-lymphocytes participate in this pro-inflammatory process. They produce CD40, a cell surface protein and IL-1. These biomolecules promote the production of enzymes in macrophages that catalyze the destruction of collagen. The CD40 ligand also stimulates macrophage production of tissue factor VII, which initiates the coagulation cascade. To prevent all these things from happening, one should try to lower the risk factors by not smoking, engaging in regular strenuous cardiovascular exercise and by consuming a healthy diet, with fewer calories, and more fresh fruits and vegetables and other sources of antioxidants" [41].

Also, hyperglycemia and diabetes often lead to heart disease. Vascular homeostasis is disrupted due to endothelial and smooth muscle cell dysfunction [41]. "This causes a further inflammation and a pro-thrombotic state. Elevated levels of RONS (ROS) lead to endothelial dysfunction, superoxide anion (O_2^-) production and NO imbalance. Peroxynitrate ($ONOO^-$) easily penetrates and crosses through cell membranes and nitrosylates proteins, which decrease the activity of endothelial NO synthase (eNOS) and antioxidant enzymes. RONS (ROS) also initiate polyol and hexosamine flux, help form AGEs, activate protein kinase C (PKC) and vascular inflammation mediated by NF-κB. Once activated, PKC increases ROS production through NADPH oxidase, which catalyzes the formation of the superoxide anion. Also, the PKC β_2 isoform catalyzes the phosphorylation of another protein, $p66^{Shc}$, causing it to move to the mitochondria, where it functions as a redox enzyme that generates more ROS and translates oxidative signals into apoptosis. Moreover, PKC leads to higher production of endothelin-1 (ET-1), which favors vasoconstriction and platelet aggregation. In the blood vessel wall, RONS (ROS) initiate vascular inflammation. They up-regulate NF-kB subunit p65 and move it to the cell nucleus, where it activates the transcription of pro-inflammatory genes encoding for monocyte chemoattractant protein-1 (MCP-1), selectins, vascular cell adhesion molecule-1 (VCAM-1), and intracellular cell adhesion molecule-1 (ICAM-1) . This helps monocytes stick to the vascular endothelium, leading to the formation of foam cells. Adhesion molecules stay up-regulated by the secretion of IL-1 and TNF-α from active macrophages. This enhances NF-kB signaling in the endothelium and also helps smooth muscle cells grow and proliferate. Also, endothelial dysfunction increases the synthesis of vasoconstrictors and prostenoids. Cyclooxygenase-2 (COX-2) is up-regulated by PKC and this is associated with an increase of thromboxane A2 and a reduction of prostacyclin (PGI2) release. As mentioned previously, RONS (ROS) also increase the level of methylglyoxal and AGEs, which led to protein nitrosylation. This includes OGlcNAcylation at the Akt binding site of the eNOS protein. This leads to reduced eNOS activity and endothelial dysfunction. Also, glycosylation of transcription factors causes up-regulation of inflammatory (TGFα, TGFβ1) and pro-thrombotic genes (plasminogen activator inhibitor-1)" [41].

Insulin resistance, caused by diabetes, develops in not just the heart, but also skeletal muscle, adipose tissue and the liver [41]. "Adipose tissue in particular produces inflammatory mediators and free fatty acids that bind to Toll-like receptors (TLRs) and activate NF-kB. They degrade the inhibitory complex formed between IkBα by IKKβ-kinase. As a result, NF-kB triggers tissue inflammation due to up-regulation of the pro-inflammatory genes IL-6 and TNF-α. Free fatty acid activation of TLRs leads to phosphorylation of the insulin receptor substrate-1 (IRS-1) by c-Jun amino-terminal kinase (JNK) and PKC. This alters its ability to activate the downstream targets PI3-kinase and Akt. This down-regulates the glucose transporter GLUT-4, leading to insulin resistance. Moreover, down-regulation of the PI3-kinase/Akt pathway leads to inhibition of eNOS and decreased NO production. Then, intracellular oxidation of free fatty acids produces ROS. This leads to vascular inflammation, synthesis of AGEs, reduced prostaglandin I2 (PGI2) synthase activity, and PKC activation. As more RONS (ROS) are formed, they can react with NO, reducing its level even more. This activates pro-inflammatory pathways that are activated by increased cytokine production. That is, TNF-α and IL-1 increase NF-kB activity and the expression of adhesion molecules. TNF-α also stimulates the expression of C-reactive protein, which down-regulates eNOS while increasing the production of adhesion molecules and endothelin-1" [41].

However, recent data has indicated that the endothelium is more important than adipocytes in causing obesity-induced insulin resistance [42]. "That is, inflammation and macrophage activation occur primarily in non-adipose tissue in obese people. Also, insulin sensitivity can be improved in patients with type-2 diabetes by restoring flow-mediated vasodilation. Still, changes in the blood lipid profile (high triglycerides, low HDL cholesterol, increased remnant lipoproteins, elevated apolipoprotein B, ApoB, and elevated LDL) can help cause atherosclerosis. When free fatty acids reach the liver, very low density lipoproteins (VLDLs) are assembled and are solubilized by making more ApoB. VLDLs are processed by cholesteryl ester transfer protein. This allows triglycerides to be transferred to LDLs, which become smaller and denser and, thus, more atherogenic. As a result, atherogenic dyslipidemia is a reliable predictor of cardiovascular risk and its pharmacological modulation reduces vascular events in subjects with type-2 diabetes and metabolic syndrome" [42].

Harmful coronary events in patients with insulin resistance are triggered by being in a prothrombotic state [42]. "Under healthy conditions, insulin inhibits platelet aggregation and thrombosis by inhibiting tissue factor and enhancing fibrinolytic action by changing the concentration of plasminogen activator inhibitor-1 (PAI-1). In contrast, insulin resistance facilitates atherothrombosis by increasing the biosyntheses of PAI-1 and fibrinogen while reducing the production of tissue plasminogen activator. In platelets, a lack of insulin leads to a down-regulation of the IRS-1/Akt pathway. This causes Ca^{2+} to accumulate. Moreover, platelet reactivity and excretion of tromboxane metabolites are increased in insulin resistant obese patients and this is reversed by losing weight or by a 3-week treatment with pioglitazone" [42].

More recently, researchers have found that microRNAs (miRNAs) are involved in the pathogenesis of vascular damage caused by hyperglycemia [42]. "That is, miRs involved in angiogenesis, vascular repair, and endothelial homeostasis are dysregulated in diabetic patients. For example, in endothelial cells exposed to excess glucose, miR-320 is highly expressed. It then targets several angiogenic factors and their receptors, including vascular endothelial growth factor and insulin like growth factor-1 (IGF-1). This is associated with decreased cell proliferation and migration. Hyperglycemia also increases the expression of miR-

221, a regulator of angiogenesis that targets the c-kit receptor, which is responsible for migration and homing of endothelial progenitor cells (EPCs). Also, miR-221 and 222 were mediate AGE-induced vascular damage. Another study showed that miR-503 is critically involved in hyperglycemia-induced endothelial dysfunction in diabetic mice and is up-regulated in ischemic limb muscles of diabetic subjects. Also, profiling of plasma miRs showed that miR-126 was down-regulated in a cohort of diabetic patients. This may be partially responsible for impaired vascular repair capacities in diabetes. Finally, miR-126 expression was reduced in endothelial progenitor cells that in diabetics" [42].

PREVENTING AND TREATMENTING HEART DISEASE

As with any disease, it is better to prevent atherosclerosis and heart disease. Even though proper diet, cardiovascular exercises and reduced emotional stress can prevent some heart attacks, sometimes these are not enough. For many people, a low dose of aspirin (81 mg, or one baby aspirin) every day is also recommended. This can also help prevent stroke. Also, statins can improve the lipid profile of the blood, preventing heart attacks. So, patients with diabetes should be treated with a statin [43]. Still, other measures are needed for diabetics. Drugs that lower the concentration of glucose in the blood can be quite effective, along with drugs that lower the blood pressure or act as diuretics. For example, metformin (Glucophage®) is the first line treatment for type-2 diabetes. It reduces the amount of glucose that is released from the liver and increases the transport of glucose into muscles, increasing insulin sensitivity and decreasing glucose absorption from the gastrointestinal tract [1]. "Metformin is also available as Janumet® in combination with the dipeptidyl peptidase (DPP-4) inhibitor, sitagliptin. Also, Glipizide (Glucotrol®) can be used as a monotherapy or combination therapy for diabetes. A similar compound called Glimepiride was introduced in 1995. It is better for patients with impaired liver or kidney function. Glipizide is a second generation, while Glimepiride is a third generation sulfonyl urea. Also, the diabetes drug, pioglitazone (Actos®) causes an increase in glucose metabolism and insulin sensitivity. It also preserves β cell function, protects cardiovascular and renal (kidney) function and reduces excess fat in the liver" [1].

Treating hypertension can also help prevent heart attacks and stroke. For some people, reducing the consumption of table salt (NaCl) can help. For non-diabetics, beta-adrenergic blockers or angiotensin-converting enzyme (ACE) inhibitors are often the first line treatment [43]. "However, for diabetics, beta-adrenergic blockers are not recommended. Instead, if the primary goal is to reduce blood pressure, a thiazide diuretic, dihydropyridine calcium-channel blocker, or angiotensin-converting enzyme inhibitor (ACE-I) are acceptable first choices. On the other hand, β-adrenergic blockers are recommended if a patient has a myocardial infarction. This is recommended for three years after myocardial infarction and when there is left-ventricular dysfunction, with or without heart failure. Moreover, β-adrenergic blockers, when titrated to full dose, are recommended for treating stable angina. If the primary goal is to prevent atherosclerotic events, antagonists of the renin–angiotensin system are preferred because they do more than just lower the blood pressure. Also, ACE inhibitors reduce recurrent strokes in diabetic patients and reduce mortality in diabetic patients after myocardial infarction. Angiotensin receptor blockers (ARBs) are similarly efficacy after myocardial infarction. For example, valsartan was as effective as captopril in decreasing total mortality in diabetic patients with myocardial infarctions complicated by left ventricular systolic dysfunction. Losartan was better than atenolol in diabetic patients with end-organ damage for the primary composite endpoint of cardiovascular death, myocardial infarction and stroke. Finally, valsartan was equivalent to amlodipine when the endpoint was cardiovascular morbidity and mortality in hypertensive diabetic patients" [43].

Inhibitors of mineralcorticoid receptors may also be useful in patients with congestive heart failure and left-ventricular dysfunction after a myocardial infarction [44]. "One of them, spironolactone, reduced mortality by 30% in patients with and without diabetes. Another one, eplerenone, reduced mortality in patients with reduced left-ventricular systolic function. It also reduced mortality in one patient who had mild heart failure and in others who had a myocardial infarction" [44].

One of the biggest problems with diabetes is that many patients die prematurely from cardiovascular disease, or heart attack. Cardiovascular disease also kills many people who are not diabetic, especially if they have hypertension, or high blood pressure. Blood pressure is the force exerted by circulating blood on the walls of blood vessels. The heart beats in cycles. The highest pressure is called the systolic and the lowest is the diastolic. Healthy blood pressures in adults are 120/80 or lower. When we go to the doctor's office, our blood pressures are measured with a device that measures the pressure at the brachial artery. This accurately measures the diastolic, but not the systolic pressure of the heart. Researchers have started using another method, which measures the central pressure in the heart, brain and kidneys. The central pressure is the pressure that each organ actually experiences. Central pressure better predicts the clinical outcome of medicines.

Some of the drugs that lower blood pressure, heart rate, chest pain and the risk of heart attack act through receptors called β-adrenergic receptors, which are part of the sympathetic nervous system (fight or flight response). This class of drugs is called the beta blockers. Atenolol (Tenormin®) is a drug that binds competitively to the β-adrenergic receptor and blocks the binding of the natural ligands, adrenaline and noradrenaline (also known as epinephrine and norepinephrine). Over 20 million people each year are treated with beta blockers [1]. "The first beta blocker (dichloroisoprenaline) was discovered in 1958 and it entered the market in 1976. In 1958, the usual way to discover new drugs was to synthesize many compounds and test them on animals. In the 1950s, James Black hypothesized that antagonists of adrenaline might reduce blood pressure. He won the Nobel Prize in medicine in 1988 for his work. Propranolol was the first clinically successful beta blocker. The family of beta blockers is so successful that many pharmaceutical companies have their own versions of them. Currently, there are over 20 of them on the market" [1]. These include carvedilol, metoprolol, propranolol and timolol.

Another class of drugs that treat high blood pressure and prevents heart attacks is the calcium channel blockers [1]. "Amlodipine (Norvasc®) is the best-selling drug in this class, with sales of $4.7 billion in 2005. It binds to calcium channels in the heart and arteries. It decreases contractions of the heart and dilates the arteries, lowering blood pressure. Amlodipine is a long acting calcium channel blocker and requires only one dose per day" [1].

Blood clots cause about 90% of all heart attacks and 80% of all strokes [1]. Blood clots form when platelets aggregate. Clopidogrel bisulfate (Plavix®) is a drug that prevents the formation of blood clots by inhibiting the binding of ADP to its platelet receptor and the subsequent activation of glycoprotein GPIIb/IIIa [1]. The modification of the platelet ADP receptor is irreversible. Millions of people have been treated with Plavix™, which had sold $74.7 billion through 2012 [45].

Even though some people do survive their first heart attack, many of them develop heart failure, for which the 5-year average survival rate is only about 50% [46]. "That is because myocardial infarctions cause cell death, followed by degradation of the associated extracellular matrix, an acute inflammatory response and the formation of a dense collagenous scar. To prevent this from happening, an injectable hydrogel derived from porcine myocardial extracellular matrix as a scaffold for cardiac repair has been developed and shown to be safe and effective in rats. Cardiac function, ventricular volumes, and global wall motion scores improved. Furthermore, a significantly larger zone of cardiac muscle was found at the endocardium in pigs injected with the matrix, compared to controls. Hemocompatibility studies with human blood indicate that exposure to the material at relevant concentrations does not affect clotting times or platelet activation. This work enables researchers to move forward in clinical studies with this cardiac-specific biomaterial that can be delivered by catheter" [46].

Another approach is to inject a self-assembling nanofiber scaffold [47]. "Currently, there are no ways to prevent the negative left ventricular (LV) remodeling process that occurs after a myocardial infarction. The only successful treatments for end-stage heart failure remain getting a heart transplant or using LV assist devices, which both have severe problems. Instead, researchers are developing a combination therapy using self-assembling peptide nanofibers and vascular endothelial growth factor (VEGF). Injecting VEGF in a self-assembling nanofiber scaffold can create a suitable microenvironment for recruiting endogenous cells to repair the heart and improve cardiac function. That is, myocardial heart failure starts with cardiomyocytes dying, followed by an inflammatory response and degradation of the extracellular matrix (ECM). It ends by forming a collagen scar. Cardiomyocyte death causes a negative LV remodeling process, which leads to increased wall stress. Ensuing molecular, cellular, and physiological events lead to LV dilation and eventual heart failure.

So, by delivering VEGF in the nanofiber scaffold, researchers were able to cause endogenous cells to be recruited, including neovasculature, circulating bone marrow cells, small cardiomyocyte-like cells, and myofibroblasts. Delivery of VEGF in the peptide nanofibers increased arteriole density in the border zone (area adjacent to the infarct). Moreover, arteriogenesis was seen to play a dominant role over angiogenesis in improving cardiac function. VEGF delivery in the nanofibers promoted higher retention of myofibroblasts 14 days after injecting them. This may be due to both the scaffold microenvironment and the newly formed arterioles. The researchers also showed that infiltration of circulating bone marrow mononuclear cells is mediated through β2-integrin adhesion" [47].

Inflammation is also an important cause of atherosclerotic plaques. This was discussed in Chapter 1 of this edition. Inflammation can be caused by obesity, high cholesterol and metabolic syndrome. So, maintaining a healthy body mass index (BMI) is an important part of good health. Although there is a genetic component to metabolic syndrome, often proper diet and cardiovascular exercise can help prevent it and the diseases that it causes.

In 2009, the FDA approved a new drug, called Prasugrel (Effient®), for preventing dangerous blood clots that can lead to heart attack or stroke. It is an ADP receptor antagonist, which irreversibly blocks the platelet receptor P2Y12 ADP, thus inhibiting platelet function [48]. During an episode of acute coronary syndrome, platelets are activated. They aggregate and adhere to each other. When plaques rupture, they expose collagen and von Willebrand factor, to which circulating platelets adhere. Thromboxane A2, ADP and other substances are released from platelet granules. This leads to platelet aggregation. It can be stopped by Prasugrel, which is a prodrug. It is converted to its active form by rapid hydrolysis catalyzed by esterases, followed by a single cytochrome P450 (CYP)–dependent step [48].

The drugs discussed in this Chapter are listed in (Table **1**).

Table 1. Drugs discussed in this chapter.

Drug	Biological Target	Clinical Use
Adalimab	TNF-α	Rheumatoid arthritis
Etanercept	TNF-α	Rheumatoid arthritis
Infliximab	TNF-α	Rheumatoid arthritis
Tocilizumab	IL6 receptor	Rheumatoid arthritis
Zolpidem	GABA	Insomnia
Eszopiclone	GABA	Insomnia
Ruboxistaurin	PKCβ isoform	Diabetic macular edema
Metformin	AMP-APK and PKC	Type-2 Diabetes
Sitagliptin	dipeptidyl peptidase	Type-2 Diabetes
Glipizide	K^+ ion channels in β cells	Type-2 Diabetes
Glimepiride	β cells in pancreas	Type-2 Diabetes
Chlorpropamide	β cells in pancreas	Type-2 Diabetes
Gliclazide	β cells in pancreas	Type-2 Diabetes
Glyburide	β cells in pancreas	Type-2 Diabetes
Pioglitazone	PPARγ	Type-2 Diabetes
Bexarotene	RXR	T-cell lymphoma
Alitretinoin	RXR	Karposi's sarcoma
Tretinoin	RXR	Acute promyelocytic leukemia
Isotretinoin	RXR	Severe nodular acne
Calcitriol	RXR	Hypocalcemia
Ergocalciferol	RXR	Hypoparathyroidism
Levothyroxine	RXR	Hypothyroidism
Rosiglitazone	RXR	Type-2 Diabetes
Gemfibrozil	RXR	Hyperlipidemia
Orlistat	Lipase	Obesity
Qnexa® phentermine plus topiramate	$5\text{-}HT^{2C}$ agonist and GABA	Obesity
Alteplase	tissue plasminogen activator	Stroke
Atenolol	β-adrenergic receptor	Heart disease
Carvedilol	β-adrenergic receptor	Heart disease
Metoprolol	β-adrenergic receptor	Heart disease
Timolol	β-adrenergic receptor	Heart disease
Benzothiadiazine	Diuretic Na^+-Cl^- cotransporter	Heart disease

(Table 1) contd.....

Chlorothiazide	Diuretic Na^+-Cl^- cotransporter	Heart disease
Hydrochlorothiazide	Diuretic Na^+-Cl^- cotransporter	Heart disease
Ramipril	ACE inhibitor	Heart disease
Simvastatin	Cholesterol biosynthesis	Heart disease
Valsartan	Angiotensin receptor	Heart disease
Losartan	Angiotensin receptor	Heart disease
Spironolactone	Mineralcorticoid receptor	Heart disease
Amlodipine	Calcium channel blocker	Heart disease
Clopidogrel bisulfate	ADP platelet receptor	Stroke
Prasugrel	ADP platelet receptor	Stroke

KEY POINTS

1. **Heart disease is the biggest killer in the world. Its major cause is obesity or metabolic syndrome, which can also lead to diabetes and stroke - the third leading killer (after cancer) and the most frequent cause of disability in the world. All of these involve imbalances in energy metabolism. More than 700,000 kcal are ingested each year by most well-fed adults, and "body weight is kept within a range of ± 1 kg, or an equivalent of 7000 kcal. That is, the regulation of energy operates with an accuracy of about 99%" [1].**

2. **Dietary carbohydrates are converted to glucose, which is used as a source of energy (especially in the brain), and as a source of intermediates for the biosynthesis of other carbohydrates, as well as nucleic acids, fats, amino acids and proteins. Fats are converted to fatty acids, which can be used for fuel. Fatty acids can react with glycerol to form triglycerides, which store energy. Dietary proteins are broken down into amino acids, which can be used to make human proteins and polypeptides [1].**

3. **Early symptoms of metabolic syndrome include excessive weight, high blood pressure, elevated blood glucose and elevated levels of lipids, especially low density lipoprotein (LDL). People who have two or more of these symptoms are at a higher risk of developing heart disease, stroke and type-2 diabetes. Childhood obesity is such a big problem that it will probably mean that this generation will be the first to have a shorter life expectancy than their parents.**

4. **Ischemic strokes need to be treated as soon as possible, so it is very important to get a stroke victim to a hospital as soon as possible, so the blockage can be removed and blood can begin to flow properly. Drugs such as aspirin, clopidrogel and dipyridamole can be given to prevent more clots from forming by preventing blood platelets from aggregating. If the patient can get to a hospital soon enough, tPA, or tissue plasminogen activator, can be given.**

5. **When blood flow to the heart is interrupted, it causes a myocardial infarction, or heart attack. The most common cause is a blockage in the coronary artery that is usually caused by the rupture of an atherosclerotic plaque. Heart attacks are the leading cause of death in men and women world-wide. Immediate treatment includes oxygen, aspirin and nitroglycerin [1].**

ABBREVIATIONS

LDL	=	Low Density Lipoprotein
VLDL	=	Very-Low-Density Lipoproteins
HDL	=	High Density Lipoprotein
tPA	=	Tissue Plasminogen Activator
acetyl-CoA	=	Acetyl-Coenzyme A
TCA	=	Cycle Tricarboxylic Acid Cycle
GTP	=	Guanidine Triphosphate
FAD	=	Flavin Adenine Dinucleotide
ATP	=	Adenosine Triphosphate
ROS	=	Reactive Oxygen Substances
AGE	=	Advanced Glycation End Product
RAGE	=	Receptor for AGE
PI3K	=	Phosphoinositide-3 Kinase
POMC	=	Pre-Proopio-Melanocortin
TNF-α	=	Tumor Necrosis Factor Alpha
HGF-1	=	Hepatocyte Growth Factor 1
VEGF	=	Vascular Endothelial Growth Factor
UVP-1	=	Uncoupling Protein 1
Fc	=	Fraction of an Antibody that is Crystallizable
IgG1	=	Immunoglobulin G1
GPCR	=	G-Protein Coupled Receptor
FDA	=	Food and Drug Administration

AMP-AKP	=	Adenosine Monophosphate-Activated Protein Kinase
NADP	=	Nicotinamide Adenine Dinucleotide Phosphorylated
RONS	=	Reactive Oxygen and Nitrogen Substances
PKC	=	Protein Kinase C
IP3	=	Inositol Trisphosphate
DAG	=	Diacyl Glycerol
DHAP	=	Dihydroxyacetone Phosphate
ECM	=	Extracellular Matrix
ATPase	=	Adenosine Triphosphatase
RAS/MAP	=	Renin-Angiotensin System/Mitogen Activated Protein
ASE	=	Angiotensin Converting Enzyme
RAS	=	Oncogene
ras	=	Oncoprotein
NAFLD	=	Non-Alcoholic Fatty Liver Disease
PPAR	=	Pattern-Recognition Receptor
NASH	=	Nonalcoholic Steatohepatitis
PPAR-γ	=	Peroxisome Proliferator-Activated Receptor-γ
PTFE	=	Polytetrafluoroethylene
RXR	=	Retionid X Receptor
KLFP	=	Krüppel-Like Factor 5
SUMO-1	=	Small Ubiquitin-Related Modifier Type 1
FXR	=	Farnesoid X Receptor
LXR	=	Liver X Receptor
RXR	=	Retinoid X Receptor
CAR	=	Constitutive Andostane Receptor
ROR	=	Retinoic Acid-Related Orphan Receptor
TAK1	=	TGFβ Activated Kinase 1
CNS	=	Central Nervous System
5-HT	=	5-Hydroxytryptamine
BMI	=	Body Mass Index
cDNA	=	Complementary DNA
CHO	=	Cells Chinese Hamster Ovary cells
VCAM-1	=	Vascular Cell Adhesion Molecule-1
NO	=	Nitric Oxide
MCSF	=	Macrophage Colony Stimulating Factor

GLUT-4	=	Glucose Transporter-4
PAI-1	=	Plasminogen Activator Inhibitor-1
miRNA	=	microRNA
IGF-1	=	Insulin Like Growth Factor-1
EPC	=	Endothelial Progenitor Cell
DPP	=	Dipeptidyl Peptidase

REFERENCES

[1] Corey EJ, Czako B, Kürti L. Molecules and Medicine. New York: John Wiley & Sons 2007.

[2] Schlosser E. Fast Food Nation. New York: Houghton-Mifflin 2001.

[3] Capra F. The Web of Life. New York: Doubleday 1996.

[4] Lazar MA. Developmental biology. How now, brown fat? Science 2008; 321(5892): 1048-9. [http://dx.doi.org/10.1126/science.1164094] [PMID: 18719271]

[5] Maratos-Flier E. The long reach of leptin. Nat Med 2008; 14(6): 604-6. [http://dx.doi.org/10.1038/nm0608-604] [PMID: 18535571]

[6] Buettner C, Muse ED, Cheng A, *et al.* Leptin controls adipose tissue lipogenesis *via* central, STAT3-independent mechanisms. Nat Med 2008; 14(6): 667-75. [http://dx.doi.org/10.1038/nm1775] [PMID: 18516053]

[7] Sun K, Wernstedt Asterholm I, Kusminski CM, *et al.* Dichotomous effects of VEGF-A on adipose tissue dysfunction. Proc Natl Acad Sci USA 2012; 109(15): 5874-9. [http://dx.doi.org/10.1073/pnas.1200447109] [PMID: 22451920]

[8] Soliman MM, Ashcroft DM, Watson KD, Lunt M, Symmons DP, Hyrich KL. British Society for Rheumatology Biologics Register. Impact of concomitant use of DMARDs on the persistence with anti-TNF therapies in patients with rheumatoid arthritis: results from the British Society for Rheumatology Biologics Register. Ann Rheum Dis 2011; 70(4): 583-9. [http://dx.doi.org/10.1136/ard.2010.139774] [PMID: 21330639]

[9] Jason M, JM Uslaner, Buettner C, *et al.* Orexin receptor antagonists differ from standard sleep drugs by promoting sleep at doses that do not disrupt cognition. Sci Trans Med 2013; 5(179ra44): 1-11.

[10] van Greevenbroek MM, Schalkwijk CG, Stehouwer CD. Obesity-associated low-grade inflammation in type 2 diabetes mellitus: causes and consequences. Neth J Med 2013; 71(4): 174-87. [PMID: 23723111]

[11] Rosengren AH, Jokubka R, Tojjar D, *et al.* Overexpression of alpha2A-adrenergic receptors contributes to type 2 diabetes. Science 2010; 327(5962): 217-20. [http://dx.doi.org/10.1126/science.1176827] [PMID: 19965390]

[12] Scully T. Diabetes in numbers. Nature 2012; 485(7398): S2-3. [http://dx.doi.org/10.1038/485S2a] [PMID: 22616094]

[13] Tabit CE, Shenouda SM, Holbrook M, *et al.* Protein kinase C-β contributes to impaired endothelial insulin signaling in humans with diabetes mellitus. Circulation 2013; 127(1): 86-95.

[http://dx.doi.org/10.1161/CIRCULATIONAHA.112.127514] [PMID: 23204109]

[14] Zhang L, Huang D, Shen D, *et al.* Inhibition of protein kinase C βII isoform ameliorates methylglyoxal advanced glycation endproduct-induced cardiomyocyte contractile dysfunction. Life Sci 2014; 94(1): 83-91.
[http://dx.doi.org/10.1016/j.lfs.2013.11.011] [PMID: 24269213]

[15] Gutterman DD. Vascular dysfunction in hyperglycemia: is protein kinase C the culprit? Circ Res 2002; 90(1): 5-7.
[PMID: 11786509]

[16] Das Evcimen N, King GL. The role of protein kinase C activation and the vascular complications of diabetes. Pharmacol Res 2007; 55(6): 498-510.
[http://dx.doi.org/10.1016/j.phrs.2007.04.016] [PMID: 17574431]

[17] Bansal D, Badhan Y, Gudala K, Schifano F. Ruboxistaurin for the treatment of diabetic peripheral neuropathy: a systematic review of randomized clinical trials. Diabetes Metab J 2013; 37(5): 375-84.
[http://dx.doi.org/10.4093/dmj.2013.37.5.375] [PMID: 24199167]

[18] Henahan S. Ruboxistaurin's future in doubt, despite promising study results. http://escrs.org/PUBLICATIONS/EUROTIMES/07May/Ruboxistaurinsfuturein.pdf 2013.

[19] Schweicher J, Nyitray C, Desai TA. Membranes to achieve immunoprotection of transplanted islets. Front Biosci (Landmark Ed) 2014; 19: 49-76.
[http://dx.doi.org/10.2741/4195] [PMID: 24389172]

[20] Grygiel-Górniak B. Peroxisome proliferator-activated receptors and their ligands: nutritional and clinical implications–a review. Nutr J 2014; 13: 17.
[PMID: 24524207]

[21] Bensinger SJ, Tontonoz P. Integration of metabolism and inflammation by lipid-activated nuclear receptors. Nature 2008; 454(7203): 470-7.
[http://dx.doi.org/10.1038/nature07202] [PMID: 18650918]

[22] Shulman AI, Mangelsdorf DJ. Retinoid x receptor heterodimers in the metabolic syndrome. N Engl J Med 2005; 353(6): 604-15.
[http://dx.doi.org/10.1056/NEJMra043590] [PMID: 16093469]

[23] Barish GD, Narkar VA, Evans RM. PPAR δ: a dagger in the heart of the metabolic syndrome. J Clin Invest 2006; 116(3): 590-7.
[http://dx.doi.org/10.1172/JCI27955] [PMID: 16511591]

[24] Jain MK. Multiple layers of metabolism. Nat Med 2008; 14(6): 603-4.
[http://dx.doi.org/10.1038/nm0608-603] [PMID: 18535570]

[25] Oishi Y, Manabe I, Tobe K, *et al.* SUMOylation of Krüppel-like transcription factor 5 acts as a molecular switch in transcriptional programs of lipid metabolism involving PPAR-δ. Nat Med 2008; 14(6): 656-66.
[http://dx.doi.org/10.1038/nm1756] [PMID: 18500350]

[26] Howe LR. Rexinoids and breast cancer prevention. Clin Cancer Res 2007; 13(20): 5983-7.
[http://dx.doi.org/10.1158/1078-0432.CCR-07-1065] [PMID: 17947457]

[27] Calkin AC, Tontonoz P. Transcriptional integration of metabolism by the nuclear sterol-activated receptors LXR and FXR. Nat Rev Mol Cell Biol 2012; 13(4): 213-24.
[PMID: 22414897]

[28] Kast HR, Goodwin B, Tarr PT, *et al.* Regulation of multidrug resistance-associated protein 2 (ABCC2) by the nuclear receptors pregnane X receptor, farnesoid X-activated receptor, and constitutive androstane receptor. J Biol Chem 2002; 277(4): 2908-15.
[http://dx.doi.org/10.1074/jbc.M109326200] [PMID: 11706036]

[29] Bays H. Phentermine, topiramate and their combination for the treatment of adiposopathy ('sick fat') and metabolic disease. Expert Rev Cardiovasc Ther 2010; 8(12): 1777-801.
[http://dx.doi.org/10.1586/erc.10.125] [PMID: 20707765]

[30] Gao J, He J, Zhai Y, Wada T, Xie W. The constitutive androstane receptor is an anti-obesity nuclear receptor that improves insulin sensitivity. J Biol Chem 2009; 284(38): 25984-92.
[http://dx.doi.org/10.1074/jbc.M109.016808] [PMID: 19617349]

[31] Huang W, Zhang J, Moore DD. A traditional herbal medicine enhances bilirubin clearance by activating the nuclear receptor CAR. J Clin Invest 2004; 113(1): 137-43.
[http://dx.doi.org/10.1172/JCI200418385] [PMID: 14702117]

[32] Tinahones FJ, Moreno-Santos I, Vendrell J, *et al.* The retinoic acid receptor-related orphan nuclear receptor γ1 (RORγ1): a novel player determinant of insulin sensitivity in morbid obesity. Obesity (Silver Spring) 2012; 20(3): 488-97.
[http://dx.doi.org/10.1038/oby.2011.267] [PMID: 21904299]

[33] Kang HS, Okamoto K, Kim YS, *et al.* Nuclear orphan receptor TAK1/TR4-deficient mice are protected against obesity-linked inflammation, hepatic steatosis, and insulin resistance. Diabetes 2011; 60(1): 177-88.
[http://dx.doi.org/10.2337/db10-0628] [PMID: 20864514]

[34] Smith BM, Smith JM, Tsai JH, *et al.* Discovery and structure-activity relationship of (1R)-8-chlo-o-2,3,4,5-tetrahydro-1-methyl-1H-3-benzazepine (Lorcaserin), a selective serotonin 5-HT2C receptor agonist for the treatment of obesity. J Med Chem 2008; 51(2): 305-13.
[http://dx.doi.org/10.1021/jm0709034] [PMID: 18095642]

[35] Hofbauer KG, Nicholson JR, Boss O. The obesity epidemic: current and future pharmacological treatments. Annu Rev Pharmacol Toxicol 2007; 47: 565-92.
[http://dx.doi.org/10.1146/annurev.pharmtox.47.120505.105256] [PMID: 17002599]

[36] Shank RP, Gardocki JF, Streeter AJ, Maryanoff BE. An overview of the preclinical aspects of topiramate: pharmacology, pharmacokinetics, and mechanism of action. Epilepsia 2000; 41 (Suppl. 1): S3-9.
[http://dx.doi.org/10.1111/j.1528-1157.2000.tb02163.x] [PMID: 10768292]

[37] Berthoud HR. Why Does Gastric Bypass Surgery Work? Science 2013 Jul 26; 341(6144): 351-2.
[http://dx.doi.org/10.1126/science.1242673] [PMID: 23888026]

[38] Hacke W, Kaste M, Bluhmki E, *et al.* ECASS Investigators. Thrombolysis with alteplase 3 to 4.5 hours after acute ischemic stroke. N Engl J Med 2008; 359(13): 1317-29.
[http://dx.doi.org/10.1056/NEJMoa0804656] [PMID: 18815396]

[39] Ptinopoulou AG, Pikilidou MI, Lasaridis AN. The effect of antihypertensive drugs on chronic kidney disease: a comprehensive review. Hypertens Res 2013; 36(2): 91-101. [http://dx.doi.org/10.1038/hr.2012.157] [PMID: 23051659]

[40] Yusef S, *et al.* The Indian Polycap Study (TIPS). Effects of a polypill (Polycap) on risk factors in middle-aged individuals without cardiovascular disease (TIPS): a phase II, double-blind, randomized trial. Lancet 2009 Apr 18; 373(9672): 1341-51. [http://dx.doi.org/10.1016/S0140-6736(09)60611-5]

[41] Libby P. Inflammation and cardiovascular disease mechanisms. Am J Clin Nutr 2006; 83(2) (Suppl.): 456S-60S. [PMID: 16470012]

[42] Paneni F, Beckman JA, Creager MA, Cosentino F. Diabetes and vascular disease: pathophysiology, clinical consequences, and medical therapy: Part I. Eur Heart J 2013 Aug; 34(31): 2436-43. [http://dx.doi.org/10.1093/eurheartj/eht149]

[43] Creager MA, Lüscher TF, Cosentino F, Beckman JA. Diabetes and vascular disease: pathophysiology, clinical consequences, and medical therapy: Part I. Circulation 2003; 108(12): 1527-32. [http://dx.doi.org/10.1161/01.CIR.0000091257.27563.32] [PMID: 14504252]

[44] Lüscher TF, Creager MA, Beckman JA, Cosentino F. Diabetes and vascular disease: pathophysiology, clinical consequences, and medical therapy: Part II. Circulation 2003; 108(13): 1655-61. [http://dx.doi.org/10.1161/01.CIR.0000089189.70578.E2] [PMID: 14517152]

[45] Forbes. The World's Ten Best-Selling Drugs http://www.forbes.com/sites/simonking/2013/07/15/ the-best-selling-drugs-since-1996-why-abbvies-humira-is-set-to-eclipse-pfizers-lipitor/.

[46] Christman KL. Treating the leading killer. Sci Transl Med 2012; 4(146): 146fs26. [http://dx.doi.org/10.1126/scitranslmed.3004565] [PMID: 22875825]

[47] Seif-Naraghi SB, Singelyn JM, Salvatore MA, *et al.* Safety and efficacy of an injectable extracellular matrix hydrogel for treating myocardial infarction. Sci Transl Med 2013; 5(173): 173ra25. [http://dx.doi.org/10.1126/scitranslmed.3005503] [PMID: 23427245]

[48] Jauregui ME, Umejei OU, Cruz MP. Prasugrel (Effient), an adenosine diphosphate receptor antagonist for the treatment of acute coronary syndrome. Pharm Therapeut 2009; 34: 417-9.

CHAPTER 4

Biology of Cancer: Genetics, Biomarkers and Clinical Approaches

Abstract: Cancer is the rapid, unregulated and pathological growth (proliferation) of abnormal cells. Even when cancers occur in the same part of the body, they can be very different diseases. When tumors become malignant, their threat depends on their ability to modify surrounding cells to form new blood vessels (angiogenesis) and other supporting cells. The typical American diet that leads to obesity can make people more susceptible to cancer. The human papilloma virus, hepatitis B and T cell leukemia virus type 1 can cause cancer of the cervix, liver and leukocytes (leukemia). Tyrosine kinases regulate many cellular processes which can contribute to cancer development and progression. KRAS is the oncoprotein that is most commonly activated in human cancer [1]. *RAS* is one of the most commonly mutated genes in human cancers. Oncogenes code for oncoproteins, which are upregulated in cancer. Another important oncogene is *PI3K*, which codes for the enzyme PI3K (phosphoinositide 3-kinase). The enzyme PTEN (phosphatase and tensin homolog) catalyzes the opposite reaction, so it is a tumor suppressor and the gene coding for it is downregulated in cancer. Human epidermal growth factor (hEGF, or HER), vascular endothelial growth factor, or VEGF, and the PI3K/Akt/mTOR (mammalian target of rapamycin) survival pathway are all important therapeutic targets in many cancers [1]. Cancer stem cells could be good targets for new drugs that will prevent the recurrence and metastasis of tumors. Also, induced pluripotent stem cells could be used to screen drugs to see if they will be effective in treating each individual patient.

Keywords: Cancer, HER, Human papilloma virus, mTOR, RAS, PI3K, Tyrosine kinase, VEGF.

INTRODUCTION

Let us look at the group of diseases called cancer, the treatment of which often involves personalized medicine. The hallmarks of cancer are inflammation, chromosomal defects, genetic instability, escape from immunosurveillance, limitless proliferative potential, self-sufficiency in growth signals, insensitivity to antigrowth signals, evasion of apoptosis, sustained angiogenesis, tissue invasion

Robert E. Smith

and metastasis. As described in Chapter 2, smoldering inflammation can lead to cancer and other diseases. Reactive oxygen substances, such as the hydroxyl radical, produced by poorly liganded iron, can cause oxidative damage to DNA, proteins and lipids, affecting the genome, epigenome, proteome and lipidome. Inflammatory cytokines can be produced. Moreover, mutations accumulate as we age, eventually leading to the formation of new cell surface antigens. These things can change a tumor's interaction with the immune system, enabling it to escape immunosurveillance. Then, it develops limitless proliferative potential as it becomes self-sufficient in producing the growth signals it needs and becomes insensitive to antigrowth signals. Finally, a tumor becomes deadly when it builds new blood vessels and metastasizes into the lymph nodes and other organs.

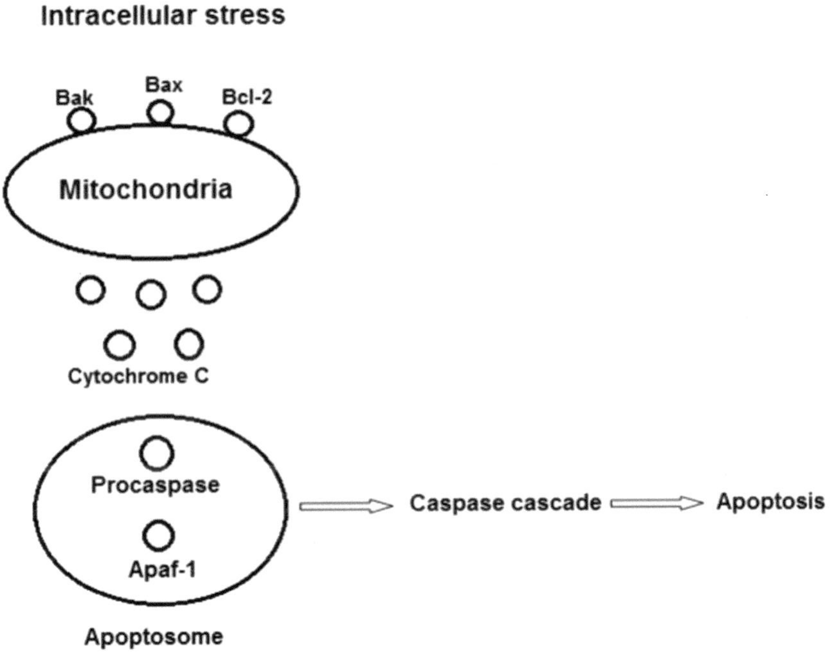

Fig. (1). Intrinsic pathway of apoptosis [1].

Most cells in our body are broken down and re-made. When a cell is damaged or just reaches the end of its normal life, it usually undergoes programmed cell death, also known as apoptosis. There is an intrinsic and an extrinsic pathway to apoptosis [1]. "The extrinsic pathway begins outside the cell, while the intrinsic pathway begins inside the cell, as shown in Figs. (**1** and **2**). In both pathways, a class of cysteine proteases called caspases catalyzes reactions that dismantle dead and dying cells. When this apoptosis is disrupted, cancer can begin. Abnormal cells should die, but they don't. The mitochondria, which regulate apoptosis in healthy cells, become defective. Not only do they not initiate apoptosis, but their metabolism changes. Despite the ready availability of O_2, glycolysis occurs as if it was in an anaerobic environment and the end product is lactate, not pyruvate. This is called the Warburg effect, after the scientist who discovered it. At first, this was thought to be a cause of cancer, but we now know that it is an effect, or a symptom. Recent data has shown that the loss of the tumor suppressor SIRT6 may be the cause of the Warburg effect" [1].

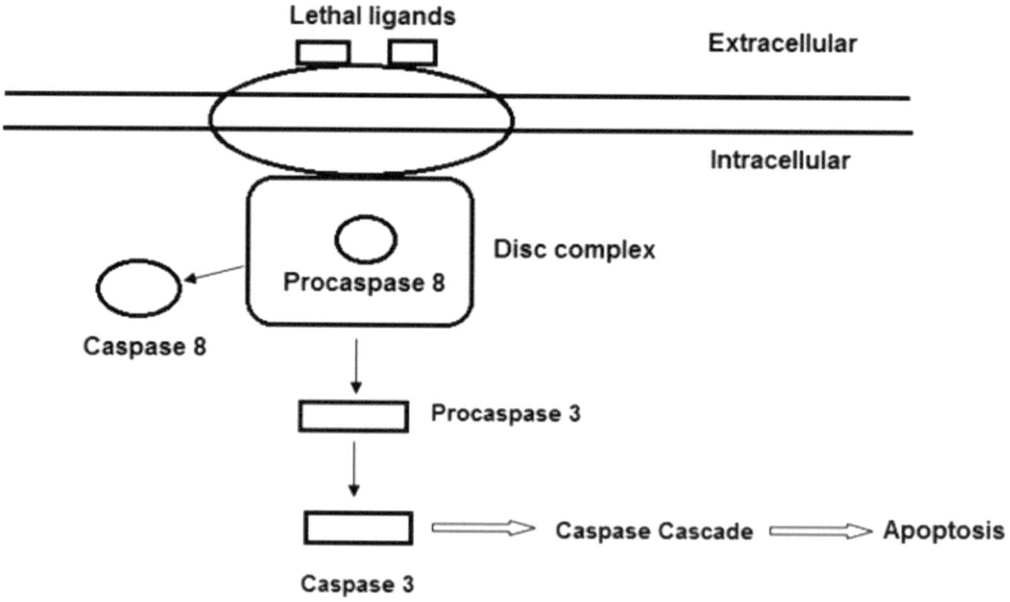

Fig. (2). Extrinsic pathway of apoptosis.

Unfortunately, there are many websites and books that promote pseudo-science and dangerous alternative cancer therapies. Some of them say that cancer can be cured if you can just get more oxygen into the cancer cells. One book (written by an accountant, not a scientist) even tells people with breast cancer to perfuse the breast with baking soda, or $NaHCO_3$, as if the oxygen in the HCO_3^- can magically convert to O_2 and stop cancer. By the way, perfusion, when done by scientists on something like an excised frog heart, means connecting a tube to one side of the heart and pumping an aqueous solution through the heart and out through another tube. The book does not go into detail on how they think perfusion of the breast should be done, but I hope that no one ever cuts into a breast and does an actual perfusion. So, let's stick to science.

An experimental hyperbaric oxygen treatment (30% O_2) has been shown to truly oxygenate tumors in rats by posttranslational modifications and restoring the normal (wild type) conformation of p53 [2]. That is, O_2 enters tumor cells, modulates the p53 pathway and induces apoptosis. Before it enters the p53 in hypoxic tumor cells is phosphorylated at serine-15, changing the conformation. Upon entering the cells, O_2 stimulates a change back to the wild type conformation of p53, which then acts as a chaperone, rescuing the mutant p53 in hypoxic tumors, resulting in tumor regression. However, side effects included inducing seizures and radiation tissue injury [2]. More recent research has shown that measuring the degree of oxygenation of tumors (using MRI) while inhaling O_2 can help guide radiation therapy [3]. This may enable personalized therapies and fewer radiation treatments while killing tumors with oxygen enhancement.

That is, oxygenation can induce apoptosis. In the extrinsic pathway, apoptosis is initiated by the binding of death ligands to death receptors on the cell membrane [1]. "Death receptors belong to the tumor necrosis factor receptor (TNFR) gene superfamily that includes TNFR-1, Fas/CD95, and the TRAIL (tumor necrosis factor-related apoptosis inducing ligand) receptors DR-4 and DR-5. They cause apoptosis mediated by caspase 8, a cysteine-aspartate protease. On the other hand, the intrinsic pathway begins when there is damage within the cell. It can be caused by oncogenes, direct DNA damage, oxidative stress and starvation. The cytoplasmic proteins, BAX and BID, bind to the outer mitochondrial membrane upon receiving appropriate stress signals. BAX and BID interact with a protein called BAK that is in the mitochondria. This causes the release of cytochrome c and other mitochondrial proapoptotic factors into the cytosol. Once in the cytosol, cytochrome c binds to apoptotic protease activating factor-1 (Apaf-1) which then forms an apoptosome that triggers the activation of the initiator procaspase-9. That is, caspases are regulated at a post-transcriptional level, so they can be rapidly activated when needed. They are first biosynthesized as inactive procaspases. Activated caspase-9 then initiates a caspase cascade, ultimately resulting in cell death. Tumor suppressor p53 protein is a sensor of cellular stress and is a critical activator of the intrinsic pathway" [1]. It initiates apoptosis by activating the transcription of proapoptotic BAX family members and repressing anti-apoptotic Bcl2 proteins.

On the other hand, p53 is not involved in the induction of apoptosis in the extrinsic pathway [1]. "Moreover, the orderly progression of the cell cycle from one stage to the next is controlled by checkpoints that prevent unwanted mitosis in response to DNA damage. Following DNA damage, the cell cycle can be arrested to enable DNA to be repaired. If the damage can't be repaired, apoptotic pathways are activated. So, there are genome stability genes that code for proteins that regulate the cell cycle or DNA repair mechanisms" [1].

The key regulators of cell cycle progression are the cyclin-dependent kinases (CDKs), cyclins, and CDK inhibitors [1]. The cyclin/CDK complexes promote cell cycle progression. On the other hand, inhibitors of the CDK protein can stop or arrest the cell cycle. Moreover, stem cells can play an important role in cancer.

Tissue-specific stem cells can be transformed into cancer stem cells, which can differentiate to form any of the many types of cancer cells that can exist in a tumor. So, cancer is the rapid, unregulated and pathological growth (proliferation) of abnormal cells [4]. That is, some cancers are organized into a hierarchy of subpopulations of tumorigenic cancer stem cells and their non-tumorigenic progeny [5]. In these cases, cancer stem cells are thought to drive tumor growth and disease progression, perhaps through resistance to therapy and metastasis. However, even in cancers that clearly have a hierarchy of tumorigenic and non-tumorigenic cells, other factors are important. Other sources of heterogeneity include clonal evolution, heterogeneity in the micro-environment and reversible changes in properties of cancer cells that can occur independently of hierarchical organization. Nowell's theory of clonal evolution proposes that cancers arise from a single cell of origin, develop genomic instability during replication and then the most aggressive clones are enriched by metastasis and the eradication of competing clones that are killed by cancer treatment [6].

Often, cancer is thought to be a process of asexual evolution driven by genomic and genetic instability [6, 7]. "There is much variation in the genotypes and phenotypes between tumors and within individual tumors, caused by genetic instability. Genetically distinct subclonal populations of cancer cells can emerge through intercellular genetic variation. This can cause selective outgrowth of clones of cells that have a phenotypic advantage within the micro-environment of the tumor. The new micro-environment at metastatic sites can help explain genetic differences between metastatic and primary sites. Moreover, there is much heterogeneity in both the patterns and extent of genomic instability in different types of tumors" [7].

Mutation, selection and adaptation were once thought to occur primarily within, and to a lesser extent, outside the primary tumor [7]. "However, cancer cells that remain after surgery can have extreme genomic heterogeneity before the metastasis begins. Later, this heterogeneity decreases due to selective clonal expansion, suggesting that the remaining cells had yet to acquire many of the important traits that fully malignant cells have. Trying to stop the cells' progression outside the primary tumor offers new challenges and opportunities for diagnosis and adjuvant therapies" [7].

Cancer is the second leading cause of death in the USA, after cardiovascular disease [7]. "There are over 100 types of clinically distinct cancers and many subsets of each type. So, cancer is actually many different diseases that occur in various parts of the body, such as the skin, stomach, mouth, throat, lungs, intestines, breasts, testicles, kidneys, prostate, pancreas, liver, ovary, bladder, blood cells and bone. Even when cancers occur in the same part of the body, they can be very different diseases. For example, the most common form of skin cancer is not very dangerous, but the rarer form, melanoma, is often fatal. Cancer in other parts of the body can have a wide range of prognoses. Cancers can be classified according to their state of advancement, rate of growth and lethality. In the beginning, abnormalities can be benign. Malignant tumors are able to form vascularized colonies, called metastases, in different parts of the body. Benign growths can't do this. They are contained and are relatively harmless, but still threatening, since the abnormal cells do have the potential to become malignant. The difference between benign and malignant lesions is associated with the emergence of genome instability" [7]. One main feature or hallmark of cancer is an ongoing production of genomic diversity, which enables gene networks to self-organize and individual cells to progress towards metastasis in a process of clonal evolution [8]. The process of self-organization can be described in network theory as the formation of attractor states (see Chapter 13).

ANGIOGENESIS

Once tumors become malignant, their threat depends on their ability to modify surrounding cells to form new blood vessels (angiogenesis) and other supporting cells. There can be cancer cells that circulate in the blood (circulating tumor cells, or CTCs) and cancer cells that disseminate into other organs (disseminated cancer cells, DCCs) [9]. "During successful dissemination, tumor cells invade the surrounding tissue of the primary tumor, intravasate into blood and lymphatic vessels, translocate to distant tissues, extravasate, adapt to the new microenvironment, and eventually seed, proliferate, and colonize to form metastases. Because dissemination occurs mostly through the blood, CTCs that have been shed into the vasculature and may be on their way to potential metastatic sites are very important. Because blood collection is simple and minimally invasive, CTCs could be used as excellent markers for predicting disease progression and survival. Identification of CTCs may also be able to guide therapeutic management and effectiveness, even in the absence of detectable metastases, and offer insights into mechanisms of drug resistance. However, CTCs have a diameter that is three to four times as large as the bores of capillaries in organs, so they can become trapped there. CTCs can be detected in the blood, but only extremely small and/or plastic CTCs can keep circulating. Also, some CTCs circulate as microclusters that may become stuck in capillaries. So, it is unclear whether actual CTCs are truly the source of metastases. Moreover, tumors can spread through the lymphatic system. So, single cell assays can be especially informative. For example, CTCs that have the human epidermal growth factor receptor 2 (HER2) on them have been found in HER2-negative breast cancers" [9].

So, cancer cells move to other parts of the body in a process called metastasis and become disseminated metastatic tumor cells (DTCs). Once DCCs become metastatic, their genomes become quite similar, indicating that an aggressive clone has expanded. So, most forms of breast cancer can almost always be cured if they are caught before they move (metastasize) into other tissues [10]. Moreover, metastasis accounts for the majority of cancer patients' deaths [11]. So, drugs that target angiogenesis continue to be developed and some are already approved. These drugs work by inhibiting angiogenesis: Bevacizumab (Avastin®), cetuximab (Erbitux®), panitumumab (Vectibix®), and trastuzumab (Herceptin®) [11].

Bevacizumab is a humanized antibody that binds to the vascular endothelial growth factor (VEGF) and neutralizes VEGF's ability to stimulate the growth of new blood vessels [11]. Cetuximab and panitumumab are monoclonal antibodies (mAbs) that are specific to the epidermal growth factor receptor, EGFR. Cetuximab is a chimeric (mouse and human), while panitumumab is a fully human mAb [11]. Trastuzumab is a mAb that is specific to the HER2 receptor. It is used in personalized medicine, as described in Chapter 1.

The following FDA approved drugs target metastasis by inhibiting tyrosine kinases: erlotinib (Tarveca®), sorafenib (Nexavar®), and sunitinib (Sutent®). Another approved drug that targets metastasis does so by inhibiting a protein called the mammalian target of rapamycin, or mTOR, an enzyme that catalyzes the phosphorylation of serine and threonine residues in proteins [11]. "The drug that targets metastasis by inhibiting mTOR is temsirolimus (Torise®). However, anti-angiogenic drugs can also enter many non-cancerous cells, causing harmful side effects. For example, fumagillin is a potent inhibitor of endothelial cell functions *in vitro* and in animal models. TNP-470 has the broadest anticancer spectrum of any known agent. However, when it was used clinically, many patients experienced neurotoxicity at the same drug doses that produced antitumor activity. So, researchers are developing ways to deliver anti-cancer drugs and inhibitors of angiogenesis to just the cancer cells, in a targeted drug delivery. At the same time, they are chemical and biologic therapies that are precisely designed to target a critical gene or signaling pathway in cancers" [11].

Recently, it was found that statins, including simvastatin, also inhibit angiogenesis and they do it in a critical tissue – the lymph nodes [12]. "Lymphangiogenesis is important because the lymph nodes are the first sites of metastasis in most epithelial cancers and malignant melanomas. Moreover, malignant tumors cause lymphatic vessels to grow. Statins were identified as inhibitors of lymphangiogenesis by doing high content screening of the database called the Library of Pharmacologically Active Compounds, or LOPAC®" [13]. They used individual human lymphatic endothelial cells along with automated image acquisition and analysis to look for a specific phenotype – the formation of lymphatic vessel sprouts [12].

So, a tumor is a collection of diverse cells, not just cancer cells that have carcinogenic mutations [14]. "Within tissues, there are epithelial cells that are surrounded by a connective framework, called stroma. The stroma consists of fibroblasts, endothelial cells, immune cells and an extracellular matrix (ECM). The ECM provides cells in tissues with information that tells them to survive. In healthy cells, the stroma is the main barrier against tumorigenesis. However, in the presence of transformed tumor cells, changes occur that support cancer progression. Fibroblasts are recruited, immune cells migrate and the ECM is remodeled and eventually vascular networks develop. Healthy fibroblasts maintain the structural framework in tissues. Quiescent fibroblasts respond differentially to damage and are activated to support repair. Although normal fibroblasts typically suppress tumor formation, cancer-associated fibroblasts (CAFs) can promote tumorigenesis. Compared with normal tissue fibroblasts, CAFs have increased proliferation, enhanced extracellular matrix production and secrete unique cytokines such as stromal cell-derived factor 1 (SDF1), vascular endothelial growth factor (VEGF), platelet-derived growth factor (PDGF) and hepatocyte growth factor (HGF). Other mesenchyme-derived cell types, such as adipocytes, can also contribute to the growth and progression of tumors. Differences in fibroblast behavior and response can cause extensive tissue remodeling that is mediated by increased expression of proteolytic enzymes such as matrix metalloproteinases. Also, pathogenic angiogenesis can occur by liberating pro-angiogenic factors within the ECM" [14].

Moreover, the tumor vasculature network is dynamic and can limit the growth of tumors [14]. "They are made by the formation of new vessels (angiogenesis). So, existing vessels within tissues, or the recruitment and differentiation of endothelial precursors from bone marrow (vasculogenesis) can contribute to vascular heterogeneity in and among tumors. Uneven vascularization and differences in vascular maturity, when combined with a lack of drainage due to poor lymphatic vessel coverage will cause complex topography and variable interstitial pressure within tumors. When tumor vasculatures are poorly organized and don't function properly, it produces areas of hypoxia and a limited supply of nutrients. A gradient develops that is based on different distances from the vascular beds. It is crucial that drugs are distributed to all cells in the tumor. These vascular networks can generate distinct micro-environments within the tumor and contribute to inter- and intratumor heterogeneity, and ultimately influence the clinical outcome. Microvessel density may also be a significant prognostic factor for indicating a poor outcome in non-small-cell lung cancer (NSCLC) as well as colorectal and breast cancers. Moreover, elevated expression of the predominant pro-angiogenic ligand VEGFA is associated with a worse prognosis than those that have a relatively low expression of VEGFA in metastatic colorectal, lung and renal cell cancers" [14].

As signals come from the ECM through cellular adhesion proteins, biochemical pathways can be activated that ensure the survival of cell. When such cells lose contact with the ECM, different pathways are activated and the cells undergo a type of programmed cell death (apoptosis) called anoikis. However, in advanced stages of cancer, this mechanism fails. Cancer cells that break free from the ECM don't die. They migrate (metastasize) into other parts of the body, where they can be especially deadly.

ROLE OF THE IMMUNE SYSTEM

Also, a healthy immune system can recognize and protect tissues from infections and damage, as described further in Chapter 7. However, both the innate and adaptive immune systems can promote and prevent tumor growth [15]. "Although the immune system has the ability to mount anti-tumor responses, immune suppression mechanisms can prevent this process. That is, many types of cancer cells can suppress the immune system. Cancer cells can escape detection by decreasing the expression of important antigenic proteins, making them invisible to cytotoxic T cells. They can also secrete proteins that inhibit effector T cells and enhance the production of regulatory T cells that suppress immune responses. Some melanomas can reorganize their surrounding cells (stroma) to form structures that are similar to those of lymphoid tissue in the immune system.

They can recruit and maintain immune cells that promote tolerance and tumor progression" [15].

Moreover, the activation of T-cells involves both stimulatory and inhibitory checkpoint signals that finely tune responses in order to prevent excessive damage and autoimmunity [14]. "One way to usurp the activation of cytotoxic T-cells in tumors is by continuously engaging inhibitory receptors on T-cells, such as cytotoxic T-lymphocyte-associated antigen-4 (CTLA-4) and programmed death 1 (PD1) by upregulating their ligands. An indirect way of preventing antitumor T-cell responses is to generate an immunosuppressive environment. A myeloid-derived suppressor cells (MDSCs, which includes neutrophils, immature dendritic cells, monocytes and early myeloid progenitors) was expanded upon upon tumor implantation. This implicates that there is tumor-initiated endocrine communication with the immune system and secretion of chemokines by tumor and cancer associated fibroblast. This includes granulocyte–macrophage colony stimulating factor, GM-CSF; or granulocyte colony stimulating factor, G-CSF. The recruitment of immunosuppressive myeloid lineages to the tumor not only suppresses adaptive immunity, but also fosters angiogenesis through the secretion of VEGFA, basic fibroblast growth factor (bFGF) and transforming growth factor β (TGF-β). MDSCs also inhibit natural killer cell function and expand the immunosuppressive regulatory T-cell population. Also, MDSCs can directly inhibit the expansion activation and migration of effector T-cells by changing the environment. B cells can suppress and support T-cell function, resulting in differential effects on tumorigenesis. However, B cells can promote tumor progression by increasing inflammation. The recruitment of mast cell has been implicated in tumorigenesis and angiogenesis. Tumor-associated macrophages (TAMs) can also drastically affect tumor progression, depending on their polarization" [14].

Moreover, different subtypes of immune cells have differential prognostic value [14]. "For example, in melanoma, colorectal and breast cancer, there is a strong correlation between high T-cell occupancy and a good clinical outcome. However, other types of T-cells, such as regulatory T cells, and helper T cells (T_H1 and T_H17) are associated with poor or good prognosis depending on the type of cancer. In breast cancer, assessment of pro- and antitumor immunity showed that a high ratio of macrophages to cytotoxic T-cells was correlated with a worse outcome than those with a low ratio. That is, a stromal signature derived from microdissected breast tumor biopsies can help predicted the prognosis, regardless of the tumor subtype. Angiogenesis, hypoxia and macrophage-mediated immune suppression signatures have been associated with poor prognosis. Also, B-cell content has been associated with a good prognosis in ovarian, as well as breast cancers. This provides a strong argument for the considering the content of the immune network as a powerful contributor to tumor development" [14].

The stroma influences not only cancer cells, but also cells that are present in a wound. The stroma can support the growth and spread of cancer, and it can help wounds heal. Scientists have examined the DNA from stroma cells surrounding tumors and found 26 genes that are called the stroma-derived prognostic predictor, SDPP [16]. Analysis of these genes can be used to help establish the proper therapy for a patient. The sequence of bases in these genes varies with the prognosis of breast cancer, because once a cancer has metastasized into other parts of the body, it is often fatal. Thus, it is very important to diagnose cancer as early as possible [17]. The interactions between tumors, stroma and the microenvironment are becoming key considerations when developing new anticancer drugs [17].

So, several drugs that target the vasculature and immune cells in the stroma have been approved by the FDA [14]. Bevacizumab, vandetanib, sunitinib, axitinib, sorafenib, pazopanib, cabozantinib and Ziv-aflibercept target the vasculature. Ipilimumab, sipuleucel-T, aldesleukin, and interferon α-2b (IFN α-2b) target immune cells.

DIAGNOSIS AND PROGNOSIS

However, cancer treatment is usually based on the diagnosis of samples obtained from tissues by biopsy [14]. "Tissues that may be cancerous are analyzed under a light microscope to evaluate histopathology. Immunostaining and selected molecular tests are used to diagnosis cancer. Treatment is based on the anatomical location and tissue of origin of the primary tumor when cancer is localized to an organ site or when cancer has metastasized and the primary site can be identified by imaging or pathological testing. Even when solid tumors re-emerge after initial treatment or progress after systemic treatment for metastatic disease, taking another biopsy to guide treatment decisions is not routine. Instead, further systemic treatment of patients with progressive metastatic disease is typically based on identifying predictive biomarkers in archived primary specimens, which may no longer represent the current disease. Examples include mutations in the BRAF gene in melanomas, HER2 (also known as ERBB2) amplification or overexpression in breast cancers, KRAS mutation in colorectal cancer and EGFR mutation in non-small-cell lung cancer. Currently, the purpose of looking for biomarkers is to look for heterogeneity between different patients. However, many people now realize that intratumor heterogeneity within the same patient is clinically relevant. That is, the status of predictive biomarkers that are used for making clinical decisions may evolve during tumor progression, especially in metastatic dissemination of the primary tumor to a distant organ or for established metastatic diseases that are under the selection pressure of treatment" [14].

CLASSIFICATION FOR CANCER DEVELOPMENT

Doctors often classify cancer according to its stage of development. In stage I, the cancer is relatively small and is contained within the organ in which it originated. In stage II, the cancer is still localized, but it is larger than in stage I and it may have spread to nearby lymph nodes. In stage III, the cancer is still larger and it has started spreading into the lymph nodes. In stage IV, the cancer has metastasized and moved into other tissues. As a cancer cells progress through these stages, they acquire more abnormalities. More of the normal constraints on cell growth are lost as a cancer grows. As mentioned already, cancer cells become surrounded by other cells which help the cancer and harm the patient. Surrounding cells include fibroblasts, inflammatory immune cells (such as macrophages), and the cells of blood vessels (smooth muscle and endothelial cells). These are all embedded in an extracellular matrix that is produced by fibroblasts, and they help the cancer cells grow and eventually metastasize [6]. So, doctors and scientists can either target the cancer cells themselves, or they can target the cells that surround and support the tumor. To identify molecular targets for therapy, one must understand the causes of cancer and the molecular mechanisms that become dysregulated.

CAUSES OF CANCER

There are many causes of cancer. Some people are born with germline genetic defects that make them more susceptible to cancer. Others have genetic defects that are caused by viruses, mutagenic environmental toxins and carcinogens. Many of the somatic mutations have been found in protein-coding regions of genes or at splice junctions. The typical American diet that leads to obesity can make people more susceptible to cancer. Certain viruses, such as the human papilloma virus, hepatitis B and T cell leukemia virus type 1 can cause cancer of the cervix, liver and leukocytes (leukemia) [14]. A type of herpesvirus, called cytomegalovirus, or CMV, is found in many people and is usually dormant and causes few clinical symptoms, but over 80% of patients with the type of brain cancer called glioblastoma multiform have CMV in their blood and brain tumors. It is the target for a vaccine that tries to slow down the progression of this fatal form of cancer. Another type of bacteria, *H. pylori*, causes stomach ulcers and predisposes a person to stomach cancer. Even the bacteria that can infect the gums in the mouth can spread to other parts of the body and increase the susceptibility to cancer. Also, many industrial chemicals can cause cancer.

Recent data has helped show how obesity can facilitate cancer [18]. That is, stromal progenitor cells from endogenous adipose tissue contribute to pericytes and adipocytes that populate the tumor microenvironment. So, obesity facilitates tumor growth in mice irrespective of their diet, suggesting a direct effect of excess white adipose tissue on cancer growth [18].

Still, the first step in starting cancer is usually one or more mutations. For example, carcinogens in tobacco products can bind to DNA and react with it, producing mutations. DNA can also be damaged by UV radiation and by reactive oxygen species, or ROS. Thus, many people feel that cancer can be prevented by consuming antioxidants, which destroy ROS and bind to Fe^{2+}, keeping it from forming the hydroxyl radical – an especially damaging ROS. Sometimes, mutations are caused by errors in copying DNA, prior to cell division. However, cells have enzymes that can proofread the copied DNA and catalyze the correction of mistakes or bases that are mis-matched. Such enzymes are often important in regulating cell growth when the cell is healthy. They control cell cycle check points and help determine whether a cell should live or die. When these DNA repair enzymes are mutated, the cells can become immortalized cancer cells. As a person gets older, the number of uncorrected mutations can accumulate, so age is an important factor in cancer.

Another factor is that the immune system becomes weaker in older people. Our immune systems produce cells that can kill some cancer cells before they develop into advanced stages. However, many cancer cells avoid attack by the immune system because they carry self-antigens on their outer membrane surface [4]. Another age-related factor is that normal human cells can only copy themselves about 60 times. This is partly due to telomeres, or tandem repeats of the following DNA base sequence: TTAGGG, located at the ends of chromosomes. These telomeres become shorter after each cell division, until they are so short that the cell can't divide any more. Cancer cells usually acquire the ability to regenerate telomeres, enabling them to copy themselves many times [4].

THE CELL CYCLE AND RTKS

When a healthy cell copies itself, it goes through a cell cycle [4]. The first stage of the cycle is the resting state, called G_0. If there are sufficient nutrients, a cell can enter a growth stage, or G_1. There is a restriction point (called R), at which the cell will decide whether to remain in G_1 or proceed to the stage of DNA synthesis (S). There are proteins called R-point gatekeepers, which allow passage into the next stage only if the proteins are phosphorylated. The level of phosphorylation is controlled by signals arriving from outside the cell. These signals activate a tyrosine kinase, which catalyzes the phosphorylation of tyrosine residues. The human genome codes for 90 proteins with tyrosine kinase domains [19] and for 518 different kinases [4]. Much research is being done to understand them better [4].

Tyrosine kinase-linked receptors (also called receptor tyrosine kinases, or RTKs) are receptors that are linked to tyrosine kinases [4]. "These receptors are also enzymes that catalyze the addition of a phosphate to tyrosine residues in proteins. RTKs are receptors for many polypeptide growth factors, cytokines and hormones. Of the 90 unique tyrosine kinase genes identified in the human genome, 58 encode receptor tyrosine kinase proteins. RTKs have been shown to be not only key regulators of normal cellular processes but also to have a critical role in the development and progression of many types of cancer. Most RTKs have only one subunit, but some, such as the insulin receptor, have more than one. Each monomer has a single domain that spans the membrane, an extracellular N-terminal region and an intracellular C-terminal region. The extracellular N-terminal region is composed of a very large domain which binds to extracellular ligands. The intracellular C-terminal region comprises domains responsible for the kinase activity of these receptors. When a growth factor binds to the extracellular domain of an RTK, its dimerization is triggered along with other adjacent RTKs. This leads to a rapid activation of the protein's cytoplasmic kinase domains, the first substrate for these domains being the receptor itself. The activated receptor then becomes autophosphorylated on multiple specific intracellular tyrosines. The phosphorylation of specific tyrosines within the activated receptor creates binding sites for proteins that contain Src homology-2 (SH2) and phosphotyrosine binding (PTB) domains. Specific proteins containing these domains include Src and phospholipase Cγ. The phosphorylation and activation of these two proteins leads to the initiation of signal transduction pathways. There are other proteins that interact with the activated receptor act as adaptor proteins and have no intrinsic enzymatic activity of their own. These adaptor proteins link RTK activation to downstream pathways, such as the MAP kinase cascade" [4].

Tyrosine kinases regulate many cellular processes which can contribute to cancer development and progression [19]. "These cellular processes include cell growth, differentiation, migration and apoptosis. There are several anti-cancer drugs that inhibit tyrosine kinases in some patients. These include imatinib (for chronic myelogenous leukemia), tratuzumab (for breast cancer) and three drugs (cetuximab, erlotinib and gefitinib) that inhibit the tyrosine kinase enzyme called the "epidermal growth factor receptor (EGFR) and are used to treat cancers of epithelial cells, such as colorectal, head and neck cancers" [19]. Cetuximab is a chimeric mAb, while erlotinib is a reversible inhibitor of EGFR, which catalyzes the phosphorylation of some of its own tyrosine residues, which activates downstream signaling cascades, including MAPK.

However, cancer cells often mutate when treated with tyrosine kinase inhibitors and survive. They become dependent on one or just a few oncogenes to maintain the malignant phenotype. Two of them are PUMA and BIM. PUMA is a member of the BCL2 family of proteins [20]. "It is activated upon inhibition of the PI3K-Akt pathway in both HER2-amplified breast cancers and EGFR mutant lung cancers. BIM is also a BCL2 family member that plays a central role in gefitinib and erlotinib-induced apoptosis in EGFR mutant lung cancers and in lapatinib-induced apoptosis in HER2-amplified breast cancers. There are distinct subsets of cancers that depend on specific driver mutations for survival, a phenomenon called oncogene addiction" [20].

ROLE OF VIRUSES IN CAUSING CANCER

In addition, some viruses produce cancer by inserting (transfecting) their genes into those of the host [4]. The cancer-causing oncogenes can code for proteins that propagate signaling cascades which are not properly regulated. For example, there are viruses that cause sarcoma, or cancer of connective tissues, such as bone or muscle. One of these is called the Harvey virus and it converts the normal, healthy RAS gene into an oncogene (HRAS) and another, called the Kirsten virus converts RAS into KRAS. KRAS is the oncoprotein that is most commonly activated in human cancer [20]. "However, oncogenic KRAS and its known downstream effectors have thus far presented intractable targets for antineoplastic drugs. Still, proteins and pathways associated with oncogenic KRAS have been identified, so they provide new therapeutic targets. Each virus was named after the person who discovered it. RAS is one of the most commonly mutated genes in human cancers" [20]. It is found in about 20% of them, but it has proved impossible to effectively target with small molecule inhibitors. The HRAS and KRAS oncogenes code for a Ras oncoprotein, which is a mutant tyrosine kinase that is always switched on. The Ras oncoprotein affects the transcription of many genes. Many R point proteins are hyperphosphorylated. Several proteins are synthesized and apoptosis is inhibited [4]. HRAS and KRAS oncogenes don't do this directly, but instead act through further (downstream) pathways, including one that involves NF-κB. Hopefully, the deadly effects of HRAS and KRAS oncogenes can be stopped by inhibiting NF-κB.

THE ROLE OF THE P53 ONCOPROTEIN

The oncogene *TP53* codes for the oncoprotein p53 that is also a target for breast cancer therapy [21]. As mentioned previously, p53 is a hub in the scale-free cellular network. We will learn more about scale-free networks, such as living cells, in Chapter 13. For now, it is sufficient to know that p53 is a tumor suppressor. "It regulates the cell cycle and helps prevent cancer. Under normal conditions, there is not much p53 in the cell, but when cells are stressed, p53 becomes activated and stabilized. It can be activated by post-transcriptional modifications. When it is mutated, p53 becomes an oncoprotein and can help cause cancer. It is the most commonly found oncoprotein in human cancers. It can activate DNA repair, stop the growth of cells, and can induce apoptosis. So, when a cell is exposed to mutagens the DNA can be damaged. The normal p53 protein can either cause the DNA to be repaired, or it can cause the defective cell to die (apoptosis), preventing its growth and differentiation into cancer cells. However, when mutated, p53 can cause cancer. So, there is a cancer vaccine undergoing clinical trials, and its target is p53" [21].

The p53 protein is a hub in the network of cellular metabolism. It interacts with many other proteins. It regulates the expression of a wide variety of genes in response to genotoxic or cellular stress [21]. The genes are involved in apoptosis, growth arrest, inhibition of cell cycle progression, differentiation and accelerated DNA repair or senescence [21]. Activated p53 can initiate apoptosis and stop cell proliferation. When p53 is mutated and no longer active, uncontrolled cell proliferation, or cancer, can occur. Progression of a healthy cell through the cell cycle is regulated by many different proteins at several check points. Information from other proteins will affect the check point proteins. The p53 protein is tetrameric. It interacts with many other biopolymers in cells. Parts of it have flexible structures, enabling it to bind to different proteins and different regions of DNA. It is a transcription factor that can activate genes to produce proteins which either halt cell cycle progression or induce apoptosis [4]. Thus, normal (or wild-type) p53 is able to sense abnormal cell function, making it an effective tumor suppressor. It can induce or suppress the expression of several genes that code for proteins involved in apoptosis, cell cycle control and senescence. It is the guardian of the genome when it is acting properly. However, a mutated form of p53 is the most frequently mutated oncoprotein found in human cancers. Transcription of the *TP53* gene is controlled by negative regulators called *MDM2* and *MDM4*. Even in cancer cells that have the normal, wild-type p53 gene, the activity of the p53 protein is often inhibited by Mdm2, another oncoprotein. Deregulation of the balance (homeostasis) of the ratio of p53/Mdm2 can lead to the malignant transformation of cells. Small molecules, or potential drugs, are being tested for their ability to prevent MDM2-p53 interaction.

The activity of the p53 protein can also be modified by phosphorylation, methylation, acetylation, glycosylation, ribosylation, summoylation, ubiquination and by interaction with other proteins, such as Mdm2 and MdmX, which are ubiquitin ligases that can also bind to the trans-activation domain of the tetrameric p53 protein [22]. "There are other proteins that can interact with Mdm2 and affect the stability and activity of p53. For many years, phosphorylation was thought to be the first step in stabilizing p53 when a cell is under stress. Phosphorylations at Ser395 and Tyr394 do inhibit Mdm2, while phosphorylations at Ser166 and Ser186 activate it. Moreover, cycles of phosphorylation-dephosphorylation and acetylation-deacetylation of Mdm2 affect its ability to inhibit p53. So, drugs that target the interactions between Mdm2 and p53 have been developed and have started clinical testing" [22].

In the second step, modified p53 binds to target genes through the conserved central, core domain, while the C-terminal region was thought to be a negative modulator that had to be modified to allow sequence-specific DNA binding [22]. "The modifications target and modify the C-terminal region of p53, thus affecting the ability of p53 to bind DNA. However, recent research indicated that both the core DNA-binding domain and the C-terminal domain of p53 possess DNA-binding activities. The core DNA-binding domain provides sequence specificity while the C-terminal domain recognizes structural features of target DNA" [22].

Acetylation is an important link between p53 and histones that regulate transcription [22]. "The same acetyl transferases, CBP/p300, Tip60 and Mof catalyze the acetylation of p53 and histones. Mof is also a key regulator of the embryonic stem cell transcriptional network, as it primes genes for developmental programs. However, ubiquitination and acetylation "are mutually exclusive. Competition between these modifications may affect the stability of p53. That is, p53 acetylation levels are markedly enhanced in response to stress, promoting p53 stabilization and activation, while ubiquination targets p53 for proteasomal degradation" [22].

ONCOGENES AND ONCOPROTEINS

So, oncogenes code for oncoproteins, which are upregulated in cancer. Before the protein is mutated, it is called a proto-oncoprotein. An important oncoprotein frequently found in some human cancers in c-Myc. As a proto-oncogene, c-Myc can regulate the expression of 15% of all genes, including genes involved in cell division, cell growth, and apoptosis [23]. "It exerts its effects on transcriptional targets through various mechanisms - there are positive effects from recruitment of histone-modifying enzymes, general transcriptional machinery, and chromatin-remodeling complexes and negative effects from recruitment of DNA methyltransferases. Often c-Myc drives cell proliferation and inhibits differentiation, as in mouse embryonic stem cells. Surprisingly, c-Myc is different in human embryonic stem cells where it induces apoptosis and differentiation" [23].

Since it has such an important role and controls so many genes, the presence of the Myc protein in cells must be carefully controlled. Thus, the transcription of the gene coding for c-Myc is tightly regulated in healthy cells. It is expressed only when cells are dividing. On the other hand, once mutated, the c-Myc gene is over-expressed, causing too much c-Myc protein to be made. "The c-Myc protein or the c-myc gene is overexpressed in a wide variety of human cancers with 80% of breast cancers, 70% of colon cancer, 90% of gynecological cancers, 50% of hepatocellular carcinomas and a variety of hematological tumors possessing abnormal myc expression. On the basis of these frequencies, it is estimated that approximately 100,000 US cancer deaths per year are associated with changes in the c-myc gene or its expression" [24].

So, the c-Myc protein is a transcription factor that can control the transcription of other genes. Myc forms a complex with another protein, called Max [24]. "When the Myc-Max complex binds to DNA, it can activate transcription. In cancer, though, the c-Myc gene is amplified, so there are many copies of it. There are over 200 copies per cell of N-myc can be found in neuroblastoma, and over 50 copies per cell of c-myc, N-myc, or L-myc may be found in small cell lung cancers. A highly regulated cell cycle permits cells to repair DNA damage before replicating, thus promoting genomic fidelity. Inappropriate cell cycle proliferation can lead to genomic instability, resulting in new mutations and abnormal chromosome number and structure. c-Myc overexpression, even transiently, can induce genomic instability that is characterized by gene amplification, aneuploidy and polyploidy. Other studies suggest that c-Myc induces the production of reactive oxygen species (ROS) by mitochondria, leading to DNA damage and genomic instability" [24].

The Myc-Max basic-helix-loop-helix leucine zipper heterodimer binds a target DNA site, termed the E-box [24]. "c-Myc regulates downstream target genes activating cell cycle regulation, apoptosis, or inhibition of cell adhesion. Examples of c-Myc target genes associated with different cellular functions include cell cycle regulation (*p19 Arf, cyclin D, cyclin E,CDK4* and *CUL1* genes), apoptosis (*ODC, LDHA, Bax, FasL* and *PRXIII* genes), metabolism (genes coding for enolase, LDHA, PRXIII, ODC, CAD, NPM, Nucleolin, and JTV-1), cell adhesion (genes coding for collagen, fibronectin and integrin) and cellular differentiation" [24].

The *c-Myc, MYCN* and *L-Myc* genes encode transcription factors that play essential roles in cell proliferation, cell growth, differentiation, and apoptosis [24]. "Myc is a key regulator of many biological activities including cell growth and division (regulation of chromatin modification and components of the biosynthetic machinery); cell-cycle progression (modulation of cyclins, cyclin dependent kinases, cyclin-dependent kinase inhibitors and phosphatases); apoptosis (p53 dependent or independent mechanisms); cell differentiation (down regulation of growth arrest genes); cell metabolism (glycolysis, amino acid biosynthesis and transport, synthesis of macromolecules and DNA metabolism); angiogenesis (upregulation of VEGF); cell adhesion and motility (control of expression of integrins)" [24]. Deregulation of Myc may result in apoptosis, genomic instability, uncontrolled cell proliferation, escape from immune surveillance, growth factor independence, and immortalization [25].

So, amplification of the c-Myc gene can cause cancer chromosomal instability (CIN). This causes an increased number of chromosomes and intratumor heterogeneity [26]. "CIN is observed in most solid tumors and is can result in poor prognosis and drug resistance. It causes an increased rate of change of chromosome number and structure and generates intratumor heterogeneity. CIN is observed in most solid tumors and is associated with both poor prognosis and drug resistance" [26]. Also, stress can mix chromosomes and cause segregation errors during cell division [27]. "Cancer has long been thought to be a disease that occurs after a series of mutations that eventually produces a cell that is well equipped for unrestricted proliferation in the proper environment. Proliferation of this cell would be expected to produce tumors containing cells with this same genome. But in reality, the cells in solid tumors have tremendous genetic variation, or intratumor heterogeneity. Understanding the molecular basis of this heterogeneity may help to understand the cellular pathways that promote tumor formation and progression, and help develop specific anticancer therapies. The genetic variation can be caused by minor DNA mutations and CIN. Structural CIN produces large chromosomal rearrangements, which may be caused by improper repair of damaged DNA. Numerical CIN produces abnormal numbers of chromosomes (called aneuploidy) and has been attributed to various defects that happen during cell division. Replication stress could promote a rapid genetic drift within a tumor, through connected rounds of changes in the structure and the number of chromosomes. Suppressing replicative stress in tumors could inhibit this genetic drift and might prevent the acquisition of resistance to therapies. Such resistance can arise when some cells of a tumor acquire mutations that confer resistance and give them a proliferative advantage Combining therapies that target replication stress with conventional therapies could help to treat CIN+ tumors" [27].

Another important oncogene is *PI3K*, which codes for the enzyme PI3K (phosphoinositide 3-kinase). It catalyzes the phosphorylation of phosphatidyl inositol (see Chapter 13 for more details on PI3K). Like other oncogenes, *PI3K* is upregulated in cancers. On the other hand, the enzyme PTEN (phosphatase and tensin homolog) catalyzes the opposite reaction, so it is a tumor suppressor and the gene coding for it is downregulated in cancer. PTEN normally blocks a PI3K (phosphoinositide 3-kinase) survival pathway, which allows cells to ignore signals that tell them to commit suicide (apoptosis). One form or isozyme of PI3K catalyzes the addition of a phosphate to phosphatidyl inositol 4,5-bisphosphate (PIP_2), to form phosphatidyl inositol 3,4,5-trisphosphate (PIP_3). The PTEN protein does the opposite – it catalyzes the dephosphorylation of PIP_3. PTEN is one of the most frequently mutated tumor suppressors that is found in human cancers [4]. So, PI3K and PTEN are part of a signaling system that can go wrong and lead to cancer.

Not only transcription factors, but also cell surface receptors can be over-expressed. Ligands that bind to the receptors can promote cell growth and division [4]. "For example, the human epidermal growth factor receptor (HER1) appears in many cancer cells. It can be either over-expressed, or it can be expressed in a mutated form that allows it to bind to several ligands that do not bind to the normal HER1. This receptor is a tyrosine kinase which catalyzes its own phosphorylation (auto-phosphorylation). The phosphorylated form binds to two proteins called Grb2 and Sos. These proteins associate with Ras, which activates three proteins on different signaling pathways. One of these pathways uses PI3K, which catalyzes the phosphorylation of the OH on carbon number 3 of inositol, to produce PIP_3. Next, PIP_3 can be hydrolyzed into inositol 1,4,5-trisphosphate (IP_3) and diacyl glycerol, which are second messengers that cause many further effects in the cell. PI_3 activates two other kinases, called Akt and GSK-3β, which lead to the phosphorylation of a protein called β-catenin which moves to the cell nucleus and activates gene transcription, leading to cell proliferation [4].

ADVANCES IN TREATING BREAST CANCER

An article in *Scientific American* described advances in treating and curing breast cancer [28]. It described the human epidermal growth factor (hEGF, or HER), which was discussed in Chapter 1. HER is an important therapeutic target in many cancers. It is associated with tyrosine kinases, which catalyze the attachment of phosphates to other proteins, which sends growth signals. One of the HER receptors, called HER2 can become locked in the on position in breast cancer cells, causing uncontrolled proliferation. The tyrosine kinase inhibitor called lapatinib is given in combination with Herceptin® to women with HER2-positive cancer [28]. "Trastuzumab (Herceptin®), pertuzumab and lapatinib are approved mAbs that act on the HER2/neu receptor. Lapatinib inhibits the tyrosine kinase activities of the two oncogenes HER2 and EGFR "(epidermal growth factor receptor, also known as HER1). Herceptin® is a mAb that binds to the HER2 receptor, reducing cancer cell proliferation. These two drugs act synergistically to inhibit the growth of breast cancer cells. However, cancer cells often develop a resistance to Herceptin®. Tumor cells lose or inactivate the tumor suppressor gene called PTEN (phosphatase and tensin homolog), which normally blocks a PI3K (phosphoinositide 3-kinase) survival pathway, which allows cells to ignore signals that tell them to commit suicide (apoptosis). PI3K and PTEN are part of a signaling system (Fig. **3**) that can go wrong and lead to cancer. The third part of this signaling system is a protein kinase called Akt or protein kinase B. Akt indirectly activates another signaling pathway, called mTOR "(for mammalian target of rapamycin). The PI3K-PTEN-mTOR pathway is aberrant in many tumors, including breast cancer" [28]. So, HER2 expression levels, PTEN activity and the amount of phosphorylated Akt are important biomarkers that help guide treatment of breast cancer.

However, when the function of PTEN or mTOR is lost, resistance to herceptin can occur. So, to prevent breast and other cancers from developing resistance to herceptin, another drug, called Afinitor® has been used. It successfully completed Phase III trials for breast cancer and is now approved for treating renal cancer, non-cancerous kidney tumors caused by the rare genetic disease tuberous sclerosis complex and for breast cancer [29]. It inhibits mTOR.

Another receptor that can be targeted is the estrogen receptor. Tamoxifen is a drug that binds to estrogen receptors on tumors and decreases DNA synthesis and competes with estrogen for binding to its receptor. Tamoxifen was a standard treatment for breast cancer in patients who have estrogen-sensitive breast cancer. That is, most breast cancers need estrogen to survive, so they can be killed by Tamoxifen, which is an estrogen receptor antagonist in breast tissue. It blocks the binding of estrogen to one of its two types of receptors (the α receptor). In another study, it was shown that the complex formed between tamoxifen and the estrogen receptor represses the transcription of a gene called ERBB2 in patients with elevated levels of ERBB2 in their breast tumors [30].

Other FDA-approved anticancer agents that are hormonal agents include: anastrozole, delta-1-testolactone, dimethyltestosterone, dromostanolone propionate, estramustine, ethinyl estradiol, exemestane, fulvestrant, letrozole, megestrol acetate, mitotane, nandrolone, raloxifene, and toremifene. Table **1** summarizes the anticancer drugs discussed so far in this chapter.

Table 1. Drugs discussed in this chapter.

Drug	Biological Target	Clinical Use
Bevacizumab	Vasculature (VEGF) in the stroma	Cancer
Vandetanib	Vasculature (VEGF) in the stroma	Cancer
Sunitinib	Vasculature (VEGF) in the stroma	Cancer
Axitinib	Vasculature (VEGF) in the stroma	Cancer
Sorafenib	Vasculature (VEGF) in the stroma	Cancer
Pazopanib	Vasculature (VEGF) in the stroma	Cancer
Cabozantinib	Vasculature (VEGF) in the stroma	Cancer
Ziv-aflib	Vasculature (VEGF) in the stroma	Cancer
Ipilimumab	Immune cells in the stroma	Cancer
Sipuleucel-T	Immune cells in the stroma	Cancer
Aldesleukin	Immune cells in the stroma	Cancer
Interferon α–2b	Immune cells in the stroma	Cancer
Lapatinib	Tyrosine kinase	Breast Cancer
Herceptin	Epidermal growth factor	Breast Cancer
Pertuzumab	Tyrosine kinase	Breast Cancer
Afinitor®	Tyrosine kinase	Breast Cancer
Tamoxifen	Estrogen receptor	Breast Cancer
Anastrozole	Aromatase	Breast Cancer
Delta-1-testolactone	Aromatase	Breast Cancer
Dimethyltestosterone	Aromatase	Breast Cancer
Dromostanolone Propionate	Aromatase	Breast Cancer
Estramustine	Aromatase	Breast Cancer
Ethinyl estradiol	Aromatase	Breast Cancer

(Table 1) contd.....

Drug	**Biological Target**	**Clinical Use**
Exemestane	Aromatase	Breast Cancer
Fulvestrant	Aromatase	Breast Cancer
Letrozole	Aromatase	Breast Cancer
Megestrol acetate	Aromatase	Breast Cancer
Mitotane	Aromatase	Breast Cancer
Nandrolone	Aromatase	Breast Cancer
Raloxifene	Aromatase	Breast Cancer
Toremifene	Aromatase	Breast Cancer
Arimidex®	Aromatase	Breast Cancer

AROMATASE

So, the enzyme aromatase (E.C. 1.14.14.1) is a target for therapy in post-menopausal women. It is in the cytochrome P450 superfamily. It catalyzes the conversion of androgens to estrogens. It catalyzes the conversion of testosterone to estradiol and androstenedione to estrone. That is, it converts the steroid A ring to an phenyl group or benzene ring. About 75% of breast cancers contain elevated levels of estrogens. In these cases, tumor cells that emerge at early stages of breast cancer over-express the gene that codes for aromatase. To treat or possibly even cure breast cancer, drugs (such as Arimidex®) that inhibit aromatase have been used. Letrozole and exemestane are also FDA-approved aromatase inhibitors. They are used as an adjuvant in hormonally responsive breast cancer. Three more aromatase inhibitors, NeuVax, dHER2 and MVF-HER-2, are in clinical trials [28].

However, Arimidex® (an aromatase inhibitor) has been shown to be superior to either tamoxifen or a combination of tamoxifen and Arimidex® in treating and curing breast cancer. It is given after surgery and radiation therapy.

VEGF

Another target for therapy in post-menopausal women is the vascular endothelial growth factor, or VEGF. Bevacizumab (Avastin®) is an FDA-approved mAb that inhibits tumor growth by blocking angiogenesis. However, it should be noted that

Avastin® may increase the risk of patients developing blood clots [31]. Monoclonal antibodies (mAbs) are antibodies (a type of protein, see Chapter 7) that are made by a single clone of white blood cells. Bevacizumab was the first clinically available angiogenesis inhibitor in the United States [32]. "It is used to treat not only breast cancer, but also metastatic colon cancer and most forms of metastatic non-small cell lung cancer. However, subsequent clinical trials that combined Avastin® with different chemotherapeutic drugs found Avastin® to be less effective than originally thought. Still, Avastin® has become a very popular anti-cancer drug, with total sales of $3.5 billion in 2007, with $2.3 billion of that in the USA. However, recent clinical studies show that it prolongs life by just a few months: colorectal cancer, 4.7 months, lung cancer 2 months, and breast cancer 1.7 months" [32].

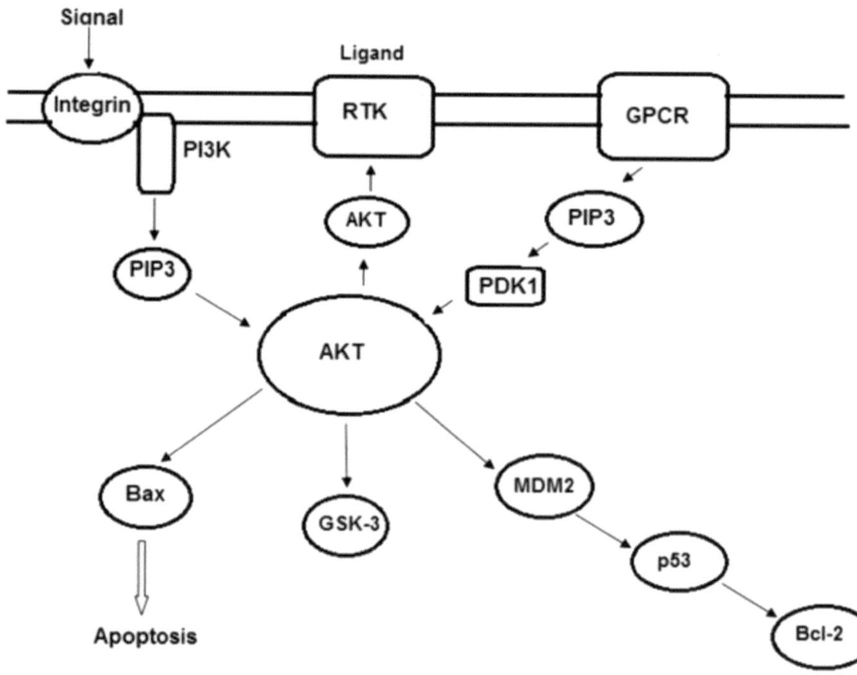

Fig. (3). PI3k and Akt signaling pathway.

The problem is the cost. It can cost up to $100,000 per year [32]. Still, how do you put a cost on life? For relatively wealthy parents with cancer, they might wonder whether the extra months are worth depleting their children's inheritance.

For the children, they might be horrified that their beloved parents would even think of worrying about the cost. Even a few extra months of life might be worth all the money in the world, even if it means a lifetime of poverty for them and their children. The problem becomes especially complicated when you consider the pain and misery that can be caused by the side effects of chemotherapy, which is still given, along with the Avastin®. If a person has health insurance, the cost may not be a factor, but if Avastin® is given to treat a different form of cancer, the cost may not be covered. Doctors are free to prescribe Avastin® or any other drug on the market for unapproved uses, at their discretion. As much as 75% of cancer drug use is of this "off label" variety [32]. "For example, one patient had gastric cancer, and this patient was given gemcitabine, a drug approved for pancreatic cancer, along with many other drugs, including Avastin®. This patient's cancer appeared to recede and Medicare paid for them. Another patient with brain cancer received Avastin® and the tumor shrank after only two infusions. The doctor began treating other brain cancer patients with Avastin™. Some insurers paid for the drug, but others, including Medicare and Medicaid, did not. A doctor at Duke University and Genentech did a type of clinical study considered less than definitive. There was no control group (standard chemotherapy without Avastin®). They found that patients lived a median of nine months, three months longer than what researchers estimated that the patients would have lived without Avastin®" [32]. One patient said, "It's really about living and not waiting to die" [32]. And what if 5 percent of Avastin patients live a lot longer than they would without the drug? I might be in that 5 percent" [32].

PI3K/AKT/MTOR PATHWAY

Another target for breast cancer therapy in post-menopausal women is the PI3K/Akt/mTOR survival pathway. As mentioned before, class I PI3K catalyzes the addition of a phosphate to PIP_2 to make phosphatidyl inositol 3,4,5-trisphosphate, $PI(3,4,5)P_3$ [33]. "There are several different isozymes of PI3K. They have been linked to many cellular functions, including cell growth, proliferation, differentiation, motility, survival and intracellular trafficking. Many of these functions depend on the ability of PI3K to activate protein kinase B, also known as Akt. PI3K activates Akt indirectly. First, PI3K catalyzes the production of PI(3,4,5)P3, which activates Akt. There are three genes in the Akt family, *Akt1*,

Akt2 and *Akt3*. Apoptosis is inhibited by the protein coded by *Akt1*, Akt. Many types of cancer are linked to Akt1. The *Akt1* gene was originally identified as an oncogene. It was found in a retrovirus called Akt8. The Akt protein activates mTOR, or mammalian target of rapamycin. The mTOR protein is a serine /threonine protein kinase. It is the catalytic component of two different signaling complexes, mTORC1 and mTORC2. They integrate signals from multiple inputs, including growth factors, amino acids, and intracellular energy supply, to regulate diverse cellular functions, including transcription, ribosome biogenesis, translation initiation, and autophagic cell death (autophagy). mTOR regulates cell proliferation, motility, and survival. It stimulates protein synthesis and the transcription of DNA into mRNA. It is a negative regulator of autophagy. Compounds that decrease the concentration of inositol or inositol 1,4,5-trisphospate (IP_3) can induce autophagy. Tuberous sclerosis 1 (TSC1) and TSC2 are upstream regulators of mTOR that form a functional complex and suppress cell growth by inhibiting mTOR activity. Stimulation by growth factors, such as insulin and the insulin-like growth factor, IGF-1, regulates mTOR signaling through the PI3K-Akt pathway. Moreover, mTORC1 controls key anabolic and catabolic processes in response to nutrient levels, energy, and growth factors. Often, mTORC1 is deregulated in human diseases, including cancer. So, there are many efforts to develop drugs that inhibit its kinase activity" [33].

The mTOR pathway is inhibited by rapamycin, which is made by the bacterium *Streptomyces hygroscopicus* [33]. "Rapamycin forms a complex with a small (12-kD) FK506-binding protein (FKBP12), which binds to the catalytic mTOR kinase domain and partially inhibits its activity. Rapamycin has been the subject of much research, not just because of its accepted clinical uses in cancer treatment and organ transplantation, but also for its capacity to prolong the lifespan of yeasts, mice and other organisms" [33]. Three drugs that are being evaluated in clinical studies to treat or cure breast cancer are BGT226, BEZ235A and RAD001 [28]. They inhibit mTORC and PtdIns 3-kinase [34 - 36]. Rapamycin itself is also in clinical trials [28]. However, it had unexpectedly weak anti-cancer activity in some early trials, possibly due to some phosphorylation sites on mTORC targets being resistant to rapamycin treatment. That is, the intrinsic capacity of a phosphorylation site on a target protein to serve as an mTORC1 substrate, a property termed substrate quality, is a major determinant of its sensitivity to modulators of the pathway [37]. So, mTORC1 effectors can respond differentially to the same signals [37].

Rapamycin itself has poor solubility, so similar compounds, called rapalogs, are being developed [38]. Rapalogs that inhibit the PI3K/Akt/mTOR pathway include sirolimus, temsirolimus and ridaforolimus [38]. Interestingly, rapamycin also significantly increases the lifespan of mice [39]. Inhibition of mTOR may also offer protection against age-related disorders, including cardiovascular disease, cancer and diet-induced obesity [39]. Healthy people should not take rapamycin, though, because it suppresses the immune system.

IGF-1

Another target for breast cancer therapy is the insulin-like growth factor type 1 receptor, or IGF-1 receptor, or IGF-1R [28]. "It is a receptor tyrosine kinase (RTK), with crucial roles in development, aging, and cancer biology. The main signaling pathways mediated by IGF-1 depend on the activation of the phosphatidylinositol 3-kinase (PI3K)-Akt and mitogen-activated protein kinase (MAPK) pathways. High levels of the IGF-1 receptor in blood have been linked with an increased risk of breast cancer. When IGF-1 binds a ligand, it regulates cellular growth and motility, and it protects cells from apoptosis. The IGF-1 receptor protects tumor cells from the effects of chemotherapy and radiation therapy. The drugs called IMC-A12, AMG 479, h7C10 and CP-751,871 are mAbs that target IGF-1 and are in clinical trials" [28]. It should be noted that IGF-1 is also found in milk. Babies and infants need to grow and develop, and IGF-1 helps this. However, children and adults in the USA and many other countries drink much milk. There are many websites on the Internet that claim that drinking milk can cause cancer because of IGF-1. However, IGF-1 is like any other protein. After eating or drinking IGF-1, it is digested and broken down in the stomach. The concentration of IGF-1 in the blood of people who drink milk is much lower than the concentration of IGF-1 in patients with breast cancer. The FDA has issued numerous statements on IGF-1, milk health and cancer research [40]. These include:

- The consumption of dietary IGF-1 plays no role in either inducing or promoting any human disease, nor does it cause malignant transformations of normal human breast cells… According to cancer research, the suggestion that IGF-1 in milk can induce or promote breast cancer in humans or premature growth stimulation in infants is scientifically unfounded.

- FDA and other scientific and regulatory bodies have thoroughly examined the safety of milk produced by rbST-treated cows and have concluded that it is safe. There is absolutely no possibility that the consumption of milk from rbST-treated cows could increase the risk of breast cancer. Insulin-like growth factor 1 (IGF-1) is a natural protein which mediates many of somatotropin's actions. It is required for normal growth and possibly health maintenance. IGF-1 is structurally and chemically similar to insulin and is normally present in most body tissues and fluids including human breast milk and saliva [40].

The following comes from the FDA website, "FDA has examined the literature and finds no definitive evidence of any direct link between IGF-1 and breast cancer. Some authors have hypothesized a link, whereas others have expressed that while IGF-1 is one of several growth factors and hormones that can contribute to an increase in cell numbers of many cell types *in vitro*, no one factor is responsible for changing normal cells into cancerous cells. Furthermore, FDA has been advised that there is no substantive evidence that IGF-1 causes normal breast cells to become cancerous [40].

Even though drinking low levels of IGF-1 in milk may not cause cancer, having a complete absence of IGF-1 may prevent cancer. In the fall of 2008 a community of small humans, or dwarfs, who have very low levels of IGF-1, were found in a remote region in Ecuador. They have no history of cancer. Similarly, mice that were genetically engineered to have very low levels of IGF-1 lived ten times longer than normal mice. This work suggests that it might be possible to reduce the amount of IGF-1, avoid cancer and live longer.

INFLAMMATORY BREAST CANCER

"The most common form of breast cancer starts in the ducts of the nipples. It spreads into the rest of the breasts and forms a lump [41]. There is a rare form of breast cancer, though, called inflammatory breast cancer, or IBC [41]. "According to the National Cancer Institute, "Inflammatory breast cancer is a rare but very aggressive type of breast cancer in which the cancer cells block the lymph vessels in the skin of the breast" [40]. "This type of breast cancer is called "inflammatory" because the breast often looks swollen and red, or "inflamed." IBC accounts for 1 to 5 percent of all breast cancer cases in the United States. Symptoms of IBC may include redness, swelling, and warmth in the breast, often without a distinct lump in the breast. The redness and warmth are caused by cancer cells blocking the lymph vessels in the skin. The skin of the breast may also appear pink, reddish purple, or bruised. The skin may also have ridges or appear pitted, like the skin of an orange (called peau d'orange), which is caused by a buildup of fluid and edema in the breast. Other symptoms include heaviness, burning, aching, increase in breast size, tenderness, or a nipple that is inverted (facing inward). Diagnosis of IBC is based primarily on the results of a doctor's clinical examination. Biopsy, mammogram, and breast ultrasound are used to confirm the diagnosis. Treatment consisting of chemotherapy, targeted therapy, surgery, radiation therapy, and hormonal therapy is used to treat IBC. Patients may also receive supportive care to help manage the side effects of the cancer and its treatment. Chemotherapy (anticancer drugs) is generally the first treatment for patients with IBC, and is called neoadjuvant therapy. Chemotherapy is systemic treatment, which means that it affects cells throughout the body. The purpose of chemotherapy is to control or kill cancer cells, including those that may have spread to other parts of the body" [40].

EARLY DETECTION

So, a very important topic in the biology of cancer is early detection. The simplest and most important first step is self-examination, especially for breast and testicular cancer. Also, please remember that men (as well as women) can get breast cancer. Many types of cancer can't be detected by self-examination alone. For breast cancer, mammograms are extremely important. Mammograms are made from low-dose X-rays. Radiologists analyze the images for any abnormal findings. During the procedure, the breast is compressed to even out the tissue and to hold the breasts still. As with any test, there are false positives and false negatives. About 7% of the tests produce false positives and about 10-20% are false negatives. At times, results are inconclusive, and further tests are required.

Ultrasound, ductography and magnetic resonance imaging (MRI) can be used as well.

Ultrasound can be used to visualize muscles, tendons, and many internal organs. It is especially useful in visualizing babies in the mother's wombs and in cleaning teeth in dental hygiene. Ultrasound is also used in ultrasound-assisted lipectomy, more commonly known as liposuction. Ultrasound can be used to image the prostate to assess the size of the gland and to help place a needle for a biopsy. When used to image breast cancer, ultrasound sends high-frequency sound waves through the breasts and converts them to images [42]. Ultrasound complements other tests. If an abnormality is seen in a mammogram, ultrasound can tell if it is solid or a fluid, but can't tell if it is malignant (*i.e.* cancer). Doctors can use ultrasound to guide a needle for a biopsy.

Ductography is used in some institutions to evaluate bloody nipple discharge. A similar approach is to aspirate some fluid through the nipple, called nipple aspirate fluid, or NAF, using a modified breast pump. Doctor E.R. Sauter at the University of Missouri - Columbia is an expert in this technique [42]. He and his colleagues found that urokinase-type plasminogen activator (uPA) and uPA receptor (uPAR) were independent predictors of the presence of breast cancer [43]. The enzyme uPA catalyzes the hydrolysis of plasminogen to produce plasmin, which is able to catalyze the break down (hydrolysis) of blood clots [44]. Also, uPA plays an important role in the processes leading to cancer invasion and metastasis [44].

MRI can be used to image breast cancer and many other things. It is a type of NMR, but is not called NMR because the letter N stands for nuclear, which means radioactivity in the medical community. There is no radioactivity in MRI or NMR. NMR measures the interaction between radio frequency electromagnetic radiation and matter that is in a strong magnetic field. In theory, the NMR signal can be produced by either changing the magnetic field strength or the radio frequencies. In old-fashioned NMR, the radio frequency was varied slowly to produce a spectrum. Once computers became available, the Fourier transform technique enabled chemists to irradiate the sample with a broad band of radio frequencies (rf) and measure the amounts of rf absorbed and the frequencies that are absorbed. In MRI, the magnetic field strength is varied. The two most commonly observed nuclei in NMR are ^{1}H and ^{13}C. The sensitivity of ^{1}H-NMR is much higher than the sensitivity of ^{13}C-NMR. Also, liquids with low viscosity produce peaks that are sharper than highly viscous liquids, gels or solids. In MRI, the ^{1}H nucleus is observed. Water is by far the most abundant source of ^{1}H in the body and in the blood. However, in tumors, the water has a higher viscosity. For this reason, MRI can be used to monitor blood flow and to produce images of tumors. Contrast agents can be injected into the patient to enhance the images. As with NMR, paramagnetic compounds can be used. For MRI, compounds containing gadolinium are often used [45, 46]. In MRI, instead of varying the rf, a static magnetic field is moved across the body, producing an image. MRI is used to image not just breast cancer, but also brain and liver cancer [45].

Another very good technique for imaging cancers is computed tomography (CT), also known as computed axial tomography (CAT) [47]. "It uses a large series of two-dimensional X-ray images to produce a three-dimensional image. In some CT machines, the patient is placed on a gantry and the X-ray source is rotated in a circular fashion around a single axis. After a time delay, the patient is moved slightly and another slice is taken. "More advanced CT machines have the X-ray source attached to a freely rotating gantry. During a scan, the patient is moved smoothly through the scanner and the X-ray source traces a helical pattern. These CT machines are called helical or spiral CTs. They do not require time delays and produce data much faster than older CT scanners. A large volume can be covered in 20-60 seconds. Often, the patient can hold his or her breath that long" [47]. Also, higher resolution scans can be acquired and better use can be made of contrast agents, which usually contain iodinated compounds. Sometimes, severe reactions to these contrast agents can occur, especially if the patient has chronic kidney disease, or is allergic to iodine.

Still, CT scans can be used to produce images of many forms of cancer, including cancer of the lungs, breast, brain, liver and colon. Early detection of lung cancer can be done with spiral CT and another imaging technique, called positron emission tomography, or PET [47].

PET scans require the injection of a short-lived radioactive compound containing a ^{18}F isotope, such as ^{18}F-fluorodeoxyglucose [47]. The radioactive nuclide emits a positron (the antiparticle to electrons). After traveling a few millimeters, the positron encounters an electron. Both particles annihilate each other and produce two photons of gamma rays, moving in opposite directions [47]. "The gamma rays are detected when they strike a scintillation material, which produces a burst of light" [47]. PET scans are often used in combination with MRI or CT scans. This provides information about the metabolism (from PET) and the anatomy (from MRI or CT). PET is used extensively in clinical oncology, for the imaging of tumors and the search for metastases. The tracer compound, ^{18}F-fluorodeoxyglucose, is rapidly taken up by cells that use glucose and it is phosphorylated in a reaction catalyzed by hexokinase, whose concentration is elevated in tumors. PET can be used for diagnosing cancer and determining its stage of development. It is particularly useful in monitoring treatments of Hodgkin's disease, non-Hodgkins lymphoma and lung cancer.

LUNG CANCER

Lung cancer is probably the leading avoidable cause of death in the USA [1]. Smoking tobacco products can cause COPD, emphysema, heart disease and lung cancer. However, not everyone who smokes dies of lung cancer and not everyone who dies of lung cancer has smoked. As with other types of cancer, the prognosis for lung cancer is better if the cancer is detected early enough. Similarly, lung cancer can be divided into the same four stages as other cancers. If caught soon enough, treatment options can include surgery, radiation therapy and/or chemotherapy [1]. Commonly used chemotherapeutic agents for non-small cell lung cancer are listed in (Table **2**).

Table 2. Commonly used chemotherapeutic agents for non-small cell lung cancer.

Brand Name	Generic Name	Molecular Target
Platinol®	Cis–platin	DNA
VP–16; VePesid®	Etoposide	Topoisomerase
Paraplatin®	Carboplatin	DNA
Taxol®	Paclitaxel	Microtubules
Taxotere®	Docetaxel	Microtubules

(Table 2) contd.....

Brand Name	Generic Name	Molecular Target
Navelbine®	Vinorelbine tartrate	Microtubules
Adriamycin®	Doxorubicin	DNA
Oncovin®	Vincristine sulfate	Microtubules
Ifex®	Ifosfamide	DNA
Gemzar®	Gemcitabine hydrochloride	Nucleoside analogue

Standard chemotherapy for lung cancer usually combines two or more drugs. Commonly used combinations include: paclitaxel plus carboplatin, cisplatin plus vinorelbine tartrate, cisplatin plus VP-16, and carboplatin plus VP-16.

In a recent study, Iressa® (gefitinib) given as a daily pill, was just as effective as Taxotere® in treating lung cancer in a phase III clinical trial [48]. Patients who took Iressa survived an average of 7.6 months, and 32% survived a year, compared to an average of 8 months and 34% for Taxotere®. However, the side effects for Taxotere® are similar to many other chemotherapeutic agents. The side effects include hair loss, numbness in the hands and feet, severe diarrhea, a decrease in blood cells and nausea. However, Iressa®, caused only skin rashes and mild diarrhea. Plus, Iressa is given as a pill, instead of intravenous injection.

More recently, the anticancer drug paclitaxel was combined with curcumin in flaxseed oil, lecithin and polyethylene glycol (PEG) to form a nanostructure [49]. The curcumin inhibits the P-glycoprotein that helps some tumors resistant to anti-cancer drugs, such as the widely used paclitaxel. Both paclitaxel and curcumin stimulate apoptosis, killing the cancer cells [49].

SKIN CANCER

Another new drug, vismodegib or ErivedgeTM, was approved in 2012 by the FDA to treat advanced cases of basal-cell skin cancer, the most common type of skin cancer [49]. It targets the hedgehog pathway, which will be discussed in Appendix 1. It is the first FDA-approved drug for metastatic basal-cell carcinoma [49]. So, there are many anticancer drugs that have a variety of targets. They are called chemotherapeutic agents. This term is also used by the FDA to describe antibiotics that are added to aquacultures to prevent infection of fish and other seafood.

MICRORNAS

MicroRNAs (miRNAs) have the potential to be excellent anticancer agents and are used now as biomarkers [50]. Many different kinds of miRNAs contribute to cancer pathogenesis. Aberrant expressions of miRNAs are implicated in cancer development and progression. For example, miR-34 (in prostate, ovary cancers) and miR-210 (in head and neck cancers) are expressed abnormally due to dysregulated transcription factors, p53 and HIF1. Epigenetic changes in promoter regions can cause aberrant expression of miR-127, which is involved in prostate and bladder cancer [50]. "Alterations in gene locus can change the expression of miR-15a/16-1, miR-29, miR-17-92, miR-125b-1, which are involved in CLL, prostate cancer, melanoma, MDS, AML, lymphomas and solid tumors. Alterations in processing steps by LIN-28 can lead to aberrant expression of let7. Alterations in processing steps by DICER can lead to aberrant expression of miR-204, miR-130b and let-7g. Many other miRNAs are potential biomarkers for many forms of cancer. Extracellular miRNAs are promising biomarkers for clinical use. Vesicles called exosomes are small intraluminal vesicles that are shed from a variety of cells. Their biogenesis is connected with endosomal sorting complex, which is required for transport in multivesicular bodies. They were originally considered to be garbage bags for discarding degraded proteins, but now they and their contents, including miRNAs, have generated great interest as intercellular communication tools. Secretory microvesicles play a crucial role in intercellular communications. Secretory miRNAs are a means of a crosstalk between tumor cells and their surrounding stromal cells, similar to cytokines, hormones and small molecules. Cancer cells may evade the attacks of T and B cells by releasing miRNAs to suppress immunoreactions, and may recruit capillary blood vessels with angiogenic miRNAs" [50].

CANCER STEM CELLS

Epithelial cancer stem cells may be good targets for anticancer therapy, especially when treating metastatic or recurring cancers. They have the ability to differentiate into cancer cells, but also to regenerate themselves and drive metastasis and multi-drug resistance. However, they express unique cell-surface proteins that can be targeted for therapy. Cancer stem cells have been identified in cancers of the blood, breast, brain, bone, skin, liver, bladder, ovary, prostate, colon and pancreas [51]. One target is the p53 protein, which mediates the effects of inhibition of the histone deacetylase, SIRT1, which is over-expressed in leukemia [52]. "Chronic myelogenous leukemia (CML) is treated with a BCR-ABL tyrosine kinase inhibitor such as imatinib, nilotinib or dasatinib. They are usually effective at first, in the chronic phase, but progress to an accelerated phase and a terminal blast crisis. SIRT1 inhibitors and knockdown using small hairpin RNA (shRNA) were shown to increase p53-induced apoptosis in leukemia stem cells and this was enhanced when combined with imatinib" [52].

The CD47-signal regulatory protein alpha (SIRPa) is another therapeutic target for cancer stem cells in solid tumors [53]. The CD47 protein is on the surface of all solid tumor cells, and it prevents phagocytic cells from attacking them. It is over-expressed, and the level of expression correlates with a decreased probability of survival in several cancers. Researchers found that mAbs targeted to CD47 showed potent anticancer activity and may prevent metastasis of human cancer that was xenografted into mice [53].

Other workers developed a method to identify agents with specific toxicity for epithelial cancer stem cells [54]. It is based on the ability of test compounds to kill immortalized nontumorigenic epithelial cells. High throughput screening of about 16,000 compounds with cells in 384 well plates identified an antibiotic, salinomycin that had been used for years in veterinary medicine [54]. More recent work has shown that salinomycin overcomes resistance to multiple drug and apoptosis in cells that model human leukemia stem cells [30]. It has also been shown that a combination of salinomycin and trastuzumab was superior to a treatment with either drug by itself [55].

Induced pluripotent stem cells (iPSCs) may also become useful in cancer therapy [56]. They can be generated by reprogramming adult cells, such as fibroblasts from the patient, avoiding immunological rejection. Potentially, patients could even grow their own tissues to transplant into themselves. For cancer treatment, iPSCs derived from T-lymphocytes could be induced to differentiate into fully active T cells with specificities for specific cancer cells. Also, iPSCs could be grown in culture and used to screen therapeutic drugs to see which ones wil be effective for each individual patient [56].

It is interesting to note that the same transcription factors that can transform differentiated cells into iPSCs can also create cancer cells by a similar process.

TELOMERASE

Another potential target for anticancer therapy is the enzyme called telomerase [57]. Telomeres are tandem repeats of TTAGGG at the ends of chromosomes. Healthy somatic cells have a limit to the number of times that they can divide, after which they become senescent. On the other hand, cancer cells become immortal as they lose their limit in cell division. This is due, in part, to the activity of the telemorase enzyme. The gene *TERT* codes for the reverse transcriptase enzyme, telomerase, which is part of a nucleoprotein complex that catalyzes the maintenance of the length of telomeres. There are drugs that inhibit telomerase that are in clinical trials for curing or treating different types of cancer [58].

Even though most of the somatic mutations found in tumors are within the protein coding regions of genes or at splice junctions, a recent study found that mutations in the core promoter of the *TERT* gene were found in 89% of the deadly melanomas examined [59, 60]. They systematically queried noncoding somatic mutations using published whole-genome sequencing data. The mutations generate new binding sites for the transcription factor E-twenty-six (ETS) and increase the transcriptional activity of the *TERT* promoter [59, 60].

MUTATIONS IN SOMATIC AND GERMLINE CELLS

So, mutations in somatic and germline cells can be important in cancer [61]. "Most tumors contain two to eight mutations that promote or drive tumorigenesis. There are also a few mutations that occur in many types of tumors and many more that only occur in a few. The driver genes can be separated into 12 signaling pathways that regulate cell fate and survival as well as maintain the genome. These include the TGF-β, MAPK, STAT, PI3K and cell cycle apoptosis in cell survival; NOTCH, Hedgehog, chromatin modification and transcriptional regulation of cell fate and DNA damage control. One of the most pressing needs of oncology is to understand these pathways better" [61]. It is also important to learn more about mutations in tumors.

In solid tumors of the colon, breast, brain, or pancreas, an average of 33 to 66 genes have subtle somatic mutations that alter the proteins which they encode [61]. "About 95% of them are single-base substitutions, while the others are deletions or insertions of one or a few bases. Of the base substitutions, 90.7% result in missense changes, 7.6% result in nonsense changes, and 1.7% result in alterations of splice sites or untranslated regions immediately adjacent to the start and stop codons. Some tumor types display many more or far fewer mutations than average. For example, melanomas and lung tumors have about 200 nonsynonymous mutations per tumor. This is due to the potent mutagens (UV light and cigarette smoke, respectively) that cause them. Tumors with defects in DNA repair form another group of outliers. For example, tumors with mismatch repair defects can harbor thousands of mutations. On the other hand, pediatric tumors and leukemias harbor far fewer point mutations. Tumors can evolve from benign to malignant by gradually acquiring a series of mutations, a process that has been studied extensively in colorectal cancer. The first (gatekeeping) mutations provide a selective growth advantage to the mutated epithelial cell, allowing it to outgrow the cells that surround it and become a microscopic clone. In the colon, gatekeeping mutations frequently occur in the *APC* gene. The small adenoma that results from this mutation grows slowly, but a second mutation in another gene, such as *KRAS*, unleashes a second round of clonal expansion. The cells with only the APC mutation may persist, but there are fewer of them than in the cells that have mutations in both genes. This process of mutation followed by clonal expansion continues, with mutations in genes such as PIK3CA, SMAD4, and TP53. This eventually produces a malignant tumor that can invade through the underlying basement membrane and metastasize to lymph nodes and distant organs. The driver mutations are the ones that confer a selective growth advantage to the tumor cells. However, each driver mutation provides only a small selective growth advantage to the cancer cell. Over time, this slight increase can expand into a large malignant tumor, containing billions of cells" [61].

At the same time, over half of the somatic mutations identified in tumors occur during the preneoplastic phase [61]. "They happen during the growth of normal cells that continuously replenish the gastrointestinal and genitourinary epithelium and other tissues. All of these pre-neoplastic mutations are passenger mutations that have no effect on tumorigenesis. It usually takes decades for to develop a metastatic cancer. Also, virtually all of the mutations in metastatic tumors are already present in many cells in the primary tumors" [61].

Nearly 20,000 protein-coding genes have been evaluated in genome-wide sequencing studies of 3284 tumors [61]. "There were a total of 294,881 mutations reported. Only 125 mutated driver genes had been discovered by 3013. Of these, 71 are tumor suppressor genes and 54 are oncogenes. About 29% of these genes were found to be mutated by unbiased genome-wide sequencing. About half of these genes encode proteins that directly regulate chromatin by modifying histones or DNA" [61].

Also, subclonal, heterogeneous mutations are important for understanding tumor evolution" [61]. "Some are intratumoral with heterogeneity among the cells of one tumor. Others are intermetastatic. They are heterogeneous among different metastatic tumors of the same patient. Still others are intrametastatic. They are heterogeneous among the cells of an individual metastasis. Each metastatic tumor is established by a single cell or small group of cells that have some founder mutations. As it grows, it acquires new mutations with each cell division. Though the founder mutations may make the cancer cells susceptible to antitumor agents, the new mutations can provide the seeds for drug resistance. Finally, there are interpatient heterogeneities among the tumors of different patients" [61]. This makes P4 medicine especially important.

There are also some important germline mutations that can make a person more susceptible to cancer. The *TP53* gene that codes for the p53 protein is one of the most important [61]. "People who have deleterious mutations in *TP53* have a nearly 90% risk of developing Li-Fraumeni syndrome. Another example is the *ATM* (ataxia telangiectasia mutated) gene that can cause pancreatic cancer when mutated. "Then there is *POLE* (polymerase epsilon, catalytic subunit) and *POLD1* (polymerase delta, catalytic subunit) that can cause multiple adenomas and/or colorectal cancer. So, whole-genome sequencing is revolutionizing medical research and may become a powerful and cost-effective diagnostic tool for managing cancer" [61].

Still, more studies are needed to apply DNA sequencing to discover mechanisms of sensitivity or resistance to drugs [62]. "There has been some success already. Sequencing of non–small cell lung cancer that was sensitive and then resistant to inhibitors of EGFR tyrosine kinase led to the routine sequencing of the *EGFR* gene to guide therapeutic choices. Whole-exome sequencing of patients treated with everolimus for advanced bladder cancer showed that a specific mutation in the tuberous sclerosis gene (*TSC1*) correlated with everolimus sensitivity. Patients with the *TSC1* mutation had a longer time until recurrence of tumor (4.1 *versus* 1.8 months). This loss-of function mutation in *TSC1* was then found in 5 of 96 advanced bladder cancers, so there may be a subgroup of patients for whom everolimus treatment might offer substantial benefit" [62].

ROLE OF CYP450 ENZYMES

Differences in CYP450 enzymes are also important [63]. "For example, the rate of metabolism of tamoxifen depends on the highly polymorphic CYP450 2D6. At one time, antidepressants were coadministered with tamoxifen to treat hot flashes until it was discovered that the antidepressants inhibited CYP2D6 activity and the production of the major metabolite, endoxifen. Coadministration was abruptly discontinued because of the risk that the cure of hot flashes was compromising the effectiveness of the antiestrogen therapy. Pharmacogenomics studies have suggested that doubling the dose of tamoxifen for intermediate metabolizers normalized plasma endoxifen concentrations to that seen in extensive metabolizers. Moreover, clinical trials of endoxifen as a treatment for breast cancer are now under way" [63].

CANCER GENOMICS

A consortium consisting of the Computational Biology Center at the Memorial Sloan-Kettering Cancer Center, the Computer Engineering Department in Bilkent University in Turkey, the Institute of Biomedicine in the Department of Medical Biochemistry and Cell Biology, the University of Gothenburg, in Sweden and Blueprint Medicines in Cambridge, MA have created a tool called the cBioPortal for Cancer Genomics" (http://cbioportal.org) [64]. "It can explore, visualize and analyze multidimensional cancer genomics data. It converts molecular profiling data from cancer tissues and cell lines into understandable genetic, epigenetic, gene expression, and proteomic events. The cBioPortal integrates different types of genomic data. This includes somatic mutations, DNA copy-number alterations (CNAs), mRNA and microRNA (miRNA) expression, DNA methylation, protein abundance, and phosphoprotein abundance. In 2013, the portal contained data sets from ten published cancer studies, including the Cancer Cell Line Encyclopedia (CCLE), and more than 20 studies that are currently in The Cancer Genome Atlas (TCGA) pipeline. It enables users to query genetic alterations and test hypotheses regarding the recurrence and genomic context of gene alterations in specific cancers" [64].

CANCER EPIGENETICS

Epigenetics are also important. For normal cells to change into cancer cells, transcription factors (TFs) must change the level of transcription of certain genes. To do this, they must contend with the underlying organization of chromatin [65]. "This includes DNA, RNA, histones, and regulatory proteins. TFs recruit chromatin regulators (CRs) that change the accessibility of target DNA and induce specific chromatin structures that are have specific histone modifications and functional roles. For example, hypermethylation of CpG islands occurs. However, TF binding depends on the accessibility of chromatin. CRs and chromatin states also serve as gatekeepers that modulate the regulatory activities of TFs "ifferentiation events in cancer and normal development often rely on promoters and enhancers that are poised for action by pioneer TFs and characteristic chromatin structures. So, TFs, cooperating CRs, and coordinated chromatin states guide successive differentiation and commitment events during the development of cancer" [65].

As a result, genetic inactivation of the DNA methyltransferase Dnmt3a occurs in leukemia and lymphoma" [65]. "Also, CRs help reprogram cells and lead to oncogenesis. Many CRs are oncogenes or tumor suppressors. They are directly affected by gain- and loss-of function mutations or by translocations. Even in the absence of direct genetic alterations, CRs may be co-opted by fusion proteins or other oncogenic factors that modulate gene expression. CRs also regulate other processes that are relevant to carcinogenesis. This includes epithelial to mesenchymal transition (EMT), senescence, genome stability, and metastasis. Also, inactive portions of the genome are partitioned between different forms of repressive chromatin in differentiated cells. Repressive structures with a compact organization resistant to regulatory activity can affect large regions of the genome. Heterochromatin marked by histone H3 lysine 9 trimethylation (H3K9me3), broad Polycomb-repressed regions enriched for H3K27me3, and highly condensed regions associated with the nuclear lamina and H3K9me2 are all repressive states. So, suppression of CRs that catalyze H3K9 methylation, including Suv39h1, Setdb1, and G9a, increases the efficiency of genetic reprogramming. As a result Suv39h1 is required for oncogene-induced senescence, and its loss is associated with decreased viability, genomic instability, and increased tumor risk. Also, when Suv39h1 is recruited to aberrant gene targets by fusion proteins, it promotes acute myeloid leukemia. Setdb1 and G9a also have a role in several malignancies. G9a has also been implicated in metastasis. MT Polycomb regulators are subject to genetic mutation or overexpression in several different types of cancer. For example, Ezh2 and Bmi1 mediate the epigenetic silencing of the proaptotic gene *CDKN2A* and are essential in cancer stem cells" [65].

Moreover, active genes and regulatory elements are associated with characteristic chromatin states [65]. "Enhancers and promoters are marked by a varying degree of H3K4 methylation and histone acetylation. The methylation mark is catalyzed by complexes that contain mixed-lineage leukemia (MLL) homologs and accessory subunits such as Wdr5. These CRs play important roles in re-programming cells and carcinogenesis. MLL fusion proteins are potent drivers of leukemia. Inactivating mutations of MLL homologs have also been identified in many types of cancer. The H3K4 demethylase Lsd1, which functions as a coactivator or corepressor in different contexts, also has a variety of roles in cancer. Also, the transition from transcriptional initiation to chain elongation is a critical regulatory step. Dot1l, an H3K79 methyltransferase that promotes this transition is required for leukemogenesis. MLL fusion proteins mediate their aberrant regulatory programs by recruiting Dot1l. As a result, leukemias with rearranged MLL are sensitive to small molecule inhibitors of Dot1l inhibitors. The multiple myeloma fusion proteins Mmset and Kdm2b catalyze the addition and removal of H3K36 methylation, respectively. This is a characteristic mark of elongating transcripts. Kdm2b contributes to the development and maintenance of leukemic stem cells" [65].

Many other CRs that are associated with active chromatin states or chromatin remodeling have been implicated in neoplastic transformation [65]. "For example, The CREB binding protein known as the (CBP)/p300 histone acetyltransferase complex is a coactivator for specific TFs at proximal and distal regulatory elements. This drives epigenetic programs that are involved in cellular proliferation, apoptosis, differentiation, and DNA stability. CBP/p300 also catalyzes the acetylation of nonhistone proteins, including p53, Rb, and c-Myc, increasing their transcriptional activity. Chromosomal rearrangements involving CBP/p300 promote hematopoietic malignancies, while loss of function mutations have been identified in colorectal and breast cancer" [65].

Dynamic chromatin regulation also requires nucleosome remodeling [65]. "The BAF remodeling complex significantly increases the efficiency of reprogramming by facilitating the binding of the Oct-4 protein to target sites on DNA.

SWI/SNF homologs, including ARID1A, are subject to recurrent genetic inactivation in many types of cancer. These and other studies show the widespread roles for CRs in human cancers" [65].

MUTATIONS

Mutations can also lead to drug resistance. That is, many chemotherapeutic agents are quite effective for some time, but then the tumor develops a resistance to the drugs. So, other drugs are being developed to prevent such resistance. One molecular target for such therapy is the enzyme poly(ADP-ribose) polymerase, or PARP [66]. "PARP1 and PARP2 facilitate DNA repair by binding single-strand breaks in DNA and catalyze the biosynthesis of branched chains of poly(ADP-ribose), which recruit DNA repair enzymes to the damaged sites. Since many chemotherapeutic drugs cause DNA damage that is repaired, due to PARP activity. Subsequently, this causes drug resistance. So, inhibitors of PARP may be able to act as chemopotentiators, improving the therapeutic potential of a chemotherapeutic drug. Moreover, they can be given after chemotherapy, to help prevent a relapse. For example, many patients with high-grade serous ovarian cancer respond well to initial chemotherapy, most relapse into incurable cancer. So, Phase III clinical trials are being planned for drugs made by three different pharmaceutical companies. In one study, a genomics company will try to find a molecular signature that can predict homologous recombination deficiency (HRD). The signature will be developed from Phase II data, and will be tested for predictive value at the end of the Phase III trial" [66].

MEDICINAL CHEMISTRY POINT OF VIEW

From the medicinal chemistry point of view, there are many drugs that are used to treat cancer. As discussed last semester, Trastuzumab or Herceptin® targets breast cancer patients who have the HER2 (human epidermal growth factor receptor 2) recepto. Gleevec® is used to treat chronic myeloid leukemia, a cancer that is caused by a definite chromosomal abnormality, called the Philadelphia chromosome [67]. This is caused by a translocation of DNA between chromosomes 9 and 22. Part of the breakpoint cluster gene (BCR) from chromosome 22 is fused with part of the ABL gene on chromosome 9. ABL stands for Abelson, the name of a virus that can cause leukemia [68]. "The fused bcr-abl fused protein speeds up cell division and inhibits DNA repair. Trastuzamab has been used for years to treat patients with metastatic breast cancer and tumors that overexpress the HER2 receptor. Even though it has been successful for many, some patients develop resistance within the first year of treatment or experience a relapse after an initial clinical response. Another way to protect drugs from rapid degradation is by forming conjugates, or covalent linkages. Conjugates can be made between chemotherapeutic drugs and monoclonal antibodies (mAb) to make antibody-drug conjugates, abbreviated as ADCs. One example is trastuzumab emtansine (T-DM1)" [68]. Trastuzamab is a monoclonal antibody that targets tumors that overexpress the human epidermal growth factor receptor 2 (HER2) [69]. "T-DM1 combines each molecule of trastuzumab (T) with up to six molecules of the antimicrotubule agent emtansine (DM1). This enables targeted delivery of the cytotoxic DM1 to cancer cells" [69]. T-DM1 was shown to be "effective in patients with HER2-positive metastatic breast cancer (MBC)" [70]. "Once T-DM1 binds to HER2, the complex undergoes receptor-mediated internalization and release of DM1 and its catabolites that cause programmed cell death, or apoptosis. Another example is an

ADC made from brentuximab and vedotin" [71].

There are also recombinant immunotoxins, in which a fusion protein is made that contains a bacterial toxin and a fragment which targets the specific cancer cells being treated. An example is anti-Tac(Fv)-PE38 or LMB-2, which contains a toxin fragment of *Pseudomonas* endotoxin and a monoclonal antibody against the cell surface protein, CD25, in cancer cells.

Several mAbs are in clinical use. They include mAbs that target soluble ligands associated with tumor growth (*e.g.*, bevacizumab), cell-surface proteins associated with hematological tumors (*e.g.*, rituximab), and cell surface proteins associated with solid tumors (*e.g.*, trastuzumab, cetuximab and panitumumab) [72].

Epoetin alfa is human erythropoietin that is is used to treat anemia that is associated with cancer chemotherapay. Chimeras made from siRNA and aptamers were able to synergistically inhibit lung cancer cell invasion, tumor growth and angiogenesis [73]. The first vaccine produced by insect cells infected with baculovirus was the HPV vaccine that protects against cervical cancer and other types of cancer. It was followed by Sipuleucel-T for prostate cancer. vaccines based on dendritic cells that are in either Phase II or II clinical trials for colorectal cancer, glioma and melanoma [74]. Recently, scientists and doctors have been able to eliminate leukemia in several patients by "converting their cytotoxic T cells into specific killers of cancer cells" [75]. "They used a lentiviral vector to express chimeric antigen receptors (CARs) in patients' cytotoxic T cells. Extracellular and intracellular domains were fused into a new, synthetic receptor for the CD8+ T cells. The gene for the CAR was delivered by a lentivirus vector, which was constructed using the backbone of the HIV-1 virus that was no longer pathogenic and can not cause AIDS. Because the natural host for HIV-1 is the T-lymphocyte, it enabled the lentivirus vector to specifically target the cancer cells and multiply in them. The patients received between 15 million and one billion cytotoxic T-lymphocytes (CTLs). These CTLs rapidly multiplied about 100-fold" [75].

There is also a liposomal daunorubicin and cytarabine, to treat and often cure cancer. More recently, non-pegylated liposomal doxorubicin (NPLD; Myocet) and liposomal daunorubicin (DaunoXome) have been approved for treating cancer [76]. Other liposomal formulations of anticancer drugs include Depocyt®, which contains the anticancer drug, cytarabine; DaunoXome®, which contains the anticancer drug, doxorubicin. There are also drugs that are prepared as polymeric micelles. These include NK911® and NK105®, which are contain doxorubicin and paclitaxel, respectively, in micelles made of poly(ethyleneglycol)-copoly (L-aspartic acid) and Genexol-PM, which contains paclitaxel in micelles made of poly(ethyleneglycol)-copoly(D,L-lactic acid). Another anticancer agent, paclitaxel, can be prepared in nanoparticles made from the common protein, albumin. It is called Abraxane® and takes advantage of albumin's natural ability to bind drugs and enter tissues to target cancer cells. So, a higher effective dose of paclitazel is delivered.

There are also allo-network anticancer drugs (adriamycin, the camptothecins, colchicine, cyclosporine, dexrazoxane, ecteinascidin 743, epothilone A, etoposide, mitoxantrone, rapamycin, paclitaxel and vinblastine) that target interfaces of biopolymers, including places where proteins interact with each other. They can bind selectively as proteins or protein-DNA complexes as they assembled in the cell [77].

Some anti-cancer drugs are also aromatase inhibitors. This includes "Letrozole (Femara®), Aminoglutethimide (Cytadren®), Anastrozole (Arimidex®), Vorozole (Rivizor®), Exemestane (Aromasin®), 4-androstene-3,6,17-trione ("6-OXO", marketed as a dietary supplement for athletes), 1,4,6- androstatrien-3,17-dione (ATD), Formestane (Lentaron®), Testolactone (Teslac®) and Fadrozole (Afema®)" [78]. Another anticancer drug that inhibits an enzyme is irinotecan (Camptosar). It inhibits the enzyme topoisomerase I, which catalyzes the unwinding of supercoiled DNA, making it available for transcription into mRNA [78].

There is also an experimental drug, called PHA-767491, which stops DNA replication in cancer cells before it can start. Many anticancer drugs interfere with DNA polymerization once it is started and can kill non-cancerous cells, but PHA-767491 acts by inhibiting a kinase, called Cdc7, which catalyzes the phosphorylation of enzymes that unwind DNA, in preparation for replication [79].

Another anticancer drug, goserelin, targets a GPCR. Other approved anticancer signaling agents include Bortezomib, Dasatinib, Erlotinib, Everolimus, Lapatinib, Nilotinib, Sorafenib, Sunitinib, and Temsirolimus. Bicalutamide acts by targeting the androgen receptor (AR). It blocks the action of androgen and testosterone, which bind to ARs. It blocks AR receptors in the brain, inhibiting the negative feedback effect of testosterone on the release of luteinizing hormone, thus increasing progesterone and estrogen levels. Bicalutamide blocks the effects of increased progesterone, but not estrogen. It is a teratogen, so women who are pregnant or could become pregnant should avoid handling it [78].

There are also selective estrogen receptor modulators (SERMs) such as Raloxifene (Evista®) that acts as an estrogen on bone, but as an antiestrogen in breast tissue. It is effective in reducing the occurrence of breast cancer, as does another SERM, tamoxifen. Raloxifene was approved by the FDA in 2002. Later, it was found that it also reduces the concentration of low density lipoproteins (LDL) [78].

There are two classes of microtubule drugs, microtubule inhibitors (vincristine and vinblastine, which are alkaloids from the Madagascar periwinkle, also known as the *Catharanthus roseus*, and once known as *Vinca rosa*), approved by the FDA in 1963 and 1965, and stabilizers Taxol (Paclitaxel) and Taxotere (Docatexel) approved in 1992, and 1996, respectively. Also, there are two semi-synthetic derivatives of vincristine, called vindesine and vinorelbine. Another class of microtubule-acting anticancer drugs contains the epothilones. Epothilones A and B are naturally occurring 16-membered macrolides, which are produced by the myxobacterium *Myxococcus xanthus* or *Sorangium cellulosum* [78]. Another natural product that acts on tubulin is called eribulin. It was isolated from the Pacific sponge and it inhibits the polymerization of tubulin. The root bark of the Combretum caffrum tree is the source of combretastatin, which is another natural inhibitor of tubulin polymerization [80]. Another drug that binds to microtubules is called Maytansine. It disrupts microtubule assembly.

Other workers have found that a new drug called pazopanib can combine synergistically with microtubule inhibitors such as paclitaxel to treat anaplastic thyroid cancer (ATC) through Aurora kinase A [43]. ATC has perhaps the worst prognosis of any cancer, with a median survival of only about 5 months regardless of stage. Pazopanib monotherapy has promising clinical activity in differentiated thyroid cancers (generally attributed to vascular endothelial growth factor receptor inhibition), yet is less effective as a single agent in treating ATC. Moreover, the mRNA coding for Aurora A (but not B or C) and Aurora A protein levels were significantly increased in ATC patients. Also, there were durable benefits in a pilot clinical trial of pazopanib plus paclitaxel therapy in a patient with metastatic ATC. Collectively, these results suggest that the combination of pazopanib plus paclitaxel is a promising therapeutic approach in ATC and that Aurora A may be a potentially viable therapeutic molecular target in ATC [81].

Some anticancer drugs target or use DNA or RNA. One of the most important of them is called either adriamycin or doxorubicin. Another, mitomycin is activated in liver cells and adds an alkyl group to bases, causing DNA to cross-link, which kills the cancer cells. Another alkylating agent is Ifosfamide, which is used to treat testicular cancer, breast cancer, lymphoma, testicular cancer, cervical cancer, bone cancer, soft tissue sarcoma, osteogenic sarcoma, and ovarian cancer [78]. Carboplatin (Paraplatin®) is a chemotherapeutic agent for treating cancer. Bleomycin (Blenoxane®) is a member of a family of glycopeptide antibiotics produced by *Streptomyces verticillus*. It damages deoxyribose in DNA, causing the strand to break. It is used to treat testicular cancer, along with head and neck cancer. There are also antimetabolites and nucleosides that are FDA-approved anti-cancer agents. They include 5-azacytidine, 5-fluorouracil, 6-mercaptopurine, allopurinol, calcium leucovorin, capecitabine, cladribine, clofarabine, cytarabine, decitabine, floxuridine, fludarabine, gemcitabine, hydroxyurea, methotrexate, nelarabine, pemetrexed, pentostatin and thioguanine [78]. Vitravene is an antisense drug that binds to mRNA that is produced by a gene coded by the cytomegalovirus (CMV), which causes CMV retinitis. DNA can also be used as a drug in gene therapy.

At the same time, surgery, mastectomy and radiation therapy are often also used to treat cancers [82]. Together with chemotherapy and immunotherapies, they have been rather successful. "Nearly 14.5 million Americans with a history of cancer were alive on January 1, 2014. This estimate does not include carcinoma *in situ* of any site except the urinary bladder, and does not include basal cell and squamous cell skin cancers. Prostate cancer (43%), colorectal cancer (9%), and melanoma (8%) are the 3 most common cancers among male cancer survivors and cancers of the breast (41%), uterine corpus (8%), and colon and rectum (8%) are the most common among female survivors. The majority of cancer survivors (64%) were diagnosed 5 or more years previously, and 15% were diagnosed 20 or more years ago" [82].

KEY POINTS

1. **When a cell is damaged or just reaches the end of its normal life, it usually undergoes programmed cell death, also known as apoptosis. When this program is disrupted, cancer can begin.**

2. **So, cancer is the rapid, unregulated and pathological growth (proliferation) of abnormal cells. There are over 100 types of clinically distinct cancers and many subsets of each type. Even when cancers occur in the same part of the body, they can be very different diseases. When tumors become malignant, their threat depends on their ability to modify surrounding cells to form new blood vessels (angiogenesis) and other supporting cells. This enables cancer cells to move to other parts of the body, a process called metastasis [1]. The following drugs work by inhibiting angiogenesis: Bevacizumab (Avastin), Cetuximab (Erbitux), Panitumumab, (Vectibix), and Trastuzumab (Herceptin).**

3. **There are many causes of cancer. Some people are born with genetic defects that make them more susceptible to cancer. Others have genetic defects that are caused by mutagenic environmental toxins and carcinogens. The typical American diet that leads to obesity can make people more susceptible to cancer. Certain viruses, such as the human papilloma virus, hepatitis B, and T cell leukemia virus type 1 can cause cancer of the cervix, liver and leukocytes (leukemia) [1].**

4. **Tyrosine kinases regulate many cellular processes which can contribute to cancer development and progression. There are several anti-cancer drugs that inhibit tyrosine kinases in some patients. These include imatinib (for chronic myelogenous leukemia), tratuzumab (for breast cancer) and three drugs (cetuxamib, erlotinib and gefitinib) that inhibit the tyrosine kinase enzyme called the epidermal growth factor receptor (EGFR) [1].**

5. **KRAS is the oncoprotein that is most commonly activated in human cancer [1]. *RAS* is one of the most commonly mutated genes in human cancers. Oncogenes code for oncoproteins, which are upregulated in cancer. An important oncoprotein frequently found in some human cancers in c-Myc. The c-Myc protein is a transcription factor that can control the transcription of other genes [1]. Another important oncogene is *PI3K*, which codes for the enzyme PI3K (phosphoinositide 3-kinase). It catalyzes the phosphorylation of phosphatidyl inositol. PTEN (phosphatase and tensin homolog) catalyzes the opposite reaction, so it is a tumor suppressor and the gene coding for it is downregulated in cancer. Another very important tumor suppressor is the p53 protein, coded for by the *TP53* gene. The p53 protein is a hub in the network of cellular metabolism.**

6. **Cancer stem cells could be good targets for new drugs that will prevent the recurrence and metastasis of tumors. Also, induced pluripotent stem cells could be used to screen drugs to see if they will be effective in treating each individual patient.**

ABBREVIATIONS:

KRAS	=	K Rat Sarcoma Oncogene
HER	=	Human Epidermal Growth Factor
VEGF	=	Vascular Endothelial Growth Factor
PI3K	=	Phosphoinositide 3-Kinase
PIP_2	=	Phosphatidyl Inositol 4,5-Bisphosphate
PIP_3	=	Phosphatidyl Inositol 3,4,5-Trisphosphate
IP_3	=	Inositol 1,4,5-Trisphosphate
PI	=	(3,4,5)P3
mTOR	=	Mammalian Target of Rapamycin
SIRT6	=	Sirtuin-6 Silent Mutating-Type Information Regulation
p53	=	Protein With a Molecular Weight of 53,000
TNFR	=	Tumor Necrosis Factor Receptor
TRAIL	=	Tumor Necrosis Factor-Related Apoptosis Inducing Ligand
caspase	=	Cysteine-Aspartate Protease
DNA	=	Deoxyribonucleic Acid
CDK	=	Cyclin-Dependent Kinase
CTC	=	Circulating Tumor Cell

DCC	=	Disseminated Cancer Cell
DTC	=	Disseminated Metastatic Tumor Cell
FDA	=	Food and Drug Administration
mAb	=	Monoclonal Antibody
CAF	=	Cancer-Associated Fibroblast
SDF1	=	Stromal Cell-Derived Factor 1
PDGF	=	Platelet-Derived Growth Factor
VEGF	=	Vascular Endothelial Growth Factor
HGF	=	Hepatocyte Growth Factor
NSCLC	=	Non-Small-Cell Lung Cancer
CTLA-4	=	T-Lymphocyte-Associated Antigen-4
PD1	=	Programmed Death 1
MDSC	=	Myeloid-Derived Suppressor Cell
GM-CSF	=	Granulocyte–Macrophage Colony Stimulating Factor
G-CSF	=	Granulocyte Colony Stimulating Factor
bFGF	=	Basic Fibroblast Growth Factor
TGF β	=	Transforming Growth Factor β
TAM	=	Tumor-Associated Macrophage
T_H1 and T_H2	=	Helper T Cells
SDPP	=	Stroma-Derived Prognostic Predictor
IFN α-2b	=	Interferon α-2b
ROS	=	Reactive Oxygen Substances
UV	=	Ultraviolet
RTK	=	Receptor Tyrosine Kinase
SH2	=	Src Homology-2
PTB	=	Phosphotyrosine Binding
EGFR	=	Epidermal Growth Factor Receptor
BCL	=	B-Cell Lymphoma
Mdm2	=	Mouse Double Minute 2
CIN	=	Cancer Chromosomal Instability
PTEN	=	Phosphatase and Tensin Homolog
MAPK	=	Mitogen-Activated Protein Kinase
mTOR	=	Mammalian Target of Rapamycin
mTORC	=	mTOR Complex

IGF-1	=	Insulin-Like Growth Factor 1
MRI	=	Magnetic Resonance Imaging
NMR	=	Nuclear Magnetic Resonance
CT	=	Computed Tomography
PET	=	Positron Emission Tomography
COPD	=	Chronic Obstructive Pulmonary Disease
PEG	=	Polyethylene Glycol
miRNA	=	microRNA
shRNA	=	Small Hairpin RNA
mRNA	=	Messenger RNA
CML	=	Chronic Myelogenous Leukemia
TSC1	=	Tuberous Sclerosis Gene
CAN	=	Copy-Number Alterations
EMT	=	Epithelial to Mesenchymal Transition
MLL	=	Mixed-Lineage Leukemia
CREB protein	=	cAMP Response Element Binding Protein
PARP	=	Poly(ADP-Ribose) Polymerase
HRD	=	Homologous Recombination Deficiency

REFERENCES

[1] Lyssiotis CA, Cantley LC. SIRT6 puts cancer metabolism in the driver's seat. Cell 2012; 151(6): 1155-6.
[http://dx.doi.org/10.1016/j.cell.2012.11.020] [PMID: 23217699]

[2] Gogna R, Madan E, Kuppusamy P, Pati U. Re-oxygenation causes hypoxic tumor regression through restoration of p53 wild-type conformation and post-translational modifications. Cell Death Dis 2012; 3: e286.
[http://dx.doi.org/10.1038/cddis.2012.15] [PMID: 22419115]

[3] Hallac , *et al.* Correlations of noninvasive BOLD and TOLD MRI with pO2 and relevance to tumor radiation response. Mag Res Med 2013.

[4] Corey EJ, Czako B, Kürti L. Molecules and Medicine. New York: John Wiley & Sons 2007.

[5] Meacham CE, Morrison SJ. Tumour heterogeneity and cancer cell plasticity. Nature 2013; 501(7467): 328-37.
[http://dx.doi.org/10.1038/nature12624] [PMID: 24048065]

[6] Bedard PL, Hansen AR, Ratain MJ, Siu LL. Tumour heterogeneity in the clinic. Nature 2013; 501(7467): 355-64.
[http://dx.doi.org/10.1038/nature12627] [PMID: 24048068]

[7] Burrell RA, McGranahan N, Bartek J, Swanton C. The causes and consequences of genetic heterogeneity in cancer evolution. Nature 2013; 501(7467): 338-45. [http://dx.doi.org/10.1038/nature12625] [PMID: 24048066]

[8] Klein CA. Selection and adaptation during metastatic cancer progression. Nature 2013; 501(7467): 365-72. [http://dx.doi.org/10.1038/nature12628] [PMID: 24048069]

[9] Plaks V, Koopman CD, Werb Z. Cancer. Circulating tumor cells. Science 2013; 341(6151): 1186-8. [http://dx.doi.org/10.1126/science.1235226] [PMID: 24031008]

[10] Marx J. Cancer biology. All in the stroma: cancer's Cosa Nostra. Science 2008; 320(5872): 38-41. [http://dx.doi.org/10.1126/science.320.5872.38] [PMID: 18388269]

[11] Folkman J. Angiogenesis: an organizing principle for drug discovery?. Nat Rev Drug Discov 2007; 6(4): 273-86. [http://dx.doi.org/10.1038/nrd2115] [PMID: 17396134]

[12] Schulz MM, *et al.* Phenotype-based high-content chemical library screening identifies statins as inhibitors of *in vivo* lymphangiogenesis. SigmaAldrich LOPAC® Library of Pharmacologically Active Compounds http://www.sigmaaldrich.com/catalog/product/sigma/sc001?lang=en&region=US. 2013; E2665-74.

[13] SigmaAldrich LOPAC® Library of Pharmacologically Active Compounds. http://www.sigmaaldrich.com/catalog/product/sigma/sc001?lang=en&region=US.

[14] Junttila MR, de Sauvage FJ. Influence of tumour micro-environment heterogeneity on therapeutic response. Nature 2013; 501(7467): 346-54. [http://dx.doi.org/10.1038/nature12626] [PMID: 24048067]

[15] Zindl CL, Chaplin D. D. Tumor immune evasion. Science 2010; 328: 697. [http://dx.doi.org/10.1126/science.1190310]

[16] Tlsty T. Cancer: whispering sweet somethings. Nature 2008; 453(7195): 604-5. [http://dx.doi.org/10.1038/453604a] [PMID: 18509432]

[17] McMillin DW, Negri JM, Mitsiades CS. The role of tumour-stromal interactions in modifying drug response: challenges and opportunities. Nat Rev Drug Discov 2013; 12(3): 217-28. [http://dx.doi.org/10.1038/nrd3870] [PMID: 23449307]

[18] Zhang Y, Daquinag AC, Amaya-Manzanares F, Sirin O, Tseng C, Kolonin MG. Stromal progenitor cells from endogenous adipose tissue contribute to pericytes and adipocytes that populate the tumor microenvironment. Cancer Res 2012; 72(20): 5198-208. [http://dx.doi.org/10.1158/0008-5472.CAN-12-0294] [PMID: 23071132]

[19] Baselga J. Targeting tyrosine kinases in cancer: the second wave. Science 2006; 312(5777): 1175-8. [http://dx.doi.org/10.1126/science.1125951] [PMID: 16728632]

[20] Adler EM. Signaling Breakthroughs of the Year. Sci Signal 2009; 2010(3): 103-.

[21] Vermeij R, *et al.* Immunological and Clinical Effects of Vaccines Targeting p53-Overexpressing Malignancies. J Biomed Biotechnol , 2011 [Epub 2011 Mar 15];2011: 702146. [http://dx.doi.org/10.1155/2011/702146] [PMID: 21541192]

[22] Kruse J-P, Gu W. Modes of p53 regulation. Cell 2009; 137(4): 609-22. [http://dx.doi.org/10.1016/j.cell.2009.04.050] [PMID: 19450511]

[23] Gardner L, Lee L, Dang C. The c-Myc Oncogenic Transcription Factor in myc oncogene. Encyclopedia of Cancer. (2nd ed.), 2002.

[24] Dang CV. MYC on the path to cancer. Cell 2012; 149(1): 22-35. [http://dx.doi.org/10.1016/j.cell.2012.03.003] [PMID: 22464321]

[25] Vita M, Henriksson M. The Myc oncoprotein as a therapeutic target for human cancer. Semin Cancer Biol 2006; 16(4): 318-30. [http://dx.doi.org/10.1016/j.semcancer.2006.07.015] [PMID: 16934487]

[26] Burrell RA. Replication stress links structural and numerical cancer chromosomal instability. Nature 2013; 494: 492-8.

[27] Janssen A, Medema RH. Cancer: Stress mixes chromosomes. Nature 2013; 494(7438): 439-41. [http://dx.doi.org/10.1038/494439a] [PMID: 23446415]

[28] Esteva FJ, Hortobagyi GN. Gaining ground on breast cancer. Sci Am 2008; 298(6): 58-65. [http://dx.doi.org/10.1038/scientificamerican0608-58] [PMID: 18642543]

[29] Afinitor Official website. http://www.afinitor.com/index.jsp?site=PC018060&source=01030&irmasrc =ONCWB0042 2014.

[30] Hurtado A, Holmes KA, Geistlinger TR, *et al.* Regulation of ERBB2 by oestrogen receptor-PAX2 determines response to tamoxifen. Nature 2008; 456(7222): 663-6. [http://dx.doi.org/10.1038/nature07483] [PMID: 19005469]

[31] Nalluri SR, Chu D, Keresztes R, Zhu X, Wu S. Risk of venous thromboembolism with the angiogenesis inhibitor bevacizumab in cancer patients: a meta-analysis. JAMA 2008; 300(19): 2277-85. [http://dx.doi.org/10.1001/jama.2008.656] [PMID: 19017914]

[32] Kolata G, Pollack A. Costly cancer drug offers hope, but also a dilemma. New York Times.http://www.nytimes.com/2008/07/06/health/06avastin.html?_r=1&oref=slogin 2008.

[33] Lee C-H, Inoki K, Guan K-L. mTOR pathway as a target in tissue hypertrophy. Annu Rev Pharmacol Toxicol 2007; 47: 443-67. [http://dx.doi.org/10.1146/annurev.pharmtox.47.120505.105359] [PMID: 16968213]

[34] Chang K-Y, Tsai S-Y, Wu C-M, Yen CJ, Chuang BF, Chang J-Y. Novel phosphoinositide 3-kinase/mTOR dual inhibitor, NVP-BGT226, displays potent growth-inhibitory activity against human head and neck cancer cells *in vitro* and *in vivo*. Clin Cancer Res 2011; 17(22): 7116-26. [http://dx.doi.org/10.1158/1078-0432.CCR-11-0796] [PMID: 21976531]

[35] Chapuis N, *et al.* Dual Inhibition of PI3K and mTORC1/2 signaling by NVP-BEZ235 as a new therapeutic strategy for acute myeloid leukemia. Clin Cancer Res 2010; 16(22): 5424-35.

[36] Medioni OS, Ayllon J, Barrascourt E, Elaidi RT, Balcaceres J, Scotte F. Everolimus (RAD001): an mTOR inhibitor for the treatment of metastatic renal cell carcinoma. Expert Rev Cancer Ther 2009; 9(6): 705-17.

[37] Kang SA. mTORC1 phosphorylation sites encode their sensitivity to starvation and rapamycin.

Science 2013 Jul 26; 341(6144): 1236566.
[http://dx.doi.org/10.1126/science.1236566]

[38] Blay J-Y. Updating progress in sarcoma therapy with mTOR inhibitors. Ann Oncol 2011; 22(2): 280-7.
[http://dx.doi.org/10.1093/annonc/mdq307] [PMID: 20591820]

[39] Kaeberlein M, Kennedy BK. Ageing: A midlife longevity drug? Nature 2009; 460(7253): 331-2.
[http://dx.doi.org/10.1038/460331a] [PMID: 19606132]

[40] U.S. Food and Drug Administration. Report on the Food and Drug Administration's Review of the Safety of Recombinant Bovine Somatotropin

[41] National Cancer Institute. Inflammatory Breast Cancer: Questions and Answers, http://www.cancer.gov/cancertopics/factsheet/sites-types/ibc , [Accessed Feb 20, 2014];

[42] Qin W, Zhu W, Wagner-Mann C, Sauter ER. Nipple aspirate fluid expression of urokinase-type plasminogen activator inhibitor-1, and urokinase-type plasminogen activator receptor predicts breast cancer diagnosis and advanced disease. Ann Surg Oncol 2003; 10: 948-53.
[http://dx.doi.org/10.1245/ASO.2003.03.518] [PMID: 14527916]

[43] Meijer-van Gelder ME, Look MP, Peters HA, *et al.* Urokinase-type plasminogen activator system in breast cancer: association with tamoxifen therapy in recurrent disease. Cancer Res 2004; 64(13): 4563-8.
[http://dx.doi.org/10.1158/0008-5472.CAN-03-3848] [PMID: 15231667]

[44] Li J, Zhao J, Yu X, *et al.* Identification of biomarkers for breast cancer in nipple aspiration and ductal lavage fluid. Clin Cancer Res 2005; 11(23): 8312-20.
[http://dx.doi.org/10.1158/1078-0432.CCR-05-1538] [PMID: 16322290]

[45] Oliva MR, Saini S. Liver cancer imaging: role of CT, MRI, US and PET. Cancer Imaging 2004; 4 Spec No A: S42-6.
[http://dx.doi.org/10.1102/1470-7330.2004.0011] [PMID: 18215974]

[46] Seo Y, *et al.* Improved prostate cancer imaging with SPECT/CT and MRI/MRSI. IEEE Trans Nucl Sci 2005; 52: 1316-20.
[http://dx.doi.org/10.1109/TNS.2005.858241]

[47] Bastarrika G, García-Velloso MJ, Lozano MD, *et al.* Early lung cancer detection using spiral computed tomography and positron emission tomography. Am J Respir Crit Care Med 2005; 171(12): 1378-83.
[http://dx.doi.org/10.1164/rccm.200411-1479OC] [PMID: 15790860]

[48] Kim ES, Hirsh V, Mok T, *et al.* Gefitinib *versus* docetaxel in previously treated non-small-cell lung cancer (INTEREST): a randomised phase III trial. Lancet 2008; 372(9652): 1809-18.
[http://dx.doi.org/10.1016/S0140-6736(08)61758-4] [PMID: 19027483]

[49] Ganta R, Amiji M. Coadministration of paclitaxel and curcumin to overcome multi-drug resistance in tumor cells. Mol Pharm 2009; 6: 928.
[http://dx.doi.org/10.1021/mp800240j] [PMID: 19278222]

[50] Iguchi H, Kosaka N, Ochiya T. Versatile applications of microRNA in anti-cancer drug discovery: from therapeutics to biomarkers. Curr Drug Discov Technol 2010; 7(2): 95-105.

[http://dx.doi.org/10.2174/157016310793180648] [PMID: 20836759]

[51] Fuchs D, Daniel V, Sadeghi M, Opelz G, Naujokat C. Salinomycin overcomes ABC transporter-mediated multidrug and apoptosis resistance in human leukemia stem cell-like KG-1a cells. Biochem Biophys Res Commun 2010; 394(4): 1098-104.
[http://dx.doi.org/10.1016/j.bbrc.2010.03.138] [PMID: 20350531]

[52] Li L, Wang L, Li L, *et al.* Activation of p53 by SIRT1 inhibition enhances elimination of CML leukemia stem cells in combination with imatinib. Cancer Cell 2012; 21(2): 266-81.
[http://dx.doi.org/10.1016/j.ccr.2011.12.020] [PMID: 22340598]

[53] Willingham SB, Volkmer JP, Gentles AJ, *et al.* The CD47-signal regulatory protein alpha (SIRPa) interaction is a therapeutic target for human solid tumors. Proc Natl Acad Sci USA 2012; 109(17): 6662-7.
[http://dx.doi.org/10.1073/pnas.1121623109] [PMID: 22451913]

[54] Gupta PB, Onder TT, Jiang G, *et al.* Identification of selective inhibitors of cancer stem cells by high-throughput screening. Cell 2009; 138(4): 645-59.
[http://dx.doi.org/10.1016/j.cell.2009.06.034] [PMID: 19682730]

[55] Oak PS, *et al.* Combinatorial treatment of mammospheres with trastuzumab and salinomycin efficiently targets HER2-positive cancer cells and cancer stem cells. Int J Cancer 2012; 131(12): 2808-19.
[http://dx.doi.org/10.1002/ijc.27595]

[56] Robinton DA, Daley GQ. The promise of induced pluripotent stem cells in research and therapy. Nature 2012; 481(7381): 295-305.
[http://dx.doi.org/10.1038/nature10761] [PMID: 22258608]

[57] Mocellin S, Pooley KA, Nitti D. Telomerase and the search for the end of cancer. Trends Mol Med 2013; 19(2): 125-33.
[http://dx.doi.org/10.1016/j.molmed.2012.11.006] [PMID: 23253475]

[58] Mocellin S, Verdi D, Pooley KA, *et al.* Telomerase reverse transcriptase locus polymorphisms and cancer risk: a field synopsis and meta-analysis. J Natl Cancer Inst 2012; 104(11): 840-54.
[http://dx.doi.org/10.1093/jnci/djs222] [PMID: 22523397]

[59] Huang FW, Hodis E, Xu MJ, Kryukov GV, Chin L, Garraway LA. Highly recurrent *TERT* promoter mutations in human melanoma. Science 2013; 339(6122): 957-9.
[http://dx.doi.org/10.1126/science.1229259] [PMID: 23348506]

[60] Horn S, Figl A, Rachakonda PS, *et al.* TERT promoter mutations in familial and sporadic melanoma. Science 2013; 339(6122): 959-61.
[http://dx.doi.org/10.1126/science.1230062] [PMID: 23348503]

[61] Vogelstein B, Papadopoulos N, Velculescu VE, Zhou S, Diaz LA Jr, Kinzler KW. Cancer genome landscapes. Science 2013; 339(6127): 1546-58.
[http://dx.doi.org/10.1126/science.1235122] [PMID: 23539594]

[62] Kilpivaara O, Aaltonen LA. Diagnostic cancer genome sequencing and the contribution of germline variants. Science 2013; 339(6127): 1559-62.
[http://dx.doi.org/10.1126/science.1233899] [PMID: 23539595]

[63] McLeod HL, McLeod HL, *et al.* Cancer pharmacogenomics: early promise, but concerted effort needed. Science 2013; 339(6127): 1563-6.
[http://dx.doi.org/10.1126/science.1234139] [PMID: 23539596]

[64] Gao J, *et al.* Integrative analysis of complex cancer genomics and clinical profiles using the cBioPortal. Sci Signal 2013; 6(269): 1-19.
[http://dx.doi.org/10.1126/scisignal.2004088]

[65] Suvà ML, Riggi N, Bernstein BE. Epigenetic reprogramming in cancer. Science 2013; 339(6127): 1567-70.
[http://dx.doi.org/10.1126/science.1230184] [PMID: 23539597]

[66] Garber K. PARP inhibitors bounce back. Nat Rev Drug Discov 2013; 12(10): 725-7.
[http://dx.doi.org/10.1038/nrd4147] [PMID: 24080684]

[67] Lozzio CB, Lozzio BB. Human chronic myelogenous leukemia cell-line with positive Philadelphia chromosome. Blood 1975; 45(3): 321-34.
[PMID: 163658]

[68] Rosenberg N, Witte ON. The viral and cellular forms of the Abelson (abl) oncogene. Adv Virus Res 1988; 35: 39-81.
[http://dx.doi.org/10.1016/S0065-3527(08)60708-3] [PMID: 2852893]

[69] Gupta M, Lorusso PM, Wang B, *et al.* Clinical implications of pathophysiological and demographic covariates on the population pharmacokinetics of trastuzumab emtansine, a HER2-targeted antibody-drug conjugate, in patients with HER2-positive metastatic breast cancer. J Clin Pharmacol 2012; 52(5): 691-703.
[http://dx.doi.org/10.1177/0091270011403742] [PMID: 21953571]

[70] Krop ie, LoRusso P, Miller KD, *et al.* A phase II study of trastuzumab emtansine in patients with human epidermal growth factor receptor 2-positive metastatic breast cancer who were previously treated with trastuzumab, lapatinib, an anthracycline, a taxane, and capecitabine. J Clin Oncol 2012; 30(26): 3234-41.
[http://dx.doi.org/10.1200/JCO.2011.40.5902] [PMID: 22649126]

[71] Shah DK, Haddish-Berhane N, Betts A. Bench to bedside translation of antibody drug conjugates using a multiscale mechanistic PK/PD model: a case study with brentuximab-vedotin. J Pharmacokinet Pharmacodyn 2012; 39(6): 643-59.
[http://dx.doi.org/10.1007/s10928-012-9276-y] [PMID: 23151991]

[72] Thompson MD, Beard DA, Wu F. Use of partition coefficients in flow-limited physiologically-based pharmacokinetic modeling. J Pharmacokinet Pharmacodyn 2012; 39(4): 313-27.
[http://dx.doi.org/10.1007/s10928-012-9252-6] [PMID: 22639356]

[73] Lai W-Y, Wang WY, Chang YC, Chang CJ, Yang PC, Peck K. Synergistic inhibition of lung cancer cell invasion, tumor growth and angiogenesis using aptamer-siRNA chimeras. Biomaterials 2014; 35(9): 2905-14.
[http://dx.doi.org/10.1016/j.biomaterials.2013.12.054] [PMID: 24397988]

[74] Geron Corporation. Geron receives FDA clearance to begin world's first human clinical trial of embryonic stem cell-based therapy. http://www.geron.com 2009.

[75] Radic M. Armed and accurate: engineering cytotoxic T cells for eradication of leukemia. BMC Biotechnol 2012; 12(6): 6.
[http://dx.doi.org/10.1186/1472-6750-12-6] [PMID: 22316161]

[76] Moghimi SM, Hunter AC, Andresen TL. Factors controlling nanoparticle pharmacokinetics: an integrated analysis and perspective. Annu Rev Pharmacol Toxicol 2012; 52: 481-503.
[http://dx.doi.org/10.1146/annurev-pharmtox-010611-134623] [PMID: 22035254]

[77] Pommier Y, Marchand C. Interfacial inhibitors: targeting macromolecular complexes. Nat Rev Drug Discov 2012; 11(1): 25-36.
[PMID: 22173432]

[78] Corey EJ, Czako B, Kürti L. Molecules and Medicine. New York: John Wiley & Sons 2007.

[79] Montagnoli A, Valsasina B, Croci V, *et al.* A Cdc7 kinase inhibitor restricts initiation of DNA replication and has antitumor activity. Nat Chem Biol 2008; 4(6): 357-65.
[http://dx.doi.org/10.1038/nchembio.90] [PMID: 18469809]

[80] Meiri D, Marshall CB, Greeve MA, *et al.* Mechanistic insight into the microtubule and actin cytoskeleton coupling through dynein-dependent RhoGEF inhibition. Mol Cell 2012; 45(5): 642-55.
[http://dx.doi.org/10.1016/j.molcel.2012.01.027] [PMID: 22405273]

[81] Isham CR, Bossou AR, Negron V, *et al.* Pazopanib enhances paclitaxel-induced mitotic catastrophe in anaplastic thyroid cancer. Sci Transl Med 2013; 5(166): 166ra3.
[http://dx.doi.org/10.1126/scitranslmed.3004358] [PMID: 23283368]

[82] DeSantis CE, Lin CC, Mariotto AB, *et al.* Cancer treatment and survivorship statistics, 2014. CA Cancer J Clin 2014; 64(4): 252-71.
[http://dx.doi.org/10.3322/caac.21235] [PMID: 24890451]

CHAPTER 5

Medicinal Chemistry and the Endocrine System

Abstract: The endocrine system consists of cells, glands and tissues that secrete hormones into the bloodstream that affect physiological and behavioral function and activities [1]. This is in contrast to the exocrine system that secretes substances into ducts. The hypothalamus connects the nervous and endocrine systems to each other through the pituitary gland, or hypophysis. It helps control body temperature, hunger, parenting and attachment behavior, thirst, fatigue, sleep, circadian rhythms and other activities of the autonomic nervous system. Darkness causes the pineal gland to secrete N-acetyl-5-methoxytryptamine, which is better known as melatonin. Melatonin is part of the system that regulates the sleep-wake cycle. It causes drowsiness and lowers the body temperature. The anterior lobe of the pituitary gland secretes growth hormone (GH), beta-endorphin, luteinizing hormone (LH), follicle stimulating hormone (FSH), melanocyte stimulating hormone (MSH), adrenocorticotropic hormone (ACTH), thyroid stimulating hormone (TSH) and prolactin (PRL). The posterior lobe stores vasopressin, also known as antidiuretic hormone (ADH) and oxytocin (OXT). The thyroid gland helps control how fast a body uses energy and makes proteins as well as influencing the sensitivity of the body to other hormones. It does this by producing T3 and T4, which are made from tyrosine and iodine. The stomach, duodenum, liver, pancreas and kidneys all secrete hormones. The kidneys secrete renin, erythropoietin, calcitrol and thrombopoietin. They regulate pH, electrolytes and blood pressure. The adrenal glands produce hormones in response to stress by synthesizing corticosteroids such as cortisol and catecholamines, such as adrenaline (epinephrine) and noradrenaline (norepinephrine). They also produce androgens in their innermost cortical layer. The adrenal glands affect kidney function by secreting aldosterone. The testes, ovarian follicle and corpus luteum are in the endocrine system, as are the placenta and uterus when a woman is pregnant. The testes secrete androgens (mostly testosterone), estradiol and inhibin. They stimulate or control the development and maintenance of male characteristics by binding to androgen receptors. This includes the activity of the male sex organs and development of male secondary sex characteristics. Androgens are also the original anabolic steroids and the precursor of all estrogens. The endocrine system also regulates the concentration of Ca^{2+}. The parathyroid gland secretes the parathyroid hormone (PTH), which stimulates Ca^{2+} release from bones, stimulates osteoclasts and Ca^{2+} reabsorption in the kidneys. Calcium regulation also occurs in the skin, which secretes the prehormone calcidiol (25-hydroxyvitamin D3), the inactive form of vitamin D. The major endocrine systems are the TRH-TSH-Y3/T4, the GnRH-

Robert E. Smith

LH/FSH-sex hormones, the CRH-ACTH-cortisol, the renin-angiotensin-aldosterone system and the leptin *vs* insulin system. The TRH-TSH-Y3/T4 system is also called [1].

Keywords: Adrenal, Endocrine system, Hypothalamus, kidneys, Parathyroid, Pituitary, Reproductive system, Thyroid.

As described in Chapter 3, adipose tissue is an important part of the endocrine and immune systems [1]. That is, the endocrine system consists of cells, glands and tissues that secrete hormones into the bloodstream to affect physiological and behavioral function and activities [1]. This includes metabolism, tissue function, sleep, mood, growth and development. On the other hand, the exocrine system secretes substances into ducts. Like the nervous system, the endocrine system sends and receives information in the form of biochemical messages. However, its effects are slower to get started and last longer than those of the nervous system. A series of glands secrete hormones. When several of them signal each other in sequence, they can be called an axis, as in the hypothalamic-pituitary-adrenal axis.

HYPOTHALAMUS

The hypothalamus is a small portion of the brain, located below the thalamus and above the brainstem [1]. "It connects the nervous and endocrine systems to each other through the pituitary gland, or hypophysis. It helps control body temperature, hunger, parenting and attachment behavior, thirst, fatigue, sleep, circadian rhythms and other activities of the autonomic nervous system. When stimulated by high amplitude oscillations (delta waves) from the thalamus or cortex in the brain, it secretes neurohormones that are often called releasing hormones, for they stimulate or inhibit the release of pituitary hormones. Secretion of GHRH and prolactin is stimulated, while thyroid releasing hormone (TRH) is inhibited. At the same time, it responds to many different signals, of internal and external origin. It responds to daylight, circadian and seasonal rhythms, odors, gonadal steroids, corticosteroids, autonomic input, stress, invading organisms, neural input from the heart, stomach and reproductive tract and biochemicals in the blood. This includes peptide hormones, leptin, gherlin, insulin, pituitary hormones, cytokines, glucose and osmolarity" [1]. So, let's look closer at this amazing part of the brain and endocrine system.

The hypothalamus has distinct nuclei and other less distinct areas. In the hypothalamus, the axons of magnocellular neurosecretory cells in the paraventricular nucleus and the supraoptic nucleus contain oxytocin and vasopressin (antidiuretic hormone), and project into the posterior pituitary gland [1]. "Much smaller neurons of the paraventricular nucleus release corticotropin-releasing hormone (CRH) and other hormones into the hypophyseal portal system, where they diffuse to the anterior pituitary. The hypothalamus also secretes orexins (also called hypocretins), ghrelin, gonadotropin-releasing hormone (GnRH), growth hormone-releasing hormone (GHRH), somatostatin, and thyrotropin-releasing hormone (TRH), along with the non-peptide hormone, dopamine (which can also act as a neurotransmitter)" [1].

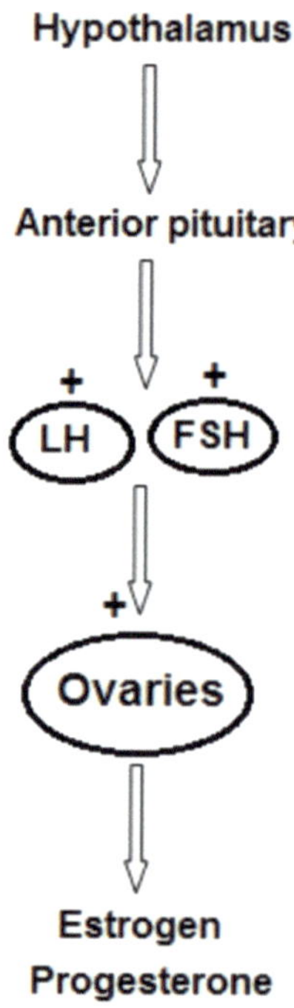

Fig. (1). Hypothalamic-adenohypophyseal (anterior pituitary) axis.

Along the hypothalamic-adenohypophyseal (anterior pituitary) axis (Fig. **1**), somatostatin is produced by the neuroendocrine cells of the periventricular nucleus. It inhibits the release of GH and TSH from the anterior pituitary [1].

"GnRH, GHRH, TRH and CRH stimulate the anterior pituitary to secrete follicle-stimulating hormone (FSH) and leutenizing hormone. Somatostatin inhibits the release of growth hormone (GH) and thyroid stimulating hormone (TSH). CRH stimulates the release of adrenocorticotropic hormone (ACTH). Dopamine neurons in the arcurate nucleus of the hypothalamus secrete dopamine, which inhibits the release of prolactin from the anterior pituitary" [1].

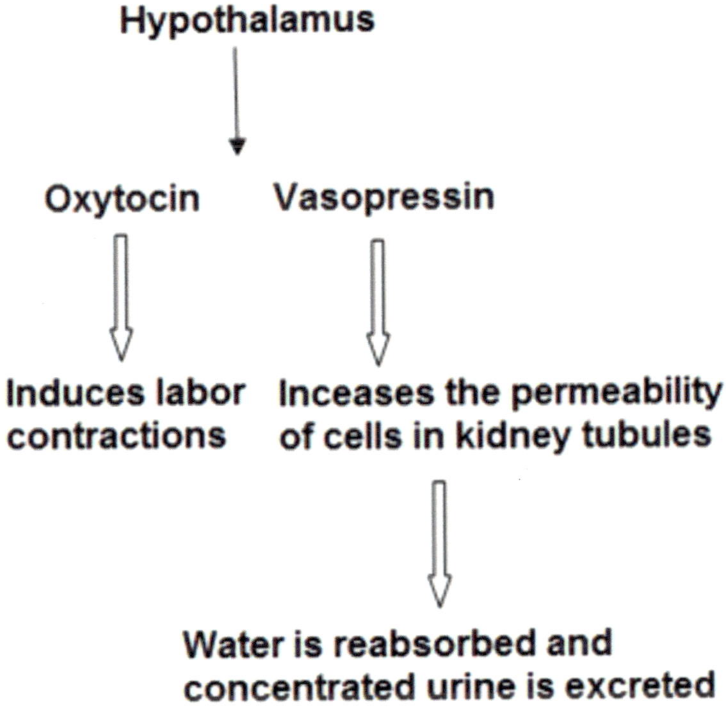

Fig. (2). Hypothalamic-neurohypophyseal (posterior pituitary) axis.

Along the hypothalamic-neurohypophyseal (posterior pituitary) axis (Fig. **2**), oxytocin (OXT) is made in the magnocellular neurosecretory cells of the paraventricular nucleus and supraoptic nucleus [1]. "OXT induces uterine contractions and lactation. Vasopressin increases the permeability of cells in the distal tubule and collecting tubule of the kidneys. This allows water to be reabsorbed and concentrated urine to be excreted. The extreme lateral part of the ventromedial nucleus helps to control food intake and satiety. When stimulated, appetite increases. Several of the hypothalamic nuclei are sexually dimorphic. They are different in males and females. There is even a sexually dimorphic nucleus in the preoptic area. Also, males and females respond differently to ovarian hormones due to estrogen-sensitive neurons that are sexually dimorphic. Male and female brains have different distributions of estrogen receptors, due to differences in neonatal exposure to gonadal steroids" [1].

Another important class of excitatory peptide hormones that helps control appetite and energy levels is called the orexins, or hypocretins. They bind to orexin receptors (members of the GPCR class) and increase the craving for food and increases wakefulness. Hypocretin-producing cells can be inhibited by leptin, but are activated by hypoglycemia and ghrelin, another peptide hormone (secreted by the hypothalamus). Even though some of it is made in the hypothalamus, much more is made by P/D1 cells in the fundus of stomach and epsilon (ε) cells of the pancreas [1]. "Ghrelin increases phosphate before meals. It stimulates food intake and fat mass and activates the reward network mesolimbic system of the brain. Ghrelin levels increase before meals and decrease after meals. It is considered to be the opposite of the hormone leptin, produced by adipose cells. It induces satiation when present at higher levels. In some bariatric procedures, the level of ghrelin is reduced in patients, thus causing satiation before it would normally occur. Ghrelin also stimulates the anterior pituitary to secrete growth hormone. The ghrelin binds a GPCR, called the growth hormone secretagogue receptor (GHSR). Ghrelin induce neurons to develop, survive and function, particularly in the hippocampus. Also, it is essential for cognitive adaptation to changing environments and learning. Finally, it activates the endothelial form of nitric oxide synthase (eNOS) in a process that depends on various kinases including Akt" [1].

The *GHRL* gene coding for ghrelin codes for a pre-propeptide that is known as the appetite-regulating hormone or growth hormone secretagogue [1]. This is hydrolyzed to produce ghrelin and obestatin, which was originally thought to have the opposite effects as ghrelin. Having a relatively low level of ghrelin is a risk factor for cardiovascular disease and for increasing the thickness of the intima-media, which is a biomarker for the early stages of atherosclerosis [2]. Low levels of plasma ghrelin might be related to early arteriosclerosis in patients with type-2 diabetes by increasing intima-media thickness level. Moreover, elevated plasma obestatin levels might protect patients with type-2 diabetes against carotid atherosclerosis to some extent.

PINEAL BODY (EPIPHYSIS)

The reddish-gray pineal body or pineal gland looks like a small (5-8 mm) pine cone, thus its name [1]. "It is in the epithalamus, near the center of the brain, between the two hemispheres and tucked into a groove where the two thalamic bodies join each other. It consists of a lobular functional part made of cells called pinealocytes that is surrounded by connective tissue. The pineal gland receives sympathetic and parasympathetic innervations. Also, some nerve fibers penetrate into the pineal gland through the pineal stalk (central innervation), while neurons in the trigeminal ganglion innervate the gland with nerve fibers containing the neuropeptide, pituitary adenylate cyclase-activating protein (PACAP). This neuropeptide stimulates adenylate cyclase, which catalyzes the biosynthesis of cAMP. It is not just a hormone but also a neurotransmitter and neuromodulator" [1]. A version of the gene that codes for PACAP (*ADCYAP1*) has been associated with post-traumatic stress disorder in women, but not men [3]. Also, there are follicles that contain a gritty material that is made of calcium phosphate, calcium carbonate, magnesium phosphate and ammonium phosphate. Increased levels of calcium and phosphorus deposits in the pineal gland have been linked with aging [1].

Darkness causes the pineal gland to secrete N-acetyl-5-methoxytryptamine, which is better known as melatonin [4]. Light inhibits the secretion of melatonin. The concentration of melatonin in the blood varies in a daily cycle or circadian rhythm. In infants, it becomes regular at about three months [5]. About 90% of it is eliminated in one pass through the liver. Melatonin levels decrease during aging [6]. Many of its physiological effects are caused by binding to its receptor, while others are caused by its wide spectrum antioxidant potential [6]. Single nucleotide polymorphisms (SNPs) in the gene coding for the melatonin MT2 receptor have been linked to an increased risk of getting type-2 diabetes [7]. "It can cross cell membranes and the blood-brain barrier, protect nuclear and mitochondrial DNA. The US FDA considers melatonin to be a dietary supplement and not a drug.

Melatonin is part of the system that regulates the sleep-wake cycle. It causes drowsiness and lowers the body temperature" [7]. Very high doses (50 mg) increased REM sleep time and dream activity [8]. Prolonged release melatonin may be able to treat insomnia in older people [9]. Prolonged release melatonin is available by prescription (Circadin®) in Europe [9]. The standard therapy for delayed sleep phase disorder (DSPD) and non-24 hour sleep-wake disorder is to take melatonin in the evening and light therapy upon awakening [10].

PITUITARY GLAND (HYPOPHYSIS)

The pituitary gland is a protrusion about the size of a pea (0.5 g) protruding off the bottom of the hypothalamus [11]. "The anterior lobe secretes growth hormone (GH), beta-endorphin, luteinizing hormone (LH), follicle stimulating hormone (FSH), melanocyte stimulating hormone (MSH), adrenocorticotropic hormone (ACTH), prolactin, thyroid stimulating hormone (TSH) and prolactin (PRL). The posterior lobe stores vasopressin, also known as antidiuretic hormone (ADH) and oxytocin (OXT) [11].

GH (also known as somatotropin) is an anabolic steroid, so it stimulates growth, cell reproduction and regeneration. The major isoform of GH has 191 amino acids and a molecular weight of 22124 Da [11]. "There are other isoforms, including one that has a lower molecular weight and is made by alternative splicing and another with a higher molecular weight that is elevated after strenuous exercise. It is a stress hormone that binds to a membrane-bound receptor and activates the MAPK/ERK pathway. This stimulates division and multiplication of chondrocytes of the cartilage. GH also stimulates the biosynthesis of insulin-like growth factor 1 (IGF-1) in the liver, through the JAK-STAT pathway. IGF-1 then stimulates growth in other tissues. More IGF-1 is produced in target tissues, so it is both an endocrine and autocrine hormone. IGF-1 also stimulates osteoblasts and bone growth" [11].

GH elevates the concentrations of free fatty acids and glucose in the blood. It increases Ca^{2+} retention, strengthening and mineralizing bones. It increases muscle mass by sarcomere hypertrophy [11]. "It promotes lipolysis, increases protein synthesis, reduces glucose uptake of glucose in the liver, while promoting gluconeogenesis. It stimulates the growth of all organs except the brain. It helps maintain the function of pancreatic islets and stimulates the immune system" [11].

It is prescribed for growth disorders in children and low GH in adults. Physicians are free to prescribe GH for off-label uses. Recently, it has been prescribed in the USA for elderly people as an anti-aging treatment. This is based partly on a study reported in 1990 in which 12 men over 60 were treated with GH [12]. The subjects who received GH experienced an increase in lean body mass and bone mineral density, unlike those who received a placebo [12]. Even though the authors did not claim that GH reversed the aging process, their results were used by people marketing and selling GH to advertize that GH is an effective anti-aging agent [13]. Actually, GH was developed to treat children afflicted by dwarfism [13]. Also, in a subsequent meta analysis of 31 articles describing 18 unique study populations, researchers found that people treated with GH were significantly more likely to experience soft tissue edema, arthralgias, carpal tunnel syndrome, and gynecomastia [14]. Moreover, they were more likely to experience the onset of diabetes mellitus and impaired fasting glucose [14]. Its use in competitive athletes is banned by the International Olympic Committee and other agencies. However, often GH is given to dairy cattle to increase milk production.

The glycoprotein hormone TSH stimulates the thyroid to produce thyroxine T_4 which is converted into triiodothyronine (T_3), which stimulates the metabolism of almost every tissue in the body [11]. "It is secreted throughout life, but is elevated during periods of rapid growth and development. About 20% of the conversion of T_4 to T_3 occurs in the thyroid. The other 80% is converted in the liver and other organs. TSH has two subunits. The alpha subunit is almost identical to that of human chorionic gonadotropin (hCG), luteinizing hormone (LH), and follicle-stimulating hormone (FSH). The α subunit is the part that responsible for stimulating adenylate cyclase, which catalyzes the production of cAMP. The β (*beta*) subunit is unique and determines the specificity for the cognate receptor" [11].

TSH concentrations are measured as part of a test of thyroid function in patients suspected of having an excess or deficiency of thyroid hormones [11]. "The range of concentrations of TSH may vary slightly, depending on the method of analysis, and do not necessarily equate to cut-offs for diagnosing thyroid dysfunction. In the UK, guidelines issued by the Association for Clinical Biochemistry suggest a reference range of 0.4-4.5 mIU/L. The National Academy of Clinical Biochemistry (NACB) stated that it expected the reference range for adults to be reduced to 0.4–2.5 μIU/mL, and the optimum concentration for hypothyroid patients is between 0.5 and 2.0 μIU/mL, TSH concentrations in children are usually higher than in adults. So, the NACB recommended reference limits are 1.3 to 19 μIU/mL for normal-term infants at birth. This decreases to 0.6–10 μIU/mL at 10 weeks, 0.4–7.0 μIU/mL at 14 months and then it slowly decreases during childhood and puberty to adult levels, 0.3 μIU/mL - 3.0 μIU/mL" [11]. Finally, there is a recombinant human TSH α (thyroid-stimulating hormone or thyrotropin-α, or rhTSH) that is used to treat thyroid cancer [15].

THYROID

The butterfly shaped thyroid gland is found on the anterior (front) of the neck, below the thyroid cartilage ("Adam's apple") [16]. It weighs 2-3 grams in neonates and 18-60 g in adults. Thyroid hormones play essential roles in the developing fetal brain. The thyroid gland has two cone-shaped lobes or wings, right and left (*lobus dexter and sinister*) that are connected by an isthmus. It is covered by a thin, fibrous sheet called the *capsula glandulae thyroidea*. Two parathyroid glands lie between the two layers of this capsule and on the posterior side of the lobes.

The thyroid gland helps control how fast a body uses energy and makes proteins as well as influencing the sensitivity of the body to other hormones. It does this by producing triiodo-and tetraiodothyronine (T_3 and T_4), which are made from tyrosine and iodine (I_2), which comes from seafood and iodized table salt. Some of it is made from free tyrosine, while the rest is made by adding iodine to tyrosyl residues in thyroglobulin (T_g). The iodine is trapped inside T_g by H_2O_2 that is made in a reaction catalyzed by thyroid peroxidase. When stimulated by TSH from the anterior pituitary, follicular cells reabsorb T_g and hydrolyze the iodinated tyrosyl residues in lysosomes, making T_3 and T_4. Once secreted into the blood, they are partly bound to thyroxine-binding globulin, transthyretin and albumin.

Only about 0.03% and 0.3% of T_4 and T_3, respectively, remain free.

Only free T_4 and T_3 can bind to one of three cognate cytosolic receptors, TR-α1 (in cardiac and skeletal muscles), TR-β1 (in the brain, liver and kidney) and TR-β2 (in the hypothalamus and pituitary glands). That is, they are like steroid hormones – they cross the cell membrane easily and enter the cytoplasm, where they bind to receptors that then migrate to the cell nucleus. The receptors bind to hormone responsive elements (HREs) in DNA either as monomers, homodimers or heterodimers with the retinoid X receptor (RXR) [1]. They regulate the transcription of responsive genes. In the absence of thyroid hormone, though, TRs form inhibitory complexes with corepressor proteins that bind to HREs and prevent transcription.

However, T_4 can also bind to a membrane-bound integrin that activates the ERK1/2 extracellular signal-regulated kinase. This affects ion transport systems, such as the Na^+/H^+ exchanger and leads to many effects, including cell proliferation. The integrins are concentrated on tumor cells and cells of the vasculature, stimulating angiogenesis and some types of cancer, including gliomas. Also, T_4 acts on the mitochondrial genome by importing nuclear TRs.

To keep the effects caused by T_4 from lasting too long, there is a negative feedback mechanism that is activated when the concentration of T_4 in the blood serum gets too high. TSH production is suppressed by excess T_4 and by other factors, such as rising levels of glucocorticoids and sex hormones (estrogen and testosterone), somatosatatin, and physiological conditions, such as cold (so thermogenesis can occur).

The thyroid also produces a peptide hormone called calcitonin that helps regulate Ca^{2+} levels in the blood serum. It protects against the loss of Ca^{2+} from the skeleton when Ca^{2+} is mobilized during pregnancy and lactation. It stimulates the movement of Ca^{2+} into bones (inhibits osteoclast activity), in opposition to the effects of parathyroid hormone (PTH). It also inhibits the absorption of Ca^{2+} in the intestines and inhibits reabsorption of Ca^{2+} by renal tubule cells, so that it can be excreted in the urine. One of its physiological effects is similar to that of PTH – they both inhibit phosphate (HPO_4^{2-} and $H_2PO_4^-$) by kidney tubules. It is made by the parafollicular cells (also called C cells) by proteolytic cleavage of a longer pre-propeptide [1]. "Calcitonin secretion is stimulated by an increase in serum Ca^{2+} and by the peptide hormones gastrin and pentagastrin. The calcitronin receptor is a GPCR and is found in the kidneys and parts of the brain. It is coupled to adenylate cyclase through G_s so it helps to make the second messenger cAMP. Calcitonin that has been extracted from salmon and purified is used to treat postmenopausal osteoporosis, hypercalcemia, Paget's disease, bone metastases and phantom limb pain. It is rapidly absorbed an eliminated. It takes only one hour to reach the peak concentration in blood plasma. It is mostly metabolized in the liver by proteolysis. It can also be used as a diagnostic marker for medullary thyroid cancer if the concentration exceeds 5 and 12 pg/mL in females and males, respectively; 40 pg/mL in children under 6 months and 15 pg/mL in children who are 6 months to 3 years old" [1].

ALIMENTARY SYTEM

The stomach, duodenum, liver, pancreas and kidneys all secrete hormones. The pancreas has endocrine and exocrine tissues. The exocrine cells secrete digestive enzymes which are delivered to the intestines. Clusters of endocrine cells (called islets of Langerhans) exist in the exocrine tissue [1]. "Within the islets, α-cells make glucagon; β-cells, insulin; δ-cells, somatostatin; and γ-cells, pancreatic polypeptide — all of which are delivered into the blood stream. Insulin and glucagon have opposite effects" [1].

The stomach secretes the hormones gastrin, ghrelin, neuropeptide Y, somatostatin, histamine and endothelin [1]. "Gastrin is made in the G cells of the antrum of the stomach (the part that is adjacent to the pyloric valve), as well as the duodenum and pancreas. It stimulates the production of gastric acid (HCl) by parietal cells. It also stimulates the growth and maturation of parietal cells, causes cells to secrete pepsinogen (the zymogen for the digestive enzyme pepsin), increases antral muscle mobility and promotes contractions. It also induces pancreatic secretions and emptying of the gallbladder. Its release is stimulated by peptides in the lumen of the stomach, in response to stomach antrum distension, vagal stimulation, partially digested proteins, amino acids, and hypercalcemia. Gastrin release is inhibited by excess stomach acid in a negative feedback mechanism. Its release is also inhibited by somatostatin, secretin, gastroinhibitory peptide, vasoactive intestinal peptide, glucagon and calcitonin. It binds to the cholecystokinin B receptors, which stimulate the release of histamine and induces the insertion of K^+/H^+ ATPase pumps into the apical membrane of parietal cells (which in turn increases H^+ release into the stomach cavity) " [1].

Ghrelin increases food intake and fat mass and activates the reward network mesolimbic system of the brain. Neuropeptide Y (NPY), like ghrelin, is a peptide that stimulates the appetite. It is a neurotransmitter in the brain and autonomic nervous system. It is a vasoconstrictor and makes adipose cells grow in fatty tissue. It is also produced in the hypothalamus and other parts of the brain. It increases food intake and the storage of energy as fat (triglycerides). It also helps to reduce anxiety and stress, reduces the perception of pain, affects the circadian rhythm, reduces voluntary alcohol consumption, lowers blood pressure and controls epileptic seizures [17]. NPY is also important in eating disorders and obesity, in which there is an increase in the concentrations of glucocorticoids in blood plasma. This causes an increase in not just glucose, but also NPY, through activation of type II glucocorticoid receptors and by blocking the negative feedback of corticotropin-releasing factor (CRF) on the synthesis and release of NPY. In the meantime, insulin resistance causes the NPYergic activity to be abolished. It should be noted that a diet high in fat and sugar stimulates the release of NPY, causing abdominal fat to increase [18]. Also, elevated concentrations of NPY may be associated with resilience against and recovery from posttraumatic stress disorder [19].

The duodenum is the first part of the small intestines. It is a hollow jointed tube about 25-38 cm long connecting the stomach to the jejunum [1]. It is responsible for the enzymatic digestion of food in the small intestine [1]. It also helps to regulate the rate of the emptying of the stomach induced by the peptide hormones secretin and cholecystekinin. These hormones are released from cells in the duodenal epithelium in response to acidic and fatty substances that are released when the pylorus opens and releases gastric chyme. These hormones also cause the liver and gall bladder to relesae bile and the pancreas to release HCO_3^- and the digestive enzymes trypsin, lipase and amylase into the duodenum.

The liver secretes insulin-like growth factor (IGF), which regulates cell growth and development, as well as having insulin-like effects [1]. IGFs have amino acid sequences similar to that of insulin [1]. The liver also affects metabolism by secreting fibroblast growth factors, FGF 21, is a hepatokine (cytokine from the liver) that acts as a global starvation signal to modulate fuel partitioning and metabolism as well as repress growth by acting on the nervous system [20]. It acts through its receptor in the brain, β-Klotho, to lower insulin, inhibit growth, alter light-dark cycle activity, increase systemic corticosterone levels and inhibit female fertility [20]. It also suppresses the expression of the neuropeptide vasopressin in the suprachiasmatic nucleus (SCN) of the brain, which also expresses receptors for ghrelin and leptin [21]. It also contributes to neuroendocrine control of female reproduction by acting on the SCN in the hypothalamus to suppress ovulation during starvation [21]. The liver also secretes angiotensin and its zymogen, angiotensinogen. They cause vasoconstriction and release aldosterone from the adrenal cortex. The liver also secretes thrombopoietin, which stimulates megakaryocytes to produce blood platelets. The liver also secretes hepcidin, which inhibits iron absorption in the intestines and iron release by macrophages.

KIDNEYS

The kidneys are shaped like kidney beans and are located at the rear of the abdominal cavity [1]. "They filter the blood and excrete wastes like urea and ammonium (NH_4^+) salts, while reabsorbing water, glucose and amino acids. They produce urine and help with homeostasis. They regulate pH, electrolytes and blood pressure. They secrete the renin, erythropoietin, calcitrol and thrombopoietin" [1].

"Renin, also known as angiotensin, is an enzyme that catalyzes the hydrolysis of the protein angiotensinogen that is produced in the liver to make angiotensin I" [1]. "Renin is secreted by the afferent arterioles from specialized kidney cells called granular cells of the juxtaglomelular apparatus in response to three stimuli: a decrease in arterial blood pressure, a decrease in NaCl in the ultrafiltrate of the urine and activity in the sympathetic nervous system. Renin circulates in the blood. After angiotensin I is produced, it is further hydrolyzed in the lungs to the vasoactive peptide called angiotensin II in a reaction catalyzed by angiotensin-converting enzyme (ACE). Angiotensin II constricts all blood vessels and acts on smooth muscles and raises the resistance that is made by these arteries to the heart. The heart works harder, increasing the blood pressure. Angiotensin II also causes the adrenal glands to release aldosterone, which causes the epithelial tubes in the distal tubule and collecting ducts of the kidneys to increase reabsorption of Na^+. It exchanges Na^+ and K^+ and water, thus maintaining the electrochemical charge while increasing the blood volume and blood pressure. The renin-angiotensin system (RAS) also acts on the central nervous system to stimulate thirst, thus increasing water consumption. The RAS also reduces the loss of urine by secreting vasopressin from the posterior pituitary gland. The normal range of concentrations of renin in the blood plasma of healthy adults is 1.98-24.6 ng/mL, when in the upright position" [1].

When the RAS system becomes over-activated it can cause hypertension [1]. "This can be treated and controlled by either ACE inhibitors (such as ramipril and perindopril) or by angiotensin II receptor blockers (ARBs, such as losartan, irbesartan or candesartan) rather than a direct oral renin inhibitor. ACE inhibitors or ARBs are also part of the standard treatment after a heart attack. Also, pediatric kidney cancer can be due to juxtaglomelular cell tumors (reninoma), Wilm's tumor and renal cell carcinoma, all of which produce renin" [1].

Erythropoietin (EPO) is an essential glycoprotein hormone that controls the production of red blood cells. About 40% of its molecular weight is due to attached sugars (glycosylation). It promotes the survival of progenitors and precursors of red blood cells by inhibiting apoptosis. Also, EPO plays an important role in the brain's response to neuronal injury and in the process of healing wounds [1]. "It helps to protect the brain under hypoxic conditions, such as stroke. Moreover, EPO may improve memory, due to its effects on the hippocampus, as well as on synaptic connectivity, neuronal plasticity, and memory-related neural networks. It binds to the erythropoietin receptor on the surface of red cell progenitors and activates the JAK2 signaling cascade (discussed further in Chapter 13). The erythropoietin receptor is also found in bone marrow as well as the peripheral and central nervous systems. EPO is also called hematopoietin and hemopoietin. It is also produced in perisinusoidal cells in the liver, primarily in the fetal and perinatal period. When exogenous EPO is used as a performance-enhancing drug it is called an erythropoiesis-stimulating agent (ESA). It can be distinguished from endogenous EPO because they undergo different post-translational modifications. Prescription erythropoietins and ESA are made by recombinant DNA in cell cultures. They include Epogen/Procrit (epoetin alfa) and Aranesp (darbepoetin alfa). They are used to treat anemia resulting due to chronic kidney disease, inflammatory bowel disease (Crohn's disease and ulcer colitis) and myelodysplasia from the cancer treatments (chemotherapy and radiation). However, the label on the container or box has warnings that there is an increased risk of death, myocardial infarction, stroke, venous thromboembolism, tumor recurrence, and other severe side effects" [1].

Calcitrol (also called 1,25-dihydroxycholecalciferol or 1,25-dihydroxyvitamin D_3) [1]. It increases the concentration of Ca^{2+} in the blood by increasing its absorption into the blood from the gastrointestinal tract [1]. It also stimulates the release of Ca^{2+} from osteoblasts in the bones, causing them to release the secondary messenger protein RANKL, which activates osteoclasts. It is sold under the names Rocaltrol, Calcijex, Decostriol and Vectical. Calcitrol can be prescribed to treat hypocalcemia, hypoparathyroidism, osteomalacia, rickets, renal osteodystrophy, and osteoporosis. It can also be used for people who are undergoing dialysis, due to kidney disease. It can also prevent corticosteroid-induced osteoporosis. The main adverse side effect is to cause hypercalcemia.

Many of the effects of calcitriol are mediated by its interaction with its receptor, also called the vitamin D receptor or VDR [1]. "When calcitrol is not bound to it, the VDR is in the cytoplasm. After binding calcitrol, it moves to the cell nucleus, where it becomes a transcription factor, promoting the expression of a gene that codes for a Ca^{2+} binding protein. As the levels of the Ca^{2+} binding protein increase, it enables the cells to actively transport more Ca^{2+} from the intestine across the intestinal mucosa into the blood. To maintain the electropotential across the cell membrane, anions (mostly HPO_4^{2-} and $H_2PO_4^-$) are also transported. As a result, calcitrol also stimulates the intestinal absorption of phosphate" [1].

Thrombopoietin is a glycoprotein protein hormone [1]. "It is also called megakaryocyte growth and development factor (MGDF). It is produced in the kidneys and liver. It regulates the production of platelets by the bone marrow. It stimulates the production and differentiation of megakaryocytes, the cells in the bone marrow that fragment into many platelets. Thrombopoietin binds to the mpl receptor (CD110) on the surface of blood platelets, thus reducing the exposure of megakaryocytes to the hormone. As a result, the increase and decrease in concentrations of platelets regulate the concentrations of thrombopoietin. Then the concentration of platelets drops too low platelets it causes more exposure of thrombopoietin to the undifferentiated bone marrow cells. This leads to differentiation into megakaryocyes and further maturation of these cells. On the contrary, high platelet concentrations lead to the reversal of this process" [1].

ADRENAL GLANDS

The adrenal glands sit on top of the kidneys [1]. "Each one has an outer cortex and an inner medulla. Both produce hormones in response to stress by synthesizing corticosteroids such as cortisol and catecholamines, such as adrenaline (epinephrine) and noradrenaline (norepinephrine). They also produce androgens in their innermost cortical layer. The adrenal glands affect kidney function by secreting aldosterone. Under normal, unstressed conditions, the human adrenal glands produce the equivalent of 35–40 mg of cortisone acetate per day. "Aldosterone acts on the distal convoluted tubule and collecting duct of the kidneys, where it causes increased reabsorption of Na^+ and increased excretion of both K^+ and H^+. The outermost layer, the zona glomerulosa is the main site for producing aldosterone, a mineralcorticoid, in a reaction catalyzed by the enzyme aldosterone synthase, also known as CYP11B2. Aldosterone is largely responsible for the long-term regulation of blood pressure. The zona fasciculata, located between the glomerulosa and reticularis, produces glucocorticoids, including 11-deoxycorticosterone, corticosterone and cortisol. The adrenal medulla is the core of the adrenal gland, and is surrounded by the adrenal cortex. It secretes about 20% noradrenaline and 80% adrenaline. The chromaffin cells of the medulla, are the body's main source of the circulating catecholamines (adrenaline and noradrenaline). Catecholamines are produced from the amino acid tyrosine. These water-soluble hormones initiate the fight or flight response. To carry out its part of this response, the adrenal medulla receives input from the sympathetic nervous system. Because it is innervated by preganglionic nerve fibers, the adrenal medulla can be considered to be a specialized sympathetic ganglion. Unlike other sympathetic ganglia, however, the adrenal medulla lacks distinct synapses and releases its secretions directly into the blood. Cortisol also promotes adrenaline synthesis in the medulla. It promotes the upregulation of phenylethanolamine *N*-methyltransferase, thereby increasing adrenaline synthesis and secretion" [1].

REPRODUCTIVE SYSTEM

The testes, ovarian follicle and corpus luteum are in the endocrine system, as are the placenta and uterus when a woman is pregnant [1]. "The testes secrete androgens (mostly testosterone), estradiol and inhibin. They stimulate or control the development and maintenance of male characteristics by binding to androgen receptors. This includes the activity of the male sex organs and development of male secondary sex characteristics. Androgens are also the original anabolic steroids and the precursor of all estrogens. The primary and most well-known androgen is testosterone, dihydrotestosterone (DHT) and androstenedione are not as well known, but are equally import in male development. DHT in the embryo life causes differentiation of penis, scrotum and prostate. Later in life DHT contributes to male balding, prostate growth and sebaceous gland activity. Circulating androgens can influence human behavior because some neurons are sensitive to steroid hormones. Androgen levels have been implicated in regulating human aggression and libido" [1].

Estradiol (also known as E2 and 17β-estradiol) is a sex hormone that readily crosses the cell membrane and binds to the estrogen receptor (ER) in the cytoplasm. The estradiol-bound receptor then enters the cell nucleus where it acts as a transcription factor [1]. "In contrast to other estrogens, estradiol binds to both the ERα and ERβ receptors. Some medications, like tamoxifen, are selective estrogen receptor modulators (SERMs). Estradiol is present in males and females. In the healthy adult male, its concentration can be between 14-55 pg/mL. In women in the follicular phase (day 5), its concentration is between 19-140 pg/mL. In preovulatory women, it can be from 110-410 pg/mL. In the luteal phase, it is between 19-160 pg/mL. In post-menopausal women, it can be from 0-35 pg/mL. That is, in the normal menstrual cycle, it acts as a growth hormone for tissues of the reproductive organs. It supports the lining of the vagina, the cervical glands, the endometrium, and the lining of the fallopian tubes. It enhances growth of the myometrium. Estradiol may also be needed to maintain oocytes in the ovary. During the menstrual cycle, estradiol is produced by the growing follicle. This triggers the events in the hypothalamic-pituitary axis that leads to the surge in luteinizing hormone, inducing ovulation. In the luteal phase, estradiol works with progesterone to prepare the endometrium for implantation. During pregnancy, estradiol increases as the placenta produces it" [1].

Estradiol also affects bones. In post-menopausal women, the lack of it can cause osteoporosis. It also plays a significant role in women's mental health, with links suggested between the hormone level, mood and well-being [1]. Sudden drops or fluctuations in, or long periods of sustained low levels of estrogen may cause significant mood-lowering. Clinical recovery from postpartum depression, perimenopause, and postmenopause can occur after estrogen concentrations were stabilized and/or restored [22]. In men, estradiol prevents apoptosis of sperm cells. Estradiol also acts as an antioxidant in the brain, so it is neuroprotective.

Estradiol can be given with progestins as part of hormone replacement therapy in post-menopausal women. When an ethinyl group is added to estradiol it becomes ethinyl estradiol. It is a major component of contraceptive drugs and devices.

Inhibin and activin are two closely related protein complexes that have almost opposite effects [1]. "Inhibin inhibits the synthesis and secretion of follicle stimulating hormone (FSH), while activin activates it. They are both part of the tumor growth factor-β (TGF-β) superfamily. In women, FSH stimulates the secretion of inhibin from the granulosa cells of the ovarian follicles, so inhibin inhibits this. There are two inhibins. Inhibin A reaches its peak in the mid-luteal phase. Inhibin B reaches a peak in the early- to mid-follicular phase, and a reaches a second peak at the onset of ovulation" [1].

The ovarian follicles and corpus luteum secrete progesterone, androstenedione, estrogens (mostly estradiol) and inhibin. Progesterone is also known as P4 (pregn-4-ene-3,20-dione [1]. "It is a steroid hormone that is involved in the menstrual cycle, pregnancy and embryogenesis. It is a member of the class of hormones called progestogens. It is the major naturally occurring progestogen. It inhibits the enzyme momoamine oxidase (MAO), which catalyzes the oxidative breakdown of the neurotransmitter serotonin. It also enhances the function of the serotonin receptor in the brain. Progesterone also plays many roles in fetus development. It converts the endometrium to its secretory stage to prepare the uterus for implantation. At the same time it affects the vaginal epithelium and cervical mucus, making it thick and impenetrable to sperm. If pregnancy does not occur, progesterone concentrations will decrease, leading to menstruation. During implantation and gestation, progesterone decreases the maternal immune network so that the pregnancy will be accepted. It also decreases the contractility of the smooth muscle in the uterus and inhibits lactation during pregnancy. The concentration of progesterone decreases after the baby is born, thus triggering milk production" [1].

Progesterone, like pregnenolone and dehydroepiandrosterone are neurosteroids. It can be biosynthesized in the CNS and is as a precursor to another major neurosteroid, allopregnanolone [1]. "Neurosteroids affect synaptic function, are neuroprotective, and affect myelination. They may also improve memory and cognitive ability. Its neurosteroid effects work by inhibiting the glycogen synthase kinase-3 beta (GSK-3 beta) pathway" [1]. GSK-3 is a serine/threonine protein that catalyzes the phosphorylation of glycogen synthase, thus reducing its activity. It is involved in type-2 diabetes, Alzheimer's disease, inflammation, cancer and bipolar affective disorder. Progesterone also affects the regulation of genes that code for proteins that induce apoptosis.

Androstenedione (also known as 4-androstenedione and 17-ketoestosterone) is a steroid hormone [1]. "It is produced as an intermediate in the biosynthesis of testosterone and the estrogens estrone and estradiol. In females, androstenedione is released into the blood by theca cells to provide a substrate for producing estrogen in granulosa cells" [1]. Also, androstenedione can increase serum testosterone levels over an eight-hour period in men who take it as a single oral dose of 300 mg per day [23]. Because androstenedione is partly converted to estrogens, people taking this as a dietary supplement may have estrogenic side-effects [1]. "In addition to its role as a precursor to testosterone, androstenedione also has androgenic properties, acting as a weak partial agonist of the androgen receptor. However, in the presence of full agonists like testosterone or dihydrotestosterone (DHT), it is antagonistic, due to its lower intrinsic activity. So, it can behave more like an antiandrogen" [1].

THE HEART AND SKELETAL MUSCLES

The heart and skeletal muscles are also endocrine organs [24 - 28]. They secrete cardiomyocytes and myocytes [24 - 28]. The atrium of the heart produces atrial natriuretic factor (ANF), brain natriuretic peptide (BNP) and C-type natriuretic peptide (CNP) that are collectively called cardiac natriuretic peptides (cNPs) [24]. "Even though BNP was first found in porcine brains, it is now known that it is expressed at higher levels in the heart. The cNPs help maintain the volume of the extracellular fluid and proper blood pressure. ANF suppresses the sympathetic outflow of the heart. Both ANF and BNP reduce vascular smooth muscle tone and peripheral vascular resistance and also decrease salt and water appetite, thus leading to a reduction in extracellular fluid volume. ANF also inhibits the secretion of vasopressin from the pituitary gland, thus decreasing water reabsorption in the kidneys. ANF and BNP exert their effects mostly through a natriuretic peptide clearance receptor (NPR-A). It is a guanyl cyclase-coupled receptor. CNP signaling occurs mostly though an NPR-B receptor. When the peptide ligands bind to their receptors, cGMP is biosynthesized and some of it enters the circulation and affects several enzymes and ion channels. This can lead to vasorelaxing effects, inhibition of reabsorption of Na^+ in the kidneys, steroidogenesis in the adrenal glands and phototransduction. There is also an NPR-C receptor that helps to remove or clear peptides and other biopolymers" [24].

All three cNPs have anti-growth and anti-proliferative properties [24]. "BNP has antifibrotic effects. It inhibits fibrosis that is induced by cGMP-dependent protein kinase and the MEK/ERK pathway. ANF antagonizes vascular smooth muscle cell growth and promotes endothelial cell function in atherosclerosis. It also inhibits the proliferation of visceral preadipocytes. The biosynthesis and secretion of BNP and ANF increase during hemodynamic overload, which occurs in chronic arterial hypertension and congestive heart failure. All three cNPs antagonize the renin–angiotensin–aldosterone system. They also induce the hydrolysis of lipids in fat cells and pre-adipocytes" [24].

The concentrations of cNPs in the blood increase during "heart failure, myocardial infarction, hypertension, left ventricular hypertrophy and pulmonary hypertension" [24]. "Since the cNPs have potent diuretic, natriuretic and hypotensive activities, they can be used to treat several cardiovascular disorders. For example, carperitide, a recombinant form of human ANF, improved congestive heart failure and acute myocardial infarction, and also protected the kidneys from nephropathy" [24].

The CNP peptide is expressed at about the same level at all times (is constitutive) and acts as a vasodilator on blood vessels while inhibiting the growth of vascular smooth muscle cells [24]. The concentration of CNP increases in patients with congestive heart failure [24].

At the same time, skeletal muscles are part of the endocrine system [25]. "Cytokines that are released by skeletal muscles are called myokines. They are released when muscles contract and then affect other tissues. Moreover, exercise causes an increase in cytokine levels, which can affect the immune system. Interleukin-6 (IL-6) is a major myokine that is released when skeletal muscles contract. Myokines like IL-6 not only help the immune system, but also help mediate changes in metabolism that occur during exercise" [25].

Skeletal muscles also produce other myokines, including members of the TNF-β superfamily of transforming growth factors that induce signaling molecules that help regulate cell growth, regeneration, differentiation, transformation, and death [26]. "One member of this superfamily, myostatin, can inhibit the repairing of damaged muscle cells. Myostatin can limit muscle mass and may become a therapeutic target. There are also activins and inhibins in the TNF-β superfamily "that help regulate the secretion of follicle stimulating hormone (FSH) from the anterior pituitary. Activins and inhibins are important in reproduction. Their concentrations in the blood can change and serve as biomarkers for ectopic pregnancy. Activin is similar to myostatin in that it inhibits muscle growth. So, activin receptors are potential therapeutic targets for neuromuscular disorders" [26].

Skeletal muscles also produce a glycoprotein called follistatin, which is not in the

TNF-β superfamily [26]. "Follistatin inhibits the synthesis and secretion of the FSH from the pituitary gland. It also inhibits the activities of activin and myostatin in the activin-myostatin-follistatin (AMF) system. Skeletal muscles produce another protein called irisin that also inhibits activin and myostatin. Irisin is made by hydrolyzing the extracellular portion of fibronectin type III domain–containing protein (FNDC) 5. Irisin is secreted into the blood" [26].

Bone morphogenic protein (BMP) is also produced by skeletal muscles [26]. "It is in the TGF-β family of proteins that are involved in skeletogenesis and differentiation as well as embryonic development. C1q/TNF-related proteins (CTRPs) are also secreted by skeletal muscles. One of them, CTRP15, is also known as myonectin. Its concentration increases after exercise, but like BMP, its physiological role is not well known. Finally, skeletal muscles also produce and secrete brain-derived neurotrophic factor (BDNF), which helps regenerate and induce myogenesis" [26].

BONES AND SKELETAL SYSTEM

Even though the skeletal system was once thought to be just connective tissue that also produced blood cells and helped regulate Ca^{2+} homeostasis, it is now known to be part of the endocrine system [29 - 39]. Bones undergo continual autopoiesis. Osteoclasts remove bone, while osteoblasts re-make it [33]. "Also, osteocytes are the last step in the osteoblast lineage. They help regulate the processes of bone remodeling and mineralization" [33]. That is, the bone marrow produces hematopoietic stem cells (HSCs) that can differentiate into myeloid cells "(monocytes, macrophages, neutrophils, basophils, eosinophils, erythrocytes/red blood cells, megakaryocytes/platelets and dendritic cells) and lymphoid cells (T-cells, B-cells and natural killer cells/NK cells) [30]. The roles of myeloid and lymphoid cells in the immune system will be discussed in more detail in Chapter 8. In addition, bones interact with adipose tissue and muscles to regulate bone density and body fat [31 - 33]. Bone marrow also produces adipocytes [32].

It was once thought that obesity can protect against osteoporosis, but recent epidemiological and clinical studies have shown the opposite [31]. "Not only obesity, but also reduces muscle mass (which often occur simultaneously in the elderly) have emerged as risk factors for osteoporosis. On the other hand, exercise increases muscle and bone mass, while decreasing fat mass. Adipose tissue secretes several pro-inflammatory cytokines such as IL-6 and TNFα that affect osteoclasts. Bones produce osteocalcin and osteopontin that help to control body weight and glucose homeostatis. So, there seems to be some feedback between the skeleton and other endocrine organs. Also, myostatin affects fat and bones. Adipokines affect the liver and skeletal muscle and affect insulin sensitivity. Adipose tissue also produces aromatase enzymes which catalyze reactions that are in the biosynthetic pathway for estrogen. That is important because estrogens have important roles in bone homeostasis. In post-menopausal women, circulating estrogen levels decrease, which leads to an increase in the number of adipocytes and decrease in osteoblasts. Ligands that bind to another nuclear hormone receptor, PPARγ, induce adipogenesis in the bone marrow, as well as inhibit osteogenesis. Parts of the Wnt signaling system can inhibit adipogenesis while promoting osteogenesis. Members of the epidermal growth factor family, such as Pref-1, or pre-adipocyte factor-1, can inhibit both adipogenesis and osteogenesis" [31].

Adipose tissue contains not only white and brown adipocytes, but also cells known as either beige fat or brite fat (from brown and white) [31]. "These beige fat cells emerge when visceral white adipose cells (VATs) transdifferentiate. VAT cells differentiate further into heat-producing brown adipose tissue (BAT) that has anti-obesity and anti-diabetic effects. These interactions may be due to the fact that adipocytes, osteoblasts and myoblasts originate from a common progenitor, pluripotent mesenchymal stem cells, or MSCs. There is an inverse relationship between MSC differentiation into either adipocytes or osteoblasts. This is mediated by cross-talk between signaling pathways that are activated by PPARγ, steroid receptors and other cytokines. Also, environmental factors and lifestyles affect adipogenesis in the bone marrow despite osteoblastogenesis. This indicates that bone marrow MSCs can undergo several types of differentiation. They can dedifferentiate and transdifferentiate in response to changes in the micro-environment. Finally, myostatin is an important link between fat, bone and muscle homeostasis. It is an important regulator of MSC proliferation and differentiation. It is expressed in the early steps in healing bone fractures. It has an affect on the proliferation and differentiation of osteoprogenitor cells, while affecting muscle mass and bone strength" [31]. MSCs can modify and influence almost all the cells of the innate and adaptive immune systems [30]. MSCs derived from the bone marrow are also important in organ repair and may be used eventually to produce new organs for people who now need to wait for an organ transplant [30].

So, it has been proposed that there is a bone-adipose axis [32]. "The number of mature osteoblasts and adipocytes in the bone marrow depends partly on the way that MSCs differentiate. So, things that increase bone marrow adipogenesis inhibit osteoblastogenesis, resulting in decreased osteoblast numbers, diminished bone formation and, potentially, inadequate bone mass and osteoporosis. In addition, mature osteoblasts and adipocytes secrete factors (including PPAR-γ2, Wnt, IGF-1, GH, FGF-2, estrogen, the glycoprotein 130 signaling cytokines, vitamin D and glucocorticoids) which can change the cell fate of MSCs and function of each other. That is, cytokines in the IL-6 family exert their effects through transmembrane receptor complexes comprised of ligand-specific receptors and the shared transmembrane receptor glycoprotein 130, or GP130. These cytokines are pro-osteogenic and anti-adipogenic. PPAR-γ2 regulates adipocyte differentiation and lipid storage in mature adipocytes. Wnts are glycoproteins that regulate many processes that occur during development. Wnt signaling plays an important role in bone formation. GH and IGF-1 correlate with bone marrow adiposity. IGF-1 is produced not only in the liver, but also in osteoblasts, independent of control by GH (which controls IGF-1 biosynthesis in the liver). As the level of IGF-1 decreases in the elderly, it can cause bone loss and osteoporosis. Also, a loss of estrogen can increase the amount of adipose cells in the bone marrow. FGF-2 is found in marrow stromal cells, osteoblasts and osteocytes and is stored in the extracellular matrix of bone. It is pro-osteogenic and anti-adipogeneic. Vitamin D promotes osteoblast differentiation and inhibits bone marrow adipogenesis. Finally, mature osteoblasts and adipocytes express some of the same genes, but the proteins that they code for have different functions in these different cells" [32].

Bone is also an energy-regulating endocrine organ [32] that also regulates mineral metabolism [33]. "That is, osteoblasts secrete the hormone osteocalcin, which helps to regulate energy metabolism. It does this by "regulating insulin secretion by pancreatic β-cells, testosterone secretion by testicular Leydig cells, fatty acid metabolism in the liver, and the insulin sensitivity of muscle and fat" [33]. There is an osteocalcin-bone-pancreas-testis axis that helps to regulate energy metabolism and the production of sex hormones [33, 36]. Osteocalcin helps to regulate male fertility [36]. "It has carboxyglutamic acid residues that can be carboxylated. When it is under-carboxylated, it is biologically active and can stimulate pancreatic β-cells to secrete insulin, while promoting insulin sensitivity in other organs and stimulates the biosynthesis of testosterone in the testis" [36].

Both osteoblasts and osteocytes secrete FGF23 [33]. It affects the kidneys, parathyroid gland, choroid plexus, and pituitary gland [33]. FGF23 helps to regulate the concentration of phosphate (HPO_4^{2-} and $H_2PO_4^-$) and 1,25-dihydroxyvitamin D_2 in the blood [35].

The bone marrow also produces early outgrowth cells (EOCs) that produce soluble hormones that have antifibrotic effects that can reduce progressive organ fibrosis in chronic kidney disease (CKD) and heart failure [38]. "That is, fibrotic injury occurs in many diseases, including CKD and the deleterious effects it can have on the heart. The EOCs secrete factors that inhibit pro-inflammatory TGF-β and collagen production. This means that these factors may be useful in cell-free treatments of heart and kidney damage in CKD" [38].

Finally, it has been shown that phytoestrogens, saponins and flavonoids can affect the osteogenic and adipogenic pathways of differentiation [39]. "The shift towards osteogenesis occurs by inhibiting PPAR-γ and C/EBPα transcription factors. So, phytoestrogens, saponins and flavonoids may help to prevent or minimize bone loss and other metabolic diseases" [39].

CALCIUM REGULATION

The endocrine system is also important in regulating the concentration of Ca^{2+} [1]. As mentioned previously, the parathyroid gland secretes the parathyroid hormone (PTH), which stimulates Ca^{2+} release from bones, stimulates osteoclasts and Ca^{2+} reabsorption in the kidneys. It also stimulates the production of activated vitamin D in the kidneys. PTH also stimulates the release of phosphate (HPO_4^{2-} and $H_2PO_4^-$) from bones and inhibits its reabsorption in the kidneys.

Calcium regulation also occurs in the skin, which secretes the prehormone calcidiol (25-hydroxyvitamin D_3), the inactive form of vitamin D [1]. However, the concentration of calcidiol is the best test for total vitamin D concentration. Also, elevated concentrations of calcidiol are associated with increasing fractional absorption of calcium from the gut

TARGETS

There are several targets for hormones. Atrial-natriuretic peptide (ANP) and brain natriuretic peptide (BNP) target cardiomyocytes in the heart. ANP is a powerful vasodilator [1]. "It reduces blood pressure by reducing systemic vascular resistance, thus reducing the amount of water, Na^+ and fats in the blood. To a lesser degree, BNP also reduces blood pressure in the same way. ANP is produced, stored, and released mainly by atrial cardiomyocytes. It is also synthesized the ventricles, brain, suprarenal glands, and renal glands. It is released in response to atrial stretch, hypervolemia (fluid overload), exercise and calorie restriction" [1]. ANP is also produced in the placenta of pregnant women by hydrolysis of pro-atrial-natriuretic peptide, catalyzed by the protease called corin [40]. "When the gene that codes for corin (*CORIN*) is mutated, it can cause preeclampsia. BNP is also known as B-type natriuretic peptide and ventricular natriuretic peptide. It is secreted by the ventricles in the heart in response to excessive stretching of cardiomyocytes. It binds to and activates the atrial natriuretic factor receptors NPRA and NPRB, like ANP, but with only about 10% of its affinity. However, the biological half-life of BNP is about twice that of ANP. Like ANP, it decreases systemic vascular resistance, central venous pressure and natriuresis (the process of excreting Na^+ in the urine)" [40].

The bone marrow is also a target of hormones. Thrombopoietin is released from the liver and kidneys. It stimulates megakaryocytes to produce platelets. Adipose tissue is also a target of hormones. Leptin and estrogens (mostly estrone) are secreted by adipocytes. Leptin decreases the appetite and increases metabolism. Finally, skeletal muscles may also be a target for vitamin D [41]. There is a vitamin D receptor in skeletal muscle cells. In older people, vitamin D is positively associated with muscle strength [40]. Clinical trials found that vitamin D supplementation in older adults with low vitamin D status improves muscle performance and reduced falls [41].

The endocrine system and immune network also interact with each other. Cortisol is immunosuppressive, while dopamine has immunomodulatory functions [1]. "On the other hand, inflammatory cytokines from the immune network activate the hypothalamus-pituitary axis. Also, cytokines stimulate the release of hepcidin from the liver, which can lead to anemia of chronic disease, a form of anemia that occurs in chronic illnesses. Hepcidin is a peptide hormone produced by the liver. It is the main regulator of iron homeostasis" [1].

DIFFUSE NEURO-ENDOCRINE SYSTEM

There is also a diffuse neuro-endocrine system that is made up of cells that secrete hormones. "They are found in the epithelium of all organs" [1].

MAJOR ENDOCRINE SYSTEMS

The major endocrine systems are the TRH-TSH-Y3/T4, the GnRH-LH/FSH-sex hormones, the CRH-ACTH-cortisol, the renin-angiotensin-aldosterone system and the leptin *vs* insulin system [1]. "The TRH-TSH-Y3/T4 system is also called the hypothalamic–pituitary–thyroid axis (HPT axis) is responsible for regulating metabolism. The hypothalamus senses the concentrations of T3 and T4 and responds by releasing TRH, which stimulates the pituitary to produce TSH. Then, TSH stimulates the thyroid to produce thyroid hormone until its concentration in the blood returns to normal. The GnRH-LH/FSH-sex hormones system is also called the hypothalamic–pituitary–gonadal axis and HPG axis. It is a critical part in the development and regulation of many of the body's systems, including the reproductive and immune systems. Fluctuations in the concentrations of these hormones cause changes in the amounts produced by each gland. These fluctuations have several widespread and local effects on the body. It also helps to control development, reproduction, and aging. The hypothalamus produces GnRH. The anterior portion of the pituitary gland produces LH and FSH. The gonads produce estrogen and testosterone. The CRH-ACTH-cortisol system is also called the hypothalamic-pituitary-adrenal axis or HPA axis. It controls reactions to stress and regulates digestion, the immune network, mood, emotions and sexuality, along with energy storage and expenditure. It is the common mechanism for interactions among glands, hormones, and parts of the midbrain that mediate the general adaptation syndrome (GAS), the body's short-term and long-term reactions to stress. The renin-angiotensin -aldosterone system regulates blood pressure and fluid (water) balance. The leptin *vs* insulin system is responsible for energy homeostasis" [1].

DISEASES

Diseases of the endocrine system include diabetes, thyroid disease and obesity [1]. "Endocrine diseases are characterized by dysregulated release of hormones in aproductive pituitary adenoma. Inappropriate signaling can cause hypothyroidism. Lack of a gland can cause type-1 diabetes and diminished erythropoiesis in chronic renal failure. Structural enlargement in a critical site such as the thyroid can cause toxic multinodal goiter. Hormones (such as estrogen) can also cause cancer" [1].

From the medicinal chemistry point of view, there are many drugs that are used to treat endocrine disorders. As discussed previously in this chapter, GH is prescribed for growth disorders in children and low GH in adults. Physicians are free to prescribe GH for off-label uses. Recently, it has been prescribed in the USA for elderly people as an anti-aging treatment. This is based partly on a study reported in 1990 in which 12 men over 60 were treated with GH [12]. Also, there is a recombinant human TSH α (thyroid-stimulating hormone or thyrotropin-α, or rhTSH) that is used to treat thyroid cancer [15]. Calcitonin that has been extracted from salmon and purified is used to treat postmenopausal osteoporosis, hypercalcemia, Paget's disease, bone metastases and phantom limb pain [1]. "It is rapidly absorbed an eliminated. It takes only one hour to reach the peak concentration in blood plasma. It is mostly metabolized in the liver by proteolysis. It can also be used as a diagnostic marker for medullary thyroid cancer if the concentration exceeds 5 and 12 pg/mL in females and males, respectively; 40 pg/mL in children under 6 months and 15 pg/mL in children who are 6 months to 3 years old" [1]. Calcitrol is sold under the names Rocaltrol, Calcijex, Decostriol and Vectical. Calcitrol can be prescribed to treat hypocalcemia, hypoparathyroidism, osteomalacia, rickets, renal osteodystrophy, and osteoporosis. It can also be used for people who are undergoing dialysis, due to kidney disease. It can also prevent corticosteroid-induced osteoporosis. The main adverse side effect is to cause hypercalcemia.

Oral contraceptives contain an estrogen-like drug, ethinyl estradiol. It is combined with any of a number of progestin derivatives. Progestin-only pills are available for women who are breast feeding and those who are not able to tolerate estrogens. There is also an emergency contraceptive to prevent pregnancy after unprotected intercourse, called Plan B. One form consists of two pills, each containing 0.75 mg of levonorgestrel. Mifepristone is a synthetic steroid that is used for the termination of pregnancy up to the 49th day of gestation. It has been tested as a morning after pill to prevent pregnancy when taken within 12 hours of unprotected intercourse. It is also used in combination with another drug called Gemeprost to terminate pregnancies between weeks 13 and 24. In the USA, levonorgestrel is the preferred emergency contraceptive. It can prevent a pregnancy up to 72 hrs after unprotected sex or contraceptive failure. Oxytocin and prostaglandin E_2 are also available to induce labor by stimulating uterine contractions. Some women take hormone replacement therapy, in which a low dose of one or more estrogens (conjugated equine estrogens) and a progestin are given. Another approach is to increase the consumption of soybeans and foods made from soybeans (such as soy milk). These will be discussed further in the next chapter.

KEY POINTS

1. **The endocrine system is now known to contain not just the adrenal thyroid and pituitary glands, parts of the brain and reproductive system, but also the kidneys, adipose tissue and alimentary tract. Cells in this system secrete hormones into the bloodstream that affect physiological and behavioral function and activities.**

2. **The hypothalamus connects the nervous and endocrine systems to each other through the pituitary gland, or hypophysis. It helps control body temperature, hunger, parenting and attachment behavior, thirst, fatigue, sleep, circadian rhythms and other activities of the autonomic nervous system [1]. Darkness causes the pineal gland to secrete N-acetyl-5-methoxytryptamine, which is better known as melatonin. Melatonin is part of the system that regulates the sleep-**

wake cycle [1].

3. The anterior lobe of the pituitary gland "secretes growth hormone (GH), beta-endorphin, luteinizing hormone (LH), follicle stimulating hormone (FSH), melanocyte stimulating hormone (MSH), adrenocorticotropic hormone (ACTH), prolactin, thyroid stimulating hormone (TSH) and prolactin (PRL). The posterior lobe stores vasopressin, also known as antidiuretic hormone (ADH) and oxytocin (OXT) [1].

4. The thyroid gland helps control how fast a body uses energy and makes proteins as well as influencing the sensitivity of the body to other hormones. It does this by producing T_3 and T_4, which are made from tyrosine and iodine.

5. The stomach, duodenum, liver, pancreas and kidneys all secrete hormones. The kidneys secrete renin, erythropoietin, calcitrol and thrombopoietin. They regulate pH, electrolytes and blood pressure.

6. The adrenal glands produce hormones in response to stress by synthesizing corticosteroids such as cortisol and catecholamines, such as adrenaline (epinephrine) and noradrenaline (norepinephrine). They also produce androgens in their innermost cortical layer. The adrenal glands affect kidney function by secreting aldosterone [1].

7. The testes, ovarian follicle and corpus luteum are in the endocrine system, as are the placenta and uterus when a woman is pregnant. The testes secrete androgens (mostly testosterone), estradiol and inhibin. They stimulate or control the development and maintenance of male characteristics by binding to androgen receptors. This includes the activity of the male sex organs and development of male secondary sex characteristics. Androgens are also the original anabolic steroids and the precursor of all estrogens [1].

8. The endocrine system also regulates the concentration of Ca^{2+}. The parathyroid gland secretes the parathyroid hormone (PTH), which stimulates Ca^{2+} release from bones, stimulates osteoclasts and Ca^{2+} reabsorption in the kidneys. Calcium regulation also occurs in the skin, which secretes the prehormone calcidiol (25-hydroxyvitamin D_3), the inactive form of vitamin D [1].

9. **There are several targets for hormones. Atrial-natriuretic peptide (ANP) and brain natriuretic peptide (BNP) target cardiomyocytes in the heart [1]. ANP is a powerful vasodilator. It reduces blood pressure by reducing systemic vascular resistance, thus reducing the amount of water, Na^+ and fats in the blood [1]. To a lesser degree, BNP also reduces blood pressure in the same way. ANP is produced, stored, and released mainly by atrial cardiomyocytes. It is also synthesized the ventricles, brain, suprarenal glands, and renal glands. It is released in response to atrial stretch, hypervolemia (fluid overload), exercise and calorie restriction.**

10. **The major endocrine systems are the TRH-TSH-Y3/T4, the GnRH-LH/FSH-sex hormones, the CRH-ACTH-cortisol, the renin-angiotensin-aldosterone system and the leptin *vs* insulin system [1]. The TRH-TSH-Y3/T4 system is also called the hypothalamic–pituitary–thyroid axis (HPT axis) is responsible for regulating metabolism [1].**

ABBREVIATIONS

GH	=	Growth Hormone
LH	=	Luteinizing Hormone
FSH	=	Follicle Stimulating Hormone
MSH	=	Melanocyte Stimulating Hormone
ACTH	=	Adrenocorticotropic Hormone
TSH	=	Thyroid Stimulating Hormone
PRL	=	Prolactin
ADH	=	Antidiuretic Hormone
OXT	=	Oxytocin
PTH	=	Parathyroid Hormone
CRH	=	Corticotropin-Releasing Hormone
TRH	=	Thyroid Releasing Hormone
TSH	=	Thyroid Stimulating Hormone
GnRH	=	Gonadotropin-Releasing Hormone
GH	=	Growth Hormone
GPCR	=	G-Protein Coupled Receptor
PACAP	=	Pituitary Adenylate Cyclase-Activating Protein
cAMP	=	Cyclic Adenosine Monophosphate
SNP	=	Single Nucleotide Polymorphism

DSPD	=	Delayed Sleep Phase Disorder
mapk	=	Mitogen Activated Protein Kinase
ERK	=	Extracellular Signal Regulated Kinase
STAT	=	Signal Transducer and Activator of Transcription
NACB	=	National Academy of Clinical Biochemistry
T_3	=	Triiodothyronine
T_4	=	T4 Tetraiodothyronine
T_G	=	Thyroglobulin
HRE	=	Hormone Responsive Elements
RXR	=	Retinoid X Receptor
ATPase	=	Adenosine Triphosphatase
NPY	=	Neuropeptide Y
CRF	=	Corticotropin-Releasing Factor
SCN	=	Suprachiasmatic Nucleus
RAS	=	Renin-Angiotensin System
ACE	=	Angiotensin Converting Enzyme
ARBs	=	Angiotensin II Receptor Blockers
RANKL	=	Receptor Activator of Nuclear Factor Kappa-B Ligand
VDR	=	Vitamin D Receptor
CYP	=	Cytochrome P450
DHT	=	Dihydrotestosterone
ER	=	Estrogen Receptor
SERM	=	Selective Estrogen Receptor Modulator
TGF-β	=	Tumor Growth Factor-β
CNS	=	Central Nervous System
GSK-3 BETA	=	Glycogen Synthase Kinase-3 Beta
DHT	=	Dihydrotestosterone
ANF	=	Atrial Natriuretic Factor
BNP	=	Brain Natriuretic Peptide
CNP	=	C-Type Natriuretic Peptide
cNP	=	Cardiac Natriuretic Peptide
cGMP	=	Cyclic Guanosine Triphosphate
IL-6	=	Interleukin-6
TNF-β	=	Tumor Necrosis Factor Beta

AMF	=	Activin-Myostatin-Follistatin
FDNC	=	Fibronectin Type III Domain–Containing Protein
BNP	=	Bone Morphogenic Protein
CTRPs	=	C1q/TNF-Related Proteins
NK cells	=	Natural Killer Cells
PPARγ	=	Peroxisome Proliferator-Activated Receptor Gamma
VATs	=	Visceral White Adipose Cells
BAT	=	Brown Adipose Tissue
MSCs	=	Mesenchymal Stem Cells
FGF-2	=	Fibroblast Growth Factor
EOCs	=	Early Outgrowth Cells
CKD	=	Chronic Kidney Disease

REFERENCES

[1] Welch K, Martini FH, Timmons MJ, Tallitsch RB. Human Anatomy. Upper Saddle River, NJ: Pearson 2014.

[2] Gu P-Y, *et al.* Relevance of plasma obestatin and early arteriosclerosis in patients with type 2 diabetes mellitus. J Diab Res 2013; 2013: 563919.

[3] Ressler KJ, Mercer KB, Bradley B, *et al.* Post-traumatic stress disorder is associated with PACAP and the PAC1 receptor. Nature 2011; 470(7335): 492-7.
[http://dx.doi.org/10.1038/nature09856] [PMID: 21350482]

[4] Reiter RJ. Pineal melatonin: cell biology of its synthesis and of its physiological interactions. Endocr Rev 1991; 12(2): 151-80.
[http://dx.doi.org/10.1210/edrv-12-2-151] [PMID: 1649044]

[5] Ardura J, Gutierrez R, Andres J, Agapito T. Emergence and evolution of the circadian rhythm of melatonin in children. Horm Res 2003; 59(2): 66-72.
[http://dx.doi.org/10.1159/000068571] [PMID: 12589109]

[6] Sack RL, Lewy AJ, Erb DL, Vollmer WM, Singer CM. Human melatonin production decreases with age. J Pineal Res 1986; 3(4): 379-88.
[http://dx.doi.org/10.1111/j.1600-079X.1986.tb00760.x] [PMID: 3783419]

[7] Peschke E, Mühlbauer E. New evidence for a role of melatonin in glucose regulation. Best Pract Res Clin Endocrinol Metab 2010; 24(5): 829-41.
[http://dx.doi.org/10.1016/j.beem.2010.09.001] [PMID: 21112029]

[8] Lewis A. Melatonin and the Biological Clock. New York: McGraw-Hill 1999; p. 23.

[9] Wade AG, Ford I, Crawford G, *et al.* Efficacy of prolonged release melatonin in insomnia patients aged 55-80 years: quality of sleep and next-day alertness outcomes. Curr Med Res Opin 2007; 23(10): 2597-605.

[http://dx.doi.org/10.1185/030079907X233098] [PMID: 17875243]

[10] Buscemi N, Vandermeer B, Pandya R, *et al.* Melatonin for treatment of sleep disorders. Evid Rep Technol Assess (Summ) 2004; 108(108): 1-7. [PMID: 15635761]

[11] Kohler M, Püschel K, Sakharov D, Tonevitskiy A, Schänzer W, Thevis M. Detection of recombinant growth hormone in human plasma by a 2-D PAGE method. Electrophoresis 2008; 29(22): 4495-502. [http://dx.doi.org/10.1002/elps.200800221] [PMID: 19003817]

[12] Rudman D, Feller AG, Nagraj HS, *et al.* Effects of human growth hormone in men over 60 years old. N Engl J Med 1990; 323(1): 1-6. [http://dx.doi.org/10.1056/NEJM199007053230101] [PMID: 2355952]

[13] Kuczynski A. Anti-aging potion or poison. NY Times 1998 April 12;

[14] Liu H, Bravata DM, Olkin I, *et al.* Systematic review: the safety and efficacy of growth hormone in the healthy elderly. Ann Intern Med 2007; 146(2): 104-15. [http://dx.doi.org/10.7326/0003-4819-146-2-200701160-00005] [PMID: 17227934]

[15] Tuttle RM, Brokhin M, Omry G, *et al.* Recombinant human TSH-assisted radioactive iodine remnant ablation achieves short-term clinical recurrence rates similar to those of traditional thyroid hormone withdrawal. J Nucl Med 2008; 49(5): 764-70. [http://dx.doi.org/10.2967/jnumed.107.049072] [PMID: 18413378]

[16] Birkholz D, Davis SF. Monitoring ENT Procedures. Principles of Neurophysiological Assessment, Mapping, and Monitoring. Springer New York 2014; pp. 207-13. [http://dx.doi.org/10.1007/978-1-4614-8942-9_16]

[17] Colmers WF, El Bahn B. Neuropeptide Y and epilepsy. Epilepsy Currents/American Epilepsy Society 2003; 2(3): 53-8.

[18] Kuo LE, Kitlinska JB, Tilan JU, *et al.* Neuropeptide Y acts directly in the periphery on fat tissue and mediates stress-induced obesity and metabolic syndrome. Nat Med 2007; 13(7): 803-11. [http://dx.doi.org/10.1038/nm1611] [PMID: 17603492]

[19] Yehuda R, Brand S, Yang RK. Plasma neuropeptide Y concentrations in combat exposed veterans: relationship to trauma exposure, recovery from PTSD, and coping. Biol Psychiatry 2006; 59(7): 660-3. [http://dx.doi.org/10.1016/j.biopsych.2005.08.027] [PMID: 16325152]

[20] Bookout AL, de Groot MH, Owen BM, *et al.* FGF21 regulates metabolism and circadian behavior by acting on the nervous system. Nat Med 2013; 19(9): 1147-52. [http://dx.doi.org/10.1038/nm.3249] [PMID: 23933984]

[21] Owen BM, Bookout AL, Ding X, *et al.* FGF21 contributes to neuroendocrine control of female reproduction. Nat Med 2013; 19(9): 1153-6. [http://dx.doi.org/10.1038/nm.3250] [PMID: 23933983]

[22] Douma SL, Husband C, O'Donnell ME, Barwin BN, Woodend AK. Estrogen-related mood disorders: reproductive life cycle factors. ANS Adv Nurs Sci 2005; 28(4): 364-75. [http://dx.doi.org/10.1097/00012272-200510000-00008] [PMID: 16292022]

[23] Leder BZ, Longcope C, Catlin DH, Ahrens B, Schoenfeld DA, Finkelstein JS. Oral androstenedione administration and serum testosterone concentrations in young men. JAMA 2000; 283(6): 779-82.

[http://dx.doi.org/10.1001/jama.283.6.779] [PMID: 10683057]

[24] Ogawa T, de Bold AJ. The heart as an endocrine organ. Endocr Connect 2014; 3(2): R31-44. [http://dx.doi.org/10.1530/EC-14-0012] [PMID: 24562677]

[25] Pedersen BK, Febbraio MA. Muscle as an endocrine organ: focus on muscle-derived interleukin-6. Physiol Rev 2008; 88(4): 1379-406. [http://dx.doi.org/10.1152/physrev.90100.2007] [PMID: 18923185]

[26] Iizuka K, Machida T, Hirafuji M. Skeletal muscle is an endocrine organ. J Pharmacol Sci 2014; 125(2): 125-31. [http://dx.doi.org/10.1254/jphs.14R02CP] [PMID: 24859778]

[27] Pedersen BK. Muscular interleukin-6 and its role as an energy sensor. Med Sci Sports Exerc 2012; 44(3): 392-6. [http://dx.doi.org/10.1249/MSS.0b013e31822f94ac] [PMID: 21799452]

[28] Jähn K, *et al.* Skeletal muscle secreted factors prevent glucocorticoid-induced osteocyte apoptosis through activation of B-catenin. Eur Cells mater 2012; 24: 197-210.

[29] Datta NS. Muscle-bone and fat-bone interactions in regulating bone mass: do PTH and PTHrP play any role? Endocrine 2014; 47(2): 389-400. [http://dx.doi.org/10.1007/s12020-014-0273-3] [PMID: 24802058]

[30] Li M, Ikehara S. Bone-marrow-derived mesenchymal stem cells for organ repair. Sten Cells Int . 2013; p. 8. Article ID 132642

[31] Migliaccio S, Greco EA, Wannenes F, Donini LM, Lenzi A. Adipose, bone and muscle tissues as new endocrine organs: role of reciprocal regulation for osteoporosis and obesity development. Horm Mol Biol Clin Investig 2014; 17(1): 39-51. [http://dx.doi.org/10.1515/hmbci-2013-0070] [PMID: 25372729]

[32] Sadie-Van Gijsen H, Crowther NJ, Hough FS, Ferris WF. The interrelationship between bone and fat: from cellular see-saw to endocrine reciprocity. Cell Mol Life Sci 2013; 70(13): 2331-49. [http://dx.doi.org/10.1007/s00018-012-1211-2] [PMID: 23178849]

[33] Pi M, Quarles LD. Novel bone endocrine networks integrating mineral and energy metabolism. Curr Osteoporos Rep 2013; 11(4): 391-9. [http://dx.doi.org/10.1007/s11914-013-0178-8] [PMID: 24193547]

[34] Shao J, Wang Z, Yang T, Ying H, Zhang Y, Liu S. Bone regulates glucose metabolism as an endocrine organ through osteocalcin. Int J Endocrinol 2015; 2015 Article ID 967673 [http://dx.doi.org/10.1155/2015/967673] [PMID: 25873961]

[35] Fukumoto S, Martin TJ. Bone as an endocrine organ. Trends Endocrinol Metab 2009; 20(5): 230-6. [http://dx.doi.org/10.1016/j.tem.2009.02.001] [PMID: 19546009]

[36] Karsenty G, Oury F. Regulation of male fertility by the bone-derived hormone osteocalcin. Mol Cell Endocrinol 2014; 382(1): 521-6. [http://dx.doi.org/10.1016/j.mce.2013.10.008] [PMID: 24145129]

[37] Colaianni G, Sun L, Zaidi M, Zallone A. Oxytocin and bone. Am J Physiol Regul Integr Comp Physiol 2014; 307(8): R970-7. [http://dx.doi.org/10.1152/ajpregu.00040.2014] [PMID: 25209411]

[38] Yuen DA, Connelly KA, Zhang Y, *et al.* Early outgrowth cells release soluble endocrine antifibrotic factors that reduce progressive organ fibrosis. Stem Cells 2013; 31(11): 2408-19. [http://dx.doi.org/10.1002/stem.1502] [PMID: 23922321]

[39] Schilling T, Ebert R, Raaijmakers N, Schütze N, Jakob F. Effects of phytoestrogens and other plant-derived compounds on mesenchymal stem cells, bone maintenance and regeneration. J Steroid Biochem Mol Biol 2014; 139: 252-61. [http://dx.doi.org/10.1016/j.jsbmb.2012.12.006] [PMID: 23262262]

[40] Zhou Y, Wu Q. Role of corin and atrial natriuretic peptide in preeclampsia. Placenta 2013; 34(2): 89-94. [http://dx.doi.org/10.1016/j.placenta.2012.11.016] [PMID: 23211473]

[41] Ceglia L, Harris SS. Vitamin D and its role in skeletal muscle. Calcif Tissue Int 2013; 92(2): 151-62. [http://dx.doi.org/10.1007/s00223-012-9645-y] [PMID: 22968766]

CHAPTER 6

Reproductive Medicine, Osteoporosis

Abstract: The fertility cycle in women is controlled by a cascade of events, which are initiated by protein and steroid hormones. These hormones create signals between the hypothalamus, the pituitary gland and the ovaries. Oral contraceptives contain an estrogen-like drug, ethinyl estradiol. It is combined with any of a number of progestin derivatives. Progestin-only pills are available for women who are breast feeding and those who are not able to tolerate estrogens. There is also an emergency contraceptive to prevent pregnancy after unprotected intercourse, called Plan B. One form consists of two pills, each containing 0.75 mg of levonorgestrel. Mifepristone is a synthetic steroid that is used for the termination of pregnancy up to the 49th day of gestation. It has been tested as a morning after pill to prevent pregnancy when taken within 12 hours of unprotected intercourse. It is also used in combination with another drug called Gemeprost to terminate pregnancies between weeks 13 and 24. In the USA, levonorgestrel is the preferred emergency contraceptive. It can prevent a pregnancy up to 72 hrs after unprotected sex or contraceptive failure. Oxytocin and prostaglandin E_2 are also available to induce labor by stimulating uterine contractions. Some women take hormone replacement therapy, in which a low dose of one or more estrogens (conjugated equine estrogens) and a progestin are given. Another approach is to increase the consumption of soybeans and foods made from soybeans (such as soy milk).

Keywords: Gemeprost, Levonorgestrel, Mifepristone, Progestin, Prostaglandin E_2, Soy.

Reproductive medicine studies fertility, sexuality, reproduction and menopause. Studying these fields may lead one to embrace systems thinking, more than reductionist thinking. The blueprint of life includes more than just DNA. In the plant kingdom, the same DNA and chromosomes can make a plant male or a female, depending on the amount of light in the environment (the length of the light and dark cycle). In amphibians, the number of males and females depends on the environment. The instructions for life also include hormones and an immune system that seems to be able to predict the future. By the time we are three years

Robert E. Smith

old we are well immunized, and our immune cells are able to distinguish self from non-self. The body's white blood cells attack and destroy cells that are foreign, but they don't bother their own cells [1]. Then, as puberty begins, we start making new cells. Somehow, our immune system is able to recognize these new cells as being self, instead of non-self. Throughout life, the immune system is able to distinguish between harmful exogenous bacteria and helpful commensal bacteria. However, there are important gender differences. Women are more susceptible to autoimmune diseases, due partly to differences in gut bacteria, which help regulate sex hormone levels [1]. The composition of the gut bacteria diverges during the onset of puberty. The bacteria in males can help protect them from type-1 diabetes [1].

Also, the innate immune system produces interferon-ε (Ifn-ε), which protects the female reproductive tract from infection [2]. "It is produced constitutively in the epithelial cells. The innate immune system senses pathogens through pattern recognition receptors that induce cytokines such as type 1 interferons, which activate effector cells. All type 1 interferons protect against HSV-2 infection, but Ifn- ε is the one that protects against *Chlamydia* infection. So, it may be important in fighting this and other sexually transmitted diseases" [2]. Machines don't do these things, living autopoietic systems do.

A group of human autopoietic systems formed a not for profit organization, the American Society for Reproductive Medicine, or ASRM [3]. They publish scientific journals and other information for patients, doctors, and anyone interested in learning more. These include topics such as adoption, age and fertility, assisted reproductive technologies, ectopic pregnancy, endometriosis, infertility, laparoscopy and hysteroscopy, multiple pregnancies and birth, pelvic pain, and others [1]. The following describes some of the basics.

The sex steroids (estrogens, progestogens and androgens) play crucial roles in the development and selection of sex in the embryo. They help masculinize or feminize the brain at birth, develop secondary sexual characteristics and control reproduction and reproductive behavior in adults [1]. All three classes of sex steroids are present in males and females, but the plasma levels of estrogens and progestogens are higher in females, while androgens are higher in males.

Endogenous human estrogens include 17β-estradiol, estriol, estrone and their conjugates. Before menopause, 17β-estradiol is the predominant estrogen in the blood. The ovaries produce it and estrone, while estriol is made by 16α-hydroxylation of estrone and estradiol [1]. Estrone is elevated after menopause, where it is made in adipose tissue. As menopause begins, there is a decrease in estrogens, which can be treated with hormone replacement therapy (HRT). So, sex steroids are important at every stage of life, and survival of the species depends on them. They bind to estrogen and progesterone receptors. They interact with other proteins to affect DNA transcription.

All three estrogens exert their activities through α and/or β-estrogen receptors. The α receptor is located in the endometrium, ovary and breast cancer cells, while the β receptor is found bones, kidneys, lungs, endothelial tissues and other organs [1]. Estriol has the weakest affinity for either receptor, estrone binds mostly to the α receptor and 17β-estradiol has the highest binding affinity for both receptors [1]. Biological responses for the short-time elevation of estrogens are similar to long-term delivery of lower doses.

Progestogens like progesterone comprise the class of steroid hormones that have pro-gestational effects. They are pro-gestational. They help maintain gestation, or pregnancy. They are also precursors of other steroids. Progestrone is the major naturally occurring progestogen [1]. It is involved in embryogenesis, pregnancy and the menstrual cycle. It is produced in the adrenal glands and ovaries, by the corpus luteum. During pregnancy, it is also produced in the placenta. It is stored in adipose tissue.

Androgens are male hormones that bind to androgen receptors and control the development and maintenance of male characteristics. They are also known as anabolic steroids. They promote the enlargement of skeletal muscle cells. They are also the precursors of estrogens. The primary androgen is testosterone, but dihydrotestosterone (DHT) and androstenedione are also quite important in male development. DHT is a metabolite of testosterone. It is made in the skin and reproductive tissue. It binds to the androgen receptor tighter than testosterone. Later in life, DHT contributes to prostate growth, male balding and sebaceous gland activity. There are also 19-carbon androgens that are made in the adrenal cortex. This includes dehydroepiandrosterone (DHEA), which is the primary source of endogenous estrogens.

So, sex steroids are important at every stage of life, and survival of the species depends on them. They bind to estrogen and progesterone receptors. For example, the endogenous ligand, 17β-estradiol, binds to the estrogen receptors ERα and ERβ with similar affinities. These nuclear hormone receptors interact with other proteins to affect DNA transcription. There is also a different type of ER that is a G-protein coupled receptor (GPCR) and is responsible for some of the faster physiological responses of estrogens.

As women age, there is a dramatic decrease in production of estrogens and menopause begins. Even though no such dramatic decrease occurs in men, serum estrogens do decrease slightly, while androgens decrease even more, causing a decrease in the ratio of testosterone to estrogens in men. Male sex steroids are androgens, including testosterone and its metabolite, 5α-dihydroxytestosterone (DHT) and estradiol. Estradiol mediates hormonal feedback at the pituitary and hypothalamus in men. A small amount of estradiol is produced in men from testosterone by a reaction catalyzed by the enzyme aromatase. This occurs mostly in adipose tissue, and partly in the Leydig cells in the testes. So, low testosterone levels that occur in some elderly men can be effectively treated with the aromatase inhibitor, anastrozole, which is also used to treat breast cancer in women [1].

Testosterone and DHT exert their effects through the androgen receptor (AR), which is a ligand-dependent transcription factor [1]. "Before binding its ligand, the AR is in the cytoplasm and is associated with a complex of heat shock proteins. When the AR binds its ligand, it undergoes a conformation change, causing the heat shock proteins to dissociate from the AR, which then dimerizes, is phosphorylated and is translocated to the cell nucleus, where it can exert its effects" [1].

On the other hand, the female reproductive system is controlled by a network that includes the hypothalamus, anterior pituitary, ovaries and reproductive tract. The fertility cycle in women is controlled by a cascade of events, which are initiated by protein and steroid hormones [4]. At the beginning of the ovulatory cycle, the concentration of progesterone is low. Progesterone increases slightly before ovulation and then again after the corpus luteum is formed. These hormones create signals between the hypothalamus, the pituitary gland and the ovaries. In the first part of the cycle, the Graafian follicle in the ovary is stimulated by the follicle stimulating hormone (FSH). The FSH is made in the anterior pituitary gland. It is released, along with luteinizing hormone (LH), by pulses of the gonadotropin-releasing hormone. FSH is a glycoprotein. It is a dimer, with each monomer containing a sugar. FSH regulates the growth, development, sexual maturation and reproductive processes. In both men and women, FSH stimulates germ cells to mature.

So, the hypothalamus secretes "gonadotropin releasing hormone (GnRH), also known as luteinizing hormone releasing hormone (LHRH) [1]. "GnRH is secreted in the hypophyseal portal bloodstream in pulses. It stimulates the release of gonadotropins, FSH and LH from the anterior pituitary gland" [1]. FSH and LH regulate the production of estrogen and progesterone. FSH stimulates the initial growth and development of immature ovarian follicles. As these follicles mature, they secrete estradiol. As estradiol levels increase, FSH production decreases and LH release is stimulated, so it peaks at the midpoint of the menstrual cycle. The dominant follicle ruptures and releases its egg (ovulation). Then LH induces luteinization of the tissue that remains, which leads to the formation of the corpus luteum, where progesterone is secreted. If the egg is not fertilized, the corpus luteum degrades, the levels of progesterone and estrogen decline, and menstruation occurs. If the egg is fertilized, a glycoprotein hormone, human chorionic gonadotropin (hCG), is released from the placenta. Many early pregnancy tests are based on the detection of hCG, which is sufficiently elevated after 14 days to indicate pregnancy. These days, though, pregnancy can be determined after only 48 hours by looking for the presence of early pregnancy factor (EPF), an immunosuppressant protein that prevents a woman's immune system from attacking the embryo.

Follicles are made in the embryo as oocytes assemble into primordeal follicles in large clusters called nests [5]. Primordial follicles remain arrested and static until they undergo a transition to primary follicles [1]. They continue to grow until they are either destroyed by a wave of apoptosis (called atresia) or ovulation. When this is not properly controlled, it can lead to premature ovarian failure or premature onset of menopause. High concentrations of maternal and fetal estradiol and progesterone inhibit early follicle development, and the decline in steroid levels that occurs later allows primary follicles to assemble and development to progress [5].

The increase in LH mid-cycle stimulates estrogen secretion from the ovaries, which inhibits release of FSH and initiates thickening of the endometrium of the uterus [4]. As the estrogen level increases, it causes a brief increase in LH, which initiates ovulation. The residual follicle, called the corpus luteum, produces progesterone, which causes further changes in the endometrium, preparing it for fertilization. Progesterone also suppresses ovulation during pregnancy [4].

In the 1950s, a scientist named Gregory Pincus screened over 200 compounds with structures similar to progesterone, to see if they could prevent ovulation. This led to the discovery of norethynodrel, a synthetic steroid and a progestin, a synthetic similar to progesterone, with similar biological activity. Although it worked well in most women, some suffered from breakthrough bleeding [4]. This did not happen when the norethynodrel contained an impurity, called mestranol, which has a methoxy ($-OCH_3$) instead of a ketone (C=O) and it has two additional C=C double bonds in the A ring. After discovering its ability to protect against bleeding, mestranol was added to norethynodrel to make the first combined oral contraceptive, Enovid, which was FDA-approved in 1960. This led to the so-called sexual revolution and many changes in society. Now, over 100 million women worldwide take oral contraceptives [4]. Instead of norethynodrel, oral contraceptives now contain an estrogen-like drug, called ethinyl estradiol. It is combined with any of a number of progestin derivatives. It binds to the human ERα and ERβ receptors with IC_{50} values of 2.34 and 9.65 nM, respectively, compares to 1.98 and 2.35 nM for 17β-estradiol [6]. Progestins can be related to either progesterone or testosterone. Progestins used in contraceptives are synthetic compounds, similar to the natural hormone, progesterone. Progestin-only pills are available for women who are breast feeding and those who are not able to tolerate estrogens.

Progesterone and synthetic progestins interact with the progesterone receptor and other steroid receptors [7]. "Some also interact with androgen receptors. Some progestins can help maintain pregnancy and transform the endometrium into the secretory phase. Some are anti-estrogenic. They down-regulate estrogen receptors and decrease the thickness of the endometrium. Others have anti-androgenic effects, in that they inhibit androgen-induced prostate growth. Anti-androgens can be either antagonists of the androgen receptor or inhibitors of the enzyme 5-α reductase, which catalyzes the conversion of testosterone into its more active form, DHT. The synthetic progestins used in the clinic are derivatives of testosterone (19-norestosterone derivatives) or progesterone (17-OH progesterone derivatives). Norethynodrel is a first generation progestin 19-norestosterone derivative. The second generation was comprised of the estrane group (norethisterone) and its metabolites and the gonane group (levonorgestrel). Norethynodrel, lynestrenol and ethynodiol acetate are prodrugs of norethisterone. They are converted into norethisterone, which is pharmacologically active. The third generation was derived from the gonane group. It includes desogestrel (DSG) and its active metabolite 3-keto-desogestrel, gestodene (GES) and norgestimate (and its active metabolite, norelgestromin). They have been used in many contraceptives available and some have androgenic activity. Drospirenone is a synthetic derivative of progesterone with anti-mineralocorticoid and proestrogenic properties not found in most other synthetic progestins. This allows it to counteract the effect of ethinyl estradiol on increasing aldosterone levels. So, water and salt retention is no longer such a problem. Drospirenone binds specifically to the progesterone receptor and not to other steroid receptors. It is used as an oral contraceptive, called Yasmin® (containing 3 mg of drospirenone plus 30 μg of ethinyl estradiol), and as a hormone replacement therapy (Angeliq®) in combination with estradiol" [7]. However, Yasmin® or Yaz® may cause a higher risk of venous thromboembolism, even though it is still quite popular and large multi-centered clinical trials seemed to demonstrate its safety, efficacy and even some weight loss [8 - 9]. Also, oral contraceptives may increase the risk of breast cancer in African-American women whose breast cancer cells that do not have the ER (they are ER^-) [10].

There are also transdermal patches and injectable hormone agents for contraception [4]. There is also an emergency contraceptive to prevent pregnancy after unprotected intercourse, the so-called Plan B. One form consists of two pills, each containing 0.75 mg of levonorgestrel.

Mifepristone is a synthetic steroid that is used for the termination of pregnancy up to the 49th day of gestation. It has been tested as a morning after pill to prevent pregnancy when taken within 12 hours of unprotected intercourse. It is also used in combination with another drug called Gemeprost to terminate pregnancies between weeks 13 and 24. It is also used to induce labor in cases of fetal death. For many years, it was known as RU-486. It was approved for use in Europe long before it was approved in the USA. In China, mifepristone is used as an emergency contraceptive, or a so-called morning after pill, also known as Plan B.

Table 1. Oral contraceptives described in this chapter.

Oral contraceptive	**Target receptor**
Enovid (mestranol plus norethynodrel)	Estrogen receptor
Ethinyl estradiol	Estrogen receptor
Norethynodrel	Progesterone receptor
Norethisterone	Progesterone receptor
Levonorgestrel	Progesterone receptor
Norethisterone	Progesterone receptor
Desogestrel (DSG)	Progesterone receptor
3-keto-desogestrel	Progesterone receptor
Gestodene (GES)	Progesterone receptor
Norgestimate	Progesterone receptor
Norelgestromin	Progesterone receptor
Drospirenone	Progesterone receptor
Mifepristone	Progesterone receptor

In the USA, the progestin called levonorgestrel is the preferred emergency contraceptive. It is given in a dose of 1.5 mg, either as two 750 μg doses 12 hours apart, or more recently as a single dose. It can prevent a pregnancy up to 72 hrs after unprotected sex or contraceptive failure. It can NOT terminate an existing pregnancy or protect against sexually transmitted diseases. It is available as an over-the-counter drug for women who are 18 or older. It can be obtained by a prescription for women who are not yet 18. Levonorgestrel is a hormone that is found in many birth control pills. When used as Plan B, it is a larger dose. It probably works by preventing ovulation. Oral contraceptives described in this chapter are summarized in (Table **1**).

Drugs are also available to induce labor by stimulating uterine contractions. Oxytocin is a natural peptide hormone that does this [4]. Oxytocin binds to G-protein coupled receptors, which initiate contraction. Also, prostaglandins can induce labor. Prostaglandin E_2 (PGE_2) (dinoprostone) is approved for the induction of labor [3]. Misoprostol is a synthetic PGE_1 derivative that can also induce labor, but it is used in the USA mainly for preventing gastric ulcers that can be caused by non-steroid anti-inflammatory drugs (NSAIDs) [2].

At the other end of a woman's reproductive cycle is menopause, which usually occurs between the ages of 48 to 55 years, with a mean age of 51"[11]. It has been defined as, "the permanent cessation of menses as a consequence of the loss of ovarian follicular activity, which can only be recognized retrospectively after 12 consecutive months of amenorrhea for which no other obvious physiological or pathological cause can be attributed" [12]. It occurs in all women and is a natural condition, unless it is caused by surgery. Some women have a hysterectomy. That is, they have their uterus removed surgically, for medical reasons. They will be infertile and have no more periods, but they will not enter menopause as long as one of the ovaries is still intact.

For some women, the effects of menopause can be severe. Hot flashes, also known as vasomotor symptoms, can be severe. Gabapentin is the only non-hormonal product shown to be as effective as low dose estrogen for treating them [13]. "Clonidine is mildly effective. However, many women take hormone replacement therapy (HRT), in which a low dose of one or more estrogens (such as conjugated equine estrogens) and a progestin are given. Progestins are synthetic proestogens. The estrogen is taken daily, with the progestin added either sequentially (cyclic regimen) or daily (continuous combined regimen), if it is needed. Also, there is tibolone, an oral synthetic steroid preparation with estrogenic, androgenic, and progestogenic actions that can also be used as HRT.

The key indications for HRT or tibolone are troublesome vasomotor symptoms (hot flushes and night sweats, with or without awakening). HRT can reduce the frequency of hot flushes by around 18 per week and in the severity of hot flushes by 87% compared with placebo. In general, women over age 60 years should not start HRT. However, women with premature (age <40 years) or early (<45 years) menopause, current guidelines recommend HRT until aged 50 for the treatment of vasomotor symptoms and bone preservation. Also, even though HRT reduces fracture risk, an increased risk of osteoporosis alone is not an indication for HRT. Similarly, although HRT may also improve mood and libido, these are not primary indications for treatment. Vaginal symptoms alone do not require systemic HRT and can be managed with local estrogens" [13].

However, there is much confusion – even among pharmacists [14]. The FDA has said that "bioidentical hormone (replacement) therapy" is a marketing term that they do not recognize, and that there is little rigorous evidence to support claims of safety and efficacy [14]. Still, a recent survey showed that most pharmacists thought that bioidentical hormone therapy (BHT) is similar or more effective than conventional hormone therapy (CHT) for treating vasomotor symptoms and preventing bone fracture caused by osteoporosis. Moreover, the majority of respondents indicated that they believed the risks of BHT and CHT were similar.

So, let's be clear. Equine estrogens used in CHT come from pregnant mares (horses). They are conjugated to a hydrophilic side group, such as a sulfate. The drug is called Premarin and was used widely considered to be quite safe. However, results from a clinical study by the Women's Health Initiative published in 2002 changed this [15]. "That is, women were given estrogen plus progestin (E+P) given as "conjugated equine estrogen (CEE) plus medroxyprogesterone acetate (MPA). There was an increased risk of breast cancer, cardiovascular disease, stroke, and thromboembolic events" [15]. These findings caused many women to discontinue taking E+P and some lost confidence in their doctors who had prescribed them [16]. Prescriptions for E+P decreased by 63% [16]. Moreover, the results of the 2002 study were generalized to all types of hormone replacement, including estrogen alone, and in women without a uterus, although only women with a uterus were used in the study [17]. Within 18 months nearly two million women in the USA who did not have a uterus stopped using estrogen therapy (ET). Subsequently, follow ups of the 2002 study were published in 2004 (intervention phase) and 2011 (long-term follow-up phase). They showed that there was no difference in mortality between women on ET and those taking placebo [18]. Moreover, there was a decrease in mortality in women aged 50-59 in the 2004 report and a reduction in mortality of 13 per 10000 women per year was reported in 2011. This was due to a decrease in coronary heart disease [19]. Despite these findings, prescriptions for ET continued to decline. So, it was estimated that avoiding ET may have shortened the lives of 18600 to 91600 women from 2002-2011 [20].

Still, the Women's Health Care Initiative has shown that there is an increased risk of cardiovascular disease in women taking HRT who have no pre-existing risk factors [21]. At the same time, estrogen may help prevent atherosclerosis in pre-menopausal women have a lower incidence of cardiovascular disease and. When activated, the ER(receptor in endothelial and myocardial cells has antioxidant and other important health effects, including improving the recovery of damaged endothelial cells, vasodilation through nitric oxide, decreasing the oxidation of low density lipoproteins (LDL), reducing platelet aggregation and increasing cyclooxygenase-2 activity. Oral contraceptives can also cause an increase in blood pressure and high density lipoprotein (HDL) levels. When post-menopausal women take estrogen, it can increase inflammatory markers, such as the C-reactive protein. Estrogen also increases cardiovascular thromboembolisms by increasing prothrombin and decreasing antithrombin III. Women also have a higher risk compared to men of sudden death caused by cardiac dysfunction, due to prolonged QT that is seen in an electrocardiogram. Healthy post-menopausal women can experience a prolonged QT interval when on HRT. So, the American College of Obstetrician and Gynecologists (ACOG) produced guidelines that say that women can take oral contraceptives if they are healthy, under 35, have well-controlled and properly monitored blood pressure and do not consume tobacco products. For healthy, non-smoking women over 35, oral contraceptives containing <50 μg of ethinyl estradiol are still safer than enduring a pregnancy and can be continued until 50 or 55 or until menopause starts, if risks and benefits are considered. However, there are insufficient data on the risks of cardiovascular disease for the newest generation of contraceptives (transdermal and vaginal).

Also, oral contraceptives can be used in the perimenopausal transition to suppress ovulation, since higher doses are needed than those required to prevent hot flashes [21]. Moreover, HRT was related to ealier-stage breast cancer in a prospective study on 49017 women [22].

According to a recent review, "no consensus has been reached on the contraindications to HRT" [23]. "However, they did recommend the following: HRT should be discontinued or avoided all together for women with a history of breast cancer, venous or arterial thromboembolic disease, including stroke and cardiovascular disease, uncontrolled hypertension, abnormal vaginal bleeding, abnormal liver function, a history of endometrial or ovarian cancer and a risk of gallbladder disease. For those women who have no contraindications, the following should be considered before starting HRT: it should be considered for perimenopausal or recently postmenopausal symptomatic women with low risk factors for cardiovascular or thromboembolic disease. Perimenopausal women should consider using cyclic HRT or (in women under 50 years) a low dose, combined oral contraceptives to minimize irregular bleeding. Women who have been postmenopausal for one to two years and wish to avoid bleeding may consider continuous combined HRT or using tibolone" [23].

So, physicians and patients should consider the nature and severity of menopausal symptoms and their impact on function and quality of life, the woman's age and health status, as well as her wishes for treatment [23]. "It is reasonable to advise younger, healthy postmenopausal women that HRT is unlikely to increase their risk of cardiovascular disease. However, HRT is not currently indicated in women at any age for preventing or treating cardiovascular disease. HRT should be considered for women who are at high risk of bone fracture if there are no contraindications. It can also be considered for women suffereing from depression and anxiety, due to menopausal symptoms, but it is not a treatment for clinical anxiety or depression. It is also important to ensure that breast and cervical screening are up to date and investigate any abnormal vaginal bleeding. Also, perimenopausal women may need contraception. Estrogen alone should be used in women after hysterectomy. For others, the progestogen component of HRT may be progesterone or a progestogen. Also, the dosage and type of HRT should be based on symptoms and possible side effects. One should start with a low dose of estrogens and consider gradually increasing the dose after four to six weeks only if troublesome vasomotor symptoms persist. It is also important to monitor the effects of HRT once it has begun. Its effectiveness should be monitored by looking for improved symptoms. Blood tests are rarely useful. If mastalgia or irregular bleeding occurs, they may respond to a lower dose of estrogens. Patients should be examined annually to evaluate indications for use, assess individuals' risk and benefit profile, and to promote lifestyle interventions to reduce or prevent chronic disease. It should also be noted that HRT does not cause weight gain. Also, the schedule for mammography and cervical smears should not be altered by HRT use. Moreover, data from large randomized controlled trials do not support the efficacy of black cohosh or other "natural remedies" for the treatment of hot flushes and "bio-identical" hormones have not been shown to be safe or effective" [23].

However, many women do use bioidenticals, partly due to their distrust of the medical establishment and popular books that have been written about them [24, 25]. So, many women ask their physicians for compounded bioidentical hormone therapy (CBHT), believing that it is safer than FDA-approved therapy. It has been estimated that CBHT is a multibillion-dollar industry, possibly affecting millions of women. The term *CBHT* refers to hormone preparations that are specifically compounded for an individual patient. However, custom CBHT is not FDA-approved for treatment of menopausal symptoms [26].

Bioidentical hormones have been called natural because they are the same hormones that are in women. They have the same chemical formula and structure. That is, 17β-estradiol, estriol and estrone from one source (like plants) are the same as 17β-estradiol, estriol and estrone in humans. Some people may not consider many bioidenticals to be "natural" because they are either extracted from plants or synthesized in a laboratory. Moreover, the word natural has no scientific meaning. It does have a popular meaning, but some people consider supernatural as being the opposite of natural. Certainly, no chemical or drug is supernatural. Unfortunately, many people consider humans to be outside of nature, so anything that is man-made (or women-made) is not natural. Such an attitude may be controversial, but still the words natural and man-made can have very emotional connotations to many people, so they make excellent words to use in marketing a product or disrespecting the competition's products. So, let's stick to science and just use the term bioidentical hormones.

One of the major objections to the use of bioidentical compounded hormones has been that there are no guidelines for doses, because no pharmacokinetic studies have been done. That changed in 2013, when a pharmacokinetic study was done in a randomized, blinded, four-arm 16-day clinical trial of forty postmenopausal women [27]. "They were assigned to one of three doses of a compounded estrogen cream (Bi-est (80:20); 2.0, 2.5, or 3.0 mg) plus compounded oral progesterone 100 mg, or to a conventional estradiol patch (Vivelle-Dot 0.05 mg) plus Prometrium 100 mg. Blood serum was analyzed for estrone, estradiol, estriol, and progesterone at multiple time intervals during the first 24-hours, and at steady-state. The results showed that the medications were tolerated well. The area under the curve of the plot of serum concentration vs time at 24 hours and at the steady state remained consistently lower for all doses of Bi-est tested relative to the estradiol patch. Also, estriol levels remained low in all study arms, but serum progesterone levels were comparable in conventional *vs.* compounded groups. So, the currently used doses of compounded hormones produced lower levels of estrogen compared to the standard-dose estradiol patch. To find comparable doses, further studies will be needed" [27].

Such studies are being done at the Mayo Clinic [28]. They wanted to find the dose of bioidentical compounded hormone therapy that is bioequivalent to conventional HT in a randomized, blinded, four-arm, phase I clinical trial. They estimated the concentrations of estrone (E1), estradiol (E2) and estriol (E3) at baseline and at steady state. They administered with two-weeks of administration of three commonly used doses of bioidentical compounded estrogen cream (Biest) and a standard dose conventional estrogen patch (Vivelle-Dot). E1, E2, and E3 values will be summarized using point estimates and 95% confidence intervals in a forthcoming report [28].

Another approach to avoid hot flashes and other symptoms of menopause is to increase the consumption of soybeans and foods made from soybeans (such as soy milk). Soybeans contain compounds called isoflavones, which may help relieve many of the symptoms of menopause. Another alternative is an herb, or dietary supplement called black cohosh, which contains a compound called deoxyactein, which may help prevent hot flashes and other symptoms of menopause. The National Center for Complementary and Alternative Medicine (a branch of the NIH) is funding several scientific studies to determine if black cohosh really does prevent symptoms of menopause and if it can help with breast cancer.

Osteoporosis is another common symptom of menopause. One isoform of the CCAAT enhancer-binding protein, or C/EBPδ, is expressed all the time (constitutively) in osteoblasts, where it regulates the transcription of the gene coding for the insulin-like growth factor-I (IGF-I) [29]. "That is, IGF-I promotes skeletal growth by stimulating bone cell replication and differentiation. Raloxifene, or Evista®, binds to the estrogen receptor and is used to prevent and treat osteoporosis. It is a selective estrogen receptor modulator. It has estrogenic effects on bone and antiestrogenic actions in the uterus and breast. Reduced levels of estrogen occur in post-menopausal women, and it is a major risk factor in osteoporosis. Raloxifene is an agonist at bone estrogen receptors. It is an estrogen receptor antagonist in breast tissue, and it is effective in reducing the occurrence of breast cancer, as does another SERM, tamoxifen. Raloxifene was approved by the FDA in 2002. Later, it was found that it also reduces the concentration of low density lipoproteins (LDL). "It is also as effective as tamoxifen in reducing the risk of breast cancer in post-menopausal women. So, it is also approved for reducing the risk of invasive breast cancer in postmenopausal women with osteoporosis and in postmenopausal women at high risk for invasive breast cancer" [29].

Alendronate (Fosamax®) is a bisphosphonate. Its structure is similar to that of phosphate, which is a natural component of bones. Also, bisphosphonates bind Ca^{2+}, another component of bone. As a result, bisphosphonates don't just inhibit the breakdown (or reabsorption) of bones, but they also are taken up by bone. So, Fosomax® is taken up by osteoclasts, where they inhibit farnesyl pyrophosphate synthase. However, it was recently reported that Fosamax® may increase the risk of osteonecrosis of the jaw and long-term use may produce giant osteoclasts.

Calcitrol is used to treat hypocalcemia, rickets and osteoporosis. It is the active form of vitamin D. It increases the absorption of calcium and phosphate from the gastrointestinal tract and kidneys. It is crucial in the maintenance of healthy bones, which contain calcium and phosphate. Calcitriol promotes the absorption of Ca^{2+} from food in the intestines and it promotes reabsorption of Ca^{2+} in the kidneys. It is used to treat hypocalcemia, rickets and osteoporosis. Also, Teriparatide (Forteo®) is a drug that contains the first 34 amino acids of parathyroid hormone, or PTH. When given once a day it stimulates the transformation of stem cells in the bone marrow to develop into osteoblasts, the cells that make new bone cells. PTH is a polypeptide with 84 amino acids. It is secreted by the parathyroid gland and it controls the concentration of Ca^{2+} in the extracellular fluid by regulating its supply from bones and from reabsorption in kidneys. PTH also stimulates the formation of calcitriol from its precursor, vitamin D_3. Sustained levels of PTH cause bone loss. On the other hand, intermittent administration of PTH stimulates bone formation. Remember that our bones (and most of the rest of our body) are being continually broken down and re-made. So, intermittent administration of PTH, or a portion of PTH, can actually increase bone mass in patients with osteoporosis.

From the medicinal chemistry point of view, there are many drugs that are used in reproductive medicine and to treat osteoporosis. Oral contraceptives contain an estrogen-like drug, ethinyl estradiol. It is combined with any of a number of progestin derivatives. Progestin-only pills are available for women who are breast feeding and those who are not able to tolerate estrogens. There is also an emergency contraceptive to prevent pregnancy after unprotected intercourse, called Plan B. One form consists of two pills, each containing 0.75 mg of levonorgestrel. Mifepristone is a synthetic steroid that is used for the termination of pregnancy up to the 49th day of gestation. It has been tested as a morning after pill to prevent pregnancy when taken within 12 hours of unprotected intercourse. It is also used in combination with another drug called Gemeprost to terminate pregnancies between weeks 13 and 24. In the USA, levonorgestrel is the preferred emergency contraceptive. It can prevent a pregnancy up to 72 hrs after unprotected sex or contraceptive failure. Oxytocin and prostaglandin E_2 are also available to induce labor by stimulating uterine contractions. Some women take hormone replacement therapy, in which a low dose of one or more estrogens (conjugated equine estrogens) and a progestin are given. Another approach is to increase the consumption of soybeans and foods made from soybeans (such as soy milk).

There are also drugs to treat osteoporosis. Raloxifene, or Evista®, binds to the estrogen receptor and is used to prevent and treat osteoporosis. It is a selective estrogen receptor modulator [1]. "It has estrogenic effects on bone and anti-estrogenic actions in the uterus and breast. Alendronate (Fosamax®) is a bisphosphonate used to treat osteoporosis. Calcitrol is used to treat hypocalcemia, rickets and osteoporosis. It is the active form of vitamin D. It increases the absorption of calcium and phosphate from the gastrointestinal tract and kidneys. It is crucial in the maintenance of healthy bones, which contain calcium and phosphate. Calcitriol promotes the absorption of Ca^{2+} from food in the intestines and it promotes reabsorption of Ca^{2+} in the kidneys. It is used to treat hypocalcemia, rickets and osteoporosis. Also, Teriparatide (Forteo®) is a drug that contains the first 34 amino acids of parathyroid hormone, or PTH. "When given once a day it stimulates the transformation of stem cells in the bone marrow to develop into osteoblasts, the cells that make new bone cells. PTH is a polypeptide with 84 amino acids" [1]. It is secreted by the parathyroid gland and it controls the concentration of Ca^{2+} in the extracellular fluid by regulating its supply from bones and from reabsorption in kidneys. PTH also stimulates the formation of calcitriol from its precursor, vitamin D_3. Sustained levels of PTH cause bone loss. On the other hand, intermittent administration of PTH stimulates bone formation. Remember that our bones (and most of the rest of our body) are being continually broken down and re-made. So, intermittent administration of PTH, or a portion of PTH, can actually increase bone mass in patients with osteoporosis.

Also, there is a drug called strontium ranelate that has been prescribed for treating osteoporosis [30 - 33]. It dissociates bone formation from bone resorption so bone production continues, while bone resorption decreases [30]. It significantly reduces the risk of all nonvertebral and in a high risk subgroup, hip fractures over a 3-yr period, and is well tolerated [30]. It stimulates the expression of the osteoblast markers ALP, BSP and OCN [31]. It increases the number of bone nodules while decreasing the number of mature osteoclasts [31]. However, it has been seen recently that there could be cardiovascular risks [32]. "That is, one study reported that patients treated with strontium ranelate have an unfavourable cardiovascular risk profile compared with users of other osteoporosis drugs. However, only the risk of death differed significantly from the rates observed in users of other osteoporosis drugs adjusted for risk factor profile. A large proportion of patients currently treated with strontium ranelate have conditions that would now be considered contraindications according to EMA" [32]. In contrast, a Clinical Practice Research Datalink (CPRD) study CPRD did not find evidence for a higher risk for cardiac events associated with the use of strontium ranelate in postmenopausal osteoporosis [33].

KEY POINTS

1. **The fertility cycle in women is controlled by a cascade of events, which are initiated by protein and steroid hormones. These hormones create signals between the hypothalamus, the pituitary gland and the ovaries. Oral contraceptives contain an estrogen-like drug, ethinyl estradiol. It is combined with any of a number of progestin derivatives. Progestin-only pills are available for women who are breast feeding and those who are not able to tolerate estrogens.**

2. **There is also an emergency contraceptive to prevent pregnancy after unprotected intercourse, the so-called Plan B. One form consists of two pills, each containing 0.75 mg of levonorgestrel. Mifepristone is a synthetic steroid that is used for the termination of pregnancy up to the 49th day of gestation. It has been tested as a morning after pill to prevent pregnancy when taken within 12 hours of unprotected intercourse. It is also used in combination with another drug called Gemeprost to terminate pregnancies between weeks 13 and 24.**

3. **In the USA, the progestin called levonorgestrel is the preferred emergency contraceptive. It is given in a dose of 1.5 mg, either as two 750 µg doses 12 hours apart, or more recently as a single dose. It can prevent a pregnancy up to 72 hrs after unprotected sex or contraceptive failure.**

4. **Drugs are also available to induce labor by stimulating uterine contractions. Oxytocin is a natural peptide hormone that does this. Oxytocin binds to G-protein coupled receptors, which initiate contraction. Also, Prostaglandin E_2 (PGE_2) (dinoprostone) is approved for the induction of labor.**

5. **For some women, the effects of menopause can be severe. Some women take hormone replacement therapy (HRT), in which a low dose of one or more estrogens (conjugated equine estrogens) and a progestin are given. Another approach is to increase the consumption of soybeans and foods made from soybeans (such as soy milk). Soybeans contain compounds called isoflavones, which may help relieve many of the symptoms of menopause.**

ABBREVIATIONS

DNA	=	Deoxyribonucleic Acid
HSV	=	Herpes Simplex Virus
Ifn	=	Interferon
ASRM	=	American Society for Reproductive Medicine
HRT	=	Hormone Replacement Therapy
CHT	=	Conventional Hormone Therapy
DHT	=	Dihydrotestosteron
ER	=	Estrogen Receptor
GPCR G	=	Protein Coupled Receptor
DHT	=	Dihydroxytestosterone
AR	=	Androgen Receptor
FSH	=	Follicle Stimulating Hormone
LH	=	Luteinizing Hormone
gnRH	=	Gonadotropin Releasing Hormone
LHRH	=	Luteinizing Hormone Releasing Hormone
EPF	=	Early Pregnancy Factor
FDA	=	Food and Drug Administration
DSG	=	Desogestrel
GES	=	Gestodene
PGE_2	=	Prostaglandin E_2
CEE	=	Conjugated Equine Estrogen
MPA	=	Medroxyprogesterone Acetate
ET	=	Estrogen Therapy
LDL	=	Low Density Lipoprotein
HDL	=	High Density Lipoprotein
CBHT	=	Compounded Bioidentical Hormone Therapy

REFERENCES

[1] Flak MB, Neves JF, Blumberg RS. Welcome to the Microgenderome. Commensal gut bacteria reinforce the gender bias observed in an autoimmune form of diabetes. Science 2013; 339: 1044-5. [http://dx.doi.org/10.1126/science.1236226]

[2] Fung KY, Mangan NE, Cumming H, *et al.* Interferon-ε protects the female reproductive tract from viral and bacterial infection. Science 2013; 339(6123): 1088-92. [http://dx.doi.org/10.1126/science.1233321] [PMID: 23449591]

[3] American Society for Reproductive Medicine. http://www.asrm.org/. 2014 March 10;

[4] Corey EJ, Czako B, Kurti L. Molecules and Medicine. New York: Wiley 2007.

[5] Kezele P, Skinner MK. Regulation of ovarian primordial follicle assembly and development by estrogen and progesterone: endocrine model of follicle assembly. Endocrinology 2003; 144(8): 3329-37. [http://dx.doi.org/10.1210/en.2002-0131] [PMID: 12865310]

[6] Harris HA, Bapat AR, Gonder DS, Frail DE. The ligand binding profiles of estrogen receptors α and β are species dependent. Steroids 2002; 67(5): 379-84. [http://dx.doi.org/10.1016/S0039-128X(01)00194-5] [PMID: 11958794]

[7] Sitruk-Ware R. Pharmacological profile of progestins. Maturitas 2008; 61(1-2): 151-7. [http://dx.doi.org/10.1016/j.maturitas.2008.11.011] [PMID: 19434887]

[8] Foidart JM, Wuttke W, Bouw GM, Gerlinger C, Heithecker R. A comparative investigation of contraceptive reliability, cycle control and tolerance of two monophasic oral contraceptives containing either drospirenone or desogestrel. Eur J Contracept Reprod Health Care 2000; 5(2): 124-34. [http://dx.doi.org/10.1080/13625180008500387] [PMID: 10943575]

[9] Huber J, Foidart JM, Wuttke W, *et al.* Efficacy and tolerability of a monophasic oral contraceptive containing ethinylestradiol and drospirenone. Eur J Contracept Reprod Health Care 2000; 5(1): 25-34. [http://dx.doi.org/10.1080/13625180008500375] [PMID: 10836660]

[10] Rosenberg L, Boggs DA, Wise LA, Adams-Campbell LL, Palmer JR. Oral contraceptive use and estrogen/progesterone receptor-negative breast cancer among African American women. Cancer Epidemiol Biomarkers Prev 2010; 19(8): 2073-9. [http://dx.doi.org/10.1158/1055-9965.EPI-10-0428] [PMID: 20647407]

[11] Barbo DM. Menopause. In: Wallis LA, Kasper AS, Reader GG, Eds. Textbook of women's health. Philadelphia: Lippincott-Raven 1998; p. 721.

[12] Cirigliano M. Bioidentical hormone therapy: a review of the evidence. J Womens Health (Larchmt) 2007; 16(5): 600-31. [http://dx.doi.org/10.1089/jwh.2006.0311] [PMID: 17627398]

[13] Hickey M, Elliott J, Davison SL. Hormone replacement therapy. BMJ 2012; 344: e763. [http://dx.doi.org/10.1136/bmj.e763] [PMID: 22344303]

[14] FDA. Bio-Identicals: Sorting myths from facts. http://www.fda.gov/downloads/ForConsumers/ConsumerUpdates/ucm049312.pdf.

[15] Rossouw JE, Anderson GL, Prentice RL, *et al.* Writing Group for the Women's Health Initiative

Investigators. Risks and benefits of estrogen plus progestin in healthy postmenopausal women: principal results From the Women's Health Initiative randomized controlled trial. JAMA 2002; 288(3): 321-33.
[http://dx.doi.org/10.1001/jama.288.3.321] [PMID: 12117397]

[16] Files JA, Ko MG, Pruthi S. Bioidentical hormone therapy. Mayo Clin Proc 2011; 86(7): 673-80.
[http://dx.doi.org/10.4065/mcp.2010.0714] [PMID: 21531972]

[17] Sarrel PM, Njike VY, Vinante V, Katz DL. The mortality toll of estrogen avoidance: an analysis of excess deaths among hysterectomized women aged 50 to 59 years. Am J Public Health 2013; 103(9): 1583-8.
[http://dx.doi.org/10.2105/AJPH.2013.301295] [PMID: 23865654]

[18] Anderson GL, Limacher M, Assaf AR, *et al.* Women's Health Initiative Steering Committee. Effects of conjugated equine estrogen in postmenopausal women with hysterectomy: the Women's Health Initiative randomized controlled trial. J Am Med Assoc 2004; 291(14): 1701-12.
[http://dx.doi.org/10.1001/jama.291.14.1701] [PMID: 15082697]

[19] LaCroix AZ, Chlebowski RT, Manson JE, *et al.* WHI Investigators. Health outcomes after stopping conjugated equine estrogens among postmenopausal women with prior hysterectomy: a randomized controlled trial. JAMA 2011; 305(13): 1305-14.
[http://dx.doi.org/10.1001/jama.2011.382] [PMID: 21467283]

[20] Sarrel PM, Njike VY, Vinante V, Katz DL. The mortality toll of estrogen avoidance: an analysis of excess deaths among hysterectomized women aged 50 to 59 years. Am J Public Health 2013; 103(9): 1583-8.
[http://dx.doi.org/10.2105/AJPH.2013.301295] [PMID: 23865654]

[21] Shufelt CL, Bairey Merz CN. Contraceptive hormone use and cardiovascular disease. J Am Coll Cardiol 2009; 53(3): 221-31.
[http://dx.doi.org/10.1016/j.jacc.2008.09.042] [PMID: 19147038]

[22] Schairer C, Byrne C, Keyl PM, Brinton LA, Sturgeon SR, Hoover RN. Menopausal estrogen and estrogen-progestin replacement therapy and risk of breast cancer (United States). Cancer Causes Control 1994; 5(6): 491-500.
[http://dx.doi.org/10.1007/BF01831376] [PMID: 7827235]

[23] Hickey M, Elliott J, Davison SL. Hormone replacement therapy. BMJ 2012; 344: e763.
[http://dx.doi.org/10.1136/bmj.e763] [PMID: 22344303]

[24] Somers S. Ageless: The Naked Truth about bioidentical hormones. New York: Three Rivers Press 2006.

[25] Somers S. I'm Too Young for This: The Natural Hormone Solution to Enjoy Perimenopause. New York: Harmony Press 2013.

[26] Files JA, Ko MG, Pruthi S. Bioidentical hormone therapy. Mayo Clin Proc 2011; 86(7): 673-80.
[http://dx.doi.org/10.4065/mcp.2010.0714] [PMID: 21531972]

[27] Sood R, Warndahl RA, Schroeder DR, *et al.* Bioidentical compounded hormones: a pharmacokinetic evaluation in a randomized clinical trial. Maturitas 2013; 74(4): 375-82.
[http://dx.doi.org/10.1016/j.maturitas.2013.01.010] [PMID: 23384975]

[28] Mayo Clinic. A pharmacokinetic evaluation of bioidentical compounded Estrogen Cream and Natural Progesterone (HRT). Mayo Clinic http://clinicaltrials.gov/ct2/show/NCT00864214?term=A+pharmacokinetic+evaluation+of+bioidentical+compounded&rank=1. 2013.

[29] Umayahara Y, Billiard J, Ji C, Centrella M, McCarthy TL, Rotwein P. CCAAT/enhancer-binding protein δ is a critical regulator of insulin-like growth factor-I gene transcription in osteoblasts. J Biol Chem 1999; 274(15): 10609-17.
[http://dx.doi.org/10.1074/jbc.274.15.10609] [PMID: 10187857]

[30] Reginster JY, Seeman E, De Vernejoul MC, *et al.* Strontium ranelate reduces the risk of nonvertebral fractures in postmenopausal women with osteoporosis: Treatment of Peripheral Osteoporosis (TROPOS) study. J Clin Endocrinol Metab 2005; 90(5): 2816-22.
[http://dx.doi.org/10.1210/jc.2004-1774] [PMID: 15728210]

[31] Bonnelye E, Chabadel A, Saltel F, Jurdic P. Dual effect of strontium ranelate: stimulation of osteoblast differentiation and inhibition of osteoclast formation and resorption *in vitro*. Bone 2008; 42(1): 129-38.
[http://dx.doi.org/10.1016/j.bone.2007.08.043] [PMID: 17945546]

[32] Abrahamsen B, Grove EL, Vestergaard P. Nationwide registry-based analysis of cardiovascular risk factors and adverse outcomes in patients treated with strontium ranelate. Osteoporos Int 2014; 25(2): 757-62.
[http://dx.doi.org/10.1007/s00198-013-2469-4] [PMID: 24322475]

[33] Cooper C, Fox KM, Borer JS. Ischaemic cardiac events and use of strontium ranelate in postmenopausal osteoporosis: a nested case-control study in the CPRD. Osteoporos Int 2014; 25(2): 737-45.
[http://dx.doi.org/10.1007/s00198-013-2582-4] [PMID: 24322476]

CHAPTER 7

The Nervous System

Abstract: The first year of life is critical in brain development, for the total brain volume doubles, as measured by MRI. This is when the brain is most susceptible to damage by genetic defects and environmental insults. It is also the time in which therapeutic intervention can have its maximum effect. A principal component of the nervous system is the neuron. Neurons are arranged in networks and circuits. The normal human brain has many local regions, or centers, and many pathways between them. The autonomic nervous system is organized into three divisions: the sympathetic, parasympathetic and enteric [1]. These maintenance activities are usually performed without conscious control or sensation. The sympathetic and parasympathetic nervous systems work to maintain a type of balance. They have opposite effects on the body. The sympathetic division is used in actions requiring quick responses. The parasympathetic division is used in actions that do not require immediate reaction [1]. Messages are sent to and from neurons in the form of primary messengers, called neurotransmitters. L-DOPA is used to treat Parkinson's disease, which affects about 1% of the population over 65. Alzheimer's disease (AD) is the most common neurodegenerative disease. Phenobarbital, carbamazepine, valproic acid and its sodium salt, gabapentin, ethosuximide, lamotrigrine, and tiagabine are anti-epileptics. Diazepam, buspirone, b-blockers, tricyclic antidepressants and monoamine oxidase inhibitors treat anxiety disorders. Currently the first-line treatment is either SSRIs or SNRIs. The four main classes of antidepressant drugs are MAOIs, TCAs, SSRIs and SNRIs.

Keywords: Alzheimer's disease, Autonomic nervous system, Neuron, Parkinson's disease.

INTRODUCTION

The nervous system can be studied using systems thinking. As mentioned before, one of the weaknesses of reductionist thinking is the inability to explain much of embryology, growth or development. The growth and development of the human brain and its cells from neural stem cells is a fascinating field [1]. Remember that stem cells have the same set of genes that differentiated cells have, but those

Robert E. Smith

genes are controlled differently in different cells. They are masters of autopoiesis. Moreover, the human nervous system is a network of molecules, ions, sub-cellular organelles, cells, tissues and organs. Chemical and electrical signals are transmitted from neuron to neuron in complex networks and circuits in the brain. This network and its circular organization are linked with other tissues and organs, including the endocrine system, digestive system, liver and immune system. As mentioned in Chapter 5, one example is the fibroblast growth factor FGF 21, which is a hepatokine (cytokine secreted by the liver) that acts as a global starvation signal to modulate fuel partitioning and metabolism as well as repress growth by acting on the nervous system [1]. It acts through its receptor in the brain, β-Klotho, to lower insulin, inhibit growth, alter light-dark cycle activity, increase systemic corticosterone levels and inhibit female fertility. It also suppresses the expression of the neuropeptide vasopressin in the suprachiasmatic nucleus (SCN) of the brain, which also expresses receptors for ghrelin and leptin [2]. It also contributes to neuroendocrine control of female reproduction by acting on the SCN in the hypothalamus to suppress ovulation during starvation [2].

Network analysis of neural systems in humans and other species show something called the small-world network phenomenon. That is, information can be transferred from one region to any other region by traversing only a few major white matter tracts [3]. So, any two neurons can communicate by the transmission of signals over just a few synapses. Networks consist of nodes or vertices, connected by line segments or edges. This will be discussed further in Appendix 2.

The human brain develops slower than the other organs, even though we are born with almost all the neurons that we will ever have. The first year of life is critical in brain development, for the total brain volume doubles, as measured by MRI [3]. This is when the brain is most susceptible to damage by genetic defects and environmental insults. It is also the time in which therapeutic intervention can have its maximum effect [3]. The brain continues to grow during the first two years of life, due to the growth and division of supporting cells (glial cells) in the white matter and there is a large increase in connections (synapses) between neurons in the gray matter. Glial cells are derived from myeloid precursors and migrate into the brain during development. Unlike the most of the rest of the body, it is mostly just these glial cells that are being continuously being broken down and re-made in the brain [4, 5]. Most adult neurons do not undergo mitosis and are not regenerated when they die. Neurogenesis probably does not occur in most regions of the brain, but some data suggest that adult neurogenesis might occur in the neocortex, striatum, amygdala, hypothalamus and brainstem [6]. There is much more evidence that adult neurogenesis occurs in the subgranular zone (SGZ) of the dentate gyrus in the hippocampus (which is involved in learning and memory) and the subventricular zone, or SVZ (in which neural stem cells reside) of the lateral ventricle [7]. Neural stem (or progenitor) cells can respond to external neural activity and differentiate into neurons [7]. This activity-dependent neurogenesis requires Ca^{2+} channels and receptors for the neurotransmitter *N*-methyl-D-aspartate (NMDA) [7]. It is called excitation-neurogenesis coupling [8]. "That is, neuronal stem cells respond to electrical signals from neighboring neurons by expressing appropriate genes and signaling pathways. This is mediated by GABA-mediated membrane depolarization, which causes an increase in intracellular Ca^{2+}. The excitation signal activates the transcription factor called NeuroD and promotes neurogenesis. Once the stem cells differentiate into neuroblasts and neurons, excitation by GABA helps the new neurons to be integrated as functional units. Moreover, neural stem cells called type 1 or radial glial-like (RGL) cells can respond to neural activity. RGLs maintain the adult neural stem cell pool in the hippocampus by remaining quiescent. The maintenance and activation of RGLs is controlled dynamically by experience and aging. RGLs that express the protein called nestin can be activated by GABA. The absence of functional $GABA_A$ receptors causes a rapid exit from quiescence and an increased production of RGLs. Interneurons in the stem cell niche that express the Ca^{2+}-binding protein parvalbumin are are a source of GABA. They are required to maintain RGLs. When GABA signaling is modulated, it affects the generation of more RGLs or causes them to remain quiescent, but not to differentiate. So, the adult brain continues to develop under the influence of electrical activity, while behavior and circuit activity control adult neurogenesis" [8].

Neurogenesis occurs throughout life in the SVZ. Neurons that are generated there migrate to the olfactory bulb, where they become interneurons [7]. Even though most neurons outside the hippocampus, SVZ, and olfactory bulb can't be regenerated, most of the proteins inside them are continuously broken down and re-made. Also, connections between neurons (synapses) are modified throughout our lives. So, it is striking that people can remember some things for many decades, even though the proteins within the neurons last for a few months or less. Consistent with systems thinking, it is the structure and organization that remains the same, not the molecules. As one thinks about basic building blocks, it is the cell and the organization of networks, and not the individual molecules that build and maintain the nervous system.

Components of the Nervous System

Still, an important part of systems thinking (and reductionist thinking) is identifying and characterizing the components of the system. So, in the science of anatomy, the nervous system is divided into the central and peripheral nervous systems [1]. The brain and spinal cord make up the central nervous system (CNS). The peripheral nervous system (PNS) is made up of all the sensory nerves [1].

A principal component of the nervous system is the neuron, which is the type of electrically excitable cell that transmits and processes information. Neurons are arranged in networks and circuits [1]. "They communicate with each other and with other types of cells through connections called synapses. Like other human cells, neurons are polarized. The inside of the cell membrane has a slight negative charge (about -70 mV), compared to the outside" [1]. When something happens to the membrane, such as electrical stimulation, the voltage difference can change. Non-neuronal cell voltages change rapidly and reach a constant value, where neuronal cells exhibit an action potential, in which the inside becomes positive for a short time and then returns to a negative value.

There are about 10-100 billion neurons in the adult brain, although brain size in humans is not related to intelligence. This is best illustrated by a disease called hydrocephalus, or water on the brain. Back in the 1940s, when a baby was born with this condition, nothing could be done except let the baby die. Now, doctors can drain the water and babies can live, grow and mature to lead ordinary lives. The disease occurs when the fluid filled spaces in the middle of the brain expand due to increased pressure, pressing the brain against the skull [9, 10]. "More than 600 individuals were examined by Dr. Lorber, and about one in ten had more than 90% of the skull volume filled with fluid, and of these, more than half have IQs over 100. A young man who was a student at Sheffield University had an IQ of 126, earned a first class honors degree in mathematics; and had virtually no brain. The neocortex of the student's brain is a layer about 1/25th of an inch (1 mm) thick, lining his skull. His brain weight was estimated to be about 2/10 of one pound (about 0.090 kg). That is only about 7% of normal brain weight. This student is one of many normal persons with almost no brain. Clearly these observational facts show that intelligence and brain size are not related. Normal brain size is not necessary to be of perfectly normal intelligence. What makes man unique is not the size of his brain" [9]. Moreover, young children with autism tend to have larger brains that do neurologically normal children. Moreover, social insects, like honeybees have only 10^6 neurons, but have complex social behaviors. Honeybees have a 'dance language' and they can learn abstract concepts, such as 'same' and 'different'. Also, some birds migrate tens of thousands of miles, performing navigational feats that humans can only do with sophisticated technology. Even the 'lowly' insects migrate to their proper locations. Unlike birds, many do not migrate back and forth between two locations in their lifetimes. Instead, many perform inter-generational migration. So, animal behavior and their brains are not like machines, in which 'bigger' means 'better' and 'smaller' means 'inferior'. One might even think of the ongoing battle between pathogenic bacteria and humans. The more antibiotics we use to try to kill them, the more resistant they become, so are we really that much smarter than the bacteria, which have been around for hundreds of millions of years? Modern *Homo sapiens* have been around for a mere 200,000 years and if we don't control our greenhouse emissions and our thermonuclear weapons, we might not survive much longer, even though bacteria will. Still, discussions about the definition of intelligence are best left to the field of psychology. For now, let us focus on human intelligence and look closer at the human brain.

It is the interaction between genetics, environment and epigenetics that make our brains unique [11]. "Studies in behavioral genetics indicate that the genetic influence on intelligence appears to be strongest later in life. That is, about 20% of individual differences in intelligence in children can be attributed to genetics, but 80% can be attributed to genetics in some studies of older adults. It is possible that genes nudge people toward choices that shape their environment in a particular way, which in turn affects their intellectual prowess" [11].

The normal human brain has many local regions, or centers, and many pathways between them. At each center, input fibers make synapses on the cell body, or soma (also known as the perikaryon) of the neuron. At the top of our brain is the cerebral cortex, also known as the neocortex [12]. It can be divided into a large neocortex and a smaller allocortex. It can also be divided into the temporal, occipital, parietal and frontal lobes, based on topography. The cerebral cortex is the part of the brain that sorts through the vast amount of information that we receive from our internal and external environments. It keeps us on time and enables the interpretation and expression of language [12]. It sorts out the information that we receive from our auditory, visual and olfactory senses. The neocortex looks gray and has six layers. It is comprised mostly of soma. The neurons send extensions called axons deeper into the brain. The axons are covered with a white myelin sheath, thus making the white matter of the brain.

The simplest neurons, though, are sensory neurons in the peripheral nervous system, or PNS [1]. They have one axon. The axons of sensory neurons are in groups in the sensory or dorsal root ganglia [13]. "The dorsal root (or posterior root) is the afferent sensory root of a spinal nerve. Afferent neurons carry signals toward the CNS. Axons conduct signals that are generated by sensory receptors in the skin or other organs. The signals go through the dorsal root and into the spinal cord or cranial nerve nuclei. Some neurons have two axons and are called bipolar. Examples include sensory cells in the retina, olfactory mucosa and auditory nerves, as well as small nerve cells in the brain that are called granule cells. Other nerve cells are multi-polar. They have only one axon, but they have a recurrent or collateral branch that feeds back onto the same type of nerve cells" [13].

Nerve cells are like all other human cells (except red blood cells), in that they have a nucleus that contains the genes and chromosomes. They also have other internal organelles, such as smooth and rough endoplasmic reticula, mitochondria and Golgi apparatus. They also have microtubules that keep the polarized shapes of the nerve cells. There is a protein called tau that helps form the rope-like structure of the microtubules, which are needed for neurons to communicate with each other and to guide nutrients and chemical messengers down the axons. Neurons are unique in that they have synaptic vesicles in their axons. The synaptic vesicles are spherical with diameters that range from 400 to 1200 Å [13], where 1 Å equals $1x10^{-10}$ m or 10 nm. They store neurotransmitters when a neuron is in the resting state. When activated, synaptic vesicles release neurotransmitters into the synaptic gap (or cleft) where they bind to their cognate receptors on the postsynaptic cell [13].

Neurons can also have several synapses with branched processes called dendrites that extend from the soma. The dendrites often have even smaller projections, called spines. Some neurons send out long axons that carry signals to other centers. These are called principal, relay, or projection neurons [13]. "Other neurons only process signals within a center and these are called intrinsic neurons, local neurons, or interneurons. The principal and intrinsic neurons, together with the incoming input fibers make up a triad of neuronal elements. This triad is the functional unit for the construction of neural circuits" [13].

A general principle in biology is that behavior depends on many levels of organization. In the brain, there are several important levels of organization. The most fundamental level is the synapse, followed by small clusters of synapses called microcircuits [13]. "These are grouped to form dendritic subunits within dendritic trees of individual neurons. The tree-like structures are patterns of outgrowths of neuronal dendrites. The whole neuron with its several dendritic subunits is the next higher level of complexity. Next, interactions between neurons form local circuits, which perform functions that are characteristic of different regions of the brain. Above this are the inter-regional pathways, columns, laminae and topographical maps. These involve several regions of the brain that mediate different types of behavior" [13]. However, all these levels of organization are inter-woven into an organization that more closely resembles circles than pyramids. That is, they are not organized in a hierarchy, with one on the top and others below them (as in a pyramid), but in an inter-dependent group in which all are important (as in a circle).

Still, scientists tend to break down the brain into smaller parts as they find that different regions of the brain that do different things. For example, we have already mentioned that the cerebral cortex is in the outer layer of the brain. The cerebral cortex can be further divided into different parts: the premotor cortex, motor cortex, primary somatic sensory cortex, parietal lobe, visual cortex and temporal lobes [1].

So, the anatomy of the brain can be divided into at least four parts and three pairs of lobes that are separated by deep, gray grooves. The parts are the right hemisphere, left hemisphere, cerebellum and brain stem [1]. "The three lobes are the frontal lobes, parietal lobes and occipital lobe. Beneath them is a fourth lobe, the temporal lobe, followed by the pons, cerebellum and brain stem. The three functions of the frontal lobes are tension, tact and tenacity. Some children with hyperactivity attention deficit disorder may have less activity in the frontal lobes. They have problems with tact, or controlling their social behavior. They also have trouble with tenacity, or the ability to stick with a task. The back edge of the frontal lobe has a cortex that controls movements in the "opposite side of the body. The left hemisphere controls movement on the right side and the right hemisphere controls movement on the left side of the body. The two hemispheres are connected by a broad band of nerve fibers called the corpus callosum" [1].

The second pair of lobes is called the parietal lobes, or association cortex. They are continuous with the other lobes of the brain. The parietal lobes are behind the frontal lobes, and in front of the occipital lobes. The parietal lobes also connect to the temporal lobe, so it can check for previous memories. The parietal lobes keep us oriented with respect to our internal and external environments.

The third lobe is called the occipital lobe, which is the visual lobe of the brain [1]. "It receives impulses from our eyes and then helps convert them into images in the brain. At the bottom of the brain are the cerebellum, pons and the brain stem. The primary somatic sensory cortex, or somatosensory system, reacts to different sensory stimuli, including pain, touch, temperature and propioception (body position). It is part of the parietal lobe, which integrates sensory information of different types, including visual, which is done in the visual cortex. The visual cortex is connected to the retina of the eye, which is also considered to be part of the brain. The temporal lobes are parts of the cerebrum. They are involved in speech, memory and hearing" [1].

The premotor cortex is often thought of as being part of the motor cortex. The premotor cortex helps guide our movements and controls some of the muscles in the body. The motor cortex works with the premotor cortex to plan and control body movements.

The part of the body that most people think of as the brain is also called the cerebrum, telencephelon or forebrain [1]. "Telencephelon refers to the embryonic structure from which the cerebrum develops. The dorsal telencephelon develops into the cerebral cortex. The ventral telencephelon develops into the basal ganglia. The brain or cerebrum is also divided into two nearly symmetric hemispheres, separated by the corpus callosum" [1].

Then there is the limbic system, which helps form the border between the cerebral cortex and subcortical layers. It is near the center of the brain [1]. The hippocampus, hypothalamus, thalamus, olfactory bulb, fornix, amygdala, anterior thalamic nuclei and limbic nuclei are all in the limbic system, which supports many functions. This includes emotion, behavior, long-term memory, and smell (olfaction).

The cerebellum is located between the brain stem and the rest of the body [1]. It helps integrate sensory perception, coordination and muscle control. At the top of the brain stem is the pons, which relays sensory information between the cerebellum and cerebrum.

Outside the central nervous system, there are clusters of nerve cells called peripheral ganglia. They are distributed throughout the body. They are part of the PNS, which is composed of two divisions, the somatic and autonomic [13]. "The somatic nervous system is composed of the axons of spinal motorneurons, which directly innervate the skeletal muscles. There are no ganglia in the somatic nervous system, so there is no information processing there. Each impulse leaving the somatic nervous system reaches the neuromuscular junction unchanged" [13].

The autonomic nervous system is organized into three divisions: the sympathetic, parasympathetic and enteric [13]. "Each of these divisions is characterized by the presence of peripheral neurons, so it can not only relay information, but also process and modify it. In the sympathetic and parasympathetic nervous systems, the peripheral neurons are located in clusters called ganglia.

In the enteric nervous system, peripheral neurons are found in sheets called plexi, found in the wall of the gut" [13].

In the sympathetic and parasympathetic nervous systems, motor neurons (also called motoneurons) located in the spinal cord send their pre-ganglionic, lightly myelinated axons to the autonomic gangion cells [13]. "The ganglion cells send their postganglionic axons out to the smooth muscles and glands. The autonomic nervous system regulates the physiological state of a person by its effects on glandular secretion, cardiovascular function, intestinal motility and many other properties" [13].

The cell membranes of neurons can be excited by electric current and the movement of ions. The most active part of the brain, the cerebral cortex, is made up of different types of neurons, each with excitable membranes. Each cortical neuron connects with 1000 to 10,000 other neurons. The brain is often thought as a self-communicating organ. The brain spends most of its resources communicating among its own neurons. This communication is done through action potentials, caused by changes in electropotential and the movement of ions.

We are born with almost all the neurons that we need. As mentioned earlier, in the first 2.5 – 3 years of life, our brains grow as our glial cells grow and divide. Neurons and their cell bodies are a major component of the gray (or grey) matter in the brain. The function of gray matter is to route sensory or motor stimuli to interneurons of the CNS. Gray matter does our mental calculations and stores our memories. As mentioned previously, the top of our brain is the cerebral cortex, which is a densely compacted area of neurons. This is where we make our decisions and where most of the gray matter of the brain is located. There is also gray matter below the surface of the brain, in compact shapes called nuclei. Examples include the thalamus, striatum, putamen, globus pallidus and caudate nucleus. The gray matter also contains capillaries and glial cells, which can be either astroglia or oligodendrocytes. The gray matter also contains axons and dendrites. Axons send information to other neurons in the central nervous system, or to muscle cells in the PNS. Dendrites receive information from other neurons. Dendrites also have smaller projections called spines. The average adult human brain contains about 10^{11} neurons and about 10^{15} synapses [14]. However, synapse is a general term for connections between not just neurons, but also between cells in the immune system [15].

So, each neuron has a cell body and there are also longer axons (up to 1 meter) extending out from the cell body. These axons conduct electrical signals away from the soma to axonal termini, called synaptic knobs. Axons usually have many branches, allowing signals to be sent to many other neurons through their dendrites and spines [14]. "Axons are the primary transmission lines of the nervous system. When packaged in bundles, they make up nerves. Individual axons are typically about 1 μm across, but may be up to one meter long.

The longest axons in the human body, for example, are those of the sciatic nerve , which run from the base of the spine to the big toe of each foot" [14].

Dendrites are long, thin, and highly-branched. They receive signals from other neurons through synapses. So, neurons send signals by way of axons and receive them by way of dendrites. Signaling between neurons occurs by transmitting chemicals (neurotransmitters) from an axonal terminus to a dendrite. It goes across a synaptic gap, or cleft, of about 0.02 μm (20 nm).

In humans and other vertebrates, the axons are coated with an electrical insulation, called the myelin sheath [16]. This allows the passage of low currents through the axons at speeds up to 120 m/s (or about 270 mph). When the myelin sheath is damaged, it can cause diseases such as multiple sclerosis, or MS. The myelin sheath is formed by glial cells, which can act like glue to hold neurons in their proper place and are the most abundant cells in the brain [16]. About 90% of the cells in the human brain are glial cells [17]. "In mammals, glial cells are classified as microglia, astrocytes, Schwann cells, or oligodendrocytes. Most of the glia and the neurons share a common origin in the neuroectoderm of the developing embryo. However, microglia are part of the immune system and enter the brain through the circulatory system in early development" [17].

The myelinated axons make up another part of the brain, called white matter, which connects various parts of gray matter and carries nerve impulses. Although scientists once thought that the brain's white matter was merely passive infrastructure, we now know that it actively affects learning and can be involved in mental illness [16]. "Although the neurons that make up the gray matter may do the brain's thinking and calculating, white matter controls the signals that neurons share, and helps different brain regions work together. A new type of magnetic resonance imaging, called diffusion tensor imaging (DTI), has shown white matter in action, revealing its under-appreciated role. Myelin is only partially formed at birth and gradually develops in different regions throughout our twenties. The timing of its growth and the degree of completion can affect learning, self-control, and mental illnesses, including schizophrenia, autism and pathological lying" [16]. Moreover, astrocytes may have an active role in brain function and information processing during development and in the adult human brain [17]. Although astrocytes are not excitable by electric stimulus (as are neurons), they are excitable by Ca^{2+}.

Astrocytes allow neurons to function by providing energy and chemicals that are needed for neurotransmission [16]. "They also act as physical barriers between synaptic connections between neighboring neurons and remove excess neurotransmitters from the extracellular space. More recently, it has been shown that astrocytes help form synapses and modulate synaptic function by communicating with neurons" [16]. Signals from astrocytes also control the action potential waveform and influence synaptic strength [18]. By shaping axonal action potential waveform, astrocytes assist in the glutamatergic transmission in the hippocampus [18]. Astrocytes also communicate with blood vessels, causing increased blood flow during enhanced neuronal activity [19]. "They also help terminate neurotransmission and recycle the neurotransmitter, glutamate, in the glutamate-glutamine cycle. Also, there are two major classifications of astrocytes: protoplasmic astrocytes in the gray matter and fibrous astrocytes in the white matter" [19]. Moreover, astrocytes communicate with each other through waves of Ca^{2+} ions, and they can communicate with neurons [16]. So, synapses are more than just pre- and post-synaptic neurons, but also an astrocyte projection that surrounds the synapse [19].

Another area of the brain is the substantia nigra, which is black and is in the midbrain. Part of the substantia nigra is the *substantia nigra compacta*, which is responsible for dopamine production in the brain, and therefore plays a vital role in reward and addiction.

Another way to look at the nervous system is by physiology, or "the role that different neural pathways perform, regardless of whether they pass through the CNS or PNS [1]. "The somatic nervous system is responsible for receiving external stimuli and regulating activities that are under conscious control, such as coordinating the body's movements. The somatic nervous system is made up of the afferent and efferent systems. The afferent system carries nerve impulses towards the CNS, so they affect the CNS. The efferent system carries nerve impulses away from the CNS to effectors such as glands and muscles, so they are the effect of nerve activity" [1].

The other part of the physiological division of the nervous system is the autonomic nervous system. As mentioned previously, in addition to the enteric system, there are the sympathetic and parasympathetic nervous systems, each with afferent and efferent nerves. The autonomic nervous system (ANS) (or visceral nervous system) is the part of the peripheral nervous system that controls and maintains homeostasis in the body [1]. "These maintenance activities are mostly done without conscious control or sensation. The ANS affects the heart rate, digestion, respiration rate, salivation, perspiration, diameter of the pupils, the discharge of urine and sexual arousal. Whereas most of its actions are involuntary, some ANS functions (such as breathing) work in tandem with the conscious mind. Its main components are its sensory system, motor system (comprised of the parasympathetic nervous system and sympathetic nervous system), and the enteric nervous system. The sympathetic and parasympathetic nervous systems work to maintain a type of balance. They have opposite effects on the body. The sympathetic division is used in actions requiring quick responses. The para-sympathetic division is used in actions that do not require immediate reaction. The main actions of the parasympathetic nervous system are summarized by the phrase "rest and repose" or "rest and digest" (in contrast to the "fight-or-flight" of the sympathetic nervous system). The parasympathetic nervous system uses acetylcholine (ACh) as a neurotransmitter. ACh has two types of receptors, muscarinic and nicotinic. The sympathetic nervous system also uses acetylcholine. The first synapse in the sympathetic chain uses nicotinic acetylcholine receptors. The target synapse is mediated by adrenergic receptors, adrenaline and noradrenaline (also known as epinephrine and norepinephrine)" [1].

Neurotransmitters

Messages are sent to and from neurons in the form of primary messengers, called neurotransmitters. There are about 50 different neurotransmitters [14]. "For a compound to be considered a neurotransmitter, the classical definition says that it must be synthesized and released presynaptically, it must mimic the action of the endogenous compound and its pharmacological identity must be made. This is different from a neurohormone, which must travel in the blood, or circulation, and act at a distinct site. There are also neuromodulators, which have no intrinsic activity, but they modulate ongoing synaptic activity. However, these definitions have become less strict as neuroscientists have learned more. Dopamine is a neurotransmitter in the striatum, but is a neurohormone when it is released from the hypothalamus, through hypophyseal circulation to the pituitary gland, where it inhibits the release of prolactin. Serotonin is a neurotransmitter in the raphe nuclei, but in the facial motor nucleus it acts as a neuromodulator and as a neurohormone. Most neuropeptides have multiple activities in the brain and gut and are neuromodulators. However, substance P is a neurotransmitter at sensory afferent nerves in the dorsal horn and spinal cord. The plethora of exceptions to the aforementioned definitions of *transmitter, modulator,* and *hormone* has generated confusion in the literature. Better to describe the activity of a neuroactive agent at a specified site rather than to give a profitless definition" [14].

According to one reference, there are five distinct classes of neurotransmitters: ACh, monoamines, amino acids, peptides and purines [14]. ACh is a neurotransmitter in the PNS, CNS, autonomic ganglia and at neuormuscular junctions in skeletal muscle. Cholinergic neurons are depleted in Alzheimer's disease, which will be discussed more in a few pages.

Monoamine neurotransmitters include dopamine, norepinephrine, epinephrine, histamine, serotonin and melatonin. Neurons make them and store them in presynaptic vesicles. When they are released into the synaptic cleft, they bind to cognate receptors on the postsynaptic side and cause a response. It is important that the response not last too long, so seizures don't occur. So, there are vesicular monoamine transporters, or VMAT, which transport the neurotransmitters back into the cell. Re-uptake of monoamines limits their concentration in the synaptic cleft and the length of the cellular response. VMAT is the main target for the deadly drug methamphetamine. It is a competitive antagonist that prevents the presynaptic cell from using VMATs to package neurotransmitters into vesicles. The result is increased release of neurotransimtter that is not dependent on the electrochemical activity of the presynaptic cell.

Norepinephrine and epinephrine are hormones as well as neurotransmitters. Norepinephrine and epinephrine are stress hormones that are involved in the fight or flight response. Epinephrine also binds to receptors in the liver, where it activates the phosphoinositide signaling system. It then causes the breakdown of glycogen to form glucose, which is released into the blood. Histamine has many roles in the body, including acting as a neurotransmitter. Antihistamines (H1 histamine receptor antagonists) produce sleep. On the other hand, destruction of histamine releasing neurons, or inhibition of histamine synthesis inhibits sleep [1]. Histamine is also important in the immune response and in allergies, as discussed in Chapter 7.

Serotonin (5-HT or 5-hydroxytryptamine) has a broad homeostatic function [12]. "Serotonin is a neurotransmitter that affects anger, aggression, body temperature, mood, sleep, aspects of sexuality, appetite and much of metabolism. Serotonin is also important in the PNS, especially in the gut, where about 80-90% of it is found. Serotonin is also found in blood platelets, where it helps mediate vasoconstriction after an injury. It is a ligand for 15 different receptors. Drugs that target these receptors are used to treat migraine headache, depression, anxiety, nausea, vomiting, and irritable bowel syndrome" [12]. The type 3 5-HT receptors are ligand-gated ion channels, but all the others are GPCRs.

As a neurotransmsitter, melatonin is important in regulating the circadian rhythm and the sleep-wake cycle. Melatonin is also an antioxidant. It helps protect mitochondria and nuclear DNA from oxidative damage. Melatonin also interacts with the immune system. It is also involved in memory and learning.

The two most abundant neurotransmitters in the human brain are glutamate and γ-amino butyric acid (GABA) [1]. A form of NMR called magnetic resonance spectroscopy, or MRS, was able to measure the concentrations of glutamate, GABA and other compounds in the brain. The concentration of glutamate in the human motor cortex was about 10 mM (10 μmol/cm^3) in younger people, but was lower in the elderly [19]. The concentration of GABA was about 1 mM [20]. Almost all the glutamate is inside the brain cells; that is the neurons and glia. Glutamate excites neurons (so it is called excitatory), while GABA inhibits them. At the same time, glutamate is a metabolic precursor of GABA. Other excitatory amino acid neurotransmitters include aspartate, cysteate and homocysteate [12]. These depolarize neurons in the CNS. Other inhibitory amino acids include glycine, taurine, and β-alanine, which hyperpolarize neurons. So, GABA is the major inhibitory neurotransmitter in the brain and glutamate is the major excitatory neurotransmitter [12]. Neurons that use glutamate and neurons that use GABA are located throughout the CNS and send projections to other neurons throughout the CNS.

Another extremely important neurotransmitter is ACh. In the brain, ACh modulates the excitability of other neurons. The cell bodies of neurons that use ACh are located in basal forebrain and send projections to the cerebral cortex [1]. "The cell bodies of other neurons that use ACh are in the pontomesencephalic region and send projections to the thalamus, cerebellum, pons and medulla. The cell bodies of other cholinergic neurons are located in the spinal cord, autonomic preganglionic nuclei and parasympathetic nuclei. They send projections to skeletal muscles, autonomic ganglia, glands, smooth muscle and cardiac muscle" [1].

Glycine is an important inhibitory neurotransmitter in the medulla, spinal cord, retina and brain stem [1]. A synthetic amino acid, N-methyl-D-aspartic acid

(NMDA), is an agonist of glutamate [1]. It binds to a subset of glutamate receptors, called the NMDA receptors. Glycine is a modulator of NMDA receptors.

There is also an important derivative of aspartate, N-acetylaspartate, which is the second most abundant molecule in the brain (after water) and it produces the strongest signal in the *in vivo* ^{1}H-NMR spectrum of the brain (when the peak due to H_2O is suppressed) and the strongest signal in the anion chromatography of brain, when suppressed conductivity detection is used. The concentration of N-acetylaspartate was found to be 8.81±0.18 mmol/kg in the white matter of the parietal cortex and 9.24±0.26 in the gray matter of the occipital lobe [20]. Other compounds present at high enough concentration to be seen by NMR in that study were creatine plus phosphocreatinine, choline, and *myo*-inositol, present at concentrations of 6.33, 1.58 and 6.68 mmol/kg in white matter and 8.04, 1.43 and 7.26 mmol/kg in gray matter [20]. The molar ratio of creatine and phosphocreatine is 4:6 because of the equilibrium in the reaction catalyzed by creatine phosphokinase. Choline is the breakdown product of acetylcholine esterase catalyzed hydrolysis of ACh. The ^{1}H-NMR peaks produced by choline also has contributions due to phosphocholine and glycerophosphocholine. The *myo*-inositol peak is due to not just inositol, but also *myo*-inositol 1-phosphate and inositol polyphosphates such as IP_3,which is an important second messenger in brain cells and in other cells throughout the body. In another study, glutamate and glutamine were also seen by ^{1}H-NMR [20].

N-acetylaspartate is a neuronal marker and is involved in maintaining the fluid balance in the brain. It is also a source of acetyl groups for the production of lipids in the myelin sheath, and it is a precursor for the synthesis of the important neurotransmitter, N-acetylaspartylglutamate, a dipeptide [20]. When the concentration of N-acetylaspartate decreases, it can be interpreted as a loss of neurons, as can occur in patients with chronic pain [21]. N-acetylaspartate is also a contributor to the production of energy in mitochondria [20]. The concentration of N-acetylaspartate is also depleted in the brains of people with brain injury, stroke and Alzheimer's disease [20].

There are many neuroactive peptides and more will probably be discovered in the future. All known neuroactive peptides are derived from larger prohormones and are active in both the CNS and PNS [12]. In the early years of neuroscience, it was thought that each type of neuron responded to only one neurotransmitter. Neurons were labeled as cholinergic, GABAergic or adrenergic. We now know that neuroactive peptides also appear in neurons in both the CNS and ANS. Table **1** shows some examples of neuropeptides and the neurotransmitters that affect them [12].

Table 1. Examples of neurotransmitters that affect neuropeptides, with their locations.

Transmitter	**Peptide**	**Location**
GABA	Somatostatin Cholecystokinin	Cortical and hippocampal neurons Cortical neurons
Acetylcholine	VIP Substance P	Parasympathetic and cortical neurons Pontine neurons
Norepinephrine	Somatostatin Enkephalin NPY Neurotensin	Sympathetic neurons Sympathetic neurons Medullary and pontine neurons Locus ceruleus
Dopamine	CCK Neurotensin	Ventrotegmental neurons Ventrotegmental neurons
Epinephrine	NPY Neurotensin	Reticular neurons Reticular neurons
Serotonin	Substance P TRH Enkephalin	Medullary raphe Medullary raphe Medullary raphe
Vasopressin	CCK, dynorphin	Magnocellular hypothalamic neurons
Oxytocin	Enkephalin	Magnocellular hypothalamic neurons

Abbreviations: VIP – vasoactive intenstinal polypeptide, NPY – neuropolypeptide Y, TRH – thyrotropin releasing hormone, CCK – cholecystokinin

Like all other neurons that release neurotransmitters, the release of neuropeptides is dependent on Ca^{2+} [12]. All known receptors for neuropeptides are G-coupled proteins. The effects of neuropeptides are mediated indirectly by second messengers, such as intracellular Ca^{2+}, cyclic nucleotides or inositol triphosphates. Like the monoamines, neuropeptides are also made by non-neuronal secretory cells in endocrine glands and cells lining the hollow viscera, such as the intestinal tract. As a result, some look at the wholeness of life and tend to think of the CNS, PNS, endocrine system, digestive system, immune system and the entire body as one interconnected network. Indeed, anyone who remembers going through puberty can attest to the fact that the endocrine system and its hormones can seem to take over the brain and cause us to make emotional decisions that the brain by itself would not make. Likewise, many of us have experienced a nervous stomach. Stress can have many different effects on the eating habits and digestion of different people. Also, excess worrying can adversely affect the immune system and the rest of the body, causing hypochondria and even real diseases. Depression is a major risk factor in developing Alzheimer's disease. Moreover, the immune system has a memory that is sometimes even better than that of the brain. Our white blood cells retain memories of pathogenic organisms that have invaded us in the past. Some of our white blood cells remember the difference between self and non-self. In a healthy immune system, our white blood cells don't attack cells identified as self, but do attack new cells that appear in the human body. However, our white blood cells are able to predict the future by being able to identify newly emerging sex cells as self when they appear during puberty. Our brains, though, have little or no ability to predict the changes that happen at the onset of puberty.

Neuropeptides

Let us look more at some different families of neuropeptides. Vasopressin and oxytocin are nonapeptides (nine amino acids) with internal 1,6-disulfide bridges [12]. "They are synthesized in the soma (or perikarya) in the relatively large neurons of the supraoptic and paraventricular nuclei and are stored in the axons of these neurons, which form the neurohypophysis, also known as the posterior pituitary. They are secreted into the pituitary portal circulation where vasopressin can act synergistically with corticotropin-releasing hormone to release ACTH (adrenocorticotropic hormone). Oxytocin is secreted during labor, delivery and nursing of babies. Vasopressin stimulates pair bonding in male prarie voles. In the kidney, vasopressin (also known as the antidiuretic hormone) facilitates water reabsortion, while oxytocin stimulates epididymal and uterine muscle contraction" [12].

Another class of neuropeptides is one called tachykinin peptides [12]. "It is a large class that is found in animals ranging from amphibians to mammals. They were named for their ability to rapidly induce contractions in the gut. The genes that produce tachykinins code for larger precursor polypeptides called preprotachykinins. There are two mammalian tachykinin genes. Different tachykinins are formed by differential slpicing of the mRNA of preprototachykinins. The least common mRNA codes only for substance P and the other two forms encode both substance P and NKA, also known as substance K. The second tachykinin gene, on human chromosome 12, encodes the neurokinin B (NKB) precursor. Substance P was the first tachykinin that was identified when scientists found it in the acetone extracts of the brain and intestine. Substance P is an undecapeptide that is found in small neuronal systems in many parts of the CNS. It is a potent depolarizing agent in the spinal cord. "Tachykinins excite neurons, evoke behavioral responses, are potent vasodilatators and contract (directly or indirectly) many smooth muscles. Tachykinins have ten to twelve amino acids. Tachykinin peptides are also involved in inflammation , and tachykinin receptor antagonists have been studied for use in treating inflammatory conditions such as asthma and irritable bowel syndrome " [12].

Another class is the VIP-related peptides [12]. "VIP (vasoactive intestinal peptide) is composed of 29 amino acids. In the neocortex, VIP-reactive neurons are quite abundant. In the PNS, VIP-active neurons innervate the gut, lungs and possibly the thyroid gland. Two other members of this family are called PHI-27 and PHM-27, and have 27 amino acids. At their N-terminus, there is a histidine (H) residue and at the carboxy terminus there is either an isoleucine (I) or a methionine (M) in PHI and PHL, respectively. Both are active in the gut. Next, there is the pituitary adenylate cyclase-activating peptide (PACAP), which is in the pituitary gland. Two other members of this class are found in the hypothalamus and they stimulate appetite and help regulate blood pressure. They are called either hypocretin-1 and -2 or orexin A and B" [12].

The next class is pancreatic polypeptide-related peptides [12]. "The first one found was given the name PPY (a neuropeptide that contains tyrosine, or Y). It was also found in the brain and named neuropeptide Y, or NPY. In many places, including the autonomic nervous system, NPY coexists with either epinephrine or norepinephrine. On the other hand, a shortened version of NPY binds to a different receptor (Y_2) and reduces the release of norepinephrine. The full NPY binds to the Y_1 receptor and increases the sensitivity of smooth muscles to norepinephrine and is one of the most potent vasoconstrictors known. During fasting, the concentration of NPY in the hypothalamus decreases" [12].

The next class is the opioid peptides [12]. "They are produced by the pituitary gland and the hypothalamus during strenuous exercise, excitement , and orgasm . They include dynorphins, endorphins and enkephalins. Endorphins have the same pharmacological profile of morphine. There are at least 12 endorphins. The major endorphin agonist is called β-endorphin. "Enkephalins are pentapeptides that are involved in regulating pain and nociception, or the activity produced in the peripheral and central nervous system by stimuli that have the potential to damage tissue. The enkephalin pentapeptides that are called Met^5-enkephalin and Leu^5-enkephalin are expressed in different neurons than the POMC neurons and are more widely distributed throughout the CNS and PNS. The prodynorphin peptides consist of forms of Leu^5-enkephalin that are extended at the carboxyl terminus. There are four major dynorphin peptides: dynorphins A, B, αand β. They are widely distributed in both the CNS and PNS" [12].

Somatostatin is another interesting neuroactive polypeptide. It inhibits the release of growth hormone [12]. "Also, it is widely distributed in the gastrointestinal tract and in the δ cells of the pancreas, where it can suppress the release of both glucagon and insulin. It is also largely concentrated in the mediobasal hypothalamus, with smaller amounts in other regions of the brain, including the dorsal root ganglia, amygdala, hippocampus and neocortex. Somatostatin hyperpolarizes neurons and can open M current channels, which muscarinic cholinergic receptors close so they can excite hippocampal and other neurons. Thus, somatostatin enhances responsiveness to acetyl choline. Somatostatin is also found in GABAergic neurons, but its effect on them was not known for a long time. A larger version of somatostatin, extended by 14 amino acids at the N terminus, also exists in the brain and gut and is equipotent with somatostatin" [12].

Cholecystokinin is another polypeptide that is a hormone in the gut, but is a neuroactive peptide in the brain [12]. "It coexists with VIP, NPY and GABA in the thalamocortical and thalamostriatal connections; with dopamine and neurotensin in the substantia nigra and "with substance P and serotonin (5-HT) in medullary neurons. Cholecystokinin is an opioid antagonist. Antagonists to one of the cholecystokinin receptors enhance opiate analgesia" [12].

Neurotensin is a tridecapeptide and one of the few that does not have an amide group at its C-terminus [12]. "Neurotensin is a potent analgesic and it induces hypothermia, accentuates barbituate and alcohol sleeping time and it increases the release of prolactin and growth hormone. Neurotensin is located primarily in the anterior and basal hypothalamus, the nucleus accumbens and septum, the mid-brain dopamine neurons and "some of the neurons in the brain stem and spinal cord" [12].

Corticotropin-releasing factor, or CRF, is found in the hypothalamus [12]. "It has 41 amino acids and it controls the secretion of corticotropin. It also increases the frequency of action potentials of neurons that fire in bursts, such as the hippocampal pyramidal cells. CRF can also induce seizure-like activity in the limbic system. It can also activate spontaneous locomotion and cause an anxiogenic response" [12].

Another neuroactive polypeptide is called the extracellular leucine-rich repeat fibronectin containing 1 (Elfn1), because it contains a repeating amino acid sequence that contains many leucine residues [21]. "It is selectively expressed by oriens–lacunosum moleculare (O-LM) interneurons. It is selectively expressed by O-LM interneurons and regulates the probability of presynaptic release so it can direct the formation of highly facilitating pyramidal-O-LM synapses. That is, even though synaptic transmission may be unidirectional, synaptic connections can be established that have bidirectional signaling. Pyramidal neurons in the hippocampus form functionally distinct synapses onto two types of interneurons. Excitatory synapses onto O-LM interneurons are facilitating and have a low release probability, but synapses onto parvalbumin interneurons are depressing and have a high probability of release of neurotransmitters" [21].

Nitric Oxide and Carbon Monoxide

On the other hand, there are two especially unusual neuroactive compounds, nitric oxide, or NO, and carbon monoxide, CO, which are gases at room temperature and are synthesized only when needed [12]. NO is synthesized from arginine in a reaction that is catalyzed by nitric oxide Synthase. All the other neurotransmitters are synthesized in the soma and are stored near the terminus of the axon in presynaptic vesicles. The neurotransmitters are released when the neuron reaches its "firing" milliVoltage, and generates an action potential. The neurotransmitter diffuses across the synaptic gap and binds to its specific post-synaptic receptor [14]. The receptor becomes activated, changes shape and triggers the opening of ion channels. This causes membrane depolarization and an action potential. This system of neurotransmission is important for mental and physical health.

NO complexes with iron in the water-soluble form of guanylate cyclase and stimulates it, so it can catalyze the production of cGMP [12]. "The rise in cGMP activates cGMP-dependent protein kinases, which catalyze the phosphorylation of other proteins. In cerebellar neurons, excitatory amino acids cause the release of NO, as did stimulation of noradrenergic peripheral neurons. NO is also thought to play a role in long term potentiation and long term depression. NO also has many effects outside the CNS and PNS. It is the endothelium-derived relaxing factor. The endothelium is the inner lining of blood vessels. When stimulated by NO, the endothelium signals the surrounding smooth muscle to relax, causing vasodilation and increased blood flow. NO is also generated by macrophages and neutrophils so they can help kill bacteria and other pathogens. NO can also contribute to ischemic and reperfusion injury" [12].

The synthesis of NO can be a target for therapeutic drugs [22]. "There are endogenous inhibitors of NO synthesis, such as asymmetric dimethylarginine (ADMA) — a naturally occurring amino acid that is a product of protein breakdown. It is released into the cytoplasm following the post-translational methylation of arginine residues within proteins and the subsequent proteolysis of these arginine-methylated proteins. ADMA inhibits all three isoforms of NO synthase. So, it can potentially produce diverse biological effects, particularly in the cardiovascular system. It is cleared in the kidneys and metabolized to L-citrulline and dimethylamine by the dimethylarginine dimethylaminohydrolase (DDAH) enzymes. So, pharmacological modification of DDAH may change the concentrations of ADMA, thus regulating the production of NO in situations where alterations in NO signaling contribute to pathophysiology" [22].

CO is generated by the conversion of heme to biliverdin, which is catalyzed by heme oxygenase [12]. Like NO, CO raises the level of cGMP. Heme oxygenase is found in the same location as guanyl cyclase and cytochrome P-450 reductase, which donates electrons for heme oxygenase [12].

Glial Cells

As mentioned before, the human brain also contains supporting cells called glial cells, or glia, from the Greek word for glue. That is, they were once thought of as the "glue" of the brain. They provide support and nutrition, form myelin and participate in signal transmission [1]. "Glial cells outnumber neurons by about 10 to 1 in the human brain. Classically, the four main functions of glial cells were thought to be to surround neurons and hold them in place, to supply nutrients and oxygen to neurons, to insulate one neuron from another, and to destroy pathogens and remove dead neurons. Glial cells also help in the development of a baby's brain. They guide neurons to their proper places and modify the growth of axons and dendrites. Unlike neurons, glial cells can undergo mitosis. More recently, it has been shown that it is incorrect to think of glial cells as merely glue to hold neurons together. Instead, they can be considered to be partners with neurons. Glial cells have a role in the regulation of repair of neurons after injury. In the CNS, glial cells can supress repair. One type of glial cell, called astrocytes, can enlarge and proliferate to form a scar and produce molecules that inhibit regrowth of a damaged or severed axon. In the peripheral nervous system (PNS), a type of glial cell called a Schwann cell makes myelin and can promote repair. After axon injury Schwann cells regress to an earlier developmental state to encourage regrowth of the axon. This difference between PNS and CNS raises hopes for the regeneration of nervous tissue in the CNS, for example a spinal cord injury or severance" [1].

Another type of glial cells, called microglia are specialized macrophages that are capable of phagocytosis, or the cellular process of engulfing solid particles by the cell membrane to form an internal phagosome, or food vacuole [23]. "Microglial cells can move around in the brain and multiply when the brain is injured. They can sense pathological changes and develop into macrophages and perform immunological functions" [23]. When present in healthy brain tissue, they are not macrophages. They help maintain the integrity of synapses. They sense defunct synapses and eliminate them. They also control the production of new synapses. When a person suffers a stroke, or develops Alzheimer's disease, reactive oxygen substances cause oxidative damage and proliferation of microglial cells [24]. Antioxidants, such as resveratrol, curcumin and apocynin, along with antioxidant-rich foods like blueberries, spinach, and spirulina can prevent the oxidative damage and brain damage caused by a temporary lack of blood flow and oxygen (ischemia), followed by reperfusion [25 - 27]. Interestingly, this damage requires the presence of O_2. It has been found that small animals, like mice, can be put into a state of suspended animation by putting them in an atmosphere of H_2S.

Diseases of the myelin, which are among the most prevalent and disabling conditions in neurology, include vascular and inflammatory demyelinating disorders of adulthood, as well as childhood leukodystrophies and cerebral palsy [28]. "These fundamentally glial disorders may be treated by glial progenitor cells (GPCs), which give rise to astroglia and oligodendrocytes that produce myelin. That is, oligodendrocytes are the sole source of myelin in the adult CNS. Their loss or dysfunction is can cause many child and adult diseases. Moreover, demyelination is involved in degenerative disorders such as Alzheimer's disease, amyotrophic lateral sclerosis and schizophrenia, as well as normal aging. Thus, demyelinating diseases may be effective targets for cell-based therapies. Axons can remyelinate after either congenital or acquired demyelination, if they are provided with myelinogenic cells. Therefore, GPCs - also referred to as either oligodendrocyte progenitor cells or NG2 cells – may become promising reagents for restoring myelin to demyelinated regions of the diseased or injured CNS" [28].

Glial cells can also be transformed into malignant cells, causing a form of brain cancer called glioblastoma multiforme. It can grow rapidly and kill a patient within a few weeks. The standard treatment (for those who choose to be treated) is a combination of brain surgery (resection), radiation and chemotherapy using the drug called temozolomide, or Temodar. The median survival for such patients is about 14 months. About 30-50% of patients express a mutant form of the epidermal growth factor receptor, called EGFRvIII. In a clinical trial, some patients received a vaccination, consisting of a peptide (PEPvIII-KLH) plus the immune stimulant granulocyte-macrophage colony-stimulating factor (GM-CSF) every 2 weeks followed by monthly booster shots [29]. "Their mean survival time increased to 18 months. More recently, the peptide vaccine called CDX-110 has been used an in currently in a phase 2/3 clinical trial, and the vaccine has been granted fast-track status by the FDA. In another clinical trial, a vaccine against cytomegalovirus (CMV) is being given. Patients will receive a surgical resection. Then, a process called leukapheresis (that separates and removes white blood cells) will be done "to obtain peripheral blood lymphocytes (PBLs) for human cytomegalovirus (CMV)-autologous lymphocyte transfer (ALT) and CMV-dendritic cell (DC) generation. Patients will then receive radiation therapy. Beginning on day one of radiation therapy, they will receive temozolomide once daily for up to 49 days. Within 3-4 weeks of receiving this treatment, patients will receive oral TMZ once daily on days 1-5. Treatment repeats every 4 weeks for up to 6 courses in the absence of disease progression or unacceptable toxicity. Beginning on day 21-23 of course 1, patients also receive an intradermal immunization and are randomized to 1 of 2 vaccine treatment arms. If the initial dose is found to be safe, another cohort of patients will receive a higher dose. If this works, then a fourth cohort of patients will receive the same treatment, except the T-cells will be labeled with ^{111}In and cultured *ex vivo* with deuterated glucose. The migration of these T-cells will be monitored by MRI and single photon emission computed tomography" [29].

In more recent work, D. Mitchell, M.D., Ph.D. reported that a vaccine against CMV has helped some patients survive for almost three years after being diagnosed with glioblastoma multiforme [30]. "CMV is a type of herpes virus that is found in almost 80% of the people in the USA. Also, more than 80% of the patients with glioblastoma multiforme have CMV in their blood and brain tumors. This does not mean that CMV causes brain cancer. It could be that the cellular changes in cancer cells and their supporting cells are better environments for CMV to grow. Still a vaccine that targets CMV will be able to kill the cells that contain it, so clinical trials are continuing" [30].

Still, scientists continue to learn more about healthy glial cells. As mentioned before, it had been thought that glial cells were mainly acting like glue and simply supported neurons. Recently, though, researchers reported that astrocytes (a type of glial cell) in the visual cortex of ferrets can process information [31]. Like neurons, astrocytes respond to visual stimuli with distinct special receptive fields and sharp tuning to visual stimuli [31]. Astrocytes that are stimulated by light also affect blood flow in blood vessels and can affect MRI and other imaging techniques that monitor blood flow by observing the water in blood [32]. Astrocytes are the most abundant type of glial cell in the brain and they communicate with neurons through neurotransmitters [32].

Pain Control

So, our neurons and glial cells interact in a complex network, in the web of life. They can help us feel good and help us feel pain when we are affected by harmful stimuli. As one book stated, "Human consciousness includes the constant awareness of an individual's state along a euphoria-dysphoria axis. At the extreme dysphoria end is pain, which in many respects is a sixth sense, no less functional than the other five. Without a sense of pain, the warning sign of illness or injury to the body, life would be far more tenuous" [14]. So, one of the main goals of neuropharmacology is the treatment of pain, both physical and psychological.

Aspirin (acetylsalicylic acid) and other non-steroid anti-inflammatory drugs, or NSAIDs inhibit the enzyme cyclooxygenase, or COX, which catalyzes the conversion of arachidonic acid [1] into the pro-inflammatory biochemical, prostaglandin G_2. Acetylsalicylic acid inhibits COX-1. Ibuprofen and naproxen reduce pain by inhibiting COX-2. It should be noted that arachidonic acid can also be produced by the hydrolysis of phospholipids that contain the arachidonoyl group on carbon number 2. Enzymes that catalyze the hydrolysis of phospholipids are called phospholipases [1]. Phospholipase-2 enzymes catalyze the hydrolysis of fatty acyl groups (like arachidonoyl) on carbon number two of a phospholipid [1]. Recent research has found that inflammatory damage in Alzheimer's disease can be caused by activation of phospholipase-2, which causes the release of arachidonic acid [33]. However, arachidonic acid and prostaglandin G_2 are not the only things that can cause pain. Sodium ion channels can cause pain, too. Lidocaine is an analgesic that blocks Na^+ channels. Lidocaine was made by "rational drug design, in which the structure of cocaine was modified [1].

A natural, but very addictive and deadly drug, cocaine, acts as an analgesic (pain killer) by inhibiting the reuptake of dopamine, noradrenaline and serotonin. So, remember that nature is not benign. Many natural substances, like cocaine and morphine, are quite deadly and addictive. So, pharmaceutical companies have developed analgesics that aren't deadly or addictive, if used properly under the care of a physician. The effort to find compounds with analgesic properties and low toxicity was rewarded in 1905 by the synthesis of procaine by the German chemist, Albert Einhorn [14]. "During the 1930s, two Swedish scientists Nils Lofgren and Bengt Lundquist, synthesized and tested several analogues of procaine and they produced lidocaine, which acts faster (5 – 10 min) and lasts about two hours. In 1948, the Swedish pharmaceutical company, Astra, began large-scale production and marketing of lidocaine as Xylocaine. It blocks the Na^+ ion channels of cell membranes. As a result, the transmission of pain signals is blocked" [14].

Another natural but deadly and addictive drug is morphine. It is the active ingredient in opium, which has been used for over a thousand years to treat pain [13]. More than 230 tons of morphine are used legally each year for the medical treatment and relief of pain [14]. It can be especially useful in relieving the pain due to cancer [34]. Providing pain relief is not just an end in itself, but it can improve the patient's prospect for survival [34]. With proper medical supervision in the treatment of extreme pain, morphine often is not addictive [34]. It acts by binding to high-affinity binding sites, aptly called opioid receptors. Morphine produces analgesia, sedation, euphoria, tranquility, along with slower respiratory, cardiovascular and gastrointestinal function [14]. "There are three classes of opioid receptors: μ, δ and κ. Mostly, morphine binds to the μ-opioid receptor. The effects of μ agonists can be reversed by the antagonist called naloxone. Partial agonists of the μ-receptor, such as methadone, can be used to treat the addiction to opiates, such as heroin. At larger doses of morphine, the μ-opioid receptor interacts with the δ and κ receptors and may change its pharmacological effects. Agonists of the (and κ receptors have limited analgesic (pain relieving) effects and many side effects. They are seldom used in medicine. There are natural ligands for these receptors. They are peptides and are called enkephalins, endorphins and dynorphins" [14].

There is a much stronger analgesic, called fentanyl (Duragesic®) that also binds to the μ-opioid receptor [14]. "It is 100 times more potent than morphine and is used as an adjuvant in surgical procedures with intravenous or epidural administration. Slow release (24 hours) transdermal (skin) patches of fentanyl, such as

Duragesic®, are widely used" [14]. Another analgesic, called remifentanil, is the shortest acting opioid (5-10 min) available for human use. Another drug, called sufentanil is 1000 times as potent as morphine and is used in heart surgery, which causes much damage to this highly innervated tissue [12]. An even more potent analgesic, called carfentanil, is 10,000 times as potent as morphine and is only used on animals [14].

A much milder and non-addictive pain reliever is acetaminophen (Tylenol®). It is the most widely used analgesic and antipyretic agent globally [14]. "Unlike aspirin and other NSAIDS, acetaminophen does not inhibit COX-1 or COX-2. Its exact mechanism of action is unclear. Acetaminophen is the preferred drug to relieve pain and fever in infants, but it has been linked to the development of asthma later in life when it is given in the first year of life. It should be reserved for children with a high fever of 38.5°C (101.3°F). Also, the toxic dose of acetaminophen is not much higher than its therapeutic dose. The toxicity increases very much when mixed with alcohol. Therefore, it is imperative that people taking acetaminophen do not drink alcohol. It can cause liver failure and death" [14]. Also, infants should never be given alcohol! In adults, the combination of 350 mg acetaminophen and 5 mg of oxycodone (a semi-synthetic opioid) is an important pain-relieving medicine that lasts about six hours, but it is available only by prescription [14]. The two drugs act synergistically. It was first marketed as Percocet, but is now available as a generic prescription medicine.

Barbituates are another class of drugs that can be used (and mis-used) as analgesics [14]. "Barbituates are derivatives of barbituric acid, which can be made by reacting diethyl malonate with urea. The first medically useful barbiturate was a sedative called barbital. It was marketed in 1904 by Bayer as a sleeping aid called Veronal. More barbituric acid derivatives were made and there are a few used in medicine today. They depress the CNS and produce sedation, sleep and even coma in some people. Abbott Laboratories made pentobarbital, sold under the name Nembutal™. Later, an oxygen was replaced by a sulfur to make thiopental, a very short-acting barbiturate. A standard intravenous dose will render a person unconscious within one minute and the duration of anesthesia is less than 10 minutes. Thus, thiopental has been used for minor surgery. However, a non-barbituate, propofol, has replaced it, because patient's recover from it much better" [14].

Thiopental and other barbiturates act by binding to the $GABA_A$ receptor [14]. "It is a ligand-gated chloride ion channel. Its endogenous agonist is GABA, the major inhibitory neurotransmitter in the brain. When GABA binds to the $GABA_A$ receptor, it causes the Cl^- ion channel to open, leading to hyperpolarization and inhibition of neuron firing and action potentials. However, an overdose of thiopental can cause death, and a high dose of thiopental is used in lethal injections when executing a death sentence in the USA" [14].

Another drug that binds to the $GABA_A$ receptor is diazepam (Valium®). By the mid-1970s, more than 8000 tons of benzodiazepines (Valium® and Librium®) were made annually [14]. "Benzodiazepines bind to the $GABA_A$ receptor in the CNS at a different site than the one at which barbiturates bind. It is called the benzodiazepine site. Binding causes a conformation change, which enhances the binding of GABA. The in late 1970s, a new benzodiazepam, called Xanax® was made. The use of benzodiazepines causes side effects such as amnesia and dependence, so long-term use is contra-indicated. However, the abrupt withdrawal of benzodiazepines after long-term use can lead to withdrawal symptoms such as anxiety, insomnia and seizures" [14].

Another application for drugs that bind to the $GABA_A$ receptor is treating insomnia [14]. "Zolpidem (Ambien®) has a rapid onset of action (usually 10 – 30 min) and extended release forms of Ambien®, called Ambien CR® lasts for at least three hours and enables deep sleep. The $GABA_A$ receptor has three subunits, called α,β and γ. GABA binds to the (subunit and opens a Cl^- ion channel. There are different subtypes of the (receptor. Benzodiazepines bind non-selectively to all α subtypes, but zolpiden binds selectively to the $(_1$ subtype, which is thought to explain its effects as a muscle relaxant and in preventing epileptic seizures. There are two other FDA-approved hypnotic agents (sleeping pills), called zaleplon and eszopiclone (Lunesta®). Although zaleplon is less effective than benzodiazepines, it can be used for long-term treatment without serious withdrawal symptoms when it is discontinued" [14].

Another use for inhibitors of neurotransmission is the treatment of epileptic seizures. Such drugs, called anticonvulsants, can affect voltage-dependent Ca^{+2} ion channels, which play an important role in the release of neurotransmitters from presynaptic vesicles [14]. "The drug called gabapentin (Neurontin®) binds to the $\alpha_2\delta$ subunit of a multiprotein complex that regulates a specific Ca^{+2} ion channel. The α_2 and δ subunits are next to each other in the cell membrane. They enhance Ca^{2+} influx through the ion channel. Gabapentin inhibits the influx of Ca^{2+} and decreases the excitability of neurons, preventing seizures and reducing the transmission of pain" [14].

Another target for hypnotic drugs is the melatonin receptor. Melatonin is a hormone that regulates the circadian rhythm, the 24 hour cycle of biochemical, physiological and behavioral processes, including sleep and wakefulness [14]. "Melatonin is formed in the brain by a series of enzyme-catalyzed steps that start with the amino acid tryptophan, which is converted to serotonin and eventually to melatonin. The synthesis of melatonin is inhibited by light, which sends signals from the retina of the eye to the pineal gland, through the suprachiasmatic nucleus in the hypothalamus. When it gets dark, melatonin synthesis begins and it reaches its peak in the middle of the night. When given as a drug, melatonin helps induce and maintain sleep, but it has a short half-life (40 min), so it can only be used to initiate sleep and ease jet lag. However, the drug called ramelteon (Rozerem®) binds to the MT_1 melatonin receptor subtype and helps people fall asleep and stay asleep. Unlike the sleeping pills that bind to the $GABA_A$ receptor, ramelteon does not influence behavior, impair cognition, memory, alertness or cause dependence or withdrawal symptoms" [14].

Another target for treating pain is the 5-hydroxytryptamine (5-HT) receptor. Note that serotonin and 5-HT are synonyms. Seven 5-HT receptor families and 14 receptor subtypes have been identified [14]. "Selective agonists of the 5-HT_{1D} and 5-HT_{1B} receptors alleviate migraine headaches. One such agonist, or drug, is sumatriptan (Imitrex®). The 5-HT_{1D} and 5-HT_{1B} receptors constrict blood vessels in the skull, unless the endogenous ligand (5-HT, or serotonin) or agonists are bound to them. Several second generation triptan drugs, called naratriptan, zolmitriptan and rizatriptan, are more potent than sumatriptan and they have been FDA-approved since 1993. However, 5-HT receptors have structures that are similar to those of other biogenic amines (like dopamine and norepinephrine), so there can be harmful side effects due to binding to off target receptors. “For example, ergot alkaloids treat migraines by binding to 5-HT_{1B} and 5-HT_{1D} receptors, but can also act on α-adrenergic receptors” [14]. Moreover, binding to the 5-HT_{1B} receptor can have opposite effects. It has been called the death receptor because of the cardiotoxic effects caused by its activation [35]. This should not be confused with death receptors that determine whether programmed cell death should occur. On the other hand, when activated by the concentrations of the serotonin precursor tryptophan in chocolate, it can cause pleasure [35].

Drugs to Help People Stop Smoking [1]

There are also drugs to help people stop smoking tobacco, which causes cancer, heart disease and chronic obstructive pulmonary disease (COPD) [1]. The alkaloid called nicotine is primarily responsible for the physiological effects of tobacco. Nicotine is efficiently absorbed through the lungs and is rapidly transported to the brain, where it causes the release of ACh, norepinephrine and dopamine [14]. “Nicotine binds to the nicotinic acetyl choline receptor (nAChR) and opens ligand-gated ion channels that allow Ca^{2+}, Na^{+} and K^{+} to pass through cell membranes. Nicotine is strongly addictive because it causes dopamine to be released in the mesolimbic pathway, which is the reward circuit of the brain. It is responsible for positive reinforcement and the feelings of pleasure. Nicotine leads to the desensitization of nAChR and the brain compensates by upregulating the number of these receptors, causing an increase in the reward pathway sensitivity. Thus, nicotine addiction is physiological, persistent and psychological. The drug called varenicline (Chantix®) binds to the α4β2 subtype of the nAChR. Binding of varenicline to this receptor causes a sustained increase of dopamine in the mesolimbic system, causing pleasure. Varenicline also prevents nicotine from binding to the nAChR receptor, since it binds to the receptor stronger than nicotine” [14].

Drugs That Treat Neurodegenerative Diseases

Let us now switch to drugs that treat neurodegenerative diseases, in which improperly folded proteins cause neuronal damage. These misfolded proteins aggregate and form solid deposits, such as the plaques of Alzheimer’s (AD) and prion (PrPD) diseases, the Lewy bodies of Parkinson's disease (PD), the nuclear and cytoplasmic inclusions of Huntington disease (HD), the Bunina bodies of familial amyotrophic lateral sclerosis (ALS), and many others [1]. In all of these diseases, proteins that are normally soluble are changed into insoluble aggregates that form insoluble, toxic deposits not only in the brain but also in skeletal and muscular tissue and in the heart and the liver [1]. Moreover, RNA-binding proteins with prion-like domains can be found in neurodegenerative diseases [15].

Although we are still learning more about human prions, some yeast prions are known to be neutral or even beneficial, depending upon genetic background and environmental conditions [1]. "An algorithm was written to detect prion domains. It identified 19 domains that can confer prion-like behavior to proteins. The human genome was searched and it found a select group of RNA binding proteins that contain a canonical RNA recognition motif (RRM) and a putative prion domain. Of 210 human RRM-bearing proteins, 29 have a putative prion domain, and 12 of these are in the top 60 prion candidates in the entire genome. These RNA-binding prion candidates now seem to be very important in the pathology and genetics of amyotrophic lateral sclerosis (ALS), frontotemporal lobar degeneration with ubiquitin-positive inclusions (FTLD-U), Alzheimer's disease and Huntington's disease. Two of the RRM-bearing prion candidates form cytoplasmic inclusions in the degenerating motor neurons of ALS patients. Mutations in them cause familial ALS. It is possible that additional RNA-binding prion candidates identified by the algorithm will soon surface as genetic modifiers or causes of many neurodegenerative conditions. Indeed, simple prion-like transfer mechanisms involving the prion domains of RNA-binding proteins could cause non-cell-autonomous neurodegenerative pathology" [36]. So, distinct prion domains were identified in yeasts [1]. Then, the human genome was searched for these domains and some of them were found in a group of RNA-binding proteins. These RNA-binding prion candidates are emerging as causes of amyotrophic lateral sclerosis (ALS), frontotemporal lobar degeneration with ubiquitin-positive inclusions (FTLD-U), Alzheimer's disease and Huntington's disease [1]. Moreover, mutations in the prion-like domains in hnRNPA2B1 and hnRNPA1 cause multisystem proteinopathy and ALS [37]. "This was discovered by using computer algorithms that were designed to identify canonical yeast prions. It predicted that about 250 human proteins, including several RNA-binding proteins associated with neurodegenerative disease, harbor a distinctive prion-like domain (PrLD) enriched in uncharged polar amino acids and glycine. PrLDs in RNA-binding proteins are essential for the assembly of ribonucleoprotein granules. Pathogenic mutations were found in PrLDs of heterogeneous nuclear ribonucleoproteins (hnRNPs) A2B1 and A1 in families with inherited degenerative diseases that attack muscle, brain, motor neuron and bone, and in one case of familial amyotrophic lateral sclerosis (ALS)" [37].

That is, wild-type hnRNPA2 and hnRNPA1 show an intrinsic tendency to assemble into self-seeding fibrils, which is exacerbated by the mutations that cause diseases [37]. "The pathogenic mutations strengthen a steric zipper motif in the PrLD, which accelerates the formation of self-seeding fibrils. They cross-seed the polymerization of wild-type hnRNP. The disease mutations promote excess incorporation of hnRNPA2 and hnRNPA1 into stress granules and drive the formation of cytoplasmic inclusions in animal models that recapitulate the human pathology. Thus, dysregulated polymerization caused by a potent mutant steric zipper motif in a PrLD can initiate degenerative diseases. So, related proteins with PrLDs should be considered as candidates for initiating and even propagating proteinopathies of muscle, brain, motor neuron and bone" [37].

Further evidence in support of the prion-like hypothesis is that a key protein in Parkinson's disease (PD), α-synuclein, can mis-fold and spread from cell-to-cell and, in a prion-like fashion, trigger the aggregation of α-synuclein in the recipient cell [38]. "An important feature of synucleinopathies is aggregated α-synuclein in intracellular inclusion bodies, called Lewy bodies (LB) or Lewy neurites (LN), depending on their location. They are the neuropathological hallmark of PD" [38].

Inflammatory Cytokines

Inflammatory cytokines are at the top of pathological signal cascades in brain diseases [39]. "That is, interleukins 1 (IL-1), IL-6, IL-17, and tumor necrosis factor (TNF) are critical for the pathogenesis of inflammation in several brain disorders. At the same time, IL-1, IL-6 or TNF can initiate fever, the fundamental immune response that is orchestrated by the brain. All three of these cytokines can cross the blood-brain barrier in the vicinity of temperature-regulating neurons in the hypothalamus. In fact, IL-1 was the first endogenous pyrogen to be recognized. Still, all three can activate independent pathways that mediate inflammatory signaling cascades and have important consequences. Also, IL-17 is produced by T helper 17 cells in the immune system. All four (IL-1, IL-6, IL-17 and TNF) are critical in several immune-mediated disorders of the CNS, including seizures and multiple sclerosis (MS). However, treatments based on IFN-β and TNF can have different effects in patients who have different autoimmune diseases" [39]. For example, blocking TNF makes symptoms of MS patients worse and improves those of patients with rheumatoid arthritis, psoriasis, or Crohn's disease [29].

Parkinson's Disease

Parkinson's disease is another important neurological disease. It affects about 1% of the population over 65 [14]. "The earliest symptoms are difficulty in movement, muscular rigidity, tremors and impairment of posture and balance. The primary defect in Parkinson's disease is the loss of dopamine-secreting neurons in the part of the brain called the substantia nigra pars compacta. These neurons help mediate neuronal circuits of the basal ganglia. They regulate the flow of information from the cerebral cortex to the motor neurons of the spinal cord. Loss of these neurons causes reduced excitation of the motor cortex. The biochemical called levodopa, or L-DOPA is used for the management of Parkinson's disease. L-DOPA is the precursor of the neurotransmitter dopamine. It is formed by the addition of a hydroxyl group to the amino acid L-tyrosine, to form L-dihydroxyphenylalanine, or L-DOPA. The reaction is catalyzed by the enzyme tyrosine-3-monooxygenase. L-DOPA is then converted to dopamine by removing the –COOH moiety. The reaction is catalyzed by L-amino acid decarboxylase. Levodopa (L-DOPA) can cross the blood-brain barrier, where it is decarboxylated to form dopamine and CO_2. To prevent this, levodopa is given in combination with an inhibitor of L-amino acid decarboxylase, called carbidopa" [14].

Alzheimer's Disease

Next, let us look at Alzheimer's disease (AD), which is the most common neurodegenerative disease [1]. Worldwide, it affects 20 million people and in the USA it affects about five million people, most of them over 70 [14, 40]. A major risk factor is age. The incidence increases from 0.5% per year at age 65 to 8% per year at age 85 years [41]. However, once people reach the age of 95, the incidence of AD actually decreases as they get older [42]. Even in healthy people, some plaques and tangles can appear in the brain [40]. However, they occur much more in AD patients and they have been seen in autopsied brains of people in their 20s [40]. Also, AD is characterized by a progressive decrease in memory, language and other cognitive functions [41]. There is a decrease in behavioral, emotional, interpersonal and social skills [14]. The neuropathological symptoms include an accumulation of extracellular amyloid plaque containing β-amyloid, and intracellular neurofibrillar tangles containing polymerized and hyper-phosphorylated tau protein [40, 41]. In AD patients, the abnormal tau protein accumulates, and tangles the microtubules that are essential for neurons to communicate with each other and to receive their needed nutrients and chemical messengers [42]. It is likely that AD is actually several diseases with several possible genetic causes [41, 42]. Still, people who are homozygous (two copies) of the ApoE epsilon-4 allele (APOE-4) have an increased risk of AD. There are three APOE genes, called APOE-2, 3 and 4 [42]. APOE-4 is linked to AD. People with one copy of APOE-4 have three times the risk of getting AD as people with only APOE-2 and 3. People who have two copies of APOE-4 (homozygous) have eight times the risk of getting AD [40]. "ApoE is a protein that influences a person's ability to develop AD. It codes for an abnormal version of a protein called β-amyloid precursor protein, or APP" [40]. APP catalyzes the hydrolysis of a large protein that is a precursor of β-amyloid. In healthy people, APP is a transmembrane glycoprotein with an unknown, but apparently essential function. Healthy β-amyloid protein is water-soluble, but improperly cut β-amyloid protein is not.

Also, about 5-10% of AD patients get AD before they turn 65. This is called early onset AD. It can be caused by a defective gene on chromosome 21, which exists as three copies (trisomy) in Down's syndrome patients. Such patients develop abnormal β-amyloid plaques. AD patients don't just make more APP, but they cut (hydrolyze) the precursor protein wrong [42]. When APP is cut wrong, it becomes sticky and clumps together, forming plaques that are neurotoxic. The plaques stimulate the immune system in the brain to mount an inflammatory response, further damaging the neurons. There are at least three genes that contribute to plaque formation, APP, presenilin 1 and presenilin 2. Also, latent herpes virus may facilitate the development of AD in people who have the defective ApoE gene [41].

However, the behavioral symptoms of AD can be prevented in some cases by staying socially connected, keeping the heart (cardiovascular system) in good shape, avoiding strokes, obtaining a higher education and avoiding depression [42]. "There is an on-going epidemiological study of retired nuns who have agreed to take intelligence tests every year while they are alive and donate their brains to AD research after they die. Strokes and untreated clinical depression emerged as important risk factors for AD. During autopsy, some of the nun's brains exhibited the amyloid plaque and neurofibrillar tangles in the entorhinal cortex and hippocampus, even though they don't show any symptoms of AD when they are alive. Doctors and scientists use a scale called the Braak scale to classify different levels of AD. If there are no neurofibrillar tangles and no plaques, there is no AD. If there are tangles in the entorhinal cortex, it is stage I or II of AD. The entorhinal cortex is located at the base of the skull and is important for memory. If there are tangles higher and deeper in the brain, and they reach the hippocampus and neighboring regions, it is stages III or IV. The hippocampus is critical to learning and memory. It processes and stores new information. If the tangles and plaques reach the neocortex, it is stages V or VI. There was one nun who stayed very connected to her sister nuns, had never suffered a stroke or depression and had a higher college education. Upon her death, an autopsy showed that she had an unusually small brain and had many tangles in her entorhinal complex and hippocampus, but not in her neocortex. She showed no symptoms of AD when she was alive" [42]. On the other hand, it is quite common for widows to develop AD after their husbands die and they are overcome with loneliness. Still, even though there is much a person can do to prevent the outwards symptoms of AD, much research continues on the biochemical and genetic components.

Also, the brains of nuns with AD had less folic acid and less lycopene in them than nuns who died with no symptoms of AD [42]. This does not necessarily mean that folic acid and lycopene prevent AD. It is possible that oxidative damage consumes the folic acid and lycopene in AD patients.

Amyloid-β is 40 – 42 amino acids long and it is produced by the hydrolysis of the much larger amyloid precursor protein, or APP, which is in the cell membrane [43]. "Amyloid-β is a secreted protein and it has only a partially understood function in healthy brains. It binds copper and other transition metals. When copper is bound, it gives Amyloid-β antioxidant properties. The N-terminus of APP is located in the extracellular domain and its C-terminus is inside the membrane. Two different enzymes catalyze the hydrolysis of the N- and C-termini. They are called β-secretase and γ-secretase, respectively. A third enzyme, called α-secretase catalyzes the hydrolysis of APP between amino acid residues 16 and 17, which prevents the formation of Amyloid-β. The gene that codes for the APP protein (called the APP gene) has been mapped to chromosome 21. Three copies of this chromosome exist in people with Down's syndrome, who also have amyloid plaques in their brains. A small percentage of people who have AD (the Volga Deutsch) have mutations in some of their genes that might help cause the disease. However, mutations of the APP gene cause only a minority of these familial AD cases. Linkage studies found a major disease locus on chromosome 14 and positional cloning led to the identification of mutations in the *presenilin-1* gene" [43]. Mutations in the *presenilin-1* gene are the most common cause of familial AD [44].

The presenilin protein is a part of an aspartyl protease complex that works with γ-secretase to catalyze the hydrolytic cleavage of APP [44]. "In preclinical cases with *presenilin-1* mutation, deposits of Amyloid-β 42 are seen. Moreover, studies with transgenic animals indicated that *presenilin-1* mutations can also lead to hyperphosphorylated tau protein even when there are no Amyloid-β plaques. Recent evidence indicated that the presenilin proteins and the normal Amyloid-β may play important roles in maintaining the proper amount of intracellular Ca^{2+} in neurons. Disease-causing mutations may form channels in the cell membrane, thus disrupting the proper regulation of intracellular Ca^{2+}" [44].

The tau protein is involved in the assembly and stabilization of microtubules [40].

"In the human brain, six isoforms of tau are produced by splicing different portions of mRNA. Filamentous tau deposits are also formed in some other rare neurodegenerative diseases, which do not have Amyloid-β plaques. Tau pathology appears first in the part of the brain called the transentorhinal region. From there it spreads to the hippocampus, amygdala and finally the cerebral cortex. Amyloid-β appears in the neocortex. Both types of inclusion happen independently, with tangles appearing first. In fact, neurodegeneration in AD starts to occur 20 to 30 years before the appearance of the first clinical symptoms. The first phase is called amnesiac mild cognitive impairment (aMCI). Its neuro-pathological features are intermediate between normal aging and AD. Tau deposits are abundant in the entorhinal cortex and hippocampus and some Amyloid-β plaques are in the neocortex. There is a chemical called Pittsburgh compound B, which has been used to visualize Amyloid-β plaques in both AD patients and normal elderly people. This may help to identify people at risk for AD and help in developing ways to prevent it" [40].

Also, one of the pathways of neuronal death in AD is mediated by free radical damage and reactive oxygen substances (ROS), produced by NADPH oxidase and derivatives of arachidonic acid [38, 41]. It is possible that Amyloid-β 42 induces ROS production and stimulates the release of arachidonic acid through NMDA receptors [38]. As a result, Amyloid-β may mediate oxidative damage in AD by acting on the NMDA subtype of ionotropic glutamate receptors. The brain is especially vulnerable to oxidative damage because it uses so much oxygen in its metabolism, and it has a relatively large amount of polyunsaturated fats that are easily damaged by ROS. In addition to aging, Down's syndrome, vascular disease, smoking tobacco and diabetes are important risk factors for AD and all of these involve oxidative damage. Antioxidants, such as NSAIDs, curcumin, apocynin, docosahexaenoic acid (DHA) and other omega-3 fats may help prevent AD [45 - 49]. However, some studies indicated that NSAIDs do not prevent AD [50].

More recently, it has been shown that the root cause of AD may be oxidative stress [51], which occurs before the buildup of amyloid plaques. There is a long dormant period of gradual oxidative damage (smoldering inflammation) that precedes and eventually leads to the seemingly sudden appearance of clinical and pathological symptoms of AD symptoms, including Amyloid-β (Aβ) deposition, neurofibrillary tangle formation, metabolic dysfunction, and cognitive decline [52]. In fact, Aβ has been shown to increase the amount of iron and oxidative stress [53]. This might be a common factor in not just AD, but also Parkinson's and Huntington's disease [54]. It is also possible that the soluble form of Aβ (amino acids 25-35) may cause neurotoxicity in the hippocampus before the appearance of amyloid plaques [55].

Recent research has shown that a peptide fragment containing six amino acids (hexamer) of Aβ can successfully treated a mouse model of multiple sclerosis [56]. "Moreover, the hexamers from the following proteins were also successful and showed anti-inflammatory properties: tau, major prion protein (PrP), heat shock protein HspB5, amylin, serum amyloid P, and insulin B chain. They were shown to be effective therapeutics in treating neuroinflammation, with no apparent toxicity" [56].

When AD can't be prevented, it must be treated [56]. Five drugs are FDA-approved for the treatment of AD, although one of them (tacrine) is not used much anymore because of hepatotoxicity [57]. Most of the drugs inhibit acetyl choline esterase. This includes galantamine, rivastigmine, tacrine and donepezil, or Aricept®[9]. Another drug, called huperzine, is a natural product extracted from the club moss *Huperzia serrata* [58]. It is a potent reversible inhibitor of acetyl choline esterase and may have fewer side effects than other FDA-approved drugs [58]. A different drug, memantine, prevents over-stimulation of the NMDA subtype of the glutamate receptor [14, 57]. However, these drugs can only slow down the progression of the disease, and they only work in some patients [57]. More recently, though, a drug called etanercept has been shown to actually improve verbal function, along with clinical, cognitive and behavioral function in a Phase 2 clinical trial [57]. Etanercept blocks the ability of TNF-α to bind to its receptor. TNF-α is a gliotransmitter that regulates synaptic function [59]. Also, a drug called Rember, or MTC (methylthioninium chloride) may halt cognitive decline in patients with mild and moderate AD [57]. "Rember is also known as methylene blue, which is used as a redox indicator in analytical chemistry. In 2009, another drug, called brain-derived neurotrophic factor, or BDNF, was shown to protect neurons and reversed synapse loss, partially normalized improper gene expression, improved cell signaling and restored learning and memory in animal models of AD. BDNF is produced in the entorhinal complex, which send signals to the hippocampus and is important in learning and memory" [57]. AD adversely affects the entorhinal complex in the early stages of AD and contributes to short-term memory loss [60]. Cardiovascular exercise and learning increase the amount of BDNF in the brain, which helps explain their ability to help prevent AD. Also, BDNF is another reminder that the nervous system is linked to other parts of the body [57]. BDNF is found in the retina, kidneys, motor neurons and prostate.

In a recent study [61], a new class of drugs (benzylideneindanone derivatives) was developed for multitarget therapy for AD. They were shown to be anti-inflammatory, capable of binding copper, prevent β-amyloid (Aβ) aggregation and inhibit monoamine oxidase. They inhibit Cu(II)-induced Aβ1−42 aggregation and disassembled the well-structured Aβ fibrils [61].

Epilepsy

Next, let us discuss epilepsy, or periodic seizures. They occur because of poorly regulated electrical and neuronal activity. There is no cure for epilepsy, but it can be suppressed, so that about 70% of patients can lead a normal life [14]. "The first anti-epileptic agent was phenobarbital. It binds to the GABAA receptor and causes a decrease in chloride current. Another early anticonvulsant, phenytoin, was discovered in 1936 and marketed in the USA until 1953 as Dilantin. It binds to sodium (Na^+) channels in the depolarized state and delays the influx of Na^+, which prevents the very fast firing of neurons that causes seizures. Another drug, carbamazepine (Tegretol) was introduced in the USA in 1968. It also acts on the Na^+ channels. Another anticonvulsant, valproic acid and its sodium salt (Depakene) acts by three different mechanisms. It not only prolongs the recovery of Na^+ channels, but also mediates voltage-gated Ca^{2+} channels and it causes an increase in the amount of GABA in the brain. Other antiepileptic drugs include gabapentin, ethosuximide (Zarontin®), lamotrigine (Lamictal®), and tiagabine (Gabitril®). Ethosuximide reduces low threshold Ca^{2+} current in the hypothalamus. Lamotrigrine mediates voltage-gated Na^+ channels. Tiagabine is a selective GABA uptake inhibitor. Tiagabine is also used to treat migraines" [14].

Anxiety Disorders

Next, let us look at anxiety disorders. These include panic disorder, obsessive-compulsive disorder (OCD), post-traumatic stress disorder (PTSD), phobias and generalized anxiety disorder (GAD) [9]. Diazepam (Valium®) was the first drug to treat anxiety. In 1986, buspirone (BuSpar®, or BMS) was approved for treating GAD. It is an agonist of postsynaptic 5-HT_{1A} receptors, so it modulates the levels and function of serotonin. Buspirone also has moderate affinity for dopamine type two receptors (DA_2). It acts as a mixture of agonist and antagonist. It also decreases the activity of norepinephrine, decreases the concentration of acetyl choline and indirectly interferes with GABA neurotransmission. When depression and anxiety occur together, a selective serotonin reuptake inhibitor (SSRI) or serotonin-norepinephrine reuptake inhibitor (SNRI) can be used [9]. "Other drugs that treat anxiety disorders include β-blockers, tricyclic antidepressants (TCAs) and monoamine oxidase inhibitors (MAOIs). TCAs include amitriptyline, nortriptyline, imipramine, paroxetine, venlafaxine and clomipramine. Serotonin reuptake inhibitors include fluoxetine and paroxetine" [9].

Currently the first-line treatment is either SSRIs or SNRIs [14]. Some SSRIs, such as fluoroxetine and paroxetine are used to treat depression. Serotonin, or 5-hydroxytryptamine (5-HT), has many diverse effects throughout the body, and is an important neurotransmitter in the central nervous system, or CNS. There are also drugs that inhibit the uptake of not only serotonin, but also norepinephrine [9]. They are called serotonin-norepinephrine reuptake inhibitors (SNRIs). Serotonin and norepinephrine play important roles in mood. When they are taken back up into a neuron, their concentrations decrease. When reuptake is inhibited, the concentrations are higher, and a person's mood can be elevated. However, it can be very dangerous for a patient to suddenly quit taking anti-depressants.

Depression

Next, we will look at depression, starting with its prevention. Common sense tells us that proper care during infancy is essential for developing positive attitudes and preventing depression and suicide. Recent research has indicated the importance of the glucocorticoid receptor in this [62]. They found decreased levels of the receptor in the hippocampus in people who had suffered from child abuse. There were important epigenetic changes. There was an increased amount of methylation on cytosine residues in the promoter for the gene coding for the receptor [62]. So, depression can occur through no fault of the patient.

Next, let us look at drugs that are used to treat depression. The four main classes of antidepressant drugs are MAOIs, TCAs, SSRIs and SNRIs [14]. Iproniazid (Marsilid®) was the first entirely synthetic antidepressant. It is a MAOI. So is tranylcypromine, a drug used for treating major depression and post-traumatic stress disorder. In 2006, another MAOI, selegiline (Emsam®), was approved for treating major depression. TCAs are inhibitors of neurotransmitter uptake. In the 1960s and 1970s, several TCAs were FDA-approved. One of them, notriptyline (Pamelor®) selectively inhibits norepinephrine uptake. Most other TCAs inhibit ion channels (Na^+, K^+ and Ca^{2+}) and can cause dangerous heart arrhymias or sudden cardiac arrest and death, so they have been replaced by SSRIs. One of them, fluoxetine, was introduced in 1988 as Prozac®. Several second-generation SSRIs were developed and approved in the 1990s, including sertraline (Zoloft®), paroxetine (Paxil®) and escitalopram (Lexapro®). Although SSRIs are safer than TCAs, they also have serious side effects. Most common are sexual side effects (anorgasmia, erectile dysfunction and decreased libido), sleep disorders (both insomnia and excess sleep) and gastrointestinal discomfort. More seriously, abrupt discontinuation of treatment with SSRIs can lead to severe withdrawal symptoms.

The sales and marketing of Paxil® is an interesting case that illustrates a major controversy in the pharmaceutical industry. It was released in 1992 and for 10 years it was advertised as being non habit-forming. In 2001, it was reported that the World Health Organization found that it was quite habit forming and had the worst withdrawal problems of any antidepressant.

SNRIs, including venlafaxine (Effexor® and duloxetine (Cymbalta®) can be effective for patients who do not respond to other medications. Bupropion is a widely used inhibitor of reuptake of dopamine and norepinephrine. It is used as both an antidepressant and an aid to stop smoking [14].

Also, lithium (as lithium carbonate, Li_2CO_3) is able to diminish abnormal mood changes, also called manic-depressive and bipolar affective disorders. It works through the inositol triphosphate (IP_3) second messenger system. Phosphoinositides act as second messengers in neurons when they are hydrolyzed to IP_3 and diacylglycerol. Lithium inhibits the enzyme inositol monophosphatase [59], which catalyzes the hydrolysis of a phosphate to produce inositol. This inhibition prevents the re-use of inositol and limits phosphoinositide signaling [63].

Antipsychotics

Next, let us look at antipsychotics. The mental illness schizophrenia affects about one percent of the population globally [14]. "Symptoms include hallucinations, antisocial and delusional behavior. It may be associated with brain abnormalities that can be detected by imaging. It also may involve defects in dopamine, GABA and acetyl choline neurotransmission. Most schizophrenics self-medicate by smoking heavily. The earliest antipsychotics were chlorpromazine and haloperidol, but they had several harmful side effects. Second generation antipsychotics are called atypical antipsychotics. They are more effective and have fewer side effects. The first of them, clozapine has been used widely, but can cause a fatal blood disorder in about 1% of the patients. Other atypical antipsychotics in use are quetiapine (Seroquel®) and ziprasidone (Geodon®). They are not as effective, but have few side effects. Olanzapine (Zyprexa®) and aripiprazole (Abilify®) are more effective, but can cause weight gain and other metabolic disorders" [14]. A recent report indicated that these atypical antipsychotics also increase the risk of stroke, especially when given to elderly patients suffering from dementia [64].

Autism

Next, let us look at autism and related disorders. There are three common symptoms that are used to diagnose autism: poor social engagement, limited verbal and nonverbal communication and repetitive behaviors [65]. When it was first reported in 1943, it was considered rare, but it is much more common now. The CDC estimated that one in 166 children has autism or a related disease [65]. In fact, many doctors use the plural, autisms, or autism spectrum disorders, because it is most likely several different diseases, with similar symptoms. In fact, some think of autism as a metabolic disorder that affects the brain, gut and immune system [66]. "Autism has become such a problem that the U.S. Congress passed the Combating Autism Act, which authorized nearly $1 billion over four years for autism research. There is a new emphasis on studying possible interactions between vulnerable genes and environmental triggers. It is thought that autisms are complicated conditions that probably have multiple causes or risk factors. There are variations of genes, such as *MET* which make children susceptible to autisms, but most people with the widely-occurring autism-susceptible variant of *MET* aren't autistic. It takes a combination of genetic and environmental factors to cause autisms. The combination is probably not additive, but rather synergistic. Some consider autisms to be examples of autoimmune diseases, which can be caused by synergistic interactions of many environmental toxins and infectious agents" [66]. However, this is controversial, since intravenous administration of immunoglobulins was not effective [67]. Still, low levels of inflammation (smoldering inflammation, Chapter 2) may be a risk factor. Children with autisms don't make as much glutathione as do neurotypical children, as reported by Jill James [66]. As discussed in Chapter 2, glutathione is a very important natural antioxidant and it is crucial for removing toxins, such as reactive oxygen substances (ROS) [66]. Dr. James has a $2.4 million grant from the NIH to try to sort out the relationships between metabolism, genes and behavior [66]. She and her colleagues are interested in looking at mitochondrial dysfunction, since this can cause more ROS and free radicals to be produced. Also, a study called the Autism Birth Cohort is being started. It will follow 100,000 pregnant women for 72 months [62]. They will study their genetics and level of metabolites in blood and urine [66]. They hope to discover dietary and other environmental factors that may increase the risk of a child becoming autistic [66, 67].

Amyotrophic Lateral Sclerosis (ALS) [1]

Another deadly disease of the central nervous system is commonly known as Lou Gehrig's disease, or Amyotrophic Lateral Sclerosis, or ALS. It is a disease of motor neurons in which death occurs two to five years after onset, and is usually due to respiratory failure [68]. It is an especially cruel disease in that the victim (like the famous baseball player, Lou Gehrig) has often lived a very healthy lifestyle, but still succumbs to the steady decline of simple motor skills, even though the cognitive skills are still excellent. As a result, the patient is fully aware of what is happening, unlike some patients who have Alzheimer's disease or brain cancer. About 10% of the ALS patients have familial ALS, meaning that there is a known genetic component [69]. Of these, about 20% have a defect in the enzyme called superoxide dismutase (SOD), which is one of body's most important antioxidants (see Chapter 3). Recently, another gene, called FUS/TLS, which is mutated in familial ALS patients, was identified [68, 69]. The normal (wild type) FUS/TLS protein is located in the nucleus of neurons, but the mutated form is in the cytoplasm and it is associated with abnormal RNA metabolism and the accumulation of cytoplasmic protein aggregates [68, 69]. The normal, healthy version of FUS/TLS is involved in DNA repair, regulation of transcription, RNA processing and transport into the cytoplasm [69].

Damaged Spinal Cords

Finally, let's finish by discussing something that is not a disease, but an injury – damaged spinal cords. When the spinal cord is severely damaged, it causes paralysis. There is no cure at this time. However, recently a company called the Geron Corporation has developed an experimental procedure that may help patients with severe spinal cord injury [70]. Embryonic stem cells will be obtained from fertility clinics which would have otherwise discarded the embryos. The embryonic stem cells have been partly developed into oligodendrocytes, which are precursors of nerve cells. The oligodendrocytes will be placed at the site of spinal cord injury. Although the main goal of this planned Phase I clinical trial is patient safety, researchers will note whether or not there is any indication of movement in the previously paralyzed regions below the damaged spinal cord [67].

Medicinal Chemistry Point of View

There are many drugs that are used to treat pain, depression, anxiety, insomnia, epilepsy and neurodegenerative disorders. The most popular analgesics target opioid receptors. Even though there are three human genes that code for them, almost all opioid analgesics bind tighter (lower k_i values) to the μ-opioid receptor (also known as mu-opioid receptor or MOR), which is a GPCR [71]. However, there are several mRNA spice variants of the μ-opioid receptor, including the full length 7-transmembrane (7-TM) version [71]. Morphine, codeine and heroin are classical ligands. Fentanyl (Duragesic®) also binds to the μ-opioid receptor and is 100 times as potent as morphine [14]. It is used to treat severe and persistent pain as well as in surgical procedures *via* intravenous or epidural administration. An analogue, remifentanil is the shortest acting opioid (5-10 min) for use in humans [12]. Hydrocodone and oxycodone (OxyContin®, Percocet®) are two others. To be effective, codeine must be converted to morphine in an *O*-demethylation reaction catalyzed by CYP-2D6 [68]. "Also, the *O*-demethylated products of hydrocodone and oxycodone, hydromorphone and oxymorphone, are more active. The EC_{50} values for codeine, hydrocodone and oxycodone acting on human μ-opioid receptors were found to be 6.59, 1.50 and 1.40 μM, respectively, while hydromorphone and oxymorphone were 0.051 and 0.049 μM, respectively. The E_{max} values were 55, 54 and 67% of the maximum stimulation for codeine, hydrocodone and oxycodone, respectively. The E_{max} values were 54 and 63% for hydromorphone and oxymorphone, respectively [72]. In a more recent study hydromorphone was also found to be much more potent than hydrocodone [73]. "The EC_{50} values for binding to the MOR were 0.047 and 2.75 μM for hydromorphone and hydrocodone, respectively. The E_{max} values were 107 and 78% for hydromorphone and hydrocodone, respectively (oxycodone was 100%)" [73].

Tramadol (Ultram®) is another analgesic that binds to MORs [74]. "There are two enantiomers: (+)-tramadol is and MOR agonist and serotonin reuptake inhibitor, while (-) tramadol inhibits norepinephrine reuptake. Tramadol is available as drops, capsules and a sustained-release formulation. Its potency is about 10% that of morphine. It provides post-operative pain relief. It produces relatively few adverse side effects and has never been a scheduled drug" [74]. So, synergistic analgesic effects are possible when a drug has multiple effects.

Synergism can also occur when drugs are combined. The combination of acetaminophen and hydrocodone (Vicodin® and other brand names) as well as acetaminophen plus oxycodone (Percocet® and other brand names) are two important examples. Hydrocodone and oxycodone can also be combined with NSAIDs and COX-2 inhibitors are also used [75]. However, NSAIDS increase the risk of bleeding and can cause kidney damage. COX-2 inhibitors increase the risk of adverse cardiac effects and acetaminophen increases the risk of severe liver damage [75]. So, the FDA lowered the maximum permissible dose of acetaminophen in opioid combination products to 325 mg [76].

On the other hand, opioids can be abused, even though this was not widely appreciated by many in the early 1990s [77]. Opioids have become more widely available, resulting in their inappropriate and illegal use. So, there are two conflicting ideas that both contain merit. Patients who are really in pain should be treated, but there is a real and urgent need to decrease the unintended and undesirable consequences that opioid diversion has on society [77]. So, it was not easy for the US FDA to decide to approve a new 12 hour extended release form of hydrocodone (Zohydro ER®), which they did in 2014, despite it not being recommended by the advisory panel [77, 78].

There is also research being done to develop new analgesics, such as iodobenzoylnaltrexamide (IBNtxA), which is an agonist of the 6-TM splice variant of the MOR [71]. It relieves acute, thermal pain about 10 times as effectively as morphine, but without the adverse effects of respiratory depression, significant constipation, reward, aversion, and dependence that morphine can cause [71].

Even though analgesics can successfully treat physical pain, other types of drugs are used to treat depression. As discussed above, the four main classes of antidepressant drugs are MAOIs, TCAs, SSRIs and SNRIs [14]. Iproniazid (Marsilid®) was the first entirely synthetic antidepressant. It is a MAOI. So is tranylcypromine, a drug used for treating major depression and post-traumatic stress disorder. In 2006, another MAOI, selegiline (Emsam®), was approved for treating major depression. TCAs include amitriptyline, nortriptyline, imipramine, paroxetine, venlafaxine and clomipramine. Selective serotonin reuptake inhibitors (SSRIs) include citalopram (Celexa®), escitalopram (Lexapro®), fluoxetine (Prozac®), paroxetine (Paxil®) and sertraline (Zoloft®) [79]. SSRIs are the most commonly prescribed antidepressants [79].

As mentioned above, in the 1960s and 1970s, several TCAs are FDA-approved. One of them, notriptyline (Pamelor®) selectively inhibits norepinephrine uptake. Most other TCAs inhibit ion channels (Na^+, K^+ and Ca^{2+}) and can cause dangerous heart arrhymias or sudden cardiac arrest and death, so they have been replaced by SSRIs. One of them, fluoxetine, was introduced in 1988 as Prozac®. Several second-generation SSRIs were developed and approved in the 1990s, including sertraline (Zoloft®), paroxetine (Paxil®) and escitalopram (Lexapro®). Although SSRIs are safer than TCAs, they also have serious side effects. Most common are sexual side effects (anorgasmia, erectile dysfunction and decreased libido), sleep disorders (both insomnia and excess sleep) and gastrointestinal discomfort. More seriously, abrupt discontinuation of treatment with SSRIs can lead to severe withdrawal symptoms.

SNRIs include venlafaxine (Effexor®), milnacipran (Savella®), duloxetine (Cymbalta®), levomilnacipran (Fetzima®), desvenlaxafine (Pristiq®) and sibutramine (Meridia®). They can be effective for patients who do not respond to other medications. Bupropion is a widely used weak inhibitor of reuptake of dopamine and norepinephrine. It is used as both an antidepressant and an aid to stop smoking [14].

Also, lithium (as lithium carbonate, Li_2CO_3) is able to decrease abnormal mood changes, also called manic-depressive and bipolar affective disorders. It works through the inositol triphosphate (IP_3) second messenger system. Phosphoinositides act as second messengers in neurons when they are hydrolyzed to IP_3 and diacylglycerol. Lithium inhibits the enzyme inositol monophosphatase, which catalyzes the hydrolysis of a phosphate to produce inositol [63]. This inhibition prevents the re-use of inositol and limits phosphoinositide signaling [63].

A very popular analgesic that binds to a receptor is diazepam (Valium®). It binds to $GABA_A$ receptors, causing a conformation change that enhances the receptor's binding of GAB_A [14]. GABA opens Cl- ion channels and hyper-polarizes neurons. Valium® is used for the treatment of anxiety, seizures, muscle spasms, insomnia and symptoms of opiate or alcohol withdrawal. However, if a patient abruptly stops taking diazepam, it can cause a drop in the efficiency of GABA receptors, which can cause withdrawal symptoms of its own [14].

Insomnia can also lead to depression. Zolpidem (Ambien®) is a drug that binds to one of the subunits (α-subunit) of the $GABA_A$ receptor in the brain, inducing sleep. Ambien® is used as a sedative and as a hypnotic agent [14].

Over-the-counter NSAIDS, such as aspirin and ibuprofen are also effective in treating pain. However, as discussed last semester, NSAIDS and antioxidants in fruits and vegetables can help prevent Alzheimer's disease (AD). Although it was once thought that they did this through their direct antioxidant effects, it is now known that most antioxidants also bind Fe^{2+}, thus preventing it from undergoing the Fenton reaction that produces very harmful superoxide anions. The polyunsaturated fats in fish oil, especially DHA, can also help prevent AD, especially if phytosterols are also consumed [80].

Another study found that the antidepressant citalopram may be useful in preventing AD [81]. "It is a selective serotonin reuptake inhibitor. It decreased the appearance of new Aβ (amyloid beta) plaque growth and arrested the growth of preexisting plaques in mice brains interstitial fluid in a dose-dependent manner" [81].

There has also been an important advance in predicting who will get AD [82]. "That is, the blood lipid profiles of 525 people over 70 with mild cognitive impairment were analyzed by LC-MS. Total lipids were extracted from 25 μL of blood plasma by adding a mixture of 25% acetonitrile, 40% methanol and 20% water containing internal standards. This denatured plasma proteins, which were removed after centrifuging the mixture. Then, the solvent was evaporated off and the residue remaining redissolved in "200 μL of buffer containing 5% methanol, 1% acetonitrile and 94% water. After centrifuging this to remove solid particles, the supernatant was injected on a C18 column. The lipids were eluted using a strong mobile phase that contained acetonitrile and isopropyl alcohol. They found that the concentrations of ten lipid metabolites decreased in those subjects who progressed from mild cognitive impairment to AD. The lipids included phosphatidyl cholines, abbreviated as PCs. They had different fatty acyls attached to the glyceride backbone. That is, PC has two acyls, abbreviated as aa. Some of those with very long chain fatty acyls, "PC diacyl (aa) C36:6, PC aa C38:0, PC aa C38:6, PC aa C40:1, PC aa, C40:2, PC aa C40:6 and PC acyl-alkyl (ae) C40:6 were depleted. Also, lysophophatidylcholine (lysoPC a C18:2), and acylcarnitines (ACs) (Propionyl AC (C3) and C16:1-OH) were depleted. This relatively simple and non-invasive blood test is much better than previously existing methods of predicting who will progress to AD" [78].

Also, drugs that target acetyl choline esterase (AchE) are used to treat AD [83]. That is, when one neuron sends a signal to another neuron, or to a muscle cell, it generates an action potential. It is important that the neuron or muscle is able to relax after being stimulated, so the acetylcholine messenger is rapidly broken down and inactivated by AchE in cholinergic neurons. Tetrohydroaminoacridine, also known as tacrine (Cognex®), donepezil (Aricept®), rivastigmine (Excelon®) and galantamine (Razadyne®) are reversible AChE inhibitors that can penetrate the blood brain barrier [83, 84]. Tacrine was the first, but it is not used much any more, due to its adverse side effects, including nausea and diarrhea [83]. Although none of them can cure AD, they can delay the worsening of symptoms by 6-12 months on average in about half the people who take them [83]. They can ease some of the memory and language deficits characteristic of this devastating disease [84]. There is also a drug (memantine, Namenda®) that blocks glutamate receptors [83]. Finally, high doses of vitamin E can help treat some of the cognitive symptoms of AD [83].

Pyridostigmine bromide, marketed under the name Mestinon or Regonol, is another AChE inhibitor [85]. It is used to treat Myasthenia gravis [85]. Other AChE inhibitors include ambenonium chloride (Mytelase), and neostigmine methyl sulfate (Prostigmin) [86]. "These drugs cannot pass the blood-brain barrier, so their action is limited to the inactivation of AChE at the neuromuscular junction. Because of this, they can be administered to treat muscle weakness, without causing the psychotropic side-effects that they might cause if they could penetrate the cholinergic pathways of the central nervous system" [86].

As mentioned previously, one of the hallmarks of AD is the accumulation of the β-amyloid peptide, also known as amyloid-β, Aβ and Ab [83]. "It can have 39-43 amino acids and is produced by the hydrolysis of the membrane-bound amyloid precursor protein. Mostly, Ab 1-40 and Ab 1-42 are produced. They self-aggregate and form large, insoluble plaques that block proteasome function and inhibit mitochondrial activity. In healthy brains, Ab activates signaling pathways that regulate the phosphorylation of the tau protein. However, when tau is hyperphosphorylated, it can no longer properly regulate axonal transport. This leads to the production of neurofibrillary tangles. Also, Ab inhibits tau degradation. Also, Ab 1-42 can bind Cu^{2+}, Zn^{2+} and Fe^{3+}, which might contribute to Ab aggregation and the formation of ROS" [83].

AD shares three things with Parkinson's and prion diseases: progressive loss of neuronal function, leading to cognitive decline, involvement of misfolded and aggregated metalloproteins (like Ab in AD) and the lack of any drug or therapy that can reverse or stop disease progression [83]. "So, in prion diseases, the prion protein (PrP) can bind metal ions like Mn^{2+} and Cu^{2+}, changing its structure and causing fibrils to form. In Parkinson's disease, a peptide called alpha-synuclein (αS) is formed. It is the main component of 5-24 μm spherical Lewy bodies that are a hallmark of the disease. It can bind Fe^{2+}, Cu^{2+}, Fe^{3+} and Mn^{2+} weakly. In healthy brains, the αS peptide is part (amino acids 61-95) of a native form of αS that has 140 amino acids. The N-terminal domain (amino acids 1-60) binds lipids, the central hydrophobic non-amyloid-β component (NAC) that can form aggregates when cleaved from the main proteins while the C-terminal region (amino acids 96-140) blocks the formation of fibrils" [83].

There are no drugs that can cure or even treat prion diseases [79]. "However, L-DOPA is used as the standard treatment for Parkinson's disease, since the concentration of dopamine in the striatum decreases to about 10% of that in healthy brains. L-DOPA is a precursor of dopamine. L-DOPA can cross the blood-brain barrier, but dopamine can not. In most cases L-DOPA can improve the symptoms, but can never cure the disease" [79]. However, there is "an atypical form of Parkinson's disease that was unusually common on the Pacific islands of Guam and New Caledonia and the French Caribbean island of Guadeloupe that did not respond to L-DOPA" [87]. It was caused by the over-consumption of graviola (*Annona muricata*) and products made from it [87, 88]. The incidence of this atypical form of Parkinson's disease decreased very much after people changed their diets [88].

Also, L-DOPA can cause adverse side effects, such as nausea, anxiety and dyskinesia [83]. "It can also cause tachycardia, because it can be converted to dopamine while it is outside the brain. The reaction can be catalyzed by DOPA decarboxylase. So, carbidopa (Lodosyn®) can be used to inhibit this enzyme. When combined with L-DOPA, the drugs are marketed as Atamet®, parcopa® and Sinemet®. There are also drugs that inhibit another enzyme that catalyzes the breakdown of L-DOPA. That is, tolcapone (Tasmar®) and entacapone (Compat®) can inhibit the enzyme catechol-*O*-methyltransferase or COMT and inhibit the metabolism of L-DOPA outside the brain. When entacapone is combined with L-DOPA, the combination is called Stalevo®). Also, MAO inhibitors such as selegiline (Eldepryl®) and rasagiline (Aziltec®) can be used to limit the destruction of dopamine catalyzed by the MAO type B enzyme in the substantia nigra and striatum. On the other hand, agonists (like bromocriptine, Parlodel®) of the dopamine D2 receptor can be used together with L-DOPA to treat Parkinson's disease. This combination is especially useful in patients who have the "OFF-ON" syndrome, in which after the usual doses of L-DOPA are no longer effective. Finally, anticholinergic drugs can help restore the balance between the activities of dopaminergic and cholinergic neurons in the brain. This includes Benzhexol®, Orphenadrine®, Procyclidene® and Bentropine®" [83].

Despite the lack of cures for neurodegenerative diseases, there is still hope [89 - 93]. For example, inhibitors of caspase-6 might prove effective in treating AD and Huntington's disease (HD) [89]. As mentioned in Chapter 4, caspases are cysteine-aspartate proteases that are involved in apoptosis. They can also be involved in inflammatory cascades [89]. "One of them, capsase-6, is a tetrameric enzyme that is involved in axon degeneration as well as AD and HD. It is widely expressed in the brain and other tissues. It can catalyze the hydrolytic cleavage of healthy APP at Asp664, which is a prerequisite for the pathology of AD" [89].

HD is characterized by progressive motor, cognitive and psychiatric problems [89]. "It is caused by an abnormal or malfunctional Huntington protein (mHtt) that is coded for by the presence of many copies of a CAG repeat in the gene coding for it. CAG codes for glutamine. The normal Htt protein serves as a scaffold to bring together several proteins that are involved in endocytosis, apoptosis, vesicle transport, cell signaling, morphogenesis and regulation of transcription. It also increases the production of brain-derived neurotrophic factor (BDNF) in striatal neurons. Unfortunately, Htt can be hydrolyzed in a reaction catalyzed by caspase-6. This produces an N-terminal fragment that contains the glutamines that relocates to the cell nucleus and forms inclusions that cause the symptoms of HD" [89].

So, inhibitors of caspase-6 may be useful in treating AD and HD [89]. "Synthetic allosteric inhibitors are being developed. Some inhibit the formation of caspase-6 tetramers and others bind to the dimer interface. Also, the protein kinase ARK5 is a potential therapeutic target since it catalyzes the addition of phosphates to caspase-6" [89].

Another class of new drugs for treating AD is based on the dibenzo [1,4,5] thiadiazepine ring structure [90]. "They are able to protect cultured neurons from mitochondrial oxidative stress and were not neurotoxic. One of them also was able to modulate Ca^{2+} channels. This is important because another hallmark of AD is an imbalance in cytosolic Ca^{2+} concentration. This family of compounds is easy to synthesize and may prove useful in treating AD in the future" [90].

Another potential target for AD, Parkinson's disease, stroke and other neurodegenerative diseases is the glucagon-like peptide 1, or GLP-1 [91]. "It is best known as an incretin hormone it facilitates the release of insulin from the pancreas during hypoglycemia. However, while monitoring patients who received GLP-1 mimetics neuroprotective and anti-inflammatory effects were seen. In animal models, the GLP-1 mimetics exendin-4, liraglutide and lixisenatide reduced Ab plaques, prevented synapse loss and memory impairment. In animal models of Parkinson's disease, exendin-4 protected dopaminergic neurons in the substantia nigra and prevented loss of dopamine in the basal ganglia, thus preserving muscle control. These results led to clinical trials, which are showing encouraging results" [91].

Drugs that inhibit both AChE and brain-specific monoamine oxidase AB (MAOAB) are also being developed for treating Parkinson's disease [92]. "The pharmacophores in rasagiline was used to develop new drugs that have several targets. Some inhibit both AChE and MAOAB while another called M30 chelates iron, scavenges free radicals and inhibits MAOAB. It can not only protect neurons from damage, but also restore damaged neurons. It does this by activating the hypoxia-inducible factor. M30 also regulates cell cycle arrest and induces not just BDNF, but also erythropoietin, and glia-derived neurotrophic factor" [92].

Rasagiline was also described in a recent review on drug development for Parkinson's disease [93]. "Also, new molecular targets such as α-synuclein, leucine-rich repeat serine/threonine protein kinase 2, and parkin have been identified. However, they might require new drug delivery systems to reach them. This includes hydrogels as well as polymeric or lipid microparticles and nanoparticles. Poly(lactic-co-glycolic acid) or PLGA and PEG can be used. In an animal model of Parkinson's disease, PLGA-encapsulated VEGF and GDNF were able to regenerate neurons".

Finally, human induced pluripotent stem cells (hiPSCs) are being developed as a possible treatment or even cure of Parkinson's disease [94, 95]. Stem cells from umbilical cords or bone marrow can be used for grafting into people as efficiently as embryonic stem cells, but only autologous stem cells (ASCs) are considered to be safe [94]. "There is no need to suppress the immune system before administering ASCs since they come from the patient himself (or herself). They can be harvested safely and easily from bone marrow. They can nestle into the part of the body into which they are grafted and differentiate into many different types of cells, including dopaminergic neurons. They can also produce chemotrophins such as BDNF, glial-derived neurotrophic factor, nerve growth factor, TGF-β, IGF-1, neurotrophin-3 and stem cell factor 3" [94]. Moreover, hiPSCs may eventually be used to treat AD, ALS and MS as well [95].

In conclusion, there are many drugs that are available for treating neurodegenerative diseases and there is much important research being done to develop new therapies.

KEY POINTS

1. **The human brain develops slower than the other organs, even though we are born with almost all the neurons that we will ever have. The first year of life is critical in brain development, for the total brain volume doubles, as measured**

by MRI [1]. This is when the brain is most susceptible to damage by genetic defects and environmental insults. It is also the time in which therapeutic intervention can have its maximum effect.

2. **A principal component of the nervous system is the neuron, which is the type of electrically excitable cell that transmits and processes information. Neurons are arranged in networks and circuits [1]. They communicate with each other and with other types of cells through connections called synapses [1].**

3. **There are about 10-100 billion neurons in the adult brain, although brain size in humans is not related to intelligence. The normal human brain has many local regions, or centers, and many pathways between them. At the top of our brain is the cerebral cortex. The simplest neurons are sensory neurons in the PNS.**

4. **The anatomy of the brain can be divided into at least four parts and three pairs of lobes that are separated by deep, gray grooves. The parts are the right hemisphere, left hemisphere, cerebellum and brain stem [1]. The three lobes are the frontal lobes, parietal lobes and occipital lobe. Beneath them is a fourth lobe, the temporal lobe, followed by the pons, cerebellum and brain stem.**

5. **The cerebellum is located between the brain stem and the rest of the body. It helps integrate sensory perception, coordination and muscle control. At the top of the brain stem is the pons, which relays sensory information between the cerebellum and cerebrum.**

6. **The limbic system helps form the border between the cerebral cortex and subcortical layers. It is near the center of the brain. The hippocampus, hypothalamus, thalamus, olfactory bulb, fornix, amygdala, anterior thalamic nuclei and limbic nuclei are all in the limbic system, which supports many functions.**

7. **The autonomic nervous system is organized into three divisions: the sympathetic, parasympathetic and enteric. It is the part of the peripheral nervous system that controls and maintains homeostasis in the body [1]. These maintenance activities are primarily performed without conscious control or sensation. The sympathetic and parasympathetic nervous systems work to maintain a type of balance. They have opposite effects on the body. The**

sympathetic division is used in actions requiring quick responses. The parasympathetic division is used in actions that do not require immediate reaction [1].

8. **Neurons and their cell bodies are a major component of the gray (or grey) matter in the brain. The function of gray matter is to route sensory or motor stimuli to interneurons of the CNS. Gray matter does our mental calculations and stores our memories.**

9. **In humans and other vertebrates, the axons are coated with an electrical insulation, called the myelin sheath. This allows the passage of low currents through the axons at speeds up to 120 m/s (or about 270 mph). When the myelin sheath is damaged, it can cause diseases such as multiple sclerosis, or MS.**

10. **Messages are sent to and from neurons in the form of primary messengers, called neurotransmitters. There are also neuromodulators, which have no intrinsic activity, but they modulate ongoing synaptic activity.**
11. **Drugs for treating pain include NSAIDs, morphine, barbituates, fentanyl (Duragesic®), acetaminophen (Tylenol®), sumatriptan (Imitrex®) and diazepam (Valium®). Zolpiden (Ambien®) is a drug for treating insomnia. Gabapentin (Neurontin®) is an anticonvulsant that is used to prevent seizures. Ramelteon (Rozerem®) is used to help people sleep.**

12. **L-DOPA is used to treat Parkinson's disease, which affects about 1% of the population over 65.**

13. **Alzheimer's disease (AD) is the most common neurodegenerative disease [1]. The neuropathological symptoms include an accumulation of extracellular amyloid plaque containing β-amyloid, and intracellular neurofibrillar tangles containing polymerized and hyperphosphorylated tau protein [1]. It has been shown that the root cause of AD may be oxidative stress, which occurs before the build up of amyloid plaques. Five drugs are FDA-approved for the treatment of AD, although one of them (tacrine) is not used much any more because of hepatotoxicity. Galantamine, rivastigmine, tacrine and donepezil, or Aricept® target acetyl choline esterase.**

14. **The first anti-epileptic agent was phenobarbital. It binds to the GABA receptor**

and causes a decrease in chloride current. Other antiepileptic drugs include carbamazepine (Tegretol), valproic acid and its sodium salt (Depakene), gabapentin, ethosuximide (Zarontin), lamotrigrine (Lamictal), and tiagabine (Gabitril).

15. **Diazepam (Valium®) was the first drug to treat anxiety. Other drugs that treat anxiety disorders include buspirone (BuSpar®, or BMS), β-blockers, tricyclic antidepressants (TCAs) and monoamine oxidase inhibitors (MAOIs) [1]. Currently the first-line treatment is either SSRIs or SNRIs. The four main classes of antidepressant drugs are MAOIs, TCAs, SSRIs and SNRIs.**

ABBREVIATIONS

MRI	=	Magnetic Resonance Imaging;
L-DOPA	=	L-Dihydroxyphenylalanine;
AD	=	Alzheimer's Disease ;
MAO	=	Monoamine Oxidase;
MAOI	=	Monoamine Oxidase Inhibitor;
TCA	=	Tricyclic Antidepressants;
SSRI	=	Selective Serotonin Reuptake Inhibitors;
SNRI	=	Serotonin-Norepinephrine Reuptake Inhibitors;
FGF 21	=	Fibroblast Growth Factor ;
SCN	=	Suprachiasmatic Nucleus ;
SGZ	=	Subgranular Zone;
SVZ	=	Subventricular Zone;
NMDA	=	N-Methyl-D-Aspartate;
GABA	=	Gamma Amino Butyric Acid;
RGL	=	Radial Glial-Like;
CNS	=	Central Nervous System;
PNS	=	Peripheral Nervous System;
DTI	=	Diffusion Tensor Imaging;
ANS	=	Autonomic Nervous System;
Ach	=	Acetylcholine;
VMAT	=	Vesicular Monoamine Transporters;
5-HT	=	5-Hydroxytryptamine;
NMR	=	Nuclear Magnetic Resonance;

NKB	=	Neurokinin B;
VIP	=	Vasoactive Intestinal Peptide;
PACAP	=	Pituitary Adenylate Cyclase-Activating Peptide;
NPY	=	Neuropeptide Y;
POMC	=	Pre-Proopio-Melanocortin;
CRF	=	Corticotropin-Releasing Factor;
O-LM	=	Oriens–Lacunosum Moleculare;
NO	=	Nitric Oxide;
cGMP	=	Cyclic Guanosine Monophosphate;
ADMA	=	Asymmetric Dimethylarginine;
DDAH	=	Dimethylarginine Dimethylaminohydrolase;
GPC	=	Glial Progenitor Cell;
GM-CSF	=	Granulocyte-Macrophage Colony-Stimulating Factor;
CMV	=	Cytomegalovirus;
PBL	=	Peripheral Blood Lymphocyte;
ALT	=	Autologous Lymphocyte Transfer;
DC	=	Dendritic Cell;
NSAID	=	Non-Steroid Anti-Inflammatory Drug;
COX	=	Cyclooxygenase;
FDA	=	Food and Drug Administration;
nAChR	=	Nicotinic Acetyl Choline Receptor;
PD	=	Parkinson's Disease;
HD	=	Huntington Disease;
ALS	=	Amyotrophic Lateral Sclerosis;
RNA	=	Ribonucleic Acid;
RRM RNA	=	Recognition Motif;
FTLD-U	=	Frontotemporal Lobar Degeneration With Ubiquitin-Positive Inclusions;
PrLD	=	Prion-Like Domain;
hnRNP	=	Heterogeneous Nuclear Ribonucleoproteins;
LB	=	Lewy Bodies;
LN Lewy	=	Neurites;
Il	=	Interleukin;
IFN	=	Interferon;

MS	=	Multiple Sclerosis;
APOE-4	=	ApoE Epsilon-4 Allele;
APP	=	β-Amyloid Precursor Protein;
aMCI	=	Amnesiac Mild Cognitive Impairment;
ROS	=	Reactive Oxygen Substances;
NADPH	=	Nicotinamide Adenine Dinucleotide Phosphate Reduced;
PrP	=	Prion Protein;
MTC	=	Methylthioninium Chloride;
BDNF	=	Brain-Derived Neurotrophic Factor;
OCD	=	Obsessive-Compulsive Disorder;
PTSD	=	Post-Traumatic Stress Disorder;
GAD	=	Generalized Anxiety Disorder;
DA2	=	Dopamine Type Two Receptor;
CDC	=	Centers for Disease Control;
NIH	=	National Institutes of Health;
SOD	=	Superoxide Dismutase

REFERENCES

[1] Bookout AL, de Groot MH, Owen BM, *et al.* FGF21 regulates metabolism and circadian behavior by acting on the nervous system. Nat Med 2013; 19(9): 1147-52. [http://dx.doi.org/10.1038/nm.3249] [PMID: 23933984]

[2] Owen BM, Bookout AL, Ding X, *et al.* FGF21 contributes to neuroendocrine control of female reproduction. Nat Med 2013; 19(9): 1153-6. [http://dx.doi.org/10.1038/nm.3250] [PMID: 23933983]

[3] Knickmeyer RC, Gouttard S, Kang C, *et al.* A structural MRI study of human brain development from birth to 2 years. J Neurosci 2008; 28(47): 12176-82. [http://dx.doi.org/10.1523/JNEUROSCI.3479-08.2008] [PMID: 19020011]

[4] Lie DC, Song H, Colamarino SA, Ming GL, Gage FH. Neurogenesis in the adult brain: new strategies for central nervous system diseases. Annu Rev Pharmacol Toxicol 2004; 44: 399-421. [http://dx.doi.org/10.1146/annurev.pharmtox.44.101802.121631] [PMID: 14744252]

[5] Bhardwaj RD, Curtis MA, Spalding KL, *et al.* Neocortical neurogenesis in humans is restricted to development. Proc Natl Acad Sci USA 2006; 103(33): 12564-8. [http://dx.doi.org/10.1073/pnas.0605177103] [PMID: 16901981]

[6] Glasper ER, Leuner B, Gould E. Adult neurogenesis finds its niche. Nat Neurosci 2008; 11: 731. [http://dx.doi.org/10.1038/nn0708-731]

[7] Ming GL, Song H. Adult neurogenesis in the mammalian central nervous system. Annu Rev Neurosci 2005; 28: 223-50.

[http://dx.doi.org/10.1146/annurev.neuro.28.051804.101459] [PMID: 16022595]

[8] Hsieh J, Schneider JW. Neuroscience. Neural stem cells, excited. Science 2013; 339(6127): 1534-5. [http://dx.doi.org/10.1126/science.1237576] [PMID: 23539589]

[9] Lewin R. Is your brain really necessary? Science 1980; 210(4475): 1232-4. [http://dx.doi.org/10.1126/science.7434023] [PMID: 7434023]

[10] Harris DH. The Man without a Brain, http://www.drtruth.org/Infinity4.html. , [Accessed Feb 20, 2014.];

[11] Why are you and your brain unique? Van Essen D Sci 2012; 338: 35-6. [http://dx.doi.org/10.1126/science.338.6103.35]

[12] Cooper JR, Bloom FE, Roth RH. The Biochemical Basis of Neuropharmacology. Oxford: Oxford Press 2003.

[13] Shepherd GM. The Synaptic Organization of the Brian. New York: Oxford University Press 1990.

[14] Corey EJ, Czako B, Kürti L. Molecules and Medicine. New York: John Wiley & Sons 2007.

[15] Yamada S, Nelson WJ. Synapses: sites of cell recognition, adhesion, and functional specification. Annu Rev Biochem 2007; 76: 267-94. [http://dx.doi.org/10.1146/annurev.biochem.75.103004.142811] [PMID: 17506641]

[16] Fields RD. White matter matters. Sci Am 2008 Feb; 298(3): 42-9. [PMID: 18357821]

[17] Allen NJ, Barres BA. Neuroscience: Glia - more than just brain glue. Nature 2009; 457(7230): 675-7. [http://dx.doi.org/10.1038/457675a] [PMID: 19194443]

[18] Debanne D, Rama S. Astrocytes shape axonal signaling. Sci Signal 2011; 4(162): pe11. [http://dx.doi.org/10.1126/scisignal.2001884] [PMID: 21364185]

[19] Kaiser LG, Schuff N, Cashdollar N, Weiner MW. Age-related glutamate and glutamine concentration changes in normal human brain: ^{1}H MR spectroscopy study at 4 T. Neurobiol Aging 2005; 26(5): 665-72. [http://dx.doi.org/10.1016/j.neurobiolaging.2004.07.001] [PMID: 15708441]

[20] Watanabe H, Takaya N, Mitsumori F. Simultaneous observation of glutamate, γ-aminobutyric acid, and glutamine in human brain at 4.7 T using localized two-dimensional constant-time correlation spectroscopy. NMR Biomed 2008; 21(5): 518-26. [http://dx.doi.org/10.1002/nbm.1223] [PMID: 18351694]

[21] Sylwestrak EL, Ghosh A. Elfn1 regulates target-specific release probability at CA1-interneuron synapses. Science 2012; 338(6106): 536-40. [http://dx.doi.org/10.1126/science.1222482] [PMID: 23042292]

[22] Leiper J, Nandi M. The therapeutic potential of targeting endogenous inhibitors of nitric oxide synthesis. Nat Rev Drug Discov 2011; 10(4): 277-91. [http://dx.doi.org/10.1038/nrd3358] [PMID: 21455237]

[23] Grachev ID, Fredickson BE, Apkarian AV. Dissociating anxiety from pain: mapping the neuronal marker N-acetyl aspartate to perception distinguishes closely interrelated characteristics of chronic pain. Mol Psychiatry 2001; 6(3): 256-8.

[http://dx.doi.org/10.1038/sj.mp.4000834] [PMID: 11326290]

[24] Graeber MB. Changing face of microglia. Science 2010; 330: 783-8. [http://dx.doi.org/10.1126/science.1190929]

[25] Wang Q, Tompkins KD, Simonyi A, Korthuis RJ, Sun AY, Sun GY. Apocynin protects against global cerebral ischemia-reperfusion-induced oxidative stress and injury in the gerbil hippocampus. Brain Res 2006; 1090(1): 182-9. [http://dx.doi.org/10.1016/j.brainres.2006.03.060] [PMID: 16650838]

[26] Lim GP, Chu T, Yang F, Beech W, Frautschy SA, Cole GM. The curry spice curcumin reduces oxidative damage and amyloid pathology in an Alzheimer transgenic mouse. J Neurosci 2001; 21(21): 8370-7. [PMID: 11606625]

[27] Lu KT, Chiou RY, Chen LG, *et al.* Neuroprotective effects of resveratrol on cerebral ischemia-induced neuron loss mediated by free radical scavenging and cerebral blood flow elevation. J Agric Food Chem 2006; 54(8): 3126-31. [http://dx.doi.org/10.1021/jf053011q] [PMID: 16608241]

[28] Goldman SA, Nedergaard M, Windrem MS. Glial progenitor cell-based treatment and modeling of neurological disease. Science 2012; 338(6106): 491-5. [http://dx.doi.org/10.1126/science.1218071] [PMID: 23112326]

[29] ClinicalTrials.gov. Vaccine Therapy in Treating Patients with Newly Diagnosed Glioblastoma Multiforme Recovering From Lymphopenia Caused by Temozolomide http://clinicaltrials.gov/ct2/show/record/NCT00627224?term=Duane+Mitchell+cytomegalovirus &rank=2. , [Accessed March 10, 2014.];

[30] Science and Reason. Human cytomegalovirus and glioblastoma, website: http://science andreason.blogspot.com/2008/05/human-cytomegalovirus-and-glioblastoma.html. , [Accessed Feb 20, 2014.];

[31] Schummers J, Yu H, Sur M. Tuned responses of astrocytes and their influence on hemodynamic signals in the visual cortex. Science 2008; 320(5883): 1638-43. http://creativecommons .org/licenses/by-nc/3.0/. [http://dx.doi.org/10.1126/science.1156120] [PMID: 18566287]

[32] Wolf F, Kirchhoff F. Neuroscience. Imaging astrocyte activity. Science 2008; 320(5883): 1597-9. [http://dx.doi.org/10.1126/science.1160122] [PMID: 18566273]

[33] Shelat PB, Chalimoniuk M, Wang JH, *et al.* Amyloid beta peptide and NMDA induce ROS from NADPH oxidase and AA release from cytosolic phospholipase A2 in cortical neurons. J Neurochem 2008; 106(1): 45-55. [http://dx.doi.org/10.1111/j.1471-4159.2008.05347.x] [PMID: 18346200]

[34] Foley KM. Controlling the pain of cancer. Sci Am 1996 Sept; 275(3): 164-5. [http://dx.doi.org/10.1038/scientificamerican0996-164] [PMID: 8701289]

[35] Palczewski K, Kiser PD. Biochemistry. As good as chocolate. Science 2013; 340(6132): 562-3. [http://dx.doi.org/10.1126/science.1238521] [PMID: 23641106]

[36] King OD, Gitler AD, Shorter J. The tip of the iceberg: RNA-binding proteins with prion-like domains

in neurodegenerative disease. Brain Res 2012; 1462: 61-80.
[http://dx.doi.org/10.1016/j.brainres.2012.01.016] [PMID: 22445064]

[37] Kim HJ, Kim NC, Wang YD, *et al.* Mutations in prion-like domains in hnRNPA2B1 and hnRNPA1 cause multisystem proteinopathy and ALS. Nature 2013; 495(7442): 467-73.
[http://dx.doi.org/10.1038/nature11922] [PMID: 23455423]

[38] Dunning CJ, George S, Brundin P. What's to like about the prion-like hypothesis for the spreading of aggregated α-synuclein in Parkinson disease? Prion 2013; 7(1): 92-7.
[http://dx.doi.org/10.4161/pri.23806] [PMID: 23360753]

[39] Steinman L. Inflammatory cytokines at the summits of pathological signal cascades in brain diseases. Sci Signal 2013; 6(258): pe3.
[http://dx.doi.org/10.1126/scisignal.2003898] [PMID: 23322904]

[40] Goedert M, Spillantini MG. A century of Alzheimer's disease. Science 2006; 314(5800): 777-81.
[http://dx.doi.org/10.1126/science.1132814] [PMID: 17082447]

[41] Praticò D, Delanty N. Oxidative injury in diseases of the central nervous system: focus on Alzheimer's disease. Am J Med 2000; 109(7): 577-85.
[http://dx.doi.org/10.1016/S0002-9343(00)00547-7] [PMID: 11063960]

[42] Snowden D. Aging with Grace: What the Nun Study Teaches Us about Leading Longer, Healthier and More Meaningful Lives. New York: Bantam Books 2001.

[43] Atwood CS, Obrenovich ME, Liu T, *et al.* Amyloid-β: a chameleon walking in two worlds: a review of the trophic and toxic properties of amyloid-β. Brain Res Brain Res Rev 2003; 43(1): 1-16.
[http://dx.doi.org/10.1016/S0165-0173(03)00174-7] [PMID: 14499458]

[44] Marx J. Fresh evidence points to an old suspect. Calcium Sci 2007; 318: 384-5.
[PMID: 17947560]

[45] Yang F, Lim GP, Begum AN, *et al.* Curcumin inhibits formation of amyloid β oligomers and fibrils, binds plaques, and reduces amyloid *in vivo*. J Biol Chem 2005; 280(7): 5892-901.
[http://dx.doi.org/10.1074/jbc.M404751200] [PMID: 15590663]

[46] Lim GP, Calon F, Morihara T, *et al.* A diet enriched with the omega-3 fatty acid docosahexaenoic acid reduces amyloid burden in an aged Alzheimer mouse model. J Neurosci 2005; 25(12): 3032-40.
[http://dx.doi.org/10.1523/JNEUROSCI.4225-04.2005] [PMID: 15788759]

[47] Hashimoto Y, Tsuji O, Niikura T, *et al.* Involvement of c-Jun N-terminal kinase in amyloid precursor protein-mediated neuronal cell death. J Neurochem 2003; 84(4): 864-77.
[http://dx.doi.org/10.1046/j.1471-4159.2003.01585.x] [PMID: 12562529]

[48] Vlad SC, Miller DR, Kowall NW, Felson DT. Protective effects of NSAIDs on the development of Alzheimer disease. Neurology 2008; 70(19): 1672-7.
[http://dx.doi.org/10.1212/01.wnl.0000311269.57716.63] [PMID: 18458226]

[49] Szekely CA, Breitner JC, Fitzpatrick AL, *et al.* NSAID use and dementia risk in the Cardiovascular Health Study: role of APOE and NSAID type. Neurology 2008; 70(1): 17-24.
[http://dx.doi.org/10.1212/01.wnl.0000284596.95156.48] [PMID: 18003940]

[50] Arvanitakis Z, Grodstein F, Bienias JL, *et al.* Relation of NSAIDs to incident AD, change in cognitive function, and AD pathology. Neurology 2008; 70(23): 2219-25.

[http://dx.doi.org/10.1212/01.wnl.0000313813.48505.86] [PMID: 18519870]

[51] Bonda DJ, Wang X, Perry G, *et al.* Oxidative stress in Alzheimer disease: a possibility for prevention. Neuropharmacology 2010; 59(4-5): 290-4. [http://dx.doi.org/10.1016/j.neuropharm.2010.04.005] [PMID: 20394761]

[52] Smith MA, Zhu X, Tabaton M, *et al.* Increased iron and free radical generation in preclinical Alzheimer disease and mild cognitive impairment. J Alzheimers Dis 2010; 19(1): 363-72. [PMID: 20061651]

[53] Wan L, Nie G, Zhang J, *et al.* β-Amyloid peptide increases levels of iron content and oxidative stress in human cell and *Caenorhabditis elegans* models of Alzheimer disease. Free Radic Biol Med 2011; 50(1): 122-9. [http://dx.doi.org/10.1016/j.freeradbiomed.2010.10.707] [PMID: 21034809]

[54] Kell DB. Towards a unifying, systems biology understanding of large-scale cellular death and destruction caused by poorly liganded iron: Parkinson's, Huntington's, Alzheimer's, prions, bactericides, chemical toxicology and others as examples. Arch Toxicol 2010; 84(11): 825-89. [http://dx.doi.org/10.1007/s00204-010-0577-x] [PMID: 20967426]

[55] Peña F, Ordaz B, Balleza-Tapia H, *et al.* Beta-amyloid protein (25-35) disrupts hippocampal network activity: role of Fyn-kinase. Hippocampus 2010; 20(1): 78-96. [PMID: 19294646]

[56] Kurnellas MP, Adams CM, Sobel RA, Steinman L, Rothbard JB. Amyloid fibrils composed of hexameric peptides attenuate neuroinflammation Sci Transl Med 2013; 5(179ra42): 1-11. [http://dx.doi.org/10.1126/scitranslmed.3005681]

[57] Roberson ED, Mucke L. 100 years and counting: prospects for defeating Alzheimer's disease. Science 2006; 314(5800): 781-4. [http://dx.doi.org/10.1126/science.1132813] [PMID: 17082448]

[58] Liang X-T, Fang W-S. Medicinal Chemistry of Bioactive Natural Products. New York: Wiley-Interscience 2006. Chapter 4 [http://dx.doi.org/10.1002/0471739340]

[59] Tobinick EL, Gross H. Rapid improvement in verbal fluency and aphasia following perispinal etanercept in Alzheimer's disease. BMC Neurol 2008; 8: 27. [http://dx.doi.org/10.1186/1471-2377-8-27] [PMID: 18644112]

[60] Binder DK, Scharfman HE. Brain-derived neurotrophic factor. Growth Factors 2004; 22(3): 123-31. [http://dx.doi.org/10.1080/08977190410001723308] [PMID: 15518235]

[61] Huang L, Lu C, Sun Y, *et al.* Multitarget-directed benzylideneindanone derivatives: anti-β-amyloid (Aβ) aggregation, antioxidant, metal chelation, and monoamine oxidase B (MAO-B) inhibition properties against Alzheimer's disease. J Med Chem 2012; 55(19): 8483-92. [http://dx.doi.org/10.1021/jm300978h] [PMID: 22978824]

[62] McGowan PO, Sasaki A, D'Alessio AC, *et al.* Epigenetic regulation of the glucocorticoid receptor in human brain associates with childhood abuse. Nat Neurosci 2009; 12(3): 342-8. [http://dx.doi.org/10.1038/nn.2270] [PMID: 19234457]

[63] Worley PF, Heller WA, Snyder SH, Baraban JM. Lithium blocks a phosphoinositide-mediated

cholinergic response in hippocampal slices. Science 1988; 239(4846): 1428-9. [http://dx.doi.org/10.1126/science.2831626] [PMID: 2831626]

[64] Douglas IJ, Smeeth L. Exposure to antipsychotics and risk of stroke: self controlled case series study. BMJ 2008; 337: a1227. [http://dx.doi.org/10.1136/bmj.a1227] [PMID: 18755769]

[65] Neimark J. Autism: it's not just in the head Discover Magzine 2007 April; 33-8. 75

[66] Nakazawa DJ. The Autoimmune Epidemic. New York: Simon & Schuster 2008.

[67] Barbaresi WJ, Katusic SK, Voigt RG. Autism, a review of the state of the science for pediatric primary health care clinicians. Arch Pediatr Adolesc Med 2006; 160(11): 1167-75. [http://dx.doi.org/10.1001/archpedi.160.11.1167] [PMID: 17088521]

[68] Kwiatkowski TJ Jr, Bosco DA, Leclerc AL, *et al.* Mutations in the *FUS/TLS* gene on chromosome 16 cause familial amyotrophic lateral sclerosis. Science 2009; 323(5918): 1205-8. [http://dx.doi.org/10.1126/science.1166066] [PMID: 19251627]

[69] Vance C, Rogelj B, Hortobágyi T, *et al.* Mutations in FUS, an RNA processing protein, cause familial amyotrophic lateral sclerosis type 6. Science 2009; 323(5918): 1208-11. [http://dx.doi.org/10.1126/science.1165942] [PMID: 19251628]

[70] Geron Corporation. Geron receives FDA clearance to begin world's first human clinical trial of embryonic stem cell-based therapy, http://www.geron.com . 2009.

[71] Wieskopf JS, Pan YX, Marcovitz J, *et al.* Broad-spectrum analgesic efficacy of IBNtxA is mediated by exon 11-associated splice variants of the mu-opioid receptor gene. Pain 2014; 155(10): 2063-70. [http://dx.doi.org/10.1016/j.pain.2014.07.014] [PMID: 25093831]

[72] Thompson CN, *et al.* Activation of G-proteins by morphine and codeine congeners: insights to the relevance of *O*- and *N*-demethylated metabolites at μ- and δ-opioid receptors. J Pharmacol Exp Ther 2004; 308(2): 547. [PMID: 14600248]

[73] Zhang Y, Wang Z, Cox DP, Civelli O. Study on the activation of the opioid receptors by a set of morphine derivatives in a well-defined assay system. Neurochem Res 2012; 37(2): 410-6. [http://dx.doi.org/10.1007/s11064-011-0627-7] [PMID: 22002663]

[74] Grond S, Sablotzki A. Clinical pharmacology of tramadol. Clin Pharmacokinet 2004; 43(13): 879-923. [http://dx.doi.org/10.2165/00003088-200443130-00004] [PMID: 15509185]

[75] Richards P, Gimbel JS, Minkowitz HS, Kelen R, Stern W. Comparison of the efficacy and safety of dual-opioid treatment with morphine plus oxycodone *versus* oxycodone/acetaminophen for moderate to severe acute pain after total knee arthroplasty. Clin Ther 2013; 35(4): 498-511. [http://dx.doi.org/10.1016/j.clinthera.2013.03.002] [PMID: 23545428]

[76] US FDA. FDA drug safety communication: Prescription acetaminophen products to be limited to 325 mg per dosage unit; boxed warning will highlight potential for severe liver failure. http://www.fda.gov/Drugs/DrugSafety/ucm239821.htm .

[77] Gould HJ III, Paul D. Hydrocodone extended-release: pharmacodynamics, pharmacokinetics and behavioral pharmacology of a controversy. Pharmacol Res 2015; 91: 99-103. [http://dx.doi.org/10.1016/j.phrs.2014.09.006] [PMID: 25301538]

[78] FDA News Release. FDA approves extended-release, single-entity hydrocodoneproduct.. US Department of Health and Human Services, Food and Drug Admin-istration. Protecting and Promoting Your Health http://www.fda.gov/newsevents/newsroom/pressannouncements/ucm 372287.htm. 2013.

[79] Mayo Clinic. Selective serotonin reuptake inhibitors (SSRIs). http://www.mayoclinic.org /diseases-conditions/depression/in-depth/ssris/art-20044825.

[80] Koivisto H, Grimm MO, Rothhaar TL, *et al.* Special lipid-based diets alleviate cognitive deficits in the APPswe/PS1dE9 transgenic mouse model of Alzheimer's disease independent of brain amyloid deposition. J Nutr Biochem 2014; 25(2): 157-69. [http://dx.doi.org/10.1016/j.jnutbio.2013.09.015] [PMID: 24445040]

[81] Sheline YI, West T, Yarasheski K, *et al.* An antidepressant decreases CSF Aβ production in healthy individuals and in transgenic AD mice. Sci Transl Med 2014; 6(236): 236re4. [http://dx.doi.org/10.1126/scitranslmed.3008169] [PMID: 24828079]

[82] Mapstone M, Cheema AK, Fiandaca MS, *et al.* Plasma phospholipids identify antecedent memory impairment in older adults. Nat Med 2014; 20(4): 415-8. [http://dx.doi.org/10.1038/nm.3466] [PMID: 24608097]

[83] Rowinska-Zyrek M, Salerno M, Kozlowski H. Neurodegenerative diseases – Understanding their molecular bases and progress in the development of potential treatments. Coord Chem Rev 2015; 284: 298-312. [http://dx.doi.org/10.1016/j.ccr.2014.03.026]

[84] Liang X-T, Fang W-S. Medicinal Chemistry of Bioactive Medicinal Products. New York: Wiley-Interscience 2006.

[85] Breyer-Pfaff U, Maier U, Brinkmann AM, Schumm F. Pyridostigmine kinetics in healthy subjects and patients with myasthenia gravis. Clin Pharmacol Ther 1985; 37(5): 495-501. [http://dx.doi.org/10.1038/clpt.1985.78] [PMID: 3987173]

[86] Flacke W. Treatment of myasthenia gravis. N Engl J Med 1973; 288(1): 27-31. [http://dx.doi.org/10.1056/NEJM197301042880107] [PMID: 4347296]

[87] Caparros-Lefebvre D, Elbaz A. Caribbean Parkinsonism Study Group. Possible relation of atypical parkinsonism in the French West Indies with consumption of tropical plants: a case-control study. Lancet 1999; 354(9175): 281-6. [http://dx.doi.org/10.1016/S0140-6736(98)10166-6] [PMID: 10440304]

[88] Badrie N, Schauss AG. Soursop (*Annona muricata* L.): Composition, nutritional value, medicinal uses, and toxicology. In: Watson RR, Preddy VR, Eds. Bioactive Foods in Promoting Health Fruits and Vegetables. Oxford: Academic Press 2009; pp. 621-43.

[89] Wang X-J, Cao Q, Zhang Y, Su X-D. Activation and regulation of caspase-6 and its role in neurodegenerative diseases. Annu Rev Pharmacol Toxicol 2015; 55: 553-72. [http://dx.doi.org/10.1146/annurev-pharmtox-010814-124414] [PMID: 25340928]

[90] González-Muñoz GC, Arce MP, Pérez C, *et al.* Dibenzo[1,4,5]thiadiazepine: a hardly-known heterocyclic system with neuroprotective properties of potential usefulness in the treatment of neurodegenerative diseases. Eur J Med Chem 2014; 81: 350-8. [http://dx.doi.org/10.1016/j.ejmech.2014.04.075] [PMID: 24858540]

[91] Hölscher C. Central effects of GLP-1: new opportunities for treatments of neurodegenerative diseases. J Endocrinol 2014; 221(1): T31-41.
[http://dx.doi.org/10.1530/JOE-13-0221] [PMID: 23999914]

[92] Youdim MBH, Kuperschmidt L, Amit T, Weinreb O. Promises of novel multi-target neuroprotective and neurorestorative drugs for Parkinson's disease. Parkinsonism and Related Disorders 2014; 20S1: S132-6.
[http://dx.doi.org/10.1016/S1353-8020(13)70032-4]

[93] Garbayo E, Ansorena E, Blanco-Prieto MJ. Drug development in Parkinson's disease: from emerging molecules to innovative drug delivery systems. Maturitas 2013; 76(3): 272-8.
[http://dx.doi.org/10.1016/j.maturitas.2013.05.019] [PMID: 23827471]

[94] de Munter JPJM, Melamed E, Wolters EC. Stem cell grafting in parkinsonism - Why, how and when Parkinsonism Rel Disord 2014; 20S1: S150-3.

[95] Payne NL, Sylvain A, O'Brien C, Herszfeld D, Sun G, Bernard CC. Application of human induced pluripotent stem cells for modeling and treating neurodegenerative diseases. N Biotechnol 2015; 32(1): 212-28.
[http://dx.doi.org/10.1016/j.nbt.2014.05.001] [PMID: 24815224]

Medicinal Chemistry - Fusion of Traditional and Western Medicine, **2015, 323-394**

CHAPTER 8

The Immune System and Immune Network

Abstract: The immune system recognizes and defends us against internal threats caused by invading organisms and pathogens. The innate immune system recognizes bacteria, fungi and other organisms, breaks them down, identifies a characteristic protein on them (an antigen) and attaches it to the surface of specific cells, which present it to the adaptive immune system for destruction. The adaptive or acquired immune system is activated after being stimulated by the innate system. The adaptive immune response is initiated by specific interactions between antigen-bound, mature dendritic cells and naïve $CD4^+$ T cells in the lymph nodes. The adaptive or acquired immune system acts once it is stimulated by the innate system. The five families of immune cells are: phagocytes, granulocytes, natural killer (NK) cells, lymphocytic T-cells and lymphocytic B-cells. The four major classes of immune system mediators are chemotactic agents, cytokines, C-reactive protein and antibodies. A fifth class of immune network mediators are the small molecules, including neurotransmitters, such as L-DOPA and catecholamines. Risk factors for autoimmune diseases include exposure to man-made chemicals. Benlysta (belimumab) is approved for treating lupus erythematosus. AIDS is caused by the retrovirus HIV. Currently, a mixture, or cocktail of antiretroviral drugs are given in what is often called highly active antiretroviral therapy, or HAART. Microbes in the intestines and lungs (acquired from the environment) keep rare invariant natural killer immune cells from triggering autoimmune diseases.

Keywords: Adaptive, Autoimmune diseases, HAART, HIV, Immune system, Innate.

Just as the nervous system can be studied by systems thinking so can the immune system or immune network. In fact, the immune and nervous systems are closely linked. Some of the molecules that are produced by our immature immune systems help regulate the growth and development of our nervous systems. There are other molecules that are used to communicate information in both the immune and nervous systems. Both systems are used in making decisions. The brain helps us recognize and defend ourselves against external threats, while the immune

Robert E. Smith

system recognizes and defends us against internal threats caused by invading organisms and pathogens. Regular cardiovascular exercise (endurance training) and proper nutrition are important for the nervous and immune systems. Both are essential in helping to establish the boundaries between self and the environment and the difference between food and poison. It is essential that we can distinguish between self and non-self. This is important because we are constantly being exposed to many man-made chemicals, viruses and organisms that are potentially very dangerous. From the outside of our skin to the inside of our bodies, we have many ways to defend ourselves. Similarly, all organisms have defenses. Both invertebrates and vertebrates have an innate immune system, but only vertebrates have an acquired (or adaptive) immune system. In fact, it is often described as an immune network.

Our first line of defense is composed of physical and chemical barriers, such as antimicrobial peptides and enzymes (like lysozyme) that are secreted by saliva, tears, the respiratory tract and the skin (a physical barrier). In the gut, commensurate (or commensal) bacteria help prevent infection by pathogenic bacteria. The ecosystem that comprises the human organism co-evolved with commensurate bacteria that have established a niche for themselves, in which they have a competitive advantage over other bacteria. The highest density of commensurate bacteria in the human body is found in the gastrointestinal tract [1]. There is a mucus tract that covers it and influences the function of antigen presenting cells (APCs) and epithelial cells so that our dendritic cells (DCs) can tolerate food and commensurate bacterial antigens [1]. This mucus tract contains glycoproteins called mucins, including MUC2. It prevents inflammation by forming a non-attached outer mucus layer that is inhabited by bacteria. MUC2 also forms an inner mucus layer that adheres to intestinal epithelial cells. It mitigates inflammatory responses to DCs by generating signals that make them tolerant to antigens [2]. DCs lie just beneath the epithelial layer of cells and presents foreign antigens which we should not try to tolerate to other cells in the immune system.

When these first lines of defense fail to block pathogenic bacteria or other immunogens, the innate and adaptive immune systems are activated. Both use white blood cells (leukocytes), which are a diverse group of cell types that mediate the body's immune response. They circulate through the blood and lymphatic system and are recruited to sites of tissue damage and infection [1].

Acute inflammation is made up of antibodies and humoral factors in the cell-free blood serum and other bodily fluids, once known as humors. This includes the complement system, coagulation system, iron-binding proteins (lactoferrin and transferrin), interferons, lysozyme and a protein called interleukin-1, or IL-1. Antibodies are made in B cells, which are lymphocytes that are made in the bone marrow [1]. "The innate immune system recognizes pathogen-associated molecular patterns (PAMPs), which include double-stranded RNA, lipo-polysachharides and other molecules that are common in viruses, bacteria and fungi, but not in healthy human cells. The innate immune system senses the presence of microbial pathogens by detecting cytoplasmic DNA. However, erroneous detection of dinucleotides can help cause some autoimmune diseases" [1].

The innate immune system also recognizes damage- or danger-associated molecular patterns (DAMPs) released from damaged cells, such as uric acid crystals or ATP [1]. "It protects the host from infection by other organisms in a relatively non-specific way. The human innate immune system recruits immune cells to the site of infection, activates the complement cascade, identifies and removes foreign substances and activates the adaptive immune system through antigen presentation. That is, the innate immune system recognizes bacteria, fungi and other organisms, breaks them down, identifies a characteristic protein on them (an antigen) and attaches it to the surface of specific cells, which present it to the adaptive immune system for destruction. The innate immune system does not have a memory and does not offer specific resistance against organisms that have invaded the host in the past. For this, the adaptive immune system is needed. It is activated by the innate immune system. It is made of many parts" [1].

Invading cells are killed by phagocytosis using a subgroup of leukocytes that circulate in the blood as monocytes [3]. "They are converted into macrophages, which enter tissues during inflammation. There is much heterogeneity in phenotype, homeostatic turnover and function when they are present in different tissues. Dendritic cells (DCs) are a distinct lineage of mononuclear phagocytes, specialized in presenting antigens to T cells while initiating and controlling immunity. DCs are crucial for activating and conditioning virus-specific T cells. They bridge innate and adaptive immunity and have additional roles in shaping the immune response to pathogens, vaccines, and tumors" [3].

Phagocytic white blood cells have several cell surface protein receptors. This include F_c receptors that bind to the F_c (crystallizable fraction) of IgG antibodies; complement receptors, which bind complement proteins; scavenger receptors, which bind a variety of polyanions on bacterial surfaces; and toll-like receptors, which recognize PAMPs on infectious agents [1].

There is also a cell-based immunity that does not include complement or B cells. Instead, cell-mediated immunity works through macrophages, natural killer (NK) cells, cytotoxic T-lymphocytes (made in the thymus) and many different cytokines that are released in response to an antigen, bacteria, fungi, protozoans and cancer cells. Both the innate and adaptive immune systems use cell-mediated and humoral components [1].

The adaptive or acquired immune system acts once it is stimulated by the innate system [1]. "Mature dendritic cells in the innate immune system form immunological synapses with T cells that have the CD4 cell surface glycoprotein, or $CD4^+$ T-cells. There are two main types of T cells: $CD4^+$, which oversee the immune response and $CD8^+$ cells which do most of the actual killing. The adaptive immune response is initiated by specific interactions between antigen-loaded, mature dendritic cells and naïve $CD4^+$ T cells in the lymph nodes. Dendritic cells link the innate and adaptive immune systems. They engulf exogenous pathogens and toxins, chop them into pieces and present them to T-cells, which come from the thymus gland. To do this properly, they must be able to distinguish between self and non-self. Although our brains may establish our personal identities to the outside world, the immune system establishes our internal identities" [1].

As the name implies, the adaptive immune system adapts to toxic challenges from viruses, bacteria, protozoa, and other organisms. Adaptive immunity uses helper T cells (T_H) that produce cytokines. Naïve T_H cells are stimulated by antigen presenting cells that have cognate antigens on their surface [1]. "The naïve cells differentiate into two different types of T_H cells, T_H1 and T_H2. The T_H1 cells secrete interferon-γ and promote cellular immunity. The $CD4^+$ T_H2 cells produce the interleukin (IL) cytokines IL-4, IL-5, IL-10 and IL-13, while producing humoral immunity. A small percentage of the cells acquire and retain a memory of one or more pieces (antigens) of the invading viruses and micro-organisms. This enables us to mount rapid and effective responses to the invading pathogens when they invade us more than once" [1].

The adaptive immune system uses the type of white blood cells that are called lymphocytes [1]. "The human body has about 2 trillion lymphocytes, constituting 20-40% of all the white blood cells. The peripheral blood contains 20-50% of them as circulating lymphocytes; the rest move within the lymphatic system. Natural killer cells are large and granular. Smaller lymphocytes are T and B cells. There are three general classifications of $CD8^+$ cytotoxic T cells: naïve cells, central memory and effector memory cells. Naïve T-cells have not yet seen their cognate antigen, but memory cells have. Memory T cells can recognize bacteria, viruses and cancer cells. Central memory cells express the cytokine CCR7 and interleukin IL-2. Effector memory T cells (T_{EM}) express the effector cytokines IFNγ and IL-4. The T_{EM} cells can reside permanently (or with low turnover rates) in peripheral tissues and migratory cells that can move from the periphery and into the blood. Moreover, T_{EM} cells can rapidly recognize and kill infected target cells. They can facilitate early containment of nascent infections. Human T lymphocytes can be genetically engineered to express almost any gene, so they can have the desired specificity. They can be used in adoptive immunotherapy for cancer" [1]. Many protocols transfer genes into bulk peripheral blood mononuclear cells (PBMCs), but recent studies have shown that effector cells derived from naïve cells may be the preferred cells for adoptive immunotherapy [4].

IMMUNE NETWORK

One of the most important features of autopoiesis is self. Living, autopoietic systems have a boundary, such as a skin or a membrane, which separates self from the environment. The boundary and almost everything inside it is being continuously broken down and re-made. Inside that boundary, our immune network distinguishes between self and non-self. Chemical messengers (chemokines, cytokines, C reactive protein and products of the complement system) mediate the immune system and link the different cells and organs, which are the nodes of the network. For example, transforming growth factor-β (TGFB) regulates the development and maturation of T_{Reg} cells and immune homeostasis. There is also a network of antibodies. They are connected to other antibodies and antigens. Antibodies are always being made, even in the absence of infection. They help define self. Every day, about 100 precursors of immune cells enter the thymus and undergo several cell divisions to make about $10^7 - 10^8$ T-cells [5].

Immune cells make cell surface proteins called immunoglobulins, or Igs. The Igs have structures that look a little bit like the letter Y when projected onto a flat surface. Igs have two heavy chains and two light chains that are joined by disulfide bonds [1]. Each chain is made up of structural domains called Ig domains. These domains of 70-100 amino acids are classified as variable, joining or constant. They are made from three types of genes, which code for the constant, diversity and joining regions. That is, part of an Ig molecule has a structure that is very similar to other Igs in its class. This is called the constant region. There are five types of mammalian heavy chains: α,δ, e, γ and μ. They make the IgA, IgD, IgE, IgG and IgM antibodies. The constant region is joined with the variable, or diversity region of the Ig, which is on the outer part of the Y-shaped arm. It has a unique structure that gives the antibody molecule its antigenic individuality and is coded for by diversity genes. These unique regions are called complementarity determining regions (CDRs). They define the idiotypes, or antigen binding specificity of the antibodies. More than a million idiotypes are made. Those that bind to self antigen are marked for destruction. Those that bind foreign antigen survive and make memory B and T cells. Only about 5% of the T cells survive the selection process that eliminates the cells that recognize self-antigens and bind them tight enough. There is a selection-driven tolerance (self-tolerance) that helps establish a type of balance or homeostasis. If the elimination process becomes too lenient, some of the T-cells that recognize self would survive and kill some of the body's own cells, causing an autoimmune disease. If the elimination process becomes too severe, some of the T-cells that recognize invading pathogens would be killed and the body could succumb to disease [3]. This homeostasis is controlled, in part, by regulatory T cells (T_{Reg}) [1], whose development and maturation is controlled by transforming growth factor-β (TGF β). So, altered TGF β signaling can predispose people to allergic responses to antigens that may be present in the environment and foods.

So, one of the most important roles of our immune system is to define self and non-self [1]. "This is done by looking at specific proteins on the cell surface. The outer membranes of the human cells in our body (except red blood cells) have proteins and glycoproteins that are unique to the individual. These proteins are called human leukocyte antigens (HLA) and they are coded for by the major histocompatibility complex (MHC) genes" [1]. Often, the terms HLA and MHC are used as synonyms [3]. They are antigens that define self. Class I HLAs are found on all human cells (except red blood cells, or erythrocytes), while class II HLAs are only found on immunocompetent cells, such as B lymphocytes and macrophages. Class I HLAs (or MHCs) display fragments of intracellular proteins on the cell surface, so they will be identified as self and not be attacked by T-lymphocytes. Class II HLAs present extracellular protein antigens to helper T cells, activating them and triggering an immune response [1]. "Antigens bind to antibodies, which are proteins produced by immune cells in vertebrates. So, every cell that produces an HLA antigen also produces an image of itself that is recognized by antibodies in other cells. If an immune cell recognizes the self antigens, chemokines bind to it and initiate programmed cell death (apoptosis) of the immune cell. If an immune cell recognizes non-self antigens, the response depends on the source of the antigen. If it comes from food, the immune cells should undergo self-destruction. If it comes from a piece of a parasite, virus or bacteria, signaling molecules will cause the immune cells to grow, differentiate and make more antibodies to kill the invading organisms" [1]. At the same time, our immune system must be able to tell the difference between pathogenic bacteria and the good bacteria that we need in our digestive system, to help us digest our food [5 - 6]. So, our immune network makes decisions, similar to our brain and central nervous system. If knowledge and intelligence are thought of as the ability to make correct decisions, then both the immune and nervous systems are part of the same whole body that makes us smart.

Throughout our lives, our immune systems go through a series of different states of homeostasis. When we are in our mother's womb, we have an underdeveloped immune system that mostly uses a different set of helper T cells (type 2, or T_H2) than what we use as adults (type 1, T_H1). We are born with a certain repertoire of immune cells, which make many antibodies and antigens. After we are born and are exposed to different antigens from the environment, our adaptive (acquired) immune systems adapt by making different repertoires of antibodies. Chemical messengers communicate between cells and tissues, to help make necessary changes, in response to changes in our internal environments. So, there is a network in which antigens make antibodies, which make more antibodies. The adaptive immune system does this, but it takes some time. So, as mentioned earlier, there is also an innate immune system that responds faster and which activates necessary changes in the adaptive immune system. Toll-like receptors, NLR proteins and dendritic cells link the two together [7]. So, we can think of the immune network as consisting of nodes (cells and tissues) connected by chemical messengers (chemokines, cytokines and receptors). The reductionist thinking of the 20th century identified the nodes and the chemical messengers. Systems thinking of the 21st century added an emphasis on the relationships between the components, how they communicate and how they are organized. A person's immune response must be kept in balance to maintain health. This is done through a network of co-signaling molecules. It is coordinated by a network of ligand-receptor interactions on the cell surface, with both co-stimulatory and co-inhibitory capacities [7]. The direction and outcome of immune responses are decided by the interplay of these complicated and often counterbalancing network interactions [7].

COMPONENTS OF THE IMMUNE SYSTEM

So, let us start by looking at the different parts of the immune system, which consists of layers of increasing selectivity. They are listed in Table **1**. The outermost layer can be skin, fur, feathers, exoskeletons, egg shells, or even the oily outer layer of leaves. This layer acts as a physical barrier to toxins and pathogens. The next layer includes lungs, intestines, commensurate bacteria, mucus, urinary tract, lymph glands and the blood-brain barrier. Our lungs can expel dust, pathogens and other irritants by sneezing and coughing. Our eyes make tears that wash. The urinary tract flushes out water-soluble toxins. Mucus in the respiratory and gastrointenstinal tracts traps and entangles microorganisms. Mucosal immunity helps protect against pathogens that invade through the respiratory, gastrointestinal and reproductive tracts. Infections at mucosal sites elicit immune responses that are focused on the site of exposure. Still, vaccines administered to mucosal surfaces elicit strong immune responses not only at the application site but also at distal sites, due to crosstalk between mucosal compartments. This permits immunization at one mucosal surface to establish immunity at distal mucosal sites. So, needle-free vaccination strategies such as aerosol or intranasal delivery might be capable of protecting against not only airway or systemic infections but also gastrointestinal or reproductive tract pathogens. Existing mucosal vaccines include FluMist and the oral polio vaccine. Moreover, a new pulmonary vaccine is being developed that is based on nanoparticles that contain antigens and Toll-like receptor agonists [8]. It elicited long-lasting antigen-specific effector memory T cell responses at several mucosal sites. Nanocapsules primed 13-times more T cells than did equivalent soluble vaccines [8].

Table 1. Components of the immune system.

Components of the Immune System
Outermost layer: skin, fur, feathers, exoskeletons and egg shells
Lungs, intestines, commensurate bacteria, mucus, urinary tract, lymph glands and the blood-brain barrier
Chemical barriers: Antimicrobial peptides (β-defensins and cathelicidins), lysozyme and phospholipase A2
Biological barriers: Commensurate bacteria
Inner defenses: the innate and acquired (or adaptive) immune systems

There are also chemical barriers. The skin and respiratory tract secrete antimicrobial peptides (β-defensins and cathelicidins) [9]. "They are also called host defense peptides, or HDPs. They are found in almost all forms of life. They kill bacteria, stimulate the innate and acquired immune systems, help heal wounds, move cells around (cell trafficking) and stimulate the growth of blood vessels. They recruit "immature dendritic cells and T-cells. They produce glucocorticoids. They stimulate macrophage phagocytosis, mast cell degranulation, complement activation and interleukin production by epithelial cells. Host defense peptides can also act as adjuvants for vaccines" [9].

There are also enzymes such as lysozyme and phospholipase A2 in saliva, tears, and breast milk that kill bacteria [1]. Vaginal secretions provide a chemical barrier following menarche, when they become slightly acidic, while semen contains defensins and zinc to kill pathogens [1]. In the stomach, gastric acid and proteases are powerful chemical defenses against ingested pathogens. There are also biological barriers. We have good bacteria that compete with pathogenic bacteria for food, space and even environmental conditions, such as pH and iron. However, antibiotics can reduce the amount of good bacteria and lead to the growth of fungi. On the other hand, one can increase the amount of good bacteria by eating yogurt, which has probiotic *Lactobaciili*. They help restore a healthy balance of microbial populations. Most probiotics are fermented milk or yogurt. Recently, many of these have appeared in stores around the world, while dietary supplements made from them are aggressively marketed. Still, no probiotic has been approved by the FDA or other government agencies for therapeutic use, even though some are under clinical trials and may be approved in the future [10].

The FDA has no definition of probiotics [10]. "It regulates them based on whether they are drugs, biologics or foods. However, regulatory approaches for such products may not cover some unique features of probiotics. They are living, dynamic organisms and are likely to lose viability and degrade over time. Manufacturing them involves a more variables than other substances. These include the effect of storage temperature on their viability and their interactions with the human genome and microbiome. Without stringent manufacturing procedures and quality controls, some probiotics may lose some important properties. So, it is interesting to note that most probiotics that would be defined as dietary supplements in the USA are regulated in Canada as natural health products. As described in Health Canada's probiotics monograph, all probiotic natural health products require premarket assessment. Moreover, to get approval, there must be evidence of strain-specific safety and efficacy under the recommended conditions of use. However, the US FDA's OTC drug monographs do not specify or require pre-market approval or a demonstration of efficacy [10].

If viruses, bacteria, fungi, parasites or toxins make it past the outer layers of our defense, they must face our inner defenses, the innate and acquired (or adaptive) immune systems [1]. "The innate immune system is coded for by germ cells. It is the first line of defense. It is made up of immune cells that have receptors that bind to microbial biochemicals. The immune cells are white blood cells, or leukocytes. They include natural killer cells, mast cells, eosinophils and basophils. They also include phagocytic cells (macrophages, dendrocytes and neutrophils) that engulf, or ingest (phagotocytize) bacteria, dead or dying cells and cell particles. Phagocytes can be labeled professional or non-professional, depending on whether they have receptors on their surfaces that bind to pieces of viruses, parasites, dead host cells, bacteria or other foreign particles. Toll-like receptors recognize bacterial biochemicals, including pieces of DNA, while RIG-I-like receptors recognize viral RNA. "After phagocytosis, macrophages and dendritic cells present antigens to the adaptive immune system. The innate immune system is predetermined and deals with a wide variety of pathogens and toxins that have a limited number of structures" [1]. Things like lipopolysaccharides, dinucleotides and other relatively small molecules are recognized by the innate system, which is often described as being non-specific, even though it shows much specificity [7]. The innate immune system provides immediate, but short term defense against infections.

The innate immune system also activates the acquired immune system, which is always called specific, because it has a memory and is long-lasting [1]. Membrane-bound toll-like receptors on innate immune cells surfaces and internal organelles provide crucial links between the innate and acquired immune systems [1]. Lectins can recognize carbohydrates made by pathogens. NLRs (nucleotide-binding domain, leucine rich) proteins can bind to molecular patterns on bacteria that make it into the cytoplasm by endocytosis. The acquired immune system is coded for by somatic cells and it is not predetermined. There is a lag time between infection and activation of the adaptive immune system, which consists of B- and T-lymphocytes. The responses of these cells are highly specific. They can remember molecules (antigens) from invading organisms, so that we can launch a fast and powerful response to future infections. It also remembers molecules produced by our own cells so that we don't kill them. The innate and acquired immune systems have opposite and complementary properties. The innate system responds fast and causes inflammation, redness and swelling and it generates heat. It also helps activate the adaptive immune system. These actions are mediated by four classes of mediators and five families of cells.

TYPES OF IMMUNE CELLS

The five families of immune cells are: phagocytes, granulocytes, natural killer (NK) cells, lymphocytic T-cells and lymphocytic B-cells [5]. "Phagocytes, granulocytes and natural killer (NK) cells are in the innate immune system, while lymphocytic T- and B-cells are in the acquired immune system. "Like erythrocytes and thrombocytes (the red cells), white blood cells (leukocytes) are made from precursors in the bone marrow. The $CD4^+$ T cells can be activated by a variety of pathogens. Once activated, they can differentiate into subsets of helper T cells which secrete the cytokines that are needed to clear pathogens and induce inflammatory responses" [5]. Each helper T cell subtype is also critical for helping B lymphocytes produce pathogen-specific antibodies [8]. Table **2** shows the types and roles of immune cells.

Phagocytes are immune cells that recognize, engulf and destroy viruses, microorganisms and organic compounds [5]. "They can be classified as either professional or non-professional, based on their ability to perform phagocytosis. The professional phagocytes are neutrophils, mast cells, monocytes, macrophages, and dendritic cells. About 5% of the circulating white blood cells are monocytes. They have a lifetime of just a few days, after which they develop into macrophages, which are in body tissues. Monocytes selectively destroy cells and materials that have been tagged with surface labels, such as antibodies, complement fragments and C-reactive protein. Monocytes circulate in the blood, bone marrow, and spleen and do not proliferate in a steady state. They are immune effector cells, equipped with chemokine receptors and adhesion receptors that mediate migration from blood to tissues during infection. They produce inflammatory cytokines and remove cells and toxins. They can also differentiate into inflammatory dendritic cells or macrophages during inflammation. Inflammation and pathogen-associated pattern-recognition receptors trigger the migration to tissues and differentiation to inflammatory dendritic cells and macrophages" [5]. Macrophages reside in lymphoid and nonlymphoid tissues and help maintain tissue homeostasis by removing cells that have undergone apoptosis, and they produce growth factors [3].

Table 2. Types and roles of immune cells.

Types of immune cells	Rolls of these cells
Phagocytes	Recognize, engulf and destroy viruses, microorganisms and organic compounds
Granulocytes	Consist of neutrophils, eosinophils, basophils and mast cells [1]
Natural killer (NK) cells	Interact with many different types of cells and influence their fate.
Lymphocytic T-cells	Bind antigens that are presented by "major histocompatibility complex (MHC) molecules that are expressed by diseased cells [1] and protect us from infection
Lymphocytic B-cells	recognize antigens in their native form and make antibodies
Macrophages	Engulf and kill microorganisms. Secrete factors that affect tissue repair and remodeling [1]
Monocytes	Recruited to inflamed tissues and become macrophages
Dendritic cells	Link the innate and adaptive immune systems; ingest and break down the various constituents of the cell [1].
Mast cells	Shape the inflammatory milieu and control the activation state of many cells that are crucial for adaptive immunity [1]
Eosinophils	Kill parasites, help regulate other immune cells, help destroy tumor cells, and promote the repair of damaged tissue [1].
Neutrophils	Kill invading microorganisms
Basophils	Produce interleukin-5 (IL5)

Macrophages are among the immune system's most important line of innate defense [10]. "They engulf and kill microorganisms and secrete factors that give them key roles in tissue repair and remodeling. Macrophages are found at low amounts in all resting tissues, but they increase in response to inflammatory signals. In many chronic diseases including heart disease, asthma, and cancer, smoldering inflammation leads to disease. Especially in cardiovascular disease, accumulated macrophages are quite harmful, and suppressing this build-up has therapeutic potential. Monocytes that circulate in the bloodstream are recruited to inflamed tissues and become macrophages. So, therapies to control inflammation often focus on interrupting this recruitment. However, the accumulation of certain macrophages, called M2 macrophages or alternatively activated macrophages (AAMs) occurs not through coaxing their precursors from the bloodstream, but by local proliferation of macrophages" [10].

Macrophages engulf and destroy foreign cells and fragments. The liver has specialized macrophages, called Kupffer cells. About 80% of the fixed macrophages in the body are located in the liver, as Kupffer cells. They ingest and degrade particulate matter and they can act as antigen-presenting cells. Kupffer cells are also a source of cytokines.

Macrophages can live for months or years. They produce many toxic chemicals and can cause collateral damage to surrounding healthy cells. This causes inflammation and their own death. Macrophages are activated by mediators that are secreted by other immune cells and they produce their own mediators that stimulate other immune cells. Macrophages can also be reservoirs for HIV, the virus that causes AIDS. Once inside a macrophage, the HIV virus is protected from attack by the immune system [5].

Dendritic cells are cells with many thin, long arms. Like red blood cells and other white blood cells, they are made from progenitor "cells in the bone marrow [1]. They are located throughout the body and are particularly abundant in interfaces between the external and internal environments, such as the skin and the lining of the nose, lungs and gastro-intestinal tract, where they are ideally placed to encounter invading pathogens. Dendritic cells can migrate into the blood and reach infected tissues. They help link the innate and adaptive immune systems. When they sense a tagged cell that is harmful, they are activated so they ingest and break down the various constituents of the cell. They contain a multi-protein complex called the proteasome, which can catalyze the hydrolysis of proteins that are tagged with ubiquitin. The small peptide fragments produced by the hydrolysis are transported to the surface of the dendritic cells, where they are held in place by a human leukocytes antigen (HLA) protein, also called a major histocompatibility complex (MHC) protein. The tagged dendritic cells then migrate to a lymph node where they encounter T-cells that have a matching receptor that binds the HLA-peptide complex [5]. The dendritic cells bind to $CD4^+$ T cells. The dendritic cell then functions as an antigen presenting cell (APC), which activates the $CD4^+$ T-cells [5].

For T cells to be activated, co-stimulatory signals must be delivered. For example, when the cluster of differentiation 28 (CD28) protein binds its ligand (CD80 and

CD86), it works with the T cell receptor to induce T cells to produce interleukins, especially IL-6, a pro-inflammatory cytokine and anti-inflammatory myokine [1] (secreted by myocytes – muscle cells). CD28 induces Bcl-2 through the mTOR pathway and enhances IL-2 transcription through protein kinase C. Other proteins (ITk and Tec) with a SRC3 domain also bind to CD28 to enhance IL-2 production. This is just one example of the protein network that regulates immune responses.

Classical DCs (cDCs) process antigens and present cells. They have high phagocytic activity as immature cells and high cytokine-producing capacity as mature cells [3]. "They are highly migratory cells that can move from tissues to the T cell and B cell zones of lymphoid organs. cDCs regulate T cell responses both in the steady state and during infection. They are generally short-lived and replaced by blood-borne precursors" [3].

Plasmacytoid DCs (PDCs) differ from cDCs in that they are relatively long-lived and some of them carry characteristic immunoglobulin rearrangements [3]. "They are present in the bone marrow and all peripheral organs. PDCs are specialized to respond to viral infection with a massive production of type I interferons (IFNs); however, they also can act as antigen-presenting cells and control T cell responses. The development of the mononuclear phagocyte system is controlled by cytokines: small secreted proteins that promote cell-cell communication and can act as growth and differentiation factors" [3].

Dendritic cells are deficient both in number and function in patients with cancer [3]. They have been used in the immunotherapy of cancer for more than a decade. Dendritic cell products have been delivered to more than 1000 cancer patients in over 150 clinical trials [11]. Cancer cells from patients can be isolated and cancer-specific antibodies can be isolated. Dendritic cells can be treated with these antibodies, which are presented on the cell surface. The activated, mature dendritic cells can activate tumor-specific cytolytic T lymphocytes, which can kill the cancer cells in the patient [10]. In 2010, the FDA approved Sipuleucel-T (Provenge®; Dendreon Corp., Seattle, WA), a novel immunotherapeutic for the treatment of asymptomatic metastatic castrate-resistant prostate cancer [11]. "It consists of dendritic cells from peripheral blood mononuclear cells. They are pulsed *ex vivo* with PA2024, a recombinant fusion protein consisting of granulocyte macrophage colony-stimulating factor and prostatic acid phosphatase (a prostate antigen), as an immunogenic agent" [11].

Recently, scientists and doctors have been able to eliminate leukemia in several patients by converting their cytotoxic T cells into specific killers of cancer cells [12]. "They used a lentiviral vector to express chimeric antigen receptors (CARs) in patients' cytotoxic T cells. Extracellular and intracellular domains were fused into a new, synthetic receptor for the CD8+ T cells. The gene for the CAR was delivered by a lentivirus vector, which was constructed using the backbone of the HIV-1 virus that was no longer pathogenic and can not cause AIDS. Because the natural host for HIV-1 is the T-lymphocyte, it enabled the lentivirus vector to specifically target the cancer cells and multiply in them. The patients received between 15 million and one billion cytotoxic T-lymphocytes (CTLs). These CTLs rapidly multiplied about 100-fold" [12].

There are similar vaccines based on dendritic cells that are in either Phase II or II clinical trials for colorectal cancer, glioma and melanoma [11]. "However, the need for leukapheresis and cell cultures limits the production of vaccines. So, peptides from tumor associated antigens are in clinical, trials, too. They need an adjuvant and/or immune modulator, but are more cost-effective. Still, they target only one or a few epitopes. For example, Stimuvax contains epitopes for both CD4 and CD8, to induce $CD8^+$ cytolytic "T cells and $CD4^+$ T helper cells" [11].

Another approach is to use viral vectors to deliver the epitopes [11]. "Poxviruses that infect birds and are harmless to humans are used. They can accept large insertions of foreign DNA consisting of many genes. Clinical trials have shown that when the transgene for a tumor-associated antigen is inserted into an appropriate virus, it becomes more immunogenic, due to its ability to induce the Toll-like receptors. Bacteria and yeasts can also be used as vectors. They can be used to activate dendritic cells and present epitopes of transgenes. This stimulates both the innate and adaptive immune systems. Still, whole tumor–cell vaccines are doing better in clinical trials. They present the patient's immune systems with a range of both known and undefined TAAs as immunogens. An even better approach may be to use whole tumor–cell vaccines. They have the advantage of presenting the patient's own tumor-associated antigens to their immune system. This may consist of point mutations and/or fused genes. In one protocol, dendritic cells and autologous tumor cells are fused together before immunizing the patient. The other approach used in clinical trials is to use allogeneic whole tumor–cell vaccines. Typically, they contain two or three established and characterized human tumor cell lines of a given tumor type. The GVAX vaccine platforms contain allogeneic pancreatic, prostate, or breast tumor cells and are a testament to the ability to provide such a vaccine for multicenter evaluation. Common targets for all these types of vaccines include oncoproteins, oncofetal antigens, differentiation-associated proteins, and viral proteins. For example, the FDA approved the Gardasil vaccine, which targets the human papilloma virus (HPV)" [11].

Cancer stem cells and/or the epithelial–mesenchymal transition (EMT) process are potential targets for anti-cancer vaccines [11]. "Both of them are associated with drug resistance. EMT is also associated with tumor extravasation and intravasation into the site of metastasis. EMT and cancer "stemness" are associated with the key proteins involved in stem-cell properties. Preclinical studies have also shown that a vaccine combined with immune stimulants, or inhibitors of immune suppression, can enhance antitumor responses. Another useful combination therapy is to use inhibitors of checkpoint proteins and immunosuppressive entities. Ipilimumab has been approved by the FDA for treating metastatic melanoma. It binds to the cell surface protein CTLA4 (T lymphocyte antigen 4), a CD28 homologue that is induced on activated T cells. CTLA4 is a co–inhibitor to keep T cell responses in check. It can enhance T cells and antitumor effects when combined with vaccines such as CEA-TRICOM recombinant poxviruses. A recent clinical trial of the PROSTVAC vaccine in combination with ipilimumab showed evidence of enhanced survival in patients. Ongoing clinical studies with anti-PD-1 and anti-PD-1 ligand 1 monoclonal antibodies as monotherapies have also shown promising results" [11].

Another advance is to combine vaccines and chemotherapy, which was previously thought to be a bad idea [11]. "However, recent data has shown that multiple forms of therapy can be used concurrently with cancer vaccines or after vaccine therapy, with additive or synergistic effects. For example, oxaliplatin and anthracyclines such as doxorubicin will induce immunogenic tumor cell death, which results in enhanced cross-priming of tumor-associated antigens by dendritic cells and subsequent activation of T cells. Clinical trials are being done using the PROSTVAC vaccine plus the taxane, docetaxel, in patients with prostate cancer. A trial that uses PROSTVAC vaccine in combination with a bone-seeking chelated radionuclide in patients with prostate cancer metastatic to the bone is also in progress. It has also been shown that a low-dose of paclitaxel (another taxane) can enhance dendritic cell function in preclinical studies. Also, hormonal therapy that is used in the treatment of several different stages of prostate cancer has been shown to induce thymic regeneration and naïve T cells. It is at this interval that vaccine therapy may be most effective" [11].

The second family of white blood cells is known as granulocytes or polymorphonuclear leukocytes. The subclasses are neutrophils, eosinophils, basophils and mast cells. They have granules that contain enzymes that catalyze the hydrolysis of DNA, RNA, triglycerides, lipids, complex carbohydrates and proteins. The granules also have an enzyme called lactoferrin, which binds iron and keeps it away from bacteria. Since iron is an essential nutrient, the bacteria die. The granules also have enzymes that catalyze the breakdown of bacterial cell walls and the contents of the cells. They also produce reactive oxygen substances (ROS), such as superoxide (O_2^-) and hypochlorite (OCl^-) [5].

Mast cells are dispersed throughout most tissues but are crucially located at the host's interface with the environment, such as the skin and mucosa [13]. "They support a role in the recognition of pathogens and other signs of infection. They function not only as sentinels, but also as modulators of innate and adaptive immune responses, ultimately influencing disease outcomes. At the earliest stages of infection, mast cells are important for communicating the presence of a pathogen to many cell types located nearby in the site of infection and distally in draining lymph nodes. To facilitate these interactions, mast cells are strategically located at the host–environment interface, proximal to blood vessels and lymphatic vessels, as well as nerve fibers and tissue-resident immune cells, including dendritic cells. Armed with granules that contain mediators, mast cells have the potential to be the first responders (within seconds to minutes) following recognition of an invading pathogen. Activation of mast cells by pathogens can cause degranulation and *de novo* synthesis of cytokines. Degranulation involves the rapid release of pre-packaged, insoluble mediators into the surrounding tissue, a strategy that gives mast cell-derived products a temporal advantage over those produced by other immune surveillance cells. Mast cells are most often associated with T_H2-type inflammatory responses. Mast cells also make products that have direct bactericidal activity. After stimulation, mast cells shape the inflammatory milieu and control the activation state of many cells that are crucial for adaptive immunity" [13].

Neutrophils are the most abundant phagocytes. They circulate in the blood and migrate to sites of infection by cytokines, chemokines and other mediators [5]. They arrive within an hour after infection and have half-lives of only 7-8 hours. They kill not only invading microorganisms, but also the body's own cells, so it is important that they don't live long [5]. "Eosinophils are only about 5% of the white blood cell population, but they are important because they have enzymes that catalyze the hydrolysis of proteins, DNA and RNA. Eosinophils can kill parasites because their granules contain a unique, toxic basic protein called cathepsin. Eosonophils are also professional antigen presenting cells. They help regulate other immune cells, are involved in the destruction of tumor cells, and promote the repair of damaged tissue. The chemokine interleukin-5 (IL5) is produced by basophils and interacts with eosinophils, causing them to grow and differentiate. Eosinophils can regulate local immune and inflammatory responses, and they accumulate in the blood and tissue of patients who have any of several inflammatory and infectious diseases. So, therapies that target eosinophils may help control asthma and allergy, as well as diseases that are not usually associated with eosinophils, such as autoimmunity and cancer. Eosinophil-targeted therapeutic agents that try to block specific steps in eosinophil development, migration and activation have recently entered clinical testing and are producing encouraging results. Corticosteroids are the most effective drugs for reducing the number of eosinophils in the blood and tissue of patients suffering from hypereosinophilic syndrome. They inhibit eosinophil survival and promote their clearance from tissues" [5]. Also, the tyrosine kinase inhibitor imatinib (Gleevec®, Novartis) is the first and only drug approved for the treatment of hypereosinophilic syndrome [14]. It can reduce blood eosinophil levels, but is only for patients with genetic alterations involving fusion genes that cause hypereosinophilic syndrome [14].

Basophils and mast cells are similar to each other. They both have receptors for IgE and granules that contain mediators of inflammation. This includes histamine, leukotrienes B and C, prostaglandin D_2 (PGD_2) and various cytokines [4]. Mast cells are found in most tissues and are more abundant than basophils. They are activated when IgE-bound receptors are cross-linked by an antigen. The IgE is provided by a B-cell. Mast cells are important in asthma, rhinitis and many types of allergies [5].

The third family of white blood cells is known as natural killer (NK) cells [5]. "They are very cytotoxic and are carefully regulated in a healthy immune system. They "are crucial components of the innate immune system, and they also influence adaptive immune responses. Through combinations of activating and inhibitory receptor-engagement, NK cells interact with many different types of cells and influence their fate. The outer portions of their cell membranes contain different types of receptors. Some activate and others deactivate NK cells. Cytokines such as TNF-α, viral double-stranded RNA and interferons can activate them. "NK cells contain granules that release perforin and proteolytic enzymes, which form pores in the membranes of invading cells and kill them by catalyzing the hydrolysis of their proteins. NK cells also remove the body's own cells that become malignant or infected with viruses, or any other cells that are tagged with IgG" [5]. NK cells have been shown to both promote and regulate autoimmune diseases like Sjörgen's syndrome [16]. "In this disease, chemokines in the salivary glands activate the innate immune system and cause an initial wave of NK cells and dendritic cells (DCs) to infiltrate. The DCs pick up apoptotic debris and migrate to the local draining lymph nodes, where they interact with NK cells that express a receptor, NKp3. This coupling triggers maturation of DCs followed by activation of autoreactive T cells. The ensuing T cell–B cell interaction generates autoantibodies and a systemic autoimmune response" [16].

The fourth family of white blood cells is lymphocytic T-cells, which originate in the thymus. They protect us from disease throughout our lives [5]. "Several hundred billion of them reside in lymphoid tissues and circulate in the blood. T cell receptors bind antigens that are presented by major histocompatibility complex (MHC) molecules that are expressed by diseased cells. About 100 million (10^8) are made each day and each population contains millions of different clones, with their own type TCR. They recognize and bind to HLA-peptide complexes on APCs. T-cells that are bound to non-self HLA-peptide complexes mature and proliferate. When the TCR binds to antigens, a cascade of molecular signaling occurs that includes phosphorylation of tyrosine residues in various proteins, activation of phospholipase C-γ1, production of diacylglycerol (DAG) and IP_3, and mobilization of Ca^{2+}" [5]. Signaling pathways that are activated include mitogen-activated protein kinases (MAPKs) or stress activated protein kinases, or SAPKs, also known as c-Jun N-terminal kinases (JNKs). The cytoskeleton is also reorganized [15]. The actin cytoskeleton organizes and maintains signaling complexes and the pathways that are initiated from cell surface receptors in T and B lymphocytes [12]. Also, the Rap1–GTP–interacting adaptor molecule plays a critical role in reorganizing actin and activating integrins in lymphocytes and platelets [15].

There is an exponential multiplication of those highly competent T-cells and an elimination of incompetent T-cells [1]. "A negative selection occurs in the thymus for any T-cells that bind strongly to HLA proteins that have a self peptide attached. Also, T-cells are distinguished by two different cell surface proteins, CD4 and CD8. The T-cells that have CD4 receptors bind class II HLAs and those having CD8 bind class I HLAs. The $CD8^+$ T-cells are cytotoxic and the $CD4^+$ cells are helper T-cells, which either activate other immune cells or release cytokines. As mentioned previously, dendritic cells bind to $CD4^+$ T-cells, activating them. Antigen-specific $CD4^+$ cells then interact with antigen-specific B cells, which proliferate and eventually differentiate into long-lived antibody-secreting cells" [1].

Cytotoxic T-cells destroy cells that are non-self and are sources of inflammation. The two largest subsets of helper T-cells are called T_H1 and T_H2 [1]. "The T_H1 cells produce interferon gamma (IF-γ) and stimulate phagocytes. The T_H2 cells produce interleukins 4, 5 and 13 (IL-4, IL-5 and IL-13) and stimulate mast cells, basophils and eosinophils" [1]. This activates antiparasitic and allergic responses.

Another type of helper T-cell, T_H17 makes IL-17 and plays a role in autoimmune diseases [5].

Most T-cells that have the CD4 cell surface protein ($CD4^+$ cells) belong to either the T_H1 or T_H2 subsets [13]. However, about 10% of the $CD4^+$ cells do not. Instead they are called regulatory T-cells (T_r or T_{Reg}) and they suppress the immune response, rather than activate it [15]. The development of T_{Reg} cells is regulated by transforming growth factor – β (TGFβ). The two most important T_{Reg} cell subsets express the transcription factor FOXP3 and develop in the thymus or can be induced in peripheral sites such as the mucosa-associated lymphoid tissue (MALT) [16]. A third important type of T_{Reg} cell secretes the immunosuppressive cytokine interleukin-10 (IL-10) and may develop from conventional $CD4^+$ T cells by activation in the presence of IL-10 or may develop from T_H1 or T_H2 cell subsets [1]. Other T cell subpopulations, including natural killer T cells, γδ T cells and $CD8^+$ T cells, can also exert potent suppressor functions in certain settings. $FOXP3^+$ T_{Reg} cells have been shown to influence the outcome of immune responses in several tissues [16]. "For example, in the intestine T_{Reg} cells have a key role in maintaining tissue homeostasis by inhibiting the overactivation of dendritic cells (DCs) and effector T cells. In the case of autoimmunity, T_{Reg} cells can have a beneficial effect by short-circuiting the inflammatory loop of T cells and antigen-presenting cells (APCs). This same general mechanism can have negative consequences in the setting of a tumor, in which T_{Reg} cells can inhibit the anti-tumor immune response, thereby preventing tumor clearance. During infection, T_{Reg} cells carry out a delicate balancing act. Preventing immunity would lead to the inability to clear the pathogen, whereas unrestrained immunity would lead to unwanted autoimmune-mediated tissue destruction. In each of the examples shown, the role of $FOXP3^+$ T_{Reg} cells was emphasized. Still, interactions between multiple immune cell types and different types of regulatory T cell are also important in the regulation of immune responses" [16].

Moreover, T_{Reg} cells may be able to be manipulated to treat autoimmune diseases and cancer [17]. "T_{Reg} cells that express the Forkhead box P3 (FOXP3) protein is crucial for regulating immune responses and in maintaining immunological self-tolerance. They can control inflammation by contacting and inhibiting antigen-presenting cells and effector T cells or by releasing anti-inflammatory cytokines such as IL-10 or transforming growth factor-β (TGFβ). They have been shown to maintain their regulatory functions for a long period of time - even in the absence of antigens that induced their generation - and are stable and transferable, thereby generating these cells that prevent and suppress unwanted immunity. There are strategies to enhance the generation of $FOX3P^+$ T_{Reg} cells and to manipulate their activation, expansion, survival or suppressive functions by interfering with intracellular signaling pathways or microRNAs (miRNAs). Moreover, the epigenetic regulation of the *FOXP3* gene can be targeted" [17].

For example, the serine/threonine kinase mTOR has a key role in regulating the immune response [17]. "The PI3K–Akt–mTOR signaling pathway suppresses the induction of FOXP3 expression *in vitro* and *in vivo*. Efficient upregulation of FOXP3 expression depends on the E3 ubiquitin protein ligase CBLB74 or phosphatase and tensin homolog (PTEN). They antagonize the function of PI3K and inhibit downstream signaling through Akt. Conversely, constitutive activation of Akt suppresses the induction of FOXP3 expression. The activity of the PI3K–Akt–mTOR network can be manipulated by pharmacological agents, and some are already approved for clinical use. Rapamycin, an allosteric inhibitor of mTOR that was approved as an immunosuppressant in 1999, has potent immunosuppressive characteristics. The mTOR inhibitor rapamycin and its analogue everolimus (Afinitor®) promotes $FOXP3^+$ T_{Reg} cell induction. Also, they increased the stability of FOXP3 expression and enhanced the survival of T_{Reg} cells. The induction of T_{Reg} cells induction is caused by the drugs interfering with the activation of the PI3K–mTOR signaling pathway, limiting cellular proliferation and dampening T cell activation. Moreover, rapamycin or everolimus may affect activated conventional T cells more than T_{Reg} cells by inhibiting the proliferation or survival of the former and thus enriching the latter" [17].

The PI3K–Akt–mTOR signaling pathway can also be targeted with drugs that affect downstream factors such as FOXO proteins and support their translocation to the cell nucleus [17]. "For example, derivatives of pyrazolopyrimidine are potent FOXO nuclear relocators and inhibit PI3K. Virtual screening and molecular modelling were used to develop a structure–activity relationship. This led to the synthesis of ETP-45658, which is a potent and selective inhibitor of

PI3K. The generation and stability of T_{Reg} cells can also be directly controlled by the epigenetic regulation of the *FOXP3* gene. It has four conserved non-coding regions that are sites of epigenetic regulation of *FOXP3* transcription. There is a demethylated region (CpG-rich intronic enhancer region) that has a T_{Reg}-specific pattern of methylation. This region is completely demethylated in human T_{Reg} cells but is fully methylated in resting conventional T cells. When CpG was demethylated by short-term treatment with the DNMT1 inhibitor 5-aza-2-deoxycytidine (5-Aza) and mTOR inhibitors rapamycin and everolimus, it stabilized the expression of FOXP3 T cells" [17].

More recently, the adenosine A2A receptor was found to be able to inhibit T cell responses in part by upregulating the expression of FOXP3 in $CD4^+$ T cells [17]. "So, the A2A receptor was blocked by using antibodies or adenosine analogues. These drugs have been used in clinical trials to treat Parkinson's disease, but they have not yet been tested on patients who have cancer" [17].

Two classes of regulatory T-cells, Tr1 and Tr3, require the anti-inflammatory IL-10 for their formation. Once the Tr1 and Tr3 cells are mature they continue to secrete large amounts of IL-10 and another anti-inflammatory cytokine called tumor necrosis factor beta, or TNF-β [11]. "Both Tr1 and Tr3 cells are located in the intestine and help us tolerate food, even though it is non-self. Another class of T_{Reg} cells is called "natural" regulatory T-cells. They express not only CD4, but also CD25, CD125 and the transcription factor FOXP. TNF-β helps produce T_{Reg} cells and causes TNF-β to be secreted in the mature T_{Reg} cells" [11].

The fifth family of white blood cells is known as lymphocytic B-cells, which originate in the bone marrow [1]. "They recognize foreign matter, including microorganisms and viruses. They produce high affinity immunoglobulin antibodies that attach to foreign matter and cooperate with other immune cells to eliminate such threats. These immunoglobulins are made up of two heavy chain proteins that are linked non-covalently to two light chain polypeptides. Humans have about 10^{12} B-cells and about 10^9 clones. Each clone produces its own antibody, which can be in the IgG, IgM, IgA, IgD or IgE class of immunoglobulins. Those B-cell clones that produce Igs which recognize non-self antigens are the ones that proliferate, or multiply. An important difference between B and T cells is that B cells recognize antigens in their native form, while T cells recognize antigens that have been processed into peptide fragments that are complexed with MHC on the surface of APCs. T cells recognize this complex by using their T-cell receptors (TCRs). On the other hand, the interaction with antigen activates B cell receptor signaling, resulting in the internalization, processing, and presentation of antigen to T lymphocytes in the context of cell surface proteins of the MHC class II family" [1]. Recognition of such processed antigen by the T cell receptor induces the formation of a stable association, or synapse, between the two cell types, resulting in the transmission of signals required for regulating the B cell response [19].

To make so many different clones, or types of immunoglobulins, B cells mix and match three different proteins by choosing three different gene segments, called the variable, diversity and joining segments, or V, D and J [18]. "The genes that code for the heavy chain undergo V(D)J recombination. The genes coding for the light chain undergo VJ recombination, to form the V(D)J exon, which codes for the part of the immunoglobulin called the antigenic determinant, or complementarity determining region (CDR). This is the part of the immunoglobulin (or antibody) that makes it bind very specifically to an antigen. The recombination starts with a recombination activating gene, called RAG1/2. It codes for an endonuclease, which catalyzes the hydrolysis (or cleavage) of V, D and J segments. V(D)J recombination first occurs in progenitor B cells, called pro-B cells. They assemble the IgH (heavy chain) V, D and J exons and μ chains are produced" [18]. Next, the IgL (light chain) V and J exons are assembled in precursor B cells, producing immature B cells that express μ plus IgL chains as IgM immunoglobulins on the cell surface [19]. "These cells (called IgM$^+$ B cells) migrate to peripheral lymphoid tissues, such as the spleen and lymph nodes, where they participate in antigen-dependent responses, including class switch recombination. This changes the function of the antibody to generate different classes of antibodies" [19].

When a B-cell encounters an APC, it ingests the antigen, displays it with an HLA II protein and binds to helper CD4$^+$ T-cells that also recognize the antigen and secrete cytokines, which stimulate B-cell proliferation [5]. "Some B-cells have large nuclei and make many copies of their antibodies. Other B-cells are long-lived memory cells. Another subtype of B-cells produces lower affinity IgM antibodies. These form pentamers and bind to the surface of bacteria, forming aggregates that are attacked and destroyed by macrophages. B-cells are distributed in tissues throughout the body, including the lymph nodes, like T-cells. A smaller fraction circulates through the blood vessels. Activated B-cells are cleared in the liver and spleen, so they survive just a few days. Naïve B-cells, which have not been exposed to antigens and memory B-cells are protected and live longer" [5].

The diversity among memory B cells was reviewed recently [20]. "When someone is infected by a virus or pathogenic organism, one type of B cells produces antibodies that may persist for months or years, depending on how long protection against the antigen-producing pathogens is required. Next, pathogen-specific B cells persist in a resting state. They circulate in the body and can be reactivated to produce more antibodies if a person is infected again. However, these two types of B cells produce antibodies that are different from each other. "The class or isotype usually changes and is no longer the IgM class produced initially and the affinity of the antibody for the pathogenic antigen is much stronger in the initial immune response. The immune memory includes not just the antibody that is made, but also in the B and T cells that combine rapidly to make more antibodies. This reproduces the previously successful immune responses" [20].

Previously, it was thought that naïve B cells expressing IgM and IgD and naïve T cells are activated by an antigen in the primary antibody response, either directly or after being processed by a dendritic cell [20]. "Depending on how they are primed, activated T cells become a type of helper T (T_H) cell types, such as T_H1, T_H2 or T_H17. Each of these has produces a distinct profile of cytokines. T_H 1 secretes IFN-γ, T_H2 secretes IL-4 and T_H17 secretes TGF-β. Independent of their interaction with B cells, activation of T cells leads to T cell memory (T_M). Signals from T cells induce the proliferation of B cells and class switching. Activated B and T_H cells can also up-regulate the transcription factor Bcl-6 and establish places in which the affinity of the antibody for the antigen increases. T_H cells within germinal centers (T_{FH} cells) are different than early subsets of T_H cells. They secrete IL-21 in addition to other, priming-specific cytokines. Affinity-matured long-lived plasma cells and memory B cells are produced from the germinal centers. They express isotypes of immunoglobulins that reflect the T_H type in the initial priming. It now appears that the persistence of switched memory B cells depends on the continuous expression of the transcription factors needed for their induction. Different classes of antibodies appear in the memory compartments. They are specialized in clearing specific types of pathogens. This can be traced back to the initial interactions between dendritic and T cells" [20].

CHEMICAL MEDIATORS OF THE IMMUNE SYSTEM

The four major classes of immune system mediators are chemotactic agents, cytokines, C-reactive protein and antibodies [5].

A group of chemotactic agents called chemokines attract immune cells to the site of infection or inflammation. Chemokines are biochemicals that attract cells. They bind to receptors on target cells, stimulating movement towards areas of increasing chemokine concentration. Chemokines include products of the complement system, which are fragments from the degradation of microorganisms and viruses, and other small molecules [5]. Chemokines are relatively small proteins (molecular weights of 8000 – 10,000) that are secreted by several types of immune cells [1]. They attract monocytes, macrophages and dendritic cells of the innate immune system and the T- and B-cells of the adaptive immune system [1]. There are about 50 chemokines and 20 receptors for them [21]. "Chemokines have been called, "small cytokines with selective chemoattractive properties, coordinating the homeostatic circulation of leukocytes as well as their movement to sites of inflammation or injury. They are classified into four families, based on the pattern of the first two cysteine (CySH) residues of the chemokine. The large CC family has its first two CySHs adjacent to each other. Many of them have older names, such as MCP (monocyte chemotactic protein), MIP (macrophage inflammatory protein), and RANTES (Regulated upon Activation, Normal T-cell Expressed, and Secreted)" [21]. They are all ligands that bind to receptors, so the new names are CCL1 to CCL28 [20]. There are also chemokines that have one amino acid between the CySHs (CXC) and some don't have two CySHs in their binding site (SC). All of them bind to G protein-coupled receptors, or GPCRs [21]. When they bind, two of the subunits, $G\alpha_1$ and $G\beta\gamma$, dissociate, leading to an influx of Ca^{2+} and activation of the phosphatidyl inositol 3-kinase and small Rho

GTPase signaling pathways [22]. "When chemokines are released by endothelial cells, chemokines tend to remain concentrated and localized to the original site of infection. The concentrations of cytokines (such as IL-6) can vary from about 10^{-12} M (pM) to 10^{-9} M (nM) during infection or trauma. When the chemokines bind to their receptors on leukocytes, they induce signaling cascades that cause a rapid increase in integrin binding, which causes the leukocytes to adhere firmly to the endothelium. When they are inside tissues, chemokines direct the migration of leukocytes. Neutrophils and monocytes are also attracted to sites of inflammation by CXC and CC chemokines. When chemokines and their GPCRs are not properly regulated, it can lead to autoimmune diseases, chronic inflammation and associated diseases, such as cancer. When properly controlled, chemokines can attract white blood cells to lymphoid tissues and sites of infection and help activate neutrophils and monocytes. "Adaptive immunity starts with the encounter between naïve antigen-specific T cells and dendritic cells that are loaded with cognate antigens. The encounter between these cells is driven by two CC chemokines (CCL15 and CCL21) and their receptor (CCR7). The dendritic cells help link the innate and adaptive immune systems" [22]. The process is activated by the presence of pathogens.

Most pathogens enter through the mucosal surfaces. Examples include toxic strains of *E. coli*, rotavirus, *Mycoplasma* bacteria and the influenza virus. Mucosal associated lymphoid tissues (MALT) are the largest immune compartment in the body [22]. "MALT protects mucosal membranes against infection. They must tolerate antigens derived from food and airborne matter and they must initiate immune responses against harmful antigens that might break through the mucosal lining. When a vaccine is administered to a mucosal site, it is taken up by M cells, dendritic cells or epithelial cells and it is passed on to antigen presenting cells, which migrate to specialized tissues, such as Peyer's patches, lymph nodes and tonsils. They affect immune cells, which migrate to the site of the mucosal infection. The cells are attracted by chemokines, such as CCL25, which binds to the chemokine receptor CCR9" [22].

Many chemotactic agents are the products of the complement system, the large family of proteins that is made in the liver and some immune cells [5]. "These proteins catalyze the hydrolysis of other proteins. Chemotactic agents provide reactive fragments that can bind to foreign and abnormal cells, tagging them for destruction. They generate small peptide fragments (called C3a, C5a and the tripeptide, N-fMet-Leu-Phe). These peptide fragments attract immune cells to sites of infection or inflammation" [5].

Fragments from the degradation of microbes, viruses, blood clots or fibers are also chemotactic agents. For example, bacteria contain CpG dinucleotides that stimulate both the innate and adaptive (acquired) immune system [5]. They promote mostly immune responses that are helped by T_H1 cells, which is especially important for newborn babies (neonates). Another type of fragment produced by bacteria are lipopolysaccharides (LPS), which bind to Toll-like receptor number four (TLR4). LPS can cause sepsis and cholestasis [23]. TLR4 also responds to LPS. Other fragments include lipoproteins and glycolipids, which are recognized by TLR2. Many bacteria contain flagella, which can be broken into the protein called flagellin, which is recognized by TLR5 [23].

The fourth class of type of chemotactic agents is called other small molecules [5]. This includes leukotrienes B_4 and C_4, which attract immune cells. These are derivatives of arachidonic acid, and can cause inflammation.

The second class of mediators is cytokines. These are proteins that are secreted by immune cells [5]. "They are growth factors that regulate and coordinate the immune response, and play a role in cell proliferation and inflammation. subtypes include interferons, monokines (made by monocytes), lymphokines (made by lymphocytes), chemokines and interleukins (made by leukocytes to act on other leukocytes). Examples include tumor necrosis factor alpha (TNF-α), which activates macrophages and T-cells; IL-1, which activates T-cells; IL-2, which activates T, B and natural killer cells; TGF-β, which controls T_{Reg} development and immune homeostasis; and interferons, which activate antiviral defenses" [5]. The interleukin-1 (IL-1) family of cytokines is made up of 11 proteins (IL-1F1 to IL-1F11) encoded by 11 distinct genes in humans and mice. IL-1–type cytokines are major mediators of innate immune reactions [24]. "Their main function is to control proinflammatory reactions in response to tissue injury by PAMPs or damage- or danger-associated molecular patterns released from damaged cells (DAMPs, such as uric acid crystals or ATP). Thus, they are major mediators of innate immune reactions, and their actions are tightly balanced" [24].

Macrophages and monocytes are sentinel cells of the innate immune system. They are a major source of IL-1α and IL-1β, but many other cell types, including epithelial cells, endothelial cells, and fibroblasts, can also produce IL-1α and IL-β [24]. "IL-1α is anchored to the cell membrane and signals through autocrine or juxtracrine mechanisms, whereas IL-1β is secreted by an unconventional protein secretion pathway. It can act in a paracrine manner or systemically. IL-1α and IL-1β rapidly induce the expression of hundreds of genes in many different types of cells, including monocytes or macrophages, epithelial and endothelial cells, and chondrocytes or fibroblasts. At the same time, IL-1α and IL-1β induce expression of their own genes, which is a positive-feedback loop that amplifies the IL-1 response in an autocrine or paracrine manner. Stimulation of transcription occurs within 30 min of exposure to IL-1α or IL-1β and can be sustained for many hrs" [24].

Type-I interferons fight different types of cancers, viral infections and autoimmune diseases. Interferons are glycoproteins that are released in response to viruses, bacteria, parasites and tumor cells. They activate natural killer cells and macrophages. They up-regulate antigen presentation to T-cells and they increase the ability of uninfected cells to resist viral infection. Also, interferons are used to treat cancer. They are given in combination with chemotherapy and radiation.

The third class of mediators is the C-reactive protein (CRP). The concentration of CRP increases about two orders of magnitude (at least 100-fold) after an infection. CRP binds to the C3b protein fragment of complement and stimulates monocytes and macrophages. CRP is also an important indicator of risk for a heart attack. Some people with relatively high CRP are more likely to have a heart attack. The concentration of CRP can be kept low by the Mediterranean diet that uses monounsaturated fats (such as those found in olive oil, including oleoyl or oleic acid). It should be added that the terms fatty acids is usually used in place of fatty acyls in the medical and popular literature, even though olive oil, salmon and other healthy foods contain triacylglycerides, which contain fatty acyls. However, if doctors and nutritionists were to tell non-chemists that they should eat triglycerides in order to lower their triglyceride levels, people would not believe them. Instead, they have to be told to eat healthy fats or fatty acids. So, in most of this book, the terms omega-3 fats, polyunsaturated fats, *trans* fats and saturated fats are used. However, in the literature on food chemistry, the term fatty acid is used, even though the authors are really referring to fatty acyls in triglycerides. In fact, free fatty acids are formed in oils when they turn rancid, due to the partial hydrolysis of triglycerides.

The fourth class of mediators is the antibodies [5]. They are Y-shaped proteins that recognize and bind to foreign agents, such as microorganisms, cell fragments and foreign molecules. There are five classes of antibodies, or immunoglobulins, IgG (the most abundant in blood), IgM, IgA, IgD and IgE. These are all made by B-cells. As mentioned before, IgEs mainly target mast cells and basophils and cause allergic responses. The stem of the Y-shaped Igs has a relatively fixed structure that binds to a specific receptor for each class. The uppermost part of the Y is highly variable, and is often called the antigen binding region, also called the complement determining region, or CDR. Each B-cell clone produces only one antibody that is specific for that clone. When an antigen is bound to an Ig that is attached to the surface of a B-cell, it is internalized and broken down (hydrolyzed) [1].

More recently, a fifth class of immune network mediators was identified. These are the small molecules, including neurotransmitters, such as L-dopa and catecholamines. This is another example of a link between the immune and central nervous systems. Another immune network mediator is serotonin, which is usually thought of as a neurotransmitter. However, it was shown that naïve T cells express a receptor, 5-HT$_7$ that binds serotonin [25]. The 5-HT$_7$ receptor is expressed more when the T cell is activated. Activation also causes two other types of serotonin receptors to be expressed, 5-HT$_{1B}$ and 5-HT$_{2A}$. When exogenous serotonin (also known as 5-HT) binds to a T cell, it causes a kinase (ERK1/2) to be phosphorylated and signals to be generated. These signals activate T-cells. In another study, it was shown that serotonin is transmitted between dendritic cells and T cells [26]. Thus, serotonin may play an important role in activating the immune system [26]. This should be remembered when giving or taking SSRIs to treat depression, especially if allergies should develop.

The immune system can also be stimulated by anticancer drugs [27]. "This includes conventional anticancer drugs that were designed to target one of the classic hallmarks of cancer: genetic instability, limitless proliferative potential, self-sufficiency in growth signals, insensitivity to antigrowth signals, evasion of apoptosis, sustained angiogenesis, tissue invasion and metastasis. Gemcitabine, a nucleoside analogue that is used to treat several carcinomas and some forms of non-Hodgkin's lymphoma, has increases the expression of class I human leukocyte antigen (HLA) on malignant cells, enhance the cross-presentation of tumor antigens to $CD8^+$ T cells and selectively kill myeloid-derived suppressor cells (MDSCs), both *in vitro* and *in vivo*. This facilitates T cell-dependent anticancer immunity. Oxaliplatin (used to treat colorectal and pancreatic cancer) and cyclophosphamide (used in combination regimens to treat some forms of lymphoma, leukemia and solid tumors) produce similar effects on class I HLA expression. Both of them also inhibit the expression of programmed death ligand 2 (PDL2), thus limiting immunosuppression by both dendritic and tumor cells. Although a high dose of cyclophosphamide exerts immunosuppressive effects and is used to treat autoimmune diseases, at low doses it selectively suppresses inhibitory cells like MDSCs and T_{reg} cells (which facilitate IFNγ- mediated anticancer immunity). Low doses favor the differentiation of $CD4^+$ T helper cells that produce IL-17-producing and sustain a T_H1 antitumor response, restores T cell and NK cell effector functions in patients with end-stage cancer, stimulates the expression of type I IFNs, expands $CD8\alpha^+$ DCs (to enable cross-presentation) and inhibits the generation of immunosuppressive cytokines, while stimulating antitumor innate immunity" [27].

Also, the anthracyclines doxorubicin and daunorubicin stimulate the immune system [27]. "Doxorubicin enhances the proliferation of tumor antigen-specific $CD8^+$ T cells in lymph nodes, while promoting tumor infiltration, first by γδ T cells that secrete IL-17, and then by CD8+ T cells that secrete IFNγ. Daunorubicin, which is now being used to treat acute myeloid leukaemia and neuroblastoma, lowers antigen expression by cancer cells. This promotes IL-2 and IFNγ synthesis by local T cells and facilitates the development of a tumor-specific immune response. Also, oxaliplatin (but not the other platinum-based compounds cisplatin and carboplatin), cyclophosphamide and anthracyclines such as doxorubicin and mitoxantrone have been shown to trigger immunogenic cell death, or ICD. ICD is when cancer cells succumb to lethal stimuli (like doxorubicin) in a process similar to apoptosis. First, the chaperone protein calcireticulin moves to the outer part of the cell membrane, exposing itself on the surface of a dying cell. This enables dendritic cell to take up tumor antigens, while releasing the non-histone chromatin-binding nuclear protein known as the high-mobility group protein B1. This stimulates antigen processing and presentation to T cells, while ATP is secreted. This activates inflammasomes and produces pro-inflammatory cytokines" [27].

Paclitaxel - the taxane that is currently most widely used in the clinic - impairs cytokine production and decreases the viability of $FOXP3^+$ T_{reg} cells but not in FOXP3–$CD4^+$ effector cells [27]. "Also, similarly to cisplatin and doxorubicin, paclitaxel increases the permeability of tumor cells to granzyme B, thereby rendering them susceptible to CTL-mediated lysis even if they do not express the antigen recognized by CTLs (cytotoxic T cells). This bystander effect perhaps explains why small numbers of CTLs can mediate potent anticancer responses when combined with some types of chemotherapy. Docetaxel, which is used to treat anthracycline-resistant breast and lung cancers, has been shown to decrease splenic MDSC levels in mice with 4T1-Neu cancers, leading to increased antitumor cytotoxic responses. Other microtubule inhibitors that affect the immune system during cancer chemotherapy include vincristine (which is currently used to treat non-Hodgkin's lymphoma) and vinorelbine (which for breast and non-small-cell lung cancers). Vincristine, in combination with doxorubicin and glucocorticoids, increases the abundance of specific DC subsets in patients with multiple myeloma and stimulates antigen presentation by them. Conversely, cancer cells succumbing to vinorelbine treatment appear to suppress antitumor immunity by promoting the bystander death of immune cells" [27].

The folate antagonist methotrexate was originally developed as an anticancer agent because it inhibits the production of tetrahydrofolate, an essential precursor for nucleotide synthesis [27]. "Because they preferentially target rapidly dividing cells, methotrexate and the second-generation folate antagonist pemetrexed (Alimta®) have immunosuppressive effects. So, methotrexate is used to treat several autoimmune disorders, including rheumatoid arthritis. Moreover, low concentrations of methotrexate may boost the maturation of DCs and enable them to stimulate T cells. So, some conventional chemotherapeutics may owe their efficacies - at least in part - to cellular cascades that stimulate antitumor immunity" [27].

More recently, targeted agents have been developed that modulate the immune system [27]. "For example, the EGFR-specific compound erlotinib exerts antitumor effects in patients with acute myeloid leukemia, a type of cancer in which EGFR is not expressed. Instead, it affects several cell-intrinsic signaling pathways in cancer. More recently, EGFR inhibition (by erlotinib or by the mAb cetuximab (Erbitux®) has been associated with the increased expression of class I and class II major histocompatibility complex (MHC) molecules, possibly promoting cancer-directed immune and/or inflammatory responses. Still, they can have side effects. Erlotinib and gefitinib cause a broad spectrum of adverse effects on skin and hair" [27].

Intriguingly, the therapeutic effects of several mAbs targeted agents appear to rely partly on off-target mechanisms, some of which are mediated by the immune system [27]. "Moreover, an antibody against mouse EGFR mediates anticancer effects that are superior to the anticancer effects of AG1478, a chemical inhibitor of EGFR, because it induces immunogenic cell death and elicits tumor-specific $CD4^+$ and $CD8^+$ T cell responses. These observations suggest that EGFR-targeted antibodies (but not relatively unspecific chemicals such as erlotinib) may efficiently activate the immune system against tumors that express EGFR" [27].

Most patients with CML who receive imatinib and are in remission exhibit robust antileukemic immune responses [27]. "Also, some imatinib-treated patients with CML make B lymphocytes that produce anti-carbohydrate antibodies with antitumor activity. In patients with gastrointestinal stromal tumors, the therapeutic outcome of imatinib treatment is influenced by the level of IFNγ produced by circulating NK cells and the expression pattern of the activating NK receptor NKp30. Moreover, at clinically relevant concentrations, imatinib inhibits STAT3 and STAT5 signaling, decreases the frequency of T_{Reg} cells and impairs their immunosuppressive function *in vivo*" [27].

At the same time, immunostimulatory antibodies are attracting much attention [27]. "They are targeted therapeutics that promote antitumor immunity by modulating immune functions. This includes ipilimumab (Yervoy®), an FDA-approved drug that antagonizes cytotoxic T lymphocyte antigen 4 (CTLA4) on the surface of helper T cells. This inhibits the development of immune tolerance. There is also MDX-1106, which specifically targets programmed death protein 1 (PD1), a transmembrane receptor that mediates immunosuppressive functions in activated T cells. There are also several antibodies that block T cell immunoglobulin and mucin domain-containing protein 3 (TIM3), a trans-membrane protein that identifies antigen-specific dysfunctional $CD8^+$ cells. Other antibodies activate CD40, in turn stimulating antigen-presenting cells for tumor-specific T cell priming and activation, alone or in combination with other anticancer interventions (including the conventional chemotherapeutics gemcitabine and dacarbazine as well as the targeted agent AZD8055). That is, these immunostimulatory antibodies reportedly reverse immune tolerance and induce consistent antitumor responses" [27].

Several other targeted anticancer agents have immunostimulatory functions [27]. "Nitrogen bisphosphonates (such as zoledronate, which is commonly prescribed for osteoporosis) have been shown to trigger inflammatory and/or innate immunity, delay the progression of breast cancer, and facilitate the expansion of patient-derived DCs for the generation of efficient vaccines against hepatocellular carcinoma. Chemical agents that antagonize proteins that inhibit apoptosis not only help kill cancer cells by tumor cell-intrinsic mechanisms, but also enhance co-stimulation. This lowers the threshold for activating the immune system in response to physiological stimuli and hence increases the efficacy of both prophylactic and therapeutic antitumor vaccine. Similarly, besides facilitating tumor cell death, several histone deacetylase inhibitors (such as vorinostat (Zolinza®), sodium butyrate and MS-275) increase the expression of receptor ligands on the surface of cancer cells that activate NK cells, thereby facilitating tumor cell recognition by them" [27].

In summary, it appears that targeted anticancer agents often affect the immune system, thereby exerting unwanted immunosuppressive functions or stimulating therapy-relevant anticancer immune responses [27]. Table **3** summarizes drugs that affect the activation of immune system.

Table 3. Drugs that affect the activation of immune system.

Drug	Effect
Gemcitabine	Increases the expression of class I human leukocyte antigen (HLA) on malignant cells, enhance the cross-presentation of tumor antigens to $CD8^+$ T cells and selectively kill myeloid-derived suppressor cells (MDSCs) [1].
Oxaliplatin	Increases the expression of class I human leukocyte antigen (HLA) on malignant cells and inhibits the expression of programmed death ligand 2 (PDL2), thus limiting immunosuppression by both dendritic cells and tumor cells. Also, it triggers immunogenic cell death [1].
Cyclophosphamide	Increases the expression of class I human leukocyte antigen (HLA) on malignant cells and inhibits the expression of programmed death ligand 2 (PDL2), thus limiting immunosuppression by both dendricites and tumor cells [1].
Doxorubicin	Stimulates the immune system by enhancing the proliferation of tumor antigen-specific $CD8^+$ T cells in lymph nodes, while promoting tumor infiltration, first by $\gamma\delta$ T cells that secrete IL-17, and then by CD8+ T cells that secrete IFNγ [1].
Daunorubicin	Stimulates the immune system by enhancing the proliferation of tumor antigen-specific $CD8^+$ T cells in lymph nodes, while promoting tumor infiltration, first by $\gamma\delta$ T cells that secrete IL-17, and then by CD8+ T cells that secrete IFNγ [1].
Paclitaxel	Impairs cytokine production and decreases the viability of $FOXP3^+$ T_{reg} cells but not in FOXP3–$CD4^+$ effector cells and increases the permeability of tumor cells to granzyme B, thereby rendering them susceptible to CTL-mediated lysis [1].
Cisplatin	Increases the permeability of tumor cells to granzyme B, thereby rendering them susceptible to CTL-mediated lysis.
Methotrexate	Immunosuppressive effects.
Erlotinib	Affects several cell-intrinsic signaling pathways in cancer and increases the expression of class I and class II major histocompatibility complex (MHC) molecules [1].
AG1478	Chemical inhibitor of EGFR.
Ipilimumab	Immunostimulatory monoclonal antibody that antagonizes cytotoxic T lymphocyte antigen 4 (CTLA4) on the surface of helper T cells.
MDX-1106	Immunostimulatory monoclonal antibody that targets programmed death protein 1 (PD1) [1].
Nitrogen bisphosphonates such as zoledronate	Triggers inflammatory and/or innate immunity
Vorinostat	Increase the expression of receptor ligands on the surface of cancer cells that activate NK cells, thereby facilitating tumor cell recognition by them. Also used to treat HIV infection by acting as a histone deacetylase inhibitor [1]
Sodium butyrate	Increase the expression of receptor ligands on the surface of cancer cells that activate NK cells, thereby facilitating tumor cell recognition by them [1].

(Table 3) contd.....

Drug	Effect
MS-275	Increase the expression of receptor ligands on the surface of cancer cells that activate NK cells, thereby facilitating tumor cell recognition by them [1].
Ipilimumab	Inhibits the immune checkpoint receptor cytotoxic T lymphocyte antigen 4 (CTLA4) [1].
Dulanermin	Stimulates the immune system by targeting the TRAIL-TRAILR signaling pathway [1].
AMG 951	Stimulates the immune system by targeting the TRAIL-TRAILR signaling pathway [1].
Mapatumumab	Stimulates the immune system by targeting the TRAIL-TRAILR signaling pathway [1].
Conatumumab	Stimulates the immune system by targeting the TRAIL-TRAILR signaling pathway [1].
Lexatumumab	Stimulates the immune system by targeting the TRAIL-TRAILR signaling pathway [1].
Belimumab	Inhibits the immune system by targeting the B cell activation factor (BAFF) [1]
Avonex	Immunosuppressant, based on interferon 1b
Betaseron	Immunosuppressant, based on interferon 1b
Extavia	Immunosuppressant, based on interferon 1b
Gilenya	Immunosuppressant that prevents lymphocytes from crossing the blood-brain barrier, reducing inflammatory damage to nerve cells.
Novantrone	Immunosuppressant
Teriflunomide	Immunosuppressant that reversibly inhibits the enzyme dihydroorotate dehydrogenase
Natalizumab	Immunosuppressant that targets the cell adhesion protein α4-integrin
Rituximab	Immunosuppressant that targets CD40 cell surface receptors
Maraviroc	Prevents spread of HIV in the immune system by targeting binding, fusing and entry of HIV into the host cell.
Enfurvitide	Prevents spread of HIV in the immune system by targeting binding, fusing and entry of HIV into the host cell.

The immune system can also be modulated by monoclonal antibodies (mAbs) that target signaling molecules cell surface of immune cells. These mAbs can cause improved immune responses [28]. Some are immunological biologics. Unlike conventional mAbs like HER2 that have a specific target, immunomodulatory biologics bind to signaling proteins on cell surfaces of host immune cells and manipulate them [28]. This, in turn modulates antigen-specific T cell receptor (TCR) and B cell receptor (BCR) signals that control the direction and magnitude of immune responses by lymphocytes [28]. Some inhibit hyperactive immune responses, while others stimulate hypoactive immune responses in cancer as well as in chronic bacterial and viral infections [28].

Both co-stimulatory and co-inhibitory signaling molecules are being studied as potential targets of mAbs, as are membrane-bound receptors that are involved in cellular migration and cell-cell adhesion [27]. “This includes the immunoglobulin (Ig) superfamily and the tumor necrosis factor (TNF) – TNF receptor superfamily. For example, the protein B7–CD28 modulates TCRs and BCRs. This influences the outcome of lymphocyte-mediated immune responses. TNF and TNFR produce divergent signals that drive hematopoiesis, the development of secondary lymphoid organs, innate and adaptive immune responses and bone absorption. Moreover, TNF and TNFR are needed for T and B lymphocytes to differentiate and make a population of memory T cells” [27].

The first superfamily of Ig co-stimulating proteins was the ligand and receptor pair B7–CD28 [28]. “Members include CD28, CTLA4, inducible co-stimulator (ICOS), programmed cell death protein 1 (PD1) and its ligand PDL1, also known as B7 homolog 1 and B7H1. Actually, these ligand-receptor pairs are counter-receptors, since they can switch their roles as ligand and receptor. CD28 is expressed all the time (constitutively) in naïve T cells, but when it binds the B7 ligand, the pair primes the naïve cells. CD28 then activates the phosphoinositide 3-kinase (PI3K) and Akt signaling pathways, along with the MAPK cascase (see chapter 11). As IL-2 and other cytokine levels rise, the survival factor B cell lymphoma XL (BCL-XL) is upregulated and the cell cycle inhibitor cyclin-dependent kinase inhibitor 1B (CDKN1B) is downregulated. This promotes T cell activation, differentiation and memory T cell formation. Conversely, CTLA4, a CD28 homologue is induced on activated T cells. It is a co-inhibitor. It inhibits T cell responses after binding a B7 ligand” [28].

Immunotherapies that target the PD1–PDL1 pair are being tested and produced encouraging results [28]. “For example, in 2011 the FDA approved ipilimumab, an inhibitor of the immune checkpoint receptor cytotoxic T lymphocyte antigen 4 (CTLA4), for the treatment of metastatic melanoma. Data from clinical trials using three different mAbs (Phase III nivolumab, Phase II lambrolizumab and

Phase I MPDL3280A) have been encouraging. Three other PD1- or PDL1-targeting agents, including one fusion protein, are also in clinical trials. Like CTLA4, PD1 regulates the activity of T cells. When PD1 is not bound to its ligand, the T cells on which it is expressed can find and kill target cells, but when bound to its ligands PDL1 and PDL2, it inhibits this. However, whereas CTLA4 inhibits T cells in the lymph nodes and does not have any tumor specificity, PD1 acts locally and its ligands are overexpressed directly on cancer cells. Their interaction with PD1 prevents tumor-infiltrating T cells from recognizing the cancer cells" [28]. The mAb nivolumab binds PD1 and caused promising response rates for patients with non-small-cell lung cancer (NSCLC; 18%), melanoma (28%) and renal cell cancer (27%) [28, 29]. Longer-term follow-up of more patients presented at ASCO showed that so far the majority of responders have responses that last longer than 1 year (28 out of 54 responders with at least 1-year follow-up data) [28, 30].

The TNF-TNFR superfamily is the other target for immunomodulatory biologics [28]. "TNFRs can be divided into three groups: death domain-containing TNFRs, decoy TNFRs and TNFR-associated factor (TRAF)-recruiting TNFRs. CD95 (also known as FAS), TNFR1, death receptor 3 (DR3), DR4 (also known as TRAILR1), DR5 (also known as TRAILR2) and DR6 are death domain-containing TNFR family members. When TNF ligands crosslink TNFRs that have a death domain, it activates caspase and apoptosis. Several therapeutic agonistic mAbs that target death domain-containing TNFRs expressed on tumors are currently in clinical trials. Decoy receptor 1 (DCR1; also known as TRAILR3), DCR2 (also known as TRAILR4), DCR3 and osteoprotegerin (OPG) are decoy TNFRs. They have no intracellular signaling domains. They bind to TNF ligands and neutralize them, blocking downstream signaling" [28].

The rest (and the majority) of TNFRs are in the third family: TRAF-recruiting TNFRs [28]. "Upon binding their TNF ligands, they recruit specific TRAF family adaptor proteins through intracellular domains to differentially activate MAPK signaling cascades, promote nuclear factor-κB (NF-κB) activation and enhance cellular proliferation and function. In addition to the successful neutralizing biologics against TNF for treating TNF-mediated inflammation, therapeutic biologics targeting other members of the TNF superfamily are promising potential treatments of cancer, bacterial and viral infections as well as autoimmune diseases" [28].

Several therapeutic biologics that target the TRAIL-TRAILR pathway are currently in clinical cancer trials [28]. "Two recombinant human TRAIL proteins, dulanermin and AMG 951, in combination with rituximab, bevacizumab (Avastin; Roche/Genentech), cetuximab (Erbitux; Bristol-Myers Squibb/Lilly) or chemotherapy, are being evaluated for treating metastatic colorectal cancers, non-Hodgkin's lymphoma and NSCLC. Mapatumumab (HGS-ETR1), a human mAb that is a TRAILR1 agonist is being used to treat solid tumors alone or in combination with chemotherapy. In completed Phase I/II trials treating non-Hodgkin's lymphoma, colorectal cancer and NSCLC, mapatumumab was well tolerated without hepatotoxicity seen in other drugs. In a clinical trial of non-Hodgkin's lymphoma, 2.5% of the patients enjoyed complete responses and one other patient had a partial response. Similarly, conatumumab, lexatumumab and CS-1008 are mAbs that target TRAILR2. They are in clinical trials for multiple cancers, alone or in combination with other mAbs or chemotherapeutic regimens. Conatumumab and lexatumumab were well tolerated and had no dose-limiting toxicity in the dose range tested. So, biologics targeting TRAIL and TRAILRs appear to have promising antitumor efficacy with limited hepatic toxicity" [28].

Another target is B cell activation factor (BAFF; also known as BLYS and TNFSF13B) [28]. "It was first identified as a TNF family ligand expressed on monocytes. It promotes B cell activation and is one of the most important costimulatory ligands for B cell development and function. BAFF and its homologue APRIL (a proliferation-inducing ligand) bind to B cell maturation antigen (BCMA) as well as transmembrane activator and CAML interactor (TACI) to promote B cell development, class switching and survival. BAFF, but not APRIL, also binds to BAFF receptor (BAFFR) to help B2 cells and marginal zone B cells develop and survive. In 2011, a human IgG1 mAb targeting BAFF, belimumab (Benlysta®), was the first new drug to be approved by the FDA for the treatment of systemic lupus erythematosus (SLE) in 50 years" [28].

The immune system is also modulated by cytokines and chemokines. They activate the acquired system, which responds only after a lag time, but it is very effective. This acquired immune system produces antibodies that bind to antigens [1]. "The cells of the adaptive (acquired) immune system are found in the bone marrow and the thymus and are called B and T cells. They are both made from stem cells in the bone marrow" [1].

B cells are involved in the humoral immune response, whereas T cells are involved in cell mediated immune response [1]. "The humoral immune response is activated by external pathogens and is mediated by antibodies that are secreted from B-cells. Cell-mediated immunity is an immune response that does not involve antibodies but instead involves the activation of macrophages, natural killer cells (NK), antigen-specific cytotoxic T-lymphocytes, and the release of various cytokines in response to an antigen. Cellular immunity eliminates infected cells and it makes helper T cells (T_{H1} and T_{H2}). Both B cells and T cells carry receptor molecules that recognize specific targets. T cells recognize targets that are non-self (pathogens) only after antigens (small fragments of the pathogen) have been processed and presented in combination with a self receptor–MHC protein complex" [1].

There are two major subtypes of T cells: the killer cells and the helper T cells [1]. "Killer T cells only recognize antigens coupled to Class I MHC molecules, while helper T cells only recognize antigens coupled to Class II MHC molecules. These two mechanisms of antigen presentation reflect the different roles of the two types of T cell. In contrast, the antigen-specific receptor on a B recognizes whole pathogens without any need for antigen processing. Each lineage of B cell expresses a different antibody, so the complete set of B cell antigen receptors represents all the antibodies that the body can manufacture" [1].

Killer T cells are a sub-group of T cells that kill cells infected with viruses (and other pathogens), or are otherwise damaged or dysfunctional" [1]. "As with B cells, each type of T cell recognizes a different antigen. Killer T cells are activated when their TCR binds to this specific antigen in a complex with the MHC Class I receptor. Recognition of this MHC:antigen complex is aided by a co-receptor on the T cell, CD8. The T cell then travels throughout the body in search of cells where the MHC I receptors bear this antigen. When an activated T cell contacts such cells, it releases cytotoxins, such as perforin, which form pores in the target cell's plasma membrane, allowing ions, water and toxins to enter. The entry of another toxin called granulysin (a protease) induces the target cell to undergo apoptosis. T cell killing of host cells is particularly important in preventing the replication of viruses. T cell activation is tightly controlled and usually requires a very strong MHC/antigen activation signal, or additional activation signals provided by helper T cells. They regulate both the innate and adaptive immune responses and help determine which types of immune responses the body will make to a particular pathogen. These helper cells have no cytotoxic activity and do not kill infected cells or clear pathogens directly. Instead, they control the immune response by directing other cells to perform these tasks. Two classes of helper T cells, T_{H1} and $T_{H2,}$ bind to B cells and activate them" [1].

B cells identify pathogens when antibodies on their surfaces bind to specific foreign antigens [1]. "The antigen-antibody complex is taken up by B cells and is processed by proteolysis to produce smaller peptides. The B cell displays the antigenic peptides on its surface, together with MHC class II molecules. This combination of MHC and antigen attracts a matching $CD4^+$ helper T cell, which releases lymphokines and activates the B cell. The activated B cell then begins to divide, and its offspring secrete millions of copies of the antibody that recognizes this antigen. These antibodies circulate in the blood plasma and lymph nodes. They bind to pathogens that express the antigen and mark them for destruction. Antibodies can bind to bacterial toxins or interfere with the receptors that viruses and bacteria use to infect cells. When B cells and T cells are activated and replicated, some of their offspring become long-lived memory cells. Throughout a person's lifetime, these memory cells will remember each specific pathogen that it encounters and will be able to mount a strong response if the pathogen is detected again. This process is called "adaptive" because it occurs during the lifetime of an individual as an adaptation to infection and it prepares the immune system for future challenges. So, immunological memory can be either passive short-term memory or an active long-term memory" [1].

Passive immunity is usually short-term [1]. "It lasts somewhere between a few days and several months. However, newborn infants have no prior exposure to foreign microorganisms and are particularly vulnerable to infection. Mothers provide several layers of passive protection. During pregnancy, IgG is transported from the mother to her baby through the placenta, so human babies have high levels of antibodies even at birth, with the same range of antigen specificities as their mother. Also, breast milk also contains antibodies that are transferred to the gut of the infant. These maternal antibodies protect against pathogenic bacterial infections until the baby can synthesize its own antibodies" [1].

On the other hand, long-term active memory is acquired by activation of B and T cells [1]. "Active immunity can also be generated artificially, through vaccination. The principle behind vaccination is to introduce an antigen from a pathogen to stimulate the immune system and develop a specific immunity against a particular pathogen without causing the disease that is associated with that organism. Infectious disease is one of the leading causes of death, so vaccination is extremely important for public health. Most viral vaccines are based on so-called live attenuated viruses. On the other hand, many bacterial vaccines are based on pieces of microorganisms, including harmless components. However, many antigens that are derived from acellular vaccines do not induce the adaptive response strongly, so most bacterial vaccines are provided with additional adjuvants that activate the APCs of the innate immune system and maximize immunity" [1].

The immune system also identifies and eliminates cancer cells [1]. "Many scientists think that we have a few cancer cells throughout much of our lives. While we remain healthy and free from malignant tumors, our immune system is able to eliminate the relatively few cancer cells that it encounters. They are able to do this because the transformed cancer cells express antigens that are not found on normal cells. To the immune system, these antigens are foreign, and their presence causes immune cells to attack the transformed cells. If the immune system is not working as well as it should, some tumor cells evade the immune system and go on to become malignant cancer cells that have a reduced number of MHC class I antigens on their surface, thus avoiding detection by killer T cells. Some tumor cells also release proteins such as the cytokine TGF-(that suppresses the activities of macrophages and lymphocytes. Also, the immune system can develop a tolerance for tumor antigens, so the immune system no longer attacks the tumor cells. Still, some macrophages can promote tumor growth by generating cytokines and growth factors that help tumors develop further. Also, a combination of hypoxia in the tumor and a cytokine produced by macrophages can induce tumor cells to decrease production of proteins that block metastasis and help the spread of cancer cells" [1]. So, maintaining a healthy immune system is important for preventing cancer morbidity and mortality.

Part of maintaining a healthy immune system is avoiding autoimmunity and autoimmune diseases. So far, we have only discussed the importance of keeping the immune system active and able to kill cancer cells, along with foreign, pathogenic organisms and viruses. However, it is important that the immune system does not become too active and begin to attack the healthy parts of the body. Autoimmune diseases and allergies are caused by the failure of the immune system to properly distinguish between self and non-self and to distinguish between food and toxins [1]. For example, in type-1 diabetes, the immune system destroys the pancreas. When a person develops an allergy to a type of food, the immune system has mistakenly identified the food, or certain molecules in the food as being harmful and an inappropriate inflammatory, allergic response is produced.

A recently published book describes the autoimmune epidemic that is occurring in the USA and other western nations [31]. The increase in the prevalence in autoimmune diseases continues to be noted [32]. About 23.5 million people in the USA suffer from one or more of the nearly 100 autoimmune diseases, including type-1 diabetes. The rate of increase in these diseases is large. The rate of increase in type-1 diabetes is about 6% per year in children who are under four years of age and 4% in children aged 10 – 14. However, there is hope. Doctor Denise Faustman was able to cure type-1 diabetes in mice by giving them TNF-α [31]. "Normally, our bone marrow and thymus produce millions of T cells (and B cells), which have different receptors. Many T cells are born "faulty" in the bone marrow, in that they react to the body itself, rather than to outside pathogens. Usually, >99% of these are recognized as auto-reactive and are killed. Still, a group of dangerous T cells can mistakenly escape from the bone marrow. However, they are more susceptible to the action of TNF-α. Faustman administered TNF-α, which only killed the aberrant T cells. It not only stopped the disease progression, but also allowed the body to regenerate islet-producing cells. Three other labs confirmed the results" [31]. However, it is one thing to cure a disease in mice and a completely different thing to cure it in people.

Other autoimmune diseases include antiphospholipid antibody syndrome (APS), multiple sclerosis (MS), rheumatoid arthritis, lupus, transverse myelitis (TM), Guillain-Barré syndrome, juvenile arthritis, psoriasis, and many others [31]. "About 80% of the people with autoimmune diseases are women. The sex hormone balance is crucial in helping to regulate one's immune and inflammatory responses. The amount of interferon gamma that is produced is a key variable in understanding who gets MS. Gamma interferon signals the immune system to start its response. Men are less likely to have a defective gene coding for increased secretion of interferon-gamma, and less of the gamma-interferon variant" [31].

Another disease, Loeys-Dietz syndrome, LDS, has shed light on the importance of TGF-β signaling on autoimmune diseases [32, 33]. "TGF-β is a cytokine that plays diverse roles in many physiological processes. One of them is regulating the maturation of regulatory T cells (T_{reg}). Patients with LDS are more susceptible to autoimmune diseases due to mutations in genes coding for receptors for TGF-β. This also implies that other people with deleterious mutations in TGF-β signaling might be more susceptible to autoimmune diseases. TGF-β also plays a crucial role in directing immune responses to potential antigens in food and the environment. So, it might be useful to test drugs that are currently approved by the FDA to see if they can affect TGF-β signaling and treat allergies and autoimmune diseases" [32].

Risk factors for autoimmune diseases include exposure to man-made chemicals, such as PCBs, dioxins, PBDEs, perfluorooctanoic acid (PFOA), phthalate esters, trichloroethylene (TCE), organochlorine pesticides, bisphenol A (BPA) and many others [30]. For example, PBDEs are used as flame retardants [31]. "Many of the common PBDEs were banned by the European Community in 2006. Phthalate esters are plasticizers used to make many man-made polymers and are added to cosmetics to make them creamier. PCBs and dioxins were part of Agent Orange, which was used in the war in Vietnam. PFOA is a plasticizer that is used in making Teflon. The EPA determined that even low-level exposure to PFOA poses "a potential risk of developmental and other adverse effects. In a recent paper from Stockholm University's Unit for Biochemical Toxicology, investigators found that all doses (even the lowest) altered the function of our immune cells at every major step in the immune response. In a study at the University of Florida in Gainesville, organochlorine pesticides caused lupus in female mice. They were exposed a dose that was four times lower than the EPA's acceptable limit. In another study, it was found that farmers who were exposed to pesticides were more likely to die of an autoimmune disease. They had a higher antinuclear antibody titer, or ANA count, which is diagnostic of autoimmune diseases" [31]. More recently, a positive association was found between the concentration of PFOA in blood and the incidence of cardiovascular and peripheral arterial diseases [34]. Doctor Allen Silverstone studies how endocrine disruptors such as BPA and PCBs disturb the regulation of the immune system [31]. "The aryl hydrocarbon receptor is in almost all human tissues and it binds dioxins and PCBs and leaves the receptor turned on way too long. This causes other biochemical and cellular effects that are unhealthy, and can lead to autoimmune diseases and altered immunity. At the University of Tokyo, investigators reported that endocrine disruptors, such as BPA, promote an increase in autoantibodies. Others found that endocrine disruptors suppress the function of some immune cells and over stimulate others. BPA has been found in 95% of the urine samples tested. The EPA has set a daily safe limit of BPA of 0.05 mg/kg. This limit can be exceeded by handling the paper on which many receipts are made that contain mg quantities of BPA. If infants were to put such paper in their mouths and suck, they could ingest several times the daily limit of BPA – at a time when they are most susceptible to developmental toxicity. More than six billion pounds of BPA have been used each year to produce resins that are used to line metal cans and hot beverage cups, to package foods, and to blend (copolymerize) with other plastic products. TCE has been widely used to clean grease and oil off of metal parts in manufacturing. The major source of environmental exposure is water, especially when taking a shower, or in a steam room or hot tub in a gym or a health club" [31].

Moreover, 160 man-made chemicals (including lead) have been detected in drinking water, mother's milk and the blood of several people [31]. It should be added that the amounts of chemicals found were very low, and may or may not be clinically relevant. As our instrumental methods become more sensitive, we are able to detect even lower amounts of many chemicals. That does not necessarily mean that we are being exposed to many more bad chemicals. Maybe we are simply able to see lower amounts of them. Our ability to detect lower and lower amounts of chemicals exceeds our ability to tell whether they are toxic to humans at such low doses. Remember that the National Toxicology Program in the USA tests the toxicities of chemicals in rodents. Such studies show that almost all of the chemicals found in drinking water, mother's milk and the blood are not toxic to rodents at very low doses. However, they are toxic at doses that are much higher than the amount to which any human would ever be exposed. Still, toxicologists assume that toxicological data on such high doses of chemicals will provide useful information on the toxicity to the small fraction of the human population that might be highly susceptible to the chemicals.

Still, something is causing the alarming increase in autoimmunity. Nakazawa also wrote that eating organic foods and foods containing antioxidants may help prevent autoimmune diseases [31]. Unfortunately, the author also questions the wisdom of vaccinating babies and young children, even though there are no data to support this. It also ignores the fact that diseases like polio can be much worse than autoimmune diseases and that the campaign to stop immunizing children has caused polio (and other deadly diseases) to re-emerge, after being nearly wiped out. In fact, a polio epidemic has emerged in Nigeria [35].

It has been suggested by others that the possible link between autism and immunization may actually be due to the fact that acetaminophen is given to treat fevers that may arise after immunization [36]. Some people do not metabolize acetaminophen as well as others, so it may be possible that the concentration of acetoaminophen in the blood could reach neurotoxic levels, even if it is not high enough to cause the much more common liver toxicity that can occur in adults who consume excess alcohol while taking acetaminophen. However, it should be added that it is still given to autistic children and there has been no indication of an increase in toxicity in those children. Still, it is striking that the incidence of autism is much lower in Cuba than in Miami, Florida. Because of the embargo imposed by the USA, acetaminophen is very difficult to obtain in Cuba, unlike Miami, which has a large population of Cuban immigrants and refugees.

The toxicities of very few of these new chemicals are ever tested by the EPA. Only if evidence of human toxicity emerges after the chemicals are introduced into the market are they ever tested. Also, it is quite possible that something as complex as autoimmunity has many possible causes, including our tendency to live over-fed, undernourished, stressful lifestyles.

Environmental toxins can mess with the immune system, causing test animals to produce autoantibodies. When mice were given TCE-contaminated water, their CD44 molecules activated T cells in the absence of viral infection [31]. Not only were more cells expressing CD44, but these $CD44^+$ cells produced more cytokines. Also, TCE can help cause lupus [31].

One way that autoimmune diseases can occur is when immune cells mistake an amino acid sequence from one's own healthy proteins for a similar sequence in a stomach flu virus. Such a mistake is made in lupus, when eight amino acids, called the Sm B amino acid sequence, were bound by antibodies of lupus patients. This same sequence is in the EBV (Epstein-Barr virus), which is a common virus, which can cause a low-grade fever in children, or mononucleosis in teenagers and adults. Young adults with high levels of antibodies against the EBV are more likely to get multiple sclerosis in 15-20 years [37]. Type-1 diabetes is an autoimmune disease in which the body attacks its own pancreas and becomes incapable of making insulin [31]. Possibly, enteroviruses cause some cases of type-1 diabetes. Coxsackie B viruses may help cause type-1 diabetes. One should note that coxsackie viruses prefer warm weather, so they may flourish in global warming [31].

Also, there is a possible link between autism and autoimmunity [31]. "Aberrant antibodies have been found in autistic children and in mothers with more than one autistic child. There is a researcher named Fairweather who said that the adaptive immune system is controlled by the innate immune system" [31]. A gene, called Tim-3 puts the brakes on the immune system, telling it to stop releasing cytokines [38]. When mast cells are stimulated, they release cytokines. When it does its job correctly, Tim-3 acts as the master brake of the immune system. As in the rest of the body, there must be a proper balance in the immune system. However, when the immune system is constantly bombarded with new man-made chemicals, called autogens, the balance is disrupted [31].

Under stress, our adrenal glands produce several hormones, including adrenaline and cortisol [1]. "The process starts in the hypothalamus, which regulates body temperature, respiration, hunger, sleep cycle, sexual function and blood pressure. When stress occurs, it causes the hypothalamus to activate the adrenal glands and pituitary gland, which secretes cortisol. When the stress ends, the adrenal glands send a message back to the hypothalamus, telling it to stop producing excess cortisol and adrenaline" [1]. This doesn't happen in chronic stress. Cortisol mobilizes the immune system. If this happens too often, the immune cells get a constant barrage of cortisol, making us less able to respond to truly stressful pathogens. Prolonged cortisol stimulation can cause an improperly functioning immune response and stimulate an autoimmune response [31].

So, let us look at some of the autoimmune diseases discussed in the book [31], starting with antiphospholipid antibody syndrome, or APS, also known as sticky blood, or Hughes syndrome. In APS, the immune system mistakenly disables the proteins needed to prevent excessive clotting. So, excess blood clotting occurs, which can cause stroke, as well as pregnancy-related complications such as miscarriage, pre-term delivery, or severe preeclampsia. APS occurs due to the autoimmune production of antibodies against cardiolipin and β_2 glycoprotein I. Cardiolipin is an important component of the inner mitochondrial membrane, where it constitutes about 20% of the total lipid [31]. "It is typically present in metabolically active cells of the heart and skeletal muscle. Also, apolipoprotein H (Apo-H) is a multifunctional apolipoprotein. One of its functions is to bind cardiolipin" [31].

Another important autoimmune disease is lupus, which affects 1.5 million Americans, or about one in two hundred people [31]. "Immune cells are tricked into acts of self-sabotage by a combination of genetic predisposition and environmental triggers. They can turn against almost any organ, including the joints, kidneys, heart, lungs, brain, blood, or skin, causing severe pain, inflammation and cellular damage. ANAs, or antinuclear antibodies, are biomarkers of lupus. These include anti-nRNP, anti-Sm, anti-double -stranded DNA, anti Ro and anti-La antibodies. They also occur in people who don't have lupus. If caught soon enough (before too much damage is done), doctors can treat lupus. To diagnose lupus, look for ANA, or antinuclear antibodies, which are able to bind to the nuclei of cells and inflict damage. These include anti-nRNP, anti-Sm, anti-double -stranded DNA, anti Ro and anti-La antibodies. However, they also occur in people who don't have lupus. Steroids and immunosuppressive drugs are given as treatment" [31]. In 2011, the FDA approved the first new drug, belimumab (Benlysta), for treating lupus in 56 years [39]. It is a human monoclonal antibody that neutralizes the B-cell survival factor, B-lymphocyte stimulator (BLyS), a member of the tumor necrosis factor (TNF) ligand family [40]. In lupus, there B cells are hyperactive. BLyS is a vital B-cell survival factor. Autoreactive B cells have a greater dependency on BLyS for their survival than do nonautoreactive B cells [40]. Interestingly, belimumab is an example of the principle that moderate inhibition is better than total inhibition. Another experimental drug for lupus, rituximab, depleted B cells more than belimimab, but it failed the clinical trials [40].

In an interesting twist, a lupus antibody has been shown to have the potential to treat cancer [41]. It can penetrate living cells and bind to their DNA. This antibody and its single-chain variable fragment have the same cytotoxic activity of known drug inhibitors of DNA repair, but with the potential therapeutic advantages and specificity of a monoclonal antibody [41]. Also, Hansen [42] found an autoantibody (3E10) that binds to single-stranded DNA tails and inhibits its repair. "It can sensitize cancer cells and human tumor xenografts to DNA-damaging therapy caused by doxorubicin and radiation" [42]. "This may be particularly applicable to malignancies related to BRCA2 including breast, ovarian, and prostate cancers. It also suggested that lupus autoantibodies may be partly responsible for the intrinsic deficiencies in DNA repair and the unexpectedly low rates of breast, ovarian, and prostate cancers observed in people who have systemic lupus erythematosus (SLE)" [42].

Another autoimmune disease is transverse myelitis, or TM, in which demyelination and injury to bundles of axonal nerves occurs in the spinal cord [31]. Embryonic mouse stem cells were used to regenerate damaged axonal nerves and myelin sheaths in paralyzed mice [43]. Growth dependent nerve factor (GDNF) binds to a receptor on the surface of an axon and causes the neuron to grow in that direction.

Another important autoimmune disease is multiple sclerosis, or MS. The immune system attacks the central nervous system, leading to demyelination of axons [1]. "When the myelin sheath is lost, the axons are no longer insulated and signals can't be sent properly. The name multiple sclerosis refers to scars or scleroses in the white matter of the brain and spinal cord. MS can be treated by intravenous injection of corticosteroids, such as methylprednisolone. The goal of this kind of treatment is to end the attack sooner and leave fewer lasting deficits in the patient. Nine disease-modifying treatments have been approved for treating MS. They include Avonex, Betaseron, Extavia and Refib, based on interferons 1a and b; Gilenya (fingolimod), Novantrone (mitoxantrone), teriflunomide and Tysabri (natalizumab). Gilenya is a modulator of the sphingosine1-phosphate receptor. It prevents lymphocytes from crossing the blood-brain barrier, reducing inflammatory damage to nerve cells. Novantrone is an immunosuppressant that is also used in cancer chemotherapy. It suppresses the activity of T cells, B cells, and macrophages that are thought to lead the attack on the myelin sheath. Teriflunomide reversibly inhibits the enzyme dihydroorotate dehydrogenase, which catalyzes a key step in the biosynthesis of pyrimidines needed for DNA synthesis. So, it reduces the activation, proliferation and function of T- and B-cells" [1]. At the same time, the replication and function of slowly dividing cells that can use dietary sources of pyrimidines through a salvage pathway is preserved [44]. "It is used to treat relapses of MS. Tysabri is a monoclonal antibody that was designed to hamper movement of potentially damaging immune cells from the bloodstream, across the blood-brain barrier and into the brain and spinal cord. All six medications are modestly effective at decreasing the number of attacks and slowing progression to disability, although their efficacy rates differ, and studies of their long-term effects are still lacking" [44].

Rheumatoid arthritis is a chronic inflammatory autoimmune disease that is characterized by swelling in the joints, tenderness and destruction of synovial joints, leading to severe disability and premature mortality [1]. Rituximab (Rituxan®) is approved for the treatment of rheumatoid arthritis, and non-Hodgkin lymphomas that are resistant to other chemotherapy treatments [1]. It works on non-Hodgkin lymphomas that have CD40 cell surface receptors on them. It is a monoclonal antibody. It is a genetically engineered chimeric protein, made from mouse and human IgG1. It binds to the CD4 antigen. It is manufactured by Chinese Hamster Ovary (CHO) cells grown in a suspension culture.

Another disease involving the immune system that is in need of a cure is AIDS, or acquired immune deficiency syndrome [1]. The AIDS pandemic has been going on for over 25 years and killed an estimated 2.1 million people in 2007, while 33.2 million people lived with AIDS. Although affected individuals may show no symptoms for a few years, eventually the concentration of T-lymphocytes that have the CD4 receptor (that is, $CD4^+$ lymphocytes) decreases from slightly over 1000 $TD4^+$ cells per μL (or per mm^3) to 200 or less. This leaves the patient with a deficient immune system. Their immune systems are compromised and highly susceptible to infections that can cause death.

As mentioned before, AIDS is caused by the retrovirus called HIV, or human immunodeficiency syndrome virus. As a retrovirus, it contains RNA, not DNA. Its RNA must be converted into DNA in the host cell in a reaction that is catalyzed by a reverse transcriptase – thus the name retrovirus. This enzyme is the molecular target for reverse transcriptase inhibitors, which were among the first approved drugs for treating AIDS. Currently, a mixture, or cocktail of antiretroviral drugs are given in what is often called highly active antiretroviral therapy, or HAART [1]. "This includes two nucleoside analogue reverse transcriptase inhibitors. HIV protease is an enzyme target for treating AIDS. It is needed for the premature HIV to mature and become infective. So, HAART treatment includes either a protease inhibitor or a non-nucleoside reverse transcriptase inhibitor. Without HAART treatment, the progression from HIV infection to AIDS occurs at a median time of nine to ten years, while HAART treatment is thought to extend that by many decades" [1].

So, HAART treatment includes three or more drugs, such as protease inhibitors, along with nucleoside and non-nucleoside reverse transcriptase inhibitors. Without HAART treatment, the progression from HIV infection to AIDS occurs at a median time of nine to ten years, while HAART treatment is thought to extend that by many decades. However, it was reported recently that a baby that was HIV positive at birth was cured after receiving antiretroviral treatment (ART) starting at just 30 hrs of age [45]. Actually it was called a functional cure. Even though no HIV could be detected in the patient after 29 days, it is possible that some HIV remained dormant and was undetectable. Still, this is the first case of a patient remaining HIV negative after stopping HAART [45].

So, HAART treatment includes three or more drugs, such as protease inhibitors, along with nucleoside and non-nucleoside reverse transcriptase inhibitors [45]. Without HAART treatment, the progression from HIV infection to AIDS occurs at a median time of nine to ten years, while HAART treatment is thought to extend that by many decades [45].

However, about 1 in 500 people who are HIV positive do not get AIDS. They are called long-term nonprogressors (LTNPs) or elite controllers. Many of them have been infected with the fully virulent form of HIV for 30 years without progressing to the point where they need medication for AIDS. There are some genetic traits that may affect progression into AIDS. They include mutations in the *FUT2* gene that codes for a protein (Galactoside 2-alpha-L-fucosyltransferase 2) that affects the progression of HIV infection, different mitochondrial DNA haplotypes that generate less ATP and reactive oxygen substances (ROS), genes that increase the production of a protein called APOBEC3G (or just A3) that disrupts reverse transcription of RNA from the HIV virus, and mutations in the CCR5 receptor that HIV needs to enter T cells [45].

The drugs maraviroc and enfurvitide interfere with binding, fusing and entry of HIV into the host cell and are called entry inhibitors. Maraviroc was discovered using high-throughput screening of CCR5 antagonists. The drug raltegravir inhibits the HIV enzyme integrase, which catalyzes the reaction needed for viral DNA to integrate into the host cell's DNA. Also, the drug interferon alpha inhibits the maturation of the viral capsid polyprotein. It was used for years to treat hepatitis B before it was approved for treating HIV/AIDS.

Although treatment is working for many, it also fails with some. In fact, the Presidential Advisory Council on HIV/AIDS indicated that we will not be able to treat our way out of this AIDS pandemic [46]. In fact, there are few complete cures for any diseases that are caused by viruses. To prevent and eventually eliminate such diseases, they must be prevented by immunization, which completely eradicated smallpox and nearly eradicated polio, until many people stopped getting their children immunized. So, there is much research on developing an AIDS vaccine and a recent Phase III trial in Thailand produced a 31% reduction in infection, compared to placebo, and the results are being analyzed by systems biology to identify molecular signatures that can predict the adaptive immune response [47]. "At the same time there have been important advances in identifying especially good epitopes that are hundreds of times more immunogenic than the broadly neutralizing antibodies that are also in development. Investigators at the Scripps Research Institute evaluated the antibodies in 1800 HIV-infected people from around the world and found two antibodies that are able to neutralize many strains of HIV. Moreover, a team at the Oregon Health and Science University has been able to use the cytomegalovirus (CMV) as a vector to deliver a vaccine against the simian version of AIDS (SIV) to HIV-infected monkeys that brought the infection under complete control in 13 of 24 them. The advantage of a vaccine delivered by CMV is that the virus remains in the body, so only one injection of the vaccine would be required" [47]. Also, a recently developed vaccine called MVA-B triggered important "effector memory T cell responses in a Phase I clinical trial" [48]. It was made from recombinant modified vaccinia virus (MVA) that expressed monomeric gp120 and the fused Gag-Pol-Nef (GPN) polyprotein of clade B of HIV [49].

In another approach, $CD8^+$ cells were targeted [50]. "This is being aided by studying the rare cases in which some people who are infected with HIV don't get AIDS – the elite controllers. Most of them express the "*HLA-B*57* or *HLA-B*27* alleles. A vaccine was developed that targeted an analogue of the HLA-B*27 protein in an animal model of AIDS. This increased the level of $CD8^+$ T cells that targeted the epitopes" [50].

So, some are trying other approaches. Remarkably, doctors in Germany were able to completely eliminate HIV in a patient who had AIDS and leukemia [51]. "To eliminate HIV, peripheral blood stem cells were transplanted to the patient from a healthy donor whose cells had compatible HLA surface antigens, to prevent rejection. The healthy donor had a key mutation. Unlike most people who have at least one copy of "a gene coding for a chemokine receptor called CCR5, the donor was homozygous for a 32 base pair deletion of the *CCR5* gene, resulting in an inactive protein. It turns out that HIV needs not only the CD4 cell surface protein to infect host cells, it also needs CCR5. As a result, the donor (and others like her who have the same mutation on both copies of their *CCR5* genes) was resistant to HIV infection. So, the adult stem cells proliferated in the patient and produced T-cells that no longer had the CCR5 receptor and were no longer susceptible to HIV infection. Following the stem cell therapy, radiation therapy was given to treat the leukemia, and it worked. The leukemia was reported to be in remission after 20 months" [51].

It should also be noted that the FDA recently approved a test for HIV that can be done at home. Since detection is crucial in preventing the spread of HIV, this could save many lives.

Another new approach uses the histone deacetylase, vorinostat, which is already approved for treating several types of cancer. It catalyzes the removal of acetyl groups from histones, thus activating DNA expression. DNA transcription can be prevented in cells by attaching acetyl groups to histones. Since HIV remains dormant in $CD4^+$ cells, antiretroviral therapy can't reach them. They can emerge after many years and cause AIDS to emerge. In a clinical trial, the proof of principal of vorinostat was demonstrated [52 - 54]. At a dose of 400 mg, it increased the expression of HIV RNA. As a result, the $CD4^+$ cells that contain HIV can be seen by the patients' immune systems, which can kill the $CD4^+$ cells and eliminate the HIV.

There have been other recent advances that indicate that there are previously unexpected connections regarding autoimmune diseases. For example, blood platelets have been shown to bind to the pathogenic bacteria *Listeria monocytogenes* and transport it to $CD8\alpha^+$ dendritic cells, for destruction [52]. It has also been shown that the commensural bacteria in the gut cooperate with the myelin autoantigen to trigger autoimmune demyelination in multiple sclerosis [55]. Another study showed that the polio virus and cancer-causing viruses can interact with bacteria in the gut and coat themselves with molecules from the bacteria [56]. This enables the viruses to interact with toll-like receptor 4 (TLR4), avoiding destruction by the immune system and enabling viral infection. Also, the structure of the microbial community changes in people with inflammatory disorders [57]. Obligate anaerobic bacteria are depleted while facultative anaerobic Enterobacteriaceae increase [57]. Nitrate that is produced as a by-product of inflammation gave commensal *E. coli* a growth advantage in the large intestines by generating electron acceptors for anaerobic respiration [58].

Another paper addresses the controversy over what is causing the increase in autoimmune diseases in children. In her book, Nakazawa said that she thinks it is due to the increased exposure to human-made chemicals, such as BPA [31]. She did mention that others claim that it is due to the relatively clean environment in which children live. This is called the hygiene hypothesis and is based, in part, on the fact that there is a lower incidence of autoimmune diseases in children who live on farms [59]. They are exposed to more bacteria than children who do not live on farms [59]. They are exposed to much less bacteria than previous generations, due to increased sanitation and cleanliness. Also, it was shown that microbes in the intestines and lungs (acquired from the environment) keep rare invariant natural killer immune cells from triggering autoimmune diseases [59]. However, other aspects of lifestyle may be important, too. Periodic expansions of the lungs through exercise are good for the bronchial smooth muscles. The modern sedentary lifestyle limits this. Also, since children stay indoors and watch TV or play video games, they are exposed to far more dust mites, which can lead to asthma [60].

Another thing that has changed in recent years is that solid food is introduced to infants at six months instead of four months. Instead of rice, infants are started with oats. New foods are introduced gradually, one at a time. It has also been found that childhood allergies can be prevented by introducing a tiny, controlled amount of food, such as peanuts [61]. Other research has found that not only the well-known T_{H2}, but also the cytokine IL-17 are involved in asthma [62]. This might explain why some people do not respond to the usual steroids that are used to treat for asthma and provide a new target for new drug development. Another paper describes a new approach to preventing asthma [63]. "Currently, this is done by injecting children with extracts obtained from allergens. Instead, a company called Circassia is preparing epitopes, or just smaller portions of the allergens, to prepare a second generation of allergy shots. Another group is developing an allergen library made from pieces of DNA from a variety of antigens. These are placed on an electronic device (a chip), which can rapidly detect which allergens to which a patient is allergic. All they need is to put a drop of the patient's blood on the microchip to do the analysis" [63].

Other advances have also been made in the way vaccines might be developed. To get the antigens in the vaccine into the body a vector must be used. Adenoviruses are convenient, but they are not effective in people who have a natural immunity to them. Researchers recently completed a Phase I trial using two rare serotype adenoviral vectors that expressed a hepatitis C viral protein [64]. "Vaccination primed T cells to respond to multiple HCV strains at a level consistent with protective immunity. Advances continue to be made for developing cancer vaccines" [64]. A recent paper reported that adding bacterial components as adjuvants provides a robust immune response [65].

Polysaccharides can also be used as an adjuvant [66]. "The 2009 H_1N_1 influenza pandemic highlighted the need for vaccines to be produced faster and more efficiently. Traditional vaccines cultivate adapted influenza viruses in eggs and take three to four months to establish. "Early in the 2009 pandemic, the initial influenza A/H_1N_1/California/04/2009 strain distributed by the Center for Disease Control (CDC) provided unsatisfactory yields. A strain that produced higher yield was required, delaying vaccine availability. Moreover, some children have severe egg allergies. So, an alternative vaccine substrate is being studied - recombinant hemagglutinin (rHA), since HA is the primary target of protective neutralizing antibodies after natural infection or vaccination. Recombinant proteins can be produced rapidly and predictably, making it an attractive option for pandemic vaccines. So, in response to the Public Health Emergency Medical Countermeasures Enterprise Review several companies are developing recombinant influenza vaccines. Insect cell-derived rHA produced using the baculovirus expression system has been evaluated in clinical trials. It protected against seasonal influenza in a study of 4648 subjects. At the same time, since antigen production is a major limiting factor in supplying vaccines, adjuvant-based dose-sparing strategies are being prioritized. Advax™ is a novel polysaccharide adjuvant based on particles of semi-crystalline delta inulin. Advax™ enhances the immunogenicity and protection of vaccines in a several animal models. Advax™ particles bind to human monocytes directly and enhance their co-stimulatory powers. In an influenza challenge study, Advax™ adjuvant significantly enhanced H5N1 vaccine protection, with 100% survival of ferrets receiving adjuvanted vaccine *versus* only 66% survival with standard H5N1 vaccine. Advax™ adjuvant significantly reduced neurological disease and H5N1 viral shedding while enabling the dose to be reduced over 3-fold" [66].

The speed and utility of the rHA approach in a real pandemic setting was tested in the H_1N_1/2009 outbreak [66]. "A clinical study performed on the first rHA vaccine to be developed during an actual pandemic addressed two main questions: whether it was possible to design, manufacture and release a recombinant vaccine within 12 weeks of identifying a new pandemic influenza strain and, whether Advax™ adjuvant would improve the immunogenicity of the recombinant antigen. The study confirmed the ability to design, manufacture and release a recombinant vaccine in such a short time and that the Advax™ adjuvant can significantly enhance the immunogenicity of rHA" [66].

In another study, it was found that gut bacteria can convert acetylcholine (lecithin) into choline and then to the toxic compound trimethylamine *N*-oxide [67]. "Lecithin is a dietary supplement that is taken by many people who do not get sufficient amounts of it in their diet. The best sources of dietary lecithin are beef liver and other meats. So, vegetarians and vegans might want to take lecithin. A lack of dietary choline can lead to non-alcoholic fatty-liver disease and muscle damage. However, if one takes too much, it can cause toxicity, due to compound trimethylamine *N*-oxide" [67].

New things continue to be discovered about immune cells. It was found that the immune system is also involved in regulating body temperature [68]. Macrophages are essential for brown fat to be able to produce heat [68]. It has also been shown that some T-cells do not originate in the thymus [69]. So, research continues to advance and new therapies are continually being developed.

From the medicinal chemistry point of view, there are many drugs that are used to treat diseases of the immune system. As discussed above, Rituximab (Rituxan®) is approved for the treatment of rheumatoid arthritis [1], and non-Hodgkin lymphomas that are resistant to other chemotherapy treatments. It works on non-Hodgkin lymphomas that have CD40 cell surface receptors on them. It is a monoclonal antibody. It is a genetically engineered chimeric protein, made from mouse and human IgG1. It binds to the CD4 antigen. It is manufactured by Chinese Hamster Ovary (CHO) cells grown in a suspension culture.

Nine disease-modifying treatments have been approved for treating MS. They include Avonex, Betaseron, Extavia and Refib, based on interferons 1a and b; Gilenya (fingolimod), Novantrone (mitoxantrone), teriflunomide and Tysabri® (natalizumab). Gilenya is a modulator of the sphingosine1-phosphate receptor. It prevents lymphocytes from crossing the blood-brain barrier, reducing inflammatory damage to nerve cells. Novantrone is an immunosuppressant that is also used in cancer chemotherapy [1]. "It suppresses the activity of T cells, B cells, and macrophages that are thought to lead the attack on the myelin sheath. Teriflunomide reversibly inhibits the enzyme dihydroorotate dehydrogenase, which catalyzes a key step in the biosynthesis of pyrimidines needed for DNA synthesis. So, it reduces the activation, proliferation and function of T- and B-cells" [1]. At the same time, the replication and function of slowly dividing cells that can use dietary sources of pyrimidines through a salvage pathway is preserved [44]. It is used to treat relapses of MS. "Tysabri is a monoclonal antibody that was designed to hamper movement of potentially damaging immune cells from the bloodstream, across the blood-brain barrier and into the brain and spinal cord. All six medications are modestly effective at decreasing the number of attacks and slowing progression to disability, although their efficacy rates differ, and studies of their long-term effects are still lacking. Currently, a mixture, or cocktail of antiretroviral drugs are given in what is often called highly active antiretroviral therapy, or HAART. This includes two nucleoside analogue reverse transcriptase inhibitors. HIV protease is an enzyme target for treating AIDS. It is needed for the premature HIV to mature and become infective. So, HAART treatment includes either a protease inhibitor or a non-nucleoside reverse transcriptase inhibitor. Without HAART treatment, the progression from HIV infection to AIDS occurs at a median time of nine to ten years, while HAART treatment is thought to extend that by many decades" [44].

So, HAART treatment includes three or more drugs, such as protease inhibitors, along with nucleoside and non-nucleoside reverse transcriptase inhibitors. Without HAART treatment, the progression from HIV infection to AIDS occurs at a median time of nine to ten years, while HAART treatment is thought to extend that by many decades. However, it was reported recently that a baby that was HIV positive at birth was cured after receiving antiretroviral treatment (ART) starting at just 30 hrs of age [45]. Actually it was called a functional cure. Even though no HIV could be detected in the patient after 29 days, it is possible that some HIV remained dormant and was undetectable. Still, this is the first case of a patient remaining HIV negative after stopping ART.

So, HAART treatment includes three or more drugs, such as protease inhibitors, along with nucleoside and non-nucleoside reverse transcriptase inhibitors. Without HAART treatment, the progression from HIV infection to AIDS occurs at a median time of nine to ten years, while HAART treatment is thought to extend that by many decades.

However, about 1 in 500 people who are HIV positive do not get AIDS. They are called long-term nonprogressors (LTNPs) or elite controllers. Many of them have been infected with the fully virulent form of HIV for 30 years without progressing to the point where they need medication for AIDS. There are some genetic traits that may affect progression into AIDS. They include mutations in the *FUT2* gene that codes for a protein [1] (Galactoside 2-alpha-L-fucosyltransferase 2) that affects the progression of HIV infection, different mitochondrial DNA haplotypes that generate less ATP and reactive oxygen substances (ROS), genes that increase the production of a protein called APOBEC3G (or just A3) that disrupts reverse transcription of RNA from the HIV virus, and mutations in the CCR5 receptor that HIV needs to enter T cells.

The drugs maraviroc and enfurvitide interfere with binding, fusing and entry of HIV into the host cell and are called entry inhibitors. Maraviroc was discovered using high-throughput screening of CCR5 antagonists. The drug raltegravir inhibits the HIV enzyme integrase, which catalyzes the reaction needed for viral DNA to integrate into the host cell's DNA. Also, the drug interferon alpha inhibits the maturation of the viral capsid polyprotein. It was used for years to treat hepatitis B before it was approved for treating HIV/AIDS.

Although treatment is working for many, it also fails with some. In fact, the Presidential Advisory Council on HIV/AIDS indicated that we will not be able to treat our way out of this AIDS pandemic [46]. In fact, there are few complete cures for any diseases that are caused by viruses. To prevent and eventually eliminate such diseases, they must be prevented by immunization, which completely eradicated smallpox and nearly eradicated polio, until many people stopped getting their children immunized. So, there is much research on developing an AIDS vaccine and a recent Phase III trial in Thailand produced a 31% reduction in infection, compared to placebo, and the results are being analyzed by systems biology to identify molecular signatures that can predict the adaptive immune response [47]. "At the same time there have been important advances in identifying especially good epitopes that are hundreds of times more immunogenic than the broadly neutralizing antibodies that are also in development. Investigators at the Scripps Research Institute evaluated the antibodies in 1800 HIV-infected people from around the world and found two antibodies that are able to neutralize many strains of HIV. Moreover, a team at the Oregon Health and Science University has been able to use the cytomegalovirus (CMV) as a vector to deliver a vaccine against the simian version of AIDS (SIV) to HIV-infected monkeys that brought the infection under complete control in 13 of 24 them. The advantage of a vaccine delivered by CMV is that the virus remains in the body, so only one injection of the vaccine would be required" [47]. Also, a recently developed vaccine called MVA-B triggered important effector memory T cell responses in a Phase I clinical trial [48]. It was made from recombinant modified vaccinia virus (MVA) that expressed monomeric gp120 and the fused Gag-Pol-Nef (GPN) polyprotein of clade B of HIV [49].

In another approach, $CD8^+$ cells were targeted [50]. "This is being aided by studying the rare cases in which some people who are infected with HIV don't get AIDS – the elite controllers. "Most of them express the *HLA-B***57* or *HLA-B***27* alleles. A vaccine was developed that targeted an analogue of the HLA-B*27 protein in an animal model of AIDS. This increased the level of $CD8^+$ T cells that targeted the epitopes" [50].

So, some are trying other approaches. Remarkably, doctors in Germany were able to completely eliminate HIV in a patient who had AIDS and leukemia [51]. "To eliminate HIV, peripheral blood stem cells were transplanted to the patient from a healthy donor whose cells had compatible HLA surface antigens, to prevent rejection. The healthy donor had a key mutation. Unlike most people who have at least one copy of a gene coding for a chemokine receptor called CCR5, the donor was homozygous for a 32 base pair deletion of the *CCR5* gene, resulting in an inactive protein. It turns out that HIV needs not only the CD4 cell surface protein to infect host cells, it also needs CCR5. As a result, the donor (and others like her who have the same mutation on both copies of their *CCR5* genes) was resistant to HIV infection. So, the adult stem cells proliferated in the patient and produced T-cells that no longer had the CCR5 receptor and were no longer susceptible to HIV infection. Following the stem cell therapy, radiation therapy was given to treat the leukemia, and it worked. The leukemia was reported to be in remission after 20 months" [51].

It should also be noted that the FDA recently approved a test for HIV that can be done at home. Since detection is crucial in preventing the spread of HIV, this could save many lives.

Another new approach uses the histone deacetylase vorinostat, which is already approved for treating several types of cancer. It catalyzes the removal of acetyl groups from histones, thus activating DNA expression. DNA transcription can be prevented in cells by attaching acetyl groups to histones. Since HIV remains dormant in $CD4^+$ cells, antiretroviral therapy can't reach them. They can emerge after many years and cause AIDS to emerge. In a clinical trial, the proof of principal of vorinostat was demonstrated [52 - 54]. At a dose of 400 mg, it increased the expression of HIV RNA. As a result, the $CD4^+$ cells that contain HIV can be seen by the patients' immune systems, which can kill the $CD4^+$ cells and eliminate the HIV.

Other advances have also been made in the way vaccines might be developed. To get the antigens in the vaccine into the body a vector must be used. Adenoviruses are convenient, but they are not effective in people who have a natural immunity to them. Researchers recently completed a Phase I trial using two rare serotype adenoviral vectors that expressed a hepatitis C viral protein [64]. Vaccination primed T cells to respond to multiple HCV strains at a level consistent with protective immunity [64]. Advances continue to be made for developing cancer vaccines. A recent paper reported that adding bacterial components as adjuvants provides a robust immune response [65].

Polysaccharides can also be used as an adjuvant [66]. "The 2009 H_1N_1 influenza pandemic highlighted the need for vaccines to be produced faster and more efficiently. Traditional vaccines cultivate adapted influenza viruses in eggs and take three to four months to establish. Early in the 2009 pandemic, the initial influenza A/H_1N_1/California/04/2009 strain distributed by the Center for Disease Control (CDC) provided unsatisfactory yields. A strain that produced higher yield was required, delaying vaccine availability. Moreover, some children have severe egg allergies. So, an alternative vaccine substrate is being studied - recombinant hemagglutinin (rHA), since HA is the primary target of protective neutralizing antibodies after natural infection or vaccination. Recombinant proteins can be produced rapidly and predictably, making it an attractive option for pandemic vaccines. So, in response to the Public Health Emergency Medical Countermeasures Enterprise Review several companies are developing re-combinant influenza vaccines. Insect cell-derived rHA produced using the baculovirus expression system has been evaluated in clinical trials. It protected against seasonal influenza in a study of 4648 subjects. At the same time, since antigen production is a major limiting factor in supplying vaccines, adjuvant-based dose-sparing strategies are being prioritized. Advax™ is a novel polysaccharide adjuvant based on particles of semi-crystalline delta inulin. Advax™ enhances the immunogenicity and protection of vaccines in a several animal models. Advax™ particles bind to human monocytes directly and enhance their co-stimulatory powers. In an influenza challenge study, Advax™ adjuvant significantly enhanced H5N1 vaccine protection, with 100% survival of ferrets receiving adjuvanted vaccine *versus* only 66% survival with standard H5N1 vaccine. Advax™ adjuvant significantly reduced neurological disease and H5N1 viral shedding while enabling the dose to be reduced over 3-fold" [66].

The speed and utility of the rHA approach in a real pandemic setting was tested in the H_1N_1/2009 outbreak [66]. "A clinical study performed on the first rHA vaccine to be developed during an actual pandemic addressed two main questions: whether it was possible to design, manufacture and release a recombinant vaccine within 12 weeks of identifying a new pandemic influenza strain and, whether Advax™ adjuvant would improve the immunogenicity of the recombinant antigen. The study confirmed the ability to design, manufacture and release a recombinant vaccine in such a short time and that the Advax™ adjuvant can significantly enhance the immunogenicity of rHA" [66].

KEY POINTS

1. **The immune system recognizes and defends us against internal threats caused by invading organisms and pathogens. The immune system is essential in helping to establish the boundaries between self and the environment and the difference between food and poison.**

2. **Our first line of defense is composed of physical and chemical barriers. When these fail to block bacteria or other immunogens, the innate and adaptive immune systems are activated.**

3. **Granulocytes have granules that contain enzymes that catalyze the hydrolysis of DNA, RNA, triglycerides, lipids, complex carbohydrates and proteins. Natural killer cells are very cytotoxic and are carefully regulated in a healthy immune system. Lymphocytic T-cells recognize and bind to HLA-peptide complexes on APCs. T-cells that are bound to non-self HLA-peptide complexes mature and proliferate. When the TCR binds to antigens, a cascade of molecular signaling occurs.**

4. **Lymphocytic B-cells recognize foreign matter, including microorganisms and viruses. They produce high affinity immunoglobulin antibodies that attach to foreign matter and cooperate with other immune cells to eliminate such threats.**

5. **In 2011, the FDA approved the first new drug, Benlysta (belimumab), for treating lupus erythematosus in 56 years.**

6. **Doctors in Germany were able to completely eliminate HIV in a patient who had AIDS and leukemia. To eliminate HIV, peripheral blood stem cells were transplanted to the patient from a healthy donor whose cells had compatible HLA surface antigens, to prevent rejection.**

7. **Microbes in the intestines and lungs (acquired from the environment) keep rare invariant natural killer immune cells from triggering autoimmune diseases.**

ABBREVIATIONS

CD4+ T	=	Cell Cluster of differentiation 4 on the surface of a T cell
DOPA	=	Dihydroxyphenylalanine
NK cells	=	Natural Killer Cells
AIDS	=	Acquired Immunodeficiency Syndrome
HIV	=	Human Immunodeficiency Virus
HAART	=	Highly Active Antiretroviral Therapy
APCs	=	Antigen Presenting Cells
DCs	=	Dendritic Cells
PDCs	=	Plasmacytoid DCs
IL-1	=	Interleukin-1
PAMPs	=	Pathogen-Associated Molecular Patterns
DAMPs	=	Danger-Associated Molecular Patterns
ATP	=	Adenosine Triphosphate
T_H cells	=	Helper T Clls
IFN	=	Interferon
PBMCs	=	Blood Mononuclear Cells
TGFB	=	Transforming Growth Factor-β
Ig	=	Immunoglobulin
HLA	=	Human Leukocyte Antigen
MHC	=	Major Histocompatibility Complex
AAMs	=	Alternatively Activated Macrophages
EMT	=	Epithelial–Mesenchymal Transition
PGD_2	=	Prostaglandin D_2
DAG	=	Diacylglycerol
MAPKs	=	Include Mitogen-Activated Protein Kinases
SAPKs	=	Stress Activated Protein Kinases
T_{reg}	=	Regulatory T-Cell
TGFβ	=	Transforming Growth Factor–β
MALT	=	Mucosa-Associated Lymphoid Tissue
FOXP3	=	Forkhead Box P3

PI3K	=	Phosphoinositide 3-Kinase
mTOR	=	Mammalian Target of Rapamycin
APC	=	Antigen-Presenting Cells
TCRs	=	T-Cell Receptors
BCRs	=	B Cell Receptors
V, D and J	=	Variable, Diversity and Joining segments of immunoglobulins
CDR	=	Complementarity Determining Region
MCP	=	Monocyte Chemotactic Protein
MIP	=	Macrophage Inflammatory Protein
RANTES	=	Regulated upon Activation, Normal T-cell Expressed, and Secreted
GPCR	=	G-Protein Coupled Receptor
GTPase	=	GTP Phosphatase
LPS	=	Lipopolysaccharide
TLR	=	Toll-Like Receptor
TNF-α	=	Tumor Necrosis Factor Alpha
TRAF	=	TNFR-Associated Factor
CRP	=	C-Reactive Protein
MDSC	=	Myeloid-Derived Suppressor Cell
ICD	=	Immunogenic Cell Death
CTLs	=	Cytotoxic T Cells
CML	=	Chronic Myelogenous Leukemia
PD1	=	Programmed Death Protein 1
DCR1	=	Decoy Receptor 1
OPG	=	Osteoprotegerin
NF κB	=	Nuclear Factor-κB
BAFF	=	B Cell Activation Factor
BCMA	=	B Cell Maturation Antigen
SLE	=	Systemic Lupus Erythematosus
APS	=	Antiphospholipid Antibody Syndrome
PCBs	=	Polychlorinated Biphenyls
PBDEs	=	Polydibrominated Ethers
PFOA	=	Perfluorooctanoic Acid
TCE	=	Trichloroethylene
BPA	=	Bisphenol A

EPA	=	Environmental Protection Agency
EBV	=	Epstein-Barr Virus
Apo-H	=	Apolipoprotein H
ANAs	=	Antinuclear Antibodies
FDA	=	Food and Drug Administration
GDNF	=	Growth Dependent Nerve Factor
rHA	=	Recombinant Hemagglutinin

REFERENCES

[1] Belkaid Y, Grainger J. Immunology. Mucus coat, a dress code for tolerance. Science 2013; 342(6157): 432-3.
[http://dx.doi.org/10.1126/science.1246252] [PMID: 24159036]

[2] Shan M, Gentile M, Yeiser JR, *et al.* Mucus enhances gut homeostasis and oral tolerance by delivering immunoregulatory signals. Science 2013; 342(6157): 447-53.
[http://dx.doi.org/10.1126/science.1237910] [PMID: 24072822]

[3] Geismann F, Manz MG, Jung S, Sieweke MH, Merad M, Ley K. Development of monocytes, macrophages, and dendritic cells. Science 2010; 327: 656.
[http://dx.doi.org/10.1126/science.1178331]

[4] Hinrichs CS, Borman ZA, Gattinoni L, *et al.* Human effector $CD8^+$ T cells derived from naive rather than memory subsets possess superior traits for adoptive immunotherapy. Blood 2011; 117(3): 808-14.
[http://dx.doi.org/10.1182/blood-2010-05-286286] [PMID: 20971955]

[5] Corey EJ, Czako B, Kürti L. Molecules and Medicine. New York: John Wiley & Sons 2007; pp. 112-9.

[6] Iijima H, Takahashi I, Kiyono H. Mucosal immune network in the gut for the control of infectious diseases. Rev Med Virol 2001; 11(2): 117-33.
[http://dx.doi.org/10.1002/rmv.307] [PMID: 11262530]

[7] Akira S, Takeda K, Kaisho T. Toll-like receptors: critical proteins linking innate and acquired immunity. Nat Immunol 2001; 2(8): 675-80.
[http://dx.doi.org/10.1038/90609] [PMID: 11477402]

[8] Awasthi A, Kuchroo VK. Immunology. The yin and yang of follicular helper T cells. Science 2009; 325(5943): 953-5.
[PMID: 19696339]

[9] Yang D, Chertov O, Oppenheim JJ. Participation of mammalian defensins and cathelicidins in anti-microbial immunity: receptors and activities of human defensins and cathelicidin (LL-37). J Leukoc Biol 2001; 69(5): 691-7.
[PMID: 11358975]

[10] Randolph GJ. Immunology. No need to coax monocytes. Science 2011; 332(6035): 1268-9.
[http://dx.doi.org/10.1126/science.1208480] [PMID: 21659591]

[11] Schlom J. Therapeutic cancer vaccines: current status and moving forward. J Natl Cancer Inst 2012; 104(8): 599-613.
[http://dx.doi.org/10.1093/jnci/djs033] [PMID: 22395641]

[12] Radic M. Armed and accurate: engineering cytotoxic T cells for eradication of leukemia. BMC Biotechnol 2012; 12(6): 6.
[http://dx.doi.org/10.1186/1472-6750-12-6] [PMID: 22316161]

[13] Abraham SN, St John AL. Mast cell-orchestrated immunity to pathogens. Nat Rev Immunol 2010; 10(6): 440-52.
[http://dx.doi.org/10.1038/nri2782] [PMID: 20498670]

[14] Fulkerson PC, Rothenberg ME. Targeting eosinophils in allergy, inflammation and beyond. Nat Rev Drug Discov 2013; 12(2): 117-29.
[http://dx.doi.org/10.1038/nrd3838] [PMID: 23334207]

[15] Patsoukis N, Lafuente EM, Meraner P, *et al.* RIAM regulates the cytoskeletal distribution and activation of PLC-γ1 in T cells. Sci Signal 2009; 2(99): ra79.
[http://dx.doi.org/10.1126/scisignal.2000409] [PMID: 19952372]

[16] Dong C, Martinez GJ. T cells: The usual subsets. Nat Rev Immunol 2012; 12. [Noticeboard no page number].

[17] von Boehmer H, Daniel C. Therapeutic opportunities for manipulating T(Reg) cells in autoimmunity and cancer. Nat Rev Drug Discov 2013; 12(1): 51-63.
[http://dx.doi.org/10.1038/nrd3683] [PMID: 23274471]

[18] Deshmukh US, Bagavant H. When killers become helpers. Sci Transl Med 2013; 5(195): 195fs29.
[http://dx.doi.org/10.1126/scitranslmed.3006850] [PMID: 23884464]

[19] Harnett MM. Immunology. B cells spread and gather. Science 2006; 312(5774): 709-10.
[http://dx.doi.org/10.1126/science.1127652] [PMID: 16675691]

[20] Tarlinton D, Good-Jacobson K. Diversity among memory B cells: origin, consequences, and utility. Science 2013; 341(6151): 1205-11.
[http://dx.doi.org/10.1126/science.1241146] [PMID: 24031013]

[21] Viola A, Luster AD. Chemokines and their receptors: drug targets in immunity and inflammation. Annu Rev Pharmacol Toxicol 2008; 48: 171-97.
[http://dx.doi.org/10.1146/annurev.pharmtox.48.121806.154841] [PMID: 17883327]

[22] Wang JH, Gostissa M, Yan CT, *et al.* Mechanisms promoting translocations in editing and switching peripheral B cells. Nature 2009; 460(7252): 231-6.
[http://dx.doi.org/10.1038/nature08159] [PMID: 19587764]

[23] Li N, Choudhuri S, Cherrington NJ, Klaassen CD. Down-regulation of mouse organic anion-transporting polypeptide 4 (Oatp4; Oatp1b2; Slc21a10) mRNA by lipopolysaccharide through the toll-like receptor 4 (TLR4). Drug Metab Dispos 2004; 32(11): 1265-71.
[PMID: 15483194]

[24] Weber A, Wasiliew P, Kracht M. Interleukin-1 (IL-1) pathway. Sci Signal 2010; 3(105): cm1.
[PMID: 20086235]

[25] León-Ponte M, Ahern GP, O'Connell PJ. Serotonin provides an accessory signal to enhance T-cell activation by signaling through the 5-HT7 receptor. Blood 2007; 109(8): 3139-46. [http://dx.doi.org/10.1182/blood-2006-10-052787] [PMID: 17158224]

[26] O'Connell PJ, Wang X, Leon-Ponte M, Griffiths C, Pingle SC, Ahern GP. A novel form of immune signaling revealed by transmission of the inflammatory mediator serotonin between dendritic cells and T cells. Blood 2006; 107(3): 1010-7. [http://dx.doi.org/10.1182/blood-2005-07-2903] [PMID: 16223770]

[27] Galluzzi L, Senovilla L, Zitvogel L, Kroemer G. The secret ally: immunostimulation by anticancer drugs. Nat Rev Drug Discov 2012; 11(3): 215-33. [http://dx.doi.org/10.1038/nrd3626] [PMID: 22301798]

[28] Yao S, Zhu Y, Chen L. Advances in targeting cell surface signalling molecules for immune modulation. Nat Rev Drug Discov 2013; 12(2): 130-46. [http://dx.doi.org/10.1038/nrd3877] [PMID: 23370250]

[29] Topalian SL, Hodi FS, Brahmer JR, *et al.* Safety, activity, and immune correlates of anti-PD-1 antibody in cancer. N Engl J Med 2012; 366(26): 2443-54. [http://dx.doi.org/10.1056/NEJMoa1200690] [PMID: 22658127]

[30] Sznol M, *et al.* Survival and long-term follow-up of safety and response in patients (pts) with advanced melanoma (MEL) in a phase I trial of nivolumab (anti-PD-1; BMS-936558; ONO-4538). J Clin Oncol Abstr 2013; 31: S3002.

[31] Nakazawa DK. The Autoimmune Epidemic, Touchstone, a division of Simon & Schuster. New York 2008.

[32] Frischmeyer-Guerrerio PA, Guerrerio AL, Oswald G, *et al.* TGFβ receptor mutations impose a strong predisposition for human allergic disease. Sci Transl Med 2013; 5(195): 195ra94. [http://dx.doi.org/10.1126/scitranslmed.3006448] [PMID: 23884466]

[33] Van Hemelrijk C, Renard M, Loeys B. The Loeys-Dietz syndrome: an update for the clinician. Curr Opin Cardiol 2010; 25(6): 546-51. [http://dx.doi.org/10.1097/HCO.0b013e32833f0220] [PMID: 20838339]

[34] Shankar A, Xiao J, Ducatman A. Perfluorooctanoic acid and cardiovascular disease in US adults. Arch Intern Med 2012; 172(18): 1397-403. [http://dx.doi.org/10.1001/archinternmed.2012.3393] [PMID: 22945282]

[35] Roberts L. Public health. Type 2 poliovirus back from the dead in Nigeria. Science 2009; 325(5941): 660-1. [http://dx.doi.org/10.1126/science.325_660] [PMID: 19661388]

[36] Schultz ST. Can autism be triggered by acetaminophen activation of the endocannabinoid system? Acta Neurobiol Exp (Warsz) 2010; 70(2): 227-31. [PMID: 20628445]

[37] DeLorenze GN, Munger KL, Lennette ET, Orentreich N, Vogelman JH, Ascherio A. Epstein-Barr virus and multiple sclerosis: evidence of association from a prospective study with long-term follow-up. Arch Neurol 2006; 63(6): 839-44. [http://dx.doi.org/10.1001/archneur.63.6.noc50328] [PMID: 16606758]

[38] Monney L, Sabatos CA, Gaglia JL, *et al.* Th1-specific cell surface protein Tim-3 regulates macrophage activation and severity of an autoimmune disease. Nature 2002; 415(6871): 536-41. [http://dx.doi.org/10.1038/415536a] [PMID: 11823861]

[39] Navarra SV, Guzmán RM, Gallacher AE, *et al.* BLISS-52 Study Group. Efficacy and safety of belimumab in patients with active systemic lupus erythematosus: a randomised, placebo-controlled, phase 3 trial. Lancet 2011; 377(9767): 721-31. [http://dx.doi.org/10.1016/S0140-6736(10)61354-2] [PMID: 21296403]

[40] Stohl W, Hilbert DM. The discovery and development of belimumab: the anti-BLyS-lupus connection. Nat Biotechnol 2012; 30(1): 69-77. [http://dx.doi.org/10.1038/nbt.2076] [PMID: 22231104]

[41] Ford JM. Lupus antibody tops cancer cells. Sci Transl Med 2012; 4(157): 157fs38. [http://dx.doi.org/10.1126/scitranslmed.3004955] [PMID: 23100623]

[42] Hansen JE, Chan G, Liu Y, *et al.* Targeting cancer with a lupus autoantibody. Sci Transl Med 2012; 4(157): 157ra142. [http://dx.doi.org/10.1126/scitranslmed.3004385] [PMID: 23100628]

[43] Myckatyn TM, Mackinnon SE, McDonald JW. Stem cell transplantation and other novel techniques for promoting recovery from spinal cord injury. Transpl Immunol 2004; 12(3-4): 343-58. [http://dx.doi.org/10.1016/j.trim.2003.12.017] [PMID: 15157926]

[44] O'Conner PO, *et al.* Randomized trial of oral teriflunomide for relapsing multiple sclerosis. N Engl J Med 2011; 364: 1293-303. [PMID: 21470008]

[45] Persaud D, *et al.* http://www.retroconference.org/2013b/Abstracts/47897.htm

[46] Presidential Advisory Council on HIV/AIDS. Achieving an HIV Generation 2004.

[47] Sekaly R, Pulendran B. Systems biology in understanding HIV pathogenesis and guiding vaccine development. Curr Opin HIV AIDS 2012; 7(1): 1-3. [http://dx.doi.org/10.1097/COH.0b013e32834e0667] [PMID: 22134343]

[48] Maurice J. Quest for an effective AIDS vaccine takes a new tack. Lancet 2011; 378(9787): 213-4. [http://dx.doi.org/10.1016/S0140-6736(11)61109-4] [PMID: 21770060]

[49] Gómez CE, Nájera JL, Perdiguero B, *et al.* The HIV/AIDS vaccine candidate MVA-B administered as a single immunogen in humans triggers robust, polyfunctional, and selective effector memory T cell responses to HIV-1 antigens. J Virol 2011; 85(21): 11468-78. [http://dx.doi.org/10.1128/JVI.05165-11] [PMID: 21865377]

[50] Mudd PA, Martins MA, Ericsen AJ, *et al.* Vaccine-induced $CD8^+$ T cells control AIDS virus replication. Nature 2012; 491(7422): 129-33. [http://dx.doi.org/10.1038/nature11443] [PMID: 23023123]

[51] Hütter G, Nowak D, Mossner M, *et al.* Long-term control of HIV by *CCR5* Delta32/Delta32 stem-cell transplantation. N Engl J Med 2009; 360(7): 692-8. [http://dx.doi.org/10.1056/NEJMoa0802905] [PMID: 19213682]

[52] Deeks SG. HIV: Shock and kill. Nature 2012; 487(7408): 439-40.

[http://dx.doi.org/10.1038/487439a] [PMID: 22836995]

[53] Archin NM, Liberty AL, Kashuba AD, *et al.* Administration of vorinostat disrupts HIV-1 latency in patients on antiretroviral therapy. Nature 2012; 487(7408): 482-5. [http://dx.doi.org/10.1038/nature11286] [PMID: 22837004]

[54] Verschoor A, Neuenhahn M, Navarini AA, *et al.* A platelet-mediated system for shuttling blood-borne bacteria to CD8α+ dendritic cells depends on glycoprotein GPIb and complement C3. Nat Immunol 2011; 12(12): 1194-201. [http://dx.doi.org/10.1038/ni.2140] [PMID: 22037602]

[55] Berer K, Mues M, Koutrolos M, *et al.* Commensal microbiota and myelin autoantigen cooperate to trigger autoimmune demyelination. Nature 2011; 479(7374): 538-41. [http://dx.doi.org/10.1038/nature10554] [PMID: 22031325]

[56] Pennisi E. Microbiology. Gut bacteria lend a molecular hand to viruses. Science 2011; 334(6053): 168. [http://dx.doi.org/10.1126/science.334.6053.168] [PMID: 21998362]

[57] Winter SE, Winter MG, Xavier MN, *et al.* Host-derived nitrate boosts growth of *E. coli* in the inflamed gut. Science 2013; 339(6120): 708-11. [http://dx.doi.org/10.1126/science.1232467] [PMID: 23393266]

[58] Jostins L, Ripke S, Weersma RK, *et al.* International IBD Genetics Consortium (IIBDGC). Host-microbe interactions have shaped the genetic architecture of inflammatory bowel disease. Nature 2012; 491(7422): 119-24. [http://dx.doi.org/10.1038/nature11582] [PMID: 23128233]

[59] Leslie M. Immunology. Gut microbes keep rare immune cells in line. Science 2012; 335(6075): 1428. [http://dx.doi.org/10.1126/science.335.6075.1428] [PMID: 22442453]

[60] Maxmen A. Breathing new life into research. Nature 2011; 479: S20-1. [http://dx.doi.org/10.1038/479S20a]

[61] Kessler R. Picky eaters. Nature 2011; 479: S8-9. [http://dx.doi.org/10.1038/479S8a]

[62] Wong CK, Ho CY, Ko FW, *et al.* Proinflammatory cytokines (IL-17, IL-6, IL-18 and IL-12) and Th cytokines (IFN-γ, IL-4, IL-10 and IL-13) in patients with allergic asthma. Clin Exp Immunol 2001; 125(2): 177-83. [http://dx.doi.org/10.1046/j.1365-2249.2001.01602.x] [PMID: 11529906]

[63] Gravitz L. In search of a booster shot. Nature 2011; 479: S17-9. [http://dx.doi.org/10.1038/479S17a]

[64] Barnes E, Folgori A, Capone S, *et al.* Novel adenovirus-based vaccines induce broad and sustained T cell responses to HCV in man. Sci Transl Med 2012; 4(115): 115ra1. [http://dx.doi.org/10.1126/scitranslmed.3003155] [PMID: 22218690]

[65] Reed SG, Bertholet S, Coler RN, Friede M. New horizons in adjuvants for vaccine development. Trends Immunol 2009; 30(1): 23-32. [http://dx.doi.org/10.1016/j.it.2008.09.006] [PMID: 19059004]

[66] Gordon DL, Sajkov D, Woodman RJ, *et al.* Randomized clinical trial of immunogenicity and safety of a recombinant H_1N_1/2009 pandemic influenza vaccine containing Advax™ polysaccharide adjuvant.

Vaccine 2012; 30(36): 5407-16.
[http://dx.doi.org/10.1016/j.vaccine.2012.06.009] [PMID: 22717330]

[67] Wang Z, Klipfell E, Bennett BJ, *et al.* Gut flora metabolism of phosphatidylcholine promotes cardiovascular disease. Nature 2011; 472(7341): 57-63.
[http://dx.doi.org/10.1038/nature09922] [PMID: 21475195]

[68] Whittle AJ, Vidal-Puig A. Physiology: Immune cells fuel the fire. Nature 2011; 480(7375): 46-7.
[PMID: 22101432]

[69] Josefowicz SZ, Niec RE, Kim HY, *et al.* Extrathymically generated regulatory T cells control mucosal TH2 inflammation. Nature 2012; 482(7385): 395-9.
[http://dx.doi.org/10.1038/nature10772] [PMID: 22318520]

CHAPTER 9

Infectious Diseases

Abstract: Infectious diseases can be caused by worms, protozoa, fungi, bacteria, viruses and even proteins (prions). Organisms, viruses and prions can be classified by their infectivity, or their ability to enter, survive and multiply in a host. There are seven classes of viruses, based on their DNA or RNA. By number, 90% of the cells in the human body are bacteria. Even though our lives depend on symbiotic bacteria, it is important that they stay in their proper places in our human bodies, or ecosystems. By the mid-1980s strains of *S. aureus* emerged which were resistant to common antibiotics. Multicellular parasites include four species of *Schistosoma*, a flatworm that causes schistosomiasis, which is second in importance only to malaria, with hundreds of millions infected worldwide. In addition to schistosomiasis, helminths can cause ascariasis, dracunculiasis, elephantiasis, hookworm, lymphatic filiaruasis, oncho-cersiasis, and trichuriasis.

Keywords: Bacteria, Fungi, Infectious diseases, Mycobacterium , *S. aureus*, Tuberculosis, Viruses.

Next, let us look at the biggest killers throughout most of human history – infectious diseases. They can be caused by parasitic worms, fungi, protozoa, bacteria, viruses and even infectious proteins, called prions. They can exist in a self-propagating state that is biologically accessible, but rarely form spontaneously [1]. "They also can replicate themselves by acting on their non-prion substrate protein, spread to naive hosts and find new substrate pools for replication. Finally, prions also cause phenotypic changes in the host. In mammals, they propagate after post-translational conversion of the host's normal, detergent-soluble, protease sensitive PrP (PrP^C or PrP-sen) to a less soluble and more proteinase-K resistant state (PrP^{RES} or PrP^{Sc}). The most infectious particle per unit protein is an oligomer containing 12-24 PrP monomers. The prototypical prion PrP^{Sc} causes Mad Cow disease, also known as bovine spongiform encephalopathy (SGE). It is also called the SGE prion. It also causes Creuzfeldt-Jakob disease in humans. Once PrP^{RES} is made, it can accumulate on the cell

Robert E. Smith

surface, in intracellular lysosomal vesicles, or in extracellular deposits such as amyloid plaques" [1].

Infectious diseases are usually contagious. They can be transferred from one person or species to another. Organisms, viruses and prions can be classified by their infectivity, or their ability to enter, survive and multiply in a host. Most biologists do not consider viruses to be alive, and none consider prions to be alive. Prions are proteins, or pieces of proteins that can cause mad Cow disease and Creuzfeld-Jakob disease. The natural, healthy prion, PrP^C, is soluble, but anchored to cell membranes by glycophosphatidyl inositol (GPI) [1]. It is coded by the *Prnp* gene [2]. The function of PrP^C is not well known, but it may act as a copper-dependent antioxidant. It is also involved in maintaining the brain's white matter, regulating the innate immune system, and forming new neurons. It may also function in long term memory, or the ability to remember things that happened hours, days, months, years or decades earlier. There are also two mammalian PrP paralogs, Doppel and Shadoo [2]. Normally, Doppel is mostly expressed in testes, but its ectopic expression can cause neurodegeneration in the central nervous system (CNS). Shadoo is expressed in the CNS. Both Doppel and Shadoo share neuroprotective properties [2].

The infectious form of PrP, PrP^{Sc}, is able to convert PrP^C proteins into an infectious form by post-translational modifications that make them less soluble and more resistant to degradation catalyzed by proteinase-K. The modified, pathogenic prions also lose their N-linked glycans and GPI anchor to the cell membrane. Aggregations of these abnormal forms can make amyloid fibers. Similar pathogenic amyloids, amyloid beta, or Aβ, form plaques and cause severe brain damage in Alzheimer's disease. The exact amyloid form of Aβ that may help cause Alzheimer's disease remains a matter of debate, but recent evidence suggests that smaller sub-amyloid Aβ oligomers are being more neurotoxic [1].

The next in size are viruses, which are more than just proteins, but still, they don't make their own surface coat or internal components – their hosts make them. Viruses are not cells, nor are they made up of cells. They don't have their own metabolism, even though they do evolve. Viruses use enzymes and structures within their host to make their surface or capsid, and to make copies of themselves. Most biologists don't consider viruses to be alive, but almost all talk about ways to kill them. Though not truly alive, viruses can infect their hosts and cause potentially fatal diseases. It is also quite likely that the human genome contains many remnants of viruses, in the form of retrotransposons. Viruses can also mutate and form new strains that can infect different species. This has happened when bird flu and swine flu infected people. This was alarming, because nobody was immune to the new strain, and the annual flu shot did not protect it, although recent research may develop better vaccines. Fortunately, the most recent strain of the flu virus had relatively low virulence, but it could mutate into a new strain that is more virulent. However, one should remember that even the different strains of influenza virus that have been around for decades kill about 36,000 people in the USA each year.

Even though viruses are not alive, they can be classified the same way as are plants and animals. They can be assigned to an order, family, subfamily, genus and species. In another classification system, viruses are classified by the way that their mRNA is made. The seven classes of viruses are: dsDNA, ssDNA, dsRNA, (+)-sense ssRNA, (-) -sense ssRNA, RNA reverse transcribing and DNA reverse transcribing. Viruses can also be classified by different types, subtypes and strains. For example, the influenza virus is divided into Types A, B and C. The natural or most common hosts for Type A viruses are birds. However, Type A viruses can also infect humans and other animals. Type B viruses are normally only found in humans. Although they have caused epidemics, they have not caused pandemics. Type C viruses cause mild illnesses and do not cause epidemics or pandemics. Types B and C are not further divided into subtypes. Because they can be so deadly, Type A influenza viruses are classified further. They have different proteins on their surfaces. There are 15 different subtypes with different hemagglutanin (HA) proteins and nine different neuraminidase (NA) proteins. So, the H_1N_1 virus has HA protein number 1 and NA protein number 1 on its surface. Especially dangerous virus subtypes can also be classified further into different strains, based on differences in their DNA or RNA, which can be caused by mutations. The recent strain of H_1N_1 swine virus that emerged in Mexico, in April 2009 has "genes" (RNA) from human and avian influenza, as well as swine. Even though most people infected in the USA with this strain of swine flu did not develop serious disease symptoms, it is very troubling, because it killed young, healthy people in the USA, Mexico and other countries.

The relative ability of viruses to cause a disease is called its virulence. Virulence is also a term used to describe the severity of the disease that the virus (or any infectious agent) can cause. Viruses cause HIV, polio, the common cold, influenza (the flu), chickenpox, cold sores and many other diseases. Many people do not use the word flu properly in non-scientific conversations. We use the term "stomach flu" to describe the illness that is caused by bacteria and is accompanied by vomiting and diarrhea, when the correct medical term is gastroenteritis. On the other hand, influenza, or the seasonal flu, is characterized by fever, sore throat, congestion, muscle aches and fatigue. It can kill people directly, or by causing pneumonia. When a disease (such as the flu or AIDS) spreads throughout the world, it is called a pandemic. Flu pandemics occurred in 1918-1919, 1957-1958 and 1968-1969 and 2009 [3]. In "normal flu seasons" in the USA, more than 36,000 people die from influenza each year and 200,000 more are hospitalized [3]. Worldwide, annual epidemics kill 250,000 to 500,000 each year [4]. Serious diseases such as Ebola, AIDS, bird flu and severe acute respiratory syndrome (SARS) are also caused by viruses. AIDS is caused by the human immunodeficiency virus, or HIV. SARS was caused by a coronavirus (SARS-CoV) in an epidemic in 2002-2003. In 2012, another human coronavirus (HCoV-EMC) was identified in autopsy specimens from a man in Jeddah, Saudia Arabia. There have been more cases since then and this virus may be even more pathogenic than SARS-CoV [5]. Another virus, called rotavirus, is the leading cause of infant mortality in the world. It kills by causing severe diarrhea and dehydration. In the USA, this is not such a serious threat, since hospitals are readily available in most communities, so an infant can be easily re-hydrated intravenously.

Some viruses can also cause cancer. The type 2 strain of Herpes simplex virus (HSV-2) causes cervical cancer and a vaccine is now available for preventing the disease. Viruses also cause hepatitis, polio, measles, mumps and even the common cold (caused by rhinoviruses). There are 99 serotypes of human rhinoviruses, which cause the common cold. This is more than just a nuisance.

When infants get the cold, it can lead to asthma. So, prions and viruses can be infectious, even though they are not life forms. The smallest organisms that can cause infectious diseases are bacteria, which are prokaryotes.

Prokaryotes do not have a cell nucleus or any other internal membrane-bound organelles. Most prokaryotes are unicellular, but a few are multicellular, and even the unicellular bacteria can combine to form a protective slime layer, or plaque. By number, 90% of the cells in the human body are bacteria. Bacteria live in symbiotic relationships with fungi, plants, animals and other bacteria. In the modern form of the theory of evolution, life is said to have evolved by cooperation, not competition. Species evolved in cooperation with each other, not by conquest and destruction. Humanity is absolutely dependent on our symbiotic bacteria. All life is dependent on bacteria. Even though healthy human fetuses are sterile in utero, they are exposed to bacteria from their mothers and the environment when they are born. So, we started our lives isolated from our mother's endogenous bacteria that helped her digest her food and extract nutrients. She also existed in an ecosystem permeated with bacteria. Bacteria in the atmosphere help form clouds that bring us rain and snow. Cyanobacteria are a type of phytoplankton, which are part of the primary foundation of the ocean's food chain, converting CO_2 into sugars and other organic compounds through photosynthesis.

The US government is funding many research projects concerning bacteria. The genomes of over 1000 bacteria have been solved. Unfortunately, many of the bacteria that are part of the ecosystem that comprises the human body can't be grown in laboratory culture, so not very much is known about many of them. To help us learn more, the NIH has funded the human microbiome project, "with the mission of generating resources enabling comprehensive characterization of the human microbiota and analysis of its role in human health and disease" [6]. Reductionist thinking and limited technology have forced most previous research to focus on one species or strain of bacteria at a time, grown in an artificial two-dimensional culture. Most bacteria in the human ecosystem depend on their microenvironment for survival, so systems thinking and improved technologies are required to study them. Advances in DNA sequencing technology has enabled researchers to study entire microbial communities, including species that can't be cultivated (cultured) individually. The field of metagenomics allows the determination of genomes of complete microbial communities harvested from natural environments, including the open ocean and human tissues [6].

Another project to study bacterial genomes is being conducted by J. Craig Venter. During a two-year cruise from Halifax, Nova Scotia, to Florida he took 200-liter samples of seawater every 200 miles and captured bacteria in filters [6]. He discovered about six million genes, more than doubling all the genes known previously to science [6]. One of his goals was to learn how to produce H_2 gas from photosynthetic organisms. He and his coworkers are trying to put genes from one or more types of bacteria into another, to produce completely new organisms that will be able to produce H_2 gas and many fine chemicals and medicines, which are currently made from much fossil fuel. In the process, he has discovered picoplankton, previously unknown forms of bacteria that are smaller than 1 μm, which was once thought to be the size of the smallest bacteria. It had been thought that most of the world's ocean was devoid of life and only had isolated oases packed with phytoplankton and larger organisms. We now know that pico-planktons are distributed throughout the upper layers of the world's oceans, but at relatively low concentration. Still, the world's oceans are huge, so this discovery added significantly to the estimated total bacterial biomass, consumption of CO_2 and production of O_2.

Even though our lives depend on symbiotic bacteria, it is important that they stay in their proper places in our human bodies, or ecosystems [6]. For example, when the bacteria in a patient's mouth are aspirated into his or her lungs, it can cause pneumonia. If the skin bacteria, *S. epidermidis* gets into the bloodstream, it can cause sepsis and endocarditis [6]. If *E. coli* from the gut or *S. aureus* from the skin and mucosal surface find their way to other parts of the body, life-threatening infections can occur [6]. The risk of acquiring these infections is greatest in hospitals, but antibiotic resistant bacteria are being found in people and even Arctic birds that have never been in a hospital and never been given an antibiotic [7]. This occurs because bacteria can transfer their DNA and genes to unrelated bacteria of different species or strains in what we call horizontal transfer. They can transfer genes in plasmids, which are circular loops of DNA that are independent of bacterial chromosomes. They can also transfer genes as transposable groups, or cassettes [6]. "When penicillin was first introduced, it was able to kill *S. aureus*, but soon, some strains of the bacteria developed the ability to make an enzyme called penicillinase, that catalyzed the breakdown of the β-lactam ring of penicillin. In the 1950s, scientists developed semisynthetic penicillins, such as methicillin, that were not broken down by penicillinase. Within two years of the introduction of methicillin, resistant strains of *S. aureus* emerged. These strains had acquired a new gene, called mecA, which coded for a binding protein that penicillin needed to bind to the bacterial cell membrane. The mecA gene was transferred in a gene cassette to *S. aureus* by a skin bacterium found in wild and domestic animals, called *S. sciuri*" [6].

Through the 1960s, 1970s and mid-1980s, methicillin-resistant *S. Aureus*, or MRSA, was just a minor nuisance. Only 1 – 5% of all *S. aureus* isolates were methicillin resistant [6]. "By the mid-1980s, though, strains of *S. aureus* emerged that were resistant to other common antibiotics. Today, 60 – 70% of all *S. aureus* isolates are multi-drug resistant MRSA. In the late 1990s, the problem got worse. In 1999, the Centers for Disease Control (CDC) reported four deaths in Minnesota and North Dakota, which could not be traced to a hospital. Genetic analysis showed that these strains of MRSA did not migrate out of hospitals, but instead acquired their mecA genes in the community outside the hospital. In 2002 isolates of MRSA were found that were fully resistant to vancomycin, which had been considered the treatment of last resort for MRSA. They had acquired a gene called vanA, from enterococci that normally reside in the intestines. The enterococci developed resistance to vancomycin in the 1980s, and were able to transfer it to MRSA. By 2008, about six community-acquired MRSA clones have spread across the world, and 50% of all *S. aureus* isolates outside hospitals in San Francisco are now resistant to methicillin" [6]. In 2010, the FDA approved an advanced cephalosporin, Ceftaroline Fosamil (Teflaro®) to treat MRSA.

Prokaryotes are divided into two domains, the bacteria and archaea. Bacteria are found throughout the biosphere, from the clouds in the sky, to the water and soil, the human gut and in seemingly hostile environments, such as oil and sludge. Archaea are found in even more extreme environments, at the bottom of the ocean in hydrothermal vents. Some are also found in humans.

Bacteria can be distinguished by their ability to be stained dark blue or violet by Gram staining with the dye called crystal violet [6]. "The stain is able to reach the membranes of both Gram positive cells and Gram negative cells, which have an additional protective layer, or outer membrane, made of lipopolysaccharides. Gram negative bacteria are harder to kill, because of this extra layer of protection. The gram stain contains a dye called crystal violet, which is the chloride (Cl^-) salt of a cation that is abbreviated CV^+. Both the CV^+ and Cl^- penetrate the cell wall and membrane and bind to negatively charged iodide anions within the inner and outer layers of the cell wall and form a complex, sometimes abbreviated as C-I. Then a de-colorizer (such as ethanol) is added. It interacts with the lipids of the cell membrane. A Gram-negative cell will lose its outer membrane and the peptidoglycan layer is left exposed. The CV-I complexes are washed from the Gram-negative cell along with the outer membrane. In contrast, a Gram-positive cell becomes dehydrated from an ethanol treatment. The large CV-I complexes become trapped within the Gram-positive cell due to the multilayered nature of its peptidoglycan" [6].

The most common bacterial disease is tuberculosis, caused by the Gram positive bacterium *Mycobacterium tuberculosis*, which kills about 2 million people a year [8]. Tuberculosis is emerging as an important disease, since some strains have become resistant to antibiotics. In Dec, 2012, the FDA approved a new drug, bedaquiline for treating multi-drug resistant tuberculosis [9]. It affects the H^+ pump for the bacterial ATP synthase that is needed to produce energy for reproduction. Pathogenic bacteria contribute to other globally important diseases, such as pneumonia, which can be caused by bacteria such as *Streptococcus* and *Pseudomonas*, and foodborne illnesses, which can be caused by bacteria such as *Shigella*, *Campylobacter* and *Salmonella*. *Salmonella* is named after a veterinary surgeon named Daniel E. Salmon. It has nothing to do with salmon (the fish). A rare serotype of *Salmonella* caused salmonellosis and food poisoning in 2008, due to contamination of fresh jalapeño peppers. In Jan., 2009, *Salmonella* was found in peanut butter-based products such as crackers, energy bars, and peanut butter cookies, as well as in pet food [9]. Salmonellosis (*Salmonella* infection) is usually caused by ingesting raw or undercooked eggs. It can also be caused by poultry, pork and cattle, pet reptiles (such as turtles) and pet rodents. "*Pseudomonas aeruginosa* is increasingly recognized as an emerging opportunistic pathogen of clinical relevance" [9]. Interestingly, an article published in the journal *Science* in 2008 showed that *Pseudomonas* may be the most common nucleator of ice crystals in clouds, thereby being of utmost importance to the formation of snow and rain around the world [10].

The *Enterobacteriaceae* have developed resistance to broad-spectrum antibiotics [11]. "Carbapenem antibiotics have been used to kill these organisms and, until recently, resistance to carbapenems has been uncommon among *Entero-bacteriaceae* in the United States. However, new β-lactamases have emerged. "They catalyze the hydrolysis of carbapenem and have contributed to an increased prevalence of carbapenem resistant *Enterobacteriaceae* (CRE). CREs are very dangerous since they often cause infections, which are often fatal. Potentially, they can become widespread through mobile genetic elements. Even though CREs occur mostly in health care settings, but they may eventually spread further into the community. These issues, combined with the limited therapeutic options available to treat patients infected with these organisms, have made CRE of epidemiologic importance nationally" [11].

So, medical science is especially interested in pathogenic bacteria and their control. So, we are very concerned that antibiotic resistance is increasing in clinical isolates of bacteria [11]. In addition to tuberculosis and pneumonia, pathogenic bacteria also cause infections such as tetanus, typhoid fever, diptheria, syphilis, and leprosy [11].

About six new antibiotics that work against Gram positive bacteria have either been approved or are in later phases of clinical investigation [6]. "There are no such new antibiotics against Gram negative bacteria, "such as *P. aeruginosa*, *A. baumannii* and *K. pneumonia*. One of the demons that come with war is pestilence. *A. baumannii* infects many veterans of the war in Iraq, and there are few antibiotics that are effective against it. Many Gram negative bacteria have become resistant to antibiotics. To make the problem worse, few pharmaceutical companies are trying to develop new antibiotics, which aren't very profitable. Also, the better an antibiotic works, the less health experts are apt to approve its use, so that bacteria don't develop a resistance to it. They often feel that such a good new antibiotic should be saved in case we need it in the future" [6].

There is also a research project to determine the bacterial resistome, or the set of genes in soil bacteria that confer antibiotic resistance to them [12]. Soil bacteria such as *Actinomyces* produce their own natural antibiotics and resistance to antibiotics. By learning about the parts of their DNA that confer antibiotic resistance to them, scientists hope to learn more about the process and hopefully find new molecular targets for new antibiotics [12].

Still, necessity remains the mother of invention. Researchers recently reported a whole new approach to killing pathogenic bacteria. They are targeting bacterial virulence and not bacterial growth [13]. Virulence is the relative ability of a pathogenic virus or organism to cause a disease. Bacteria have specific genes that make them virulent. Many pathogenic bacteria recognize host cells by recognizing a membrane-bound kinase, called QseC, which has a histidine residue in its active site [13]. When QseC senses either adrenaline, noradrenaline or a bacterial signal, it catalyzes the attachment of a phosphate to itself (autophosphorylation). This initiates a cascade of events that leads to the transcription of several bacterial virulence factors [13]. These scientists found a molecule that they called LED209 that inhibits the binding of signals to QseC and inhibits bacterial virulence. This offers a strategy for developing a whole new class of wide spectrum antibiotics [13].

Still, multi-drug resistant bacteria remain a problem. Not only MRSA, but also drug-resistant *Neisseria gonorrhoeae*, *Mycobacterium tuberculosis*, extended-spectrum β-lactamase (ESBL) or pan-resistant *E. coli*, *Klebsiella pneumonia*, *Pseudomonades* and *Acinetobacter* species, and the NDM-1 Superbug have emerged. A novel approach to making better antibiotics that could be effective in treating these is to use antisense inhibition of key bacterial genes [14]. "This includes antisense oligodeoxynucleotides (ODNs) that are based on modified structures, such as peptide nucleic acids (PNAs) and phosphorodiamidate morpholino oligomers (PMOs). Antisense oligonucleotides are designed to bind to target mRNA to prevent translation or bind to DNA to prevent gene transcription. They have demonstrated potency in the expression of genes in a sequence-specific and dose dependent manner at low concentrations. Also, delivery systems based on cell penetrating peptides has enabled the effective display of intracellular antisense molecules and the inhibition of targeted genes both *in vitro* and *in vivo*" [14].

Prokaryotes are not the only organisms that can cause disease. Fungi, protozoa and multicellular organisms are eukaryotes and some of them are pathogenic. They have a cell nucleus which is bound by a nuclear membrane. Eukaryotic cells have other internal organelles that have their own membranes.

Fungi are heterotrophic eukaryotic organisms that have a cell wall made out of chitin, a carbohydrate polymer that is external to the cell membrane [14]. "Most species grow as multicellular filaments. Yeasts, molds and mushrooms are examples of fungi. Many fungi produce toxins that are called mycotoxins. The FDA analyzes foods for mycotoxins, such as aflatoxins, which are highly carcinogenic, causing liver cancer. Mycotoxins can occur in spoiled grains. "Other notable mycotoxins include ochratoxins, patulin, ergot alkaloids, trichotecenes and fumonisins. Fungi with the potential to cause serious human diseases are in the genera *Aspergillus, Candida, Cryptoccocus, Histoplasma* and *Pneumocystis*. Several pathogenic fungi are also responsible for relatively minor human diseases, such as athlete's foot and ringworm" [14].

Protozoa are unicellular eukaryotes [14]. "Some of them are parasites, which can exist only in or on a host, including animals and humans. For example, the genus *Plasmodium* causes malaria, the infectious disease that kills the most people in the world. The World Health Organization (WHO) estimated that there were 247 million cases of malaria in 2006" [14]. There are nine species of *Plasmodium*, four of which cause malaria in humans. Human and mosquito hosts are both essential for survival of the parasite.

Malaria causes millions of deaths each year [15]. The most infectious species of *Plasmodium* is *P. falciparum*. Its life cycle starts with an asexual phase and progresses to later phases that are sexual. The sporozite forms enter human hosts through the mosquito's saliva, when the mosquito penetrates the skin and withdraws blood. Sporozites reach the liver within 30 minutes and infect liver cells. This starts the hepatic stage, which lasts 6 – 15 days. During this time, thousands of daughter parasites, called merozoites, are formed. The infected liver cells burst and release the merozoites, which infect red blood cells, called erythrocytes. The erythrocytes burst, releasing more merozoites, which cause the biosynthesis of cytokines, which cause fever and chills, the usual symptoms of malaria. Some merozoites can enter a sexual phase, in which gametocytes are formed. The gametocytes can be transormed only in the gut of mosquitoes. After several more stages, they are ready to infect more humans. There are drugs that act on these stages. Atovaquone acts on the hepatic stage. Chloroquine acts on the erythrocyte stage. It keeps a heme-iron complex in solution, so it can kill the parasites. Chloroquine may also interfere with other enzymes and may even bind to the parasite's DNA. Recently, it was shown that chloroquine might also be effective against Nipah and Hendra, two viruses that recently jumped from animals to humans.

Artemisinin also acts on the erythrocytic stage of malaria. Its mode of action has not been proved, but is thought to form reactive oxygen substances (ROS), which can kill the parasite.

In a recently reported clinical study, a new anti-malaria vaccine was proven to be safe and effective [16, 17]. The vaccine, is called RTS,S. It prevents the malaria parasite from infecting and surviving in liver cells [16]. It contains a fragment of the circumsporozite protein that is on the outer surface of the parasite's cell membrane [16, 17]. The protein is fused with a hepatitis B protein that helps trigger a more active immune response [16, 17]. The vaccine includes an adjuvant called AS01, which helps stimulate the immune system. AS01 is a liposome that contains immunostimulants called monophosphoryl lipid A (MPL) and a saponin-derived molecule called QS21. The RTS,S vaccine started Phase III testing in Tanzania in 2009. On May 5, 2009, five infants aged 5 – 17 months were immunized and 16,000 more will be immunized at 11 sites in 7 African countries as the study continues.

Pathogenic eukaryotes include the protozoans that are called trypanosomes. This includes *Tyrpanosoma*, which causes trypanosomiasis (African sleeping sickness and South American Chagas disease) [8]. "A drug called ivermectin (Stromectol®)

is the most effective broad spectrum antiparasitic agent. It acts by opening glutamate-gated Cl^- channels in invertebrates. It is used to prevent sleeping sickness, also known as onchocerciasis. Another parasitic protozoan is *Leishmania*, which causes leishmaniasis. It is spread by the bite of the female sandfly. The two main forms of the disease are infection of internal organs and infection of the skin. Infection of internal organs is usually fatal, and affects about 500,000 people each year. The drug miltefosine (Impavido®) is effective against several stages in the life cycle of *Leishmania* parasites. Miltefosine is thought to inhibit phospholipid metabolism and may interfere with membrane signal transduction in the parasite. Two other important protozoan parasites are *Cryptosporidium parvum*, which causes diarrhea and *Giardia intestinalis*, which causes intestinal infections. The drug Nitazoxanide (Alinia®) treats both of these. It acts by inhibiting the parasitic enzyme pyruvate-ferredoxin oxidoreductase, which plays an important role in energy metabolism. After absorption, nitazoxanide is rapidly converted to its active form by hydrolysis of an acetyl group" [8].

Multicellular parasites include four species of *Schistosoma*, a flatworm which causes schistosomiasis, which is second in importance only to malaria, with hundreds of millions infected worldwide [8]. Schistosomiasis is treated with the drug called Praziquantel (Biltricide). Its mode of action is thought to involve increasing the permeability of the *Schistosoma* cell membrane to Ca^{2+} ions.

There are other species of helminths, or parasitic worms. Those that live in the intestines are called intestinal parasites. All worm species are passed onto humans by poorly cooked meat, contaminated water, feces and mosquitoes, especially in areas of the world that don't have an adequate supply of fresh water. In addition to schistosomiasis, helminths can cause ascariasis, dracunculiasis, elephantiasis, hookworm, lymphatic filiaruasis, onchocersiasis, and trichuriasis. Ascariasis can be treated by several drugs, including mebendazole (Vermox®), which blocks the uptake of glucose in the worm. Piperazine is another anti-helminthic agent, and it is an acetylcholine antagonist. Albendazole is a broad-spectrum antihelminthic agent that decreases ATP production in the worm [8]. Elephantitis is treated with albendazole, in combination with either ivermectin or diethylcarbamazine. It can also be treated with the antibiotic doxycycline, which kills symbiotic bacteria in the worm that causes elephantitis. Albendazole binds to tubulin and prevents it from forming microtubulin in the cells in the worm that causes elephantitis. Ivermectin and the related drug avermectin are macrocyclic lactones that bind and activate glutamate chloride channels. Hookworm infection can be treated with local cryotherapy when it is still under the skin, or by albendazole. Albendazole plus ivermectin is also used to treat lymphatic filariasis. Onchocerciasis can be treated with ivermectin. Onchocerciasis can be treated with mebendazole, by blocking the uptake of glucose by pinworms, roundworms, tapeworms, hookworms and whipworms.

KEY POINTS

1. **Infectious diseases can be caused by worms, protozoa, fungi, bacteria, viruses and even proteins (prions). The prototypical prion causes Mad Cow disease, also known as bovine spongiform encephopathy (SGE).**

2. **Organisms, viruses and prions can be classified by their infectivity, or their ability to enter, survive and multiply in a host. There is also a natural, healthy prion, called PrP^{C}, which is found in cell membranes. The infectious form of PrP called PrP^{Sc}, is able to convert PrP^{C} proteins into an infectious form by post-translational modification that makes them less soluble and more resistant to degradation.**

3. **The seven classes of viruses are: dsDNA, ssDNA, dsRNA, (+)-sense ssRNA, (---sense ssRNA, RNA reverse transcribing and DNA reverse transcribing. Because they can be so deadly, Type A influenza viruses are classified further. There are 15 different subtypes with different hemagglutanin (HA) proteins and nine different neuraminidase (NA) proteins. So, the H_1N_1 virus has HA protein number 1 and NA protein number 1 on its surface. The relative ability of viruses to cause a disease is called its virulence. Some viruses can also cause cancer. The type 2 strain of Herpes simplex virus (HSV-2) causes cervical cancer and a vaccine is now available for preventing the disease.**

4. **The smallest organisms that can cause infectious diseases are bacteria, which are prokaryotes. By number, 90% of the cells in the human body are bacteria. All life is dependent on bacteria. Most bacteria in the human ecosystem depend on their microenvironment for survival. Even though our lives depend on symbiotic**

bacteria, it is important that they stay in their proper places in our human bodies, or ecosystems. By the mid-1980s strains of *S. aureus* emerged that were resistant to common antibiotics. Today, 60 – 70% of all *S. aureus* isolates are multi-drug resistant MRSA.

5. **Fungi, protozoa and multicellular organisms are eukaryotes and some of them are pathogenic. Many fungi producc toxins that are called mycotoxins. Protozoa are unicellular eukaryotes. The genus *Plasmodium* causes malaria, the infectious disease that kills the most people in the world. Pathogenic eukaryotes include the protozoans that are called trypanosomes. Multicellular parasites include four species of *Schistosoma*, a flatworm which causes schistosomiasis, which is second in importance only to malaria, with hundreds of millions infected worldwide. In addition to schistosomiasis, helminths can cause ascariasis, dracunculiasis, elephantiasis, hookworm, lymphatic filiaruasis, onchocersiasis, and trichuriasis.**

ABBREVIATIONS

PrP	=	Prion Protein
SGE	=	Spongiform Encephalopathy
GPI	=	Glycophosphatidyl Inositol
Aβ	=	Amyloid Beta
dsDNA	=	Double Stranded DNA
ssDNA	=	Single Strand DNA
dsRNA	=	Double Strand RNA
(+)-sense ssRNA	=	Positive Sense Single Strand RNA
(-)-sense ssRNA	=	Negative Sense Single Strand RNA
HA	=	Hemagglutanin
NA	=	Neuraminidase
SARs	=	Severe Acute Respiratory Syndrome
MRSA	=	Methicillin-Resistant S. Aureus
CDC	=	Centers for Disease Control
FDA	=	Food and Drug Administration
CV	=	Crystal Violet
CRE	=	Carbapenem Resistant Enterobacteriaceae

ESBL	=	Extended-Spectrum B-Lactamase
ODNs	=	Oligodeoxynucleotides
PNAs	=	Peptide Nucleic Acids
PMOs	=	Phosphorodiamidate Morpholino Oligomers
WHO	=	World Health Organization
ROS	=	Reactive Oxygen Substances
MPL	=	Monophosphoryl Lipid A

REFERENCES

[1] Caughey B, Baron GS, Chesebro B, Jeffrey M. Getting a grip on prions: oligomers, amyloids, and pathological membrane interactions. Annu Rev Biochem 2009; 78: 177-204. [http://dx.doi.org/10.1146/annurev.biochem.78.082907.145410] [PMID: 19231987]

[2] Passet B, Halliez S, Béringue V, Laude H, Vilotte JL. The prion protein family: looking outside the central nervous system. Prion 2013; 7(2): 127-30. [http://dx.doi.org/10.4161/pri.22851] [PMID: 23154632]

[3] Doshi P. Trends in recorded influenza mortality: United States, 1900-2004. Am J Public Health 2008; 98(5): 939-45. [http://dx.doi.org/10.2105/AJPH.2007.119933] [PMID: 18381993]

[4] National Institute of Allergy and Infectious Diseases. Influenza Research at NIAID, website: www3.niaid.nih.gov/topics/Flu/. , [Accessed March 20, 2014];

[5] Human Microbiome Project. website: nihroadmap.nih.gov/hmp/. , [Accessed Feb 20, 2014];

[6] Taubes G. The bacteria fight back. Science 2008; 321(5887): 356-61. [http://dx.doi.org/10.1126/science.321.5887.356] [PMID: 18635788]

[7] Sjölund M, Bonnedahl J, Hernandez J, *et al.* Dissemination of multidrug-resistant bacteria into the Arctic. Emerg Infect Dis 2008; 14(1): 70-2. [http://dx.doi.org/10.3201/eid1401.070704] [PMID: 18258081]

[8] Corey EJ, Czako B, Kürti L. Molecules and Medicine. New York: John Wiley & Sons 2007.

[9] Jones D. Tuberculosis success. Nat Rev Drug Discov 2013; 12(3): 175-6. [http://dx.doi.org/10.1038/nrd3957] [PMID: 23449293]

[10] Christner BC, Morris CE, Foreman CM, Cai R, Sands DC. Ubiquity of biological ice nucleators in snowfall. Science 2008; 319(5867): 1214. [http://dx.doi.org/10.1126/science.1149757] [PMID: 18309078]

[11] Gupta N, Limbago BM, Patel JB, Kallen AJ. Carbapenem-resistant *Enterobacteriaceae*: epidemiology and prevention. Clin Infect Dis 2011; 53(1): 60-7. [http://dx.doi.org/10.1093/cid/cir202] [PMID: 21653305]

[12] D'Costa VM, McGrann KM, Hughes DW, Wright GD. Sampling the antibiotic resistome. Science 2006; 311(5759): 374-7. [http://dx.doi.org/10.1126/science.1120800] [PMID: 16424339]

[13] Rasko DA, Moreira CG, Li R, *et al.* Targeting QseC signaling and virulence for antibiotic development. Science 2008; 321(5892): 1078-80.
[http://dx.doi.org/10.1126/science.1160354] [PMID: 18719281]

[14] Bai H, Xue X, Hou Z, Zhou Y, Meng J, Luo X. Antisense antibiotics: a brief review of novel target discovery and delivery. Curr Drug Discov Technol 2010; 7(2): 76-85.
[http://dx.doi.org/10.2174/157016310793180594] [PMID: 20836761]

[15] Mackinnon MJ, Marsh K. The selection landscape of malaria parasites. Science 2010; 328(5980): 866-71.
[http://dx.doi.org/10.1126/science.1185410] [PMID: 20466925]

[16] Enserink M. Malaria. Vaccine comes another step closer. Science 2008; 322(5908): 1622-3.
[http://dx.doi.org/10.1126/science.322.5908.1622b] [PMID: 19074320]

[17] Bejon P, Lusingu J, Olotu A, *et al.* Efficacy of RTS,S/AS01E vaccine against malaria in children 5 to 17 months of age. N Engl J Med 2008; 359(24): 2521-32.
[http://dx.doi.org/10.1056/NEJMoa0807381] [PMID: 19064627]

CHAPTER 10

Vaccines

Abstract: An extremely important part of disease prevention is vaccination, which improves the immune response to a particular disease. Vaccines save lives and prevent deadly diseases that used to take millions of lives, especially the lives of infants. Vaccines can be made from dead or inactive organisms or viruses. Vaccines can also contain "live" attenuated viruses. The tetanus and diphtheria vaccines contain inactivated toxic compounds. Children should be vaccinated against hepatitis B, hepatitis A, diphtheria, tetanus, pertussis, polio, *Pneumococcus*, measles, mumps, rubella, rotavirus, human papilloma virus (HPV), Meningococcus, *Orthomyxoviridae* (flu virus) and varicella (chicken pox).

Keywords: (HPV), Chicken pox, Diphtheria, Flu virus, Hepatitis A, Hepatitis B, Human papilloma virus, Measles, Meningococcus, Mumps, Orthomyxoviridae, Pertussis, Pneumococcus, Polio, Rotavirus, Rubella, Tetanus, Varicella, Vaccination.

INTRODUCTION

An extremely important part of disease prevention is vaccination, which improves the immune response to a particular disease. Vaccines save lives and prevent deadly diseases that used to take millions of lives, especially the lives of infants. They have led to the eradication of smallpox, the near eradication of polio and have prevented billions of deaths worldwide. The idea of vaccination (or variolation) originated in Chinese traditional medicine in which a person was deliberately infected with cowpox. It is consistent with the idea in traditional Chinese medicine that a very low dose of a toxin can help prevent an illness. Since vaccination with weakened smallpox virus is much safer, variolation was banned in England in 1849, but it was used longer in China and it caused an outbreak of smallpox in the 1960s. Edward Jenner first popularized the use of cowpox to immunize humans against smallpox. The word vaccine comes from the Latin word for cow, vacca. In France, Louis Pasteur and others continued the

Robert E. Smith

work. The earliest vaccines were made from partially attenuated virus (smallpox and rabies) or inactivated bacteria (pertussis). Since the 1930s, many vaccines have been prepared by injecting viruses into fertilized chicken eggs. The eggs are kept warm as the viruses multiply for a few days. The eggs are cracked open and the viruses are inactivated using chemicals that chop them into pieces.

In time, more advanced methods were used. This includes chemical treatment of a protein toxin to form a toxoid (such as tetanus, diphtheria), development of a purified and inactivated virus (hepatitis A), development of virus-like particles (such as hepatitis B, human papillomavirus) and the use of purified polysaccharides (pneumococcal vaccines) [1]. "Cell cultures have been used ever since they were first used to develop the polio vaccine. Salk's inactivated polio vaccine was grown in primary monkey kidney cells. There are now Vero cells derived from African green monkey kidneys, Madin–Darby Canine Kidney (MDCK) cells, and PBS-1 cells (HepaLife Technologies, Boston, MA) that are grown either adherently in roller bottles, NUNC cell factories (Thermo Scientific, Roskilde, Denmark), or on microcarriers, or in suspension cultures in stirred-tank bioreactors or disposable wave bioreactors. There are also proprietary human cell lines such as PER.C6® (Johnson & Johnson, Leiden, the Netherlands), AGE1.CR® (ProBiogen, Berlin, Germany), and EB14® (Vivalis, Nantes, France) which are usually grown in suspension cultures in serum-free media. The Vero cell line is currently the most widely accepted by regulatory authorities for vaccine development, and it has been used for over 30 years. The smallpox vaccine ACAM20001 (Sanofi Pasteur, Lyon, France), the pediatric rotavirus vaccines Rotateq1 (Merck, Whitehouse Station, NJ), Rotarix1 (GSK, Brentford, Middlesex, UK), the H5N1 pandemic influenza vaccine Preflucel1 (Baxter, Deerfield, IL) and the Japanese encephalitis vaccine Ixiaro1 (Intercell, Vienna, Austria) are all produced from Vero cell lines grown in containers as large as 7000 L" [1]. By the end of the 20th century, most of the vaccines that could be developed by mimicking natural infection with live attenuated or inactivated viruses or bacteria had been developed [2]. "However, there are still diseases that cause much morbidity and mortality for which there are no immunizations. This includes infections by bacteria in the genuses *Campylobacter and Chlamydia*, the bacteria *Helicobacter pylori* and *Streptococcus* groups A and B. It also includes the following diseases: dengue fever, leishmaniasis, malaria, schistosomiasis, shigella, tuberculosis and urinary tract infections. It also includes diseases caused by the following viruses: cytomegalovirus, HIV, Epstein-Barr virus (causes mononucleosis), hepatitis C, herpes simplex, and respiratory syncytial virus, rhinovirus. Moreover, a universal flu/influenza vaccine is needed to replace the annual vaccine. Finally, vaccines are being developed to prevent or cure allergies, autoimmune diseases and many forms of cancer" [2].

VIRAL VECTORS

To immunize against diseases caused by viruses, viral vectors are used instead of whole pathogenic viruses [1]. "Adenovirus vectors deliver DNA to target cells, can incorporate cDNA expression cassettes and have a low oncogenic potential because they do not insert their genome into the host DNA. Also, recombinant adenoviruses replicate well in the specific host cell, but do not replicate after being injected into the patient. Viral vaccines can be purified by using polyethyleneglycol (PEG) to precipitate them. Also, conjugated vaccines are made by combining an antigen with a carrier protein that increases the immunogenicity of the antigen. For example, three vaccines against encapsulated bacterial pathogens, *Neisseria meningitides*, *Streptococcus pneumoniae*, and *Haemophilus influenzae* type b (Hib), were significantly improved by covalently attaching (conjugating) the polysaccharides (PS) in the original vaccines to carrier proteins. The five main carrier proteins used in vaccines today are: tetanus toxoid (TT), diphtheria toxoid (DT), crossreactive material 197 (CRM197), N. meningitides outer membrane protein (OMP), and non-typeable *H. influenza* derived protein D (PD)" [1].

There are also virus-like particles that are used to make vaccines [1]. "They are structural proteins that do not contain a genome and are incapable of a spreading infection. Virus-like particles (VLPs) have a repetitive antigenic structure that is capable of efficiently stimulating both cellular and humoral immune responses. Recombivax HB1 for prevention of hepatitis B infection was the first recombinant protein vaccine for human use. It is made by inserting the gene for a hepatitis B surface antigen (HBsAg) into genetically engineered strains of *S. cerevisiae*. Two VLP-based vaccines for HPV have since been approved by the FDA. They are

Gardasil® (Merck) and Cervarix® (GSK), both used for the prevention of human papilloma virus (HPV) and cervical cancer" [1]. They target the L1 major capsid protein that self-assembles spontaneously to form capsomers. Purified capsomers are converted to capsid viral-like particles *in vitro* to make the prophylactic HPV vaccine. Gardisil® contains L1 particles from four different types of HPV, and Cervarix® contains two. They protect against the two most prevalent types, HPV16 and HPV18. They causc about 70% of all the cases of cervical cancer, but they cost between $4 and $6 to vaccinate each girl with three injections over six months [3]. "However, they do not protect against 30% of the cases of cervical cancer, they must be kept refrigerated and there is a risk that other HPV types will proliferate in place of HPV 16 and HPV 18. Moreover, they are useless against someone who is already infected. So, next generation HPV vaccines are being developed. So, some researchers are trying to develop a vaccine that targets the L2 protein, which stimulates a weaker response, but is active against more types of HPV. One advantage of this approach is that parts of L2 proteins from 12 different types of HPV can be fused to make one polypeptide that can be grown in the bacterium *E. coli*, which is not possible with Gardasil® or Cervarix®. Using bacteria to produce a vaccine simplifies the manufacturing process and enables it to be done in many local laboratories, significantly decreasing costs. In another approach, researchers are developing a vaccine that will contain whole E7 proteins from the five most common cancer-causing HPV types, as well as poly(I:C) — a double-stranded RNA molecule that signals the presence of a virus. Together, these molecules stimulate a swift build-up of $CD8^+$ immune cells. Because whole proteins are used instead of parts of proteins this vaccine (called Penatrix, is more stable and does not need to be refrigerated" [3].

CANCER VACCINES

That is, cancer can be caused by viruses or bacteria, including *Helicobacter pylori*, hepatitis B and C viruses, Epstein–Barr virus, HIV, herpes virus, and HPV. About 1.9 million cases of cancer in 2002 can be attributed to infections. So, a new paradigm shift in preventing cancer is vaccination against infectious agents. There are six types of cancer vaccines in development: antigen/adjuvant vaccines, DNA vaccines, vector-based vaccines, tumor cell vaccines, dendritic cell vaccines, and anti-idiotype vaccines [1]. "The first FDA-approved therapeutic cancer vaccine was Dendreon's patient-specific prostate cancer vaccine Provenge (Sipuleucel-T). It has antigen-presenting cells (APCs) that present prostatic acid phosphatase (PAP) to T-cells. The PAP is linked to granulocyte-macrophage colony stimulating factor (GM-CSF), a protein that induces a powerful immune response. *In vitro* generation of dendritic cells (DC) loaded with tumor associated antigens (TAAs) has also been tested on glioblastoma multiforme, an aggressive and deadly primary brain tumor. There is also OncoVEXGM-CSF, a therapeutic oncolytic vaccine for melanoma and head and neck cancer that is in Phase III clinical trials. It is a version of herpes simplex virus which has been engineered to replicate selectively in tumors, leaving surrounding healthy tissues unharmed. It is also engineered to express GM-CSF. This enhances the anti-tumor immune response to tumor antigens that are released following the replication of lytic viruses. Still, an estimated 2.5 million people die each year from diseases that can be prevented by available vaccines that are not being fully utilized. To help address this concern, manufacturers are using total quality manufacturing (TQM), Process Analytical Technology (PAT), Quality by Design (QbD), and Real-time Release (RTR) to ensure safety and quality" [1].

Often, the protein given in an immunization has low immunogenicity, so an adjuvant is added to boost immune responses [1]. "Tested adjuvants include synthetic emulsions (oil-in-water), alum (aluminum hydroxide gel or aluminum phosphate), lipid formulations, and cytokine-based adjuvants like type-one interferon, with alum being the most widely used. Adjuvants increase inflammation and activate the compliment cascade. For example, a stable form of Gardasil® (Merck) is prepared by adding nonionic surfactants and adjusting the salt concentration and pH. It is also important to note that many manufacturers of vaccines have no longer use thimerosal as a preservative and, in general, no preservative is used for single dose vials or syringes" [1].

VACCINES FROM DEAD OR INACTIVE ORGANISMS OR VIRUSES

Vaccines can be made from dead or inactive organisms or viruses, which are not free-living organisms. Still, many people refer to killed viruses. Examples of vaccines containing previously killed "organisms" include vaccines against flu (influenza virus, as opposed to the stomach flu), cholera (the bacterium *Vibrio cholerae*), bubonic plague (the bacterium *Yersenia pestis*), and the hepatitis A virus. Vaccines containing "live" attenuated viruses include the rotavirus, yellow fever, measles, rubella and mumps. There are two forms of H_1N_1 vaccines. One is an injection, made from pieces of killed influenza virus. The other is a nasal spray that contains live attenuated virus.

On the other hand, the tetanus and diphtheria vaccines contain inactivated toxic compounds. Hepatitis and human papilloma (HPV) vaccines contain fragments of a virus. The pneumococcal conjugate vaccine (PCV7) is made of the sugars from the cell membranes of seven serotypes of the bacterium *Streptococcus pneumoniae* conjugated with diphtheria proteins. The varicella vaccine prevents the disease commonly known as chicken pox, caused by the varicella-zoster virus. The same virus remains in the body and can cause shingles in adults. So, a shingles vaccine is available. The meningococcal vaccine (MCV4) prevents infection by the bacterium *Neisseria meningitidis*, also known as *meningococcus*, which causes meningitis. It is made from the outer polysaccharide capsule of the bacterium. The *Haemophilus influenza* type B (Hib) vaccine is made from a polysaccaharide from the *H. influenza* bacteria. The polysaccharide is conjugated to a protein, which increases a child's immune system's ability to recognize it and mount an appropriate immune response.

CHILDHOOD IMMUNIZATIONS

Thanks to state laws requiring children to be vaccinated before they start kindergarten, 77% of the children are up to date in their immunizations [4]. Table **1** lists vaccines that are given in the first 12 years of life. It was constructed from information in an article in Time Magazine [4].

Table 1. Vaccines given in the first 12 years of life.

		Age in months								Age in years	
Vaccine	**0**	**1-2**	**4**	**6**	**6-18**	**12-15**	**12-23**	**15-18**	**≥18**	**4-6 yrs**	**11-12 yrs**
Hepatitis B	x	x			x						
DTaP		x	x	x				x		x	
Hib		x	x	x		x					
Polio (IPV)		x	x		x					x	

(Table 1) contd.....

		Age in months								Age in years	
Vaccine	**0**	**1-2**	**4**	**6**	**6-18**	**12-15**	**12-23**	**15-18**	**≥18**	**4-6 yrs**	**11-12 yrs**
PCV7		x	x	x		x					
Rotavirus (RV)		x	x	x							
Influenza*					x	.	.	.	.	.	....
MMR						x				x	
Varicella						x				x	
Hepatitis A							x		x		
Tdap											x
MCV4											x
HPV											x x x

* Vaccinations for influenza are recommended throughout life, once per year, and can be especially important for the elderly.

Abbreviations: DTaP – Diphtheria, tetanus and pertussis (whooping cough);
Hib – *Haemophilus influenza* type B vaccine
PCV7 – Pneumococcal conjugate vaccine; MMR – Measles, mumps and rubella;
Tdap – Diphtheria, tetanus and pertussis, with a reduced dose of the diphtheria and pertussis vaccines, compared to DTaP
MCV4 – Meningococcal vaccine; HPV – Human papilloma virus

Also, cholera, meningococcal meningitis, the plague, typhoid fever, hepatitis B, Japanese encephalitis, rabies, tick-born encephalitis, varicella zoster and yellow fever are being prevented by vaccines [4].

Recently, the issue of required immunizations made it into prime time televised news. A parent who did not immunize his children asked why other parents were so concerned. That is, “If your child is immunized, he (or she) is protected, so why should you care whether my children are immunized? They can’t infect your children”. Unfortunately, this is wrong. People who are not immunized can catch communicable diseases. These diseases can infect infants who are too young to have received their first vaccines. Also, immunocompromised people can become infected. This includes people who are HIV positive, those who are undergoing chemotherapy, radiation therapy or large doses of corticosteroids. It also includes people who have leukemia. Also, there is always a very small population of people whose vaccines did not “take”. They can become infected, too. If everyone were to be immunized, the number of people not protected would be so small that it would be almost impossible for them to infect each other. However, when a sizeable portion of the population is not immunized, they can catch serious diseases and infect the small number of people in whom the immunizations did not "take", or were not effective. It is also likely that unimmunized children will be taken to a doctor when they get sick. When they are taken to a doctor, he or she will probably prescribe antibiotics. Since the parents don't trust the medical establishment, they probably will not give their children the full dose of antibiotics. They will stop when symptoms of the illness go away. As we have already learned, this leads to antibiotic resistant bacteria. So, let us look closer at the different types of vaccines that prevent deadly diseases.

The vaccine for hepatitis B uses one of the proteins (hepatitis B surface antigen, HBsAg) that is in the outer lipid viral envelope. It was originally prepared from patients who were already infected, but the vaccine is now usually prepared by recombinant DNA, in which DNA coding for the viral protein is inserted into bacteria and is expressed there.

The DTaP (or DPT) vaccine immunizes against diphtheria, tetanus, pertussis and polio. The DPT (or DTaP) vaccine comes in two forms, containing either whole cell pertussis (DTwP) or acellular pertussis (DTaP). The acellular version is considered to be safer and contains fewer antigens than the cellular form. Older versions of the DTP vaccine contained a preservative called thiomersol (also known as thimerosal), which contained mercury. The usual protocol for immunization is five doses between ages 2 months and 15 years.

Diphtheria is a contagious disease of the upper respiratory tracts that affected an estimated 100,000 to 200,000 people in the USA in the 1920s and killed 13,000 to 15,000 people, mostly children. It is caused by the aerobic Gram-positive bacterium *Corynebacterium diphtheria*. It causes the gradual deterioration of the myelin sheath in the CNS and PNS. Tetanus is a medical condition that is characterized by the prolonged contraction of skeletal muscles. It is also known as lockjaw, since it caused the jaws to lock, making eating and drinking very difficult or impossible. It is caused by the obligate anaerobic bacterium *Clostridium tetani*. Polio, or poliomyelitis, is an acute viral infection caused by the poliovirus that is usually spread by the fecal-oral route. About 90% of those infected by the poliovirus exhibit no symptoms, but the unfortunate 10% who do can suffer from paralysis and death. The first polio vaccine was developed by Jonas Salk and consisted of three strains of attenuated (dead) poliovirus that had been "killed" with formalin, an aqueous solution of formaldehyde. It was administered by injection. The second polio vaccine was (and is) administered orally, so it is called the OPV, or oral polio vaccine. It is produced by the passage of the virus through non-human cells at a sub-physiological temperature, which produces spontaneous mutations in the viral genome. The use of OPV was discontinued in the USA in 2000, but it is still used in other parts of the world. In 2002, a pentavalent, or five-component vaccine, was developed. It contains acellular pertussis, purified and inactivated diphtheria and tetanus protein toxins and the inactivated (mutated) poliovirus. This is the DTaP vaccine.

The PCV7 vaccine, also called Prevnar, is a Pneumococcal conjugate vaccine that prevents infection by the Gram-positive alpha hemolytic bacterium *Streptococcus pneumonia*. It can cause many diseases, including pneumonia, septicemia and meningitis. In the USA, a heptavalent vaccine has been recommended since 2000 for all children between two and 23 months old and in at risk children between 24-59 months old. It protects against the seven most common forms of pneumonia. Like other conjugate vaccines, it is prepared by forming a covalent bond between a poor antigen and a carrier protein. In 2010 a a more broadly protective version of Prevnar, Prevnar 13 was approved.

The rotavirus is a double-stranded RNA virus in the family of viruses called Rotaviradae. It is the leading cause of diarrhea in infants and young children and the leading cause of infant mortality (due to dehydration). It was not described until 1973. When visualized under an electron microscope, it looks like a wheel (rota in Latin). In 1998 a rotavirus vaccine was approved for use in the USA. Clinical trials found it to be 80-100% effective in preventing severe diarrhea, but it was withdrawn from the market after it was discovered that it may have contributed to an increased risk for a type of bowel obstruction in a small fraction of vaccinated infants. There are seven "species" of rotavirus, called A through G. Rotavirus A causes about 90% of the rotavirus infections and gastroenteritis in humans. In 2006, two new vaccines against rotavirus A infection were shown to be safe and effective in children. Doses were recommended at ages two and four months, and at six months, when needed. The first dose can be given as early as six weeks, but should be given before 14 weeks, six days. RotaTeq is a live, oral pentavalent vaccine against rotavirus, but the FDA had received 28 reports on infants getting intussusception, since it issued a report on Feb 15, 2007 [5]. On April 3, 2008, it was reported that the FDA approved a second rotavirus vaccine, called Rotarix [6]. It is a monovalent (G1P1A) human-attenuated version of the RIX4414 strain of rotavirus and is given in two doses [7, 8].

The influenza (flu) vaccine is designed to protect against the highly variable influenza virus, or *Orthomyxoviridae,* a family of RNA viruses that includes five genera: Influenza viruses A through C, Thogotovirus, and Isavirus. From the Centers for Disease Control, or CDC [9]: “The single best way to protect against the flu is to get vaccinated each year. There are two types of vaccines: The "flu shot"— an inactivated vaccine (containing killed virus) that is given with a needle, usually in the arm. The flu shot is approved for use in people older than 6 months, including healthy people and people with chronic medical conditions. The nasal-spray flu vaccine — a vaccine made with live, weakened flu viruses that do not cause the flu (sometimes called LAIV for “live attenuated influenza vaccine” or FluMist®). LAIV (FluMist®) is approved for use in healthy people 2-49 years of age who are not pregnant. Each vaccine contains three influenza viruses-one A (H_3N_2) virus, one A (H_1N_1) virus, and one B virus. The viruses in the vaccine change each year based on international surveillance and scientists' estimations about which types and strains of viruses will circulate in a given year. Antibodies that provide protection against influenza virus infection develop in the body about two weeks after vaccination” [8]. However, not all A (H_1N_1) viruses are the same. The vaccine does prevent infections from the A (H_1N_1) that has existed for decades, but it is not effective against the A (H_1N_1) that emerged in 2009 and is commonly called the swine flu. On the other hand, the neuraminidase inhibitor, oselamivir (Tamiflu®), is effective against the swine flu, but the Centers for Disease Control says that Tamiflu® may not always be able to treat influenza type A, the most common influenza virus in 2008.

Even though most influenza vaccines are manufactured in chicken eggs, this does take time. Viruses are infected into embryonated hen’s eggs, where they multiply and are secreted into the allantoic fluid. They are harvested, chemically inactivated and washed with a detergent. Either the whole virion is produced or individual viral proteins (hemagluttanin, HA, and neuraminidase, NA) are purified split-product, subvirion, or subunit vaccines. There is a trivalent inactivated influenza vaccines (TIVs) contain about 15 μg each of three HAs, derived from influenza A subtype H_1N_1, H_3N_2 and influenza B. However, production is time consuming has led to shortages of vaccines.

FASTER WAYS TO MAKE VACCINES

In case of an emerging epidemic, it would be better to have a faster way to make vaccines. So, methods have been developed to make vaccines and even pharmaceutical drugs using insect cells that are grown in culture that are infected by baculoviruses [10]. They help control insect populations in nature and when present as pests in agriculture [11]. "The two baculoviruses used the most as vectors to produce recombinant proteins are *Autographa california* multiple nucleopolyhedrovirus (AcMNPV) and to a lesser extent Bombyx mori (Bm) NPV. In contrast to bacterial expression systems, insect cells can appropriately modify proteins after they are translated from mRNA. Moreover, baculoviruses are safe. They do not infect people. They can be grown in media that does not require added serum and cultures can be scaled up. Large quantities of live influenza virus are not produced, so no high bio-containment manufacturing environment is required. So, manufacturing personnel are not exposed to live influenza viruses and it is impossible for a live virus to escape during production" [11].

The first vaccine produced by insect cells infected with baculovirus was the HPV vaccine that protects against cervical cancer and other types of cancer (see Chapter 9). It was followed by Sipuleucel-T for prostate cancer (see Chapter 13) and a new influenza virus (flu) vaccine, called FluBlok [11 - 13]. It uses the AcMNPV to produce three times the HA active ingredient, derived from three different strains of the influenza virus, two of influenza A and one of B [12, 13]. Quality control of the purity, potency, quality and quantity of the final product can be ensured by using mathematical models, along with appropriate analytical methods, including process monitoring and control [14]. The mathematical models predict the effects of environmental factors such as pH, ionic strength and temperature [14]. However, there is still another vaccine (Flucelvax) that is produced in dog kidney cells. There is another in development that will be a tetravalent vaccine that will have HA from two strains of influenza B [15].

CLASSIFYING VIRUSES

Remember that viruses can be classified by their family and genus and how their mRNA is made. The influenza virus, or Orthomyxoviridae, is a (-) sense single stranded RNA virus. Negative-sense viral RNA is complementary to mRNA, so it is converted to (+) sense RNA before being translated into viral proteins. However, according to the ICTVdB Index of Viruses [16], "The genus Influenza virus A is comprised of a cluster of strains that replicate as a continuous lineage and can genetically reassort with each other. Therefore, although 15 different HA subtypes and 9 different NA subtypes are recognized among influenzaviruses A replicating in birds, separate species designations have not been accorded to these subtypes. All isolates are capable of exchanging RNA segments (reassortment)" [16]. That is, each of the different strains of the H_1N_1 (or any other subtype) can exchange their RNA and make slightly different proteins, giving the virus subtypes different virulences and abilities to be transmitted to others. There are eight individual molecules of RNA in an influenza virus. Each of them codes for one or two polypeptides. Because the RNA molecules are separate, they exchange pieces of themselves (recombine) and produce more viruses (progeny) with mixed and different (mutated) proteins.

FLU (INFLUENZA) VACCINES

Unfortunately, the currently used flu vaccine targets hemagglutinin and neuraminidase (H and N proteins) that evolve rapidly. Thus, the vaccine you receive this year may not be effective against next year's flu virus. Still, the vaccine can prevent influenza and save lives for a single season. When given to pregnant women, the flu vaccine was seen to also protect their fetuses and newborn babies [17]. The flu vaccines are available as either an injection of "killed" viruses or as a nasal spray of "live" attenuated virus. The live attenuated viral form of the vaccine is also called LAIV and FluMist. It is not recommended for people under two years of age or older than 50. The flu vaccine is usually grown in fertilized chicken eggs. Vaccines have also been formulated against avian (bird) flu and it is used in many countries to immunize chickens.

More recently, researchers at the Dana Farber Institute have developed an experimental vaccine that may be effective against many forms of the human influenza A virus [18]. It is based on the type of monoclonal antibodies called neutralization antibodies that bind to a conserved region of the various hemagglutinin (HA) proteins (surface antigens) that each type of virus needs to enter human cells and cause the flu [18]. It was effective against all types of group 1 influenza viruses that were tested, including the H5N1 that causes the bird flu and H_1N_1 that caused the Spanish flu [18]. At about the same time, researchers at the Scripps Research Institute and Burnham Institute found some human antibodies that also bind to the conserved region of HA [18]. The antibodies were tested on mice which were infected with relatively high doses of bird flu virus that would normally be fatal. One group of mice was given the antibodies before being infected and another group was given the antibodies three days after being infected. Most of the mice survived, indicating that the vaccine might prevent and cure the bird flu [19]. The next step will be to do more animal studies and if these are successful, clinical trials in humans can begin. In the meantime, clinical trials on the new vaccine against the 2009 version of influenza A, or H_1N_1 swine flu showed it to be immunogenic [20].

MMR VACCINE THE MMR

vaccine protects against measles, mumps and rubella. It is a mixture of three "live" attenuated viruses. It is given to children around the age of one year, with a second dose before starting school [21]. "The second dose is not a booster. It is a dose to produce immunity in the small fraction of children who fail to develop measles immunity after the first dose. Strains of measles and mumps viruses were grown in hen's eggs and chick embryo cell cultures. This produced strains of virus which were adapted to the hen's egg and not human cells. So, these strains are attenuated strains. The Rubella component, Meruvax, was made using a human cell line derived in 1961 from embryonic lung tissue" [21]. "The MMR vaccine is given by subcutaneous injection. As a measure of its success, the incidence of measles once about 900,000 cases per year, but dropped to about 60 cases per year in recent years, but increased markedly in 2013, largely due to parents refusing to have their children immunized" [21, 22].

VARICELLA VACCINE

The varicella vaccine protects against chickenpox [23]. It is a "live" attenuated vaccine. This means that the so-called live (more properly called active and able to cause infection) virus was modified in the laboratory to produce a non-virulent strain that can grow in the body and produce immunity, without causing chickenpox. It was approved by the FDA in 1995 and is recommended for children and other people who are 12 months or older who are susceptible to chickenpox. The vaccine that is sold by Merck, originated as a virus that came from a child who had a natural strain of varicella, called Oka/Merck [24]. Next, it was introduced into human embryonic lung cell cultures, to which they adapted and propagated in embryonic guinea pig cell cultures. Finally it was propagated in human diploid cell cultures called WI-38, followed by MRC-5 at Merck [24]. It is reconstituted into a 0.5 mL subcutaneous dose that contains a minimum of 1350 plaque forming units (or PFUs) [24]. There is also a vaccine called Zostavax® for people over 60 to prevent shingles. This is important because the varicella zoster virus that causes chickenpox can remain dormant in nerve cells for decades, but re-emerge in elderly people and cause shingles.

HEPATITIS A VACCINES

There are four hepatitis A vaccines available that prevent infection by the hepatitis A virus (HAV) that can cause liver disease. HAV is a member of the Picornaviridae family. Even though viruses are not living organisms, HAV is still considered to be a unique species. In countries where hepatitis A is endemic, almost everyone in infected during childhood without showing symptoms of liver disease. This provides immunity and effectively prevents its appearance in adults.

DTAP VACCINE

The Tdap vaccine is similar to the DTaP vaccine, except that the Tdap vaccine is a combined booster for tetanus, diphtheria and pertussis. In 2005, it replaced the next booster for tetanus and diphtheria that had been recommended previously. It is recommended for adults between the ages of 19 – 65.

MCV4 VACCINE

The MCV4 vaccine prevents infection by the bacterium called Neisseria meningitides. There are between 1400 and 2800 cases of meningococcal diseases each year in the USA. In general, meningitis is an inflammation of the meninges, which are the membranes that surround the spinal cord and brain. Bacterial meningitis is caused by N. meningitides. College freshman who live in dormitories and teenagers from 15 – 19 years of age are the most susceptible in the USA. In addition to the MCV4 vaccine, there is a polysaccharide based vaccine that goes by two names, either MPSV4 or PSV4. It is marketed as Meomune® by Aventis Pasteur. The MCV4 vaccine is also marketed by Aventis Pasteur as Menactra®. It contains polysaccharide antigens from three serogroups of N. meningitides (A, C and Y) along with antigens from W-135 strains that are conjugated to the toxic diphtheria protein. This quadrivalent conjugate vaccine was developed to produce a greater response in young children and a longer lasting immunity. The MCV4 vaccine is recommended as the preferred vaccine for people aged 2 through 55 years. The MPSV4 vaccine is for people older than 55.

MENINGITIS VACCINES

Meningitis can also be caused by the mumps virus or pneumococcal infections. Vaccines against them have already been described. Meningitis can also be caused by viruses, physical injuries and cancer. Patients with viral meningitis often improve on their own, but they are usually kept in a hospital for observation. Antibiotics don't affect viruses and are useless against them. Even though we can cure many bacterial infections with antibiotics, there are no cures for any viral infections, just treatments. If viral meningitis is caused by the herpes virus, acyclovir can be used, as it is used to help treat HIV and herpes infections. In HIV positive patients, the immune system is compromised, so meningitis can also be caused by fungi, which are treated with antifungal agents.

HUMAN PAPILLOMA VIRUS VACCINE

The last vaccine on the list is the HPV, or human papilloma virus vaccine. As the name implies, it prevents infection by the human papilloma virus (HPV), which is usually sexually transmitted and is the causative agent of cervical cancer. There are about 40 types of HPV and most of them cause no symptoms and go away on their own [25]. The vaccine is marketed under the name Gardasil. It is a qudrivalent recombinant vaccine that protects against four types (6, 11, 16 and 18) of HPV. It is given as three injections over six months [25]. Some types, though, can cause cervical cancer and less common genital cancers. According to the CDC, about 12,000 women are diagnosed with cervical cancer each year and almost 4000 die each year [16]. The HPV vaccine is recommended for all girls between the ages of 11 and 12, but it can be given to girls as young as 9 or 10. The HPV vaccine is also recommended for girls and women who are between 13 and 26 who have not yet been vaccinated [25]. More research is being done on the safety of the HPV vaccine in women over 26, and the FDA will probably approve it if it is found to be safe and effective. Research is also being done on vaccinating males [25].

More than 26 million doses of Gardasil have been given since it was introduced in June, 2006 [26]. An estimated 8 million girls in the USA have received it. Gardasil was approved after being tested on about 11,000 people [27]. Side effects were rare and mild. They included things like pain or tenderness at the injection site. Among the 8 million girls who have received Garadsil in the USA, there have been 7802 reports of adverse effects, including 15 deaths and 31 reports of Guillame-Barre syndrome, a potentially paralyzing and life threatening disease. The rate of serious effects are about half of what is reported for other vaccines [27].

RISKS TO PHARMACEUTICAL COMPANIES

This illustrates an important point that was first mentioned in Chapter 1. Pharmaceutical companies run big risks when they introduce a new drug or vaccine on the market. They can spend billions of dollars and years of effort to develop the drugs and vaccines. If they are tested on 11,000 people and there are no severe side effects, then the rate is less than 0.01%, or 100 parts per million. So, it is possible that the rate could be 0.001%, or 10 parts per million. If the vaccine is given to over 8 million people, there would be 80 cases of severe side effects. Moreover, it is very difficult to prove cause and effect, in either the legal or scientific sense. Just because a person gets a disease or dies after consuming something, it doesn't mean that that particular thing caused the disease or death. Of course, arguments like that were used by tobacco companies for decades before enough scientific and legal research was able to prove that smoking tobacco can cause cancer. At the same time, not everyone who gets lung cancer is a current or former smoker and many smokers don't die of lung cancer. Many don't live long enough to get lung cancer. They often die from heart disease. Still, there are some smokers who live long lives despite the odds. Such are the problems with epidemiology and statistics. They can tell us with almost certainty about what will happen to most individuals in a large group, but tell us almost nothing about any single individual. Moreover, human biology is complex. Problems and illnesses seldom ever have a single cause or a simple cure.

EFFORT TO DEVELOP NEW VACCINES

Still, work continues in the effort to develop new vaccines. Recent advances in molecular genetics, molecular and cellular immunology, structural biology, bioinformatics, computational biology, nanotechnology, formulation methods, and systems biology are being used [2]. "For example, genomic-based antigen discovery is being used to help design vaccines against multiple bacterial pathogens. Also, analyses of the memory B cell and antibody repertoires from virus-infected subjects have identified broadly neutralizing antibodies against HIV, influenza, and other viruses. This information is being used to design highly conserved vaccines based on epitopes. Advances in adjuvant and vector delivery technologies are improving the potentiation of vaccines. This offers new strategies for improving successful vaccine response rates in neonates and the elderly. However, translating these advances into approved vaccines is impeded by major gaps in our knowledge of human immune responses. Methods are needed that focus immune responses on subdominant protective epitopes, to elicit long-term memory responses, and to drive antibody maturation processes. These gaps are being addressed by these technological advances. This includes developing better ways to analyze immune responses at the single-cell and systems levels [2].

One approach is to sequence the entire genomes of viruses and pathogenic organisms and use bioinformatics to design vaccines to discover new antigens [2].

This was done for the bacterium *Neisseria meningitides*, also known as meningococcus, the causative agent in >50% of the cases of bacterial meningitis. Over 600 potential antigens were analyzed for antigenicity, and 30 of them are surface proteins that may be able to induce bactericidal antibodies. The complete genomes of groups A and B streptococcus, *Streptococcus pneumoniae*, *Staphylococcus aureus*, and *Chlamydia* have been sequenced. Discovering antigens can also help identify antigens that can be used to induce neutralizing antibody responses and generate vaccines based on T cells. For hypervariable pathogens like HIV, protective antigens are being identified by interrogating the antibody repertoire in HIV positive people. Broadly neutralizing antibodies are being identified by techniques that can look at single B cells. The structures of these antibodies bound to antigens are also being determined. Moreover, the structures of these antibodies when bound to carrier proteins that can act as scaffolds or vectors that will be the basis of antibodies are being determined. Broadly neutralizing antibodies against not only HIV, but also HCV and influenza are being used as templates to design antigens for vaccines [2].

DISCOVERING ANTIGENS AND EVALUATING CELL-MEDIATED IMMUNE RESPONSES

There have also been major advances in discovering antigens and evaluating cell-mediated immune responses [2]. "These responses help control many acute and chronic diseases, provide critical helper signals to elicit antibody responses. These are important components of strategies to develop vaccines against some parasites, intracellular bacteria , and viral infections. Recent technological advances include enzyme-linked immunosorbent spot (ELISPOT), intracellular cytokine staining (ICS), mass spectrometry coupled with epitope-identification algorithms, tetrameric staining reagents, genomics and proteomics of more complex pathogens such as mycobacterium and plasmodium, and advances in immune monitoring have combined to enable advances antigen discovery for T cell–based vaccines. Also, computational optimization methods can design mosaic proteins based on fragments of natural viral genetic sequences. They may be able to provide diverse coverage that is comparable to that of thousands of separate peptides. There are also algorithms that identify key epitopes" [2].

Also, signaling pathways and receptors of the innate immune system are being studied [2]. "The increased understanding of the receptors and molecules that drive innate immune responses is leading to the development and testing of new adjuvants for vaccines. Aluminum salts, oil in water emulsions, virosomes and toll-like receptor 4 agonists such as bacterially derived monophosphoryl lipid A (MPL) are some of the adjuvants currently used in licensed vaccines. Also, there are many adjuvants currently in development for increasing $CD4^+$ helper T cell, $CD8^+$ cytotoxic T cell, and humoral immune responses" [2].

On the other hand, pathogens that cause respiratory, gastrointestinal and sexually transmitted diseases initiate infection at a mucosal surface [2]. "So, it may be necessary to induce mucosal immunity to prevent and control them. Vector delivery systems designed to stimulate mucosal and systemic immunity have been developed. Some are already in clinical trials, including adenovirus, para-myxovirus, and some bacterial vectors. Dendritic cells can be targeted with antibodies that are specific for their cell surface receptors. Dendritic cells are antigen presenting cells that provide initial surveillance for exposure to pathogens. After doing this, they mature rapidly and migrate to lymph nodes, where they induce immune responses. So, they play a central role in inducing vaccine-mediated immune responses by providing antigen-specific and costimulatory signals needed to activate T cells" [2].

ADVANCES IN ANALYZING IMMUNE RESPONSES

Advances are also being made in analyzing immune responses at both the single cell and the systems level [2]. "Systems vaccinology is the coupling of systems biology with a better understanding of the pivotal role of innate immunity in increasing adaptive immune responses. This is now being applied to vaccine discovery. DNA microarrays, high-throughput DNA sequencing, mass spectrometry, proteomics, bioinformatics, and computational methods enable data integration that is the basis of systems vaccinology. For example, research on yellow fever vaccination has identified early innate signatures that correlate with the immunogenicity of the vaccines. Similar approaches are now being used to develop influenza and malaria vaccines. Deep sequencing and bioinformatics are used to examine the antibody repertoire. This is being used to develop broadly neutralizing antibodies for an HIV vaccine. In addition, multiplexed flow cytometry and intracellular cytokine staining assays now enable analyses of cell phenotype, subset identity, immune activation, intracellular signaling status and effector activities. At the same time, single cell gene expression analyses allow greater specificity in immune monitoring. Combining these tools with other new methods for analyzing antigen-specific cells with multiplexed, highly sensitive microarrays and sequencing technologies will allow responses in blood and selected tissues to be understood better. This will enable future analyses to be made in the tissues where the battle between pathogen and host largely occurs, rather than simply relying on measurements of peripheral blood. All together, these advances provide the foundation for major advances in vaccine discovery" [2].

KEY POINTS

1. **An extremely important part of disease prevention is vaccination, which improves the immune response to a particular disease. Vaccines save lives and prevent deadly diseases that used to take millions of lives, especially the lives of infants.**

2. **Vaccines can be made from dead or inactive organisms or viruses. Vaccines can also contain "live" attenuated viruses. The tetanus and diphtheria vaccines contain inactivated toxic compounds.**

3. **Children should be vaccinated against hepatitis B, hepatitis A, diphtheria, tetanus, pertussis, polio, Pneumococcus, measles, mumps, rubella, rotavirus, human papilloma virus (HPV), Meningococcus, *Orthomyxoviridae* (flu virus) and varicella (chicken pox).**

ABBREVIATIONS

HPV	=	Human Papilloma Virus
cDNA	=	Complementary DNA
PEG	=	Polyethyleneglycol
Hib	=	Haemophilus Influenzae Type b
PS	=	Polysaccharides
TT	=	Tetanus Toxoid
DT	=	Diphtheria Toxoid

CRM197	=	Crossreactive Material 197
OMP	=	Outer Membrane Protein
PD	=	Non-Typeable H. Influenza Derived Protein D
VLP	=	ProVirus-Like Particles
HBsAg	=	Hepatitis B Surface Antigen
APCs	=	Antigen-Presenting Cells
FDA	=	Food and Drug Administration
PAP	=	Prostatic Acid Phosphatase
GM-CSF	=	Colony Stimulating Factor
DCs	=	Dendritic Cells
TAAs	=	Tumor Associated Antigens
TQM	=	Total Quality Manufacturing
PAT	=	Process Analytical Technology
QbD	=	Quality by Design
RTR	=	Real-Time Release
PCV7	=	Pneumococcal Conjugate Vaccine
MCV4	=	Meningococcal Vaccine
DTaP	=	Diphtheria, Tetanus and Pertussis (whooping cough)
IPV	=	Polio
RV	=	Rotavirus
MMR	=	Mumps, Measles and Rubella
LAIV	=	Live Attenuated Influenza Vaccine
TIVs	=	Trivalent Inactivated Influenza Vaccines
AcMNPV	=	Autographa California Multiple Nucleopolyhedrovirus
PFUs	=	Plaque Forming Units
HAV	=	Hepatitis A Virus
MPL	=	Monophosphoryl Lipid A

REFERENCES

[1] Josefsberg JO, Buckland B. Vaccine process technology. Biotechnol Bioeng 2012; 109(6): 1443-60. [http://dx.doi.org/10.1002/bit.24493] [PMID: 22407777]

[2] Koff WC, Burton DR, Johnson PR, *et al.* Accelerating next-generation vaccine development for global disease prevention. Science 2013; 340(6136): 1232910. [http://dx.doi.org/10.1126/science.1232910] [PMID: 23723240]

[3] Sanderson K. Vaccination: A durable design. Nature 2012; 488(7413): S7.

[http://dx.doi.org/10.1038/488S7a] [PMID: 22932439]

[4] Time. How safe are vaccines. www.time.com/time/health/article/0,8599,1808438-1,00.html. , 2008 May 21; [Accessed June 20, 2012];

[5] Medical News Today. FDA Issues Notice On RotaTeq Rotavirus Vaccine, www.medicalnewstoday.com/articles/63074.php. , [Accessed June 20, 2012.];

[6] MSNBC. FDA clears drugmaker's new rotavirus vaccine, www.msnbc.msn.com/id/23944370/. , [Accessed June 20, 2012.];

[7] Linhares AC, *et al.* A short report on highlights of RIX4414: A Latin American experience. Vaccine 2006; 24: 3784-5.
[http://dx.doi.org/10.1016/j.vaccine.2005.07.027] [PMID: 16098636]

[8] Ruiz-Palacios GM, Pérez-Schael I, Velázquez FR, *et al.* Human Rotavirus Vaccine Study Group. Safety and efficacy of an attenuated vaccine against severe rotavirus gastroenteritis. N Engl J Med 2006; 354(1): 11-22.
[http://dx.doi.org/10.1056/NEJMoa052434] [PMID: 16394298]

[9] CDC Centers for Disease Control. Seasonal Flu. Key Facts about Seasonal Flu Vaccine, http://www.cdc.gov/FLU/protect/keyfacts.htm. , [Accessed June 20, 2012.];

[10] Cox MM. Recombinant protein vaccines produced in insect cells. Vaccine 2012; 30(10): 1759-66.
[http://dx.doi.org/10.1016/j.vaccine.2012.01.016] [PMID: 22265860]

[11] van Oers MM. Opportunities and challenges for the baculovirus expression system. J Invertebr Pathol 2011; 107 (Suppl.): S3-S15.
[http://dx.doi.org/10.1016/j.jip.2011.05.001] [PMID: 21784228]

[12] Cox MM, Hashimoto Y. A fast track influenza virus vaccine produced in insect cells. J Invertebr Pathol 2011; 107 (Suppl.): S31-41.
[http://dx.doi.org/10.1016/j.jip.2011.05.003] [PMID: 21784229]

[13] Cox MMJ, Hollister JR. FluBlok, a next generation influenza vaccine manufactured in insect cells. Biologicals 2009; 37: 182e-9.
[http://dx.doi.org/10.1016/j.biologicals.2009.02.014]

[14] Roldão A, Vicente T, Peixoto C, Carrondo MJ, Alves PM. Quality control and analytical methods for baculovirus-based products. J Invertebr Pathol 2011; 107 (Suppl.): S94-S105.
[http://dx.doi.org/10.1016/j.jip.2011.05.009] [PMID: 21784235]

[15] Willyard C. Cell-based vaccines yield only modest advances for seasonal flu. Nat Med 2013; 19(1): 4.
[http://dx.doi.org/10.1038/nm0113-4] [PMID: 23295991]

[16] ICTVdB. Index of Viruses, http://www.ncbi.nlm.nih.gov/ICTVdb/Ictv/fs_ortho.htm.

[17] Zaman K, Roy E, Arifeen SE, *et al.* Effectiveness of maternal influenza immunization in mothers and infants. N Engl J Med 2008; 359(15): 1555-64.
[http://dx.doi.org/10.1056/NEJMoa0708630] [PMID: 18799552]

[18] Sui J, Hwang WC, Perez S, *et al.* Structural and functional bases for broad-spectrum neutralization of avian and human influenza A viruses. Nat Struct Mol Biol 2009; 16(3): 265-73.
[http://dx.doi.org/10.1038/nsmb.1566] [PMID: 19234466]

[19] Ekiert DC, Bhabha G, Elsliger MA, *et al.* Antibody recognition of a highly conserved influenza virus epitope. Science 2009; 324(5924): 246-51. [http://dx.doi.org/10.1126/science.1171491] [PMID: 19251591]

[20] Greenberg ME. Response after One Dose of a Monovalent Influenza A (H_1N_1) 2009 Vaccine - Preliminary Report. N Engl J Med 2009; 361: 1-10. [http://dx.doi.org/10.1056/NEJMoa0907413] [PMID: 19528190]

[21] D'Argenio DA, Wilson CB. A decade of vaccines: Integrating immunology and vaccinology for rational vaccine design. Immunity 2010; 33(4): 437-40. [http://dx.doi.org/10.1016/j.immuni.2010.10.011] [PMID: 21029955]

[22] Centers for Disease Control. Measles Outbreaks. cdc.gov/measles/outbreaks.html.

[23] Vázquez M, LaRussa PS, Gershon AA, Steinberg SP, Freudigman K, Shapiro ED. The effectiveness of the varicella vaccine in clinical practice. N Engl J Med 2001; 344(13): 955-60. [http://dx.doi.org/10.1056/NEJM200103293441302] [PMID: 11274621]

[24] Merck&Co, Inc.. Varivax® Varicella live vaccine. merck.com/product/usa/pi_circulars/v/varivax/varivax_pi.pdf. , [Accessed March 20, 2014];

[25] CDC. HPV vaccine information for young women. www.cdc.gov/hpv/. , [Accessed March 20, 2014.];

[26] Gardasil. www.gardasil.com/. , [Accessed March 20, 2014];

[27] Web MD. Gardasil safety: Questions and answers. www.webmd.com/cancer/cervical-cancer/news/20080708/gardasil-safety-questions-and-answers. , [Accessed March 20, 2014.];

CHAPTER 11

Preventing Diseases by Proper Nutrition

Abstract: The five major food groups: cereals, vegetables, fruits, dairy, meat (and meat substitutes), and fats, oils and sweets. A lack of folic acid causes birth defects, such as *spina bifida*, which leaves the victim severely disabled. Folic acid is also found in multi-vitamin supplements, and these are recommended for pregnant women. It is better to eat many types of fish, than to eat red meat. If at all possible, mother's should be encouraged to breast feed their babies. The best-selling, most interesting and controversial dietary supplements are multi-vitamins. The American Medical Association (AMA) does not recommend them. Instead, the AMA recommends getting your vitamins and minerals from a healthy, balanced diet. The National Institutes of Health maintains several pages on their website that have fact sheets on many dietary supplements. The NIH has an office of dietary supplements. It provides information on the use and safety, nutrient requirements, database resources, news and research. However, it is the FDA that has regulatory responsibility for dietary supplements, as dictated by the dietary supplement health and education act, or DSHEA, passed in 1994. The DSHEA indicated that the dietary supplement manufacturer is responsible for ensuring that a dietary supplement is safe before it is marketed, but the FDA is responsible for taking action against any unsafe dietary supplement product after it reaches the market [1]. The popular dietary supplement, açaí, will make you gain weight – not lose weight. Another supplement, *myo*-inositol, may help prevent lung cancer in smokers.

Keywords: Açaí, AMA, American Medical Association, DSHEA, FDA, Folic acid, Myo-inositol, National Institutes of Health, NIH.

NUTRITION

Modern medicine now recognizes the importance of nutrition in maintaining good health and managing diseases. Although hard science and reductionist thinking were used to determine which nutrients are essential, systems thinking is also needed. The needs of each individual must be considered. For example, in the USA and many countries that emphasize western medicine, over-consumption of calories is a major concern, but in much of the world malnutrition and starvation

Robert E. Smith

are problematic. Malnutrition can be a problem in both cases. The popular fast food diet in the USA may provide plenty of calories, but is often lacking in vitamins and dietary fiber that is found in fruits and vegetables. However, multivitamins seldom do any good to well-fed people and can actually do much harm, especially if they contain iron and are consumed by men over 50. In contrast, vitamins save lives and prevent blindness in many disadvantaged communities.

Water is at least as important as food. Unfortunately, billions of people don't have access to clean water. In parts of the world that have plenty of clean water, many say that it is important to drink at least eight 8-ounce glasses of water every day to maintain good health. Caffeinated beverages, sweetened fruit juices and soft drinks are often considered to be dehydrating, even though usually they are not [1]. The USDA recommends different amounts of water for people of different age and sex, such as 3.7 L/day for adult men and 2.7 L/day for women. Even though sweetened beverages may not be dehydrating, they are a major contributor to obesity in the USA and other countries.

Still, food is essential, and billions of people don't get enough to eat. However, in so-called "developed countries" like the USA, obesity is a major health problem. It can cause diabetes and other diseases. In this chapter, we will learn that proper diet and nutrition can prevent obesity and many diseases. We will learn about the five major food groups: cereals, vegetables, fruits, dairy, meat (and meat substitutes), and fats, oils and sweets. Cereals include bread, oatmeal, rice and pasta. They contain complex carbohydrates. It is better to eat whole grain foods. This includes whole grain bread, as opposed to white bread. It also includes brown and wild rice, as opposed to white rice. Similarly, whole grain pasta is better than white pasta. The USDA recommends eating three or more ounce-equivalents of whole grain products each day [2]. Two cups of fruit and 2.5 cups of vegetables per day are recommended for a 2000 calorie per day diet, suitable for a sedentary man 51 – 70 years old, or a sedentary woman 19 – 30 years old [2]. This does not include the two most popular vegetables in the USA, iceberg lettuce and French fries. It does include all the colorful fruits and vegetables that contain vitamins, minerals and antioxidants.

Antioxidants are nutrients that prevent damage that can be caused by oxidants, that is, reactive oxygen and nitrogen substances, or RONS. The five vegetable subgroups are: dark green, orange, legumes, starchy vegetables and other vegetables. The USDA also recommends consuming 3 cups of fat-free or low-fat milk or equivalent milk products, such as soy milk or other soy products. The USDA recommends 5.5 ounce equivalents from the meat and bean group. This can be lean meat, fish, beans, nuts or seeds. The USDA recommends 24 grams (6 tsp) of oils. There is also a discretionary Calorie allowance of 267 Calories, which can be such things as 18 grams of solid fat (saturated fat, from red meat), or 8 tsp of sugar, in sweetened foods or beverages [2].

Of course, the number of Calories and servings of food should be higher for active people. The estimated daily Caloric intake for active women aged 19 – 30 is 2400 Calories and for active men 19 – 30 years old it is 3000 Calories [2]. Note that there are two different definitions of Calories and calories. In a science class (especially physics), one calorie (lower case) is defined as the amount of energy needed to raise the temperature of one gram of deionized water in the liquid phase by 1°C. Actually, calories and kilocalories are not SI (metric) units and are somewhat imprecise. That is, the exact amount of energy needed to raise the temperature of liquid water by 1°C depends on the starting temperature of the water. Still, the difference in the amount of energy needed to heat liquid water from 1°C to 2°C is very close to the amount of energy needed to heat liquid water from 10°C to 11°C, so the original definition of a calorie is close enough to be used for most practical purposes. One kilocalorie is equal to 1000 calories. One kilocalorie is the amount of energy needed to increase the temperature of 1 kg (1000 g) of water by 1°C. Labels on containers of foods and beverages are based on a different definition. They use the commonly accepted definition that one Calorie (upper case) in food is the same as one kilocalorie as defined by scientists. So, theoretically, the 100 Calories in a small apple have enough stored energy (as carbohydrates) to increase the temperature of 100 kg of liquid water 1°C, if it is burned up completely in a bomb calorimeter.

However, it should be noted that the caloric content of food is not based on an artificial experiment, in which a small sample of food is burned up completely in a bomb calorimeter in the presence of much oxygen. The calorimeter contains a thermometer to measure the increase in temperature that was caused by the combustion. However, not all calories are equal when considering the effect on health and obesity. It is more important to consider the effect of different foods on the glucose concentration in the blood, which is defined by the glycemic index.

To measure the glycemic index, a measured amount of food or beverage (usually 50 g) was given to ten subjects, and the changes in amount of glucose in the blood were measured over a two hour time period. The change in glucose concentration *vs* time, for 120 min was plotted. The area under the curve was the glycemic index and the average for ten subjects was used. Two different benchmarks can be used. Sometimes glucose is used and is assigned a glycemic index of 100. Others use white bread to define a glycemic index of 100. By this measure, glucose has a glycemic index of about 140. Thus, a 100 Calorie (or kilocalorie) candy bar has a higher glycemic index that a 100 Calorie apple, so the apple is much healthier and less likely to lead to obesity. Foods such as most fruits and vegetables have relatively low glycemic indices of 55 or less. Whole wheat products, brown rice and table sugar have intermediate glycemic indices of 55 – 69. Corn flakes, white rice, baked potato, watermelon, white bread, cereal and candy have high glycemic indices of over 69. This is all based on glucose having a glycemic index of 100. So, the caloric content of food is based on the ingredients. Protein and carbohydrates are assigned the value of 4 Cal/g and fat 9 Cal/g, regardless of the source. Bomb calorimetry is not used in the assignment of caloric content.

A recent study found that a low-glycemic index diet was better than a low-carbohydrate diet [3]. It clearly showed that not all calories are the same. Foods with a high glycemic index should be avoided, while whole grain foods and unprocessed fruits and vegetables should be encouraged. As a result, some of the older advice to avoid excess fat and salt is deceptive. Dietary sugar, be it glucose, sucrose, fructose, white rice and processed foods containing added sugar should be avoided.

In one study, after subjects lost 10-15% of their weight on different diets, the resting and total energy expenditures were greatest with a low-fat diet, intermediate with the low-glycemic index diet and lowest with the very low-carbohydrate diet [3]. So, a combination of whole grain foods, fruits, vegetables, omega-3 fats and polyunsaturated fats should be a part of a healthy diet. Saturated fats and *trans* fats should be avoided, along with most processed foods.

LIMITS TO USA RECOMMENDATIONS

The USDA recommendations have their limits, though. There is no distinction made between different types of meat or fish. It is now known that there is a big difference between good fats (in fish) and bad fats (in meat). Good fats are unsaturated, polyunsaturated and omega-3 fats. They are especially important for the developing brains of babies and infants. In adults, they help prevent diseases of inflammation, including heart disease, stroke, cancer, arthritis, Alzheimer's disease, and others. On the other hand, saturated fats and *trans* fats that are found in meat and partly hydrogenated oils are bad.

Also, there is no USDA recommendation regarding organic foods, or genetically modified (GM) crops. Also, the dangers of eating mass-produced meat in the USA are not indicated. The advantages of vegetarianism are not described. The advantages of eating or drinking soy, rice or almond "milk" instead of cow's milk or other dairy products, are not mentioned. The importance of antioxidants is not stressed.

It is better to eat many types of fish, than to eat red meat. Fish and vegetable oils are better than butter, margarine, or solid fat from pork or beef. Cold water oily fish, such as salmon, anchovies, herring and mackerel, are good sources of omega-3 fats. Some of the omega-3 fats, such as alpha linoleic acid (ALA), eicosapentaenoic acid (EPA) and docosahexaenoic acid (DHA) are essential and may prevent heart disease. However, farm-raised Atlantic salmon has been found to have relatively high levels of toxic pollutants, called PCBs and dioxins [4]. Unsaturated fats, including polyunsaturated fats and omega-3 fats, are liquids and are much better than saturated or *trans* fats, which are solid at room temperature. Even though cookies and crackers taste good, they contain *trans* fats and should be avoided.

RISKS IN EATING MASS-PRODUCED MEAT

There are other risks from eating mass-produced beef, pork or chicken, especially in the USA. Cows, pigs and chickens are mass produced by keeping them confined in small spaces, and packed closely together in pens or cages. They are fed antibiotics, which prevent disease, but causes diarrhea. They are fed growth hormones, to make them grow bigger and faster. They spend their lives knee-deep in their own waste. They are butchered in meat-packing plants, in which many undocumented workers are forced to work close to each other, with sharp knives and high quotas that need to be met. Many workers are missing parts of their fingers. Because many of the workers fear being deported, they do not speak up when there is a problem. Instead of total quality management, it is management by fear. If a worker makes a mistake when dissecting a cow, the fecal material in the cow's guts can be mixed with the rest of the beef. Workers are tempted to say nothing, for fear of being fired and deported. The fecal material contains the bacteria *E. coli*, which can cause food poisoning. If a worker slices off a piece of his or her finger, it can end up in the beef. Cattle are also fed high protein concentrates that come from the parts of other cattle that people won't eat, including the brain. This can cause Mad Cow disease. Similar problems exist in the mass production of pork and chicken. There is a huge stench caused by the waste from hog farms, especially. For more details, a book called Fast Food Nation, by Eric Schlosser, is recommended [5].

Adding to the problem is the lack of sufficient money to hire USDA employees to inspect our beef. In her book, Animal, Vegetable, Miracle, Barbara Kingsolver tells us that only 0.1% of all slaughtered cows need to be inspected for Mad Cow disease, in a policy of "don't look, don't find" [6]. Most meat that is produced in the USA comes from confined animal feed organizations, or CAFOs. Over 1000 chickens can be kept in a room that is equivalent to a small bathroom. All CAFO animals are fed antibiotics, with the hope of preventing bacterial infection. Unfortunately, it doesn't work very well. Most of the chickens produced by CAFOs contain bacteria. To survive, these bacteria must be resistant to antibiotics. So, the major source of antibiotic resistant bacteria is not people who have been given too many antibiotics, but animals from CAFOs [6]. Antibiotics tend to give animals (and people) diarrhea. So, CAFO animals, especially cattle, spend their entire lives knee deep in their own fecal material [5]. So, beef is often contaminated with *E. coli* from the gut [5]. The fecal material (especially from hogs) is a major source of environmental pollution. CAFO animals produce much more fecal material than all the humans on our planet!

So, if you have to eat beef, pork, turkey or chicken in the USA, it is better to eat animals grown on the free range. That is, the animals should be allowed to roam freely, eat (and defecate) in the open range, or pastures. When they defecate, the fecal material enters the soil instead of our rivers and drinking water (as in CAFOs). Sometimes meat produced in this way is labeled organic, meaning the producer has spent some extra money to have his or her organization inspected. In her book, Animal, Vegetable, Miracle, Barbara Kingsolver tells us that it is just as good (and usually better) if you buy your meat and other foods from local producers who you know, even if it isn't certified as organic [6]. That is, organic foods that have to travel hundreds or thousands of miles to reach you are not as good for the environment as locally produced foods. It takes much fossil fuel to transport foods over a long distance, which has a negative impact on the environment. Free range meat will not have growth hormones, antibiotics, fecal material, or human flesh. Similarly, there can be advantages to eating locally produced grains, cereals, fruits, and vegetables.

It is also much better for the environment to be a vegetarian. The international meat industry generates about 18% of the world's greenhouse gas emissions (more than transportation) and switching to vegetarianism can reduce your carbon footprint by up to 1.5 tons of CO_2 per year [7]. In an article in *Scientific American*, it was reported that beef production generates greenhouse gases that contribute to 13 times more global warming than chickens and 57 times as much as potatoes [8]. Still, in her book, Animal, Vegetable, Miracle, Barbara Kingsolver tells us that it is almost impossible to be a complete vegetarian, since there are almost always a few pieces of small insects that are parts of the foods we eat [6]. She points out that Hindus are often strict vegans and think that they eat no animal products at all. However, when Hindus began immigrating to England in the early 1900s, many of them became anemic because of a lack of protein. The English farmers used different farming methods that kept the amount of insects much lower on their on their fruits and vegetables than in India, so the protein intake of the immigrants decreased, causing anemia [6]. I also remember my son's high school biology teacher telling his class that if you sleep in the basement of your house and snore, you probably swallow an occasional spider or two.

FISH AND SEAFOOD

Still, many people recommend eating fish instead of meat. There are advantages and disadvantages to this. First, not all fish (or seafood) is the same. As mentioned before, oily fish, such as salmon and mackerel, are high in omega-3 fats, so they might prevent heart attack, stroke, dementia, cancer and even attention deficit hyperactivity disorder (ADHD). However, they tend to have relatively high levels of toxic pollutants, such as heavy metals (like mercury) and dioxins (which were used in Agent Orange in the Vietnam War). So, some recommend not eating farm raised Atlantic salmon and eating no more than two servings a week of wild salmon. Please note that most grocery stores sell fresh salmon at a higher price than frozen salmon, even though the fresh salmon was farmed raised and the frozen was caught wild and has much less contamination from pollutants [4]. So, when the USDA recommends meat or fish, there is no distinction between farm raised and wild salmon, or between shrimp, which has almost as much cholesterol as lean meat. However, there are websites that list seafood that should or should not be eaten. One of them recommends eating farmed raised catfish, caviar, clams, mussels, oysters, scallops, striped bass, American or Canadian shrimp, and sturgeon, along with wild salmon, tilapia, crab, anchovies, crab, halibut and others [9]. They list several as being bad for your health and/or the environment. Thus includes farm-raised Atlantic salmon, orange roughy, grouper, bluefin tuna, swordfish, imported shrimp and others. Also, the National Resources Defense Council has a website that warns pregnant women about the dangers to their babies of eating too much seafood that is contaminated with mercury. They say, "Children under six, as well as women who are pregnant or planning to become pregnant, are the most vulnerable to mercury's harmful effects. They should restrict or eliminate certain fish from their diet, including ahi or bigeye tuna, tilefish, swordfish, shark, king mackerel, marlin, orange roughy and fish caught in any waters that are subject to a mercury advisory. Women with elevated mercury levels should ideally begin avoiding or restricting their consumption of mercury-laden fish as much as a year before they become pregnant" [10]. They list the following as being low in mercury: Anchovies, Butterfish, Catfish, Clam, Crab (Domestic), Crawfish/Crayfish, Croaker (Atlantic), Flounder, Haddock (Atlantic),

Hake, Herring, Mackerel (N. Atlantic, Chub), Mullet, Oyster, Perch (Ocean), Plaice, Pollock, Salmon (Canned), Salmon (wild), Sardine, Scallop, Shad (American), Shrimp, Sole (Pacific), Squid (Calamari), Tilapia, Trout (Freshwater), Whitefish, and Whiting. They listed these as being ones to avoid, since they have the highest in mercury: Mackerel (King), Marlin, Orange Roughy, Shark, Swordfish, Tilefish, Tuna and Bigeye. Similarly, the US Department of Health and Human Services, along with the EPA (Environmental Protection Agency), has a website that recommends the following: [11]. "By following these 3 recommendations for selecting and eating fish or shellfish, women and young children will receive the benefits of eating fish and shellfish and be confident that they have reduced their exposure to the harmful effects of mercury. Do not eat Shark, Swordfish, King Mackerel, or Tilefish because they contain high levels of mercury. Eat up to 12 ounces (2 average meals) a week of a variety of fish and shellfish that are lower in mercury. Five of the most commonly eaten fish that are low in mercury are shrimp, canned light tuna, salmon, pollock, and catfish. Another commonly eaten fish, albacore ("white") tuna has more mercury than canned light tuna. So, when choosing your two meals of fish and shellfish, you may eat up to 6 ounces (one average meal) of albacore tuna per week. Check local advisories about the safety of fish caught by family and friends in your local lakes, rivers, and coastal areas. If no advice is available, eat up to 6 ounces (one average meal) per week of fish you catch from local waters, but don't consume any other fish during that week" [11].

In 2004, the FDA and Environmental Protection Agency (EPA) advised pregnant and nursing mothers and their young children to limit their consumption of fish that are low in mercury to no more than 12 ounces (about two portions) per week [11]. This was challenged by another group of experts, the Maternal Nutrition Group and the National Healthy Mothers, Healthy Babies (HMHB), who recommended that such women and children should eat at least 12 ounces of fish per week. They recommended fish like salmon, tuna, sardines and mackerel [12]. "The Group found that eating fish is the optimal way to gain the benefits of long-chain omega-3 fatty acids, docosahexaenoic acid (DHA) and eicosapentaenoic acid (EPA). Seafood is the richest dietary source of DHA and EPA in Americans' diets. The Group also recognized that selenium, an essential mineral found in certain ocean fish, accumulates and appears to protect against the toxicity from trace amounts of mercury" [12]. However, that website has changed. Instead, one can refer to a report from Purdue University [13]. Please note that fish like salmon actually contain omega-3 fatty acyls and not omega-3 fatty acids, but the fatty acyls in triglycerides are converted to free fatty acids when the triglycerides are digested.

OTHER SOURCES OF OMEGA-3 AND POLYUNSATURATED FATS

However, it is possible to consume protein, omega-3 fats and polyunsaturated fats without eating seafood or fish oil. Some people recommend beans, nuts and soy products as meat and fish substitutes as sources of protein. DHA and EPA can be obtained as dietary supplements, derived from seaweed, which is the source being used in a clinical study to see if DHA can prevent ADHD in children whose mothers took DHA as a dietary supplement when they were pregnant. This avoided the problems with mercury or dioxin contamination, and it avoided possible allergies that some people have to fish or fish oils. Also, it is important to know that the healthy polyunsaturated fats can react with light and oxygen in the air and turn rancid. Thus, olive oil and flaxseed oil should be kept in dark bottles and flaxseed oil should be kept in the refrigerator. So, there is much more to know about the details of a healthy diet, and not all the foods in the so-called meat and meat substitute group are the same.

SOYBEANS AND SOY PRODUCTS

One member of the meat, fish, nuts and bean food group is especially healthy. It is soybeans and soy products. They can provide the necessary protein that meat, fish, nuts and beans provide, but soy offers other unique benefits. Soy does not have any cholesterol, antibiotics, growth hormones or blood, unlike mass-produced cow's milk. Moreover, soy protein may be much better than animal protein and be able to prevent cancer. Soy also has antioxidants, called isoflavones, which also prevent cancer and heart disease. The isoflavones are a type of antioxidant. They prevent damage that can be caused by reactive oxygen substances (ROS), which can cause diseases of inflammation, including cancer and heart disease. The isoflavones are also phytoestrogens. That is, they are plant-derived compounds that bind to isolated estrogen receptors *in vitro*, similar to tamoxifen, which is used to treat breast cancer [14]. Unfortunately, this has caused some people to mistakenly think that soy formula should not be given to infants [15]. Hundreds of millions of babies and infants around the world are given soy formula as their major source (or only source) of food. Many of them are allergic to cow's milk, and many of their parents refuse to eat meat or dairy products for religious reasons, and most of these babies grow up to have healthy children and grandchildren.

TURMERIC

Turmeric is a popular spice that has been used for over a thousand years to add flavor to foods and to help treat and prevent many types of illnesses [16, 17]. That and its continued use by billions of people indicate that it is safe and non-toxic. It is made from the mature rhizomes of *Curcuma liga* Linn [17]. "The rhizomes are steeped in boiling water, dried in the sun and polished to obtain turmeric sticks. Turmeric contains oils, small amounts of carbohydrates, fats and 2-9% curcuminoids. It is also has omega-3 fats" [17]. There have even been clinical studies that showed that as much as 12 g/day can be consumed without any adverse side effects [16]. It has anti-hyperglycemic and insulin sensitizing effects [18]. Like many of the most effective prescription drugs, curcumin has many biochemical effects [16]. "It affects the transcription and activities of lipoxygenase, cyclooxygenase, glutathione-*S*-transferase and CYP450 enzymes. Curcumin also modulates transcription factors and growth factors and the metabolic pathways that are linked to them. This includes the receptor for the epidermal growth factor (EGFR), fibroblast growth factor 2, activator protein-1 (AP-1), nuclear factor κB and nuclear transcription factor E2-related factor 2 (Nrf2). Curcumin also affects the expression of cytokines as well as their activities. It can also change the proliferation of cells by influencing the actions of cyclin proteins needed for apoptosis. It is also antiproliferative, since it influences signaling by oncogenes and tumor suppressor genes. It can also help patients with arthritis, neurological disorders and diabetes" [16]. Finally, it can modulate the BRCA1 protein and the induction of apoptosis in triple negative breast cancer [19].

The most biologically active compound in turmeric is the polyphenol called curcumin [20]. It is a β-diketone that is quite hydrophobic. It dissolves readily in olive oil and flaxseed oil, but not in water. When mixed with water and heated, curcumin can be hydrolyzed [20 - 22]. It is hydrolyzed faster in neutral and alkaline aqueous solutions, but much slower in acidic solutions [20]. It has poor bioavailability and is not absorbed well from the gut [16]. However, its bioavailability can be increased by dissolving it in olive oil [22] or divalent cations in a mixture of glycerol and water [23]. Its bioavailability and effectiveness against drug-resistant forms of cancer can be improved by encapsulating curcumin in polymeric nanoparticles [24 - 26].

GINGER ROOT

Ginger root is another popular and very healthy food (or spice) from China, Malaysia and the Indian subcontinent [27, 28]. Like turmeric, it contains curcumin [28]. Its other main bioactive components are gingerols and shogaols, including 6-gingerol and 6-shogaol [28]. Ginger root also has a relatively high antioxidant potential [28]. The pungent aroma of fresh ginger root is due to gingerols, which are phenolic compounds [27]. "The pungent aroma of dry ginger root is due to shagaols, which are formed from gingerols during heating. The pharmacokinetics of 6-gingerol have been reviewed. A methanolic extract of ginger roots may be able to lower body weight, blood glucose, blood lipids and hyperinsulinemia. Ginger may also help to lower blood pressure, as well as have anti-inflammatory and analgesic effects. It is also widely recognized as being able to alleviate disorders of the gastrointestinal tract caused by motion sickness and cancer chemotherapy. Ginger root and/or its extracts can also protect against radiation and chemical-induced tissue damage [27]. Others have described the ability consuming two 1-g capsules containing fresh ginger root powder to lower fasting blood sugar, hemoglobin A1c, apolipoprotein B, apolipoprotein A-I and malondialdehyde in patients with type-2 diabetes [29]. It may also be effective in treating osteoarthritis [30].

ANTIOXIDANTS IN FRUITS AND VEGETABLES

Antioxidants are also found in many fruits and vegetables. All of them share the common property of reacting with ROS, to prevent oxidative damage. Highly colored fruits and vegetables contain antioxidants. Some antioxidants are vitamins. Vitamin A is found in carrots, spinach, kale, cantaloupe, sweet potatoes and other yellow vegetables. Vitamin C is found in spinach, tomatoes, kale, citrus fruits, broccoli, Brussel sprouts, cantaloupe, red peppers and others. Vitamin E is found in many vegetable oils, fish oil, nuts, green leafy vegetables and fortified cereals. So, eating fruits and vegetables can prevent many diseases in the USA and other industrialized nations. On the other hand, in much of the world, obesity is not the biggest problem. Instead it is hunger and malnutrition.

FIGHTING STARVATION AND MALNUTRITION

Unfortunately, there are many poor people, especially women and children, who suffer from hunger, malnutrition and starvation. Long before people lived long enough to die of heart disease or cancer, the struggle to find enough food was more than a full time job. Humans were hunters and gatherers, long before the beginning of agriculture. Civilizations arose near rivers, as people learned to cultivate wheat, barley, millet, rice and squash. Calendars and ancient observatories were developed to help farmers know when rivers (like the Nile) would flood, when crops should be planted and when they should be harvested. Eventually, humans learned to make metal tools, which helped them plow and cultivate fields and irrigate crops. Then there was a scientific revolution during the Renaissance and Age of Enlightenment. This encouraged trial and error experiments in plant breeding and it produced improved crops. This helped produce the Guernsey cow, which produces much milk. It also produced the idea of crop rotation, or rotating the planting of grains and legumes. Other developments included adding limestone to soil, and many inventions, such as a seed drill, cotton gin, and the mechanical reaper. By the late 1800s, steam power began to replace animal power on the farm. The demand for food for urban workers and raw materials for industry brought about major changes in world trade. Large agribusinesses emerged in capitalist economies, but central planning emerged in Communist countries. Mass starvation was used as a means of control and terror by Stalin in the Soviet Union and by Mao Zedong in the People's Republic of China. This caused the death by starvation of tens of millions of people, especially in Ukraine and rural China. Fortunately, this did not happen in the USA and other non-Communist countries. Large agribusiness has not caused starvation, even though it has displaced most small farmers in the USA.

GREEN REVOLUTION

In 1968, the director of US Agency for International Development (USAID) coined the term Green Revolution to describe the worldwide transformation of agriculture that led to significant increases in food production between the 1940s and 1960s. The Green Revolution started in 1943 with the establishment of the Office of Special Studies, which was a collaboration between the Rockefeller Foundation and the Mexican president, Manuel Avila Camacho. The main initiative was the development of high-yield varieties of maize and wheat. The Mexican government invested heavily in rural infrastructure and new varieties of seeds were widely adopted. Mexico became self-sufficient in wheat production by 1951 and has been exporting wheat ever since. The Green Revolution spread to India, where the Ford Foundation was involved. The Minister of Steel and Mines, C. Subramaniam worked with the International Maize and Wheat Improvement Center (CIMMYT) to improve plant breeding, irrigation, development and financing of agrichemicals.

By the late 1970s, yields of rice increased by 30 percent. Next, the Rockefeller and Ford Foundations jointly established the International Rice Research Institute (IRRI) in the Philippines in 1960. High-yield varieties spread throughout the Philippines, followed by Indonesia, Pakistan, Sri Lanka, and many other countries throughout Latin America, Asia and North Africa. Another organization (USAID) became involved in subsidizing the development of rural infrastructure and shipping fertilizer. This was followed by the Consulting Group on International Agricultural Research (CGIAR) in 1971, which responded to some earlier criticism of the Green Revolution. They developed a more holistic view of agriculture.

Farmers and scientists need a good understanding of soil chemistry and how it affects agriculture. They study chemical reactions in the soil that affect the growth of plant life. They have developed herbicides and pesticides that have dramatically increased productivity and decreased hunger and malnutrition.

Pesticides have saved many lives. For example, DDT suppressed a typhus outbreak in Naples, Italy in 1944-1945. It also controlled river blindness in the tropics by killing the insect vector (black fly). Several insecticides have helped control the mosquitoes that cause malaria in Africa, the Middle East and Asia. In the past decade, biotechnology has produced herbicide-resistant and insect-resistant crops. In 2006, ten million farmers in 22 countries planted more than 100 million hectares with genetically modified crops that were developed by biotechnology (one hectare is 10000 square meters). An herbicide, called Roundup™ or glyphosate, is the best-selling herbicide in the world. Many people consider it to be relatively safe and can kill hundreds of types of weeds and 80% of the genetically modified crops (also called genetically modified organisms, or GMOs) have genes inserted into them to make them resistant to glyphosate. Much of modern agriculture depends on this one chemical. In 2005 in the USA, glyphosate-resistant crops were planted on 90% of the soybean acreage, 60% of the cotton and 18% of the corn [31]. Unfortunately, weeds that are resistant to glyphosate have emerged. Instead, agribusiness is using a broad-leaf herbicide called dicamba. Scientists have developed crops that are resistant to dicamba, which is becoming an environmentally friendly alternative to glyphosate. Dicamba does not persist in soil and it is non-toxic to humans and other mammals [31]. They put a bacterial gene into plants to metabolize excess dicamba before it can accumulate to toxic levels [31].

POLLUTION, SOIL CONTAMINATION, ECOLOGICAL RISKS AND HUMAN HEALTH

Farmers and scientists are also concerned with pollution, soil contamination, ecological risks and human health. They want to be able to predict the fate, mobility and toxicity of contaminants. They want to measure the cation exchange capacity of the soil. They are also very interested in emergent problems that threaten our food supply. For example, colonies of honeybees have been disappearing in a Colony Collapse Disorder in the USA, Brazil, Canada and parts of Europe. The US Department of Agriculture (USDA) has called this the biggest threat to our food supply. It may be caused by a virus [32], a commonly used pesticide [33], a decrease in biodiversity in flowering plants [34] and a parasitic fly [35].

CHEMICAL COMPOSITION OF SOIL AND THE CHANGES INVOLVED IN THE PRODUCTION OF CROPS AND LIVESTOCK

Agricultural chemistry studies the chemical composition of soil and the changes involved in the production of crops and livestock. To improve yields, livestock in the USA often spend their entire lives in confined spaces, living in their own waste. As mentioned in Chapter 10, these are called CAFOs. To help prevent disease, animals grown in CAFOs are given antibiotics, which give them diarrhea, which contaminates the meat with fecal material containing pathogenic strains of E. coli. The meat that is consumed in fast food restaurants is often contaminated with fecal material [5]. Also, waste from hog farms can be especially noxious to surrounding communities. Cattle are often fed growth hormones to increase the amount of meat or milk that is produced. In meat packing plants, workers are often afraid of being fired, even though they work in close quarters under hazardous conditions. Many of them have had fingers cut off and the human flesh makes its way into the beef or pork. Cattle feed in the USA may also contain remains from dead cows, including cow brains. This can lead to mad cow disease, also known as spongiform encephalopathy and BSE. Also, it takes much more energy to produce meat than beans or nuts, which can provide needed protein. Producing the needed extra energy increases the production of greenhouse gases and contributes to global warming.

GENETICALLY-MODIFIED ORGANISMS (GMOS)

Agricultural chemists work with producers of food and herbs to improve yields and quality. They study the causes and effects of biochemical reactions that are related to growth and production. They are also interested in producing disease-resistant crops and livestock. Sometimes, this involves the use of genetically-modified organisms (GMOs), which are controversial. GMOs are organisms whose DNA has been modified by the tool of genetic engineering that is called recombinant DNA technology. Pieces of DNA from one organism can be recombined with the DNA of another organism. This was done when sequencing human DNA. Human chromosomes were broken into smaller pieces, which were recombined (or spliced) into the DNA of a yeast artificial chromosome. Also, smaller pieces of DNA can be spliced, or recombined, into bacteria. Some GMOs contain a gene from a jellyfish that is green and fluoresces. Other genes can code for other dyes. GloFish™ are the first GM animal that is sold as a pet. Bioengineering departments in colleges and universities are producing new life forms that never existed before, such as *E. coli* that smells like bananas instead of fecal material. GMOs can have very practical benefits. J. Craig Venter and Institute for Biological Energy Alternatives are trying to put genes from one or more types of bacteria into another, to produce completely new organisms that will be able to produce H_2 gas and many fine chemicals and medicines, which are currently made at the cost of much fossil fuel.

Soybean, corn, maize, canola, rice, cotton seed oil and wheat have been genetically modified. Even though many say that genetically modified crops that are resistant to herbicides are safer and more environmentally friendly than other crops [36], others, such as the European Union, disagree. The first GMO introduced to Europe in 1996 was a tomato paste made from a GM tomato that was first developed in the USA. However, eventually public confidence in government inspectors and regulators decreased. Protests began over the introduction of Monsanto's "Roundup-Ready" soybeans. Other crops to emerge were insect-protected cotton and herbicide tolerant soybeans. GM crops have been widely accepted in the USA, Brazil, Argentina, South Africa, India and China. Genetically modified cotton in China that is resistant to the bollworm does not need to be sprayed with pesticides. Between 1996 and 2005, the total surface area of land cultivated with GMOs had increased by a factor of 50, from 17,000 km^2 (4.2 million acres) to 900,000 km^2 (222 million acres), of which 55 percent were in the United States. In the USA, by 2006 89% of the planted area of soybeans, 83 percent of cotton, and 61 percent maize was genetically modified varieties [37]. In fact, in Japan, Mexico, much of Europe, and in some parts of the USA and Canada, GMOs are not allowed. Until 2005, GMOs were illegal in Brazil, even though many were grown illegally. Now, the Brazilian government is funding research on GM sugar cane, for the production of sugar and ethanol [36]. This continues to be a source of controversy. There is the possibility of cross pollination with wild-type crops, which do not contain the genetic modification. Pollen can be dispersed over large areas by wind, animals, and insects. There is a concern that genes may recombine almost anywhere in a host. The properties of a gene, such as the degree to which it is expressed, can be very dependent on its location on a recombined chromosome. This can cause serious adverse health effects, as has been seen in genetically modified humans, or gene therapy. There is also a fear of newly emerging viruses as genetically modified pigs are developed to produce organs for human transplantation. Pigs are much easier to clone than primates, and pig organs closely resemble those of humans.

A relatively non-controversial application of genetically modified organisms is the creation of new types of bacteria for industrial applications. They are already being used in carefully controlled laboratory environments to mass produce medicines, vitamins and other dietary supplements. Research is being done to create genetically modified bacteria that will be able to replace chemical refineries. They will convert CH_4 and CO_2 into organic chemicals and pharmaceutical compounds that are now being made from oil and petroleum. Other research is trying to develop bacteria that can mass produce H_2, for the upcoming hydrogen economy that will replace the current petroleum-based world economy. An article in Scientific American, Engineering Life: Building a FAB for Biology [38] goes into more details on how bacteria can be modified for industrial applications. Although they are not really alive, viruses can also be made artificially. The complete chemical synthesis of the polio virus has been reported [39].

NUTRIENTS IN FOODS

In the Department of Agriculture, inspectors check meat for quality and scientists focus on the environmental fate of pesticides. Agricultural chemists also study ways to increase the amount of nutrients in food. For example, organic acids, such as citric, tartaric, malic and ascorbic acids all enhance the bioavailability of Ca^{2+} [40]. It has also been found that diets rich in fruits and vegetables help prevent diseases of inflammation, because they contain antioxidants. Also, there is a website that lists the nutrients in some of the healthiest foods [41]. Table **1** contains a partial list.

Also, please note how difficult it can be to get enough potassium. Most people think that eating one or two bananas a day will provide sufficient potassium.

Actually, each banana provides only 11% of the recommended daily allowance (RDA). Much better sources of potassium are tomato paste and orange juice. Still, it is important to control the number of Calories you consume. So, one can of tomato paste is not healthy if you have to eat a whole pizza to get the tomato paste. Similarly, almost all the orange juice in the USA is sweetened. So, it should be consumed only in moderation, depending on how many other Calories you consume. An excellent source of potassium is salt substitute. It tastes almost the same as table salt, or sodium chloride. Instead, it contains potassium chloride and calcium phosphate. Table **2** shows the amounts of calcium and potassium in some foods.

Table 1. Vitamins in some of the healthiest foods.

	Vit A	5000 IU[a]	Vit B1	1.2 mg[a]	Vit B2	1.4 mg[a]
	IU	%RDA	mg	%RDA	mg	%RDA
Asparagus, 1 cup frozen	1451	29.0	0.117	9.8	0.185	13.2
Red Pepper, 1 cup 149 g	4665	93.3	0.064	5.3	0.127	9.1
Broccoli, frozen 1 cup	1860	37.2	0.101	8.4	0.149	10.6
Brussel sprouts, frozen 1 cup	1435	28.7	0.167	13.9	0.125	8.9
Cauliflower, cooked 1 cup	782	15.6	0.067	5.6	0.095	6.8
Garlic, 1 clove 3 g	0	0.0	0	0.0	0	0
Onions, raw 1 cup	0	0.0	0	0.0	0	0
Parsley, 10 g	842	16.8	0.032	2.7	0.038	2.7
Spinach, raw 1 cup 30 g	2813	56.3	0.171	14.3	0.295	21.1
Tomatoes, raw 123 g	1025	20.5	0.046	3.8	0.034	2.2
Tomato paste 1 cup	3996	79.9	0.157	13.1	0.401	28.6
Carrots, frozen 1 cup	26571	531.4	0.026	2.2	0.044	3.1
Kale, frozen 1 cup	19115	382.3	0.069	5.8	0.091	6.5
Oranges, 180 g 1 cup	405	8.1	0.157	13.1	0.072	5.1
Grapes, raw 50 g	33	0.7	0.11	9.2	0.112	8
Cantaloupe	5411	108.2	0.066	5.5	0.03	2.1
Bananas, 1 cup	94	1.9	0.034	2.8	0.11	7.9
Salmon, sockeye 155g	321	6.4	0.02	1.7	0.086	6.1
	Vit B6	2.0 mg[a]	Vit C	60 mg[a]	Vit K	80 μg[a]
	mg	%RDA	mg	%RDA	mcg	%RDA

(Table 1) contd.....

Asparagus, 1 cup frozen	0.036	1.8	43.9	73	144	180
Red Pepper, 1 cup 149 g	0.434	21.7	190	317	7.3	9
Broccoli, frozen 1 cup	0.239	11.95	101.2	169	220.1	275
Brussel sprouts, frozen 1 cup	0.448	22.4	70.8	118	299.9	375
Cauliflower, cooked 1 cup	0.158	7.9	56.3	94	15.5	19
Garlic, 1 clove 3 g	0.037	1.85	0.9	2	0.1	0
Onions, raw 1 cup	0.192	9.6	11.8	20	0.5	1
Parsley, 10 g	0.053	2.65	13.3	22	164	205
Spinach, raw 1 cup 30 g	0.059	2.95	8.4	14	1027	1284
Tomatoes, raw 123 g	0.098	4.9	16.9	28	9.7	9
Tomato paste 1 cup	0.566	28.3	57.4	96	29.9	37
Carrots, frozen 1 cup	0.164	8.2	5.6	9	21.4	27
Kale, frozen 1 cup	0.179	8.95	53.3	89	1147	1434
Oranges, 180 g 1 cup	0.108	5.4	95.8	160	0	0
Grapes, raw 50 g	0.138	6.9	5.1	9	7.3	9
Cantaloupe	0.05	2.5	58.7	98	160	200
Bananas, 1 cup	0.433	21.65	10.3	17	0.8	1
Salmon, sockeye 155g	0.236	11.8	0	0	0	0

[a] These are the amounts of each that are recommended each day (RDA).

Salmon also has 103% of the recommended daily allowance (RDA) of tryptophan, 84% of the RDA of omega-3 fatty acids, of the RDA of 76% Se, and 262 cal per 4 oz serving. Shrimp has 103% of the RDA of tryptophan, 14% of the RDA of omega-3 fatty acids, 64% of the RDA of selenium (Se), 230 mg of the RDA of cholesterol and112 cal per 4 oz. Olive oil has 9 cal per g (120 cal per tbsp). Walnuts (0.25 cup) have 91% of the RDA of omega-3 fats, 42% of the RDA of manganese, 20% of the RDA of copper, 15% of the RDA of tryptophan and 163 Calories. Cashews (0.25 cup) have 38% of the RDA of copper, 22% of the RDA of magnesium and 22% of the RDA of tryptophan and 17% of the RDA of phosphorus (P).

Also, bananas are an excellent source of potassium and other nutrients. However, almost all of the bananas sold in the USA and Europe are a single culture – the Cavendish. It is being threatened by fungal disease. So, research is being done to develop disease-resistant cultivars [42].

Another fruit that can be very healthy is grapefruit. It is low in calories and is a good source of vitamin C and other nutrients. However, it can cause sever interactions with some prescription drugs [43]. This is because the pulp and juice from grapefruits contain a class of compounds called furanocoumarins that can inhibit the CYP (cytochrome P450) enzymes that metabolize drugs [44 - 48]. As a result, some very important and popular drugs, like atorvastatin (Lipitor®) may not be metabolized as fast as they should, so they can cause toxicity and death. As a result, patients are warned not to eat grapefruit or drink grapefruit juice when taking any of 43 prescription drugs, including Alfentanil (oral) Amiodarone, Apixaban, Atorvastatin, Buspirone, Clopidogrel, Crizotinib, Cyclosporine, Darifenacin, Dasatinib, Dextromethorphan, Domperidone, Dronedarone, Eplerenone, Erlotinib, Erythromycin, Everolimus, Felodipine, Fentanyl (oral), Fesoterodine, Halofantrine, Ketamine (oral), Latatinib, Lovastatin, Lurasidone, Maraviroc, Nifedipine, Nilotinib, Oxycodone, Pazopanib, Pimozide, Primaquine, Quinine, Quetiapine, Quinidine, Rilpivirine, Rivaroxaban, Silodosin, Simvastatin, Sirolimus, Solifenacin, Sunitinib, Tacrolimus, Tamsulosin, Ticagrelor, Triazolam, Vandetanib, Venurafenib, Verapamil and Ziprasidone [44]. It should be noted that there are different CYP isozymes, some of which are especially susceptible to inhibition by furanocoumarins [45]. Even though grapefruit and its juice are especially noteworthy, black raspberry, black mulberry, plum, and wild grapes also inhibit CYPs [46]. Moreover, some vegetables, including carrots, celery, dill, cilantro, parsnips and parsley, inhibit CYP1A2 activity in humans [47] Also, there are some dietary supplements that can cause severe reactions and toxicity when taken with certain drugs and some drugs can interact with other drugs [48]. This includes *Hypericum perforatum* (St. John's wort),

Table 2. Calcium and potassium content of some healthy foods.

	Calcium	1000 mg	Potassium	4700 mg
	mg	%RDA	mg	%RDA
Asparagus, 1 cup frozen	32	3.2	124	2.6
Red Pepper, 1 cup 149 g	10	1	314	6.7
Broccoli, frozen 1 cup	41	4.1	457	9.7
Brussel sprouts, frozen 1 cup	40	4	495	10.5
Cauliflower, cooked 1 cup	22	2.2	176	3.7

(Table 2) contd.....

Garlic, 1 clove 3 g	5	0.5	12	0.3
Onions, raw 1 cup	46	4.6	234	5.0
Parsley, 10 g	14	1.4	55	1.2
Spinach, raw 1 cup 30 g	30	3	167	3.6
Tomatoes, raw 123 g	12	1.2	292	6.2
Tomato paste 1 cup	94	9.4	2657	56.5
Carrots, frozen 1 cup	37	3.7	352	7.5
Kale, frozen 1 cup	94	9.4	417	8.9
Oranges, 180 g 1 cup	52	5.2	326	6.9
Grapes, raw 50 g	16	1.6	306	6.5
Cantaloupe	14	1.4	427	9.1
Bananas, 1 cup	8	0.8	537	11.4
Salmon	9	0.9	149	3.2

Ginkgo biloba, Angelica, Cortex acanthopanacis (Acanthopanax bark), *Glycyrrhiza uralensis* (licorice), ginseng, *Salvia miltiorrhiza* (red sage) and *Allium satirum* (garlic extract) [48].

DIETARY SUPPLEMENTS

Dietary supplements are big business in the USA. Note that one of them is garlic extract. There is a school of thought that says that an extract can never be as effective as the whole food or herb, so eating a clove of garlic is thought to be better than consuming an extract. It is possible that there are things in the whole garlic that are important for maximum absorption of the nutrients and bioactive ingredients. It is also more likely that one could consume too high a dose of such concentrated ingredients, where it is doubtful that anyone would like garlic so much that they would eat themselves to death with it.

There are recommended daily allowances (RDA) of vitamins and minerals. Some of the most important, like folic acid, are added to enriched breads, cereals, flours, corn meals, pastas, rice, and other grain products. A lack of folic acid causes birth defects, such as *spina bifida*, which leaves the victim severely disabled. Folic acid is also found in multi-vitamin supplements, and these are recommended for pregnant women. However, not all pregnancies are planned and not everyone can afford to buy multi-vitamins. So, not just the US FDA, but also the governments of many other nations fortify foods with folic acid, also known as folate and vitamin B_9.

Good nutrition starts in the womb and continues with mother's breast milk. Folic acid is an essential part of pre- and post-natal nutrition. Almost all doctors tell their pregnant patients and clients to take folic acid, as part of their pre-natal vitamins. It is so important that it is added to bread, pasta and some cereals. In addition to folic acid and multivitamins, pregnant women should take two 500 mg doses of DHA three times a day, preferably with meals or snacks. It is an important nutrient for the developing human brain. Another super food or spice for pregnant women is cinnamon. Cinnamon is an excellent source of antioxidants. Continue eating cinnamon and turmeric after your baby is born. When your child gets old enough to eat cookies, consider adding one or both of these spices to the batter.

BREAST FEEDING

If at all possible, mother's should be encouraged to breast feed their babies. During the first few months, mother's milk is loaded with antibodies, to protect the baby from infection. It is also rich in the amino acid asparagine, which is important in brain development. After a few months, babies need more sulfur-containing amino acids to grow hair, fingernails and toenails, and the mother responds appropriately. For further healthy development, the milk needs to contain more Calories – and it does. Human milk also contains a type of carbohydrate called oligosaccharides, which feed good bacteria in the gut. When a baby is first born, its gut is rapidly populated by bacteria from its mother. Babies born transvaginally receive their pioneering strains of bacteria from their mother's vagina, which is less diverse than the lower intestinal tract [49]. On the other hand, babies born by Caesarian section are initially colonized by bacteria that live in the epidermis and are more susceptible to allergies and asthma than babies born transvaginally. The pioneering bacteria could have a lasting effect on the immune system. As the baby is fed human milk, a special mixture of nutrients and antimicrobial proteins influences the neonatal microbiota. Mother's breast milk also contains secretory IgA immunoglobulins (sIgAs) as a layer of defense. The sIgAs exert selective pressure that promotes the colonization of the baby's alimentary tract by innocuous, but metabolic useful microbes. The maternal sIgAs also shield the baby and infant's immune system from its own microbiota. Microbial antigens that are bound by sIgAs are identified and treated in a tolerogenic mode, partly due to sIgAs poor ability to fix complement [49].

Throughout pregnancy and during nursing, women should be encouraged to take two 500 mg gel capsules of DHA three times a day, for a total of 3000 mg, or three grams. Women should also be encouraged to do abdominal exercises after delivering a baby. During pregnancy, a woman's abdominal muscles tend to atrophy, or become weak from not being used as much as before. This can lead to constipation, which can lead to other health problems. It is also recommended that women take dietary fiber if they have problems with constipation. This is also useful for anyone who has had abdominal surgery or any other surgery or injury that keeps you inactive. As in pregnancy, the abdominal muscles become week, due to inactivity. It is also useful for the elderly, who often suffer from constipation. So, dietary fiber can be very important in preventing constipation. The best sources of dietary fiber are whole grain food, fruits and vegetables. Dietary supplements that contain dietary fiber include pectin, psyllium, and beta-glucan.

VITAMINS

It is also important to get the proper amounts of vitamins. Table **3** summarizes their names and properties:

The daily values (DVs) have replaced the former recommended daily allowances (RDAs) and adequate intakes (AIs). Some vitamins are needed in such low amount that they are better measured in international units (IUs), which are based on their biological activities. Moreover, some, like vitamin A, can be in different forms. The IU system allows them to be compared. That is, one IU of vitamin A equals 0.3 μg of retinol, 0.6 μg of β-carotene or 1.2 μg of other carotenoids.

The best-selling, most interesting and controversial dietary supplements are multi-vitamins. The American Medical Association (AMA) does not recommend them. Instead, the AMA recommends getting your vitamins and minerals from a healthy, balanced diet. Moreover, much research has shown that excess, unliganded (free) iron and copper can form deadly hydroxyl radicals that can cause cancer, heart, Alzheimer's, Huntington's and Parkinson's diseases, as well as diseases caused by prions, bacteria and environmental toxins [50 - 52]. For this reason, iron was removed decades ago from many multivitamins for men and women over 50, but almost all contain copper. So, some recommend not taking them [52].

Table 3. Olive harvested areas in the world.

Vitamin	Daily Value[1]	Chemical Name	Biological Function	Effect of Deficiency
	5000 IU	Retinoids	Vision	Night blindness
B_1	1.5 mg	Thiamin	Metabolism, cofactor	Beriberi
B_2	1.7 mg	Riboflavin	Metabolism, cofactor	Ariboflavinosis
B_3	20 mg	Niacin	Metabolism, cofactor	Pellagra
B_5	10 mg	Pantothenic Acid	Metabolism, cofactor	Paresthesia
B_6	2 mg	Pyridoxine	Metabolism, cofactor	Anemia
B_7	300 μg	Biotin	Metabolism, cofactor	Dermatitis
B_9	400 μg	Folic Acid, Folate	Metabolism, cofactor	Birth defects
B_{12}	6 μg	Cyanocobalamin	Metabolism, cofactor	Anemia
C	60 mg	Ascorbic Acid	Metabolism, antioxidant	Scurvy
D	400 IU	Ergocalciferol	Bone growth, strength	Rickets, bone disease
E	30 IU	Tocopherol	Antioxidant	Anemia
K	80 μg	Phylloquinone	Blood clotting, cofactor	Bleeding won,t stop

[1] USDA, based on a 2000 Calorie/day diet http://www.crnusa.org/about_recs.html.

Vitamin A is required for vision and, like vitamins C and E, it is a powerful antioxidant. Antioxidants are important, because reactive oxygen and nitrogen substances (RONS, also known as reactive oxygen and nitrogen species) can be deadly. Sometimes that is good, sometimes it is bad. When bacteria invade your body, white blood cells make RONS, which kill the bacteria. Unfortunately, other cells in our body also make RONS. Fortunately, our cells make some natural antioxidants, such as the enzyme called superoxide dismutase (SOD). We supplement this by consuming foods and beverages that contain vitamins A, C and E, and other antioxidants. If left undisturbed, RONS can cause diseases of inflammation, such as arthritis, heart disease, stroke, Alzheimer's disease and cancer. So, vitamins and a diet rich in fruits, vegetables and green tea can prevent many diseases, including cancer.

The B vitamins are needed for proper metabolism. They are needed for some enzymes to be active. They are called cofactors, since the enzymes are sometimes called factors. The cofactors help the enzymes catalyze some of the reactions in metabolism. Many of the B vitamins are more commonly called by their chemical names, such as thiamin, niacin, riboflavin and folic acid. Fruits and vegetables are good sources of all the vitamins in the table above, except vitamin B_{12}, which is in meat. Thus, people who eat no fish, meat, eggs, or dairy products should take vitamin B_{12} as a dietary supplement, available in soy milk and other foods. Another nutrient that is sometimes grouped with the B vitamins is called choline, which is part of the important neurotransmitter acetylcholine and is part of a phospholipid called phosphatidyl choline, which is also called lecithin. In fact, choline can be taken as lecithin in a dietary supplement. The Institute of Medicine recommends 425 to 550 mg daily, for women and men, respectively as an adequate intake of choline, and a tolerable upper intake level (UL) for adults of 3.5 g/day [53]. Beef, eggs, chicken, turkey and salmon are good sources of choline [1]. However, it should be noted that "The critical adverse effect from high intake of choline is hypotension, with corroborative evidence on cholinergic side effects (*e.g.*, sweating and diarrhea) and fishy body odor" [53]. Also, choline can be converted to trimethylamine N-oxide (TMAO) and betaine by commensural bacteria in the gut [54]. These metabolites were shown to predict risk for cardiovascular disease [54].

Vitamin C is also an antioxidant and a cofactor for the biosynthesis of many important molecules. This includes the protein collagen, essential to the development and maintenance of scar tissue, blood vessels and cartilage. Vitamin C is also a cofactor for the biosynthesis of hormones, such as adrenaline and noradrenaline.

Vitamin D is needed for proper metabolism of calcium. Calcium is needed for the growth and strength of bones and teeth. Dietary supplements that contain calcium often contain vitamin D, which helps the body absorb the calcium properly. Vitamin D also affects the immune system and helps keep it inbalance. As mentioned before, white blood cells in our immune system are needed to kill bacteria. Vitamin D helps this by stimulating the immune system when it needs stimulation. However, the immune system can get carried away and become too active. When we have an allergic reaction to a bee sting, poison ivy, pollen, mold, or other things, our immune system is too active. We can become inflamed, as the affected part of our body turns red and feels pain. Vitamin D and medicines (like the asthma medicine in Advair®) that mimic vitamin D can suppress the immune system and relieve the pain and irritation and open the airways, so we can breathe.

Vitamin E is an antioxidant and it is needed for proper fat metabolism. Vitamin E deficiency can be seen in people who can't absorb dietary fat very well. This includes premature babies (weighing less than 1500 grams, or 3.5 pounds) and people with other rare conditions, such as Crohn's disease, liver disease, or pancreatic deficiency. Vitamin E is like vitamins A and D, in that they are fat soluble. That is, they accumulate in the fatty cells and tissues of your body, unlike the water soluble vitamins (B and C), which stay in the water, in your blood, the cytoplasm of your cells and eventually your urine. So, it is almost impossible to overdose on the B vitamins or vitamin C. Your body can only absorb so much, and the rest ends up in your urine. On the other hand, you can kill yourself by taking too much vitamin A, D or E (or too many multi-vitamins that contain these). These fat-soluble vitamins are not excreted in the urine, and can bio-accumulate to toxic levels in your fat. However, you would have to eat many pills all at once to do this.

There are some people who think that very large doses of vitamins, called mega-doses, might help prevent cancer and other diseases. This is wrong, and taking too much of some vitamins or minerals can help cause diseases. It is a common mistake to think that if something is good for you, then lots of its must be really good for you. This is not true. As just mentioned, if you take many vitamin pills at once, you could die. Still, some might recommend taking ten to 100 times the recommended daily allowance (RDA) of some vitamins, especially vitamin C. In fact, a Nobel Prize winning chemist, Linus Pauling, did much research and founded a research institute to test the hypothesis that mega-doses of vitamin C can prevent the common cold, and maybe even cancer. Unfortunately, Linus Pauling died of prostate cancer at the age of 83. Though his research institute continues to do much useful research in biochemistry, his hypothesis regarding cancer was not supported in clinical studies done by others [55]. Moreover, in 2007, the prestigious *Journal of the American Medical Association* published a review article that said that supplements containing vitamins A, E and carotene may increase deaths due to cancer and other diseases, but more research is needed for vitamin C and an important mineral, selenium [56]. In 2008, the same journal published a clinical study in which vitamins C and/or E were given for ten years to 14,641 physicians who were all over 50 and were in good health [57]. The results indicated that neither vitamin C nor E (given separately or together at doses of 500 mg vitamin C and/or 400 IU of vitamin E) prevented cardiovascular disease [58]. In 2011, the *British Journal of Nutrition* published a prospective study that indicated that several vitamins and minerals, including copper and iron were linked to increased mortality in people 65 years old and over [58].

The National Institutes of Health maintains several pages on their website that have fact sheets on many dietary supplements [59]. The pages describing vitamins C and E also say that there is no convincing evidence that these vitamins can prevent cardiovascular disease or cancer [59]. Other pages, such as the one on calcium, indicate that the typical American diet usually doesn't provide enough Ca^{2+}. The RDA is different for different age groups, ranging from 210 mg/day for infants from 0 – 6 months to 1200 mg/day for men and women who are over 50 years old. On the other hand, the tolerable upper limit (TUL) for Ca^{2+} is 2500 mg/day for children over 1 and all adults. No TUL has been established for infants. So, Ca^{2+} (and almost everything else) must be carefully controlled. Insufficient Ca^{2+} can help cause osteoporosis (see Chapter 5), and can be a risk factor for other diseases [59]. However, too much Ca^{2+} can be hypercalcimia, impaired kidney function, and is a risk factor for prostate cancer [59].

So, vitamins are carbon-containing (organic) molecules that our bodies can't make for themselves, so we must obtain them in our diet. Minerals, like selenium, sodium, potassium, iron, calcium, magnesium, and zinc and many others are metal cations, do not contain carbon, and we must have some (but not too much) in our diet. We need so little selenium that it would take a microscope to see the microgram amounts that are present in multi-vitamin supplements. If there is enough selenium to see with the naked eye, though, it could be toxic. Doctors knew that selenium was toxic long before they found out that it was actually an essential nutrient. It took the invention of a powerful analytical technique (inductively coupled plasma, or ICP) to enable the detection of the very low amounts of selenium in the body. Subsequently, some enzymes were found that require selenium for their activity.

Minerals We need other minerals in the right amount, too. Table **4** lists the daily values recommended by the USDA.

Table 4. Amounts of minerals recommended per day and upper levels.

Micronutrient	Daily Value (DV)[1]	Upper Level (UL)
Calcium	1000 mg	2000 mg
Iron	18 mg	45 mg
Phosphorus	1000 μg	4 g
Iodine	150 μg	1000 μg
Magnesium	400 mg	350 mg
Zinc	15 mg	40 mg
Selenium	70 μg	400 μg
Copper	2 mg	10 mg
Manganese	2 mg	11 mg
Chromium	120 μg	ND
Molybdenum	75 μg	2000 μg
Chloride	3400 mg	3600 mg

[1] USDA, based on a 2000 Calorie/day diet http://www.crnusa.org/about_recs.html.

There is no DV for sodium on the micronutrient web page, but another (Electrolytes and Water) says that adults need 1.5 g/day and there is no maximum tolerable upper limit (UL).

www.iom.edu/Global/News%20Announcements/~/media/442A08B899F44DF9AAD083D86164C75B.ashx

Sodium it is in all foods and beverages, except distilled and deionized water. We get sodium from table salt (sodium chloride, or NaCl), and too much salt can cause bloating and even high blood pressure. This is because the cells in our body need to keep the amount of sodium under close control. When we eat too much salt it enters our cells. To keep the cells from dying, they must take in water to dilute the salt and keep it at the right concentration. Thus, our cells and our bodies swell, make us look fatter than we are, and can raise our blood pressure and cause a heart attack. Still, compared to selenium or other minerals, sodium has a relatively low toxicity – it takes much more sodium to kill you than selenium.

For people with high blood pressure that is caused by eating too much salt, there is a salt substitute. It is potassium, or potassium chloride. In some ways, potassium can almost be thought of as the opposite of sodium when it comes to the human body. The daily value is 4.7 g/day for adults. When you eat sodium or salty foods, you get thirsty and drink more water, and urinate less. When you eat fruits and vegetables like pomegranate juice, bananas, kale and asparagus, you get potassium and it makes you urinate more. They are natural diuretics. However, it can be very dangerous to take too much potassium as a pill or dietary supplement, because it is easy to overdose and die. Potassium is toxic at a much, much lower dose than sodium. While sodium might kill some people if they have high blood pressure, too much potassium will kill anybody. Fortunately, multi-vitamins only provide 1% of the daily value of potassium, and bottles of potassium chloride are clearly labeled with a warning to consult a physician before taking any. So, even though most Americans don't get enough potassium in their diets that are low in fruits and vegetables, potassium chloride supplements are almost never a good remedy, since the dose in a single pill is so low. However, there is plenty in tomato paste, tomato soup, tomatoes, asparagus, kale, pomegranate juice or bananas to be in any danger of illness of death.

Other minerals must be kept in balance, too. Calcium is the most abundant mineral in the body, and 99% of it is found in the teeth and bones. It is also an important messenger in cell signaling, as it affects the constriction and relaxation of blood vessels, muscle contraction, nerve impulses and the secretion of hormones. Its concentration in the blood is carefully controlled. When the concentration decreases too much, the parathyroid gland sends signals that cause parathyroid hormone (PTH) to be secreted. PTH stimulates the conversion of vitamin D to its active form, calcitrol, which increases the absorption of dietary calcium in the small intestines. When the concentration of calcium in the blood increases enough, the parathyroid stops secreting PTH and the kidneys excrete excess calcium. However, in chronic kidney disease, calcium levels can get too high, so such patients should not take calcium or drugs like bisphosphonates that are used to treat osteoporosis. Calcium supplements often contain zinc and magnesium and vitamin D, which is required for optimum calcium absorbance. Sometimes, zinc is recommended as a treatment for sore throats and colds. Throat lozenges containing vitamin C and zinc are popular. However, the Linus Pauling research institute mentioned a medical study that found that too much zinc can increase the chance of getting prostate cancer [60].

Over 60% of the magnesium in the body is in the bones and 27% is in the muscles. It is involved in over 300 metabolic reactions, including energy production, ion transport across cell membranes, cell signaling and the synthesis of nucleic acids. Magnesium toxicity can occur if too much is consumed in supplements, such as magnesium-based laxatives or antacids or in people who suffer from chronic kidney disease. The USDA says that the daily maximum tolerable upper limit of magnesium (350 mg) in dietary supplements is lower than the recommended daily value (400 mg). This is because no adverse effects have been reported from dietary sources of magnesium, but have been reported for dietary supplements. Also, calcium and magnesium are found in hard water. It has been found that the incidence of cardiovascular disease is lower in cities that have relatively hard water [61].

Manganese is a similar mineral that is an important part of many enzymes, including superoxide dismutase, an important natural antioxidant. It catalyzes the destruction of highly reactive superoxide anions. However, one should be careful about manganese in some dietary supplements for osteoarthritis, which can exceed the tolerable upper limit of 11 mg/day.

Chromium is an essential trace element when it is present in the +3 oxidation state, but is a known human carcinogen when present in the +6 state. When it is present in food or dietary supplements, it is Cr^{3+}. It enhances the action of insulin and is involved in carbohydrate, fat, and protein metabolism. Broccoli is an excellent source, containing 11 μg per one cup serving. Still, this provides only 9% of the recommended daily value, so it is present (as Cr^{3+}) in most multivitamins.

Molybdenum is an important cofactor for at least four enzymes. Dietary molybdenum deficiency has never been seen in humans. We know that is essential from studies on people with rare inborn errors of metabolism that result in a deficiency of sulfite oxidase, which requires molybdenum.

Iron is also an essential mineral. It is a redox-active metal that can exist in oxidation states ranging from -2 to +6. In aqueous solutions, iron precipitates as a hydroxide unless the pH is sufficiently low (as in gastric fluid), or unless it is tightly bound (liganded or chelated) to a water-soluble molecule like hemoglobin. So, iron is quite versatile. It carries O_2 and is involved in many electron transfer reactions. Almost two-thirds of it is bound to heme groups in hemoglobin (in circulating erythrocytes). A readily mobilizable store holds another 25%, while most of the remaining 15% is in myoglobin (in muscle cells). As such, it can bind O_2 and carry it through the circulatory system (as part of hemoglobin) or transfer O_2 from capillaries and into muscle cells. Iron is also carried from one tissue to another as part of the protein transferritin and it is stored inside cells in ferritin or hemosiderin. These proteins keep iron in the Fe^{3+} state, while iron in hemoglobin can be either Fe^{2+} or Fe^{3+}, depending on whether the hemoglobin is oxygenated (Fe^{3+}) or deoxygenated (Fe^{2+}). Actually, when iron is in the +3 oxidation state the protein is called methemoglobin, whereas hemoglobin has Fe^{2+}. The recommended daily allowance (RDA) of iron for boys, most men and postmenopausal women is 8 mg/day. The RDA for premenopausal women is 18 mg/day. The tolerable upper intake level (UL) is 45 mg/day, with gastrointestinal upset being the main symptom [62].

The needs for iron are quite diverse. Iron deficiency is one of the leading risk factors for disability and death worldwide, affecting an estimated 2 billion people [48]. However, an excess of dietary iron can cause atherosclerosis, heart disease, stroke, neurodegenerative diseases and many forms of toxicity [51, 52]. Improperly liganded iron can react with hydrogen peroxide that is produced in mitochondria to make the highly reactive and damaging hydroxyl radical. That is, humans are unable to actively excrete excess iron, so it must be stored. Human diet can provide both organic (heme) and inorganic (non-heme) iron. Specific protein transporters exist for both types. The intestinal heme iron transporter (HCP1) is upregulated by hypoxia and iron deficiency. Non-heme iron is transported from the intestinal lumen and into enterocytes by the divalent metal transporter 1 (DMT1).

TOXICITY OF POORLY CHELATED IRON

Even though iron deficiency may be a leading cause of disability and disease word-wide, an excess of iron may be a leading cause of disease and death in the elderly. That is, the iron hypothesis states that differences in the amounts of stored iron can explain why men are more prone to heart disease than women. Iron stores in men rise after adolescence, but remain low in women and only begin to rise after the age of about 45. The maximum sex difference in heart disease is also at age 45. Moreover, medicines like aspirin that cause gastrointestinal blood loss may protect against heart disease by decreasing iron stores. This may also partly explain the protective effect of NSAIDs on Alzheimer's disease. That is, a systems biology analysis showed that poorly liganded iron and copper can help cause not just Alzheimer's, but also Parkinson's, Huntington's, prions, bacteriocides and many forms of chemical toxicity [52]. That is, there is a continuous autocatalytic production of hydroxyl radicals involving poorly liganded iron, leading to cell death *via* apoptosis. That is, once Fe^{3+} is produced in the Fenton reaction, eq (**1**), it can react with superoxide anions, $O_2^{\cdot -}$, (also produced by mitochondria) to regenerate Fe^{2+} eq (**2**).

$$Fe^{2+} + H_2O_2 \rightarrow Fe^{3+} + OH^- + OH^{\cdot} \quad (\mathbf{1})$$
$$Fe^{3+} + O_2^{\cdot -} \rightarrow Fe^{2+} + O_2 \quad (\mathbf{2})$$

In a related reaction, the superoxide anion can react with nitric oxide (NO) to make the highly reactive peroxynitrite anion, $ONOO^-$, which can react with tyrosine or cysteine in proteins (to make nitrosylated proteins), with DNA (to make 8-hydroxy-20-deoxyguanosine) and with fatty acyls to make nitrosylated fats. So, to prevent the formation of excess hydroxyl radicals and superoxide anions, it is important to keep Fe^{2+} fully liganded. So, it has been recommended that people over 50 eat little or no meat and no iron-fortified cereals. They should also stop taking multivitamins that contain either iron or copper. In fact, most multivitamins for people over 50 have not contained iron for the past 20-30 years, so the major dietary sources of iron are iron-fortified cereals and red meat. Moreover, one study showed that several commonly used multivitamin and mineral supplements that contain iron may increase mortality in older women (average age of 61.6 years), but calcium is associated with lower risk [66, 67]. However, a recent meta analysis showed that multivitamin-multimineral treatment has no effect on mortality risk in the general population [68].

HAZARDOUS DIETARY SUPPLEMENTS

Some dietary supplements are so hazardous that they have made the consumer report's list of twelve to be avoided. Some of these may seem obscure and of little concern to most Americans, but one of them, lobelia, is wide-spread in the USA. It is more commonly known as puke weed. It contains an alkaloid, called lobeline sulfate, which is the active ingredient, or active toxin. supposedly, it can be used to get a person to quit smoking, and it can be used as an emetic, to cause vomiting. However, ipecac can induce vomiting without risk of toxicity, and this is what pediatricians recommend, not lobelia. The May 1983 issue of the *FDA Consumer* contained a list of unsafe plants, including lobelia [69]. The article said that "Overdoses of the plant or extracts of the leaves or fruits produce vomiting, pain, sweating, paralysis, depressed temperatures, rapid but feeble pulse, collapse, coma, and even death." The article had no references, but the May 1991 issue of the same magazine repeated the claim that lobelia can cause "breathing problems, convulsions, and even coma and death when used in large amounts," and cited a magazine called Tyler's *Honest Herbal* as a source. Despite the above assertion in an unofficial publication, the FDA allowed lobeline sulfate to be sold over-th--counter without prescription in doses of 6 mg per day as recently as 1990, as a smoking deterrent [69]. Remember that the dose is the poison. If an individual is able to vomit before consuming very much lobeline, the toxic substance will not reach the target organs in sufficient concentration, and there won't be any toxicity. However, if the individual does not vomit soon enough, or if he is taking other medicines, the result can be fatal.

So, remember that nature is not benign. Just because something is "all natural", it does not mean that it is harmless. Still, there are many healthy foods, dietary supplements and herbal remedies that doctor's and medicinal chemists should know about. Almost 50 million people in the USA don't have health insurance, so they can't afford to see a doctor regularly or get prescriptions. Worldwide, far more people are uninsured and don't buy prescription drugs. For them, dietary supplements are the only option. Even people who do see doctors and get prescription medicines still have health problems. They may not respond to traditional medicines. Many are fatigued, suffer constant pain, have serious allergies and other untreated health problems. Others simply want a more natural approach to health care. For all of them, dietary supplements may be appropriate.

SOME OF THE MOST IMPORTANT AND POPULAR DIETARY SUPPLEMENTS

So, let's look at some of the most important and popular dietary supplements that have been analyzed by Consumer Labs and others. These include multi-vitamins, individual vitamins, fish oil, omega-3 fatty acids, EPA, DHA, DHEA (De-hydroepiandrosterone), flaxseed oil, black cohosh, red yeast rice, coenzyme Q10, caffeine, green tea, red yeast rice, ginseng, ginkgo, curcumin, and soy isoflavones.

First, let's sort out some of the confusing acronyms and abbreviations, starting with EPA. Usually, that stands for the Environmental Protection Agency, but in the previous paragraph, it stands for eicosapentaenoic acid, or eicosapentaenoyl, when part of a triglyceride. That is the name that chemists and biochemists use for the fatty acid (fatty acyl) that has 20 carbons, five double bonds, and is an omega-3 fat. That is, there is a carbon-carbon double bond three carbons from the omega (terminal) carbon. So, it is designated n-3, omega-3 or ω-3. It is also abbreviated as 20:5(n-3). It is very similar to DHA, docosahexaenoyl or docosahexaenoic acid, which has 22 carbons, six double bonds and is an omega-3 fat, with the abbreviation 22:6(n-3). This is not to be confused with DHEA, or 5-dehydroepiandrosterone, a steroid hormone. It has been called the mother of all hormones. It is taken by some people to help them build muscle mass and to help handle stress. It has nothing to do with the omega-3 fat called DHA. There is also an acronym, DSHEA, which stands for the dietary supplement health and education act, passed by Congress in 1994. The DSHEA indicated that "the dietary supplement manufacturer is responsible for ensuring that a dietary supplement is safe before it is marketed, but the FDA is responsible for taking action against any unsafe dietary supplement product after it reaches the market". The FDA has contracts with the Association of Official Analytical Chemists

(AOAC) to develop standardized methods for analyzing supplements. The FDA also has a contract with the University of Mississippi National Center for Natural Products Research. The title of the contract is "botanical dietary supplements: science-based authentication". Like DHEA, DSHEA has nothing to do with DHA. Also, please note that most popular and scientific literature avoid the use of the words triglyceride and fatty acyl. That is probably good, for most people would think doctors and scientists are crazy if they tell you that one needs to take triglycerides (containing DHA and/or EPA) to lower the triglyceride level in the blood.

The amounts of EPA and DHA are very different in different brands of fish oil. The total EPA plus DHA can be anywhere from 8-90%, according to Consumer Labs. For example, Jarrow Formula's Max DHA with 500 mg DHA plus 72 mg EPA in every dose of two soft gel capsules. It is purified by distillation (which removes PCBs and dioxins). On the other hand, GlaxoSmithKline sells an FDA-approved prescription fish oil product called Lovaza®. It is the only prescription "drug" that is FDA approved for treating very high triglycerides. Each one gram capsule contains 38% DHA, 47% EPA, and 17% other fish oils. It is made up of omega-3 acid ethyl esters. The ethyl esters can be digested and are converted to free omega-3 fatty acids, which can be re-attached to glycerol in human cells as fatty acyls to make mono-, di- and triglycerides, which can be used in the body and are especially important in the brain. However, strict vegetarians might prefer to buy DHA that has been prepared from algae instead of from fish.

Also, many people like to eat flaxseed oil, which contains a much higher amount of omega-3 fats. It should also be added that the fats in flaxseed oil can't be used directly by human cells. They need to be converted by human metabolism and the efficiency is only about 10%. So, the 7.7 g of omega-3 fats per serving is equal to about 770 mg of usable omega-3 fats. It is also important to mention that much of the original research on flaxseed oil was done by Doctor Johanna Budwig, in Germany. She suggested mixing flaxseed oil with cottage cheese or a German and Swiss delicacy called Quark (which is similar to yogurt). The reason she gave was that cottage cheese and Quark are relatively high amount of sulfur, which would supposedly bind the omega-3 fats in flaxseed oil and increase their absorption through the stomach and into the blood. Onions and garlic also have sulfur-containing compounds, so if this really helps the absorption of omega-3 fats, they should work as well. However, sulfur-containing compounds don't bind omega-3 fats tightly (if at all), so they probably don't make a difference.

Doctor Budwig claimed that she had cured several people of cancer by giving them flaxseed oil. She was criticized by some for not following strict scientific methods in her research and praised by others who nominated her for a Nobel Prize. The truth is probably somewhere in between. People are still doing research to see if flaxseed oil can prevent breast cancer and other diseases. By itself, it almost certainly can't cure cancer. Complex problems have complex causes. There are many things that need to be done to prevent or cure cancer. We are not simple machines. Omega-3 fats and many other things can help lower your odds of getting cancer or any other disease of inflammation.

It should be added that flax seeds are almost useless. They are small and have very hard shells. So, the only way to get the oil out of the seeds is to crush them with a lot of pressure. This usually requires a machine. The seeds are too small for you to be able to effectively crush them when you chew. So, you will never get the oil. The seeds will be like insoluble fiber and just pass through you when you defecate. This is fortunate, since flaxseeds are usually found in baked bread. So, the seeds are heated in the oven with the dough. This causes the healthy omega-3 fats to be oxidized, forming very unhealthy oxidation products. The C=C bonds in omega-3 fats are easily oxidized when exposed to heat or light, so flaxseed oil should never be used as a cooking oil. It should be in a dark bottle and be stored refrigerated.

Two other words that are often misused are anise and licorice. Licorice is the name of a plant (*Glycyrrhiza glabra*) and a type of candy made from licorice extract, sugar and a binder. Licorice extract has many properties, including sweetness, due to glycyrrhizin, which is sweeter than sugar (sucrose, glucose or fructose). Licorice also contains a similar compound, glycyrrhizic acid. In Japan, it is used to treat viral hepatitis and in China, it is used for tuberculosis. Some types of licorice candy also have anise oil, which contains the fragrant chemical, anethole. So, the smell of licorice (or liquorice) is due to anise or anise seed which comes from a flowering leguminous plant, *Pimpinella anisum*. The same fragrance is in the spices, fennel and tarragon.

On the other hand, these are quite different than Chinese and Japanese star anise. Chinese star anise is used to prepare a tea that is given to infants to treat colic. Chinese star anise is an entirely different genus and species (*Illicium verum*) than *P. anisum* (anise). It is often sold as a dehydrated powder, since it is a tea. As such, it is difficult to distinguish between it and Japanese star anise, *Illicium religiosum*, which is to be used only as incense and never to make a tea. It contains several potentially deadly neurotoxins, including anisatin. A few years ago, there were cases of seizures in infants who were given tea made from mislabeled star anise. Since then, analytical methods have emerged to identify anisatin and distinguish Chinese from Japanese star anise.

Another important supplement is black cohosh, or Remifemin® from Glaxo-SmithKline. Black cohosh is the common name for an herb that has the botanical name *Actaea racemosa* [70], formerly known as *Cimicifuga racemosa*. The active ingredients are triterpene glycosides, including 23-epi-26-deoxyactein [71] which was formerly named 27-deoxyactein. Black cohosh has been shown to be effective in treating menopausal symptoms [72].

Red yeast rice is a dietary supplement that contains lovastatin (also known as monacolin K and mevinolin) and other statins. It has been shown to decrease the concentration of low density lipoproteins (LDL) and total cholesterol in patients who could not tolerate prescription statins very well [73, 74]. In the first study, three 600 mg capsules containing 1.02 mg lovastatin per capsule of red yeast rice were given twice daily [73]. In the second study, 2400 mg were given twice daily [74]. Both studies emphasized that people wanting to take red yeast rice should consult with their physicians first, since some people have liver and/or muscle toxicities when taking statins.

Even though statins lower blood cholesterol, they may decrease the amount of an important healthy lipid, called coenzyme Q10, or CoQ10. Coenzyme Q refers to coenzymes that are needed to produce energy in the mitochondria of cells. Cells in different organisms can have six to ten repeat units (called isoprenoids). Mammalian and humans have the form of CoQ that has 10 repeat units, so it is called CoQ10. It is important for energy metabolism, but all of our cells can produce it, including muscle cells, especially in the heart. If there were not enough CoQ10, we could be vulnerable to heart attacks. As a result, dietary supplements containing CoQ10 are popular, but controversial, especially if you are not taking statins. It is not that CoQ10 is harmful; it's just that it might be a waste of money for most people. However, there has been a book written about CoQ10 that claims that CoQ10 supplements should be taken, since they can (supposedly) prevent and even cure cancer [75].

CoQ10 has been used to treat cardiovascular diseases, high blood pressure, diabetes, cancer and even infertility [76]. Although CoQ10 is found in all human cells, its concentration is highest in the heart, pancreas, liver and kidneys. Healthy people don't need to take CoQ10 supplements, since human cells can make their own CoQ10. However, when taking statins, suffering from heart disease and when we become very old, we may not be able to make enough of our own and require CoQ10 supplements. In clinical studies, CoQ10 has been shown to improve two markers of cardiac health in people suffering from cardiomyopathy [77]. In a different study, CoQ10 helped the performance of world-class endurance athletes. It helped Finnish cross-country skiers [78]. It has been used to treat muscular dystrophy [79]. It helped cause a complete regression of tumors in breast cancer patients when given along with a conventional protocol of therapy. In one patient, the cancer had spread, but CoQ10 helped completely eliminate the metastatic cancer [80].

In a letter to the FDA, a group of scientists and doctors, called the International Coenzyme Q10 Association issued a position statement on the adverse effects related to the use of statins because they block the biosynthesis of CoQ10 [81]. Moreover, patients who already have had congestive heart failure may be especially susceptible to damage due to the decrease in CoQ10 caused by statins. The letter went on to say, "It is possible that the recently reported statin-related deaths are the tip of a side effect iceberg and the magnitude of the potential problem cannot be overstated. It is urgently incumbent upon the scientific community, the pharmaceutical industry and the United Sates Food and Drug Administration to be certain that we are not inadvertently creating a life-threatening deficiency of an essential co-factor in many millions of patients".

Even though statin therapy has been shown to save lives, the long-term effects on heart disease may offset some of the gains. So, people taking prescription statins and/or red yeast rice may want to take CoQ10 supplements.Also, CoQ10 is a critical addition to conventional treatments for patients with congestive heart failure, even when the traditional medicine seems to be successful.

Unfortunately, many people also take dietary supplements that contain large amounts of caffeine. Some of them are energy drinks and others are for weight loss. When taken in moderation (in coffee and tea), caffeine stimulates the central nervous system, temporarily preventing drowsiness and restoring alertness. It is also the primary treatment for breathing disorders caused by apnea of prematurity and may be effective in preventing bronchopulmonary dysplasia in premature infants. It is the world's most widely consumed psychoactive drug. It is found in the seeds, leaves, and fruit of some plants, where it acts as a natural pesticide that paralyzes and kills some insects that feed on them, as well as enhancing the reward memory of pollinators. It is most commonly consumed as an infusion extracted from coffee beans or tea leaves. These can be decaffeinated by extracting with water or by supercritical fluid extraction. It is also in yerba maté, guaraná, guayusa and yaupon holly. When consumed in moderation (500 mg or less daily), it is usually safe and may even protect against some diseases, including Parkinson's disease [82]. However, it can be toxic at doses over 10 grams per day for adults. Even though energy drinks contain much less than this, they can still be quite harmful – especially to children and adolescents [83]. Common symptoms of caffeine toxicity include nervousness, anxiety, restlessness, insomnia, gastrointestinal upset, tremors, tachycardia, psychomotor agitation and in rare cases, death. Moreover, they are not well regulated and are widely advertised as being useful [82]. For example, the list of ingredients on some energy drinks lists guaraná as one of the ingredients, along with caffeine. Few people outside of Brazil will know that guaraná itself contains much caffeine.

One healthy source of moderate amounts of caffeine is green tea. It is made from the leaves of *Camellia sinensis* that have been through a minimal oxidizing process. It is not the caffeine, but the antioxidants that make green tea so healthy. The leaves used to make the green tea are 20-45% polyphenols, of which 60-80% are catechins [84]. Epigallocatechin (EGCG) is the most abundant catechin. The

US FDA has approved a topical ointment (sinecatechins 15%, Verege®) for the treatment of external genital and perianal warts caused by the human papillomavirus. The active ingredients are catechins that are extracted from green tea. They are immunomodulators that inhibit major viral functions. Also, meta-analyses of previous publications found that a relatively high consumption of green tea was associated with a 20% and 18%reduction of the risks of breast cancer and colorectal cancers, respectively. In a prospective cohort study of >40000 Japanese adults found that green tea was inversely associated with cardiovascular mortality. Moreover, participants who consumed >5 cups per day had a significantly reduced incidence of stroke [84]. However, many people consume very large quantities of catechins in dietary supplements containing extracts of green tea. So, there has been a spike in the number of people admitted to emergency rooms due to liver failure, caused by the over-consumption of green tea extracts and other dietary supplements, including bodybuilding supplements that contain steroids and pills that supposedly cause weight loss [85].

Ginseng is another popular dietary supplement. It can be any one of 11 species in the genus *Panax*. The dried roots and rhizomes of ginseng have been used in Asia for at least 2000 years [86] and by Native Americans long before the arrival of Europeans [87]. In four recent years, ginseng has accounted for 15-20% ($300 million annually) of the market share of dietary supplements in the United States [88]. There has been some confusion about the terminology and the distinction among Asian (Chinese or Korean) ginseng (*Panax ginseng* C.A. Meyer, from the family Aralaceae), North American ginseng (*Panax quinquefolius* Aralaceae), and so-called Siberian ginseng (*Eleutherococus senticosus*). The pharmacologic properties of ginseng are usually attributed to its triterpene glycosides, called ginsenosides [89], although bioactive polyacetylenes, sesquiterpenes, peptidoglycans and polysaccharides have also been identified [90]. It has been used for thousands of years to maintain homeostasis and enhance vital energy, especially in Asian countries. Several studies have been published showing its potential health effects, including improving systemic immune function, glucose metabolism, sexual function and mental capacity. Oxidative stress has been linked to all of these, so a clinical study was done on healthy volunteers. It found that

P. ginseng extracts (1 or 2 g/day) enhanced the antioxidant defense mechanism in healthy people [90].

Ginkgo biloba is another popular dietary supplement. It is a highly revered tree that has been used for over 1500 years in China [91]. It has been called a living fossil, since it first appeared on Earth before the Age of the Dinosaurs – during the Permian epoch, 270 million years ago. The medicinal parts of the tree are the fresh or dried leaves. A dry, standardized extract is prepared by mixing the dried leaves with acetone plus water, followed by purification, which includes removing the water and acetone. The active ingredients are divided into five categories: flavonoids (0.5 – 1.8%), biflavones (0.4 – 1.9%), proanthocyanins (8 – 12%), trilactonic diterpenes (0.06 – 0.23%), ginkgolides A, B, C and trilactonic sesquiterpene bilabolids (0.04 – 0.2%) [91]. Ginkgolide B is a potent inhibitor of platelet activating factor, also known as PAF. It inhibits the binding of PAF to its receptor. This prevents PAF from causing bronchioconstriction or constriction of the airways in the lungs. This causes coughing, wheezing and shortness of breath. PAF also induces inflammation and changes in vascular permeability. So, ginkgolide B has important anti-inflammatory activities because it inhibits PAF.

Ginkgo biloba protects against damage caused by the temporary loss of blood flow, which is also called ischemia. It is the most common form of stroke. *Ginkgo* reduces unwanted neutrophils from infiltrating the brain after ischemia. It increases blood flow to prevent further ischemic damage that can lead to dementia. *Ginkgo biloba* has additional antioxidant properties due to the flavonoids in it. However, most people take *Ginkgo* to improve their memory. It might work for some people. Indeed, many people swear by it. Each individual must decide for himself or herself if this is true for them. However, there is clear clinical evidence that *Ginkgo* does NOT improve the memories of people suffering from mild cognitive impairment or Alzheimer's disease [92].

However, *Ginkgo* might be helpful in treating vertigo, dizziness or both. A clinical study on 44 patients was done. Half took 80 mg of a standardized *Ginkgo* extract (called Egb 761) twice daily and the other took a standard medicine called beta-histidine dihydrochloride. The group taking the *Ginkgo* extract enjoyed the same relief of dizziness and vertigo as the other group [93].

The indications and uses of *Ginkgo* are: symptomatic relief of organic brain dysfunction, including intermittent claudication (ischemic muscle pain, limping), vertigo and tinnitus (of vascular origins). Claudication is one of the most common things caused by peripheral vascular disease, which is caused by atherosclerosis. Claudication is caused by physical activity and relieved by a short rest. Unproven uses of *Ginkgo* are: brain dysfunctions that result in dizziness and headaches with emotion lability (volatility) and anxiety. *Ginkgo* has been proven to improve memory and concentration caused by peripheral occlusive disease, also known as peripheral arterial disease (PAD). In Chinese medicine, *Ginkgo* is used to treat asthma, tinnitus (ringing in the ear), hypertonia (abnormal increase in muscle tension and a reduced ability of a muscle to stretch) and angina pectoris (chest pain).

Some people are allergic or sensitive to *Ginkgo* preparations, and they should not take it. So, if you want to take *Ginkgo* (or any other new food or supplement), first try just a very small amount to see if you might develop an allergic reaction. If all is well after about 5 min, you can try taking the complete recommended dose. On the other hand, patients with intracranial hemorrhage or systemic arterial hypertension should not take *Ginkgo* due to a report of subarachnoid hemorrhage [94]. There have also been occasional gastrointestinal complaints. Spontaneous bilateral subdural hematomas, subarachnoid hemorrhage and an increase in bleeding have been associated with excess, chronic *Ginkgo* consumption [94].

Ginkgo is available in liquid and solid forms for oral intake. It is supplied as capsules or tablets containing anywhere from 30 to 500 mg total weight. It is also supplied as concentrated liquid extract. *Ginkgo biloba* should be standardized to contain 24% flavone and 6% terpene lactones. A standard dose of this is 40-80 mg, three times a day. Some studies have shown that it is effective in treating dementia, peripheral arterial occlusive disease and equilibrium disorders (vertigo and tinnitus).

Soy isoflavones are also popular. They can be consumed as part of soy products or as extracts from soybeans. They are estrogen mimetics, but unlike environmental toxins like bisphenol A, they can be quite healthy. Their biological effects are strongly influenced by their metabolism, which depends on the types of bacteria present in the intestines [95]. They can be metabolized into equol, which has a higher estrogenic activity than the naturally occurring isoflavone called daidzein. Moreover, soy isoflavones can be either estrogenic or anti-estrogenic, depending on the tissue. That is, there are estrogen receptors in not just reproductive organs, but also in bones, liver, heart and the brain. Soy isoflavones also have effects that do not depend on their estrogenic activity. This includes being able to inhibit tyrosine kinases, which are important causative agents in cancers. Interestingly, the incidence of breast cancer in Asia, where average isoflavone intakes from soy foods range from 25-50 mg/day [96], is lower than breast cancer rates in the Western countries where average isoflavone intakes in non-Asian women are less than 2 mg/day [97, 98].

Two important organizations have websites that contain much information on dietary supplements and other subjects. They are the National Standard [99] and the National Institute of Environmental Health Science (NIEHS) [100]. The National Standard has databases on foods, herbs and supplements and four other subjects and interactive tools. The NIEHS has information on environmental health and toxicology. One of the choices on the left side of the NIEHS home page is National Toxicology Program (NTP). If you click on it, the home page for the NTP [101] will be uploaded. Under the section called Testing Information, you can click on testing status of agents at NTP, and it will let you search the database for chemicals or herbal remedies, such as grape seed extract, and you will be able to download the NTP report on the rodent toxicity of a standardized grape seed extract. There will also be recent reports, such as the report on bisphenol A. Also on the NTP home page is a database of substances that have been nominated for study at the NTP. It shows that four substances were recommended for study in 2008.

The NIH has an office of dietary supplements [102]. It provides information on the use and safety, nutrient requirements, database resources, news and research. The NIH has also funded several grants to study the possible efficacy of some supplements. However, it is the FDA that has regulatory responsibility for dietary supplements, as dictated by the dietary supplement health and education act, or DSHEA, passed in 1994. The DSHEA indicated that "the dietary supplement manufacturer is responsible for ensuring that a dietary supplement is safe before it is marketed, but the FDA is responsible for taking action against any unsafe dietary supplement product after it reaches the market". The FDA has contracts with the AOAC International to develop standardized methods for analyzing supplements.

The DSHEA also provided a definition of dietary supplements. They included products that contain vitamins and minerals, amino acids, enzymes, herbs, organs, tissues, glandular extracts and metabolites. Many people have the mistaken idea that if something is good for you, then more of it must be even better. This can lead people to over-fertilize their garden and over-consume substances that may be helpful or even essential at lower concentrations. It is well known in the field of toxicology that the dose is the poison. All substances are toxic, if given at a high enough doses. Even water is toxic if too much is consumed, often as an initiation rite in a sorority or fraternity. Fat soluble vitamins can be toxic if too high a dose is taken. Water-soluble vitamins are excreted in the urine when the dose is too high, so some substances are not toxic because a high dose can't reach the cells, short of an intravenous injection. Similarly, therapeutic or dietary substances can't do any good if they don't reach their target organs. Many substances can't pass through the blood-brain barrier, for example, and won't reach the brain.

Dietary supplements are big business in the USA. According to the Nutritional Business Journal, total sales of dietary supplements were $23.7 billion in 2007, with vitamin supplements being the biggest seller. Weight loss supplements sales were $1.7 billion and sports supplements were $2.5 billion. In December 2008, the FDA issued a warning that some weight loss supplements contained active pharmaceutical ingredients, including sibutramine, fenproporex, phenytoin, bumetanide, furosemide, fluoxetine, phenolph thalein, rimonabant, and cetilistat [103].

So, one should be careful when buying weight loss supplements. One might think that he or she is buying a natural product when in reality they could be buying a prescription drug or a drug that has been banned.

One dietary supplement whose name has become quite popular in the USA is the tropical fruit called acai or açaí in Portuguese. It grows on palm trees in the Amazon rain forest in Brazil and other countries. It has been consumed by undernourished natives for hundreds of years. It is an oleaginous fruit like olives or avocados. When squeezed, it produces an oil, not a juice. The berry surrounds a relatively large seed that is about 95% of its weight. In the Amazon, fruits are picked and the berries are scraped off and made into a pulp that can be sold as thick açaí, or water can be added to it to make medium or thin açaí. Depending on the amount of water added, it is labeled in Brazil as açaí grosso (if it has 14-95% total solids), açaí médio (10-14% total solids) or açaí fino (<10% solids). Açaí grosso is not exported out of Brazil, so you can only find açaí médio or fino in the USA and over the Internet. You can also find many fraudulent products labeled açaí that claim that it is a super food and can help you lose weight and even turn dark hair blonde. However, the açaí berry is 50% fat, as triglycerides and will stain your hands, clothes and hair dark. So, there is a website that shows a picture of an overweight woman with dark hair in a picture labeled before and supposedly the same woman who is thin and blonde labeled after. In Brazil and other knowledgeable countries the "before" and "after" labels would be reversed. In Brazil, everybody knows that açaí will make you gain weight and stain your hair dark. Also, it is an acquired taste that most people find disagreeable – if it is genuine. However, there are many products in the USA labeled açaí that taste like grape or raspberry juice, because that is what they are. Until recently there was no good way to tell if açaí is genuine.

However, methods are available for analyzing açaí to see if it is genuine. The simplest test is a visual inspection, to see if it is dark purple or black. The next easy test is to measure the total solids by weighing the sample of açaí, let the water evaporate off and weigh the solid residue. The next test is to add hot water to the residue and see if it all dissolves. It should not. There should be a significant amount (about 24%) of insoluble solids. The next test that requires no analytical equipment except for a balance for weighing is to extract the açaí with hexane or a mixture of methanol plus chloroform (1:2, v/v), which will extract out the fats. Genuine acai should be at least 40% fat. If an NMR is available, the fat can be analyzed to show that it is made of triglycerides containing about 68% oleoyl, 7% linoleoyl, 1.5% linolenoyl and 23% saturated fatty acyls. The insoluble solids should produce a characteristic NMR spectrum. Finally, açaí should contain cyanidin-O-glucoside and cyanidin-O-rutinoside, when analyzed by HPLC or LC-MS/MS. It should also have a relatively high oxygen radical absorbance capacity, or ORAC value. It is the high ORAC value of 1027 μmol of Trolox equivalents per gram (TE/g) that was reported by one group [104] makes many people think it is a super food. It should be noted that another species of *Euterpe*, *E. precatoria*, is also called acai in Spanish speaking regions of the Amazon rain forest. It has a higher ORAC value than *E. oleracea* [105]. Also, a type of sorghum bran was found to have 3124 μmol TE/g. The spices cinnamon, cloves, oregano and turmeric have been reported to have higher values than açaí, too.

It should be added that there are many products sold in the USA and over the Internet that are labeled açaí juice and some say it is 100%, with no water added. This is impossible. As mentioned before, açaí produces an oily pulp, not a juice. It is possible to add water to açaí berries, filter them and produce a purple liquid that will contain cyanidin-O- glucoside and cyanidin-O-rutinoside. However, these account for less than 10% of the total ORAC value that is obtained for genuine, unfiltered açaí when it is extracted with a mixture of acetone and water [105]. So, the so-called "juice" does have water added to it, making it at the most 14% açaí, not 100%. Also, most of the "juices" taste like grape or raspberry juice. To see if any juice contains grape juice, it can be analyzed by ion chromatography to look for tartrate, which is found in grapes, but not açaí or pomegranate juice [106, 107]. So, açaí is an excellent food for people who need more calories in their diets, but it will not help people lose weight. Açaí does contain many antioxidants, so if it is consumed in moderation and if a person does not consume too many total calories, it could be a healthy addition to the diet and may help provide energy to endurance athletes. For a more detailed and very readable discussion about açaí, see an article written in The New Yorker magazine, which talks about a genuine frozen açaí pulp sold by Sambazon™ [108]. The details on analytical methods for verifying the authenticity and quality of acai have been described [109 - 111].

Another dietary supplement that might be good is *myo*-inositol. In a recent study of people who smoke tobacco products, it was found that the phosphatidylinositol 3-kinase (PI3K) pathway is over-activated in the respiratory tract of smokers with lung cancer of pre-cancerous abnormal dysplastic lesions [112]. Even though tobacco smoke is a huge risk factor for lung cancer, not all smokers get it and we didn't know why. It was already known that several signal transduction pathways are deregulated in lung cancer, including not just the PI3K, but also the ras and p53 pathways. The authors looked at the types of genes expressed in smokers. They found that the PI3K pathway was the one that was clearly activated. They looked at tissues from patients with dysplasias who had been successfully treated with *myo*-inositol, an inhibitor of the PI3K pathway, and found that the PI3K pathway was effectively down-regulated [112].

KEY POINTS

1. **Another important way to prevent diseases and maintain good health is through proper nutrition. The five major food groups: cereals, vegetables, fruits, dairy, meat (and meat substitutes), and fats, oils and sweets. The USDA recommends eating three or more ounce-equivalents of whole grain products each day. Two cups of fruit, 2.5 cups of vegetables, and 5.5 ounce equivalents from the meat and bean group per day are recommended for a 2000 calorie per day diet. It is more important to consider the effect of different foods on the glucose concentration in the blood, which is defined as the glycemic index.**

2. **A lack of folic acid causes birth defects, such as *spina bifida*, which leaves the victim severely disabled. Folic acid is also found in multi-vitamin supplements, and these are recommended for pregnant women. In addition to folic acid and multivitamins, pregnant women should take two 500 mg doses of DHA three times a day, preferably with meals or snacks. It is an important nutrient for the developing human brain.**

3. **It is better to eat many types of fish, than to eat red meat. Fish and vegetable oils are better than butter, margarine, or solid fat from pork or beef. Cold water oily fish, such as salmon, anchovies, herring and mackerel, are good sources of omega-3 fatty acids. Some of the omega-3 fatty acids (ALA, EPA and DHA) are essential and may prevent heart disease. However, farm-raised Atlantic salmon has been found to have relatively high levels of toxic pollutants, called PCBs and dioxins [9]. Unsaturated fats, including polyunsaturated fats and omega-3 fats, are liquids and are much better than saturated or *trans* fats, which are solid at room temperature. Even though cookies and crackers taste good, they contain**

trans fats and should be avoided.

4. If at all possible, mother's should be encouraged to breast feed their babies. During the first few months, mother's milk is loaded with antibodies, to protect the baby from infection. It is also rich in the amino acid asparagine, which is important in brain development. After a few months, babies need more sulfur-containing amino acids to grow hair, fingernails and toenails, and the mother responds appropriately. For further healthy development, the milk needs to contain more Calories – and it does. Human milk also contains a type of carbohydrate called oligosaccharides, which feed good bacteria in the gut. When a baby is first born, its gut is rapidly populated by bacteria that can cause disease. As he (or she) is fed human milk, healthy bacteria replace the unhealthy bacteria.

5. The best-selling, most interesting and controversial dietary supplements are multi-vitamins. The American Medical Association (AMA) does not recommend them. Instead, the AMA recommends getting your vitamins and minerals from a healthy, balanced diet. Vitamin A is required for vision and, like vitamins C and E, it is a powerful antioxidant. The B vitamins are needed for proper metabolism. Vitamin C is also an antioxidant and a cofactor for the biosynthesis of many important molecules. Vitamin D is needed for proper metabolism of calcium. Calcium is needed for the growth and strength of bones and teeth. Dietary supplements that contain calcium often contain vitamin D, which helps the body absorb the calcium properly. Vitamin E is an antioxidant and it is needed for proper fat metabolism.

6. The National Institutes of Health maintains several pages on their website that have fact sheets on many dietary supplements. For people with high blood pressure that is caused by eating too much salt, there is a salt substitute. It is potassium, or potassium chloride.

7. Many people like to eat flaxseed oil, which contains a much higher amount of omega-3 fatty acids (or fatty acyls). It should also be added that the fats in flaxseed oil can't be used directly by human cells. They need to be converted by human metabolism and the efficiency is only about 10%. So, the 7.7 g of omega-3 fatty acids per serving is equal to about 770 mg of usable omega-3 fatty acids. Flax seeds are almost useless. They are small and have very hard shells. So, the

only way to get the oil out of the seeds is to crush them with a lot of pressure.

8. **Red yeast rice is a dietary supplement that contains lovastatin (also known as monacolin K and mevinolin) and other statins. People wanting to take red yeast rice should consult with their physicians first, since some people have liver and/or muscle toxicities when taking statins. Even though statins lower blood cholesterol, they may decrease the amount of an important healthy lipid, called coenzyme Q10, or CoQ10. Coenzyme Q refers to coenzymes that are needed to produce energy in the mitochondria of cells.**

9. **The NIH has an office of dietary supplements. It provides information on the use and safety, nutrient requirements, database resources, news and research. However, it is the FDA that has regulatory responsibility for dietary supplements, as dictated by the dietary supplement health and education act, or DSHEA, passed in 1994. The DSHEA indicated that the dietary supplement manufacturer is responsible for ensuring that a dietary supplement is safe before it is marketed, but the FDA is responsible for taking action against any unsafe dietary supplement product after it reaches the market [1].**

10. **The popular dietary supplement, açaí, will make you gain weight – not lose weight. Another supplement, *myo*-inositol, may prevent lung cancer in smokers.**

ABBREVIATIONS

AMA	=	American Medical Association
FDA	=	Food and Drug Administration
DSHEA	=	Dietary Supplement Health and Education Act
NIH	=	NationalInstitutes of Health
USDA	=	US Department of Agriculture
GM	=	Genetically Modified
GMOs	=	Genetically Modified Organisms
ALA	=	Alpha Linoleic Acid, DHA Docosahexaenoic Acid
EPA	=	Eicosapentaenoic Acid
PCBs	=	Polychlorinated Biphenyls
CAFOs	=	Confined Animal Feed Organizations
ADHD	=	Attention Deficit Hyperactivity Disorder
EPA	=	Environmental Protection Agency

ROS	=	Reactive Oxygen Substances
USAID	=	US Agency for International Development
CIMMYT	=	International Maize and Wheat Improvement Center
CGIAR	=	Consulting Group on International Agricultural Research
GFP	=	Green Fluorescent Protein
RDA	=	Recommended Daily Allowance
CYP	=	Cytochrome P450
TMAO	=	Trimethylamine N-Oxide
TUL	=	Tolerable Upper Limit
DV	=	Daily Value
ICP	=	Inductively Coupled Plasma
PTH	=	Parathyroid Hormone
HCP1	=	Heme Iron Transporter
DMT1	=	Divalent Metal Transporter 1
DHEA	=	Dehydroepiandrosterone
AOAC	=	Association of Official Analytical Chemists
CoQ10	=	Coenzyme Q10
EGCG	=	Epigallocatechin
PAF	=	Platelet Activating Factor
PAD	=	Peripheral Arterial Disease
NIEHS	=	National Institute of Environmental Health Science
NTP	=	National Toxicology Program
TE/g	=	Trolox Equivalents Per Gram
ORAC	=	Oxygen Radical Absorbance Capacity
NMR	=	Nuclear Magnetic Resonance
PI3K	=	Phosphoinositide 3-Kinase
p53	=	Protein with a Molecular Weight of 50,000.

REFERENCES

[1] Grandjean AC, Reimers KJ, Bannick KE, Haven MC. The effect of caffeinated, non-caffeinated, caloric and non-caloric beverages on hydration. J Am Coll Nutr 2000; 19(5): 591-600. [http://dx.doi.org/10.1080/07315724.2000.10718956] [PMID: 11022872]

[2] USDA National Nutrient Database for Standard Reference. Nutrient List http://www.ars.usda.gov/Services/docs.htm?docid=22769. , [Accessed Feb 15, 2014];

[3] Ebbeling CB, Swain JF, Feldman HA, *et al.* Effects of dietary composition on energy expenditure

during weight-loss maintenance. JAMA 2012; 307(24): 2627-34.
[http://dx.doi.org/10.1001/jama.2012.6607] [PMID: 22735432]

[4] Foran JA, Carpenter DO, Hamilton MC, Knuth BA, Schwager SJ. Risk-based consumption advice for farmed Atlantic and wild Pacific salmon contaminated with dioxins and dioxin-like compounds. Environ Health Perspect 2005; 113(5): 552-6.
[http://dx.doi.org/10.1289/ehp.7626] [PMID: 15866762]

[5] Schlosser E. Fast Food Nation. New York: Houghton-Mifflin 2001.

[6] Kingsolver B. Animal, Vegetable, Miracle. New York: Harper Perennial 2007.

[7] Park A. Time Magazine 51 Things we can do to save the environment. http://www.time.com/time/specials/2007/environment/. , 2007 May; [Accessed Feb 20, 2014];

[8] Fiala N. The greenhouse hamburger. Sci Am 2009 Feb; 300(2): 72-5.
[http://dx.doi.org/10.1038/scientificamerican0209-72] [PMID: 19186709]

[9] Oceans Alive. website: http://www.oceansalive.org/eat.cfm?subnav=bestandworst. , [Accessed Feb 21, 2014.];

[10] National Resources Defense Council. Mercury contamination in fish. website: http://www.nrdc.org/health/effects/mercury/protect.asp. , [Accessed Feb 21, 2014.];

[11] U.S. Department of Health and Human Services and U.S. Environmental Protection Agency. What You Need to Know about Mercury in Fish and Shellfish, website: http://www.epa.gov/hg/advisories.htm. , [Accessed Feb 21, 2014.];

[12] National Healthy Mothers. Healthy Babies Coalition, Brainy babies, healthy kids. Website: http://blog.syracuse.com/family/2007/10/_washington_in_a.html. , [Accessed March 11, 2014.];

[13] Purdue University. Fish for Your Health Website: http://www.fish4health.net. , [Accessed Feb 18, 2014.];

[14] Piersen CE. Phytoestrogens in botanical dietary supplements: implications for cancer. Integr Cancer Ther 2003; 2(2): 120-38.
[http://dx.doi.org/10.1177/1534735403002002004] [PMID: 15035899]

[15] Setchell KD, Zimmer-Nechemias L, Cai J, Heubi JE. Isoflavone content of infant formulas and the metabolic fate of these phytoestrogens in early life. Am J Clin Nutr 1998; 68 (6 suppl): 1453S-61S.
[PMID: 9848516]

[16] Turmeric SK. An overview of potential health benefits. Nutr Today 1010(45): 216-25.

[17] Krishnaswamy K. Traditional Indian spices and their health significance. Asia Pac J Clin Nutr 2008; 17 (suppl. 1): 265-8.
[PMID: 18296352]

[18] Ghorbani Z, Hekmatdoost A, Mirmiran P. Anti-hyperglycemic and insulin sensitizer effects of turmeric and its principle constituent curcumin. Int J Endocrinol Metab 2014; 12(4): e18081.
[http://dx.doi.org/10.5812/ijem.18081] [PMID: 25745485]

[19] Rowe DL, Ozbay T, O-Regan RM, Nahta R. Modulation of the BRCA1 protein and induction of apoptosis in triple negative breast cancer cell lines by the polyphenolic compound curcumin. Breast Cancer Clin Res 2009; 2: 61-75.

[20] Food and Agriculture Organization. Curcumin Chemical and Technical Assessment. ftp://ftp.fao.org /es/esn/jecfa/cta/CTA_61_Curcumin.pdf. 2004.

[21] Tiwari V, Shanker R, Srivastava J, Vankar PS. Changes in antioxidant activity of spices – turmeric and ginger on heat treatment. Electron J Environ Food Chem 2006; 5: 1313-7.

[22] Jäger R, Lowery RP, Calvanese AV, Joy JM, Purpura M, Wilson JM. Comparative absorption of curcumin formulations. Nutr J 2014; 13: 11.
[http://dx.doi.org/10.1186/1475-2891-13-11] [PMID: 24461029]

[23] Zebib B, Mouloungouli Z, Noirot V. Stabilization of curcumin by complexation with divalent cations in glycerol/water system. Bioinorg Chem Appl 2010; 292760. Epub 2010 Jun 16
[http://dx.doi.org/10.1155/2010/292760]

[24] Bisht S, Feldmann G, Soni S, *et al.* Polymeric nanoparticle-encapsulated curcumin ("nanocurcumin"): a novel strategy for human cancer therapy. J Nanobiotechnology 2007; 5: 3.
[http://dx.doi.org/10.1186/1477-3155-5-3] [PMID: 17439648]

[25] Gou Q, Liu L, Wang C, *et al.* Polymeric nanoassemblies entrapping curcumin overcome multidrug resistance in ovarian cancer. Colloids Surf B Biointerfaces 2015; 126: 26-34.
[http://dx.doi.org/10.1016/j.colsurfb.2014.12.012] [PMID: 25543980]

[26] Scarano W, de Souza P, Stenzel MH. Dual-drug delivery of curcumin and platinum drugs in polymeric micelles enhances the synergistic effects: a double act for the treatment of multidrug-resistant cancer. Biomater Sci 2015; 3: 163-74.
[http://dx.doi.org/10.1039/C4BM00272E]

[27] Ali BH, Blunden G, Tanira MO, Nemmar A. Some phytochemical, pharmacological and toxicological properties of ginger (*Zingiber officinale* Roscoe): a review of recent research. Food Chem Toxicol 2008; 46(2): 409-20.
[http://dx.doi.org/10.1016/j.fct.2007.09.085] [PMID: 17950516]

[28] Yeh H-Y, *et al.* Bioactive components of two various gingers (*Zingiber officinalei* Roscoe) and antioxidant effect of ginger extracts. LWT Food Sci Technol 2014; 55: 329-34.
[http://dx.doi.org/10.1016/j.lwt.2013.08.003]

[29] Khandouzi N, Shidfar F, Rajab A, Rahideh T, Hosseini P, Mir Taheri M. The effects of ginger on fasting blood sugar, hemoglobin a1c, apolipoprotein B, apolipoprotein a-I and malondialdehyde in type 2 diabetic patients. Iran J Pharm Res 2015; 14(1): 131-40.
[PMID: 25561919]

[30] Selga G, *et al.* Pharmacological and clinical effectiveness of *Zingiber officinale* and *Alpinia galangal* in patients with osteoarthritis. In: Atroshi F, Ed. Pharmacology and Nutritional Intervention in the Treatment of Disease. Rijeka, Croatia: Intech 2014.
[http://dx.doi.org/10.5772/58417]

[31] Behrens MR, Mutlu N, Chakraborty S, *et al.* Dicamba resistance: enlarging and preserving biotechnology-based weed management strategies. Science 2007; 316(5828): 1185-8.
[http://dx.doi.org/10.1126/science.1141596] [PMID: 17525337]

[32] Cox-Foster DL, *et al.* A metagenomic analysis of microbes in honeybee colony collapse disorder. Science 2008; 318: 283-7.

[http://dx.doi.org/10.1126/science.1146498]

[33] Johnson RM, Ellis MD, Mullin CA, Frazier M. Pesticides and honey bee toxicity – USA. Apidol 2010; 41: 312-31.
[http://dx.doi.org/10.1051/apido/2010018]

[34] Levy S. What is best for bees? Nature 2011; 479: 164-5.
[http://dx.doi.org/10.1038/479164a] [PMID: 22071743]

[35] Core A, Runckel C, Ivers J, *et al.* A new threat to honey bees, the parasitic phorid fly *Apocephalus borealis*. PLoS One 2012; 7(1): e29639.
[http://dx.doi.org/10.1371/journal.pone.0029639] [PMID: 22235317]

[36] Academics Review. academicsreview.org/reviewed-content/genetic-roulette/section-6/6-2-new-herbicide-tolerant-crops/. , [Accessed Feb 20, 2014];

[37] GMO Compass, website: www.gmo-compass.org/eng/news/stories/273.gmos_fuelling_brazilian _future.html . , [Accessed Feb 20, 2014];

[38] Baker D, Church G, Collins J, *et al.* Bio FAB Group. Engineering life: building a fab for biology. Sci Am 2006 June; 294(6): 44-51.
[http://dx.doi.org/10.1038/scientificamerican0606-44] [PMID: 16711359]

[39] Cello J, Paul AV, Wimmer E. Chemical synthesis of poliovirus cDNA: generation of infectious virus in the absence of natural template. Science 2002; 297(5583): 1016-8.
[http://dx.doi.org/10.1126/science.1072266] [PMID: 12114528]

[40] Shiowatana J, *et al.* Enhancement effect study of some organic acids on the calcium bioavailability of vegetables. J Agric Food Chem 2006; 54: 9010-6.
[http://dx.doi.org/10.1021/jf062073t] [PMID: 17117785]

[41] The World's Healthiest Foods website: http://www.whfoods.com/foodstoc.php?PHPSESSID= 049812100195a7976e242b1fd2e10fc7. , [Accessed Feb 20, 2014.];

[42] Alves VC, *et al.* Sensory and volatile compounds profile of Thap Maeo banana: An option for the replacement of pest-susceptible cultivars. Nat Prod J 2012; 2: 205-11.

[43] Bailey DG, Dresser G, Arnold JM. Grapefruit-medication interactions: Forbidden fruit or avoidable consequences? CMAJ 2013; 185(4): 309-16.
[http://dx.doi.org/10.1503/cmaj.120951] [PMID: 23184849]

[44] Bozina N, Bradamante V, Lovrić M. Genetic polymorphism of metabolic enzymes P450 (CYP) as a susceptibility factor for drug response, toxicity, and cancer risk. Arh Hig Rada Toksikol 2009; 60(2): 217-42.
[http://dx.doi.org/10.2478/10004-1254-60-2009-1885] [PMID: 19581216]

[45] Kim H, Yoon YJ, Shon JH, Cha IJ, Shin JG, Liu KH. Inhibitory effects of fruit juices on CYP3A activity. Drug Metab Dispos 2006; 34(4): 521-3.
[http://dx.doi.org/10.1124/dmd.105.007930] [PMID: 16415112]

[46] Kang A-Y, Young LR, Dingfelder C, Peterson S. Effects of furanocoumarins from apiaceous vegetables on the catalytic activity of recombinant human cytochrome P-450 1A2. Protein J 2011; 30(7): 447-56.
[http://dx.doi.org/10.1007/s10930-011-9350-0] [PMID: 21847668]

[47] Liu C-X, Yi XL, Si DY, Xiao XF, He X, Li YZ. Herb-drug interactions involving drug metabolizing enzymes and transporters. Curr Drug Metab 2011; 12(9): 835-49. [http://dx.doi.org/10.2174/138920011797470083] [PMID: 21619518]

[48] Huang S-M, Lesko LJ. Drug-drug, drug-dietary supplement, and drug-citrus fruit and other food interactions: what have we learned? J Clin Pharmacol 2004; 44(6): 559-69. [http://dx.doi.org/10.1177/0091270004265367] [PMID: 15145962]

[49] Maynard CL, Elson CO, Hatton RD, Weaver CT. Reciprocal interactions of the intestinal microbiota and immune system. Nature 2012; 489(7415): 231-41. [http://dx.doi.org/10.1038/nature11551] [PMID: 22972296]

[50] de Valk B, Marx JJ. Iron, atherosclerosis, and ischemic heart disease. Arch Intern Med 1999; 159(14): 1542-8. [http://dx.doi.org/10.1001/archinte.159.14.1542] [PMID: 10421276]

[51] Altamura S, Muckenthaler MU. Iron toxicity in diseases of aging: Alzheimer's disease, Parkinson's disease and atherosclerosis. J Alzheimers Dis 2009; 16(4): 879-95. [PMID: 19387120]

[52] Kell DB. Towards a unifying, systems biology understanding of large-scale cellular death and destruction caused by poorly liganded iron: Parkinson's, Huntington's, Alzheimer's, prions, bactericides, chemical toxicology and others as examples. Arch Toxicol 2010; 84(11): 825-89. [http://dx.doi.org/10.1007/s00204-010-0577-x] [PMID: 20967426]

[53] The National Academies Press. Choline. www.nap.edu/openbook.php?record_id=6015&page=390. , [Accessed June 22, 2013];

[54] Wang Z, Klipfell E, Bennett BJ, *et al.* Gut flora metabolism of phosphatidylcholine promotes cardiovascular disease. Nature 2011; 472(7341): 57-63. [http://dx.doi.org/10.1038/nature09922] [PMID: 21475195]

[55] Barrett S. High doses of vitamin C are not effective as a cancer treatment Website: www.quackwatch.org/01QuackeryRelatedTopics/Cancer/c.html. , [Accessed June 22, 2013];

[56] Bjelakovic G, Nikolova D, Gluud LL, Simonetti RG, Gluud C. Mortality in randomized trials of antioxidant supplements for primary and secondary prevention: systematic review and meta-analysis. JAMA 2007; 297(8): 842-57. [http://dx.doi.org/10.1001/jama.297.8.842] [PMID: 17327526]

[57] Sesso HD, Buring JE, Christen WG, *et al.* Vitamins E and C in the prevention of cardiovascular disease in men: the Physicians' Health Study II randomized controlled trial. JAMA 2008; 300(18): 2123-33. [http://dx.doi.org/10.1001/jama.2008.600] [PMID: 18997197]

[58] Bates CJ, Hamer M, Mishra GD. Redox-modulatory vitamins and minerals that prospectively predict mortality in older British people: the National Diet and Nutrition Survey of people aged 65 years and over. Br J Nutr 2011; 105(1): 123-32. [http://dx.doi.org/10.1017/S0007114510003053] [PMID: 20807458]

[59] National Institutes of Health. Office of Dietary supplements, Dietary supplement Fact Sheets website: http://ods.od.nih.gov/Health_Information/Information_About_Individual_Dietary_Supple ments.aspx.

, [Accessed June 20, 2013];

[60] Leitzmann MF, Stampfer MJ, Wu K, Colditz GA, Willett WC, Giovannucci EL. Zinc supplement use and risk of prostate cancer. J Natl Cancer Inst 2003; 95(13): 1004-7. [http://dx.doi.org/10.1093/jnci/95.13.1004] [PMID: 12837837]

[61] Schroeder HA. Relation between mortality from cardiovascular disease and treated water supplies: variations in states and 163 largest municipalities of the United States. JAMA 1960; 172: 1902-8. [http://dx.doi.org/10.1001/jama.1960.03020170028007] [PMID: 14443614]

[62] USDA. Chapter 9 www.nal.usda.gov/fnic/DRI/DRI_Vitamin_A/290-393_150.pdf.

[63] Zimmermann MB, Hurrell RF. Nutritional iron deficiency. Lancet 2007; 370(9586): 511-20. [http://dx.doi.org/10.1016/S0140-6736(07)61235-5] [PMID: 17693180]

[64] Pizarro F, Yip R, Dallman PR, Olivares M, Hertrampf E, Walter T. Iron status with different infant feeding regimens: relevance to screening and prevention of iron deficiency. J Pediatr 1991; 118(5): 687-92. [http://dx.doi.org/10.1016/S0022-3476(05)80027-7] [PMID: 2019922]

[65] Yip R, Binkin NJ, Fleshood L, Trowbridge FL. Declining prevalence of anemia among low-income children in the United States. JAMA 1987; 258(12): 1619-23. [http://dx.doi.org/10.1001/jama.1987.03400120069027] [PMID: 3625969]

[66] WHO/UNICEF/UNU. Iron Deficiency Anemia Assessment, Prevention, and Control. Geneva: World Health Organization 2001.

[67] Mursu J, Robien K, Harnack LJ, Park K, Jacobs DR Jr. Dietary supplements and mortality rate in older women: the Iowa Women's Health Study. Arch Intern Med 2011; 171(18): 1625-33. [http://dx.doi.org/10.1001/archinternmed.2011.445] [PMID: 21987192]

[68] Macpherson H, Pipingas A, Pase MP. Multivitamin-multimineral supplementation and mortality: a meta-analysis of randomized controlled trials. Am J Clin Nutr 2013; 97(2): 437-44. [http://dx.doi.org/10.3945/ajcn.112.049304] [PMID: 23255568]

[69] Bergner P. Medical herbalism J Clin Practitioner , 2001 [Accessed Dec., 2013]; http://medherb.com/Materia_Medica/Lobelia_-_Is_lobelia_toxic_.htm.

[70] Compton JA, Culham A, Gibbings JG, Jury SL. Phylogeny of Actaea including Cimicifuga (Ranunculaceae) inferred from nrDNA ITS sequence variation. Biochem Syst Ecol 1998; 26: 185-97. [http://dx.doi.org/10.1016/S0305-1978(97)00102-6]

[71] Chen S-N, Li W, Fabricant DS, *et al.* Isolation, structure elucidation, and absolute configuration of 26-deoxyactein from *Cimicifuga racemosa* and clarification of nomenclature associated with 27-deoxyactein. J Nat Prod 2002; 65(4): 601-5. [http://dx.doi.org/10.1021/np010494t] [PMID: 11975513]

[72] Borrelli F, Ernst E. Black cohosh (Cimicifuga racemosa) for menopausal symptoms: a systematic review of its efficacy. Pharmacol Res 2008; 58(1): 8-14. [http://dx.doi.org/10.1016/j.phrs.2008.05.008] [PMID: 18585461]

[73] Becker DJ, Gordon RY, Halbert SC, French B, Morris PB, Rader DJ. Red yeast rice for dyslipidemia in statin-intolerant patients: a randomized trial. Ann Intern Med 2009; 150(12): 830-839, W147-9. [http://dx.doi.org/10.7326/0003-4819-150-12-200906160-00006] [PMID: 19528562]

[74] Halbert SC, French B, Gordon RY, *et al.* Tolerability of red yeast rice (2,400 mg twice daily) *versus* pravastatin (20 mg twice daily) in patients with previous statin intolerance. Am J Cardiol 2010; 105(2): 198-204.
[http://dx.doi.org/10.1016/j.amjcard.2009.08.672] [PMID: 20102918]

[75] Bliznakov EG, Hunt GL. The Miracle Nutrient Coenzyme Q10. New York: Bantam Books 1987.

[76] Gaby AR. The role of coenzyme Q10 in clinical medicine. Altern Med Rev 1996; 1: 168-72.

[77] Langsjoen PH, Langsjoen PH, Folkers K. Long-term efficacy and safety of coenzyme Q10 therapy for idiopathic dilated cardiomyopathy. Am J Cardiol 1990; 65(7): 521-3.
[http://dx.doi.org/10.1016/0002-9149(90)90824-K] [PMID: 2305693]

[78] Ylikoski T, Piirainen J, Hanninen O, Penttinen J. The effect of coenzyme Q10 on the exercise performance of cross-country skiers. Mol Aspects Med 1997; 18 (suppl.): S283-90.
[http://dx.doi.org/10.1016/S0098-2997(97)00038-1] [PMID: 9266538]

[79] Folkers K, Simonsen R. Two successful double-blind trials with coenzyme Q10 (vitamin Q10) on muscular dystrophies and neurogenic atrophies. Biochim Biophys Acta 1995; 1271(1): 281-6.
[http://dx.doi.org/10.1016/0925-4439(95)00040-B] [PMID: 7599221]

[80] Lockwood K, Moesgaard S, Yamamoto T, Folkers K. Progress on therapy of breast cancer with vitamin Q10 and the regression of metastases. Biochem Biophys Res Commun 1995; 212(1): 172-7.
[http://dx.doi.org/10.1006/bbrc.1995.1952] [PMID: 7612003]

[81] Littaru GP. A Letter from the Institute of Biochemistry 2001.

[82] Ascherio A, Zhang SM, Hernán MA, *et al.* Prospective study of caffeine consumption and risk of Parkinson's disease in men and women. Ann Neurol 2001; 50(1): 56-63.
[http://dx.doi.org/10.1002/ana.1052] [PMID: 11456310]

[83] Reissig CJ, Strain EC, Griffiths RR. Caffeinated energy drinks--a growing problem. Drug Alcohol Depend 2009; 99(1-3): 1-10.
[http://dx.doi.org/10.1016/j.drugalcdep.2008.08.001] [PMID: 18809264]

[84] Schneider C, Segre T. Green tea: potential health benefits. Am Fam Physician 2009; 79(7): 591-4. [PMID: 19378876]

[85] O'Conner A. Spike in harm to liver is tied to dietary aids www.nytimes.com/2013/12/22/us/ spike-i--harm-to-liver-is-tied-to-dietary-aids.html?pagewanted=all&_r=1&. 2013. NY Times

[86] Hou JP. The chemical constituents of ginseng plants. Comp Med East West 1977; 5(2): 123-45. [PMID: 608333]

[87] Li W, Gu C, Zhang H, *et al.* Use of high-performance liquid chromatography-tandem mass spectrometry to distinguish *Panax ginseng* C. A. Meyer (Asian ginseng) and *Panax quinquefolius* L. (North American ginseng). Anal Chem 2000; 72(21): 5417-22.
[http://dx.doi.org/10.1021/ac000650l] [PMID: 11080895]

[88] Gillis CN. *Panax ginseng* pharmacology: a nitric oxide link? Biochem Pharmacol 1997; 54(1): 1-8.
[http://dx.doi.org/10.1016/S0006-2952(97)00193-7] [PMID: 9296344]

[89] Tyler V. Rational Phytotherapy. (4th ed.), Berlin, Germany: Springer 1999.

[90] Kim H-G, Yoo SR, Park HJ, *et al.* Antioxidant effects of Panax ginseng C.A. Meyer in healthy subjects: a randomized, placebo-controlled clinical trial. Food Chem Toxicol 2011; 49(9): 2229-35. [http://dx.doi.org/10.1016/j.fct.2011.06.020] [PMID: 21699953]

[91] Gruenwald J, *et al.* PDR for Herbal Medicines. Montvale, NJ: Medical Economics Company 2000.

[92] DeKosky ST, *et al. Ginkgo biloba* for treatment of dementia. JAMA 2008; 300(19): 2253-62. [http://dx.doi.org/10.1001/jama.2008.683] [PMID: 19017911]

[93] Cesarani A, Meloni F, Alpini D, Barozzi S, Verderio L, Boscani PF. *Ginkgo biloba* (EGb 761) in the treatment of equilibrium disorders. Adv Ther 1998; 15(5): 291-304. [PMID: 10345150]

[94] Vale S. Subarachnoid haemorrhage associated with *Ginkgo biloba*. Lancet 1998; 352(9121): 36. [http://dx.doi.org/10.1016/S0140-6736(05)79516-7] [PMID: 9800751]

[95] Rowland I, Faughnan M, Hoey L, Wähälä K, Williamson G, Cassidy A. Bioavailability of phyto-oestrogens. Br J Nutr 2003; 89 (suppl. 1): S45-58. [PMID: 12725656]

[96] Messina M, Nagata C, Wu AH. Estimated Asian adult soy protein and isoflavone intakes. Nutr Cancer 2006; 55(1): 1-12. [http://dx.doi.org/10.1207/s15327914nc5501_1] [PMID: 16965235]

[97] Van Erp-Baart MA, Brants HA, Kiely M, *et al.* Isoflavone intake in four different European countries: the VENUS approach. Br J Nutr 2003; 89 (suppl. 1): S25-30. [http://dx.doi.org/10.1079/BJN2002793] [PMID: 12725653]

[98] de Kleijn MJ, van der Schouw YT, Wilson PW, *et al.* Intake of dietary phytoestrogens is low in postmenopausal women in the United States: the Framingham study(1-4). J Nutr 2001; 131(6): 1826-32. [PMID: 11385074]

[99] The National Standard. website: http://www.nationalstandard.com/. , [Accessed June 21, 2012];

[100] NIEHS. website: (www.niehs.nih.gov/). , [Accessed March 11, 2014];

[101] NTP. website: (www.niehs.nih.gov/). , [Accessed March 11, 2014];

[102] NIH Office of Dietary supplements. website: ods.od.nih.gov. , [Accessed March 11, 2014];

[103] U.S. Food and Drug Administration. Questions and Answers about FDA's Initiative Against Contaminated Weight Loss Product website: www.fda.gov/Drugs/ResourcesForYou/Consumers/QuestionsAnswers/ucm136187. , [Accessed Feb 21,2014]; Last Updated: 01/27/2011

[104] Schauss AG, Wu X, Prior RL, *et al.* Antioxidant capacity and other bioactivities of the freeze-dried Amazonian palm berry, *Euterpe oleraceae* mart. (acai). J Agric Food Chem 2006; 54(22): 8604-10. [http://dx.doi.org/10.1021/jf0609779] [PMID: 17061840]

[105] Kang J, *et al.* Bioactivities of açaí (*Euterpe precatoria* Mart.) fruit pulp, superior antioxidant and anti-inflammatory properties to *Euterpe oleracea* Mart. Food Chem 2012; 133: 671-7. [http://dx.doi.org/10.1016/j.foodchem.2012.01.048]

[106] Zhang Y, Krueger D, Durst R, *et al.* International multidimensional authenticity specification (IMAS)

algorithm for detection of commercial pomegranate juice adulteration. J Agric Food Chem 2009; 57(6): 2550-7.
[http://dx.doi.org/10.1021/jf803172e] [PMID: 19249817]

[107] Liao H, Shelor CP, Chen Y, Sabaa-Srur AU, Smith RE, Dasgupta PK. Anion composition of açaí extracts. J Agric Food Chem 2013; 61(25): 5928-35.
[http://dx.doi.org/10.1021/jf4014185] [PMID: 23772604]

[108] Colapinto J. Strange fruit. The rise and fall of açaí. The New Yorker 2011; 87: pp. 37-41.

[109] Smith RE, *et al.* Proposed benchmark methods for analyzing açaí (*Euterpe oleraceae* Mart.). Nat Prod J 2012; 2: 76-85.

[110] Luo R, *et al.* Distinguishing components in Brazilian açaí and in products obtained in the USA by using NMR. Nat Prod J 2012; 2: 86-94.

[111] Smith RE, *et al.* Insoluble solids in Brazilian and Floridian açaí. Nat Prod J 2012; 2: 95-8.

[112] Gustafson AM, Soldi R, Anderlind C, *et al.* Airway PI3K pathway activation is an early and reversible event in lung cancer development. Sci Transl Med 2010; 2(26): 26ra25.
[http://dx.doi.org/10.1126/scitranslmed.3000251] [PMID: 20375364]

CHAPTER 12

New Problems and Solutions

Abstract: Many new phenomena emerge at higher levels of organization. For example, it is possible that the virus causes the flu in birds or pigs will mutate to a form that can easily infect and kill people. In addition, as the global climate continues to change, several tropical diseases could appear as cooler regions warm up as the habitats of the carriers of tropical diseases like dengue and yellow fevers spread. It is also possible that the poliovirus or smallpox virus could re-emerge, even though they have been almost eradicated. On the other hand, scientists are creating new, genetically altered life forms, in which some consider an emergent solution, while others consider genetic engineering to be a terrible problem. Scientists use genetic modification (GM), biotechnology, gene splicing and recombinant DNA technology. At the same time, nanotechnology has the potential to turn relatively inactive molecules into potent drugs. Stem cell technology may aslo be able to provide many medical benefits. More recently, there has been a shortage of some important prescription drugs. This can continue to happen if a major producer has problems in manufacturing and has to stop production for a while, or if there is not enough profit to be made and they shut down production. As a result, patients may experience unacceptable delays in receiving 210 different medicines for cancer, anesthesia Parkinson's disease, schizophrenia, osteoporosis and organ failure. Finally, the ebola virus has emerged as a new problem, so therapies and vaccines are being developed.

Keywords: Ebola virus, Emergent properties, Genetic modification, Global warming, Nanotechnology.

EMERGENT PROBLEMS

Many new phenomena emerge at higher levels of organization. Although the word emergent may not be very popular with reductionist thinkers, it has been used by physicians for decades to refer to problems, diseases or epidemics that emerge unexpectedly. We have already discussed two important examples: antibiotic resistant bacteria and multidrug resistant cancers. Another frightening example of a possible emergent problem is the likelihood that the virus that causes the flu in birds or pigs will mutate to a form that can easily infect and kill people. One

Robert E. Smith

doesn't have to go too far back in history to see that a flu virus mutated into a form that caused the deadly influenza pandemic from 1918-1920 [1]. It was caused by an unusually severe and deadly strain of the H1N1 subtype of the influenza A virus [1]. It killed anywhere from 40 – 100 million people, which is more than the number of people killed in World War I or in the pandemic in the Middle Ages that was caused by the Black Death, also known as the bubonic Plague. However, the Black Death killed a larger percentage of people and it changed all of society. Some people died within 24 hours of showing their first symptoms, but others recovered. Entire villages were wiped out, while villages next to them were unaffected. Similarly, some people in affected cities and villages did not get sick and some who got sick did recover. Still, it changed everything. It interrupted the 100 year war between England and France. Peasants started living in abandoned castles. Fields full of crops were left unattended, since nobody was left to harvest them. It is possible that history could repeat itself if new pandemics, such as a contagious bird flu or a virulent form of swine flu, were to emerge. There is already a severe shortage of health care professionals, hospital emergency rooms and mortuaries. If a new pandemic emerges, there won't be anywhere near enough doctors and nurses to care for the patients, and society could become overwhelmed with corpses. It could change everything, just as the Black Death did.

In 2009, the influenza virus in pigs (or swine) began infecting people, but so far it has not been very virulent. Most people infected by it recovered. Still, many people were infected, indicating that the virus had penetrated the human population. Note that the term swine flu was first used, but then it was identified as the H1N1 2009 flu virus. In contrast, the bird flu virus did not pass easily into people, but it was quite virulent in those who became infected, with almost half of them dying.

However, the worst pandemic to emerge in the last part of the 20th century was AIDS. It has killed about 25 million people since it was first recognized in 1981, and about 33 million people are affected by HIV worldwide [2]. It can be especially bad in poorer countries and people who can't afford to buy drugs to treat AIDS that have been patented by major pharmaceutical companies. In the absence of antiretroviral therapy, the average time from HIV infection to AIDS was nine to ten years [2] but can be much longer with HAART therapy. The average survival time after developing AIDS is about 9.2 months, but can range from two weeks to 20 or more years [2]. With recent developments in antiretroviral therapy, it is hoped that people can live even longer with HIV and research continues to develop better drugs and possible vaccines. Moreover, people who are homozygous (have two copies) of an altered version of the CCR5 white blood cell receptor are resistant to some strains of HIV [2]. Still, HIV is capable of mutating into forms that are resistant to antiretroviral therapy, so extensive research continues. Also, some countries (Brazil and India) are by-passing patent laws, so that they can obtain much cheaper generic versions of AIDS drugs and save millions of lives.

Other diseases have the potential to emerge and cause an enormous loss of life and human suffering. The strain of bird flu, called Influenza A virus subtype H5N1 [3] is endemic in many types of birds, especially in Southeast Asia. It has directly killed tens of millions of birds and caused the culling of hundreds of millions of other birds, in an attempt to keep it from spreading [3]. Some people have caught this form of influenza by being in contact with affected birds, especially chickens. About 60% of those infected have died. So far, there is no evidence that H5N1 can be transmitted from one person to another. However, if H5N1 ever mutates into a strain that can be transmitted between people, it could kill up to 150 million people [3]. Billions of dollars are being spent to develop influenza vaccines that might be able to prevent a pandemic [3]. Also, a strain of H5N1 that can be transmitted to ferrets was genetically engineered [4]. It can be used to study human adaptation to the virus [4].

GLOBAL CLIMATE CHANGE

Other diseases could emerge, too. As global warming continues, several tropical diseases could appear in cooler regions [5]. "Disease-carrying mosquitoes are spreading as climate shifts allow them to survive in formerly inhospitable areas" [5]. "The year 2012 was the worst year ever in Texas for illness caused by the West Nile virus. Mosquitoes that can carry dengue fever viruses were previously limited to elevations of 3,300 feet but recently appeared at 7,200 feet in the Andes Mountains of Colombia. Malaria has been detected in new higher-elevation areas in Indonesia. Global climate change is projected to significantly increase the range conducive to the transmission of both dengue and yellow fevers. The global incidence of dengue was estimated to be 390 million people and a new serotype of the virus that causes it was recently discovered" [5]. There is no vaccine to prevent or drug to cure dengue, so it strikes much fear in the hearts of public health workers.

"At least 279 species of plants and animals are already responding to global warming" [6]. "Species' geographic ranges are shifting toward the North and South poles and spring events have shifted earlier by an average of two days per decade. In Washington's Olympic Mountains, sub-alpine forest has invaded higher elevation Alpine meadows" [6]. Also, there could be many viruses, bacteria and other parasites that have been buried in remote jungle areas that may have affected ancient civilizations, but are unknown to modern man (or woman). As civilization continues to encroach upon such places, new diseases could emerge as these pathogenic organisms are uncovered.

Another possible problem is the re-emergence of the poliovirus or smallpox virus, even though these diseases have almost been eradicated. To the best of our knowledge, smallpox has been eradicated, but small amounts of the smallpox virus still exist in laboratories in the USA and Russia. Some people think that these supplies of the virus should be destroyed, but others think they shouldn't, because we might need them to produce vaccines should smallpox re-emerge. Similarly, polio has been eradicated in most countries, including the USA. However, it is now possible to synthesize poliovirus DNA, as indicated in a recent controversial publication [7].

BIOTECHNOLOGY-BASED SOLUTIONS

Scientists are creating new, genetically altered life forms, in what some consider an emergent solution, while others consider genetic engineering to be a terrible problem. It is also known as genetic modification (GM), biotechnology, gene splicing and recombinant DNA technology. Regardless of what it is called, the structure and characteristics of genes in an organism are modified by either molecular cloning or transformation. Many people point to the potential of this growing technology to improve human life and health, and its many accomplishments to date. Others point to the potential for GM and GMOs (genetically modified organisms) to cause death and destruction, as well as some of its proven failures. Let us try to look at both points of view, starting with the positive aspects of GM and biotechnology.

Technology earned a good reputation when it helped convert Mexico from a country that could not feed itself, to a net exporter of food through the development of high-yield varieties of maize and wheat. The Mexican government invested heavily in rural infrastructure and new varieties of seeds were widely adopted. Mexico became self-sufficient in wheat production by 1951 and has been exporting it ever since. The Green Revolution spread to India, where many people worked to improve plant breeding, irrigation, development and financing of agrichemicals. By the late 1970s, yields of rice increased by 30 percent. The Nobel peace prize was awarded in 1970 to Norman Borlaug, who started the research that led to this green revolution that has saved hundreds of millions of lives and prevented wars that could have started between countries that might have been desperate to obtain arable land from other countries so they could feed their people.

Next, the Rockefeller and Ford Foundations jointly established the International Rice Research Institute (IRRI) in the Philippines in 1960. High-yield varieties spread throughout the Philippines, followed by Indonesia, Pakistan, Sri Lanka, and many other countries throughout Latin America, Asia and North Africa. Another organization (USAID) became involved in subsidizing the development of rural infrastructure and shipping fertilizer. This was followed by the Consulting Group on International Agricultural Research (CGIAR) in 1971, which responded to some earlier criticism of the Green Revolution. They developed a more holistic view of agriculture. They used methods like Rapid Rural Appraisal and Participatory Rural Appraisal adopted to help scientists understand the problems faced by farmers and even give farmers a role in the development process.

However, some parts of advanced technology can be controversial. For example, even though herbicides and pesticides can dramatically increase productivity while decreasing hunger and malnutrition, they can be toxic to birds, fish and people. So, many of them were banned in the USA, but not in all countries. Also, biotechnology was used to produced antibiotics and growth hormones for livestock as well as herbicide-resistant and insect-resistant crops. This has increased the amount of food, corporate profits, tax revenue and the controversy. The basis for this concern is that genes do not always go where we want them to go or stay where they are originally put. That is, for genes to be transcribed into hnRNA and then processed into mRNA, a piece of DNA called a promoter is needed to activate the process. They are adjacent to the site at which DNA transcription is initiated. Enhancers are also regions in the genome that can activate transcription, but they are located further from the site at which mRNA synthesis is initiated. Unfortunately, the promoter can activate the transcription of other genes, besides the one being introduced into the target organism, such as a GMO.

Even if the gene does remain in the GMO, the seeds that are produced can blow in the wind and settle into farms that are trying to grow only organic crops that are not genetically modified. The GM crops may be able to cross-pollinate organic crops and spread the inserted gene. Unlike chemical pollutants that are not able to reproduce, these genes can be reproduced and spread through the environment. This has led companies to develop crops and livestock (like farm-raised salmon and turkeys) that are sterile. This has been called the Terminator Technology. It is profitable, because farmers have to buy more of the expensive GM seed each year. However, it supposedly protects the biotechnology industry from the criticism that GMOs will mate with normal, wild-type organisms and spread the gene pollution. Supposedly, if the GMOs can't reproduce sexually, their DNA can't spread. In the short term of a few years, this might be true, but in the longer time frame, it may be false. Throughout evolution, genes have been transferred from humans to enteric bacteria, like *E. coli*. Once a gene is in bacteria, it can spread to other, unrelated bacteria, and back to humans and other eukaryotic organisms. However, this can take thousands or even millions of years to occur. The more important concern is that people and livestock who eat GM food might get some of genes into their own cells and tissues. It is hoped that all of the DNA that we eat is digested and broken down in the stomach so that it can never spread throughout the body. This may not happen completely. If genes or genetic elements containing promoters were to make it into human or animal cells and tissues, they could cause undesirable consequences. Remember that a gene can code for different proteins in different cells and different organisms. So, a gene that causes the desired modification in one organ or organism might cause a completely different and possibly fatal effect in other organs and possibly even an entire ecosystem. So, there is much concern about eating GMOs and some countries forbid growing GMOs within their borders. Many more require all GM foods be properly labeled, so people can choose whether or not to buy them and eat them. Still, starvation and malnutrition are terrible problems that can be partly solved by producing more food at lower cost, so many countries grow or import GM foods, which saves many lives.

So, many scientists point to the many benefits of biotechnology, especially those aspects of biotechnology that don't involve GMOs or GM foods. For example, genetic engineering and biotechnology were used to determine the sequence of all the three billion base pairs in the human genome and the billions of other base pairs in many other organisms. This helped scientists understand the basic science better and to identify potential molecular targets for new drugs. Moreover, biotechnology has been used to produce medicines that have saved millions of lives. Biotechnology is also used to determine the genetic changes that occur in each individual patient's cancer, making personalized treatments possible. Biotechnology is also creating some organisms that are not sold as food, but have important environmental benefits, such as bacteria that can metabolize harmful persistent environmental pollutants, such as PCBs and dioxins. Also, research is being done to develop genetically modified bacteria and algae to produce oil, hydrogen and many important chemicals. So, many people are in favor of using some organisms that have been altered genetically as well as some aspects of biotechnology, especially when they don't have to eat any of the products.

At the same time, one can consume only organic foods or foods that people grow themselves or are grown locally by people we can trust. However, many people can't afford that luxury. Instead, if more food can be produced at lower cost, it can save millions from malnutrition and starvation. However, if the long term costs are even greater, then it might be better to develop sustainable farming practices. So, genetic engineering will probably continue to be part of the solution, and also part of the problem. It is up to us scientists and educators to learn as much as we can, teach others and get involved in important discussions about genetically modified food and other emergent problems.

Although it may not seem to be part of medicinal chemistry, global climate change is probably an even more serious emergent problem for human health. It may or may not be caused partly by natural variation, but human activities also help cause it. In reductionist thinking and in management by fear, it is more important to fix the blame than to fix the problem. Many so-called conservative American politicians say that humans did not cause global warming, so we don't need to do anything to try to stop it, since it was supposedly not caused by us. Still, many evangelical Christians and Catholics (led by the Pope) encourage their followers to understand global climate change and work to limit greenhouse gas emissions.

Systems thinkers and total quality leaders are not concerned with assigning blame. They try to fix the problem, instead of the blame. Global climate change could increase extreme poverty, flooding, lack of clean water and food and could cause international conflicts, including wars. All of these could increase disease and human suffering. So, many doctors, scientists and politicians are trying to find ways to limit the production of greenhouse gases like CO_2 and methane (CH_4) that can cause global warming.

NANOTECHNOLOGY

There is a bright side. There are emerging technologies that could help cure diseases. For example, nanotechnology has the potential to turn relatively inactive molecules into potent drugs. Nanotechnology is the use of molecules and/or atoms that are the size of nanometers (10^{-9} meters, or 10^{-3} μm). For example, carbon nanotubes (also known as Buckyballs) are made from individual carbon atoms (about 1 nm in diameter) that combine with each other to make nearly perfect (defect-free) structures with very useful properties. In fact, completely new properties emerge when carbon and other materials are reduced to the size of nanometers. Carbon nanotubes have high electrical and thermal conductivity, strength, stiffness and toughness. Normally opaque copper becomes transparent.

Inert materials (like platinum and gold) become catalysts. Stable materials (like aluminum) become combustible. Solids (like gold) turn to liquids at room temperature. Insulators (like silicon) become conductors. So, it may be possible to make much stronger materials for construction, better electronic devices and more powerful computers, and unique materials to provide alternatives to fossil fuels for our energy and chemical needs.

It has been suggested that nanotechnology could produce a new industrial revolution that could produce trillions of dollars of new wealth, create millions of new jobs and save many lives. Solar panels made from nanometer-sized silicon could provide clean and cheap electricity. Nanoparticles made from tiny proteins called aptamers can be used to encapsulate an anticancer drug and treat prostate cancer [8]. An aptamer is any small synthetic nucleic acid ligand, such as a short DNA or RNA fragment, that can bind to a specific target molecule, such as a protein or metabolite. Quantum dots (or qdots) that are made from tiny crystals which have pieces of protein attached to their surfaces may be used to bind to cancer cells and produce multicolored signals for better and earlier detection of cancer. Nano-gold particles may be used to enhance radiation therapy for cancer. Eventually, it may be possible to make nanoparticles or nanomachines that have specific antibodies on their surface, which are able to bind to specific cancer cells. Once bound, the nanomachine will synthesize or simply release a drug that will kill the cancer cells to which it is bound.

Scientists are developing nanodots and nanobots (very small robots) that could be part of the next generation of medicine and could start a medical revolution. Starting materials that are the size of atoms can be arranged in such a way that they automatically self-assemble and form structures with amazing properties. These can be programmable self-replicating manufacturing systems that could start a new Industrial Revolution. There are two central elements: a universal computer and a universal constructor [9]. "The universal computer contains a program that directs the behavior of the universal constructor. The computer is not an electronic device, but a molecular sensor or probe that can tell when the constructor should be activated. The universal constructor, in turn, is used to manufacture both another universal computer and another universal constructor. Once construction is finished, the program contained in the original universal computer is copied to the new universal computer and program execution is started. The computer and constructor both shrink to the molecular scale, while the constructor takes on additional detail that is consistent with the desire to manipulate molecular structures with atomic precision. Research is being done to develop nano-structures that can identify a cancer cell and kill it. The device would have something that acts like a small molecular computer, along with several molecular building blocks and a supply of doxorubicin or other medicine. The device would circulate freely throughout the body, and would periodically sample its environment by determining whether the binding sites were or were not occupied. Occupancy statistics would allow determination of the concentration of molecules produced by cancer cells. The nano-device could incorporate many binding sites and could monitor the concentrations of many molecules. The computer could determine if the profile of concentrations fit a pre-programmed "cancerous" profile and if so, would release the poison that kills the cancer cells" [9].

Nanotechnology may also help cure AIDS. Several copies of the therapeutically inactive drug called SDC-1721 were bound to gold nanoparticles [10]. Single copies of SDC-1721 were already known to bind weakly to the receptor (CCR5) that HIV uses to gain entry into white blood cells. When several copies were bound to gold nanoparticles that were about the size of a protein, several SDC-1721 molecules were able to bind to CCR5 and prevent HIV from infecting cells. The very low concentration of 10 nM (10^{-8} mol/L) of the SDC-1721-nanoparticle conjugate inhibited HIV replication [10].

Nanotechnology has already produced clinical advances. The anti-cancer drug doxorubicin can be encapsulated into liposomes, which significantly lowers its cardiotoxicity. Newer nanoparticles called IT-101 are made from polymers that are bound to the anticancer drug camptothecin. The encapsulated drug lasts much longer in the blood stream, giving it time to reach cancer cells and be gradually released into them. In a phase I study, camptothecin encapsulated into IT-101 nanoparticles proved safe and showed signs of being effective [11].

Another approach that is being tested in a Phase 1 trial is the use of encapsulated small inhibitory RNA (siRNA). Proteins on the surface of the nanoparticles target receptors on the surface of cancer cells [11]. Once the nanoparticles bind to the cancer cells (which have many of these receptors), the siRNA is released and will specifically inhibit the production of a critical oncoprotein from the targeted gene [11].

More recently, nanosized interbilayer-crosslinked multilamellar vesicles (ICMVs) containing antigen along with two Toll-like receptor agonists were made. They are used as adjuvants in a new vaccine that can be delivered to the lungs [12].

So, nanotechnology continues to be an important emerging technology. Moreover, it is an important tool in P4 medicine.

P4 MEDICINE

As described in Chapter 1, personalized, predictive, preventive, and participatory (P4) medicine is revolutionizing modern medicine. DNA sequencing is becoming so rapid and reliable that the genome of each individual can be determined. That is, mathematics, the foundation of reductionist thinking, is used to quantify huge datasets from patients, while physicists, chemists, biologists and engineers develop the analytical tools needed to generate the data. All of this can be linked through the internet and used in mobile healthcare applications. Data about biomarkers from many people and many different tissues in an individual can be used to help everyone work with their physicians to make their own medical decisions. This will tell the individual and his or her doctor what his or her probable health will be and what disease susceptibilities exist. Next, it is becoming relatively cheap and easy to analyze very small samples of an individual's blood for many proteins that are biomarkers of potential diseases. Nanotechnology is being used to develop biochips that can do the analyses. This will enable the individual and the doctor the ability to assess regularly the individual's health, so dis-eases can be prevented by proper changes in diet, lifestyle or administration of preventive medicines. However, one problem should be mentioned. Some companies offer the supposed ability to analyze a person's DNA to tell them something about their so-called "race". Supposedly, one can find out what percentage of their DNA is of African, European or Asian origin. However, different companies report different results on the same DNA taken from the same individual. Another potential problem with sequencing a person's DNA is that it can indicate that a person might be susceptible to some diseases. Some people might mistake this for indicating that they will definitely get the disease. Remember the nocebo effect. When people are told that they are susceptible to a disease, they are often more likely to become ill than if they did not know they were susceptible. Also, if a person finds out that they are not susceptible to a disease such as lung cancer or alcoholism, they might feel free to drink and smoke.

STEM CELL TECHNOLOGY

Another technology that could provide many medical benefits is stem cell technology. Hopefully, these stem cells will help people who have become paralyzed by spinal cord injuries. So, let us look closer at stem cells and their possible uses in medicine. Stem cells can be thought of as masters of autopoiesis, since they control the making (poiesis) of the body [13]. They are unique in that they can not only self-renew, or make more stem cells, but also make differentiated cells. When other highly differentiated human cells undergo mitosis, they produce two identical cells. When a stem cell divides, it can do so in two different ways. It can produce two identical stem cells, to maintain a reservoir of stem cells for the future. On the other hand, they can divide to produce one stem cell plus a more differentiated cell. Embryonic stem (ES) cells can make any human cell type. They can divide to produce one ES cell plus a slightly differentiated adult stem cell that is specific for each tissue. The ES cells are often called totipotent, since they can make any human cell type in the body, while adult stem cells are pluripotent, because they can make many (but not all) human cell types.

Embryo-like, pluripotent stem cells are also present in adult organisms and circulate in the organism [14]. Pluripotent stem cells can differentiate into many different types of cells. For example, pluripotent stem cells in the liver can differentiate into all the different cell types that exist in the liver. The same is true for other tissues. Also, the gonads may contain a reservoir of embryo-like, pluripotent stem cells in adult organisms. On the other hand, many ES cells develop into organ-specific stem cells that can produce different cell types in the organs. Such organ-specific stem cells are senescent [15]. That is, they can remain largely unchanged for many years, until they are needed. For example, slightly differentiated stem cells, such as neural stem cells, differentiate further in the developing fetus, baby and child's nervous system to make the many different cells types in the brain and spinal cord. Moreover, throughout a person's lifetime, hematopoietic stem cells can differentiate into all the different types of blood cells.

Hematopoietic stem cells can also be transplanted from someone else when a patient has certain types of cancer, blood diseases, bone marrow failure syndromes, or congenital immune deficiencies [16] These are called allogeneic stem cells, where allogeneic means that they come from another person. However, it is often impossible to find a perfect match, so graft *versus* host disease (GVHD) can occur. Both T and B cells play roles in the pathogenesis of chronic GVHD, suggesting that there is a general loss of immune tolerance. Since $CD4^+Foxp3^+$ regulatory T cells (T_{regs}) contribute to the maintenance of peripheral tolerance, it is not surprising that patients with chronic GVHD have a relative deficiency of T_{regs}. Moreover, Interleukin-2 (IL-2) is a critical homeostatic cytokine for T_{regs} and a low dose of it restored the homeostasis of $CD4^+$ T cell subsets and promoted the reestablishment of immune tolerance. There was a selective expansion of functional T_{regs} and clinical improvement of chronic GVHD [16].

A dynamic systems view of stem cell biology has been proposed [17]. They considered cellular states to be points in multidimensional state space. Each axis in that space represented the abundance of genes, proteins and metabolites. Pertubations of the interactions between them can cause the state of a cell to shift from a stem cell to a differentiated cell. Stem cells and differentiated cells not only must be able to remain in the proper state when perturbed slightly by noise-generated fluctuations and be able to differentiate further when perturbed by proper signals. This balance must be maintained to ensure proper development during childhood and good health in adults. This is done by fluctuating and oscillatory gene expressions [17].

Both ES and adult stem cells can be grown artificially in culture. All existing ES cell cultures were obtained from in the inner cell mass of four to five-day old preimplantation embryos, which had to be destroyed afterwards [18]. More recently, scientists have been able to produce stable ES cell lines from younger embryos (three days old) that had only eight cells, called blastomeres [18]. By taking just one blastomere, the embryo survived. It is hoped that ES cells will be useful in research and in clinical applications. It is also possible that pluripotent stem cells can be produced from adult cells.

In 2006, scientists in Japan were able to convert mouse skin cells into ES cells by inserting just four genes that forced the expression of the transcription factors Oct4, Sox2, Klf4, and c-Myc into the adult cells, using retroviruses as transduction vectors [19]. This procedure also worked using human skin cells, both in Japan and in Wisconsin. This produced induced pluripotent stem (iPS) cells, or iPSCs. This could be an excellent way to mass produce one's own stem cells, which could be further stimulated to produce different organs for organ transplants from one's self. This would eliminate the current difficult problem of organ rejection when organs are donated by someone else. Moreover, adult mice have been generated from iPS [20].

The iPS cells may also be very useful in research. Douglas Melton, the co-director of the Harvard Stem Cell Institute (HSCI) was able to switch one type of adult pancreatic cell, which does not produce insulin, to a type that does - without using stem cells at all [21]. This could help researchers better understand type-1 diabetes, which is starting to affect children as young as six months old [21]. Other scientists have been able to convert fibroblasts into liver cells [22], neurons [23] and heart cells [24]. In 2012 scientists described the production of functional thyroid in mice from ES cells [25]. They overexpressed the transcription factors NKX2-1 and PAX8. The transcription factor NKX2-1, is also known as NK2 homeobox 1 and the thyroid transcription factor. It helps regulate the transcription of genes in the thyroid, lungs and diencephalon. It is known to interact with PAX8 (PAX8), which is a transcription factor that contains a paired box domain, an octapeptide, and a paired-type homeodomain. It is involved in thyroid follicle cell development. Mutations in the PAX8 gene can lead to thyroid diseases. When the transcription factors NKX2-1 and PAX8 were overexpressed, it caused the ESCs to differentiate into thyroid follicle cells that functioned well when transplanted into athyroid mice [25]. The potential medical applications of iPS and a detailed review of them were published recently [25].

Until recently, it was not known whether iPS can ever be able to be reprogrammed so that they can make any cell in the human body, like ES cells can [26]. However, Japanese scientists were able to reconstitute female germ cell development and produce fertile offspring. They intend to do the same thing with other mammals, including humans [26].

Another potential application for iSPCs is to make new donor organs. Ideally, cells could be taken from the patient, and then be converted into the organ that is needed. This would eliminate the threat of organ rejection that occurs after transplantation. Recently, a vascularized and functional human liver was made from liver buds made from iPSCs [27].

SHORTAGE OF MEDICINES

Another emergent problem is a shortage of certain medicines. This can happen if a major producer has problems in manufacturing and has to stop for a while, or if there is not enough profit to be made. As a result, patients are experiencing unacceptable delays in receiving 210 different medicines for cancer, anesthesia Parkinson's disease, schizophrenia, osteoporosis and organ failure [28]. Many of the drugs that are hard to get are no longer protected by patents that have expired, and their generic replacements, though effective, are not very profitable. This makes important drugs affordable to more people, but makes the manufacturers less money. Some people think that the answer is less regulation by the FDA [29], but it is important to remember that insufficient regulation would allow companies to make their medicines in dirtier environments with less motivation to ensure quality control.

In fact, even with current regulations, contaminated steroids (methylprednisolone acetate) were contaminated with a fungus in 2012 that caused meningitis and killed people. This was due to an unclean work environment. So, even if a cGMP organization manufactures drugs in accordance with standard SOPs and documents everything clearly, it is of no avail if the work environment is filthy.

Also, many people who don't have prescription coverage with their health insurance are tempted to buy medicines over the Internet made in unknown locations. Medicines that are fraudulent or are past their expiration date or contain insufficient active ingredients are readily available to a desperate public. Moreover, at a time when many people want lower taxes and less government involvement in their lives should remember that the cost of a poorly regulated health care industry would be much more. Finally, there are some politicians who would cut funding to the EPA or eliminate it completely. This would mean not just more pollution, but also less (or no) toxicology testing for industrial chemicals. Also remember that completely unregulated farming before 1863 led to the evils of slavery – even though it meant that everybody had a job. In fact, conditions approximating slavery still exists in the USA – especially in the meat processing industry and picking fruits and vegetables by undocumented workers [30].

TESTING FOR PROSTATE SPECIFIC ANTIGEN

PSA Another emerging issue is whether it is beneficial to test men for their blood serum levels of prostate specific antigen, PSA. It is a secreted protein that is coded for by the prostate-specific protein kallikrein 3 and is usually present at low concentrations in the blood. It can become elevated in prostate cancer, benign prostate hyperplasia (BPH) and even as a result of bacterial infection. In 1986, the FDA approved the PSA test as a way to detect prostate cancer in men over 50. Its use has helped diagnose men with prostate cancer, but most have indolent disease, which is of little danger. Unnecessary treatments can cause severe side effects, including sexual dysfunction, incontinence and depression. As a result, the U.S. Preventive Services Task Force (USPSTF) deemed the evidence to be insufficient to recommend routine use of PSA as a screening test among men younger than age 75 [31]. So, new biomarkers are being investigated [32]. These include prostate cancer antigen 3 (PCA3), a long noncoding RNA that is elevated in >90% of prostate cancer tissues, but not in normal or BPH tissues. It can be detected in a urinalysis. Another biomarker is a TMPRSS2-ERG gene fusion, which is common in prostate cancer and accounts for 90% of prostate cancer fusions. Another biomarker was found by profiling the expression of RNA. It is the enzyme α-methylacyl–coenzyme A racemase (AMACR). Low levels of its expression have been correlated with metastasis and recurrence of prostate cancer. There have also been more than 50 single nucleotide polymorphisms (SNPs) that are risk factors for prostate cancer. Another potential biomarker is to measure the amount of circulating tumor cells and the number of copies of critical genes such as PTEN (phosphatase and tensin homolog), a tumor suppressor.

OTHER IMPORTANT RECENT DISCOVERIES

There have also been some other important discoveries. For example, it was found that the hypoxia inducible factor HIF-1α works with a long non-coding RNA, lincRNA-p21 to modify the Warburg effect in cancer cells [33]. That is, cancer cells are characterized by enhanced glycolysis and reduced oxidative phosphorylation, even in the presence of oxygen. HIF-1 is a key mediator of this. It is a heterodimeric transcription factor with HIF-1α and HIF-1β subunits. The HIF-1α subunit responds to O_2. It not only regulates genes expression, but also induces the expression of a microRNA-210, which decreases the expression of the iron-sulfur cluster assembly proteins ISCU1/2, thus repressing oxidative phosphorylation. long non-coding RNA, (lncRNAs) have diverse functions, including being involved in the integrity of the nuclear structure, inactivating the extra X chromosome, and regulating gene expression at different levels, such as chromatin remodeling, transcription, and posttranscriptional processing. So, the recent study is important because it also showed that lincRNA-p21 a previously identified p53-inducible lncRNA plays an important role in the regulation of hypoxia-enhanced glycolysis. So, it may be a good therapeutic target for cancer [33].

Recent research has also discovered a weakness in cancer cells that become resistant to chemotherapy by targeting the fuel that cancer stem cells need for metabolism [34]. “That is, there is metabolic symbiosis that is known as the “Reverse Warburg Effect” in which tumor cells induce oxidative stress, which causes reactive oxygen substances to be produced. This leads to autophagy (mitophagy) and aerobic glycolysis. Due to mitochondrial dysfunction, L-lactate and ketone bodies are released into the tumor microenvironment and act as energy-rich fuels for the mitochondria. This induces the biogenesis of more mitochondria and oxidative phosphorylation that promotes tumor growth. So, there are two opposing metabolic pathways, in which stromal cells undergo aerobic glycolysis and provide fuel to cancer cells. Moreover, two concentrations of two proteins were found to change when the Warburg effect occurs. They are Caveolin-1 (Cav-1) and monocarboxylate transporter 4 (MCT-4), which exports lactate and other monocarboxylates out of the cell. Cav-1 is downregulated, while MCT-4 is upregulated in stromal cells of breast, head and neck cancers" [34].

Another recent discovery shed light on the aging process [35]. "The protein called plasminogen activator inhibitor-1 (PAI-1) was shown to play a critical role in aging. It is a serine protease inhibitor that inhibits extracellular proteolysis, which contributes to accelerated aging. Since advanced age contributes to the development of frailty and disease in humans, slowing it down could be the most important emerging advance in medicine. That is, cellular senescence keeps cells from proliferating. It is associated with the production and secretion of a set of proteins called the "senescence- messaging secretome" (SMS). It includes the insulin-like growth factor-binding proteins (IGFBPs), interleukins (ILs), transforming growth factor type β (TGF-β), and plasminogen activator inhibitor-1 (PAI-1). The recent study showed that PAI-1 plays an important role in tissue calcification in aging. Moreover, an experimental new drug (TM5441) that inhibits PAI-1 was given to an experimental model of accelerated aging in mice. It quadrupled their life span and kept their organs healthy and functioning. The experimental drug could potentially be used to treat human diseases that cause accelerated aging such as chronic kidney disease, diabetes and HIV infection as well as the effects of cigarette smoking" [35].

RECENTLY APPROVED DRUGS

There have been some recently approved drugs. For example, mipomersen sulfate (Kynamro®), is a second generation antisense oligonucleotide that is used to treat homozygous familial hypercholesterolemia [36]. "It is a second generation antisense oligonucleotide that binds the mRNA that codes for the protein called apoB, a ligand for the low density lipoprotein (LDL) receptor. This prevents it from being translated into ApoB. This decreases the elevated concentration of LDL, which is a hallmark of this disease" [36].

The drug called nimodipine (Nymalize®) was also approved in 2013 for the treatment of subarachnoid hemorrhage [37]. It is a Ca^{2+} channel blocker. The contractile processes of smooth muscle cells are dependent upon Ca^{2+}, which enters these cells during depolarization as slow ionic transmembrane currents. Nimodipine inhibits calcium ion transfer into these cells and thus inhibits contractions of vascular smooth muscle [37].

Macitentan (Opsumit®) was approved for treating pulmonary arterial hypertension [38]. "Macitentan and its metabolite are antagonists of the endothelin receptors ETA and ETB. Endothelins are vasocontrictive proteins that help raise blood pressure. ET-1 and its two receptors, ETA and ETB comprise the endothelin system that can change the structure of tissues and cause fibrosis. Macitentan was designed to be lipophilic and nonionic at a physiological pH so that it could target the lungs. Its major metabolite is also pharmacologically active" [38].

Polidocanol (Varithena®) was approved for treating varicose veins [39]. "It was originally developed as a local anesthetic with properties similar to lidocaine. It is safe, effective, and a relatively painless sclerosing solution with minimal toxicity. As a detergent, it disrupts cell membrane lipids, ultimately leading to sclerosis of the underlying vein" [39].

A cream containing 1% luliconazole (Luzu®) was approved as an antifungal agent that decreases the synthesis of ergosterol by inhibiting the enzyme lanosterol demethylase [40]. "It is indicated for the topical treatment of interdigital tinea pedis, tinea cruris, and tinea corporis caused by the organisms *Trichophyton rubrum* and *Epidermophyton floccosum*, in adults 18 years of age and older. Luzu is supplied as a cream for topical administration" [40].

Trametinib (Mekinist®) was approved for treating unresectable or metastatic melanoma with BRAF V600E or V600K mutations [41]. "It inhibits mitogen-activated protein kinase kinases 1 and 2 (MEK-1 and MEK-2). They are threonine/tyrosine kinases that are often upregulated in various types of cancer. They play a key role in the activation of the RAS/RAF/MEK/ERK signaling pathway that regulates cell growth" [41].

Brimonidine (Mirvaso®) was approved for treating the facial erythema of rosacea

[42]. "It is an agonist of the α2-adrenergic receptor with potent vasoconstrictive activity. It is effective because rosacea I caused by abnormal cutaneous vasomotor activity and inflammation. By binding to the α2-adrenergic receptor, it prevents the vasoactive effects of adrenaline in the cutaneous facial tissue. Brimonidine had already been approved and used to treat glaucoma, so it is quite safe" [42].

Dabrafenib (Tafinlar®) was approved for treating unresectable or metastatic melanoma with BRAF V600E mutation [43]. It inhibits some mutated forms of BRAF kinases, as well as wild-type BRAF and CRAF kinases. Some mutations in the BRAF gene, including those that result in BRAF V600E, can result in constitutively activated BRAF kinases that may stimulate tumor cell growth [43].

A gel containing mechlorethamine (Valchlor®) has been approved for treating Stage IA/IB mycosis fungoides-type cutaneous T-cell lymphoma [44]. It is an alkylating agent that inhibits the growth of rapidly proliferating cells [44].

A first in its class anticancer drug, VLX-600 was given approval by the FDA to enter phase 1 and 2 trials. It kills populations of sleeping cells in regions of solid tumors that are metabolically compromised due to poor vascularization. These areas cannot be reached by conventional cancer therapeutics. That is, there are areas with poor vascular supply in solid tumors, where cancer cells divide more slowly due to a lack of O_2 and nutrients. Cancer cells in these areas enter something that is called a sleeping state. These cells start proliferating after being treated with radiotherapy or conventional chemotherapy, causing the tumors to regrow. Such regrowth between therapy cycles affects morbidity and mortality. VLX600 kills sleeping cancer cells by inducing mitochondrial dysfunction [45].

A combination of fluticasone furoate, an inhaled corticosteroid (ICS), and vilanterol, a long-acting β2-adrenergic receptor agonist (LABA) was approved for treating COPD [46]. "It is called Breo Ellipta®. It is used once daily, as opposed to the twice daily dose of the previous treatments that used a long-acting β2-agonist and inhaled corticosteroids. It was shown to be a safe and effective bronchodilator with a rapid onset of action" [46].

A low dose of paroxetine mesylate, a selective serotonin reuptake inhibitor, was approved as the drug named Brisdelle® for treating vasomotor symptoms of menopause [47]. Duavee ® was also approved for treating vasomotor symptoms of menopause [48]. It is a mixture of conjugated estrogens and bazedoxifene, a selective estrogen receptor modulator (SERM) designed to mimic the beneficial qualities of estrogens, while blocking estrogen in tissues where it can be harmful [48].

Canagliflozin (Invokana®) is an inhibitor of sodium-glucose co-transporter 2 that was approved for treating type-2 diabetes [49]. It reduces concentrations of glucose in the blood glucose levels by increasing the amount of glucose excreted in the urine. It is specifically indicated as an adjunct to diet and exercise to improve glycemic control in adults [49]. Alogliptin (Nesina®) was also approved for treating type-2 diabetes [50]. It is a dipeptidyl peptidase IV (DPP IV) inhibitor. DPP-4 inhibitors slow the inactivation of incretin hormones GLP-1 (glucagon-like peptide-1) and GIP (glucose-dependent insulinotropic peptide), both of which play a role in regulating blood glucose levels [50].

Ospemifene (Osphena®) was approved for treating dyspareunia and vulvar and vaginal atrophy due to menopause [51]. "It binds to estrogen receptors, resulting in the activation of estrogenic pathways in some tissues and the blockade of estrogenic pathways in other tissues. For postmenopausal woman with a uterus, the addition of a progestin should be considered to reduce the risk of endometrial cancer. A woman without a uterus does not need a progestin" [51].

Tocilizumab (Actemra®) was approved for treating polyarticular juvenile idiopathic arthritis [52]. "It is a humanized anti IL-6 receptor monoclonal antibody. It binds specifically to both soluble and membrane-bound IL-6 receptors and has been shown to inhibit IL-6-mediated signaling through these receptors. IL-6 is a pleiotropic pro-inflammatory cytokine produced by a variety of cell types including T- and B-cells, lymphocytes, monocytes and fibroblasts. IL-6 is also produced by synovial and endothelial cells leading to local production of IL-6 in joints affected by inflammatory processes such as rheumatoid arthritis" [52]. Canakinumab (Ilaris®) was also approved for treating polyarticular juvenile idiopathic arthritis in patients two years old or older [52]. It is a human monoclonal anti-human IL-1ß antibody [53].

There is also a new drug called LCZ696 that may provide a huge advance in preventing heart failure [54]. "It is a combination of two drugs that lower blood pressure (antihypertensives), valsartan and sacubitril. It was shown to be better than enalapril, which is an inhibitor of angiotensin converting enzyme (ACE inhibitor) that has been the standard therapy. It inhibits neprylisin, an endopeptidase that catalyzes the hydrolytic breakdown of several endogenous vasoactive peptides. By inhibiting neprylisin, the sacubitril in LCZ696 increases the concentrations of these peptides, thus countering vasoconstriction, sodium retention, and maladaptive tissue remodeling. The valsartan blocks the angiotensin receptor. The combination of valsartan and sacubitril as LCZ696 reduced heart-related deaths by 20% compared to enalapril and reduced the risk of being hospitalized for heart failure by 21%" [54].

There have also been drugs that were approved recently for treating bacterial infections. delamanid (Deltyba®) was approved for treating multidrug-resistant pulmonary tuberculosis. It was designated an orphan drug in 2008 for treating a rare disease. It inhibits the biosynthesis of mycolic acid [54]. "When taken twice daily at a dose of 100 mg and combined with an optimized background regimen, it increased sputum culture conversion (SCC) after two months when compared placebo. SCC is used to determine when a patient is no longer infectious" [55].

Another antibiotic, dalbavancin (Dalvance®), was approved for treating acute bacterial skin and skin structure infections (ABSSSI) caused by some Gram-positive bacteria, such as methicillin-resistant Staphylococcus aureus (MRSA) [56, 57]. It is the first and only IV antibiotic approved for treating ABSSSI with a two-dose regimen of 1000 mg followed one week later by 500 mg, each administered over 30 mins. Patients treated with dalbavancin had fewer adverse events than those treated with vancomycin-linezolid. Dalvance was granted priority review as a Qualified Infectious Disease Product (QIDP), in accordance with the Generating Antibiotics Incentives Now (GAIN) Act, which was passed by Congress in 2012 to help make antibiotic development processes smoother. It is a lipoglycopeptide that has a long half-life in blood plasma, enabling it to be given only once a week [57].

The FDA also approved Debio 1450 as a QIDP for treating MRSA [58]. It is rapidly a metabolite (Debio 1452, which inhibits the enoyl-acyl carrier protein reductase (FabI) that catalyzes the final step in each cycle of fatty acid biosynthesis in MRSA [58].

A combination of 1.2% clindamycin phosphate and 3.0% benzoyl peroxide was also approved by the FDA for treating acne. Clinamycin is a topical antibiotic that is available by prescription, while gels containing 10% benzoyl peroxide are sold over the counter, due to its bacteriocidal, anti-inflammatory and keratolytic activities [59]. "However, it can cause dose-dependent skin irritation. The combination was shown to be safe and effective. The combination acts synergistically to successfully treat acne – even in people with very sensitive skin" [59].

Research also continues on the role of the gut microbiome in obesity and disease. It was shown that a change in diet from all animal to all plant products rapidly changed the gut microbiome [60]. "The animal-based diet increased the abundance of bile-tolerant microorganisms (*Alistipes*, *Bilophila* and *Bacteroides*) while decreasing the levels of Firmicutes (*Roseburia*, *Eubacterium rectale* and *Ruminococcus bromii*) that metabolize dietary plant polysaccharides. Also, increases in the abundance and activity of *Bilophila wadsworthia* on the animal-based diet support a link between dietary fat, bile acids and the outgrowth of microorganisms capable of triggering inflammatory bowel disease" [60].

At the same time, new ways of treating AIDS are being developed. One study showed that HIV-1 can induce a type of cell death called pyroptosis, which uses caspase-1 after being activated by inflammasomes [61]. So, an inhibitor of caspase-1 named VX-765 was shown to inhibit pyroptosis and HIV-1 induced death of $CD4^+$ T cells [61].

A new drug called ibalizumab was also approved for treating HIV. It is a humanized IgG4 monoclonal antibody that binds to human CD4 and blocks the entry of HIV into $CD4^+$ cells [62]. "The recent demonstration of effective preexposure prophylaxis (PrEP) with tenofovir vaginal gel and combination oral tenofovir/emtricitabine has renewed hope for novel biomedical prevention strategies" [63]. "However, even if effective, PrEP based on daily use of antiretroviral drugs is not always easy to implement. Therefore, an alternative PrEP strategy requiring less frequent dosing is needed. Ibalizumab may be such a strategy since it is safe and effective" [63].

There have also been dramatic breakthroughs in curing hepatitis C (HCV) [64]. The drug called ABT-450 was approved for treating it [65]. It inhibits an HCV protease [65]. When combined with the non-nucleoside polymerase inhibitor dasubavir for 12 weeks, the cure rate was almost 100% and there were none of the severe side effects that can occur when using interferon [64]. "All 25 patients with genotype 1b infection who were treated without ribavirin had undetectable HCV RNA levels 24 weeks after the end of therapy" [65].

There has also been an important advance in treating COPD. The drug umeclidinium bromide (also known as UMEC and INCRUSE ELLIPTA) is an antimuscarinic inhalation powder that was approved for helping prevent the flow of air into the lungs [66]. It is given at a dose of 125 μg, in combination with 25 μg of the long-acting β_2 agonist vilanterol (also known as VI) for once-daily long-term maintenance of air flow [66].

On the other hand, the FDA named the therapy called CTL019 as the breakthrough therapy of the year [67]. "It is a personalized cellular therapy for treating acute lymphoblastic leukemia (ALL). That is, a patient's T-cells are removed by aphereis and are genetically modified. They have a chimeric antigen receptor that couples an anti-CD19 single chain F_v domain with intracellular T-cell signaling domains of the T-cell receptor and put back into the patient's blood. It resulted in a 90% remission rate for up to two years in the 30 children and adults who received it" [67].

At about the same time, the European Union Committee for Medicinal Products for Human Use (CHMP) recommended that the EU approve obinutuzumab (Gazyvaro®) in combination with chlorambucil for the treatment of adult patients with previously untreated chronic lymphocytic leukemia (CLL) who have comorbidities making them unsuitable for a certain type of chemotherapy (full-dose fludarabine) [68, 69]. It is a glycoengineered humanized monoclonal antibody that binds to the CD20 cell surface protein and prevents B-cell cancers [69].

"The CD20 protein is a good therapeutic target because it is only expressed on B-cell precursors and mature B-cells, but not on hematopoietic stem cells or blood plasma cells. This prevents liver toxicity and helps maintain sufficient concentrations in the blood to be effective in causing CD20 proteins to reorganize into lipid rafts in the cell membranes of B-cells. This activates the complement-mediated cell death" [69]. Gazyvaro® is called Gazyva® in the United States and the rest of the world.

Another important advance in treating cancer was the recent description of a bioengineered protein to prevent metastasis [70]. "It is an altered version of a receptor that binds a protein called Axl, which is a receptor tyrosine kinase. Altered signaling through the Axl receptor has been linked to metastatic cancer. It is on the surface of cells. When its ligand (a protein called Gas6) binds to it, metastasis can emerge. So, researchers used high-throughput screening and X-ray crystallography to evaluate over 10 million variants of Axl to find one that binds Gas6 the tightest. Then, they were able to design a version of Axl that acts as a decoy that can bind Gas6 in a way that does not lead to metastasis. In the process, Gas6 proteins are all used up in binding to this altered version of Axl, so none are available to bind to the wild-type Axl protein. When given as an intravenous dose to mice with aggressive breast and ovarian cancers. It caused a 78% and 90% reduction in metastatic nodules in breast and ovarian cancer, respectively" [70].

Another recent study identified a protein called suppressor of cytokine signaling 3 (SOCS3) as a potential therapeutic target for treating triple negative breast cancer breast cancer, in which the genes coding for the estrogen, progesterone and human epidermal growth factor 2 (Her2) receptors [71]. "The study found that SOC3 mediates the regulation of inflammatory cytokines that are produced when PTEN and p53 proteins are suppressed. That is, when PTEN and p53 were knocked down in human mammary cells, this activated signaling by the pro-inflammatory interleukin-6 (IL6)/Stat3/nuclear factor κB system. This produced highly cancer stem cells and tumors whose gene expression profile mimicked that found in triple negative breast cancers. During this process, the SOC3 protein is degraded by proteolysis. So, SOC3 could be an excellent new therapeutic target for treating triple negative breast cancer. This is quite significant because it indicated that over-the-counter drugs like aspirin and Naproxen should be tried in clinical studies of patients with triple negative breast cancer" [71].

A different study found more about the mechanism by which the anti-cancer drug Taxol works by binding to microtubules [72]. "It showed exactly where Taxol binds to microtubules, freezing them in place and preventing them from separating chromosomes when a cancer cell tries to divide. That is, "Taxol inserts itself into the tubulin protein and prevents compaction of the alpha and beta subunits, so that it remains intact unable to depolymerize and carry out its normal function". These results may help to develop improved microtubule inhibitors" [72].

It has also been shown recently that high dose parenteral ascorbate can work synergistically with paclitaxel and carboplatin to treat ovarian cancer [73]. "Previous studies using oral doses ascorbate were not successful, but the bioavailability of parenteral doses is much higher, making them effective. So, the concentration of ascorbate after intravenous administration can be as high as 10 mM, while oral doses result in concentrations almost never exceed 0.2 mM (200 μM) in the blood. "When patients have normal renal function and glucose-- -phosphate dehydrogenase (G6PD) activity, toxicity is minimal even with intravenous doses as high as 1.5 g/kg, equivalent to 105 g for a 70-kg person". At such high doses, ascorbate acts as a pro-oxidant, producing H_2O_2 at concentrations in extracellular fluids that are toxic to cancer cells, but not normal cells. It induced DNA damage and decreased ATP concentrations, while activating the ataxia telangiectasia mutated (ATM)/AMP–activated protein kinase (AMPK) pathway. This caused inhibition of mTOR and death in ovarian cancer cells" [73]. "The combination of parenteral ascorbate with the conventional chemotherapeutic agents carboplatin and paclitaxel synergistically inhibited ovarian cancer in mouse models and reduced chemotherapy-associated toxicity in patients with ovarian cancer" [73].

In another important advance, the first biological (Cyramza) for treating gastro-esophageal junction adenocarcinoma that has progressed after the standard first-line therapy [74]. "It is marketed as Ramucirumab®. It is a humanized IgG1 mAb VEGFR-2 antagonist. That is, signaling through VEGF and its receptor, VEGFR, are important in the pathogenesis of gastric adenoma" [74].

Advances have also been made in treating pancreatic cancer, which is usually fatal [75]. "Since 1997, the standard first-line treatment was gemcitabine the 5-year survival rate is only 2%, and 1-year survival rates is anywhere from 17 to 23%. In this study, it was found that albumin-bound paclitaxel particles (nab-paclitaxel) can increase the concentration of gemcitabine in tumors and lead to an improved survival rate" [75].

Another breakthrough is that the FDA approved a new anti-obesity drug, Contreve [76]. "It targets the receptors for the incretin hormones (glucagon-like peptide-1, or GLP-1, and glucose-dependent insulinotropic polypeptide, or GIP). It also decreased the concentration of glycosylated hemoglobin A1c (HbA1c) in humans with type 2 diabetes. Incretins are secreted from the intestine in response to ingested nutrients. They lower postprandial glucose excursion. GLP-1 also suppresses appetite, resulting in reduced adiposity. Treatment with GLP-1 receptor (GLP-1R) agonists provides sizable glycemic benefits without a risk of hypoglycemia. It also decreases total body weight. It does this without harmful side effects such as nausea and vomiting that can occur with monotherapy using GLP-1R agonists" [76].

There have also been some recent discoveries involving the use of pomegranates and cinnamon as aids in preventing and/or treating neurodegenerative diseases [77, 78]. "That is, an important antioxidant in pomegranates and pomegranate juice, punicalagin, inhibits neuroinflammation in microglia by interfering with NK-κB signaling. So, it may help prevent Alzheimer's and Parkinson's diseases" [77]. "Also, cinnamoncan reverse the deleterious changes that occur in the brains of mice with Parkinson's disease (PD). Cinnamaldehyde, cinnamyl acetate and cinnamyl alcohol are metabolized to cinnamic aid, which is converted to sodium benzoate, an FDA-approved drug used in the treatment for hepatic metabolic defects associated with hyperammonemia. Sodium benzoate then enters into the brain, stops the loss of Parkin and DJ-1, protects neurons, normalizes neurotransmitter levels, and improves motor functions in mice with PD" [78].

Another drug, Tecfidera (dimethyl fumarate) was approved by the by the European Commission (EC) as a first-line oral treatment for people with relapsing-remitting multiple sclerosis (RRMS), the most common form of multiple sclerosis (MS) [79]. "It targets and activates the Nrf2-signalling pathway. This causes antioxidant response elements to be upregulated. It has also been used to treat psoriasis, so it has an excellent safety profile" [79].

Another study found that the antidepressant citalopram may be useful in preventing Alzheimer's disease [80]. "It is a selective serotonin reuptake inhibitor. It decreased the appearance of new Aβ (amyloid beta) plaque growth and arrested the growth of preexisting plaques in mice brains interstitial fluid in a dose-dependent manner" [80].

There has also been an important advance in predicting who will get Alzheimer's disease [81]. "That is, the blood lipid profiles of 525 people over 70 with mild cognitive impairment were analyzed by LC-MS. Total lipids were extracted from 25 μL of blood plasma by adding a mixture of 25% acetonitrile, 40% methanol and 20% water containing internal standards. This denatured plasma proteins, which were removed after centrifuging the mixture. Then, the solvent was evaporated off and the residue remaining redissolved in 200 μL of buffer containing 5% methanol, 1% acetonitrile and 94% water. After centrifuging this to remove solid particles, the supernatant was injected on a C18 column. The lipids were eluted using a strong mobile phase that contained acetonitrile and isopropyl alcohol. They found that the concentrations of ten lipid metabolites decreased in those subjects who progressed from mild cognitive impairment to Alzheimer's disease. The lipids included phosphatidyl cholines, abbreviated as PCs. They had different fatty acyls attached to the glyceride backbone. That is, PC has two acyls, abbreviated as aa. Some of those with very long chain fatty acyls, PC diacyl (aa) C36:6, PC aa C38:0, PC aa C38:6, PC aa C40:1, PC aa, C40:2, PC aa C40:6 and PC acyl-alkyl (ae) C40:6 were depleted. Also, lysophophatidylcholine (lysoPC a C18:2), and acylcarnitines (ACs) (Propionyl AC (C3) and C16:1-OH) were depleted. This relatively simple and non-invasive blood test is much better than previously existing methods of predicting who will progress to Alzheimer's disease" [81].

In a different development, the FDA put two investigational new drugs (nintedanib and pirfenidone) on the fast track for evaluation and (hopefully approval) for treating idiopathic pulmonary fibrosis (IPF) [82, 83]. "In IPF, lungs fibrose, or scar, rapidly. Death usually occurs three to five years after diagnosis. No drugs have helped so far. However, Nintedanib (formerly called BIBF-1120) has antifibrotic properties that are mediated by inhibiting several tyrosine kinase receptors that have been implicated in the pathogenesis of IPF. Patients receiving nintedanib enjoyed significant reductions in the rate of decline in forced vital capacity after one year of therapy" [82]. Pirfenidone has antioxidant, anti-TGF and anti-PDGF effects [84]. Both drugs received fast track, priority review, orphan product, and breakthrough designation.

The FDA also approved a drug (Ruconest) to treat acute angioedema attacks caused by mutations in the C1 esterase inhibitor (C1INH) [85]. So, Ruconest is a recombinant C1INH that is safe and effective [86].

The FDA also approved pegylated naloxegol to treat opioid-induced constipation in patients with noncancer pain [87]. "In the USA, over 240 million prescriptions of opioid pain killers are prescribed each year. Anywhere from 40-90% of the patients taking them develop constipation. Naloxegil is an antagonist of the μ-opioid receptor. After having poly(ethylene glycol) attached to it (pegylated naloxegol), it can no longer cross the blood-brain barrier. So, it doesn't block the analgesic effects of opioids. It effectively reduced constipation without changing the analgesic effects of opioids" [87].

The FDA also approved a new drug, albiglutide (Tanzeum®) for treating type-2 diabetes. It is a glucagon-like peptide-1 receptor agonist (GLP-1 RA) with a long half-life that "lowers glycosylated hemoglobin (A1C) and reduces weight by stimulating glucose-dependent insulin secretion, suppressing glucagon secretion, delaying gastric emptying, and promoting satiety" [88]. It only requires a once- a-week dose and has fewer harmful side effects than previously used drug, liraglutide [88].

A new antiviral drug, favipiravir, is being tested for treating infections by the RNA-based influenza and ebola viruses and may be effective against the norovirus [89]. "The norovirus is the most common cause of gastroenteritis in the United Kingdom and other countries. It "causes an unpleasant but relatively short-lived case of vomiting and diarrhea, but chronic infection can cause major health problems for people with compromised immune systems". Favipiravir destroys viruses by causing lethal mutagenesis by changing its RNA. Since RNA-based viruses replicate and mutate rapidly, they are susceptible to this new drug" [89].

However, in 2014, it was the ebola virus that has attracted the most attention. A report by the World Health Organization (WHO) stated that as of November, 2014, there had been 15 935 reported cases of Ebola virus disease (EVD), with 5689 deaths [90]. "The incidence of EVD was stable in Guinea, stable or declining in Liberia, but may still be increasing in Sierra Leone. To prevent the spread of EVD, patients should be isolated and great care must be taken to bury the dead bodies of victims, since they are highly contagious. As at 23 November, the case fatality rate among all patients recorded as hospitalized in the three intense-transmission countries since the outbreak began, and for whom there is a definitive outcome recorded, is 60% in Guinea, 61% in Liberia, and 60% in Sierra Leone. The success of Nigeria and Senegal in halting the transmission of EVD highlights the critical importance of preparedness. Strong political leadership, early detection and response, public awareness campaigns, and strong support from partner organizations were crucial in preventing the spread of EVD in both countries included" [90].

Still, new therapies are being developed. This includes giving favipiravir after a person is exposed to the ebola virus [91]. It was "100% protection against aerosol Ebola virus E718 infection; protection was shown in immune-deficient mice after 14 days of twice-daily dosing. T-705 was also shown to inhibit Ebola virus infection in cell culture. T-705 is likely to be licensed for use against influenza in the near future and could also be used with a new indication for filovirus infection. Members of the Filoviridae family include Ebola virus, Sudan virus, Bundibugyo virus and Marburg virus - all of which cause sporadic outbreaks of severe hemorrhagic fever with high case fatality rates. There are currently no licensed vaccines or therapies against the filoviruses and they are infectious at very low doses. These characteristics mean that the filoviruses are considered possible biowarfare/bioterrorism threat agents that we need to defend against" [91]. "Favipiravir is now under advanced development for influenza and has completed Phase II and is undergoing Phase III clinical trials" [91, 92].

Still, it would be much better if a vaccine against the ebola virus could be developed. So, it was very encouraging when it was reported that the first human trial of an experimental Ebola vaccine had promising results [93]. All 20 healthy adults who received the vaccine in a trial sponsored by the U.S. National Institutes of Health (NIH) produced an immune response and developed anti-Ebola antibodies, while causing no serious side effects [93]. The Ebola virus is one of the most feared viruses in the entire world, due to its high mortality rate [94].

The ebola virus was discovered in 1976 in Central Africa, but was not well-known until an outbreak emerged in 1995 in Kikwit, Zaire (now the Democratic Republic of Congo) [94]. "It was originally classified as a member of the *Rhabdoviridae* family, but was subsequently assigned to the newly defined *Filoviridae* family, based on the morphological, physiochemical and biological features, including its filamentous shape. Filoviruses are in the order *Mononegavirales* and are comprised of two genera: *Marburgvirus* and *Ebolavirus*. The genus *Ebolavirus* has four different species: *Ivory Coast ebolavirus*, *Reston ebolavirus*, *Sudan ebolavirus*, and *Zaire ebolavirus*. They all contain a single, negative strand of linear RNA that is transcribed in host cells to make polyadenylated mRNAs. They code for a 3'leader sequence, nucleoprotein (NP), two virion proteins (VP 24, VP 30, VP 35 and VP40), a glycoprotein (GP), polymerase (L) and 5' trailer. Translation of these mRNAs produces seven structural proteins. Filoviruses enter host cells by receptor-mediated endocytosis. The GP mediates receptor binding and subsequent fusion of the viral particles to the host cell membranes" [94].

It has attracted much attention due to the outbreak in Western Africa in 2014 [95, 96]. On Nov. 26, 2014, the WHO reported that there have been 15 935 reported cases of Ebola virus disease (EVD), with 5689 reported deaths [97]. It also said that "600 cases were reported in the three most-affected countries in the past week. The case incidence is stable in Guinea, stable or declining in Liberia, but may still be increasing in Sierra Leone. As at 23 November, the case fatality rate among all patients recorded as hospitalized in the three intense-transmission countries since the outbreak began, and for whom there is a definitive outcome recorded, is 60% in Guinea, 61% in Liberia, and 60% in Sierra Leone. The bodies of patients who have died from EVD are extremely infectious. Therefore, conducting burials in a safe and dignified manner is a crucial component of efforts to halt the transmission of the disease. The success of Nigeria and Senegal in halting the transmission of EVD highlights the critical importance of preparedness. WHO, the UN and other partners are accelerating the deployment of international preparedness strengthening teams to help countries build upon their existing work and planning. At the end of each mission, technical experts remain in country to support and maximize capacity-building efforts to prepare for public health emergencies, including EVD" [97].

Symptoms of EVD usually occur after an incubation time of four to ten days [98]. "It begins with "flu-like" symptoms, including fever, myalgia and chills, followed by diarrhea and vomiting. It can get much worse, with hemorrhaging and multiple organ failure. This often leads to death between 6 to 16 days after symptoms first appear. The patients who survive can recover completely as they develop an immune response. EVD is diagnosed by reverse transcriptase polymerase chain reaction (RT-PCR), which will detect the viral RNA (its genome). It can be detected 48 hrs after a person is infected. Usually, IgM antibodies appear in a patient's blood starting two days after the first symptoms appear. They can persist for 30-168 days. Meanwhile, IgG antibodies start to appear 6 to 18 days after EVD first begins. They can persist for years. The antibody profile is quite different between people who eventually die and those who survive. This difference is a prognostic marker for patient management. There are four ebola viruses that are pathogenic to humans. They can enter the body through mucosal surfaces, abrasions and injuries in the skin or by direct parental transmission. Infection through intact skin is considered unlikely, although it can't be excluded as a possibility. Once in the body, ebola viruses can infect immune cells (macrophages, monocytes and dendritic cells), epithelial and endothelial cells, fibroblasts, hepatocytes and the adrenal gland. Ebola virus is usually treated by fluid replacement, but "the WHO declared that, considering the magnitude and severity of the current outbreak, it is ethical to use experimental drugs for treatment and prevention of EVD". This includes favipiravir, TKM-ebola (lipid nanoparticle with small interfering RNAs, siRNAs that target three ebola virus proteins), BCX-4430 (a nucleoside analogue), AVI-6002 (a phosphorodiamidate morpholino oligomer that silences genes), and ZMapp (a mixture of three monoclonal antibodies that are specific for the ebola virus). ZMapp has been used to treat a few patients" [98]. However, supplies of it have run out, so vaccines and other drugs are being fast tracked [99].

In a completely different development, it was shown that drugs that had been banned by the U.S. FDA were found in several dietary supplements [100]. "The FDA issues class I drug recalls when a drug may reasonably cause serious adverse side effects or death. About half of the recalls since 2004 have been dietary supplements that were found to be adulterated with banned prescription drugs. So, a private lab (Flora Research Labs) analyzed 27 of the 274 dietary supplements that were recalled by the FDA between Jan. 1, 2009 and Dec. 31, 2012. The supplements were purchased an average of 34 months after they were recalled. One or more adulterants were still found in 66.7% of recalled supplements that could still be purchased. This included supplements for athletic (sports) enhancement, weight loss and sexual enhancement. Also, 63% (17/27) contained the same supplement adulterant that was identified by the FDA" [100].

Also, work continues on discovering genes that are mutated in cancer [101]. "They analyzed somatic point mutations in exome sequences from 4742 human cancers and their matched normal-tissue samples across 21 cancer types. They found that large-scale genomic analysis can identify nearly all known cancer genes in these types of tumors. They also identified 33 genes that were not previously known to be significantly mutated in cancer, including genes related to proliferation, apoptosis, genome stability, chromatin regulation, immune evasion, RNA processing and protein homeostasis. Moreover, analysis of larger sample sizes will reveal many more genes that are mutated. Their results may help to guide the next stage of cancer genomics" [101]. So, research continues in many areas – especially P4 medicine.

KEY POINTS

- **Many new phenomena emerge at higher levels of organization. One possible emergent problem is the likelihood that the virus that causes the flu in birds or pigs will mutate to a form that can easily infect and kill people.**

- **As global warming continues, several tropical diseases could appear in cooler regions. Global climate change is projected to significantly increase the range**

conducive to the transmission of both dengue and yellow fevers.

- **Another possible problem is the re-emergence of the poliovirus or smallpox virus, even though these diseases have almost been eradicated. To the best of our knowledge, smallpox has been eradicated, but small amounts of the smallpox virus still exist in laboratories in the USA and Russia.**

- **Scientists are creating new, genetically altered life forms, in what some consider an emergent solution, while others consider genetic engineering to be a terrible problem. It is also known as genetic modification (GM), biotechnology, gene splicing and recombinant DNA technology.**

- **Technology earned a good reputation when it helped convert Mexico from a country that could not feed itself, to a net exporter of food through the development of high-yield varieties of maize and wheat. The Green Revolution spread to India, where many people worked to improve plant breeding, irrigation, development and financing of agrichemicals. By the late 1970s, yields of rice increased by 30 percent. High-yield varieties spread throughout the Philippines, followed by Indonesia, Pakistan, Sri Lanka, and many other countries throughout Latin America, Asia and North Africa.**

- **There are emerging technologies that could help cure diseases. For example, nanotechnology has the potential to turn relatively inactive molecules into potent drugs. Eventually, it may be possible to make nanoparticles or nanomachines that have specific antibodies on their surface, which are able to bind to specific cancer cells. Once bound, the nanomachine will synthesize or simply release a drug that will kill the cancer cells to which it is bound. The anti-cancer drug doxorubicin can be encapsulated into liposomes, which significantly lowers its cardiotoxicity.**

- **Another technology that could provide many medical benefits is stem cell technology.**

- **Another emergent problem is a shortage of certain medicines. This can happen if a major producer has problems in manufacturing and has to stop for a while, or if there is not enough profit to be made. As a result, patients are experiencing unacceptable delays in receiving 210 different medicines for cancer, anesthesia Parkinson's disease, schizophrenia, osteoporosis and organ failure.**

ABBREVIATIONS

GM	=	Genetic Modification
GMO	=	Genetically Modified Organism
IRRI	=	International Rice Research Institute
CGIAR	=	Consulting Group on International Agricultural Research
ICMVs	=	Interbilayer-Crosslinked Multilamellar Vesicles
P4 medicine	=	Predictive, Preventive, Personalized and Participatory Medicine
ES cells	=	Embryonic Stem Cells
iSPC	=	Induced Pluripotent Stem Cells
FDA	=	Food and Drug Administration
GVHD	=	Graft *Versus* Host Disease
T_{regs}	=	Regulatory T Cells
cGMP	=	Current Good Manufacturing Practices
SOP	=	Standard Operating Procedure
PSA	=	Prostate Specific Antigen
USPSTF	=	U.S. Preventive Services Task Force
PCA3	=	Prostate Cancer Antigen 3
BPH	=	Benign Prostate Hyperplasia
lncRNA	=	Long Non-coding RNA
AMACR	=	α-Methylacyl–Coenzyme A Racemase
Cav-1	=	Caveolin-1
MCT-4	=	Monocarboxylate Transporter 4
PAI-1	=	Plasminogen Activator Inhibitor-1
SMS	=	Senescence- Messaging Secretome
IGFBPs	=	Insulin-Like Growth Factor-Binding Proteins
IL	=	Interleukin
TGF-β	=	Transforming Growth Factor Type β
LDL	=	Low Density Lipoprotein
MEK-1 and MEK-2	=	Mitogen-Activated Protein Kinase Kinases 1 and 2
ICS	=	Inhaled Corticosteroid
LABA	=	Long-acting β2-Adrenergic Receptor Agonist
COPD	=	Chronic Obstructive Pulmonary Disease

SERM	=	Selective Estrogen Receptor Modulator
DPP IV	=	Dipeptidyl Peptidase IV
GLP-1	=	Glucagon-Like Peptide-1
GIP	=	Glucose-Dependent Insulinotropic Peptide
ACE	=	Angiotensin Converting Enzyme
SCC	=	Sputum Culture Conversion
ABSSSI	=	Acute Bacterial Skin and Skin Structure Infections
MRSA	=	Methicillin-Resistant Staphylococcus Aureus
QIDP	=	Qualified Infectious Disease Product
GAIN	=	Act Generating Antibiotics Incentives Now Act
PrEP	=	Preexposure Prophylaxis
HCV	=	Hepatitis C
UMEC	=	Umeclidinium Bromide
ALL	=	Acute Lymphoblastic Leukemia
CLL	=	Chronic Lymphocytic Leukemia
SOC3	=	Suppressor of Cytokine Signaling 3
HER2	=	Human Epidermal Growth Factor 2
PTEN	=	Phosphatase and Tensin Homolog
G6PD	=	Glucose-6-Phosphate Dehydrogenase
ATM	=	Ataxia Telangiectasia Mutated
AMPK	=	AMP–Activated Protein Kinase
VEGF	=	Vascular Endothelial Growth Factor
VEGFR	=	VEGF Receptor
EC	=	European Commission
RRMS	=	Relapsing-Remitting Multiple Sclerosis
MS	=	Multiple Sclerosis
Aβ	=	Amyloid Beta
LC-MS	=	Liquid Chromatography – Mass Spectrometry
C1INH	=	C1 Esterase Inhibitor
EVD	=	Ebola Virus Disease
WHO	=	World Health Organization
NIH	=	National Institutes of Health
NP	=	Nucleoprotein
VP	=	Virion Protein
GP	=	Glycoprotein

RT-PCR = Reverse Transcriptase Polymerase Chain Reaction

siRNA = Small Interfering RNA

REFERENCES

[1] Wikipedia. Spanish Flu, website http://en.wikipedia.org/wiki/Spanish_flu , [Accessed Feb 22, 2014.];

[2] Wikipedia. AIDS, website en.wikipedia.org/wiki/AIDS , [Accessed Feb 22, 2014.];

[3] Wikipedia. Influenza A subtype H5N1, website: en.wikipedia.org/wiki/H5N1 , [Accessed Feb 22, 2014.];

[4] Morens DM, Subbarao K, Taubenberger JK. Engineering H5N1 avian influenza viruses to study human adaptation. Nature 2012; 486(7403): 335-40. [http://dx.doi.org/10.1038/nature11170] [PMID: 22722191]

[5] Normile D. Surprising new dengue virus throws a spanner in disease control efforts. Science 2013; 342(6157): 415. [http://dx.doi.org/10.1126/science.342.6157.415] [PMID: 24159024]

[6] National Resources Defense Council. Consequences of Global Warming, Website: http://www.nrdc.org/globalWarming/fcons.asp , [Accessed Feb 22, 2014.];

[7] Cello J, Paul AV, Wimmer E. Chemical synthesis of poliovirus cDNA: generation of infectious virus in the absence of natural template. Science 2002; 297(5583): 1016-8. [http://dx.doi.org/10.1126/science.1072266] [PMID: 12114528]

[8] Dhar S, Gu FX, Langer R, Farokhzad OC, Lippard SJ. Targeted delivery of cisplatin to prostate cancer cells by aptamer functionalized Pt(IV) prodrug-PLGA-PEG nanoparticles. Proc Natl Acad Sci USA 2008; 105(45): 17356-61. [http://dx.doi.org/10.1073/pnas.0809154105] [PMID: 18978032]

[9] Merkle RC. Nanotechnology and Medicine, website: www.zyvex.com/nanotech/nanotechAndMedicine.html , [Accessed Feb 22, 2014.];

[10] Bowman MC, Ballard TE, Ackerson CJ, Feldheim DL, Margolis DM, Melander C. Inhibition of HIV fusion with multivalent gold nanoparticles. J Am Chem Soc 2008; 130(22): 6896-7. [http://dx.doi.org/10.1021/ja710321g] [PMID: 18473457]

[11] Heath JR, Davis ME, Hood L. Nanomedicine targets cancer. Sci Am 2009; 300(2): 44-51. [http://dx.doi.org/10.1038/scientificamerican0209-44] [PMID: 19186705]

[12] Li AV, Moon JJ, Abraham W, *et al.* Generation of effector memory T cell-based mucosal and systemic immunity with pulmonary nanoparticle vaccination. Sci Transl Med 2013; 5(204): 204ra130. [http://dx.doi.org/10.1126/scitranslmed.3006516] [PMID: 24068737]

[13] Ramalho-Santos M. Stem cells as probabilistic self-producing entities. BioEssays 2004; 26(9): 1013-6. [http://dx.doi.org/10.1002/bies.20053] [PMID: 15351971]

[14] Nardi NB. All the adult stem cells, where do they all come from? An external source for organ-specific stem cell pools. Med Hypotheses 2005; 64(4): 811-7. [http://dx.doi.org/10.1016/j.mehy.2004.08.026] [PMID: 15694702]

[15] Miura T, Mattson MP, Rao MS. Cellular lifespan and senescence signaling in embryonic stem cells. Aging Cell 2004; 3(6): 333-43.
[http://dx.doi.org/10.1111/j.1474-9728.2004.00134.x] [PMID: 15569350]

[16] Matsuoka K, *et al.* Low-dose interleukin-2 therapy restores regulatory T cell homeostasis in patients with chronic graft-*versus*-host disease. Sci Trans Med 2013; 5(179ra43): 1-11.
[http://dx.doi.org/10.1126/scitranslmed.3005265]

[17] Furusawa C, Kaneko K. A dynamical-systems view of stem cell biology. Science 2012; 338(6104): 215-7.
[http://dx.doi.org/10.1126/science.1224311] [PMID: 23066073]

[18] Simpson JL. Blastomeres and stem cells. Nature 2006; 444(7118): 432-5.
[http://dx.doi.org/10.1038/444432a] [PMID: 17122844]

[19] Takahashi K, Yamanaka S. Induction of pluripotent stem cells from mouse embryonic and adult fibroblast cultures by defined factors. Cell 2006; 126(4): 663-76.
[http://dx.doi.org/10.1016/j.cell.2006.07.024] [PMID: 16904174]

[20] Boland MJ, Hazen JL, Nazor KL, *et al.* Adult mice generated from induced pluripotent stem cells. Nature 2009; 461(7260): 91-4.
[http://dx.doi.org/10.1038/nature08310] [PMID: 19672243]

[21] Park A. Stem cell research; The quest resumes. CNN Report 2009 Jan 29;

[22] Sekiya S, Suzuki A. Direct conversion of mouse fibroblasts to hepatocyte-like cells by defined factors. Nature 2011; 475(7356): 390-3.
[http://dx.doi.org/10.1038/nature10263] [PMID: 21716291]

[23] Kim K-S. Converting human skin cells to neurons: a new tool to study and treat brain disorders? Cell Stem Cell 2011; 9(3): 179-81.
[http://dx.doi.org/10.1016/j.stem.2011.08.004] [PMID: 21885012]

[24] Qian L, Huang Y, Spencer CI, *et al. In vivo* reprogramming of murine cardiac fibroblasts into induced cardiomyocytes. Nature 2012; 485(7400): 593-600.
[http://dx.doi.org/10.1038/nature11044] [PMID: 22522929]

[25] Antonica F, Kasprzyk DF, Opitz R, *et al.* Generation of functional thyroid from embryonic stem cells. Nature 2012; 491(7422): 66-73.
[http://dx.doi.org/10.1038/nature11525] [PMID: 23051751]

[26] Robinton DA, Daley GQ. The promise of induced pluripotent stem cells in research and therapy. Nature 2012; 481(7381): 295-305.
[http://dx.doi.org/10.1038/nature10761] [PMID: 22258608]

[27] Takebe T, Sekine K, Enomura M, *et al.* Vascularized and functional human liver from an iPSC-derived organ bud transplant. Nature 2013; 499(7459): 481-4.
[http://dx.doi.org/10.1038/nature12271] [PMID: 23823721]

[28] Hayashi K, Ogushi S, Kurimoto K, Shimamoto S, Ohta H, Saitou M. Offspring from oocytes derived from *in vitro* primordial germ cell-like cells in mice. Science 2012; 338(6109): 971-5.
[http://dx.doi.org/10.1126/science.1226889] [PMID: 23042295]

[29] The Washington Times. A regulatory drug shortage 2012 June 22;

[30] Schlosser E. Reefer Madness. Boston: Mariner Books 2003.

[31] U.S. Preventive Services Task Force. Screening for prostate cancer www.uspreventiveservices taskforce.org/prostatecancerscreening.htm , [Accessed Feb 4, 2014.];

[32] Prensner JR, Rubin MA, Wei JT, Chinnaiyan AM. Beyond PSA: the next generation of prostate cancer biomarkers. Sci Transl Med 2012; 4(127): 127rv3.
[http://dx.doi.org/10.1126/scitranslmed.3003180] [PMID: 22461644]

[33] Yang F, Zhang H, Mei Y, Wu M. Reciprocal regulation of HIF-1α and lincRNA-p21 modulates the Warburg effect. Mol Cell 2014; 53(1): 88-100.
[http://dx.doi.org/10.1016/j.molcel.2013.11.004] [PMID: 24316222]

[34] Martinez-Outschoorn U, Sotgia F, Lisanti MP. Tumor microenvironment and metabolic synergy in breast cancers: critical importance of mitochondrial fuels and function. Semin Oncol 2014; 41(2): 195-216.
[http://dx.doi.org/10.1053/j.seminoncol.2014.03.002] [PMID: 24787293]

[35] Eren M, Boe AE, Murphy SB, *et al.* PAI-1-regulated extracellular proteolysis governs senescence and survival in Klotho mice. Proc Natl Acad Sci USA 2014; 111(19): 7090-5.
[http://dx.doi.org/10.1073/pnas.1321942111] [PMID: 24778222]

[36] Page MM, Bell DA, Hooper AJ, Watts GF, Burnett JR. Best Pract Res Clin Endocrinol Metab 2013; 28(3): 387-403.
[http://dx.doi.org/10.1016/j.beem.2013.10.004] [PMID: 24840266]

[37] Macdonald RL, Cusimano MD, Etminan N, *et al.* SAHIT Collaboration. Subarachnoid hemorrhage international trial lists data repository (SAHIT). World Neurosurg 2013; 79(3-4): 418-22.
[http://dx.doi.org/10.1016/j.wneu.2013.01.006] [PMID: 23295631]

[38] Iglarz M, Binkert C, Morrison K, *et al.* Pharmacology of macitentan, an orally active tissue-targeting dual endothelin receptor antagonist. J Pharmacol Exp Ther 2008; 327(3): 736-45.
[http://dx.doi.org/10.1124/jpet.108.142976] [PMID: 18780830]

[39] Peterson JD, Goldman MP, Weiss RA, *et al.* Treatment of reticular and telangiectatic leg veins: double-blind, prospective comparative trial of polidocanol and hypertonic saline. Dermatol Surg 2012; 38(8): 1322-30.
[http://dx.doi.org/10.1111/j.1524-4725.2012.02422.x] [PMID: 22620717]

[40] Jarratt M, Jones T, Kempers S, *et al.* Luliconazole for the treatment of interdigital tinea pedis: A double-blind, vehicle-controlled study. Cutis 2013; 91(4): 203-10.
[PMID: 23763082]

[41] Falchook GS, Lewis KD, Infante JR, *et al.* Activity of the oral MEK inhibitor trametinib in patients with advanced melanoma: a phase 1 dose-escalation trial. Lancet Oncol 2012; 13(8): 782-9.
[http://dx.doi.org/10.1016/S1470-2045(12)70269-3] [PMID: 22805292]

[42] Fowler J, Jarratt M, Moore A, *et al.* Brimonidine Phase II Study Group. Once-daily topical brimonidine tartrate gel 0·5% is a novel treatment for moderate to severe facial erythema of rosacea: results of two multicentre, randomized and vehicle-controlled studies. Br J Dermatol 2012; 166(3): 633-41.

[http://dx.doi.org/10.1111/j.1365-2133.2011.10716.x] [PMID: 22050040]

[43] Hauschild A, Grob JJ, Demidov LV, *et al.* Dabrafenib in *BRAF*-mutated metastatic melanoma: a multicentre, open-label, phase 3 randomised controlled trial. Lancet 2012; 380(9839): 358-65. [http://dx.doi.org/10.1016/S0140-6736(12)60868-X] [PMID: 22735384]

[44] Lessin SR, Duvic M, Guitart J, *et al.* Topical chemotherapy in cutaneous T-cell lymphoma: positive results of a randomized, controlled, multicenter trial testing the efficacy and safety of a novel mechlorethamine, 0.02%, gel in mycosis fungoides. JAMA Dermatol 2013; 149(1): 25-32. [http://dx.doi.org/10.1001/2013.jamadermatol.541] [PMID: 23069814]

[45] Zhang X, *et al.* Induction of mitochondrial dysfunction as a strategy for targeting tumour cells in metabolically compromised microenvironments. Nat Commun 2014; 5: 3295. [http://dx.doi.org/10.1038/ncomms4295]

[46] Kerwin EM, Scott-Wilson C, Sanford L, *et al.* A randomised trial of fluticasone furoate/vilanterol (50/25 μg; 100/25 μg) on lung function in COPD. Respir Med 2013; 107(4): 560-9. [http://dx.doi.org/10.1016/j.rmed.2012.12.014] [PMID: 23352226]

[47] Weber L, Thacker HL. Paroxetine: a first for selective serotonin reuptake inhibitors - a new use: approved for vasomotor symptoms in postmenopausal women. Womens Health (Lond Engl) 2014; 10(2): 147-54. [http://dx.doi.org/10.2217/whe.14.3] [PMID: 24601805]

[48] Goldstein SR. Still another selective estrogen receptor modulator to enhance women's health: this time in combination with conjugated equine estrogens. Menopause 2014; 21(1): 1-2. [http://dx.doi.org/10.1097/GME.0000000000000174] [PMID: 24326285]

[49] Stenlöf K, Cefalu WT, Kim KA, *et al.* Efficacy and safety of canagliflozin monotherapy in subjects with type 2 diabetes mellitus inadequately controlled with diet and exercise. Diabetes Obes Metab 2013; 15(4): 372-82. [http://dx.doi.org/10.1111/dom.12054] [PMID: 23279307]

[50] Masuda K, Aoki K, Kamiko K, *et al.* Glycemic control after addition of the dipeptidyl peptidase-4 inhibitor alogliptin in patients with type 2 diabetes showing inadequate response to thrice-a-day treatment with α-glucosidase inhibitors. Expert Opin Pharmacother 2013; 14(9): 1111-8. [http://dx.doi.org/10.1517/14656566.2013.790365] [PMID: 23600363]

[51] Portman DJ, Bachmann GA, Simon JA. Ospemifene Study Group. Ospemifene, a novel selective estrogen receptor modulator for treating dyspareunia associated with postmenopausal vulvar and vaginal atrophy. Menopause 2013; 20(6): 623-30. [PMID: 23361170]

[52] Imagawa T, Yokota S, Mori M, *et al.* Safety and efficacy of tocilizumab, an anti-IL-6-receptor monoclonal antibody, in patients with polyarticular-course juvenile idiopathic arthritis. Mod Rheumatol 2012; 22(1): 109-15. [http://dx.doi.org/10.3109/s10165-011-0481-0] [PMID: 21667343]

[53] Ruperto N, Brunner HI, Quartier P, *et al.* PRINTO; PRCSG. Two randomized trials of canakinumab in systemic juvenile idiopathic arthritis. N Engl J Med 2012; 367(25): 2396-406. [http://dx.doi.org/10.1056/NEJMoa1205099] [PMID: 23252526]

[54] McMurray JJ, Packer M, Desai AS, *et al.* PARADIGM-HF Investigators and Committees. Angiotensin-neprilysin inhibition *versus* enalapril in heart failure. N Engl J Med 2014; 371(11): 993-1004.
[http://dx.doi.org/10.1056/NEJMoa1409077] [PMID: 25176015]

[55] Gler MT, Skripconoka V, Sanchez-Garavito E, *et al.* Delamanid for multidrug-resistant pulmonary tuberculosis. N Engl J Med 2012; 366(23): 2151-60.
[http://dx.doi.org/10.1056/NEJMoa1112433] [PMID: 22670901]

[56] U.S. Food and Drug Administration. FDA approves Dalvance to treat skin infections. http://www.fda.gov/NewsEvents/Newsroom/PressAnnouncements/ucm398724.htm , [Accessed Nov. 20, 2014.];

[57] Boucher HW, Wilcox M, Talbot GH, Puttagunta S, Das AF, Dunne MW. Once-weekly dalbavancin *versus* daily conventional therapy for skin infection. N Engl J Med 2014; 370(23): 2169-79.
[http://dx.doi.org/10.1056/NEJMoa1310480] [PMID: 24897082]

[58] Kaplan N, Hafkin K. Preclinical pharmacokinetics and pharmacodynamics of Debio 1450, a new Staphylococcocal-specific antibiotic. IACC 2014 abstract. Website: http://www.icaaconline.com/php/icaac2014abstracts/data/papers/2014/F-985a.htm , [Accessed Nov. 20, 2014.];

[59] Kawashima M, Hashimoto H, Sáenze ABA, Ono M, Yamada M. Clindamycin phosphate 1.2%–benzoyl peroxide 3% fixed-dose combination gel has an effective and acceptable safety and tolerability profile for the treatment of acne vulgaris in Japanese patients: a phase III, multicentre, randomised, single-blinded, active-controlled, parallel-group study. Brit J Dermatol 2014. (in print)

[60] David LA, Maurice CF, Carmody RN, *et al.* Diet rapidly and reproducibly alters the human gut microbiome. Nature 2014; 505(7484): 559-63.
[http://dx.doi.org/10.1038/nature12820] [PMID: 24336217]

[61] Gaiha GD, Brass AL. Immunology. The fiery side of HIV-induced T cell death. Science 2014; 343(6169): 383-4.
[http://dx.doi.org/10.1126/science.1250175] [PMID: 24458634]

[62] Pace CS, Fordyce MW, Franco D, Kao CY, Seaman MS, Ho DD. Anti-CD4 monoclonal antibody ibalizumab exhibits breadth and potency against HIV-1, with natural resistance mediated by the loss of a V5 glycan in envelope. J Acquir Immune Defic Syndr 2013; 62(1): 1-9.
[http://dx.doi.org/10.1097/QAI.0b013e3182732746] [PMID: 23023102]

[63] WuXi Pharma Tech. FDA Approves Ibalizumab manufactured by WuXi PharmaTech for treatment of patients on expanded access. www.wuxiapptec.com/press/detail/221/18.html , [Accessed Nov. 20, 2014.];

[64] Medical University of Vienna. New combination therapy for hepatitis C. http://www.meduniwien.ac.at/homepage/1/news-and-topstories/?tx_ttnews[tt_news]=4751&cHash=4d5fe14500c40f9fdfeca0d9c61688f2 , [Accessed Nov. 20, 2014];

[65] Ferenci P, Bernstein D, Lalezari J, *et al.* PEARL-III Study; PEARL-IV Study. ABT-450/r-ombitasvir and dasabuvir with or without ribavirin for HCV. N Engl J Med 2014; 370(21): 1983-92.
[http://dx.doi.org/10.1056/NEJMoa1402338] [PMID: 24795200]

[66] Celli B, *et al.* Once-daily umeclidinium/vilanterol 125/25 μg therapy in COPD. A randomized, controlled study. Chest 2014; 145: 981-91.

[http://dx.doi.org/10.1378/chest.13-1579]

[67] Maude SL, Frey N, Shaw PA, *et al.* Chimeric antigen receptor T cells for sustained remissions in leukemia. N Engl J Med 2014; 371(16): 1507-17.
[http://dx.doi.org/10.1056/NEJMoa1407222] [PMID: 25317870]

[68] Hofmann-LaRoche Ltd. Roche's Gazyvaro approved in Europe for patients with the most common type of leukemia . www.roche.com/media/media_releases/med-cor-2014-07-29.htm , [Accessed Nov 28, 2014];

[69] Rioufol C, Salles G. Obinutuzumab for chronic lymphocytic leukemia. Expert Rev Hematol 2014; 7(5): 533-43.
[http://dx.doi.org/10.1586/17474086.2014.953478] [PMID: 25163491]

[70] Kariolis MS, Miao YR, Jones DS II, *et al.* An engineered Axl 'decoy receptor' effectively silences the Gas6-Axl signaling axis. Nat Chem Biol 2014; 10(11): 977-83.
[http://dx.doi.org/10.1038/nchembio.1636] [PMID: 25242553]

[71] Kim G, Ouzounova M, Quraishi AA, *et al.* SOCS3-mediated regulation of inflammatory cytokines in PTEN and p53 inactivated triple negative breast cancer model. Oncogene 2014; 1-10.
[PMID: 24531711]

[72] Alushin GM, Lander GC, Kellogg EH, Zhang R, Baker D, Nogales E. High-resolution microtubule structures reveal the structural transitions in αβ-tubulin upon GTP hydrolysis. Cell 2014; 157(5): 1117-29.
[http://dx.doi.org/10.1016/j.cell.2014.03.053] [PMID: 24855948]

[73] Ma Y, *et al.* High-dose parenteral ascorbate enhanced chemosensitivity of ovarian cancer and reduced toxicity of chemotherapy. Sci Trans Med. 2014; 6: p. (222)222ra18.

[74] Fuchs CS, Tomasek J, Yong CJ, *et al.* REGARD Trial Investigators. Ramucirumab monotherapy for previously treated advanced gastric or gastro-oesophageal junction adenocarcinoma (REGARD): an international, randomised, multicentre, placebo-controlled, phase 3 trial. Lancet 2014; 383(9911): 31-9.
[http://dx.doi.org/10.1016/S0140-6736(13)61719-5] [PMID: 24094768]

[75] Von Hoff DD, Ervin T, Arena FP, *et al.* Increased survival in pancreatic cancer with nab-paclitaxel plus gemcitabine. N Engl J Med 2013; 369(18): 1691-703.
[http://dx.doi.org/10.1056/NEJMoa1304369] [PMID: 24131140]

[76] Finan B, *et al.* Unimolecular dual incretins maximize metabolic benefits in rodents, monkeys, and humans. Sci Transl Med. 2014; 5: p. (209)209ra151.

[77] Olajide OA, Kumar A, Velagapudi R, Okorji UP, Fiebich BL. Punicalagin inhibits neuroinflammation in LPS-activated rat primary microglia. Mol Nutr Food Res 2014; 58(9): 1843-51.
[http://dx.doi.org/10.1002/mnfr.201400163] [PMID: 25066095]

[78] Khasnavis S, Pahan K. Cinnamon treatment upregulates neuroprotective proteins Parkin and DJ-1 and protects dopaminergic neurons in a mouse model of Parkinson's disease. J Neuroimmune Pharmacol 2014; 9(4): 569-81.
[http://dx.doi.org/10.1007/s11481-014-9552-2] [PMID: 24946862]

[79] Orthmann-Murphy JL. A new era of therapy in multiple sclerosis: Balancing the options and

challenges ahead. Neurol Report 2014; Winter, 23-.

[80] Sheline YI, West T, Yarasheski K, *et al.* An antidepressant decreases CSF Aβ production in healthy individuals and in transgenic AD mice. Sci Transl Med 2014; 6(236): 236re4. [http://dx.doi.org/10.1126/scitranslmed.3008169] [PMID: 24828079]

[81] Mapstone M, Cheema AK, Fiandaca MS, *et al.* Plasma phospholipids identify antecedent memory impairment in older adults. Nat Med 2014; 20(4): 415-8. [http://dx.doi.org/10.1038/nm.3466] [PMID: 24608097]

[82] Hunninghake GM. A new hope for idiopathic pulmonary fibrosis. N Engl J Med 2014; 370(22): 2142-3. [http://dx.doi.org/10.1056/NEJMe1403448] [PMID: 24836311]

[83] King T, *et al.* The ASCEND study: A randomized, double-blind, placebo controlled trial of Pirfenidone in patients with idiopathic pulmonary fibrosis (IPF). Am J Respir Crit Care Med 2014; 189: A6602.

[84] Hilberg O, Simonsen U, du Bois R, Bendstrup E. Pirfenidone: significant treatment effects in idiopathic pulmonary fibrosis. Clin Respir J 2012; 6(3): 131-43. [http://dx.doi.org/10.1111/j.1752-699X.2012.00302.x] [PMID: 22697264]

[85] Manson AL, Dempster J, Grigoriadou S, Buckland MS, Longhurst HJ. Use of recombinant C1 inhibitor in patients with resistant or frequent attacks of hereditary or acquired angioedema. Eur J Dermatol 2014; 24(1): 28-34. [PMID: 24556385]

[86] Riedl MA, Levy RJ, Suez D, *et al.* Efficacy and safety of recombinant C1 inhibitor for the treatment of hereditary angioedema attacks: a North American open-label study. Ann Allergy Asthma Immunol 2013; 110(4): 295-9. [http://dx.doi.org/10.1016/j.anai.2013.02.007] [PMID: 23535096]

[87] Chey WD, Webster L, Sostek M, Lappalainen J, Barker PN, Tack J. Naloxegol for opioid-induced constipation in patients with noncancer pain. N Engl J Med 2014; 370(25): 2387-96. [http://dx.doi.org/10.1056/NEJMoa1310246] [PMID: 24896818]

[88] Trujillo JM, Nuffer W. Albiglutide: a new GLP-1 receptor agonist for the treatment of type 2 diabetes. Ann Pharmacother 2014; 48(11): 1494-501. [http://dx.doi.org/10.1177/1060028014545807] [PMID: 25136065]

[89] Arias A, Thorne L, Goodfellow I. Favipiravir elicits antiviral mutagenesis during virus replication *in vivo*. eLife 2014; 3: e03679. [http://dx.doi.org/10.7554/eLife.03679] [PMID: 25333492]

[90] WHO. WHO Ebola Response Map - Situation Report. http://www.who.int/csr/disease/ebola/ situation-reports/en/?m=20141126 , [Accessed Nov. 30, 2014.];

[91] Smither SJ, Eastaugh LS, Steward JA, Nelson M, Lenk RP, Lever MS. Post-exposure efficacy of oral T-705 (Favipiravir) against inhalational Ebola virus infection in a mouse model. Antiviral Res 2014; 104: 153-5. [http://dx.doi.org/10.1016/j.antiviral.2014.01.012] [PMID: 24462697]

[92] Furuta Y, Gowen BB, Takahashi K, Shiraki K, Smee DF, Barnard DL. Favipiravir (T-705), a novel

viral RNA polymerase inhibitor. Antiviral Res 2013; 100(2): 446-54. [http://dx.doi.org/10.1016/j.antiviral.2013.09.015] [PMID: 24084488]

[93] CNN. Ebola vaccine. http://www.cnn.com/2014/11/27/health/ebola-outbreak/index.html?hpt=hp_t3 , [Accessed Nov. 30, 2014.];

[94] Feldmann H, Jones S, Klenk H-D, Schnittler H-J. Ebola virus: from discovery to vaccine. Nat Rev Immunol 2003; 3(8): 677-85. [http://dx.doi.org/10.1038/nri1154] [PMID: 12974482]

[95] Du Troit A. Ebola virus in West Africa. Nat Rev Microbiol 2014; 12: 312. [http://dx.doi.org/10.1038/nrmicro3267]

[96] Kahrstrom CT. Ebola crisis continues. Nat Rev Microbiol 2014; 12: 792. [http://dx.doi.org/10.1038/nrmicro3392]

[97] World Health Organization. WHO Ebola Response Map - Situation Report . http://www.who.int /csr/disease/ebola/situation-reports/en/?m=20141126 2014.

[98] Goeijenbier M, van Kampen JJ, Reusken CB, Koopmans MP, van Gorp EC. Ebola virus disease: a review on epidemiology, symptoms, treatment and pathogenesis. Neth J Med 2014; 72(9): 442-8. [PMID: 25387613]

[99] Kåhrström CT. Ebola crisis continues. Nat Rev Microbiol 2014; 12: 792. [http://dx.doi.org/10.1038/nrmicro3392]

[100] Cohen PA, Maller G, DeSouza R, Neal-Kababick J. Presence of banned drugs in dietary supplements following FDA recalls. JAMA 2014; 312(16): 1691-3. [http://dx.doi.org/10.1001/jama.2014.10308] [PMID: 25335153]

[101] Lawrence MS, Stojanov P, Mermel CH, *et al.* Discovery and saturation analysis of cancer genes across 21 tumour types. Nature 2014; 505(7484): 495-501. [http://dx.doi.org/10.1038/nature12912] [PMID: 24390350]

CHAPTER 13

Communication and Signaling in Medicinal Chemistry

Abstract: To sustain life and good health, it is essential that cells and tissues can communicate with each other. Organelles within cells, cells within tissues, tissues within our body, and all of the bodies in a society must sense their internal and external environments and respond appropriately to changes. Hormones, neurotransmitters and cytokines can act as primary messengers. Secondary messengers include is Ca^{2+}, IP_3 and diacyl glycerol (produced by the hydrolysis of phosphoinositides), arachidonic acid (produced by the hydrolysis of phospholipids that have arachidonoyl on carbon number 2 of the glycerol backbone), ceramide, eicosanoids, lysophosphatidic acid, NO (nitric oxide), cAMP and cGMP. The IP_3 receptor, or IP_3R is a membrane-bound complex of glycoproteins. It is a Ca^{2+} channel that is activated by IP_3, which is a secondary intra-cellular messenger. Inter- and intracellular communication can be thought of as a network that contains many items (nodes) that have anywhere from one to thousands of connections. The most widely connected nodes are called hubs. Probably the major genetic hub in human and many other mammalian cells is the gene TP53 which codes for the protein p53. About 50% of all human cancers have one or more mutations in p53 that alter DNA transcription.

Keywords: Arachidonic acid, Calcium, Ca^{2+}, Ceramide, cAMP, cGMP, Diacyl glycerol, Eicosanoids, GPCR, IP_3, Lysophosphatidic acid, Nitric oxide.

INTRODUCTION

To sustain life and good health, it is essential that cells and tissues can communicate with each other. Biochemical messengers are made inside subcellular organelles, at the cell membrane and by symbiotic bacteria. These biochemicals can affect nearby cells in paracrine signaling or distant cells in the endocrine system. It is important that the whole process is properly controlled. This is donr through networks, which contain nodes (such as organs, tissues, cells, membranes, organelles, and biochemicals) that are linked to each other. Information flows between nodes, so that the network can adapt to changes in the

Robert E. Smith

external and internal environments and make appropriate changes in those environments. Information about the internal and external environments is continuously being sent and analyzed the messengers. Primary messengers, such as hormones and neurotransmitters bind to receptors, which can amplify the signal through secondary messengers that cause cellular and physiological responses. The primary signal is transduced, or changed, into a secondary signal, such as intracellular Ca^{2+}. To ensure ruggedness and complete transfer of information, there is much cross-talk and redundancy between different pathways. This produces a robust system that enables us to adapt to environmental changes, so homeostasis can be maintained. The whole body is involved in this carefully controlled system.

The conscious mind often leads one to assume that the brain controls everything. This is wrong. Our brains work in concert with the rest of our body to influence our activities. That is, the wholeness of life is an important concept in medical science. We might artificially divide the body into different components to study them, but everything is linked to everything else [1, 2]. For example, some receptors are found in the brain, but other forms of the same receptors exist in other parts of the body. For example, acetylcholine receptors are located throughout the body and brain. Other neurotransmitters, including glutamate, serotonin and norepinephrine, are found in other parts of the body. This is one reason why drugs that affect neurotransmitter uptake in the brain can have side effects in other parts of the body.

So, there are molecules called neurotransmitters that are one type of primary messenger. They send information from the one neuron to another at synapses, or from neurons to muscles at neuromuscular junctions. However, some of these neurotransmitters are also found in other parts of the body, such as the stomach and the tongue. The stomach can send signals to the brain to indicate hunger or satiety. Similarly, tongues have taste buds that tell our brains if the things we put in our mouths taste good. For example, the excitatory neurotransmitter glutamic acid can bind to glutamate receptors on our tongues, which adds flavor to foods that contain it. That is why Chinese food often contains the sodium salt of glutamate (monosodium glutamate, or MSG). This is just one example of how a primary messenger can start the flow of information.

So, let us look closer at the biology of information flow, signaling and receptors. Organelles within cells, cells within tissues, tissues within our body, and all of the bodies in a society must sense their internal and external environments and respond appropriately to changes, for the good of the cells, tissues, whole person and society. This includes the approximately 10^{11} neurons in the brain [3]. They are interconnected in a network that send and receive information within the brain and respond to information coming from the rest of the body. Moreover, we receive much information from the external environment and from people around us in a society. We compare this to information we have learned throughout our lives and try to make appropriate responses. At the same time, the immune network responds to stress caused by the brain's reaction to stimuli. For example, depression can be bad for the immune system. In any case, information is sent in the form of messengers. Hormones, neurotransmitters and cytokines can act as primary messengers. Most of these bind to protein receptors on the outside of a cell, but some hormones bind to intracellular receptors.

Some of these receptors amplify the primary message by making many secondary messengers, which cause important effects inside the cell. The most widely used second messenger is Ca^{2+}, which is stored in the endoplasmic reticulum (ER). Once released from the ER, Ca^{2+} can bind to some cytosolic molecular messengers directly, but can only bind to others indirectly. Instaed, it can bind to the protein calmodulin, which can then bind to and affect other protein messengers, such as the calcium/calmodulin (CaM)-dependent protein serine-threonine kinases (CaMKs). In excitable cells (such as neurons), Ca^{2+} can also enter the cell from outside of the cell by passing through voltage and ligand-gated Ca^{2+} channels on the cell membrane. In muscle cells, the hormone adiponectin causes extracellular Ca^{2+} to enter the cell, activating the adiponectin receptor 1, which activates protein kinases and changes mitochondrial activity. Decreased levels of adiponectin and its receptor help cause the mitochondrial dysfunction and insulin resistance that occurs in diabetes.

In addition, Ca^{2+} and calmodulin dependent protein kinase II (CaMKII) have important regulatory functions in the heart and brain, and when chronically activated, it can be pathological [4]. "For example, acute hyperglycemia can lead to the attachment of N-acetylglucosamine to CaMKII, which activates it. Under normal, healthy conditions, CaMKII is autoinhibited by the interaction between the regulatory and catalytic subunits. When Ca^{2+} and calmodulin bind to the regulatory domain, it activates this enzyme. It can be further modified by phosphorylation and oxidation to form a stable version of CaMKII even after the Ca^{2+}/calmodulin complex dissociates" [4]. Then, Ca^{2+} and activated CaMKII can produce many secondary messengers that can amplify some of the other important second messengers including IP_3 and diacyl glycerol (produced by the hydrolysis of phosphoinositides), arachidonic acid (produced by the hydrolysis of phospholipids that have arachidonoyl on carbon number 2 of the glycerol backbone), ceramide, eicosanoids, lysophosphatidic acid, NO (nitric oxide), cAMP and cGMP. These secondary messengers can cause a variety of effects on the cell and a person's health. One way that they can do this is for them to cause proteins to be phosphorylated. Such proteins have a modular structure, including a site to bind the primary messenger, another site that becomes phosphorylated and another that interacts with other proteins, which form complexes that affect cells.

So, intercellular communication can be divided into different steps. First, the primary messenger (ligand) binds to its receptor. Second, the signal is transduced, or converted into a second messenger, which can affect other messengers. This can be done different ways. A kinase can catalyze the attachment of phosphates to target proteins. A guanylate cyclase can make cGMP; adenylate cyclase can make cAMP. A lipase can produce arachidonic acid, IP_3 and/or diacylglycerol, all of which are second messengers. On the other hand, some primary messengers, such as steroids, can pass through the cell membrane, bind their receptors in the cytosol, then migrate with them to the cell nucleus where they bind to nuclear receptors, called ligand responsive elements, or hormone responsive elements. Fig. (**1**) shows how ligands like cytokines, survival factors, hormones, neurotransmitters, growth factors, wingless (Wnt) and death factors bind to cytokine receptors (CRs), receptor tyrosine kinases (RTKs), Frizled (FRz) and the Fas receptor (FasR) which transmit their signals through other messengers that enter the cell nucleus and affect gene transcription.

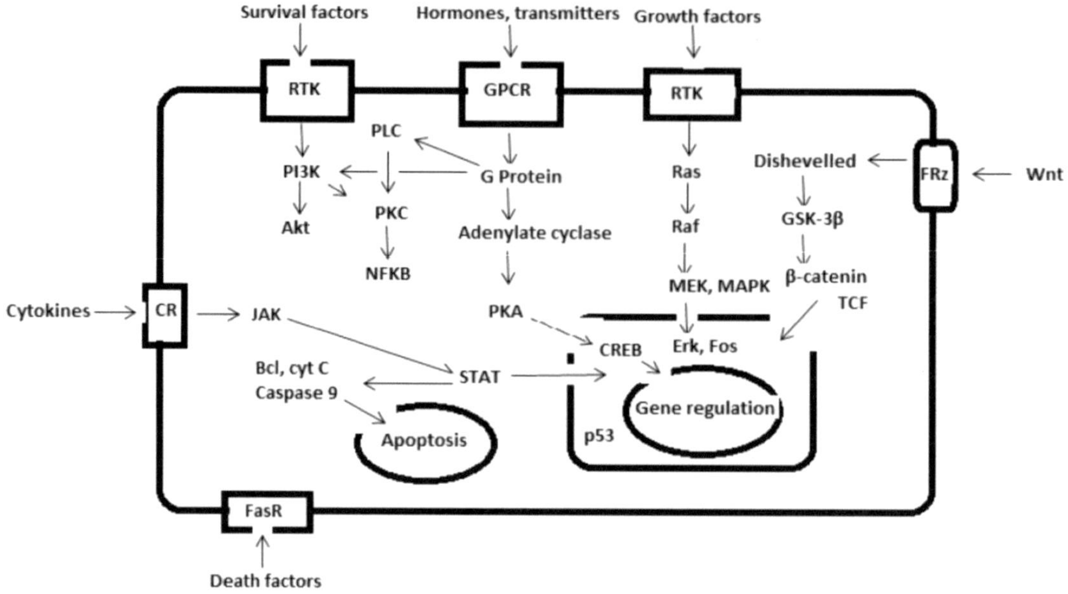

Fig. (1). Signal transduction.

Some messengers are sent to endosomes [5]. "Endosomes can amplify these signals, but they can also generate intracellular signals in the absence of messengers" [5].

Third, the message can be amplified by producing many copies of another molecule or ion, such as cAMP or Ca^{2+}. Also, messengers can activate or suppress the transcription of genes within the cell nucleus.

Those receptors that receive primary messages from the outside of a cell usually have three parts. One is on the outside of the cell membrane, another is within the membrane and spans the membrane and the other is on the inside of the cell. When a ligand binds to the outer portion of the receptor, it can cause a conformation change within the membrane-spanning region, which can cause changes in the inner portion. This can activate internal enzymes, such as adenylate cyclase, which catalyzes the formation of cAMP. After an appropriate time, the signal is turned off. In the case of cAMP, this is catalyzed by phosphodiesterase (PDE). Thus, the changes in a cell's external environment can stimulate changes in the internal environment [4].

G PROTEIN-COUPLED RECEPTORS

The most common class of receptors is the G protein-coupled receptors (GPCRs). They are glycoproteins, with sugars attached on the extracellular region. There are about 950 such receptors in human cells [4]. They are integral, membrane-spanning receptors. The conformation of a GPCR changes when it binds its cognate ligand. Once they are activated, GPCRs can couple (or bind) to heterotrimeric G proteins on the cytoplasmic side of the cell membrane.

These heterotrimeric G proteins have three subunits: α, β and γ. The α-subunit binds either GDP or GTP. Activation causes the bound GDP to dissociate from the α-subunit of the G protein and be replaced by GTP. The complex between GTP and the α-subunit can dissociate from the rest of the GPCR and migrates over to its target, such as adenylate cyclase, which catalyzes the formation of cAMP. The cAMP activates another type of enzyme called a kinase, which catalyzes the addition of a phosphate group to other proteins. The signal is terminated when another enzyme, GTPase, catalyzes the hydrolysis of one of the phosphates of GTP to form GDP and phosphate.

Based on phylogenic analysis, the International Union on Pure and Applied Chemistry (IUPAC) has divided GPCRs into three classes, based on amino acid sequence homology [6]. The protoypes for them are the rhodopsin, calcitonin and glutamate metabotropic receptors, which are classes 1, 2 and 3, respectively [6], also known as A, B and C [7]. The different classes have different structures that are formed by a network of non-covalent bonds that were discovered using a systematic analysis of high-resolution GPCR structures [8]. However, others divided GPCRs into five families: rhodopsin (701 members), adhesion (24 members), frizzled/taste (24 members), glutamate (15 members), and the secretin family (15 members) [9]. More recently, an completely different classification system was described that is based on ligands – not structure [10].

Another important class of GPCRs binds adenosine as their ligand [11]. "There are four adenosine receptors, A_1, A_{2A}, A_{2B} and A_3. They can be found throughout the body and have diverse effects on health and diseases" [11].

So, adenosine is in the generic drugs adenocard and adenoscan that are used to treat upraventricular tachycardia. Moreover, many clinically used drugs (including dipyridamole and methotrexate) act by changing extracellular adenosine concentrations and signaling [11]. "It is also important to note that caffeine is an adenosine receptor antagonist. This is important because caffeine consumption has been linked to a reduced risk of Parkinson's disease, dementia and Alzheimer's disease, type-2 diabetes and chronic liver cirrhosis. There are also several drugs in clinical trials that target adenosine receptors. However, previous trials on some drugs found some side effects due to drug interactions with adenosine receptors that are not in the organ being targeted for therapy. Still, all four types of adenosine receptors are elevated in autoimmune diseases and cancer, so they are being investigated as therapeutic targets. This includes allosteric enhancers that amplify the action of agonists by stabilizing the ternary complex formed between the agonist, the adenosine receptor and the G protein, thus minimizing the side effects of the drug. However, there can be different effects on the young and adults. That is, A_{2A} receptor signaling has a protective role in the hypoxic brain of immature mice, but a harmful role in adult mice" [11].

Recent studies have shown that high concentrations of adenosine can promote cancer metastasis [12, 13]. In addition, hypoxia-inducible factor 1α (HIF-1α) stimulates the transcription of genes coding for adenosine A_{2B} receptors in some cells. This stimulates cell scattering, the first step in metastasis [13]. "That is, in healthy tissues, cells stick to each other through structures called adherens junctions. The small GTPase called Rap1 promotes the formation and maintenance of adherens junctions. To do this, it must be at the plasma membrane and be posttranslationally modified by the attachment of a prenyl group. There is a signaling cascade that is initiated by adenosine A_{2B} receptors. This cascade suppresses the prenylation of Rap1B by phosphorylating it" [13]. So, adenosine promotes cancer by activating A_{2B} receptors on tumor cells and by activating adenosine receptors on immune cells, which suppresses antitumor immune responses [14, 15].

On the other hand, the physiological functions of some GPCRs are unknown. Such receptors are referred to as orphan GPCRs [8]. Like all the GPCRs, they have an extracellular ligand-binding domain on their amino end, seven transmembrane α-helices, and a carboxyl end inside the cell membrane. This can interact with heterotrimeric G proteins. The ligand binding region is between helices 3 and 4. Upon binding the ligand, the α subunit usually dissociates from the other two subunits (α and γ), but not always [12]. Not only the α subunit, but also the β-γ complex can catalyze the activation of other proteins in signal transduction pathways [12].

Once activated, GPCRs can cause the α subunit of the cognate G protein to break away from the β-γ complex, and can act as a signaling molecule itself. For example, the α subunit can activate L-type Ca^{2+} channels, such as phospholipase A_2 and muscarinic acetylcholine receptors. At the same time, the α subunit can also act as a messenger. Within each family, there can be more than one member. For example, in the $G_{\alpha i}$ family there is both an α_i and an α_o. The $G_{\alpha o}$ family is homologous to the $G_{\alpha i}$ family in that $G_{\alpha o}$ and $G_{\alpha i}$ can often bind to the same G protein-coupled receptors (GPCRs) but have different effectors. $G_{\alpha o}$ binds to the GTPase activating protein called Rap1GAP. This promotes the degradation of Rap1GAP by the proteasomal pathway. Once the receptor is stimulated, more $G_{\alpha o}$ is released to interact with Rap1GAP, which causes Rap1 activation [13].

So, GPCRs can activate many different signaling pathways [16]. Examples include protein kinases, cAMP/PKA (protein kinase A), Ca^{2+}/PKC, Ca^{2+}/NFAT (nuclear factor of activated T-Cells), PLC (phospholipase-C), PTKs (protein tyrosine kinases), MAPK/ERK (mitogen activated protein kinase or extracellular kinase), MAPK/p38, PI3K-Akt, nitric oxide (NO)-cGMP, Rho, NFkB (nuclear factor of activated T-Cells) and the Janus kinase (JAK/STAT) pathways. Also, Smoothened is a GPCR in the hedgehog signaling pathway, which is conserved from fruit flies to humans. In humans, the sonic hedgehog (SHH) protein can affect the cells that secrete it (paracrine) and other cells after being secreted. The drug vismodegrib (Erivedge®) that is approved for treating basal cell carcinoma, targets the smoothened receptor that is downstream of the SHH protein.

GPCRs can stimulate tyrosine kinases, which catalyze the phosphorylation of tyrosine residues on proteins. One of them is the target of the drug called Gleevec®, which has been used for 30 years to treat and cure chronic myelogenous leukemia (CML). It is used in personalized medicine to treat and cure leukemia cancer patients who have a reciprocal translocation between chromosomes 9 and

22. This produces a *bcr-able* fusion gene that codes for an abnormal tyrosine kinase that is not properly regulated. Cancer can occur when the signaling process is not properly regulated.

PROTEIN KINASES

There are many other protein kinases (PK), which also catalyze the addition of a phosphate to another protein or even to themselves and affect signal transduction. That is, they enable phosphates to be added to proteins. This can activate some proteins and inactivate others that are involved in signal transduction. They have been classified based on structural and functional properties by Hanks and Hunter [17]. The ACG group contains PKs that are regulated by cAMP (PKA) and cGMP (PKG), as well as PKs like Akt that are closely related to them. PKC, which is activated by diaceylglycerol, is also in this group. The CaMK group contains PKs that are regulated by Ca^{2+} and calmodulin. The CMGC group contains cyclin-dependent kinases, the ERK kinase family and the glycogen synthase kinase II family. The PTK group contains conventional, protein-tyrosine kinases (PTKs), some of which span the cell membrane and others that don't. The membrane spanning PTKs include the Src, Csk, Abl, EGF, PDGF, insulin receptor family, and many others. The OPK group contains other PKs. This includes the MEK, MEKK, Raf, the TGFβ receptor family and others [17].

ABBREVIATIONS

ER	=	Endoplasmic Reticulum
cAMP	=	Cyclic AMP
cGMP	=	Cyclic GMP
DAG	=	Diacylglycerol
GPCR	=	G Protein-Coupled Receptor
GTPases	=	GTP Hydrolase
NRs	=	Nuclear Receptors; IP_3 Inositol 1,4,5-Trisphosphate
CRACs	=	Ca^{2+}-Release Activated Ca^{2+}Channels
CaMKIII	=	Calmodulin Dependent Protein Kinase II;NO Nitric Oxide
PI3K	=	Phosphatidylinositol 3-Kinase
PDE	=	Phosphodiesterase
IUPAC	=	International Union on Pure and Applied Chemistry

PKA	=	Protein Kinase A, PKB Protein Kinase B (Also Known as Akt)
PKC	=	Protein Kinase C
PDK	=	Protein Kinase D
HIF-1α	=	Hypoxia-Inducible Factor 1α
GIPs	=	GPCR-Interacting Proteins
PIP_2	=	Phosphatidyl Inositol 4,5-Bisphosphate
PLC	=	Phospholipase-C
PTKs	=	Protein Tyrosine Kinases
RTK	=	Receptor Tyrosine Kinase
MAPK/ERK	=	Mitogen Activated Protein Kinase or Extracellular Kinase
NFkB	=	Nuclear Factor of Activated T-Cells
JAK	=	Janus Kinase
SHH	=	Sonic Hedgehog
CML	=	Chronic Myelogenous Leukemia
RyR	=	Ryanodine Receptor
Pln	=	Phospholamban; CRP C-Reactive Protein
CRP	=	C-Reactive Protein
PLC	=	Phospholipase-C
NFAT	=	Nuclear Factor of Activated T-Cells
Caln	=	Calcineurin
AKAp	=	A Kinase Anchoring Protein
ERK	=	Extracellular Signal-Regulated Kinase
BDNF	=	Brain-Derived Neurotrophic Factor
HSL	=	Hormone Sensitive Lipase
AMPK AMP	=	Activated Protein Kinase
sGC	=	Soluble Guanylate Cyclase
PtdIns 3-P	=	Phosphatidylinositol 3-Phosphate
PtdIns	=	3,4-P_2or PIP_2 Phosphatidylinositol-3,4-Bisphosphate
PtdIns 3,4,5-P	=	$_3$ or PIP_3 Phosphatidylinositol 3,4,5-Trisphosphate
PH domain	=	Plekstrin Homology Domain
mTORC2	=	Mammalian Target of Rapamycin Complex 2
PTEN	=	Phosphatase and Tensin Homolog
JAK-1	=	Janus Kinase-1

BCR	=	B-Cell Receptor
SYK	=	Spleen Tyrosine Kinase
BCAP	=	B-Cell Receptor Associated Protein
PDK1	=	Phosphoinositide-Dependent Kinase-1
PRC	=	Polycomb Repressive Complex
EGF	=	Epidermal Growth Factor
TNF	=	Tumor Necrosis Factor
DISC	=	Death-Inducing Signaling Complex
MAC	=	Mitochondrial Apoptosis-Induced Channel
Apaf-1	=	Apoptotic Protease Releasing Factor-1
eNOS	=	Endothelial Nitric Oxide Synthase
PDE	=	Phosphodiesterase
PFK	=	Phosphofructokinase
eIF4EBP	=	Eukaryotic Initiation Factor-4E Binding Protein
GAB2	=	GRB2-Associated Binding Protein-2
CREB	=	CAMP Response Element Binding Protein
CREs	=	CAMP Responsive Elements
AR	=	Androgen Receptor
XIAP	=	X-Linked Inhibitor of Apoptosis Protein
IGF-1	=	Insulin-Like Growth Factor-1
IPMK	=	Inositol Polyphosphate Multikinase
PIKK	=	Phosphatidylinositol 3-Kinase-Related Kinase
V-ATPase	=	Vacuolar ATPase
GEF	=	Guanine Nucleotide Exchange Factor
TCR	=	T-Cell Receptor
MMP	=	Matrix Metalloproteinase
TIMPs	=	Tissue Inhibitors of Matrix Metalloproteinases
PDGF	=	Platelet Derived Growth Factor, PDGFR Platelet Derived Growth Factor Receptor
HGF	=	Hepatocyte Growth Factor
HMBG1	=	High-Mobility Group Box 1
SOS	=	Son of Sevenless
GRB2	=	Growth Factor Receptor-Bound Protein-2
GEFs	=	Guanine Nucleotide Exchange Factors
GAPs	=	GTPase–Activating Proteins
PKR	=	RNA–Dependent Protein Kinase

bZIP	=	Basic Zipper
C/EBP CCAAT	=	Enhancer-Binding Proteins
ERs	=	Estrogen Receptors
ARs	=	Androgen Receptors
PXR	=	Pregnane X Receptor
CAR	=	Constitutive Androstane Receptor
RXR	=	Retinoid X Receptor
PPAR	=	Peroxisome Proliferator Activated Receptor
LXR	=	Liver X Receptor
FXR	=	Farsenoid X Receptor
HNF	=	Hepatocyte Nuclear Factor
NRs	=	Nuclear Receptors
MR	=	Mineralcorticoid Receptor
snoRNPs	=	Small Nucleolar Ribonucleic Acid - Protein Complexes
PAX2	=	Paired Box 2
KLF	=	Krüppel-Like Factors
iNOS	=	Inducible Nitric Oxide Synthase
nNOS	=	Neuronal Nitric Oxide Synthase
SRY	=	Sex Determining Region Y
PAX8	=	Paired Box Gene 8
YAP	=	Yes Association Protein
NFAT	=	Nuclear Factor of Activated T-Cells
eRNA	=	Enhancer RNA
PLC	=	Phospholipase-C
CalmK-II	=	Calmodulin Kinase-II
NCAM	=	Neural Cell Adhesion Molecule
APCs	=	Antigen Presenting Cells
MHC	=	Major Histocompatibility Complex
APJ	=	Apical Junction Complex
TNFR	=	Tumor Necrosis Factor Receptor
IL-1R1	=	Interleukin1 Receptor Type 1; Inhibitor of NFκ
B	=	Kinase (IKK)
NEMO	=	NF-κB Essential Modulator

REFERENCES

[1] Capra F. The Web of Life. New York: Doubleday 1996.

[2] Barabasi A-L. Linked The New Science of Networks. Cambridge, Mass.: Perseus Publishing 2002.

[3] Corey EJ, Czako B, Kürti L. Molecules and Medicine. New York: John Wiley & Sons 2007; pp. 112-9.

[4] Erickson JR, Pereira L, Wang L, *et al.* Diabetic hyperglycaemia activates CaMKII and arrhythmias by O-linked glycosylation. Nature 2013; 502(7471): 372-6.
[http://dx.doi.org/10.1038/nature12537] [PMID: 24077098]

[5] Rajagopalan S, Moyle MW, Joosten I, Long EO. DNA-PKcs controls an endosomal signaling pathway for a proinflammatory response by natural killer cells. Sci Signal 2010; 3(110): ra14.
[http://dx.doi.org/10.1126/scisignal.2000467] [PMID: 20179272]

[6] Foord SM, Bonner TI, Neubig RR, *et al.* International Union of Pharmacology. XLVI. G protein-coupled receptor list. Pharmacol Rev 2005; 57(2): 279-88.
[http://dx.doi.org/10.1124/pr.57.2.5] [PMID: 15914470]

[7] Venkatakrishnan AJ, Deupi X, Lebon G, Tate CG, Schertler GF, Babu MM. Molecular signatures of G-protein-coupled receptors. Nature 2013; 494(7436): 185-94.
[http://dx.doi.org/10.1038/nature11896] [PMID: 23407534]

[8] Kobilka BK. G protein coupled receptor structure and activation. Biochim Biophys Acta 2007; 1768(4): 794-807.
[PMID: 17188232]

[9] Lin H, Sassano MF, Roth BL, Shoichet BK. A pharmacological organization of G protein-coupled receptors. Nat Methods 2013; 10(2): 140-6.
[http://dx.doi.org/10.1038/nmeth.2324] [PMID: 23291723]

[10] Tse MT. G protein-coupled receptors: finding common threads. Nat Rev Drug Discov 2013; 12(3): 187.
[http://dx.doi.org/10.1038/nrd3959] [PMID: 23411721]

[11] Chen J-F, Eltzschig HK, Fredholm BB. Adenosine receptors as drug targets--what are the challenges?. Nat Rev Drug Discov 2013; 12(4): 265-86.
[http://dx.doi.org/10.1038/nrd3955] [PMID: 23535933]

[12] Bockaert J, Perroy J, Bécamel C, Marin P, Fagni L. GPCR interacting proteins (GIPs) in the nervous system: Roles in physiology and pathologies. Annu Rev Pharmacol Toxicol 2010; 50: 89-109.
[http://dx.doi.org/10.1146/annurev.pharmtox.010909.105705] [PMID: 20055699]

[13] Iyengar R. G alpha z-GAP (Gz-GAP) in G alpha i pathway. Sci Signal (Connections Map Component in the Database of Cell Signaling, as seen 16 September 2010), http://stke.sciencemag.org/cgi/cm/stkecm;CMN_9181 , [Accessed March 20, 2014];

[14] Linden J. Adenosine promotes tumor metastasis. Sci Signal 2013; 6(277): pe20.
[http://dx.doi.org/10.1126/scisignal.2004290] [PMID: 23716715]

[15] Ntantie E, Gonyo P, Lorimer EL, *et al.* An adenosine-mediated signaling pathway suppresses prenylation of the GTPase Rap1B and promotes cell scattering. Sci Signal 2013; 6(277): ra39.

[http://dx.doi.org/10.1126/scisignal.2003374] [PMID: 23716716]

[16] SA Biosciences. Pathways online, http://www.sabiosciences.com/pathwaysonline/ . 2012.

[17] Hanks S, Hunter T. The eukaryotic protein kinase superfamily: kinase (catalytic) domain structure and classification. FASEB J 1995; 9: 576-96.

CHAPTER 14

Systems Thinking in Medicinal Chemistry

Abstract: Systems biology is essential for P4 medicine, as are autopoiesis, network theory and the concept of emergent properties. That is, the basic, fundamental unit of life is the cell, not the atoms and molecules in the cell. Many new properties emerge when atoms, molecules and ions are organized in a living cell. The functions of a cell do not depend on just the properties of the individual molecules, but also on how these molecules interact. Autopoiesis means self-production. It is a network of production processes, in which the function of each component is to participate in the production or transformation of itself and the other components in the network. The production processes are circular. Life is a cyclic process that produces the components of a living system. Based on the autopoietic theory of life, the biosphere of Earth is often thought of as a living system. Bacteria can be thought of as the catalysts that maintain the atmosphere in its present state, far from equilibrium, but stable, like homeostasis in a cell, organ or organism. Networks permeate living systems. Living systems and the internet are examples of a type of network called a scale-free network. These networks are dominated by a few well-connected nodes, called hubs. Most nodes in the network have a few connections, but a small number of nodes have a seemingly unlimited number of connections. In the scale-free network that is in living cells, there are many levels of organization. Each of them can be viewed as a network. There is a network of genes, a metabolic network, a regulatory network and a cellular network.

Keywords: Autopoiesis, Gaia, Hubs, Nodes, Scale-free network, Systems biology.

SYSTEMS BIOLOGY

Systems biology has always been behind traditional medicine and is now an integral part of modern, westernized medicine. It is essential for P4 medicine, as are autopoiesis, network theory and the concept of emergent properties. That is, the basic, fundamental unit of life is the cell, not the atoms and molecules in the cell. This is because many new properties emerge when atoms, molecules and ions are organized in a living cell. Moreover, almost all of the atoms, molecules and ions in our cells are continuously being broken down or replaced. This

Robert E. Smith

happens while the organization and appearance of adult organisms remains nearly constant. There are slow, gradual changes as we age and much faster changes as a fetus develops, a baby is born and an infant grows. Important changes also occur during pubescence and again during and after menopause. Still, the changes are much slower than those that occur in the atoms, molecules and ions. For example, the skin cells and molecules in them are replaced every few days. So, the hands that we shake when we first meet someone look the same way when we see them a few days later, as does our face. We may feel and look the same while the organization of our cells and tissues remains nearly constant, but the molecules are different. So, one's anatomy, physiology and even personal identity don't change either. They are based on the same cells and tissues, even though the molecules have changed. We can't be reduced to a collection of fundamental subatomic or atomic particles, as suggested by reductionist thinking.

However, math - the foundation of reductionist thinking - is still essential. In reductionist thinking, quantitative mathematics is the basis for all science. Many consider it to be the only true science, since it can produce linear equations that can calculate exact answers to questions that are asked properly. That is, if a mathematician, physicist or chemist knows the position and momentum of a piece of matter, as well as all the forces acting on it, its position and momentum at any time in the future can be calculated with no uncertainty. Unfortunately, it is almost never possible to know all the forces that are acting on a piece of matter. Still, for practical purposes, many of the forces can be ignored. For example, friction can be ignored when calculating the position and momentum of a 10 kg rock that is falling off a cliff. They can be calculated by simple linear equations, such as force equals mass times acceleration. It is even possible to calculate the position and momentum of two large objects that revolve around each other (such as two stars in a binary system that has no planets).

However, there are no exact solutions to the non-linear equations that are needed to calculate the position and momentum of three objects, such as the sun, Earth and moon. Still, they can be calculated well enough to enable us to put men on the moon and send the SOSO Solar and Heliospheric Observatory to orbit the sun and study it. However, many aspects of life can only be described by non-linear equations that have no exact solution, but many approximate solutions. This can be especially important to Gaia and the survival of humanity. That is, there are many different ways that Gaia can change in response to variations in her temperature. For example, to redistribute the solar energy that reaches the tropics, ocean currents such as the Gulf Stream emerged that carry warm water to Western Europe, giving it a temperate climate, as opposed to the much colder climate in similar latitutes in Canada and Russia. This, like other ocean currents, is driven by differences in NaCl concentrations (salinity) between surface and underlying ocean water. That is, as salty ocean water freezes at in Polar regions, the ice that is produced has very little dissolved NaCl, but the liquid water that remains on the surface becomes enriched in NaCl. As more and more polar ice melts, the salinity of the water will decrease. This will decrease the force that drives the thermohaline overturning current and may eventually change or even stop the Gulf Stream. This could make the climate of England feel like that in Siberia. Similarly, an anomalous high pressure atmospheric system caused record high temperatures in the Alaska and the Western USA, while forcing cold, Polar air into the Noertheast. This caused record cold and snowfalls. So, the current set of approximate solutions to the non-linear equations that drive weather and climate can change to different solutions that may have devastating effects.

Similarly, human biology is based much more on non-linear than linear equations. There is even much debate on whether life can be completely described by mathematics, or even by physics or chemistry. For example, the autopoietic theory of life holds that cognition and self-awareness are emergent properties of life. So a corollary to this is that non-living computers and robots can never become self aware or have feelings. Others feel that it is possible that consciousness can be described by math and that computers may eventually become self aware. So far, there is probably insufficient data to know one way or the other.

This should not be confused with the Heisenberg uncertainty principle that states that both the exact position and momentum of a single electron can never be known simultaneously. It is the very nature of the electron. So, we may not be able to calculate the position and momentum of every single electron in a glass of water, we can be sure that we can drink it and satisfy our thirst. On the other hand, there are many aspects of life that can't be described by quantitative mathematics. That is, graph theory and network theory have emerged as crucial tools in learning modern biology and medicine. In graph theory, points (vertices or nodes) are connected by edges, which may or may not have a strength and direction. That is, the connections between some points may have different strengths (such as a graph of personal relationships). These are called weighted networks. Some, like afferent nerves, may have connections, or edges that go in one direction, while others (like friendships) may be bidirectional, while others (like neighbors in a neighborhood) can be undirected. This field of math focuses on relationships and looks at how things are connected. It is qualitative instead of quantitative. Still, the foundation of reductionist thinking (math) is also a fundamental part of systems thinking, even if sometimes it is qualitative instead of quantitative.

Systems biology is based on systems thinking, which is quite different from reductionist thinking. In systems thinking, the whole is often not equal to the sum of its parts. Individual parts can interact with positive synergism so that they produce a larger effect than they would if they acted individualy. They can also interact with negative synergism, in which the net effect is less than that caused by if the parts had acted independently. Moreover, whole new effects can occur as parts interact. Different levels of complexity can emerge when smaller components interact to form larger units. New and often unexpected properties emerge at higher levels of complexity. One of the best examples is in the emergence of life itself. In the past, some have considered it to be too complicated to be worth studying by reductionist methods [1]. However, systems thinkers feel that embryonic development and human behavior are worth contemplating, and studying. Even the well-known reductionist thinker Richard Dawkins said in a surprise ending to his book, The Selfish Gene, “We, alone on earth, can rebel against the tyranny of the selfish replicators” [1]. This seems to imply that we have a free will. Most systems thinkers would agree, although they would say that other organisms are also controlled by much more than just their genes (“selfish relicators” [1]).

Other things that systems thinkers might say are as follows. **There is seldom a single solution to a single problem. Everything is connected to everything else. Living organisms are not machines.** Moreover, most organisms can’t survive by themselves and none live in isolation. Plants and animals form symbiotic (or close) relationships with bacteria and other organisms. Often, one can't survive without the other. In other cases, organisms just help each other. For example, bacteria can form relationships called syntrophy with other bacteria. Each organism exhibits growth characteristics that depend on the presence of the other organism. Also, bacteria, fungi and viruses are important parts of the human organism, which is an ecosystem. As many microbiologists like to say, only 10% of the cells in our body are human, the other 90% are bacteria. Remember, though, that bacterial cells are much smaller than human cells, so our bodies are predominantly made up of human cells, by weight, if not by number. Still, until the last few years, little was known about the bacteria in our bodies. Scientists were not been able to grow most of our bacteria outside of the human body. So, another genome project, called the Human Microbiome Project was conducted by the U.S. NIH from 2008-2013. Its goal was to learn more about how changes in the human microbiome can affect human health and disease. Just as the Human Genome Project determined the base sequence of all the bases in human DNA, the Human Microbiome Project determined the base sequences of bacteria that colonize humans. Most of these bacteria can't survive or grow in cell culture (as can *E. coli*), so their DNA was sequenced by hydrolyzing the combined DNA of all the microbiome and sequencing the fragments produced. This is called metagenomics. It is the study of genes taken from organisms in their natural environment. Often it focuses on specific genes, such as the ones that code for the 16S ribosomal RNA gene. So, several species of bacteria have been identified in the human microbiome, based on their DNA. It is also important to note that some of the genes in the modern human genome appear to be ancestral genes from *E. coli*. Similarly, the human genome seems to have many segments that probably originated in viruses. So, our internal human cells do not live in isolation, just as our whole human bodies do not live in isolation from environmental bacteria and viruses. Human life emerged as an ecosystem, with many different types of cells and viruses.

EMERGENT PROPERTIES

So, emergent properties are important in systems thinking. These are properties that appear, or emerge, only when components of a system interact. Most of us are aware of emergent properties, but have seldom thought of them in that way. For example, consider geometry. A single point contains limited information. Two points connected by a line segment have new, emergent properties. Consider a traveler who is lost. He may know that he is located at a given point on a map. By itself, this is almost useless. The traveler wants to know where the destination is and how to get to it. The destination is a second point, connected by a line segment to the first point. The properties of length and direction emerge as the two points interact at a higher level. Next, let's consider chemistry. Individual hydrogen and oxygen atoms are very reactive and have very different properties than hydrogen and oxygen molecules. New properties emerge as these atoms interact to form molecules. Although some people will say that chemistry is based entirely on the principles of physics, we are only able to prove this with the hydrogen atom. The Schrödinger wave equation can only be solved for the hydrogen atom. Quantum mechanical calculations can only provide approximate solutions for all other atoms and all molecules. These calculations require some assumptions. Thus, it is possible that the hydrogen in H_2 and the oxygen in O_2 have emergent properties that can't be explained entirely from the principles of physics. That is, individual hydrogen and oxygen atoms have very different properties than they have when they combine to make a H_2 or O_2 molecule. Completely different properties also emerge when H and O react to make a water molecule. Similarly, new properties emerge when molecules interact to form living cells, which interact to form tissues, which interact in organisms, which interact with each other and with bacteria, fungi and viruses to form a human ecosystem.. Better examples occur in biology. Many biologists consider the cell to be the fundamental building block of life. The individual molecules, such as DNA, RNA, proteins, fats, sugars and small molecules (including water) have new properties when they are organized into a living cell. New properties emerge as networks grow in size and complexity.

Another example is the emergence of coordinated or synchronous activity. In theater and sporting events, performers inspire audiences to applaud. Usually, the hand clapping is random and chaotic. Sometimes, though, synchronous applause starts, as if an invisible conductor had given a signal. Similarly, fireflies in southwest Asia gather in the millions around mangrove trees. They usually flash their lights chaotically, but sometimes they start to flash their fluorescent tails off and on in synchrony. They produce light that can be seen for miles. Male crickets are another example. They chirp loudly to attract females. When there are just a few, they do not coordinate their chirping. When there are enough of them, they chirp in synchrony.

One of the best examples of emergent behavior occurs in a newly fertilized embryo. How does an egg develop and become all the different human cells in the body? It has the same DNA as every other adult cell in the body, but it has very different properties than any other cell. It has different types of proteins and RNAs that affect the choice of genes that are transcribed.

Emergent properties also appear in ecosystems. When humans began commercial farming on the coast of Australia, the runoff of fertilizers added nutrients to the sea. These nutrients helped sea stars to grow and their life cycles were accelerated. These sea stars are feeding off coral reefs, destroying them. Sadly, there are many other cases of seemingly small changes in an ecosystem can cause unexpected results. For example, intestinal bacteria can influence obesity and inflammatory bowel disease, while fungi interact with the innate immune system through the receptor called Dectin-1. There is a polymorphism in the gene coding for Dectin-1 that increases the susceptibility to ulcerative colitis. So, human, bacterial and fungal cells are part of the human body, or ecosystem that also contains viruses.

T hat is, ecology is the study of how environmental factors affect organisms and how organisms affect these factors and each other. It is quintessential systems science. It does not look at individual atoms or molecules. The individual organisms are the building blocks. It looks at how they all interact. Ecosystems are interconnected networks of organisms. For humans to survive, we must fit into a larger ecosystem that provides us with food and degrades our waste products. Although this may seem like biology, many principles of math, physics and chemistry are involved. The approach taken in this course is that of the systems thinker. To understand the chemistry, it is necessary to understand some biology, physics, math and even engineering. The math will include graph or network theory. However, this does not contradict wel-established principals of quantitative math and physics. The organizing principles that emerge in network theory complement and add to the laws of physics, they do not override them. To help understand this, there are references which provide more details on reductionist thinking [1,2], systems thinking [3,4], emergent properties [5], and the environment [6,7]. One of them [3] is especially good at describing the importance of systems thinking, non-linear mathematics and autopoiesis in biology and medicine.

Systems thinkers have also made significant contributions to industry and the world economy in the form of total quality manufacturing (TQM) and total quality leadership (TQL). They are based on the idea that a team of motivated employees can work more efficiently and competitively if they work together instead of working separately. This is another example of how the whole is more than the sum of its parts. One corollary of this is that if you want to sell a product, first survey the market (the environment) and see what the customers are demanding. You don't develop a product and hope that someone might want it; just as animals did not develop legs in hope that solid land would eventually appear on the Earth. Legs appeared only after dry land appeared. They work together with the rest of the body to provide entirely new, emergent capabilities. Similarly, not only all cells and organs in our body work together, all the employees and management work together in TQM. Our bodies are managed by total quality. If a cell or an organ develops a problem, the whole organism tries to find a cure and doesn't care about laying blame somewhere. For example, if you don't look where you are going when you are running, you might trip over a rock and break your wrist when you fall. You don't poke yourself in the eye for not seeing the rock. You don't hit yourself in the head to punish your brain for not paying better attention to your environment. Instead, you use the very parts of the body that made the mistake to fix the mistake. You might get an eye exam and a new prescription for eyeglasses. Your brain will remember to pay better attention in the future. Assigning blame is irrelevant. Similarly, in modern industry under TQM, if a production process has a problem, the whole team works together to solve it, without assigning blame. Often the person who caused the problem becomes the person who is best able to fix the problem.

Unfortunately, in organizations that don't use TQM or TQL, assignment of blame is paramount and fear rules. The motto becomes, "Don't fix the problem, fix the blame". Workers focus on keeping their own jobs, even if it costs someone else theirs. Often, the enemy is not a competing company, but coworkers who can be blamed for problems that emerge. If there is a problem, workers are encouraged to hide it. Before TQM was introduced, factory workers were often told that they are being paid to obey orders. You are not paid to think. If you think there is a problem, you are being stupid. It is the engineer's responsibility to identify problems and solve them. One example of this is former communist systems, in which workers had ever-growing quotas. It did not matter if the products were defective, if they had no use, or if the environment was damaged. If you didn't meet your quota, heads would roll. In such a system, management always reported successes, never failures. Similarly, in many companies before the 1980s, management ruled through fear. Union workers in an assembly line did not have the power or motivation to stop production if a problem emerged. They were told they were stupid, and their paycheck depended on them pretending that they believed it.

In the 1970s, Japanese industry started the ideas of TQM and TQL. Workers on an assembly line were given the power to stop production as soon as they saw a problem. Even if they caused the problem themselves, they would speak up. It no longer mattered who caused the problem. They did not waste time or resources in assigning blame. The important thing was not who caused the problem, but who identified the problem and who solved it. The element of fear was minimized. Supervisors no longer perpetuated the climate of fear. American industry noticed and adopted TQ. Now, a good supervisor (or a good teacher) is one who listens to the workers (or students), provides them what they need to do their job, and removes obstacles that may appear.

It might be good to remember the principles of TQM when addressing the biggest threat to public health – global climate change. Even though it is well established that human industrial activity is a major cause, when this fact is presented with anger and hatred, some of the biggest polluters become offended and decide to pollute even more. These same people might also enjoy watching a movie in which one of their own destroys a fictional asteroid that is heading towards Earth, thus saving humanity. It didn't matter that there was nobody to blame for the asteroid. It only mattered that a hero found the problem and fixed it. Similarly, as the debate in the USA over how to decrease our emissions of greenhouse gases grows, much of industry is realizing that they must find a way. They may be motivated more by avoiding damage to the economy than damage to the environment, but the result could be the same – a decrease in greenhouse gasses. Still, environmental friendliness may emerge in the most unlikely people.

SYSTEMS THINKING

Systems thinkers study network theory and the science of complexity. They realize that biological systems depend on their component parts, but they are somewhat autonomous. We continuously break down the components of our bodies and re-make them, but each individual person retains its identity. For example, we are all continuously making new skin cells. Within a week, all the cells on the surface of my hand are different, but it is still the same hand. When I shake your hand, I may be using different molecules this week, but you still recognize me and my hand. The organization stays the same and is an important part of the definition of life.

Systems thinkers realize that we are not machines with a set of instructions that can only be found in our DNA. Systems thinkers realize that a newly fertilized embryo has the same DNA as every other adult human cell in my body, but it is organized very differently. The embryo will divide in two. Usually, the cells continue to divide and only one person is born. Sometimes the two cells will become identical twins, which can have different personalities and different susceptibilities to diseases, even though they have the same DNA [8]. Scientists don't understand how or why this happens.

In older forms of reductionist thinking, DNA was thought to be the blueprint of life. One popular series of movies even told how pieces of DNA found in amber that was over 100 million years old could be used to make Dinosaurs. That is, all that one needs to make an organism is DNA. This is not true.

If DNA were truly the blueprint of life and if organisms were totally controlled by their DNA, then bacteria would express their DNA in the same way and they would have the same fate in life. This does not happen. Genetically identical cells and organisms demonstrate remarkable diversity, even when they are exposed to the same environment, such as the same cell culture for bacteria [9]. This is due to the stochastic nature of DNA transcription and other cellular events. That is, cell fate and behavior depend on a mixture of predictable and random elements. That is, the RNA and/or proteins that can be coded for by a gene may be predictable under a controlled experiment on a single cell at a single time. However, the interactions between promoters and inhibitors of gene transcription depend on the random movement of proteins and/or RNA. Once the proteins and/or RNA reach their DNA targets, the effects may be predictable, but getting to the target involves some randomness. Cells in the human body take advantage of their inherent stochastic variability to respond to continuous microenvironmental and intracellular variations, so they can survive at the expense of the rest of the clonal cell population.

SINLE CELL TECHNOLOGIES

This is important when studying biology and medicine. Almost all the existing literature on intra- and intercellular communication as well as the effects of drugs on cells was based on studies of many cells. For example, all the cells in a given tissue or cell culture were homogenized and analyzed as one. Some of the cells may have been affected quite differently than others. Only the average response was measured. To improve on this, scientists have developed ways to look at single cells. These single-cell analysis technologies have been able to recover, monitor and characterize population heterogeneity with single-cell resolution [10, 11]. The ability to look at individual cells can help the study of cancer cells. For example, single tumor cells can be isolated from the blood by an instrument (called DEPArray), which can isolate, move and image one metastatic tumor cell from a mixture of 100,000 cells [12].

Also, a silicon glass microwell platform has been developed for high resolution imaging and high content screening with single cell resolution [13]. "It has an exchangeable silicon glass microchip in a holder that fits in conventional microscopes. The microchips contain arrays of miniaturized wells. The size and layout of the wells can be designed for different applications, including single cell imaging. The device enables long-term assays (several days) with read-outs based on high resolution imaging or high-content screening. It was used to study the selective survival and clonal expansion of B cells that were transfected to express the human leukocyte antigen HLA Cw6, linked to green fluorescent protein (HLA-Cw6-GFP) compared to untransfected B cells that were grown under antibiotic selection for 3 days" [13].

So, when exposed to different environments, these differences in seemingly identical calls can become greater. Once an embryo has divided into 100 cells the fate of each cell depends on its location in the blastula. Even the original embryo is asymmetric. The animal pole is at one end and will become the ectoderm and mesoderm. The vegetal pole is at the other end and will become the endoderm. The cells have the same DNA, but they have different environments and will have different developmental fates. One of the most ambitious goals of network biology is to learn more embryology. Recently, systems thinkers have described a "network of human cell differentiation (NHCD) based on cell types (nodes) and differentiation steps (links) from the fertilized egg to a developed human" [14]. The network consists of 873 cell types, and not just the 407 adult cell types. So, the network includes cells, such as "placenta cells that are generated from the fertilized egg during embryo development, as well as other somatic cell types that are important to control embryo and fetus development" [14].

Synchronous activity is another emergent property that is required for life. Thousands of pacemaker cells fire (generate action potentials) in synchrony to control the contraction rate of the heart. When a newly fertilized animal embryo starts making many new cells, it starts rapid and synchronous cell cycles [15]. While the cytoplasm stays about the same size, the number of cell nuclei increases exponentially with time. Messenger RNA (mRNA) and proteins from the mother guide the development of these new cells. The embryonic genome is not activated until after the first few cell cycles. As the fertilized oocyte transforms into a totipotent zygote, no new proteins are being produced. In adult cells, proteins can control gene transcription by binding to pieces of DNA and either activating or inactivating the transcription of adjacent genes. The activation of genes in the newly fertilized embryo is controlled by RNA. Fully grown oocytes contain many RNA messages, corresponding to 20-45% of all mice genes [15].

Systems thinkers study the properties of networks. They realize that the functions of a cell do not depend on just the properties of the individual molecules, but also on how those molecules interact. They study the ways that individual components interact and influence structure and function. They look for higher levels of organization as cells interact with each other to form clusters, colonies, tissues and organs. They use engineering tools, such as control systems analysis, to organize the relationships and describe them with mathematical equations. They study the properties of networks. They learn what makes a network robust and what can make a network vulnerable. They also realize that living organisms are not machines. Machines can be understood and their actions can be predicted by linear equations, such as work equals force times displacement. All that is needed is to understand the basic building blocks of the machine. If you know the position and momentum of each component, along with Newtonian physics, you can predict exactly how the machine will be affected by different forces. If I kick a 1 kg sphere with a force of 10 N, I can predict exactly the position and momentum of the sphere, if I ignore friction and wind. If I kick it a second time with the same force, the sphere will move the same distance again. Life forms are not so predictable. If I kick my Karate sparring partner, the results are not predictable. If my instructor is watching, my partner will probably do what he has told been told to do. My partner may not change his position or momentum, but he might change his attitude. If I kick him again, the results can be unpredictable. If he is in a good mood and I don't kick very hard, he might just laugh at me. If he is in a bad mood, his position and momentum might change considerably! So, very little in life can be described by linear equations. Instead, most of life is based on nonlinear equations in a Nonlinear Dynamical Systems Theory (DST) [3, 16]. The recent book, The Systems View of Life, clearly describes the importance on nonlinear differential equations in modern biology [3]. Unlike linear equations that have one definite solution, nonlinear equations may not. Instead they can have several approximate solutions. So, a nonlinear living system that is continually changing can exist in a part of phase space in which it is rather stable. However, when environmental factors change, many aspects of the organism can change quickly and move to a different set of attractors in phase space. That is, unlike a nonliving system in which doubling a given input or force will double the effect, living systems may not be affected at all by a mee doubling of input. This is especially true in psychotherapy, in which a patient may undergo many sessions of therapy with little or no effect, then suddenly change after a single key insight [16]. That is, people have a type of emotional or psychological inertia, in which they keep on the same behavioral course as long as they are not acted upon by a powerful force [16]. However, this does not mean that people do not want or need change [3]. Many prisoners lose their sanity when kept in solitary confinement for just a couple of weeks. We all want and need change, but we want to decide for ourselves what changes we want to make. On the other hand, we seldom want change to be forced upon us – especially the changes that come with aging and disease. So, the behavior and inertia of living systems are quite different from those of nonliving objects.

There are other examples that support the idea that we are not machines. Scientists almost always use genetically identical rats to study the efficacy and toxicity of chemicals. Despite the identical genes, there is always considerable variability in the results. Some of the rats who received zero dose, or a placebo, show some effect. This placebo effect is well documented and does not exist with machines. Those of us who have raised a child know that young children can react hysterically when they get a cut or a scrape, especially if it has been caused by a sibling. We also know that a child will feel much better if we put a Band-Aid on the wound, kiss it and tell the child we love him or her. It might also help if we punish the sibling. Unfortunately, some people forget this when they are confronted with a broken machine like a car. Getting angry and kicking it will not make it better, nor will chewing gum and Band-Aids, although some used car salesmen might try to make you think otherwise (while they guarantee you that they have easy credit for you).

Systems thinkers also realize that we are not separate from our environment. Although genetics is an important factor, environment is important too. The environment affects us and we affect the environment. Our structure and that of the environment are coupled with each other. Organisms and the environment form a series of interconnected networks, called ecosystems.

Ecosystems are like other complex systems. They have many components that have many degrees of freedom and are non-linear. These complex systems are far from equilibrium. They can be described by nonlinear differential equations that have no exact solution, but have many approximate solutions, as described in chaos theory. New solutions can emerge unexpectedly, so there can be unpredictable consequences.

AUTOPOIESIS, ALLOSTASIS AND ATTRACTORS

Many systems thinkers use the term autopoiesis to describe the way living things are organized [3, 17 - 19]. Autopoiesis means self-production. That is, every organism and cell has a boundary that separates itself from the environment and interacts with the environment. The human body is covered with skin that acts as the first line of defense to protect our inner organs from environmental insults. It is covered with receptors that sense the environment. Of the five senses, touch is the most important. People can live long, happy, productive lives with dysfunctional senses of smell, taste, sight or hearing, but nobody can survive without the sense of touch. Moreover, the loving touch of parents or primary care givers is essential for the healthy development of a baby or infant. Still, it is not the atoms or molecules in the skin that are important. It is the whole person who does the touching. The skin cells themselves are continually being broken down and remade, but the person is the same. So, it has been claimed that "the human body replaces 98% of its atoms every year while maintaining its unique pattern" [17].

So, autopoiesis is a network of production processes, in which the function of each component participates in the production or transformation of itself and the other components in the network [3]. The production processes are circular. In the cell, there are metabolic cycles, such as the tricarboxylic acid cycle. The series of events between one cell division and the next is the cell cycle. Organisms have life cycles from one generation to the next. An insect starts as an egg, and becomes a larva and pupa, before becoming an adult. Ecosystems also have circular organization. It is a motif of life.

Another common feature of living things is that they form dynamic, dissipative structures that are far from equilibrium. Dynamic, continually changing systems can be described by differential equations that specify their behavior over a short period of time. They are open systems that obtain useful matter and energy from the environment and then dissipate them in different forms into the environment

[17]. They manage the flow of matter and energy in such a way that they ensure the the organization of the internal components and life itself is maintained [17]. This dissipation and the driving forces behind it settle the dynamic system into its condition of life. Even though the word homeostasis is used to describe this condition, perhaps the term allostais is more appropriate. That is, homeostasis implies that the body maintains itself within a fixed range of values. Actually, life is not fixed, but is on a trajectory. For example, thc range of concentrations of sex hormones needed to sustain normal, healthy life is quite different in a fetus, an infant, an adolescent and a post-menopausal woman. The concept of allostasis emphasizes the importance of adapting to continual changes in the environment – both internal and external [17]. Moreover, illnesses do emerge as a part of one's homeostatic (or allostatic) response to the environment. When certain parameters (like blood glucose) fall outside of healthy boundaries, the body responds by becoming ill [3]. So, one can look at all the possible states of health, including the mind and body. In fact, many scientists and doctors realize that the mental and physical aspects of health are closely linked [3]. For example, depression is extremely unhealthy and is a major risk factor for Alzheimer's disease. On the other hand, even the happiest, most positive people in the world can become depressed when they discover that they have an incurable disease. Remember the nocebo effect. Once a person is diagnosed as having a disease or even a genetic "defect" that can "cause" a disease, they are far more likely to get the disease. So, there is almost all diseases are "psychosomatic". One's attitude and behavior are important aspects of one's health.

This can be described using a tool that physicists use, called phase space. It is a graphical representation all possible states of a system. It can be a graphical representation of genes in a cell or different states of health, development and aging. In such a graph, groups or sets of points are found in different regions called attractors. These are sets of points that represent steady-state conditions to which the system can converge. For example, a graph of temperature *vs* days of the year will produce different groups of points (attractors) in different seasons. For example, during the summer in temperate climates, the high temperatures during the summer will form one attractor and the low temperatures in the winter will form another attractor. In medicine, the part of the phase space corresponding to one's typical health or behavior is called an attractor. In psychotherapy, the first sessions that seem to be unproductive may actually be forcing the patient towards a different set of attractors. For example, there may be one set of attractors in a depressed patient and a different set of attractors in a happy state.

From a medicinal chemistry perspective, there are changes in the expression of genes into proteins or different types of RNA, epigenetics, and post-translational modifications that occur throughout life. Remember, attractors are sets of points to which a dynamic system converges when stimulated by environmental changes. So, the rapid changes in the internal environment during development force an infant's attractors to change. Similarly, changes caused by infectious viruses or organisms can force a living system towards one set of attractors, while successful therapies force living systems towards a different set of attractors. Theoretically, living systems can be described graphically by such attractors that define the way that they are organized. The organization and form stay the same, even though the individual components and molecules are continuously being broken down and re-made. Living systems are organized in a closed, causal, circular process that allows for evolutionary change in how the circular process is maintained, but not in the circularity itself [3].

So, most developing cells undergo a one-way path towards attractors (such as genetic networks) as they differentiate into other types of cells [20]. "However, stem cells keep their potential to both proliferate and differentiate. So, dynamical-systems approaches have been developed to describe the transition between stem cells and differentiated cells.

They emphasized fluctuating and oscillatory levels of gene expression, since these are specific properties of stem cells" [20].

Cells have RNA, genes, proteins, and metabolites. The cellular state at a particular time can be represented as a point in multidimensional state space in which each axis represents the abundance of each component [20]. "The levels of expression of genes (or proteins) are a major aspect of these components. Interactions between the cell's components can cause the state to shift. This can be thought of as a trajectory in the state space. Changes over time in the expression of genes or proteins can restrict the cellular state to a certain region, which is defined as an attractor in dynamical-systems theory. After a slight perturbation (change in gene expression levels), a state returns to its original attractor. The attractor can be in a fixed state over time (that is, a fixed-point attractor) where the synthesis and degradation of cellular products are balanced, or a set of dynamically changing states with temporally oscillating gene expressions. A system can have many attractors with different compositions. So, each attractor can be considered to be a distinct cell type that corresponds to different valleys into which it can fall in an energy landscape. For example, when two genes, A and B, mutually suppress each other's expression (known as a toggle switch), two fixed-point attractors form: one with activated A and suppressed B, and the other with activated B and suppressed A. These two attractors can be thought of as two distinct cell types. Moreover, gene expression is affected by random fluctuations, often called noise. Attractors are robust. After a small change, the cellular state returns to its original attractor. So, attractors can explain why there are robust, distinct cell types. However, the concept of fluctuating and oscillatory gene expressions is needed to explain the differentiation of stem cells in a dynamical-systems approach" [20].

Others have looked at the attractor landscape of the p53 protein's regulatory network to obtain therapeutic hints for drugs that will target several genes or gene products simultaneously [21]. By characterizing the attractor landscape of p53, a genetic circuit was drawn around p53. That is, cells can exist in different states with different potentials for differentiation. Upon differentiation, a cell falls to a lower potential. After establishing a genetic network, it is studied by mathematical models to delineate the potential landscape of the network. So, the genetic circuit for p53 included feedback loops that respond to DNA damage. For example, p53 stimulates a phosphatase (Wip1), which inhibits the kinase ataxia telangiectasia mutated (ATM) protein, which leads to inhibition of the E3 ubiquitin ligase Mdm2 [21].

When DNA is damaged, the landscapes of both normal and cancer (MCF7) cells change, and a single cyclic attractor representing cell cycle arrest was observed for both cells [21]. This was verified experimentally. So, new drug therapies may be developed by characterizing the biological mechanisms underlying diseases with a systems-level approach [21].

At the same time, systems engineers speak in terms of cycles and hypercycles. For example, the tricarboxylic cycle is a simple cycle. The product of each reaction is a substrate for the next reaction. The catalysts for each step of the reaction come from outside the tricarboxylic acid cycle. As long as the cycle is fed with sufficient nutrients, and as long as there is a need for the products of the cycle, it will proceed at a nearly constant rate. In a hypercycle, though, the product of at least one reaction in the cycle is not a substrate, but is a catalyst for at least one reaction in the cycle. If it catalyzes its own reaction, it is an autocatalyst. For example, a mature ecosystem is a hypercycle in which all the matter is recycled. Plants and photosynthetic bacteria take in CO_2 and exhale O_2. Herbivores eat plants and fungi. Carnivores and omnivores eat herbivores and eat each other. All animals excrete waste and eventually die and decay, providing food for animals and bacteria. Bacteria have symbiotic relationships with plants and animals. The Gaia system is also a hypercycle. The autocatalytic units managing the system are mostly bacteria. Stable ecosystems and Gaia are examples of hypercycles that have nearly constant amounts of autocatalysts, so they do not grow.

Hypercycles sometimes have ever-increasing amounts of autocatalysts and show hyperbolic growth. The replication of viruses in an infected cell is an example. Once a virus is activated in a host cell, it makes copies of itself and the enzymes (RNA or DNA polymerase) needed to make more viruses. Growth continues until the cell becomes so full of viruses that it explodes. Other systems can act as an autocatalyst and growth is exponential. Doubling time remains constant. Examples include bacteria and yeasts when they divide and form two identical copies (clonal replication). Thus, cycles and hypercycles are common features of living systems [18].

One conclusion from the theory of autopoiesis is that the circular closure of the nervous system requires one to understand cognition in a new way [3]. “The nervous system is not only self-organizing, but is self-referring. Perception can’t be viewed as the accurate and objective representation of an external reality. Instead, it is the continuous creation of new relationships within the network of neurons. Cognition is not the accurate representation of an external reality, but an internal representation of a world that is brought forth by the individual. The mind is not a thing that is separate from the body or located solely in the brain. Instead, it is considered to be the process of life. Cognition is the activity of self-generation and self-perpetuation of autopoietic networks" [3].

The theory was developed by Humberto Maturana, and Francisco Varela, from the University of Santiago, Chile [22]. Its purpose was to introduce the importance of the individual in biology, at a time when most emphasis was on the species. Autopoiesis starts by recognizing that there is a boundary (cell membrane, skin, *etc.*) that separates the individual from the environment. That boundary, and everything inside it, are being continually broken down, re-made and organized, all under the control of the individual.

This is a clear distinction from a computer or a machine, which is made by someone else (human, chimpanzee, *etc*), and under the control of another. A living cell or organism and its inner processes (controlled cell death, metabolism, *etc*) are under the control of the individual and are made by that individual. However, the environment does stimulate the cell and organism, and they affect each other. The structure of the environment is changed by the cell or organism. Likewise, the structure of the cell or organism is changed by its interaction with the environment. Another way to say this is that there is structural coupling. Still, the changes in the cell or organism are made by the cell or organism, and are under its control.

Life is a cyclic process that produces the components of a living system. The average cell has 10^5 macromolecules, and each of these is destroyed and re-made 10^4 times in a cell's life time [23]. As Milan Zeleny said, "Throughout all this turnover of matter, the cell maintains its distinctiveness, cohesiveness and relative autonomy. This maintenance of unity and wholeness, while the components are being continuously or periodically disassembled and rebuilt is called autopoiesis" [23]. These components self-organize in this process and produce a system that is far from equilibrium. This all occurs at and within a boundary, such as my skin or the cell wall or membrane of a bacterium, a plant or an animal. Structure emerges from the functioning of the autopoietic organization. It also affects the maintenance, adaptability and viability of the organization in the environment. The membrane or skin separates self from the environment, but also connects self to the environment. There is a structural coupling between the organism and its environment. Moreover, complex tissues self-organize from a homogeneous population of stem cells [24].

The membrane and almost everything within it are continually being broken down and re-made in a cyclic process. The process of programmed cell death is called apoptosis, which starts when specific molecules bind to death receptors (like Fas, or APO-1) and start a cascade of events, leading to cell death. A company called Genentech has developed factors (Apo2L/TRAIL) that bind to death receptors, also known as apoptosis antigens (APO or Apo). They are being given in combination with other drugs in clinical trials on non-Hodgkins lymphoma [25].

There are many examples of things being broken down and re-made in the body. The cells of our stomach lining are replaced every three days. Skin and red blood cells live for a few days to a week. Teeth and bones are regenerated every 10-15 years. Even the proteins within long-living neurons are recycled every 90 days. Still, the form and organization of the cells and the organism stay the same and under the control of that organism or cell.

A corollary to this is that every organism is under its own self-control. You can stimulate someone, but you can't force them to do anything [26]. If the stimulus is strong enough, one might feel compelled to do something, but still only that person can do it. Through torture, an interrogator may inspire a victim to sign a confession, but only the victim can make the final decisions that bring the hand to hold the pen and sign the name. Even if an organism doesn't have a brain, it can make decisions. When a cell is exposed to a chemical gradient, it must make a decision to move towards or away from the stimulus. The process of interaction between and mutual definition of a living unit and its environment is called cognition. Moreover, innovation is natural and is the driving force behind evolution. In contrast to reductionist thinkers who may theorize that evolution is driven by the survival and propagation of the gene or the species, the autopoietic theory of life suggests that innovation is at work. We give our children much more than just our genes. Sperm and eggs are much more than just naked DNA. There are also RNA, polysaccharides, lipids, proteins and other molecules, all organized in a specific way. We give these to our children, along with a suitable organization, environment and core values. All of these things help ensure that the genes, the individual, family, society and the species will endure.

Innovation occurs as the constantly changing individual and environment interact [3]. "Because their structures are coupled, the individual affects the environment and the environment affects the individual. Genes and their translation can be affected by other (epigenetic) factors. Temperature profoundly affects the biology and behavior of all cold-blooded animals. Temperature influences their sex ratio, length of incubation, emergence from the nest, growth, activity, survival, nesting intervals, hibernation, and distribution. Cloned animals do not have identical appearance or behavior. Identical twins may have nearly identical appearance, but they have different susceptibility to diseases and (when encouraged), develop distinct personalities. Thus, genes are not strict codes that control everything in a machine-like organism. Instead, genes are part of the autopoietic system. Their structure is coupled to that of the environment. They are continuously being broken down and re-made" [3].

Another feature of autopoiesis and life is that they are dissipative structures. A dissipative structure is a structure that dissipates matter and energy in a cooperative way. It takes in matter and energy and dissipates them. It is far from equilibrium, which a characteristic of death. Living systems are far from equilibrium. They require a continuous supply of matter (nutrients) and energy, which they dissipate into the environment. Other cells or organisms in the ecosystem can use the matter that enters the environment. One organism's waste is another organism's food. Also, life is organized into a web of cycles and cycles within cycles, fueled and powered by external sources of matter and energy.

There are three criteria for defining an autopoietic system. First, the system must build its own boundary that separates itself from the environment. This construction is due to the reactions and other activities that occur at and within the boundary. This construction is determined by the organism itself [15]. The second feature of an autopoietic system is that there is structural coupling between the system and its environment. The third requirement for an autopoietic system is that the membrane and almost everything within it are continually being broken down and re-made.

The central characteristic of an autopoietic system is that it undergoes continual structural changes, while maintaining its web-like pattern of organization and its identity [3]. "For example, almost all the molecules in me and all the molecules on my face, my eyes and my brain are different than they were just a few days or months ago, but I am pretty much the same. Still, I am a dissipative structure, requiring the intake of nutrients (which are converted to energy and internal components) and the discharge of wastes, on a one way path that eventually will lead to my biological death. My molecules may change, but I retain my identity. I also believe that I can make myself change for the better if I try hard enough, regardless of what my "genetic code" might be" [3].

GAIA

Based on the autopoietic theory of life, the biosphere of Earth is often thought of as a living system [3]. The atmosphere is her boundary and it is far from equilibrium. This was first recognized by the atmospheric scientist, James Lovelock [27]. Lovelock used Gaia, the name given to the ancient Greek goddess of the Earth, as a new name for Mother Earth. He realized that our planet's atmosphere contains oxygen at a much higher concentration than methane or carbon dioxide. In contrast, the atmosphere of Mars is mostly carbon dioxide, with very little oxygen, strongly suggesting that Mars has little or no life. This is because, given enough time, methane and oxygen will react to form carbon dioxide and an equilibrium will be established. Non-living objects are usually at equilibrium, especially if they have been around for a few billion years. The atmosphere of Mars appears to be at equilibrium (although large amounts of frozen water, or ice, have been found recently on the Martian south pole). Our planet's atmosphere is far from equilibrium. Our planet has gone through drastic changes. Four billion years ago, there was little or no oxygen, but mostly methane and carbon dioxide. The sun was weaker and fewer cosmic rays reached our planet. The first life forms were obligate anaerobes. Oxygen, had it been present, would have been a deadly poison. After a couple of billion years, aerobic organisms evolved, and they caused an increase in the amount of oxygen in the atmosphere. The sun became stronger and more cosmic rays reached our planet. Eventually, some of these oxygen molecules reacted, rose to the upper atmosphere and formed an ozone layer, which blocks cosmic rays and much UV light. Our planet's temperature was stabilized. Moreover, everything beneath the atmospheric boundary is being continuously broken down and re-made. The atmosphere, hydrosphere, biosphere and even the geosphere are continually being broken down and re-made in a set of inter-connecting cycles.

Lovelock's ideas were developed further by Lynn Margulis, who explained the role of bacteria in maintaining our planet's autopoietic system [28]. Bacteria can be thought of as the catalysts that maintain the atmosphere in its present state, far from equilibrium, but stable, like homeostasis in a cell, organ or organism. Bacteria also continually create, transform and maintain the biosphere's metabolic processes. They are the catalysts of a hypercycle that links oxidation and reduction, turning O_2 into CO_2 and *vice versa*. Photosynthetic bacteria produce O_2, nitrogen fixing bacteria convert N_2 into nitrate, thermophilic bacteria oxidize H_2S and CH_4, while other bacteria produce H_2S, CH_4 and other gases. Bacteria in soil help accelerate the weathering (or break down) of rocks, which combine with rain water and carbon dioxide to form carbonates. Plants and animals recycle enormous amounts of CO_2, but the amount of CO_2 remained stable, until the industrial revolution and the more recent acceleration of the production of CO_2 by burning more and more fossil fuels.

To help us understand Gaia better, think of it as being like a tree. As the tree grows, there is only a thin layer of living cells around its perimeter, just beneath the bark. All the wood inside is dead. Similarly, Gaia is covered by a thin layer of living organisms that we call the biosphere. It reaches down to about seven miles below sea level to about five to six miles above sea level. Neither the atmosphere above it, nor the rocks, lava and molten iron in the troposphere below it, are alive. Just as the bark of a tree protects the thin layer of living cells beneath it, Gaia's atmosphere protects the biosphere with its ozone layer. Outer space and the Earth's interior also are part of Gaia's environment.

Autopoiesis can also be used when discussing ecosystems. The function of every component of a food web is to transform other components in the web. Plants take up energy, carbon dioxide, inorganic compounds and ions from the environment and produce organic and organo-metallic compounds, including macromolecules. These compounds are passed through the ecosystem, serving as food to herbivores, which serve as food to carnivores. The animals eliminate waste, which provides fertilizer and nutrients for plants. The entire network can be described as a hypercycle [17]. It is self-regulated through multiple feedback loops. Individual components (organisms and their cellular components) continually die, decompose and are replaced by the network's own process of transformation. Life makes and changes the environment, which feeds back and changes the organisms, through constant, interacting cycles.

NETWORKS

Networks permeate living systems. Brains are a network of neurons that are connected by axons and supported by glial cells. Cells are networks of ions and molecules, connected by biochemical reactions. Some molecules (ATP, H_2O) are highly connected nodes in the biochemical network. On the other hand, some compounds (like 1,3-bisphosphoglycerate) are intermediates in just one metabolic pathway and are not hubs. At the same time, people, animals and plants can not exist without symbiotic relationships. The average adult human has about 10^{15} cells, only 10% of which are human. The rest are bacteria, which help us digest our food and properly treat and excrete our wastes. In addition, societies are networks of people, connected by family, friends and work. Food webs and ecosystems are networks of organisms.

So, most organisms have symbiotic or syntrophic relationships within living networks. When one organism lives off the products of another organism it is called syntrophy. An example of syntrophy is when two different strains or types of bacteria require each other for growth. On the other hand, symbiosis is the close association of two organisms, usually for a long time. Ectosymbiosis is when one species lives on the body or surface of the host. For example, the Egyptian Plover bird enters the open jaws of the crocodile and eats parasites that feed on crocodiles. It finds food and is a safe from predators. Many symbiotic relationships are beneficial to both organisms (mutualism), some to just one (commensalisms) but some are harmful. Neutralism is when neither organism is harmed. If one species benefits and the other is harmed, it is parasitism. Competition is when both organisms are harmed. Often, symbiosis is essential. Most plants require fungi living on their roots to provide them nutrients, such as nitrogen. In hydrothermal vents and cold seeps on the ocean floor, giant clams, mussels and tube worms contain symbiotic bacteria.

Networks also pervade technology. The Internet and World Wide Web comprise a network of hardware (routers and cables) and software (web sites) that links individual computers. Electricity is provided through a network of power grids. Transportation systems are networks of airports, train stations, cities and villages, connected by airlines, railroad lines, roads and highways. There are also networks between government and industry that work together for the common good.

Just as successful governments are more than just a collection of individual people, our bodies are more than a collection of genes and building blocks. Proper organization and networks are essential for life and helps make us human. In the cell, information storage and processing, along with the execution of cellular programs, reside in distinct levels of organization. The genes may contain essential information for life, but they are not sufficient. RNA, proteins, enzymes, hormones and proper organization are also required. Genes interact with histones and other proteins, as well as with RNA and with other genes. They are also affected by the environment. Insects of the order Hymenoptera (ants, bees, wasps) have a queen, who makes one mating flight and stores the sperm for years. She rations the sperms out to her eggs, allowing some of them to be fertilized and become females. They have identical DNA, but only one of them will become a queen. It will be the one who gets the right food and care. Each female is diploid. It has a full set of queen genes and a full set of worker genes. The ones that get activated depend on the environment. There is structural coupling. Networks are important.

Whether they are natural or man-made, networks have similar properties. Living systems and the internet are examples of a type of network called a scale-free network [29]. One characteristic of scale-free networks is that a plot of the probability that a node will have k links, N, having a particular number of connections, k, follows the equation: $N(k) \sim k^{-\gamma}$ [30]. "Another characteristic is that these networks are dominated by a few well-connected nodes, called hubs. Most nodes in the network have a few connections, but a small number of nodes have a seemingly unlimited number of connections. Scale free networks have common architectures and properties and are governed by fundamental laws that seem to apply equally to cells, computers, language and society. Their organizing principles have significant implications for drug discovery, the treatment of diseases, halting the spread of epidemics, and defending the Internet from hackers" [30].

The properties of a network depend on the type of network and its topology, or structure [29]. "Though scale-free networks occur throughout the biosphere and in technology, so do random networks. An example of a random network might be a party in which 100 people attend, and none of them know each other. They will form a random network as they meet the people around them. If you serve some fruit juice and cheese, soon they will break up into groups of two or three. Suppose the juice comes in red bottles and green bottles. Unknown to the guests, the green bottles have the better tasting juice. Suppose you secretly tell one of the guests the secret and ask him or her to only tell their new acquaintances. If it takes about 10 minutes for the first group of three to break up into three new groups, how long will it take before almost everyone wants the juice in the green bottle? It would take 16 hours for the first person to meet all of the other 99 people. Mathematicians Paul Erdos and Alfred Renyi showed that it would only take 30 minutes. Even in random networks, a giant cluster emerges as the network grows and forms more links, until each node has an average of one link. Mathematicians call this the emergence of a large component. Physicists call it a percolation or phase transition. Sociologists would say that you just formed a community" [30].

Mathematicians understand that networks can be considered graphs. The nodes are points and they are connected by line segments that may or may not be directed. For example, carbon almost always moves in one direction in an ecosystem. Plants are eaten by animals. Their carbon is transferred to them. Very few plants are able to eat animals, so consumption is directed. Similarly, websites in the World Wide Web are connected in one direction. There may be a clear path from one website to another (A to D, going through B and C), but the reverse path usually doesn't exist. Ecosystems and the World Wide Web are examples of directed networks.

When the connections are not directed but are made at random, there is an average number of connections for a node. Few nodes have much more or less than the average. A plot of the number of nodes, N, having a particular number of connections, k, would look like a bell curve. An example would be a plot of the height of all the adult men in the USA. Very few are under four feet or over eight feet. Most are a little less than six feet tall.

When connections are not made at random, a scale-free network can emerge. Examples include the internet and World Wide Web. We do not connect our computers, routers, cables or web pages randomly. On the World Wide Web, some websites like Google are hubs, with many connections or links. Most, though, have just one or two links to other web pages. A plot of the probability, N, that a node will have k links (or degrees) having a particular number of connections, k, follows the equation: $N(k) \sim k^{-\gamma}$. This is called a power law. If plotted on a log-log scale, it produces a straight line. The exponent γ is called the degree exponent. It is one of the parameters that are used to characterize a scale-free network.

Scale-free networks have some other common properties or characteristics. First, they tend to grow. As they grow, new nodes are added and new links are formed between nodes. Nodes with the most links are called hubs. Scale-free networks are resilient to random attacks. For example, if a toxin or pathogen were to randomly knock out, or inhibit a reaction in the metabolic network of the cell that is not a hub, it would have little effect. Similarly, if the airport in Kansas City were shut down, it would have little effect on air traffic, as airplanes could be re-routed. In fact, several smaller airports could be shut down with little effect on air traffic over all. If a few hubs, such as New York, Los Angeles, Atlanta and Chicago, were shut down, it could shut down the whole air transportation network in the USA. Similarly, if a toxin inactivates a few key hubs in cellular metabolism, it can kill a cell. Scale-free networks are robust and resistant to random attacks, but they are quite vulnerable to attacks directed at hubs.

Another property of scale-free networks is preferential attachments of nodes. If a node (such as a website) has been around for a while it will have more connections. If a node already has many connections to other nodes, it will tend to make more connections. Another way to say this is that the rich get richer or the best-connected get even better connected. Sometimes, though, a new node can attract many links. Google in early 2000 had few links. In June, 2000, Yahoo! Fired their search engine (Inktomi) and hired Google, even though it had fewer links and had not been around as long. This is because Google was more fit, or offered more and better services. It has become one of the largest hubs in the World Wide Web. Thus, the number of links that a node has depends on its fitness, as well as the number of nodes it already has, and the length of the time the node has existed. Every network has its own fitness distribution, which tells how similar or different the nodes are.

It is also useful to measure the path length, which tells us how many links we need to pass through to travel between two nodes. There can be many alternative paths between two nodes. The shortest path is the path with the smallest number of links between the selected nodes. The mean path length,<*l*>, is the average over the shortest paths between all pairs of nodes and is a measure of a network's overall navigability [31]. The average degree <k>, average path length and average clustering coefficient all depend on the number of nodes and links (N and L) in the network [31].

The cost, κ, of a network is defined as the number of edges present in the network divided by the number of total possible edges (N(N-1)/2), where N is the total number of nodes in the network. This term is also often referred to as the sparsity of the network. The efficiency, E, of a network is equal to the inverse of that network's path-length L. The modularity, m, of a network is a measure of the degree to which the network is composed of modules. A minimum spanning tree is a subgraph of the network made up of the smallest subset of edges of the graph where all nodes are still connected together. The minimum spanning tree has been used to create a connectivity backbone of an anatomical brain network [32].

Networks can also be characterized by the clustering coefficient. That is, if one node (A) is connected to another (B), and it is connected to a third node (C), there is a possibility that A is also directly connected to C. This can be quantified by the clustering coefficient, CI, given by the equation: CI = 2nI/k(k-1), where nI is the number of links connecting the kI neighbors of node I to each other. That is, CI equals the number of triangles that go through node I, and kI(kI –1)/2 is the total number of triangles that could pass through node I, should all of node I's neighbors be connected to each other [31]. There is also an average clustering coefficient, <C>, which is the overall tendency of nodes to form clusters or groups. An important measure of the network's structure is the function C(k),which is defined as the average clustering coefficient of all nodes with k links. For many real networks C(k) is directly proportional to k^{-1}, which would indicate that a network is hierarchical [31]. A hierarchy parameter, β, is defined as the slope on a log-log plot of the clustering coefficients and the degrees (k) across all nodes in the network.

There is also a phenomenon called a small world network, in which there is a large amount of clustering and a short path length between any two nodes in the network [32]. "For example, social small world network is when one's friends are more likely to know each other than not. Another example is that the average social distance between any two people in the United States is equal to three. Neural systems in humans and other organisms also show this small world property. That is, information can be transferred from one region of the brain to another by traversing only a few major tracts of white matter. This makes anatomical sense, because human brain activity is characterized by both segregation and integration of function. This modular structure maximizes the brain's efficiency and adaptability. Moreover, the brain's individual processing units are coordinated by rhythmic electrical oscillations. The coherence of the more spatially distributed systems is mediated by lower frequency brain waves, *e.g.* β-band (12-30 Hz). On the other hand, cognition and the integration of information into a coherent perception occur over shorter physical distances. It is thought to be mediated by high frequency γ-band oscillations (≥ 40 Hz). This combination of functional segregation and integration over short and long distances as mediated by different frequency bands suggests high topological clustering and a short topological path length" [32].

The small-worldness scalar, σ, is the ratio of the clustering coefficient of a network, C, normalized by the clustering coefficient of a random graph, Cr, to the path-length of a network, L, normalized by the path length of a random graph, L_r. That is $\sigma=(C/C_r)/(L/L_r) = \gamma/\lambda$ where the normalized clustering coefficient $\gamma = C/C_r$

and the normalized path-length $\lambda = L/L_r$ [25]. The brain has this small world characteristic [32].

A human anatomical network was constructed, based on cortical thickness measurements in a large group of 124 healthy adults of both genders [33]. "The network had 104 connections out of a maximum possible number of 2862, indicating a sparsity level of 7.3%. Also, 45 regions out of the total 54 were connected. This large connected component showed robust small world properties. There was a predominance of short range connections and a power-law degree distribution, indicating that the network was broad-scale, and the presence of hubs predominantly located in the association cortex" [33]. This was demonstrated experimentally using functional magnetic resonance imaging (fMRI) [32]. "Also, the anatomical brain network has a modular structure. The anatomical brain network in healthy people was organized into six topological modules. Nodes inside a given module were highly connected to other nodes inside that module but less connected to nodes in other modules. These six modules were closely aligned with known functional domains such as auditory/language, strategic/executive, sensorimotor, visual, and mnemonic processing. This suggests that there is a direct relationship between distinct functions and highly structured anatomical connectivity. There was also a big difference in centrality, which is a measure of an index region's relative importance to communication in the network by looking at how many of the shortest paths between other regions must pass through that index region. The occipital cortices in the Alzheimer's network showed abnormally high centrality while regions of the temporal and parietal heteromodal association cortices were significantly less central than they were in the healthy brain" [32].

Network theory is useful in studying not just Alzheimer's disease, but also other neurodegenerative diseases, as well as aging itself [34 - 39]. It has shown that the brain is organized as a small-world and scale-free network with hierarchical modularity [34]. It is characterized by high clustering (local connectedness) and short path lengths between nodes. Also, the probability of any given node having a certain number of connections, N(k), is given by the power law equation: $N(k) \sim k^{-\gamma}$. This means that there are some highly connected nodes (hubs) that are called a 'rich club'. Finally, each module or cluster of nodes is made up of smaller components, while also being part of a larger unit or component. The organization is closely regulated and is essential for proper cognitive function [34].

Brain organization has also been divided into structural and functional networks [35]. Historically, neurological diseases have been classified as being either global or local [34]. However, improved understanding of neurological networks has challenged this. For example, Alzheimer's disease was once classified as global. We now know that the pathology is not distributed evenly throughout the brain, but is found mostly in hubs. Moreover, several neurodegenerative diseases are probably caused by overloading hubs, causing them to fail. This disrupts the normally healthy network structure of the brain. At first, there will be more communication with nodes that are higher in the hierarchical structure, as nodes lower in the hierarchy fail. As hubs become overloaded in the chronic phase, information will be rerouted to nodes that are lower in the hierchy. So, in the acute phase of neurological disorders, much of the information processing shifts from local to global nodes. In the chronic phase, the opposite happens. This can lead to contradictory results being reported for disorders such as Alzheimer's disease [34].

There can also be confusion due to the differences between the ways that brain networks are visualized and divided into categories. That is, they can be divided into either structural or functional networks [35]. "In either case, the brain network can be illustrated as a graph that is sometimes called a connectome, since it shows all the connections between nodes. However, nodes are not easily defined or identified. They are done in different ways in structural and functional networks. Structural networks can be cerebral cortex correlations and by tracking white fibers. Anatomical, structural networks are analyzed by calculating the correlations between cortical thickness (or volume) between all pairs of regions in a given method of creating parcels (parcellization). Structural networks can also be made from data obtained by diffusion tensor imaging (DTI), which is also called diffusion MRI. It tracks bundles of white matter fibers. This can find changes in white matter" [35].

On the other hand, functional brain networks are made from functional connectivities between pairs of brain regions [35]. "The connectivity is based on the changes over time in the patterns of neuronal activities of parts of the brain that are anatomically separated. It can be measured by functional magnetic resonance imaging (fMRI), which monitors changes in blood flow. In the past, many such graphs of functional connectivity were based on brain atlases that divided the brain into different anatomical or cytoarchitectural boundaries. More recently, though, functional atlases have been developed" [35].

So, the analysis of the brain network depends on which atlas is used to divide the brain into parcels [35]. "Atlases are needed to neuroimaging data into graphs with nodes linked by edges.

However, the small world properties (γ and λ) of the brain networks are apparent regardless of the atlas that is used. Still, the AAL and Brodmann atlases have been popular, while some have shown that the LPBA40 atlas was better [35].

Others have compared functional networks (based on fMRI while subjects are resting) and structural networks (based on DTI) [35]. "These two methods have shown that there is a close relationship between structural and functional connections. When combined, data from fMRI and DTI improve the parcellation of the data. DTI can sub-divide some functionally diverse parts of the brain. In addition, fMRI data can do the parcellation of the cerebral cortex. Still, different atlases can quantitate the network properties differently. For example, many atlases contain little or no information about the strengths of the connectivities, but this was improved by using clustering theory to produce the CC200 and CC400 atlases" [35].

More recently, a minimum spanning tree (MST) has been used to provide better analyses of brain networks [36]. It is a subgraph of a larger graph, or a subnetwork of a larger network. An example is shown in Fig. (**1**).

The MST can help improve comparisons between structural and functional networks that have produced conflicting results in the past [36]. " The MST can solve problems caused by limits in these two ways of analyzing networks. That is, previous analyses have not been able to deal with changes in the strengths of connections in weighted networks or in link density for unweighted networks. MST is not sensitive to changes in strengths or densities of connections.

Fortunately, it is sensitive to alterations in the topology of networks. So, MST is especially useful in studying brain networks, because it avoids methodological biases. Even though it does not show all the connections between nodes, it does provide an unbiased mathematical description of brain networks" [36].

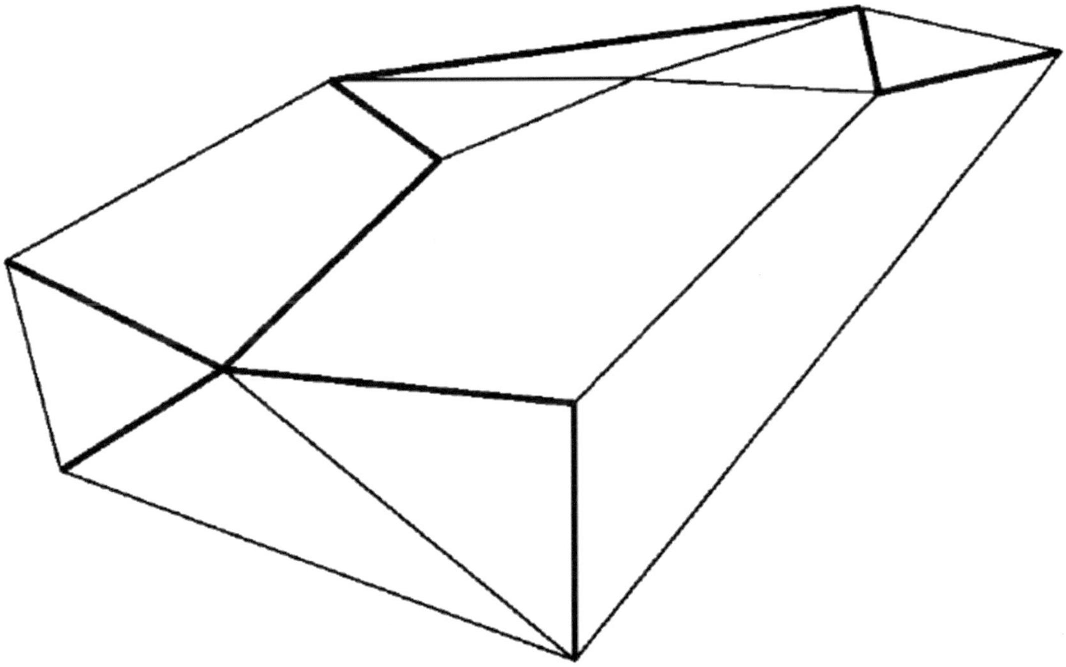

Fig. (1). Example of a minimum spanning tree (bold line segments) that shows the minimum distances betwen all the nodes (points).

NETWORKS IN AGING AS WELL AS NEURODEGENERATIVE AND PSYCHIATRIC DISEASES

The changes in the brain network as people become older and suffer from senescence has been studied [37]. "Previous studies had already established that there are important changes in the network topology during normal aging. This has been seen in both structural and functional analyses. Global efficiency and hub centrality are reduced, while local clustering increases. However, structural and functional analyses can produce conflicting results. This is because each method uses different metrics to evaluate the network topology. This can prevent proper understanding of changes in topology that occur during aging and senescence. Instead, MST used to analyze diffusion-weighted images of 382 healthy people aged 20.2 to 86.2 years. Unlike previous methods of network analysis, the MST were not heavily influenced by age-related changes in the density of the connections in the network. Instead, MST metrics illuminated important differences in linear and nonlinear correlations that agreed with known changes in the integrity of white matter in the aging brain" [37].

On the other hand, neurodegenerative and neuropsychiatric diseases can destroy the small world properties of the brain [35]. "This includes schizophrenia, attention deficit hyperactivity disorder (ADHD) and Alzheimer's disease (AD). In patients with ADHD, global efficiency can decrease while local efficiency increases in the brain metworks. Moreover, the brains of patients with AD have increased connectivity and randomization in the small world network. That is, there is a decreased clustering coefficient and increased path length in AD patients" [35].

Before AD is diagnosed, there can be mild cognitive impairment (MCI) [38]. "So, a recent study evaluated the diffusion weighted images (DWI) and MRI images of 30 subjects who had MCI. Baseline connectivity maps were produced. The mean clustering coefficient and characteristic path lengths were determined. DWI was able to find white matter fiber bundles that connect cortical and subcortical regions. The baseline results were able to predict which subjects would develop AD" [38].

In a similar study, DWI was used to study the brain connectivities of AD patients [39]. "Connectivity matrices were evaluated using both whole-brain tractography and flow-based measurements. An algorithm called support vector machines was used to learn from a training set of data to classify subjects as being healthy or suffering from MCI or AD. Measures of connectivity were obtained from the cortex and then entered into connectivity matrices. These matrices were able to quantify several characteristics of the network. The combination of fiber and flow connectivities was best able to distinguish between the brains of healthy subjects from those who had AD" [39].

Others studied 303 multiple sclerosis patients. They were divided into six groups according to the severity of their white matter lesions [32]. "The small-world properties of each group's cortical-thickness based network were measured using the global efficiency, which is defined by as the inverse of the network's average path-length. The global network efficiency was directly correlated with the white matter lesion severity in all six patient groups with significant regional inefficiencies apparent in the insula, precentral gyrus, prefrontal and temporal association cortices. This provides evidence for the disconnection hypothesis of multiple sclerosis, in which aberrant neuronal connectivity encompasses widely distributed brain regions. Regional differences have also been found in people with schizophrenia" [32].

Moreover, the networks derived from 92 patients with Alzheimer's disease were compared to those derived from 97 healthy people" [32]. "The diseased network had a larger clustering coefficient (indicating abnormally high local efficiency) and longer path length (indicating decreased global efficiency) than the healthy networks. There was also a decreased robustness to targeted attacks on the network hubs" [32].

NETWORK ANALYSIS OF MOTHER'S MILK

The importance of breast feeding was described in Chapter 11. It not only provides essential nutrients, but also helps protect babies from infectious diseases. So, it is noteworthy that a recent study evaluated interactions within the human milk microbiome [40]. "It was found that there are two disconnected sub-networks in the microbiome. One of them consisted of seven genera as the nodes and all of their pari-wise connections. Some were facilitative and others were cooperative. On the other hand, the interactions between bacteria in the second sub-network were mostly cooperative. The only non-cooperative nodes (bacteria) in this sub-network were mutually cooperative *Staphylococcus* and *Cornybacterium*, some of which were pathogenic. However, this cooperative interaction was inhibited by the other cooperative nodes in this second sub-network. This is especially important because *Staphylococcus* bacteria can cause infectious mastitis and sudden infant death syndrome. It was also found that the milk microbiome is a small world network. In this small world network, *Staphylococcus* and *Cornybacterium* form a pathological alliance that is effectively countered by the other bacteria in the milk microbiome" [40].

CONCLUSION

So, in the scale-free network that is in living cells, there are many levels of organization. Each of them can be viewed as a network. There is a network of genes, a metabolic network, a regulatory network and a cellular network. The metabolic network is the network of metabolic reactions. The regulatory network controls everything from development, to metabolism to cell death. The cellular network is a network of cellular components and organelles, connected by biochemical reactions and physical links. The web of life contains these networks, as well as all protein-protein, protein-DNA, DNA-RNA, protein-lipid, sugar-lipid (lipopolysaccharide) and sugar-protein (glycoprotein) interactions. Ecosystems contain additional links between organisms, bacteria and viruses. So, network theory and systems thinking are emerging as guiding principles in biology and medicine in the new millennium.

KEY POINTS

1. **In systems biology, the basic, fundamental unit is the cell, not the atoms and molecules in the cell. Many new properties emerge when atoms, molecules and ions are organized in a living cell.**

2. **Systems thinkers have also made significant contributions to industry and the world economy in the form of total quality manufacturing (TQM) and total quality leadership (TQL). It is based on the idea that a team of motivated employees can work more efficiently and competitively if they work together instead of working separately.**

3. **The functions of a cell do not depend on just the properties of the individual molecules, but also on how those molecules interact. We are not separate from our environment. Although genetics is an important factor, environment is important too. The environment affects us and we affect the environment. Our structure and that of the environment are coupled with each other.**

4. **Ecosystems are like other complex systems. They have many components that have many degrees of freedom and are non-linear. These complex systems are far from equilibrium. They can be described by nonlinear differential equations that have no exact solution, but have many approximate solutions, as described**

in chaos theory. New solutions can emerge unexpectedly, so there can be unpredictable consequences.

5. **Autopoiesis means self-production. It is a network of production processes, in which the function of each component is to participate in the production or transformation of itself and the other components in the network. The production processes are circular. Life is a cyclic process that produces the components of a living system.**

6. **There three criteria for defining an autopoietic system are that the system must build its own boundary that separates itself from the environment; there is structural coupling between the system and its environment; and the membrane and almost everything within it are continually being broken down and re-made. The central characteristic of an autopoietic system is that it undergoes continual structural changes, while maintaining its web-like pattern of organization and its identity.**

7. **Based on the autopoietic theory of life, the biosphere of Earth is often thought of as a living system. The atmosphere is her boundary and it is far from equilibrium. This was first recognized by the atmospheric scientist, James Lovelock who used Gaia, the name given to the ancient Greek goddess of the Earth, as a new name for Mother Earth.**

8. **Bacteria can be thought of as the catalysts that maintain the atmosphere in its present state, far from equilibrium, but stable, like homeostasis in a cell, organ or organism. Bacteria also continually create, transform and maintain the biosphere's metabolic processes. They are the catalysts of a hypercycle that links oxidation and reduction, turning O_2 into CO_2 and *vice versa*.**

9. **Networks permeate living systems. Brains are a network of nerve cells that are connected by axons. Cells are networks of molecules, connected by biochemical reactions. Some molecules (ATP, H_2O) are highly connected nodes in the biochemical network. People, animals and plants can not exist without symbiotic relationships. Organization and networks are essential for life and helps make us human.**

10. **Living systems and the internet are examples of a type of network called a scale-**

free network. These networks are dominated by a few well-connected nodes, called hubs. Most nodes in the network have a few connections, but a small number of nodes have a seemingly unlimited number of connections. Scale free networks have common architectures and properties and are governed by fundamental laws. In the scale-free network that is in living cells, there are many levels of organization. Each of them can be viewed as a network. There is a network of genes, a metabolic network, a regulatory network and a cellular network.

ABBREVIATIONS

TQM	=	Total Quality Manufacturing
TQL	=	Total Quality Leadership
HLA	=	Human Leukocyte Antigen
NHCD	=	Network of Human Cell Differentiation
RNA	=	Ribonucleic Acid
DNA	=	Deoxyribonucleic Acid

REFERENCES

[1] Dawkins R. The Selfish Gene. Oxford: Oxford Press 1976; pp. 39-214, 215.

[2] Lederman L. The God Particle, Delta; Reprint edition. McHenry, IL 1994.

[3] Capra F, Luisi PL. The Systems View of Life. Cambridge: Cambridge University Press 2014. [http://dx.doi.org/10.1017/CBO9780511895555]

[4] Davies PC. Emergent biological principles and the computational properties of the Universe. Complexity 2004; 10: 11-5. [http://dx.doi.org/10.1002/cplx.20059]

[5] Luisi PL. Emergence in chemistry: Chemistry as the embodiment of emergence. Found Chem 2002; 4: 183-200. [http://dx.doi.org/10.1023/A:1020672005348]

[6] Manahan SE. Environmental Chemistry. Boca Raton: CRC Press 2005.

[7] Anastas PT, Warener JC. Green Chemistry. Oxford: Oxford Press 1998.

[8] Qiu J. Epigenetics: unfinished symphony. Nature 2006; 441(7090): 143-5. [http://dx.doi.org/10.1038/441143a] [PMID: 16688142]

[9] Raser JM, O'Shea EK. Noise in gene expression: origins, consequences, and control. Science 2005; 309(5743): 2010-3. [http://dx.doi.org/10.1126/science.1105891] [PMID: 16179466]

[10] Mannello F, Ligi D, Magnani M. Deciphering the single-cell omic: innovative application for translational medicine. Expert Rev Proteomics 2012; 9(6): 635-48.

[http://dx.doi.org/10.1586/epr.12.61] [PMID: 23256674]

[11] Shapiro E, Biezuner T, Linnarsson S. Single-cell sequencing-based technologies will revolutionize whole-organism science. Nat Rev Genet 2013; 14(9): 618-30.
[http://dx.doi.org/10.1038/nrg3542] [PMID: 23897237]

[12] Smith C. Cancer shows strength through diversity. Nature 2013; 499(7459): 505-8.
[http://dx.doi.org/10.1038/499505a] [PMID: 23887432]

[13] Frisk TW, Khorshidi MA, Guldevall K, Vanherberghen B, Önfelt B. A silicon-glass microwell platform for high-resolution imaging and high-content screening with single cell resolution. Biomed Microdevices 2011; 13(4): 683-93.
[http://dx.doi.org/10.1007/s10544-011-9538-2] [PMID: 21465090]

[14] Galvão V, Miranda JG, Andrade RF, Andrade JS Jr, Gallos LK, Makse HA. Modularity map of the network of human cell differentiation. Proc Natl Acad Sci USA 2010; 107(13): 5750-5.
[http://dx.doi.org/10.1073/pnas.0914748107] [PMID: 20220102]

[15] Stitzel ML, Seydoux G. Regulation of the oocyte-to-zygote transition. Science 2007; 316(5823): 407-8.
[http://dx.doi.org/10.1126/science.1138236] [PMID: 17446393]

[16] Bornas X, Nogeura M, Pincus D, Buela-Casal G. Emotional inertia: psychotherapy process and outcome. Intl J Clin Psych health 2014; 14: 232-9.

[17] Forrest CB. A living systems perspective on health. Med Hypotheses 2014; 82(2): 209-14.
[http://dx.doi.org/10.1016/j.mehy.2013.11.040] [PMID: 24368035]

[18] Jantsch E. Cycles and Hypercycles. In: Zeleny M, Ed. Autopoeisis, a Theory of Living Organizations. New York: North-Holland/Elsevier 1981.

[19] Varela FG, Maturana HR, Uribe R. Autopoiesis: the organization of living systems, its characterization and a model. Curr Mod Biol 1974; 5(4): 187-96.
[PMID: 4407425]

[20] Furusawa C, Kaneko K. A dynamical-systems view of stem cell biology. Science 2012; 338(6104): 215-7.
[http://dx.doi.org/10.1126/science.1224311] [PMID: 23066073]

[21] Wang W. Therapeutic hints from analyzing the attractor landscape of the p53 regulatory circuit. Sci Signal 2013; 6(261): pe5.
[http://dx.doi.org/10.1126/scisignal.2003820] [PMID: 23386744]

[22] Maturana H, Varela F. Autopoiesis and Cognition, published originally as De Maquinas y Seres Vivos, Editorial Universitaria, Santiago, Chile, 1972, translated and re- published by D. Reidel, Dordrecht, Holland 1980

[23] Zeleny M, Ed. Autopoeisis, a Theory of Living Organizations. New York: North-Holland/Elsevier 1981.

[24] Sasai Y. Cytosystems dynamics in self-organization of tissue architecture. Nature 2013; 493(7432): 318-26.
[http://dx.doi.org/10.1038/nature11859] [PMID: 23325214]

[25] Wenner M. Profile: Learning from death. Nature 2008 May 15; 453(7193): 271-3. [http://dx.doi.org/10.1038/453271a] [PMID: 18480787]

[26] Capra F. The Hidden Connections. New York: Doubleday 2002.

[27] Lovelock J. Gaia as seen through the atmosphere. Atmosph Envi 1972; 6: 579. [http://dx.doi.org/10.1016/0004-6981(72)90076-5]

[28] Lovelock J, Margulis L. Biological modulation of the Earth's atmosphere. Icarus 1974; 21: 471. [http://dx.doi.org/10.1016/0019-1035(74)90150-X]

[29] Barabási A-L, Bonabeau E. Scale-free networks. Sci Am 2003 May; 288(5): 60-9. [http://dx.doi.org/10.1038/scientificamerican0503-60] [PMID: 12701331]

[30] Barabási A-L. Linked The New Science of Networks. Cambridge, Mass: Perseus Publishing 2002.

[31] Barabási A-L, Oltvai ZN. Network biology: understanding the cell's functional organization. Nat Rev Genet 2004; 5(2): 101-13. [http://dx.doi.org/10.1038/nrg1272] [PMID: 14735121]

[32] Bassett DS, Bullmore ET. Brain anatomy and small world networks. In: Bianchi MT, Caviness VS, Cash SS, Eds. Network Approaches to Diseases of the Brain. IL: Bentham Science Oak Park 2012.

[33] He Y, Chen ZJ, Evans AC. Small-world anatomical networks in the human brain revealed by cortical thickness from MRI. Cereb Cortex 2007; 17(10): 2407-19. [http://dx.doi.org/10.1093/cercor/bhl149] [PMID: 17204824]

[34] Stam CJ. Modern network science of neurological disorders. Nat Rev Neurosci 2014; 15(10): 683-95. [http://dx.doi.org/10.1038/nrn3801] [PMID: 25186238]

[35] Yao Z, Hu B, Xie Y, Moore P, Zheng J. A review of structural and functional brain networks: small world and atlas. Brain Informatice 2015; 2: 45-52. [http://dx.doi.org/10.1007/s40708-015-0009-z]

[36] Tewarie P, van Dellen E, Hillebrand A, Stam CJ. The minimum spanning tree: an unbiased method for brain network analysis. Neuroimage 2015; 104: 177-88. [http://dx.doi.org/10.1016/j.neuroimage.2014.10.015] [PMID: 25451472]

[37] Otte WM, van Diessen E, Paul S, *et al.* Aging alterations in whole-brain networks during adulthood mapped with the minimum spanning tree indices: the interplay of density, connectivity cost and life-time trajectory. Neuroimage 2015; 109: 171-89. [http://dx.doi.org/10.1016/j.neuroimage.2015.01.011] [PMID: 25585021]

[38] Nir TM, Jahanshad N, Toga AW, *et al.* Alzheimer's Disease Neuroimaging Initiative (ADNI). Connectivity network measures predict volumetric atrophy in mild cognitive impairment. Neurobiol Aging 2015; 36 (Suppl. 1): S113-20. [http://dx.doi.org/10.1016/j.neurobiolaging.2014.04.038] [PMID: 25444606]

[39] Prasad G, Joshi SH, Nir TM, Toga AW, Thompson PM. Alzheimer's Disease Neuroimaging Initiative (ADNI). Brain connectivity and novel network measures for Alzheimer's disease classification. Neurobiol Aging 2015; 36 (Suppl. 1): S121-31. [http://dx.doi.org/10.1016/j.neurobiolaging.2014.04.037] [PMID: 25264345]

[40] Sam Ma Z, Guan Q, Ye C, Zhang C, Foster JA, Forney LJ. Network analysis suggests a potentially

'evil' alliance of opportunistic pathogens inhibited by a cooperative network in human milk bacterial communities. Sci Rep 2015; 5: 8275.
[http://dx.doi.org/10.1038/srep08275] [PMID: 25651890]

Appendix

Structures of Small Molecule Drugs and Hormones Discussed in This Edition

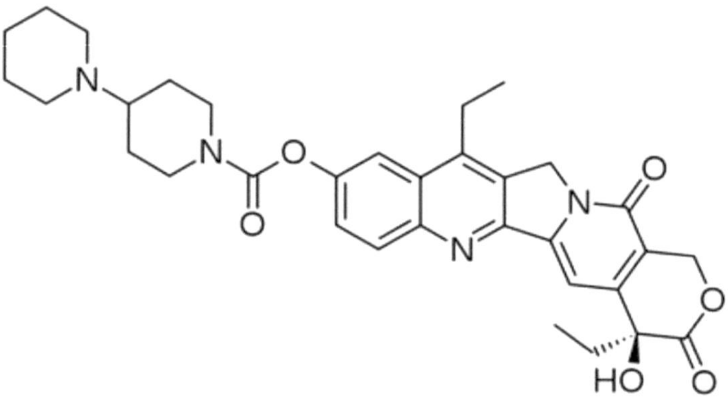

Fig. (1). Camptosar.

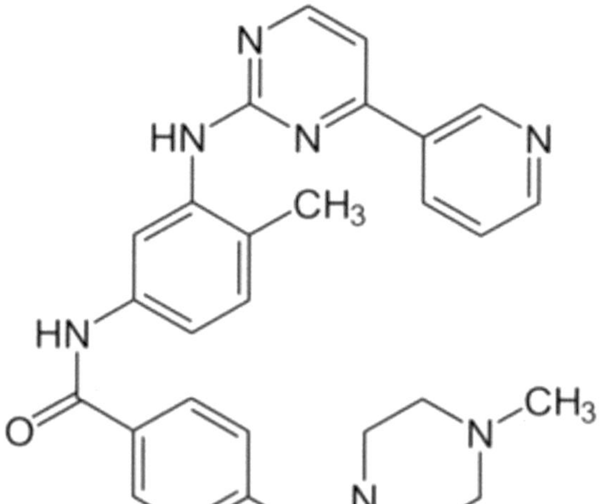

Fig. (2). Imatinib, Gleevec® for chronic myeloid leukemia.

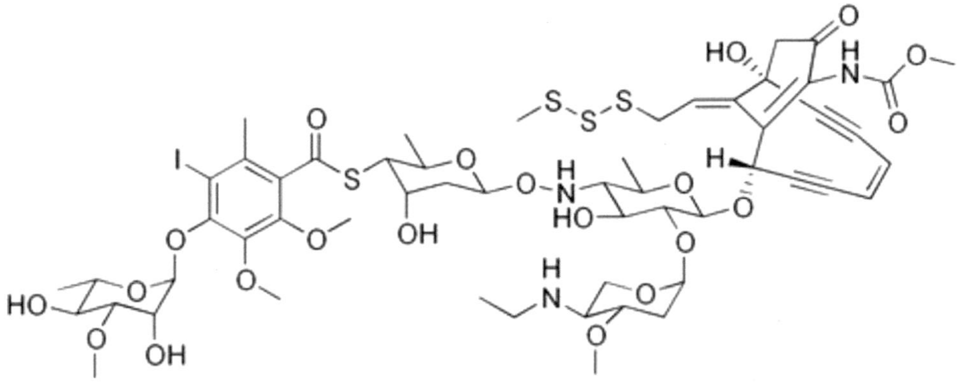

Fig. (3). Calicheamicin.

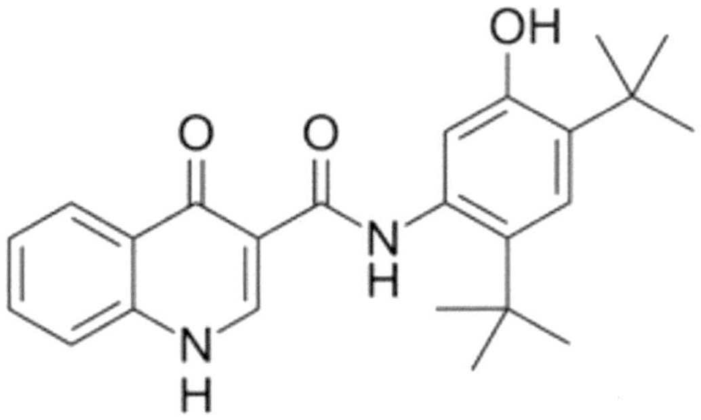

Fig. (4). Ifacaftor, Kalydeco®, for cystic fibrosis.

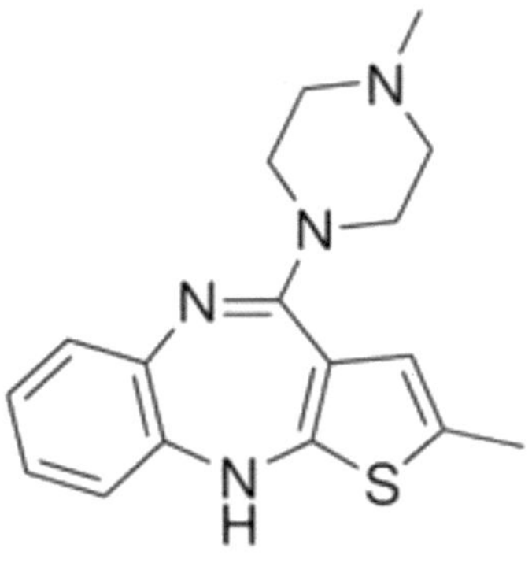

Fig. (5). Olanzapine Zyprexa®, an atypical antipsychotic.

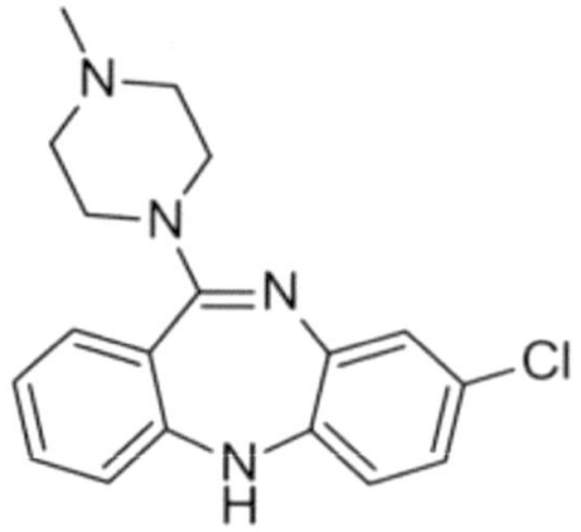

Fig. (6). Clozapine, an atypical antipsychotic.

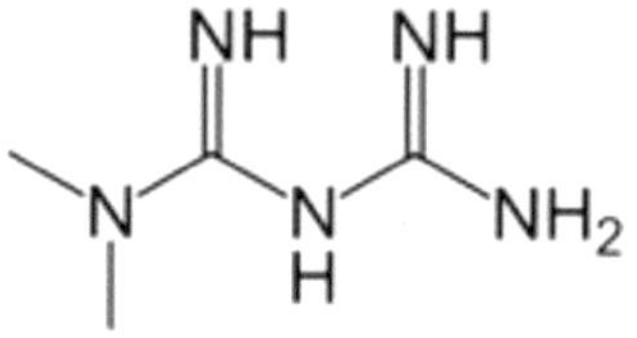

Fig. (7). Metformin (Glucophage®).

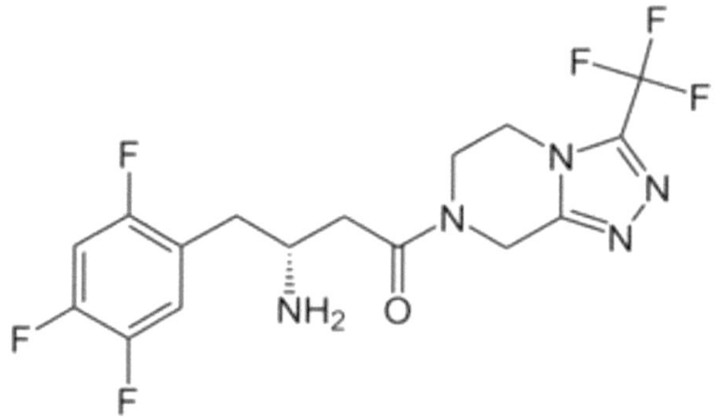

Fig. (8). Sitagliptin, Januva®, an antidiabetic that inhibits dipeptidyl peptidase-4 (DPP-4).

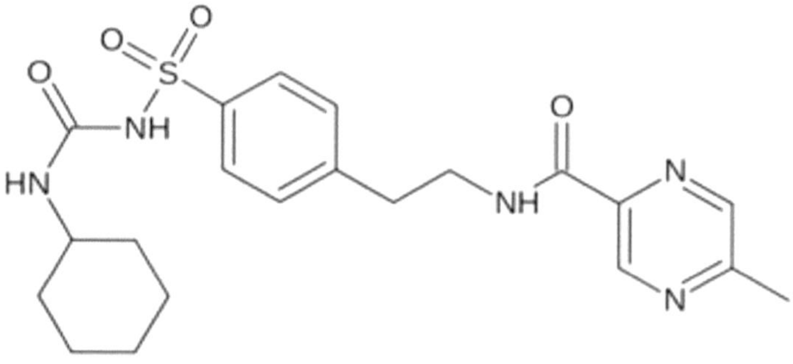

Fig. (9). Glipizide, (Glucotrol®), CYP substrate and oral antidiabetic.

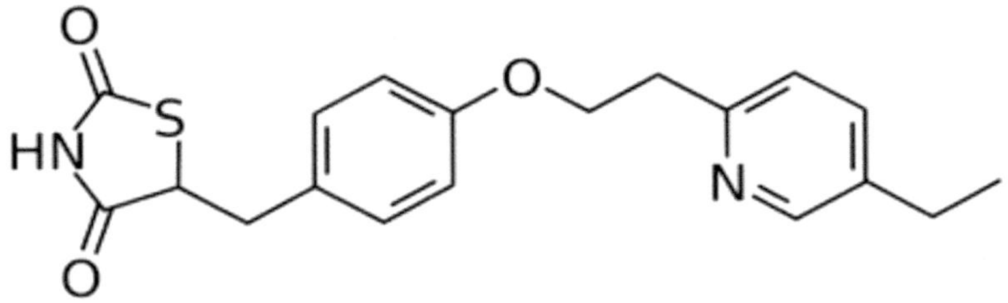

Fig. (10). Pioglitazone (Actos®), an antidiabetic.

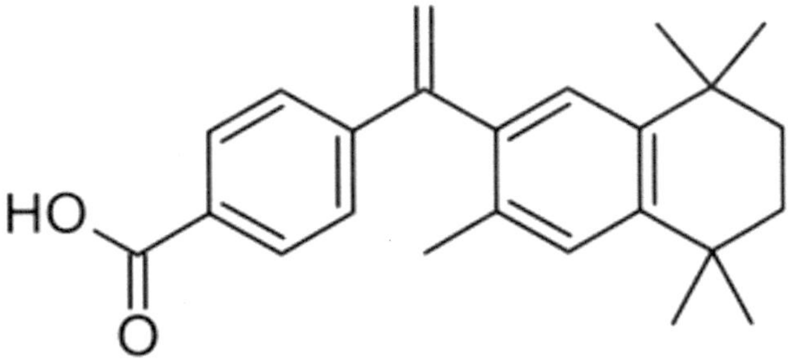

Fig. (11). Bexarotene, Targretin® (for refractory cutaneous T-cell lymphoma).

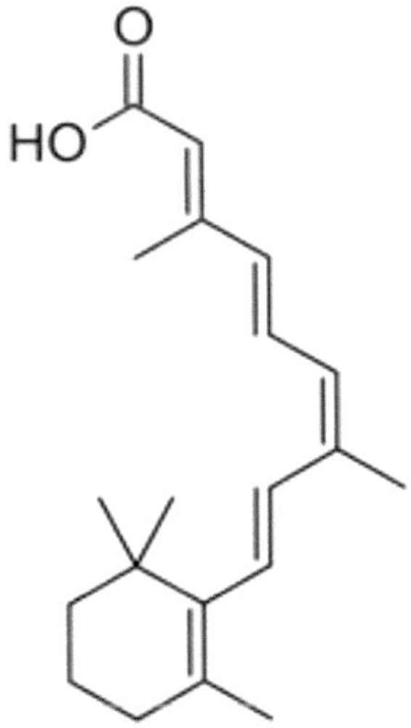

Fig. (12). Alitretinoin, 9-*cis* retinoic acid, for Karposi's sarcoma, topical only.

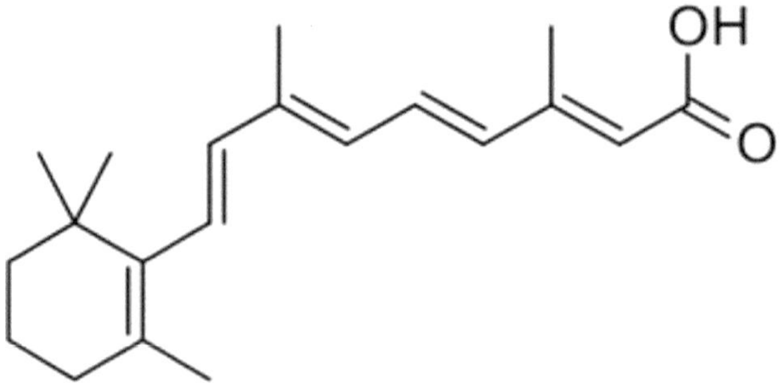

Fig. (13). Tretinoin, all *trans* retinoic acid, for acute promyelocytic leukemia and acne.

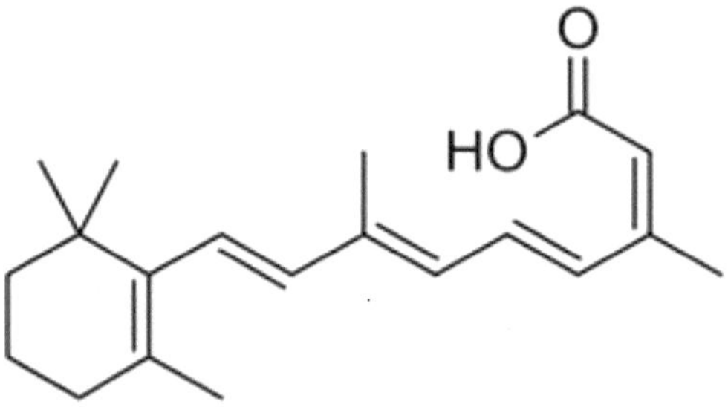

Fig. (14). Isotretinoin, 13-*cis*-retinoic acid, for severe nodular acne.

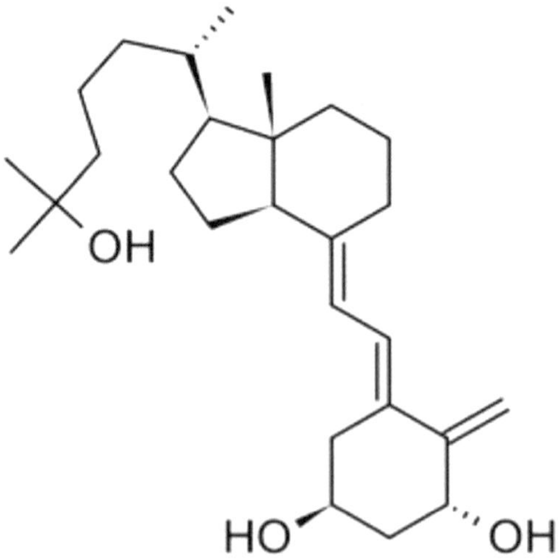

Fig. (15). Calcitriol, 1,25-dihydroxycholecalciferol, for hypocalcemia due to chronic kidney failure and hypoparathyroidism.

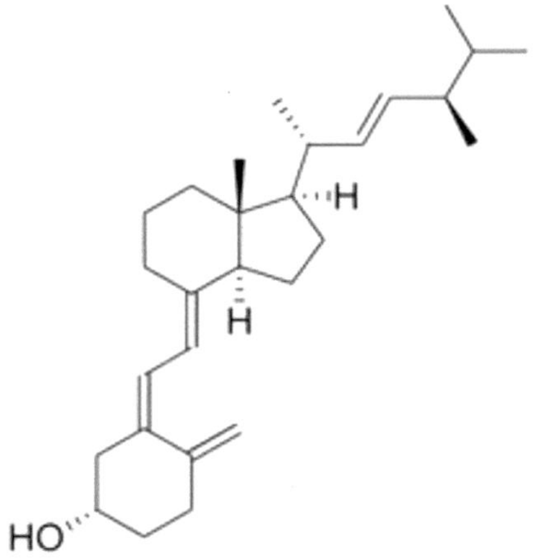

Fig. (16). Ergocalciferol, vitamin D_2, for vitamin-D resistant rickets, hypoparathyroidism and familial hypoposphatemia.

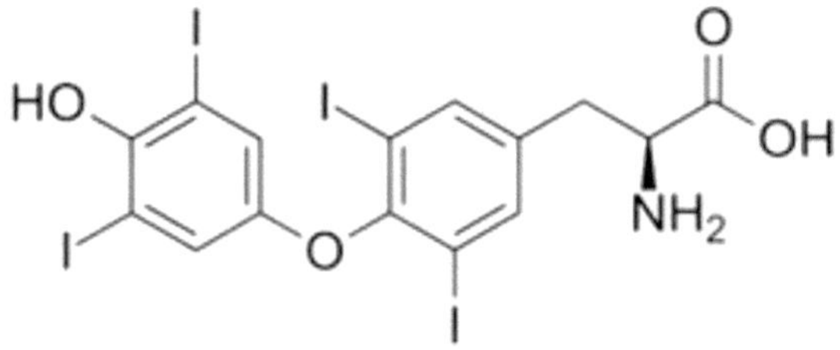

Fig. (17). Levothyroxine, L-thyroxine, for hypothyroidism, euthyroid goiters and Hashimoto's thyroiditis.

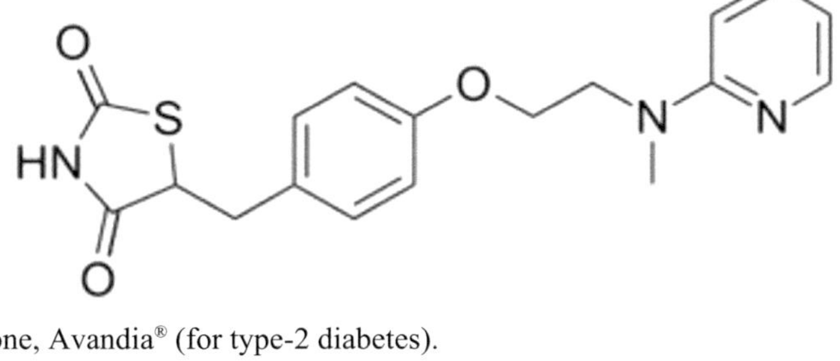

Fig. (18). Rosiglitazone, Avandia® (for type-2 diabetes).

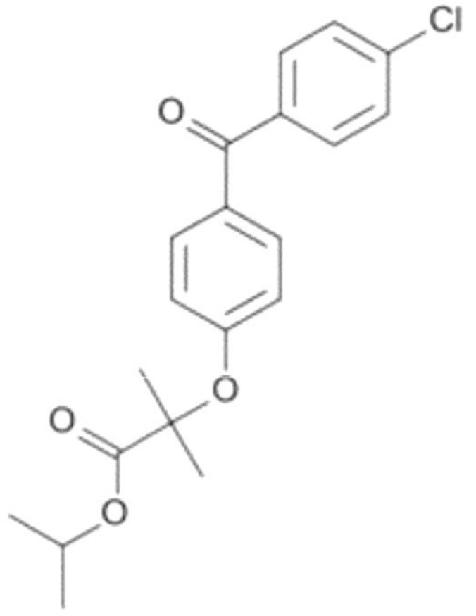

Fig. (19). Fenofibrate, for Types IIa, IIb, IV, and V hyperlipidemia.

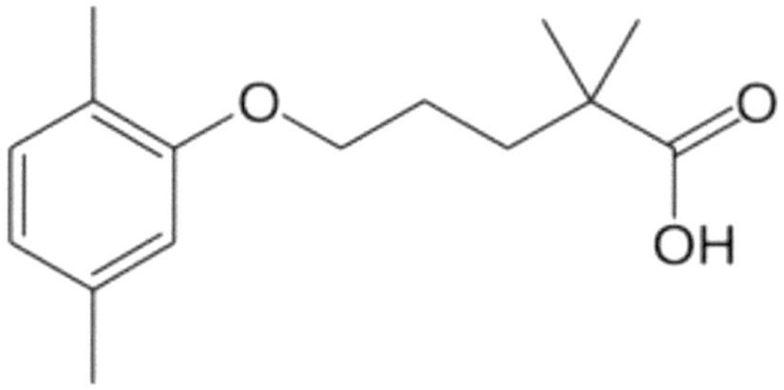

Fig. (20). Gemfibrozil, Lopid®, for dyslipidemia.

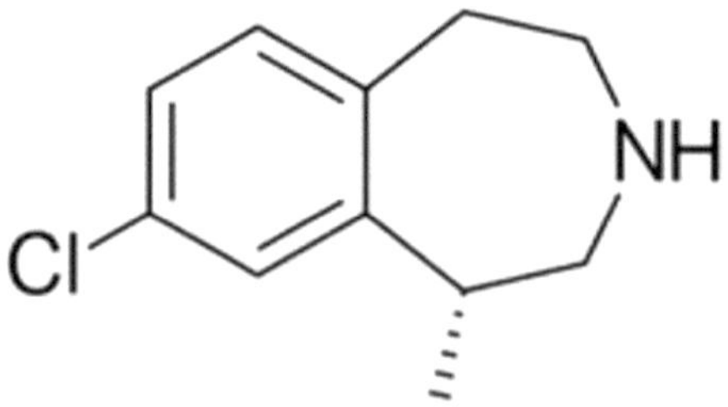

Fig. (21). Lorcaserin (Belviq®).

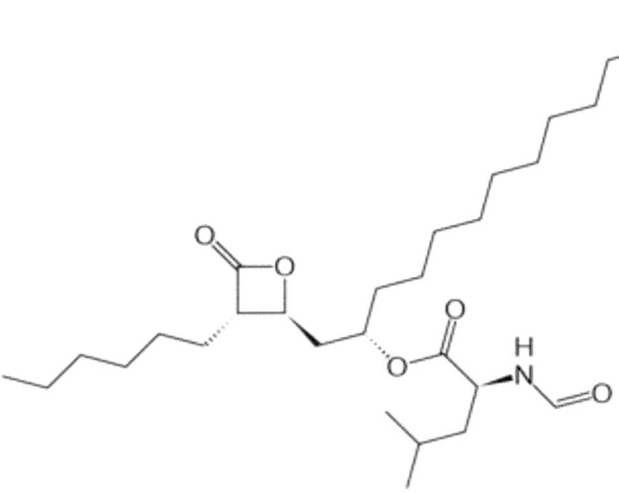

Fig. (22). Orlistat (Xenical®) is the hydrogenated derivative of lipstatin.

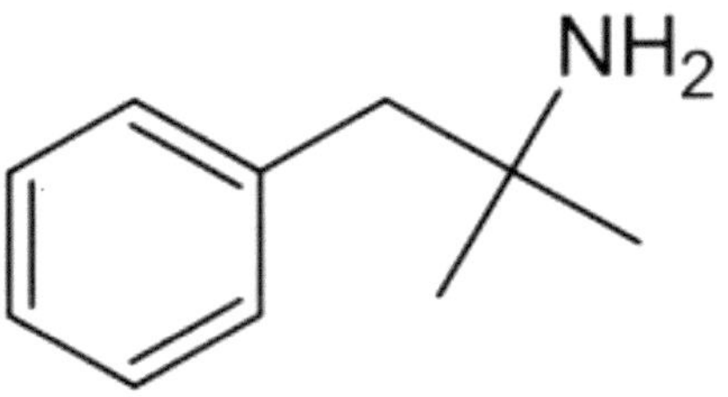

Fig. (23). Phentermine, a component of Qnexa®, also known as Qsymia™.

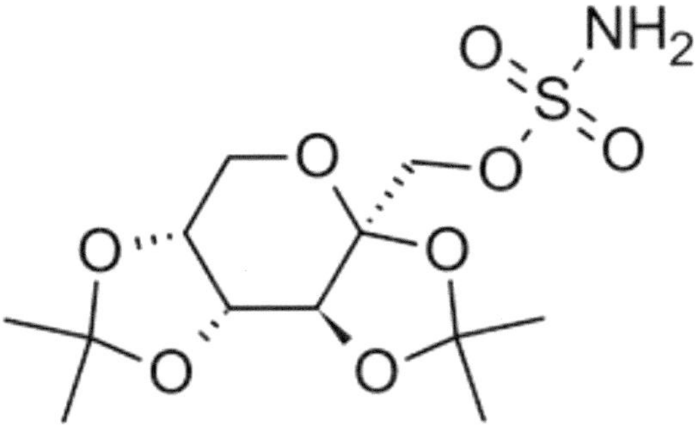

Fig. (24). Topiramate, a component of Qnexa®, also known as Qsymia™.

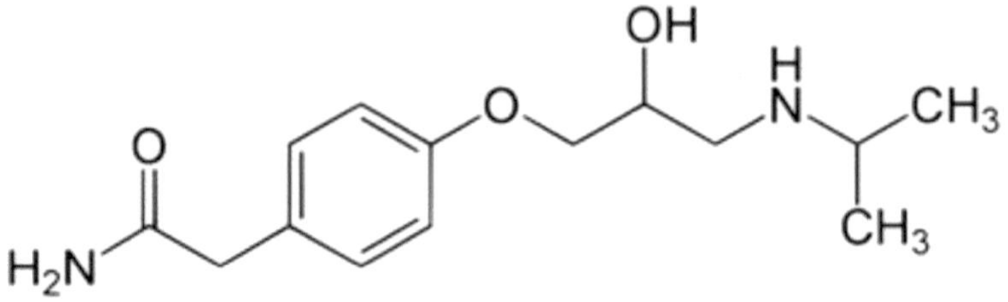

Fig. (25). Atenolol (for hypertension).

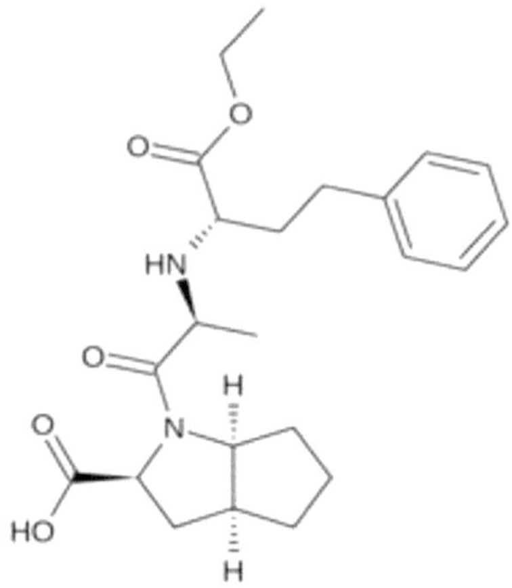

Fig. (26). Ramipril, an ACE inhibitor.

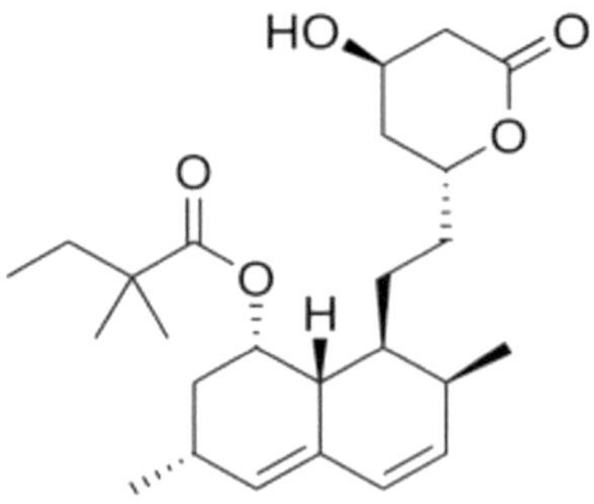

Fig. (27). Simvastatin, for dyslipidemia.

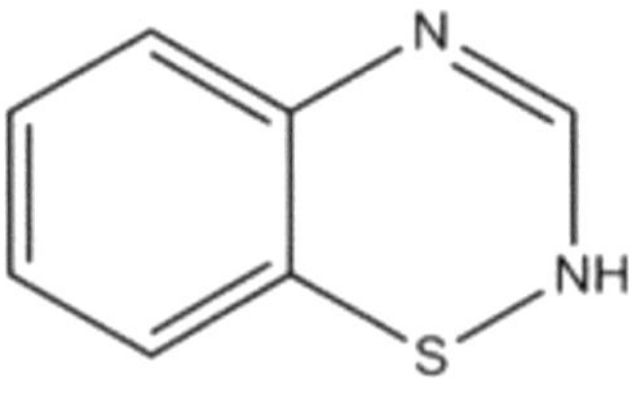

Fig. (28). Benzothiadiazine, a diuretic.

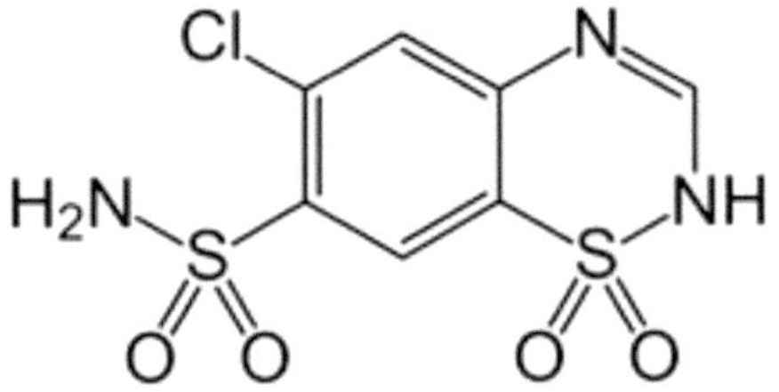

Fig. (29). Chlorothiazide, a diuretic.

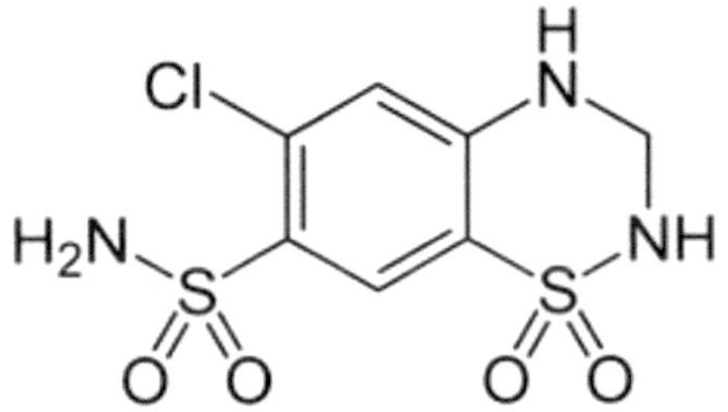

Fig. (30). Hydrochlorothiazide, a diuretic.

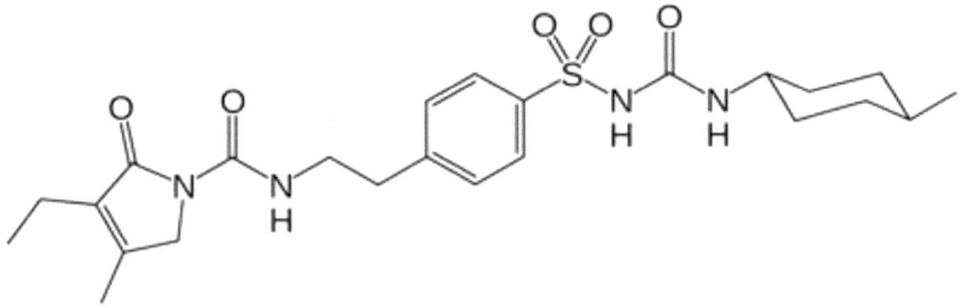

Fig. (31). Glimepiride, a sulfonylurea antidiabetic.

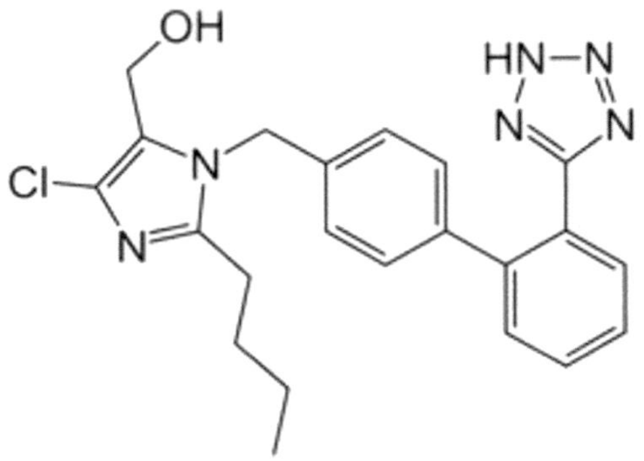

Fig. (32). Losartan, CYP substrate and antihypertensive angiotensin II receptor antagonist.

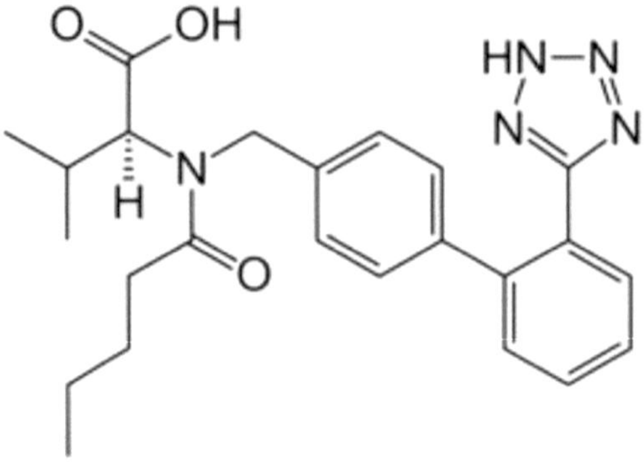

Fig. (33). Valsartan, CYP substrate and antihypertensive angiotensin II receptor antagonist.

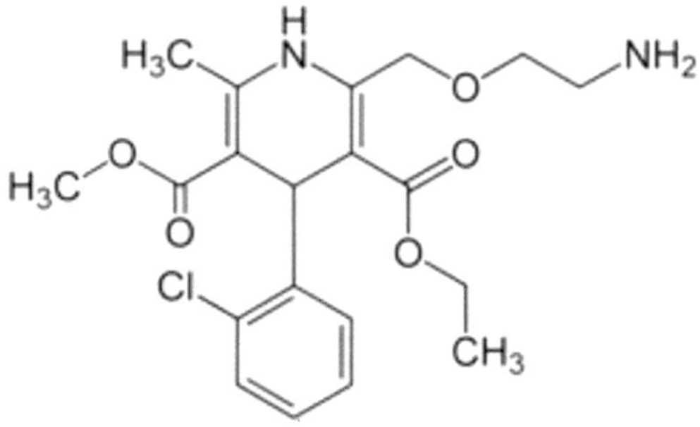

Fig. (34). Amlodipine, Calcium channel blocker.

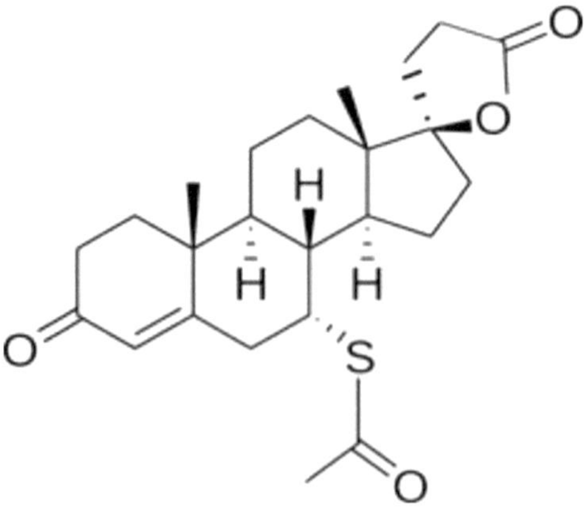

Fig. (35). Spironolactone, a diuretic.

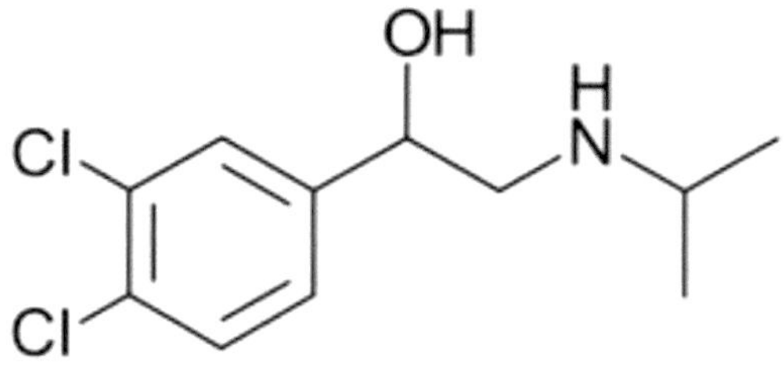

Fig. (36). Dichloroisoprenaline, a beta blocker.

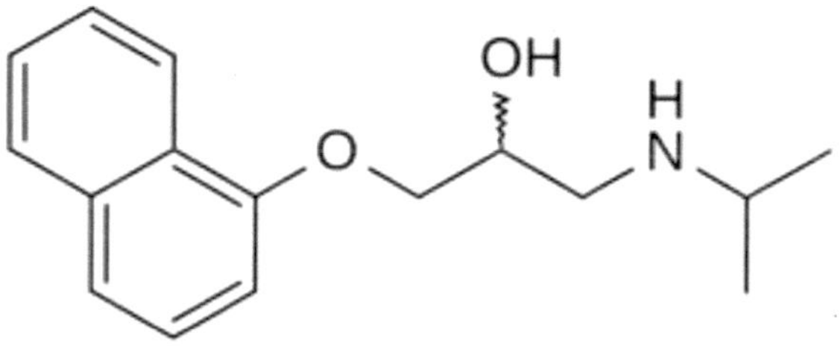

Fig. (37). Propranolol, CYP substrate and beta blocker.

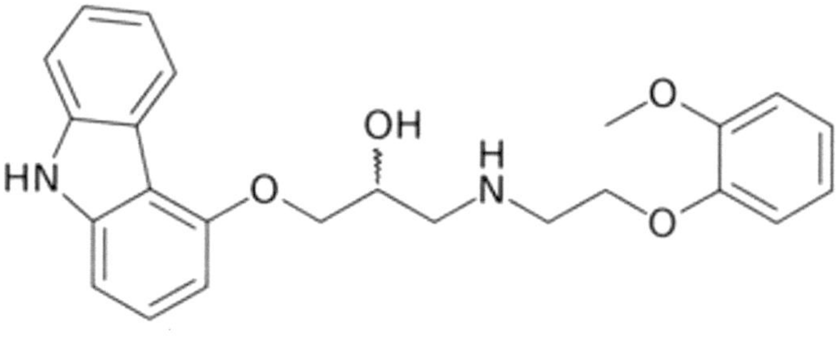

Fig. (38). Carvedilol, a beta blocker, for congestive heart failure.

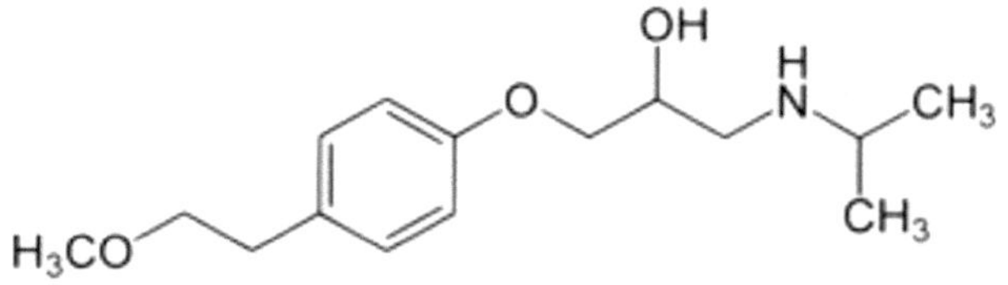

Fig. (39). Metoprolol, a beta blocker, for hypertension.

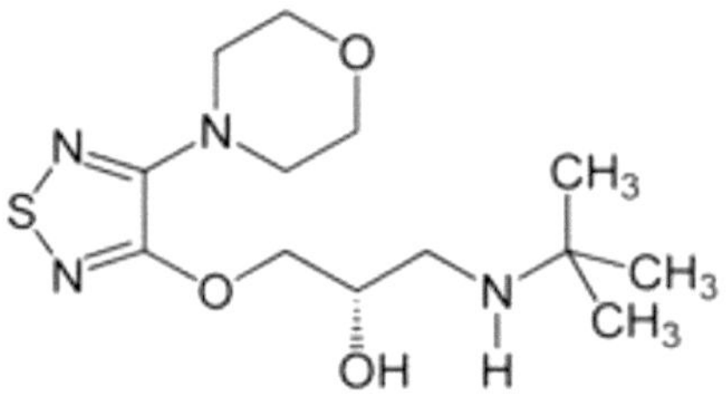

Fig. (40). Timolol, CYP substrate and beta blocker.

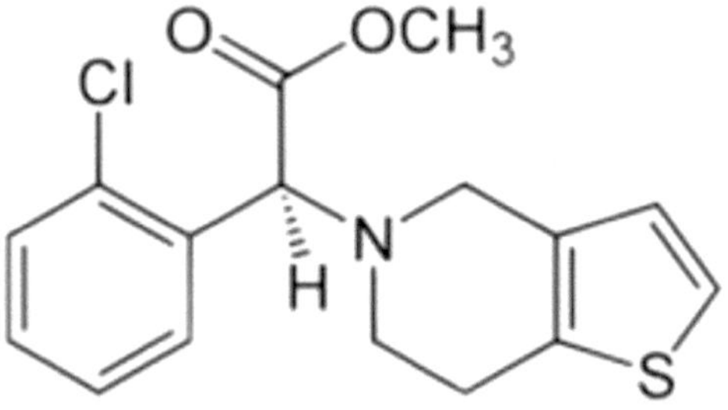

Fig. (41). Clopidogrel, inhibits formation of blood clots.

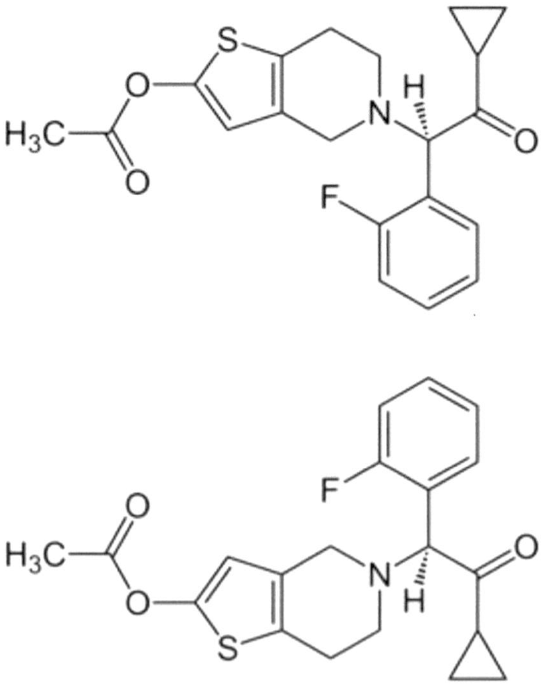

Fig. (42). Prasugrel, a platelet inhibitor, two enantiomers.

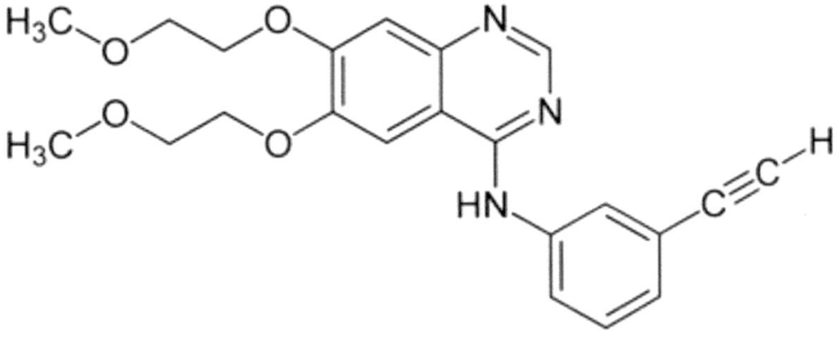

Fig. (43). Erlotinib, (Tarveca®), anticancer agent that inhibits tyrosine kinase.

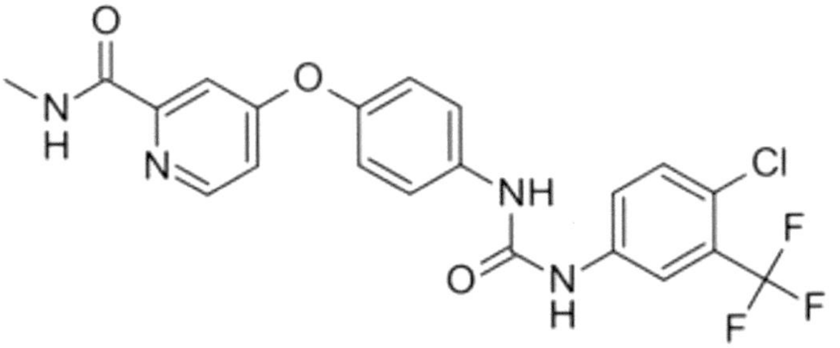

Fig. (44). Sorafenib (Nexavar®), anticancer agent that inhibits tyrosine kinase.

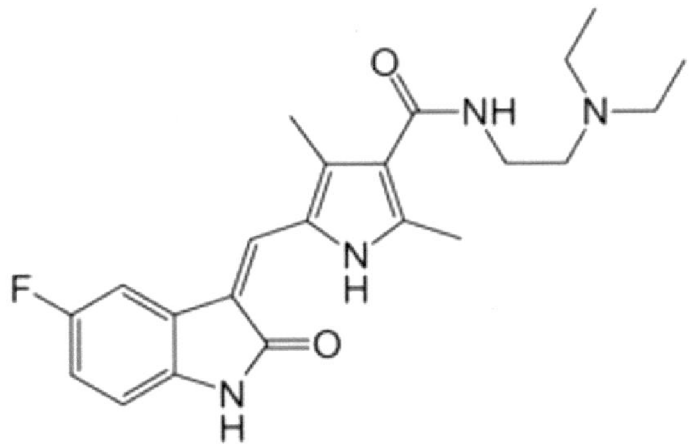

Fig. (45). Sunitinib (Sutent®), anticancer agent that inhibits tyrosine kinase.

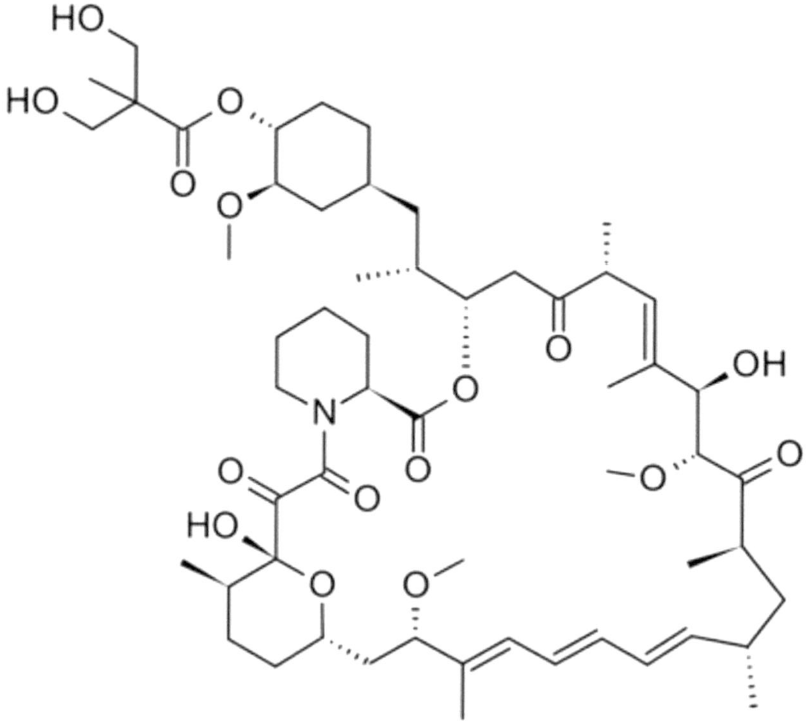

Fig. (46). Temsirolimus (Torise®) targets metastasis by inhibiting mTOR.

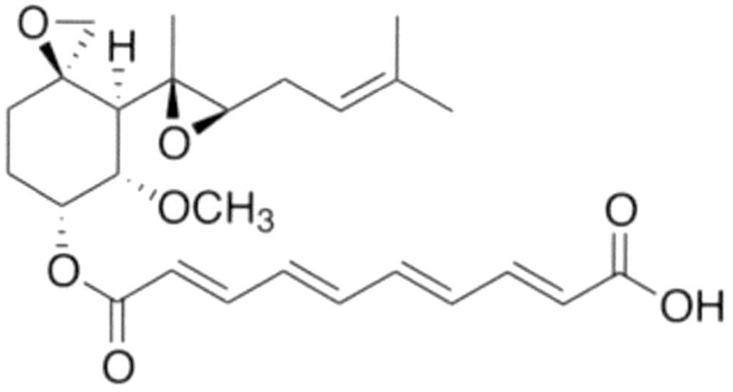

Fig. (47). Fumagillin, targets the vasculature.

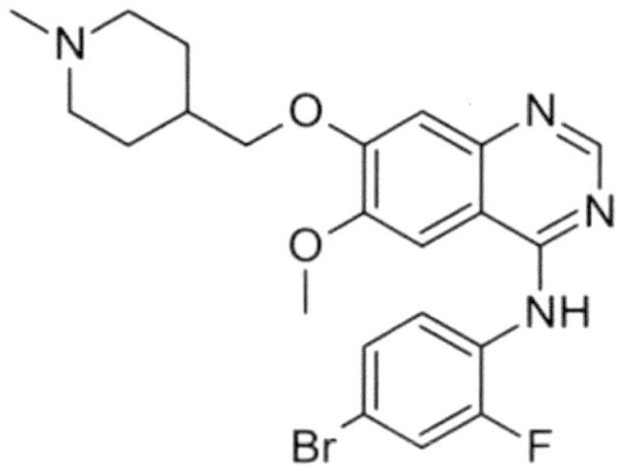

Fig. (48). Vandetanib, a tyrosine kinase inhibitor that targets the vasculature.

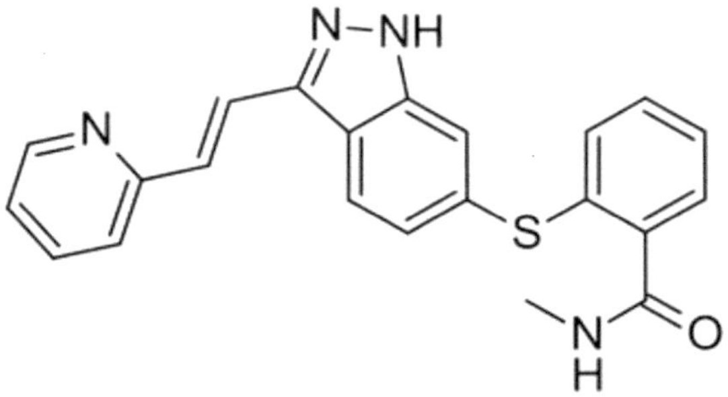

Fig. (49). Axitinib, a tyrosine kinase inhibitor that targets the vasculature.

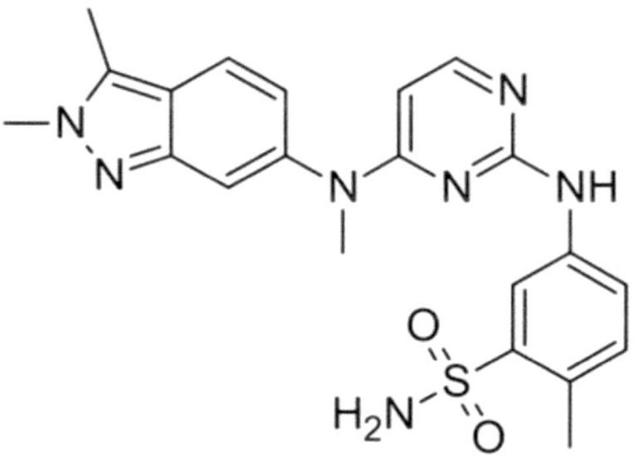

Fig. (50). Pazopanib, a tyrosine kinase inhibitor that targets the vasculature.

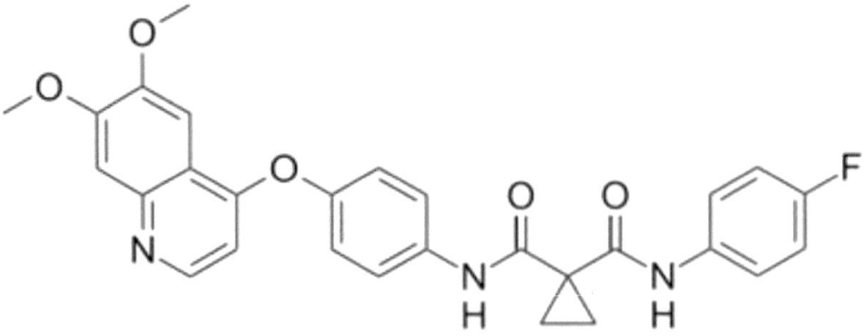

Fig. (51). Cabozantinib, a tyrosine kinase inhibitor that targets the vasculature.

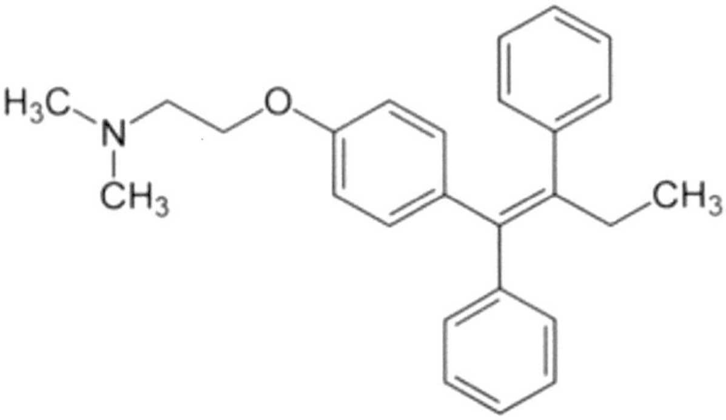

Fig. (52). Tamoxifen, an anticancer agent.

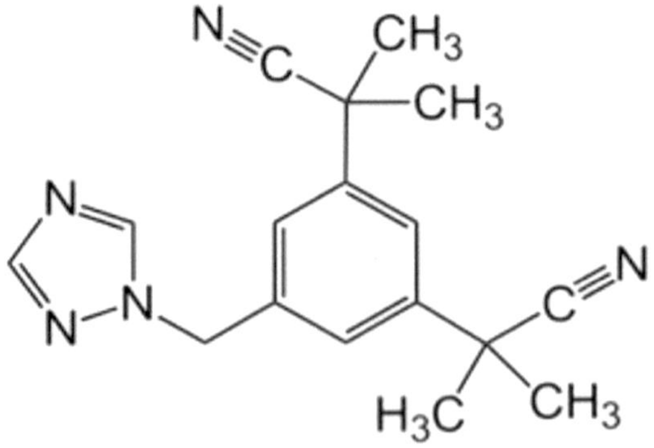

Fig. (53). Anastrozole, Arimidex®, a hormonal anticancer agent.

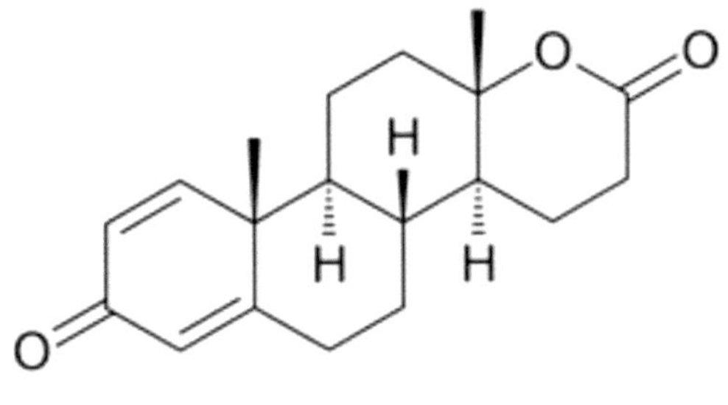

Fig. (54). Delta-1-testololactone, a hormonal anticancer agent.

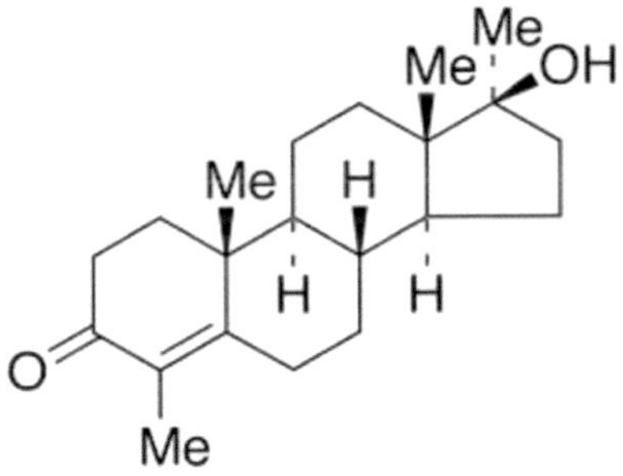

Fig. (55). Dimethyltestosterone, a hormonal anticancer agent.

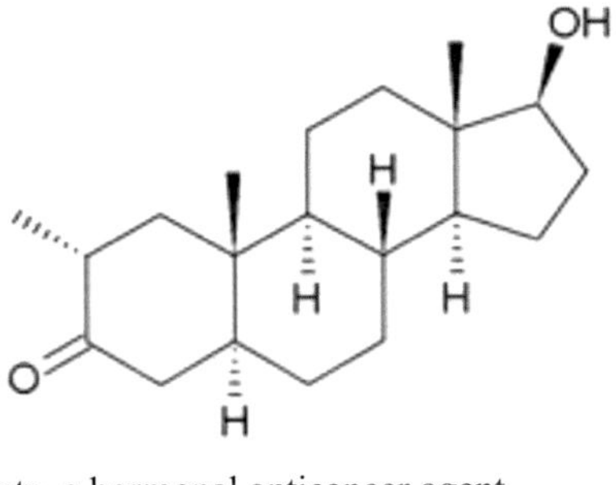

Fig. (56). Dromostanolone propionate, a hormonal anticancer agent.

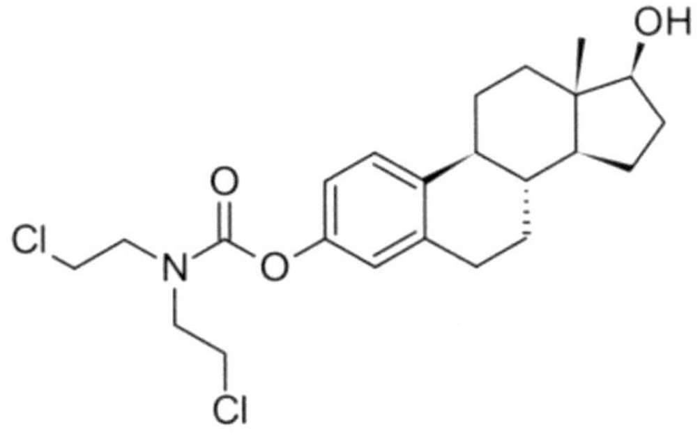

Fig. (57). Estramustine, a hormonal anticancer agent.

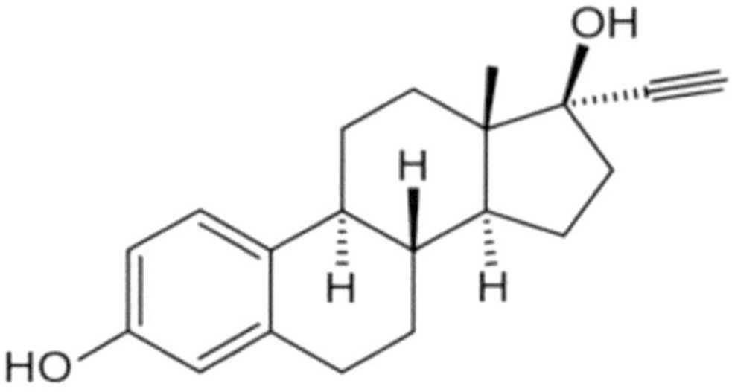

Fig. (58). Ethinyl estradiol, a hormonal contraceptive.

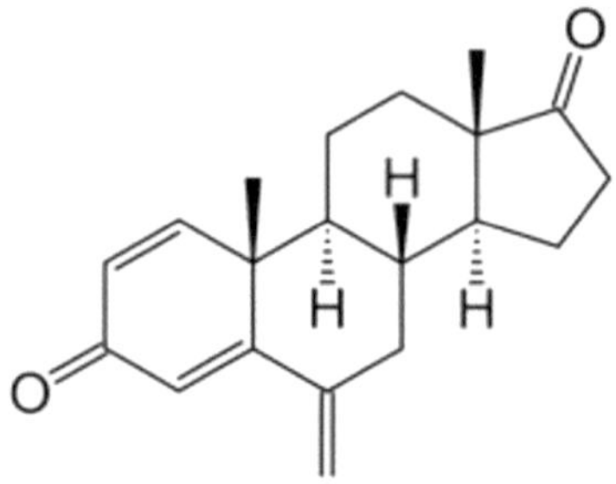

Fig. (59). Exemestane, an aromatase inhibitor.

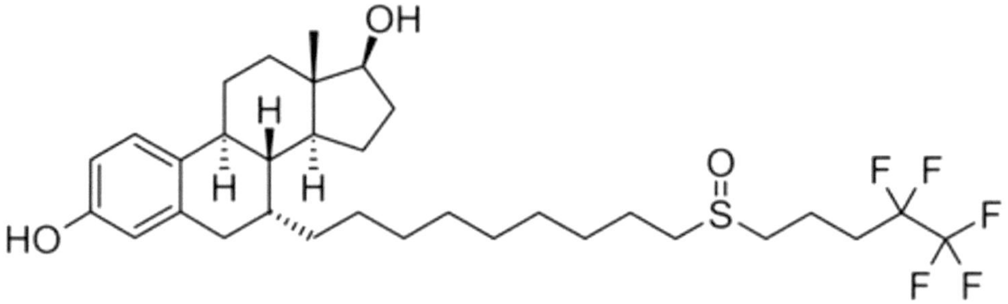

Fig. (60). Fulvestrant, an aromatase inhibitor.

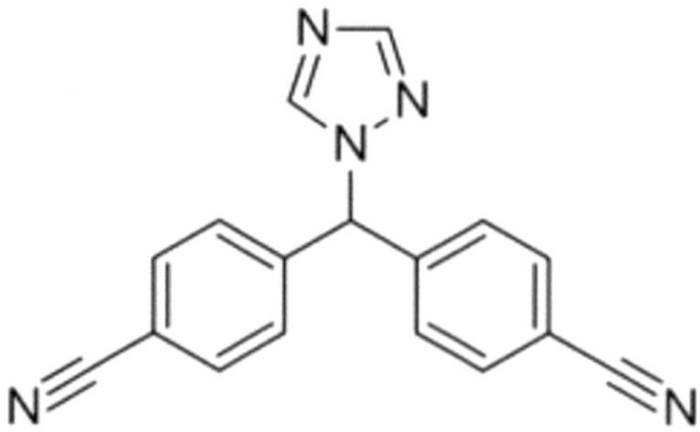

Fig. (61). Letrozole, an aromatase inhibitor.

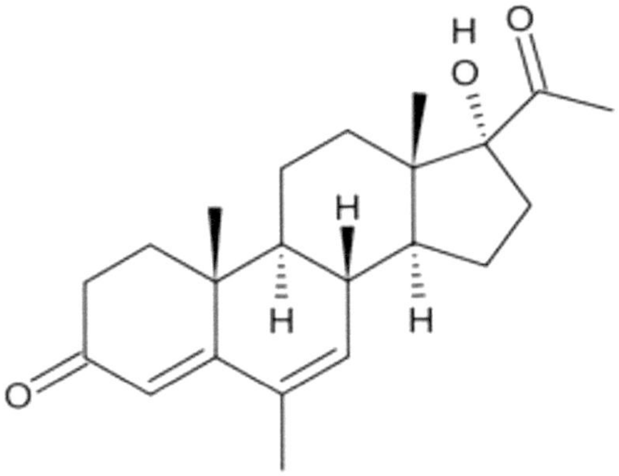

Fig. (62). Megestrol acetate, an aromatase inhibitor.

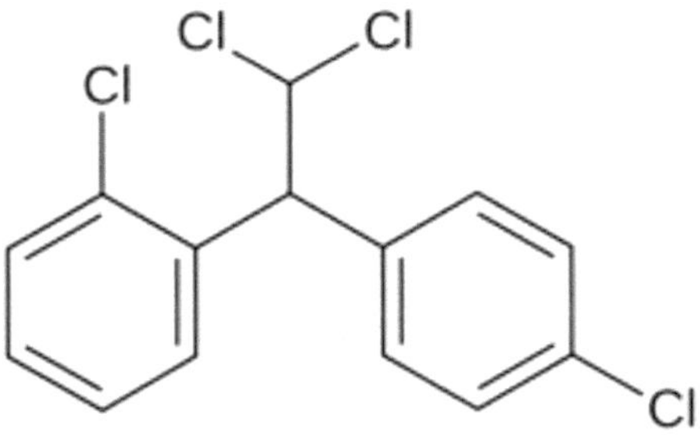

Fig. (63). Mitotane, an aromatase inhibitor.

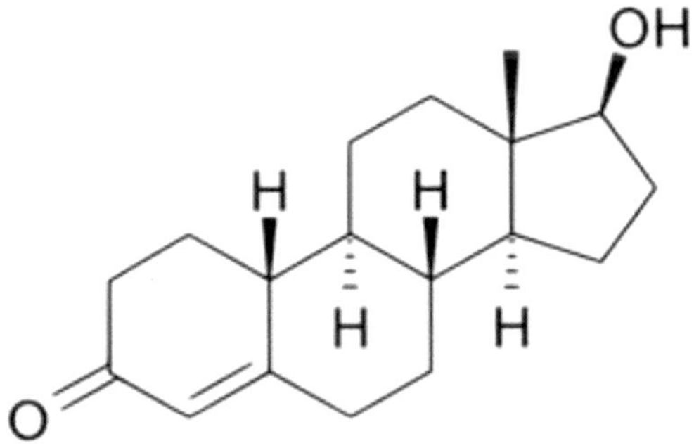

Fig. (64). Nandrolone, an aromatase inhibitor.

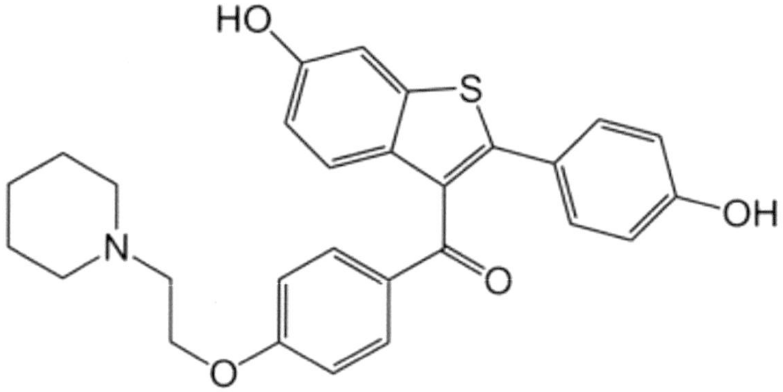

Fig. (65). Raloxifene, an aromatase inhibitor.

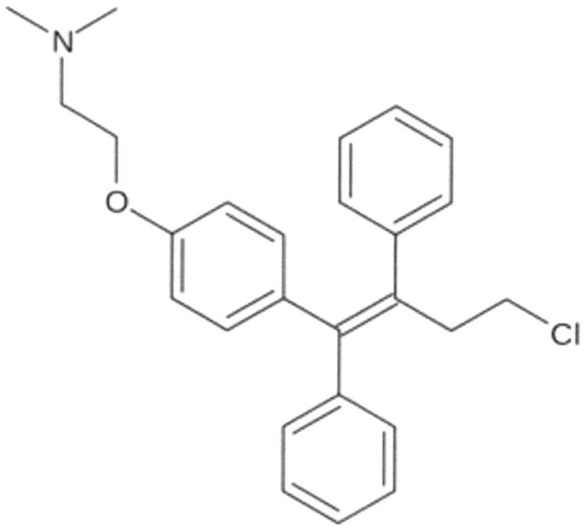

Fig. (66). Toremifene, a selective estrogen receptor modulator.

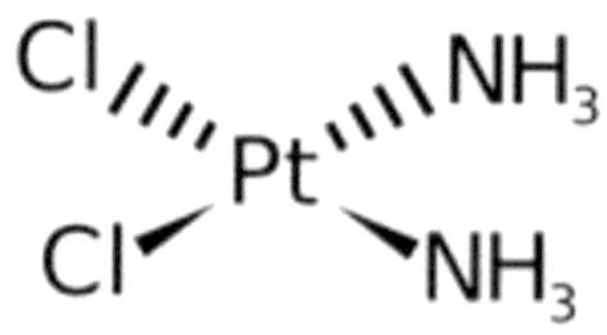

Fig. (67). Cis-platin, an anticancer drug.

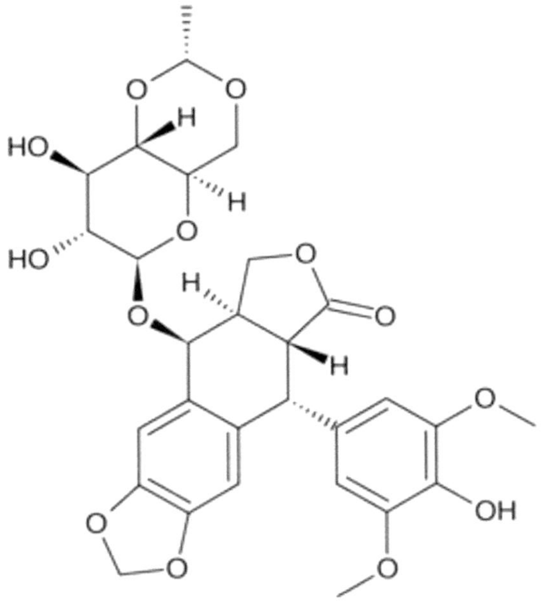

Fig. (68). Etoposide, a topoisomerase inhibitor and allo-network drug.

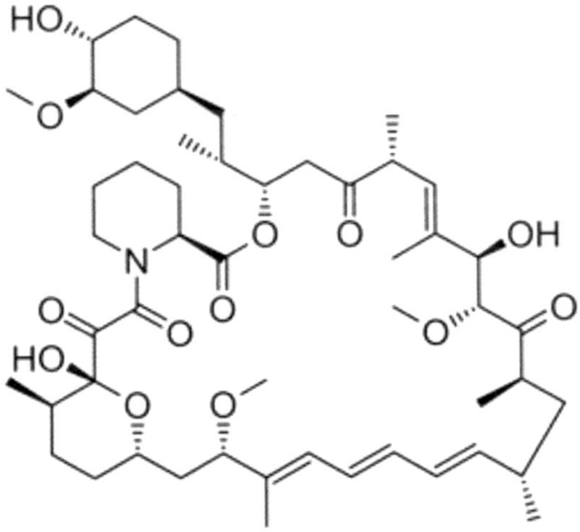

Fig. (69). Rapamycin (Sirolimus), an allo-network drug.

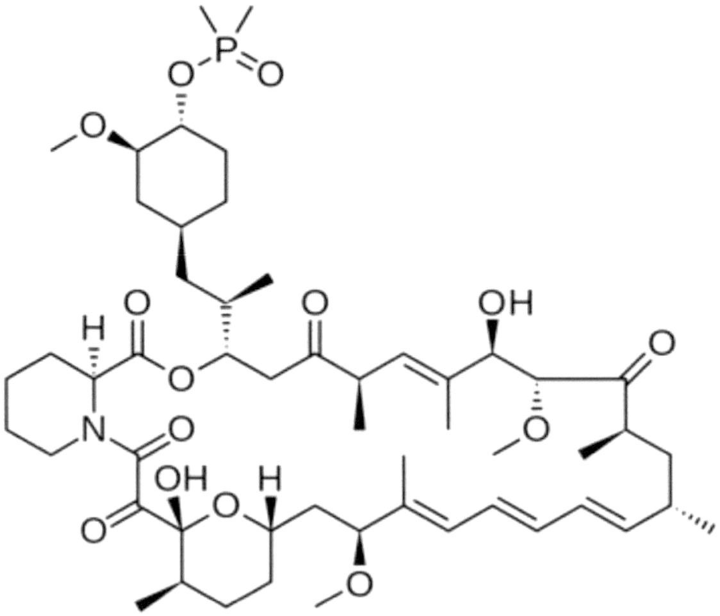

Fig. (70). Ridaforolimus, a rapalogs that inhibits the PI3K/Akt/mTOR pathway.

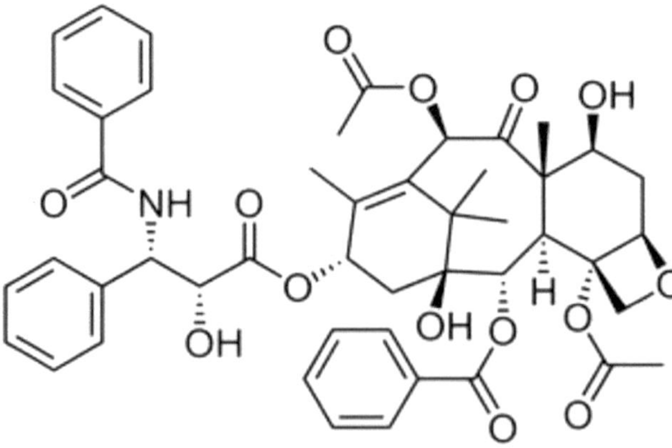

Fig. (71). Paclitaxel anticancer agent.

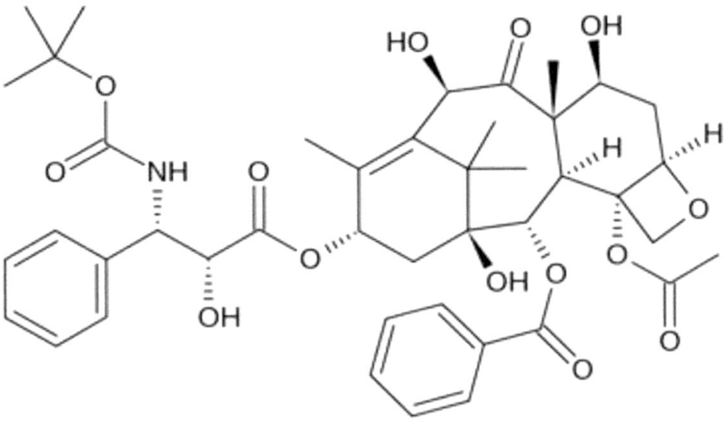

Fig. (72). Docetaxel, Taxotere®, a microtubule inhibitor that affects the immune system during cancer chemotherapy.

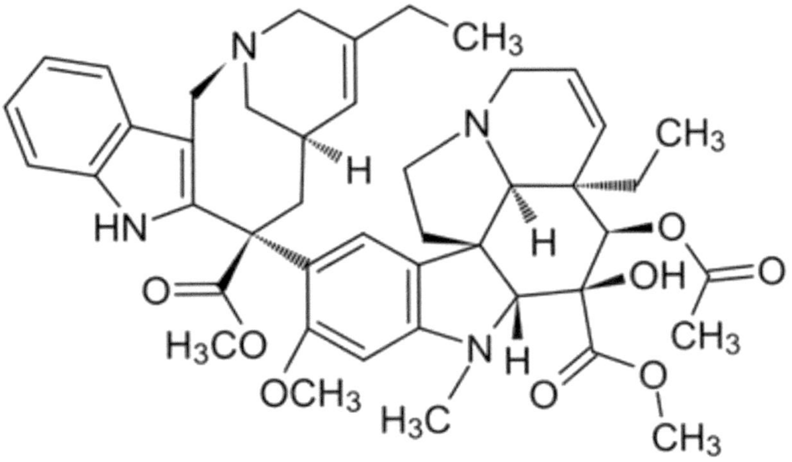

Fig. (73). Vinorelbine, anticancer.

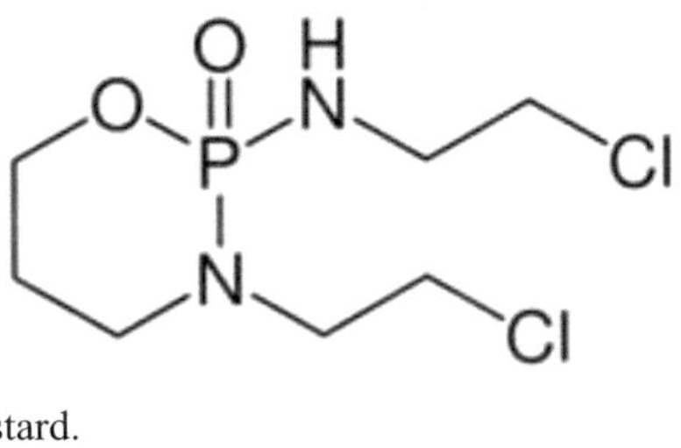

Fig. (74). Ifosfamide, nitrogen mustard.

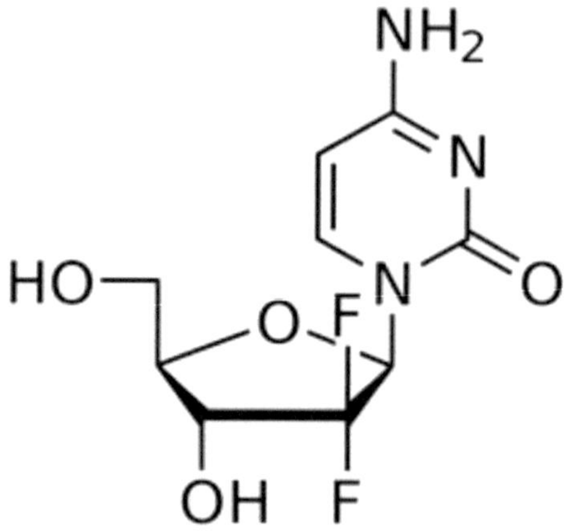

Fig. (75). Gemcitabine, Nucleoside analogue anticancer drug.

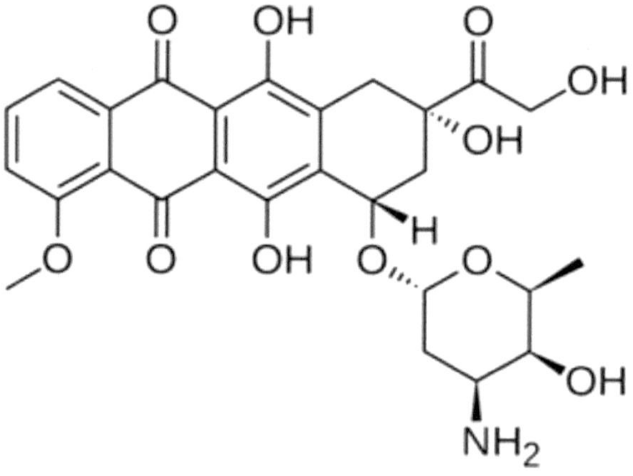

Fig. (76). Adriamycin (Doxorubicin) anticancer agent.

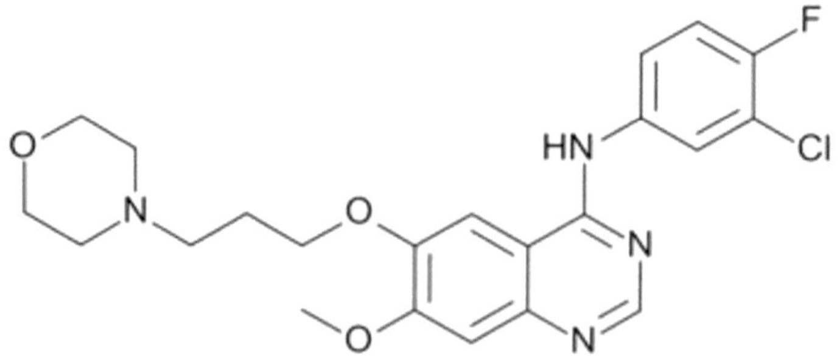

Fig. (77). Gefitinib, Iressa®, EGFR inhibitor and anticancer drug.

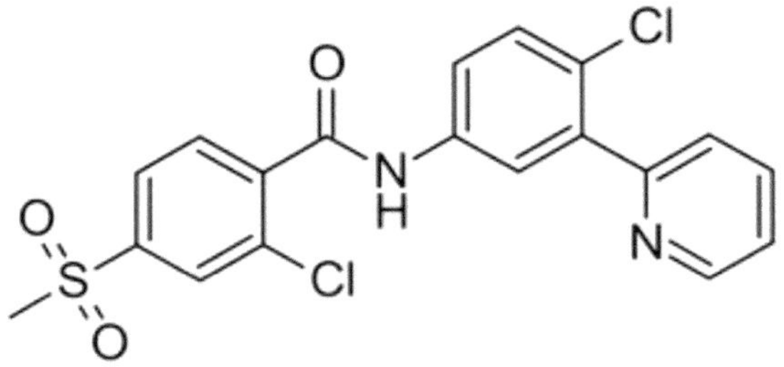

Fig. (78). Vismodegib or Erivedge®, anticancer drug.

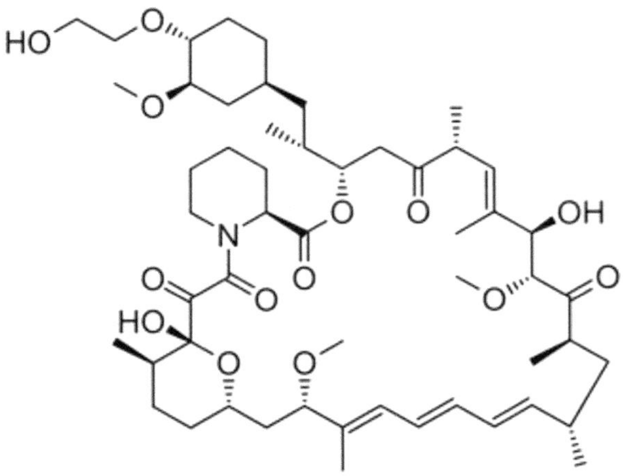

Fig. (79). Everolimus, Zortress®, an mTOR inhibitor.

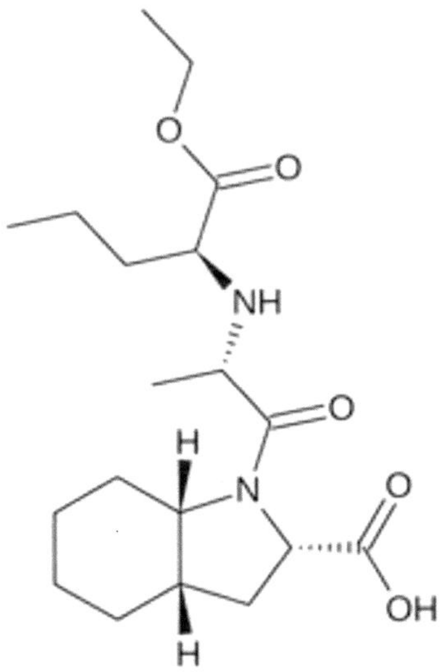

Fig. (80). Perindopril, an ACE inhibitor.

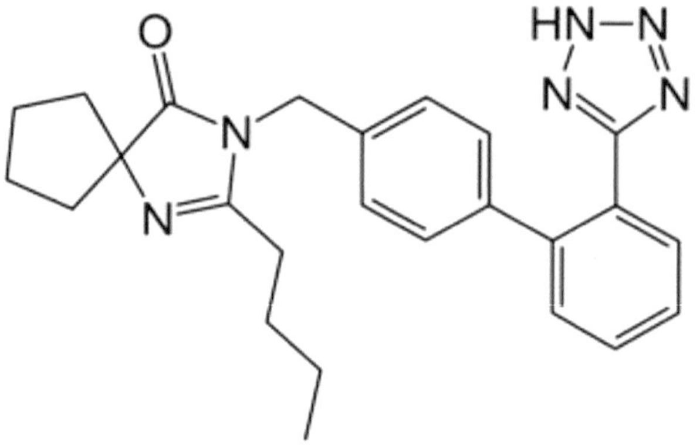

Fig. (81). Irbesartan, Aprovel®, an angiotensin II receptor antagonist.

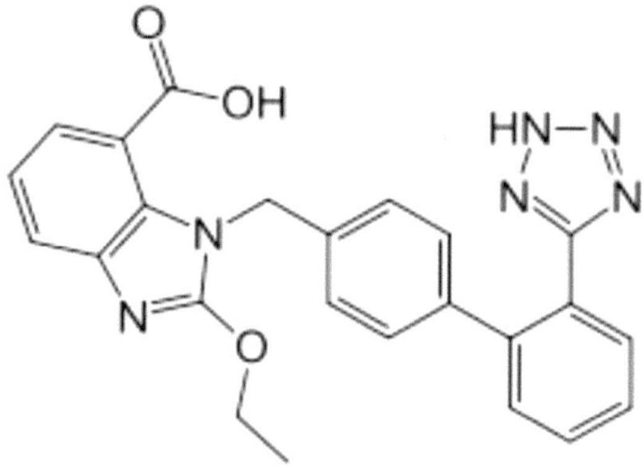

Fig. (82). Candesartan, Blopress®, an angiotensin II receptor antagonist.

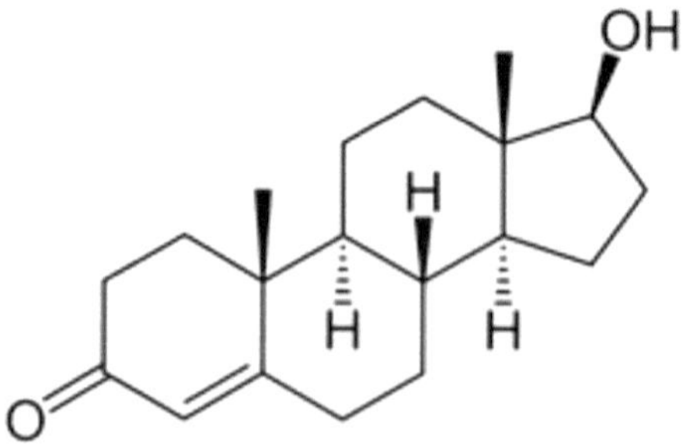

Fig. (83). Testosterone.

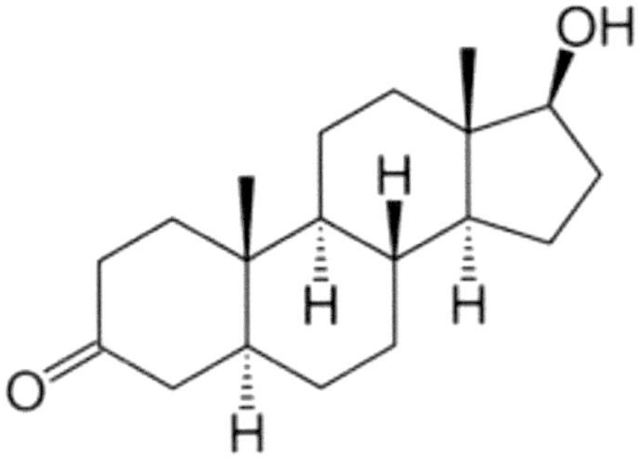

Fig. (84). Dihydrotestosterone (DHT), a sex steroid.

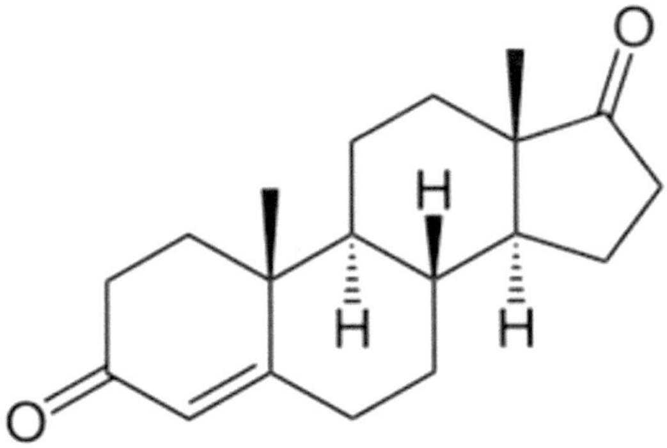

Fig. (85). Androstenedione, a steroid hormone.

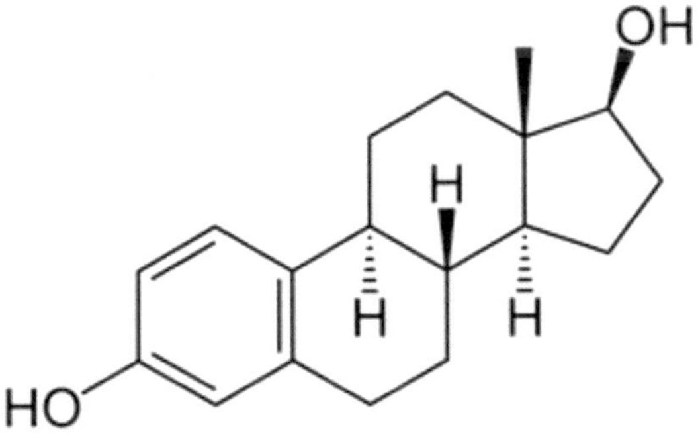

Fig. (86). 17β-estradiol, for vasomotor symptoms associated with menopause.

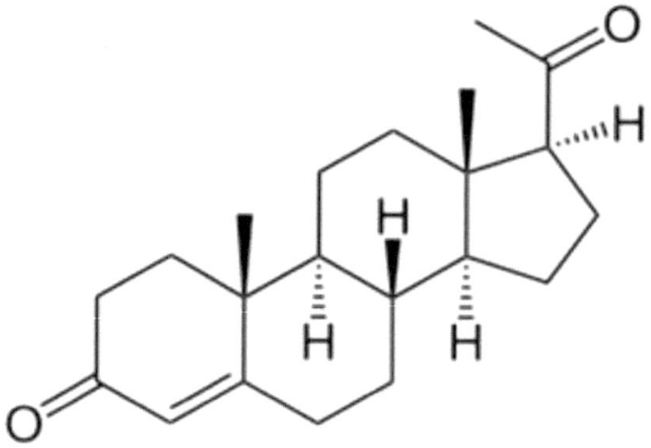

Fig. (87). Progesterone, a sex steroid hormone.

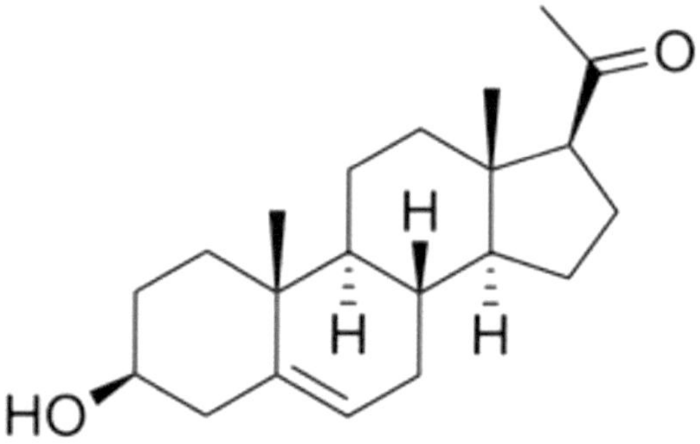

Fig. (88). Pregnenolone, a sex steroid hormone.

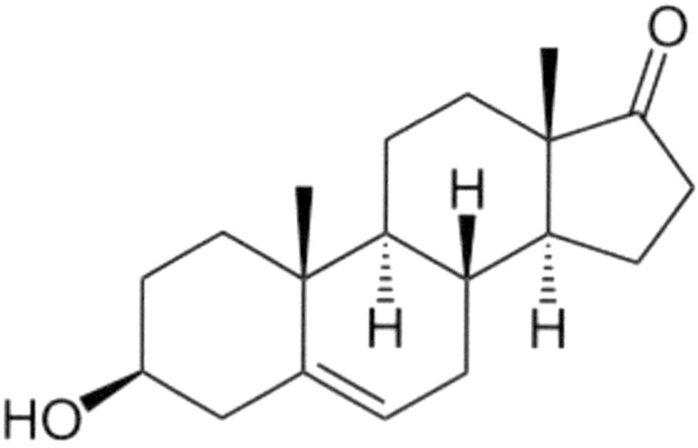

Fig. (89). Dehydroepiandrosterone, a sex steroid hormone.

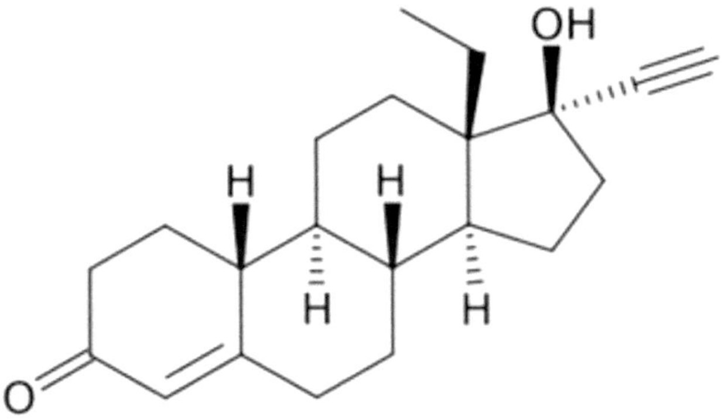

Fig. (90). Levonorgestrel, or "plan B" is the preferred emergency contraceptive. It can prevent a pregnancy up to 72 hrs after unprotected sex or contraceptive failure.

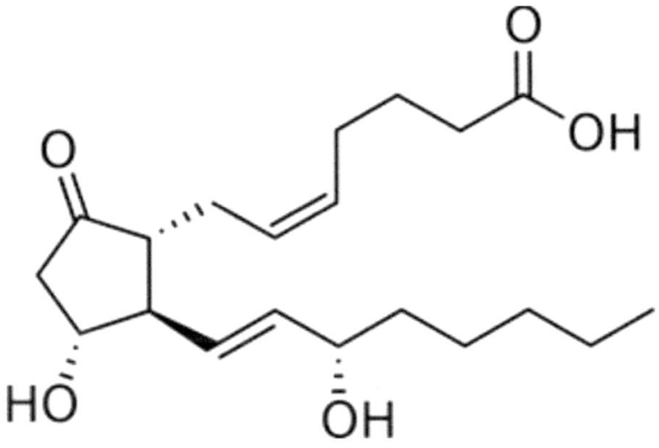

Fig. (91). Prostaglandin E_2 (PGE_2) (dinoprostone) is approved for the induction of labor.

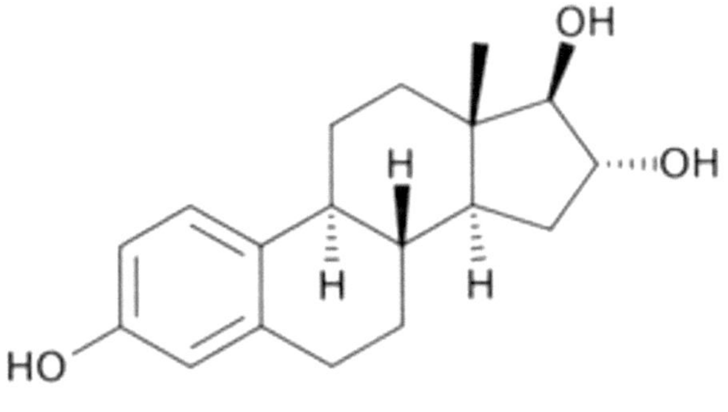

Fig. (92). Estriol, a sex steroid hormone.

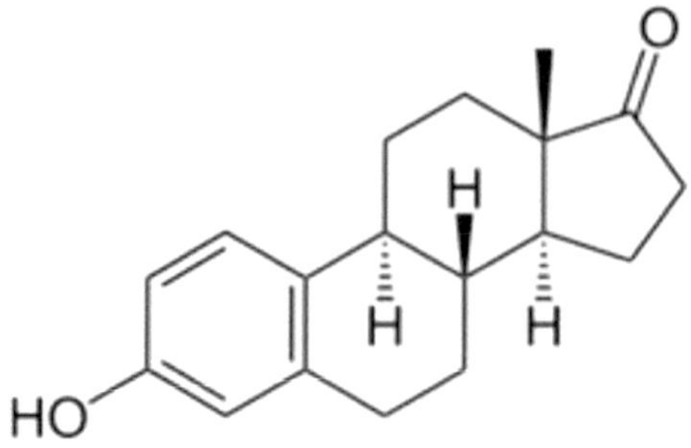

Fig. (93). Estrone, a sex steroid hormone.

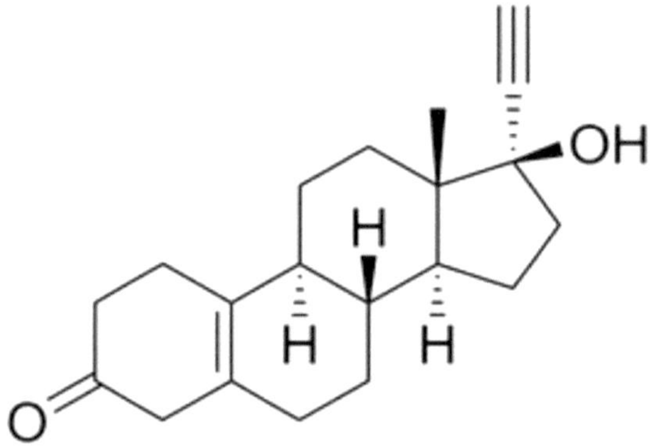

Fig. (94). Norethynodrel, a steroidal progestin.

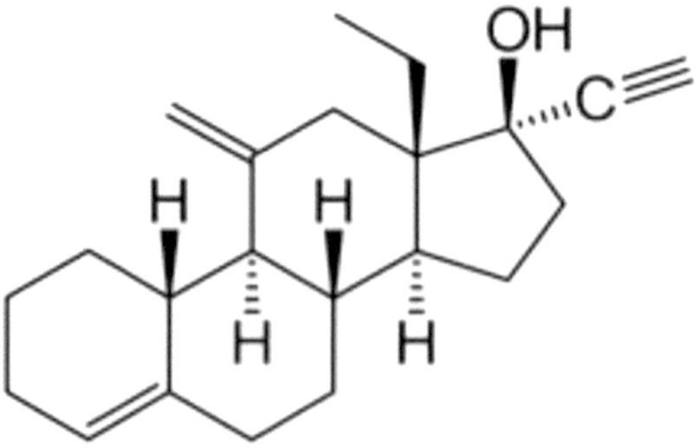

Fig. (95). Desogestrel (DSG) is used in third generation oral contraceptives.

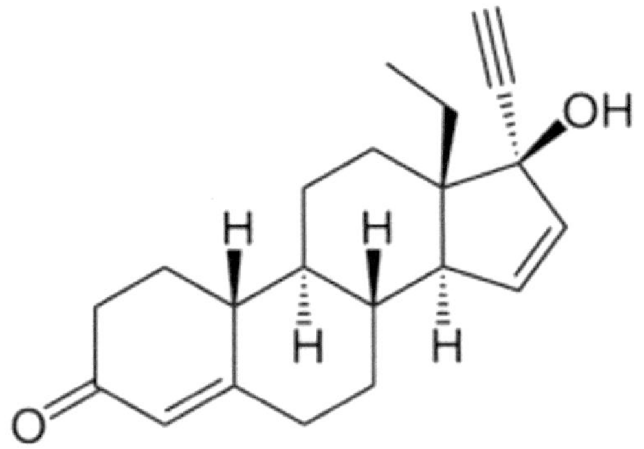

Fig. (96). Gestodene (GES) is used in third generation oral contraceptives.

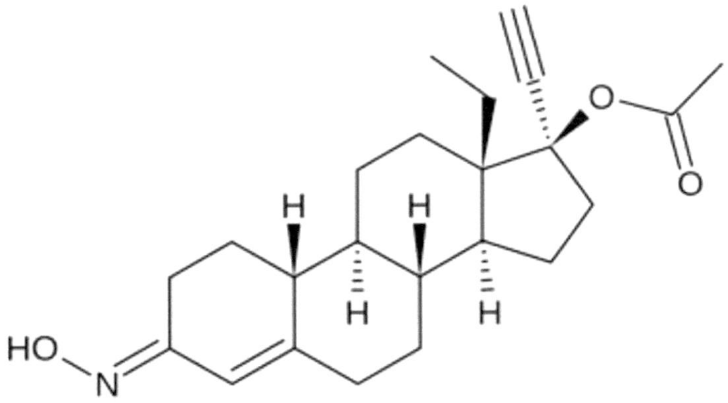

Fig. (97). Norgestimate is a progestin that is used in third generation oral contraceptives.

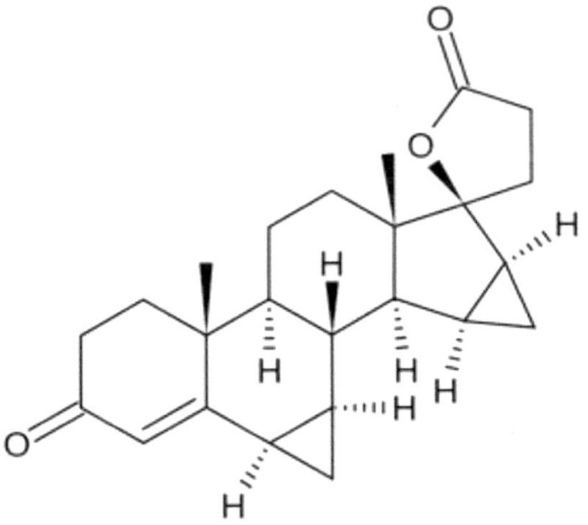

Fig. (98). Drospirenone is a synthetic derivative of progesterone with anti-mineralocorticoid and pro-estrogenic properties. It is used in the oral contraceptive called Yasmin.

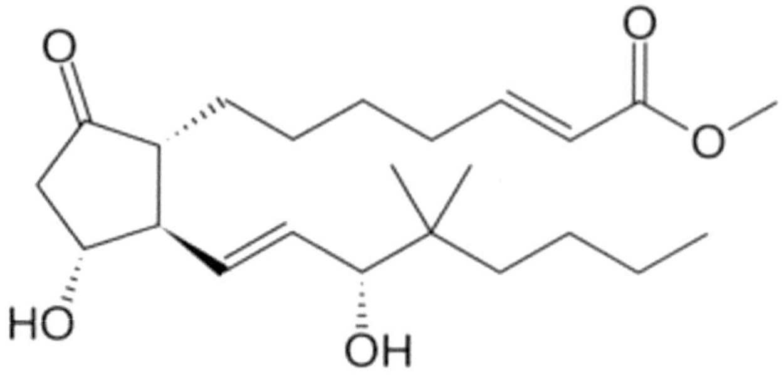

Fig. (99). Gemeprost can terminate pregnancies between weeks 13 and 24. It is also used to induce labor in cases of fetal death. For many years, it was known as RU-486.

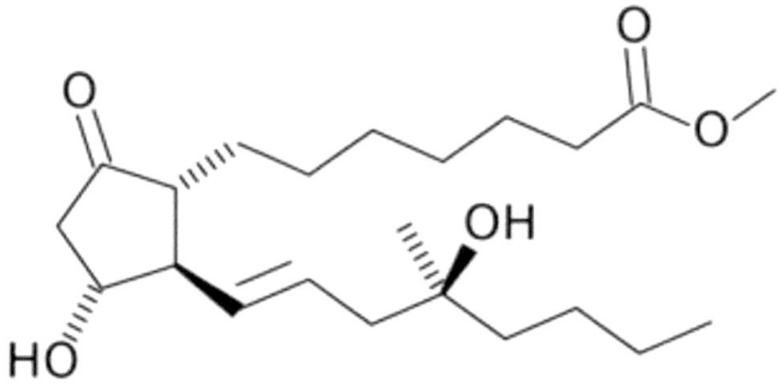

Fig. (100). Misoprostol is a synthetic PGE_1 derivative that can also induce labor, but it is used in the USA mainly for preventing gastric ulcers that can be caused by non-steroid anti-inflammatory drugs (NSAIDs).

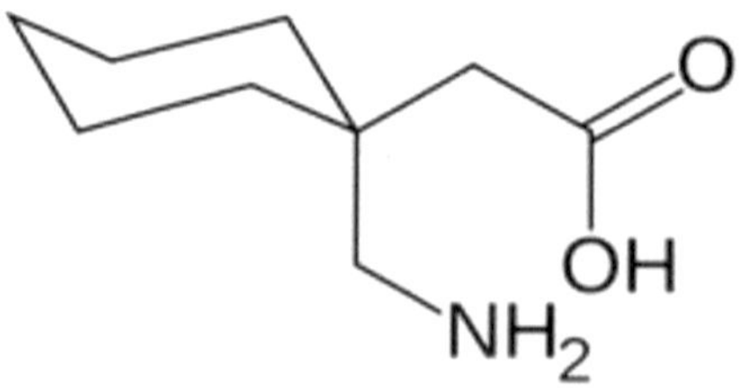

Fig. (101). Gabapentin (Neurontin®) is a pain killer that binds to the $\alpha_2\delta$ subunit of a multiprotein complex that regulates a specific Ca^{+2} ion channel. Also, gabapentin is the only non-hormonal product shown to be as effective as low dose estrogen for treating hot flashes.

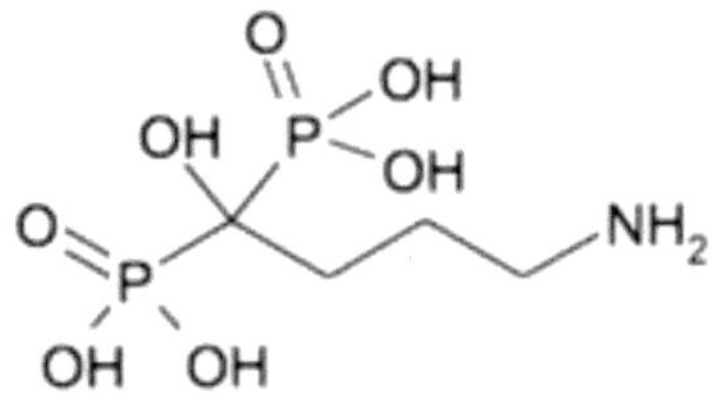

Fig. (102). Alendronate (Fosamax®), a bisphosphonate for osteoporosis.

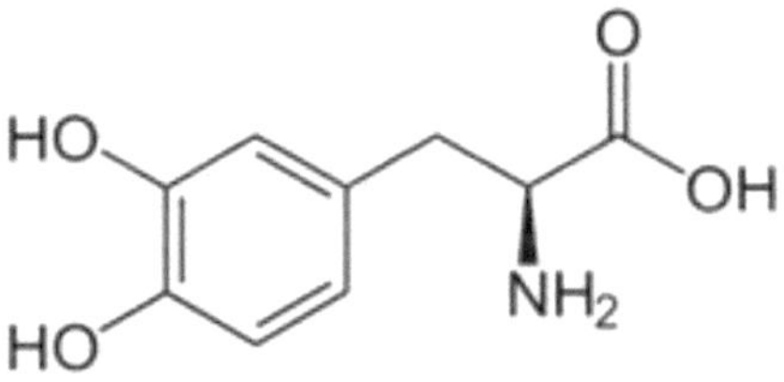

Fig. (103). L-DOPA Parkinson's disease.

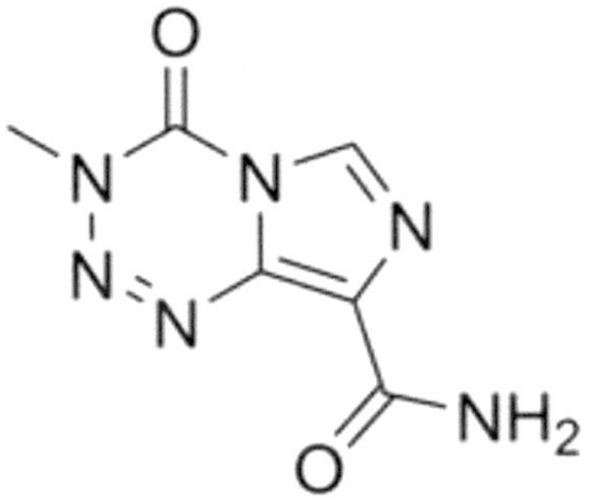

Fig. (104). Temozolomide is an alkylating agent that is used for glioblastoma multiforme.

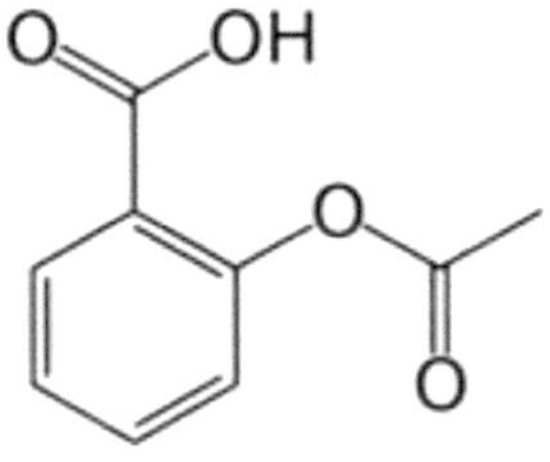

Fig. (105). Acetylsalicylic acid (NSAID).

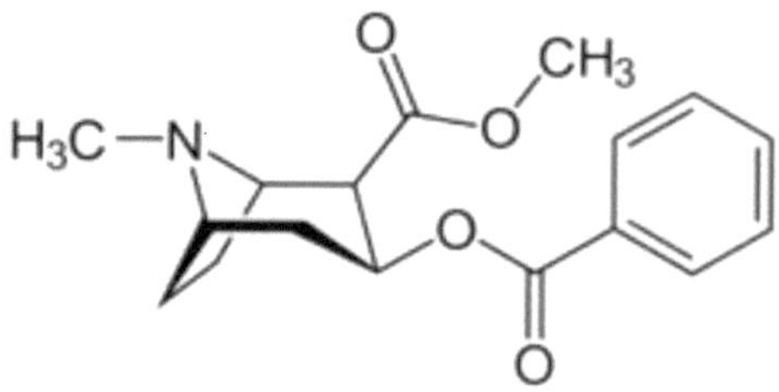

Fig. (106). Cocaine.

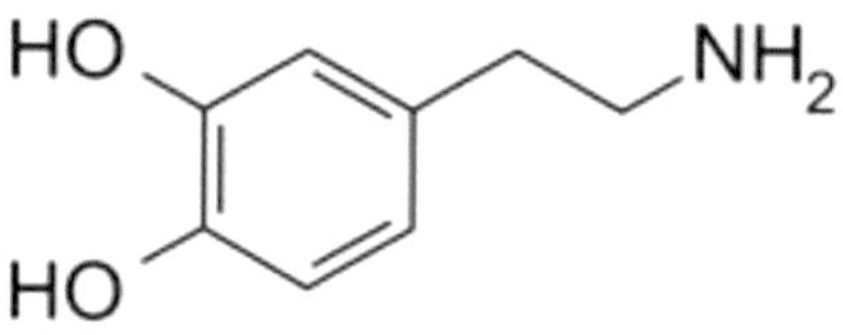

Fig. (107). Dopamine.

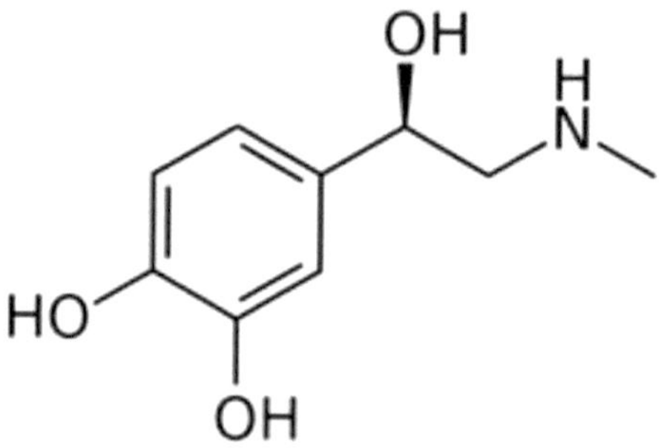

Fig. (108). Adrenaline (Epinephrine).

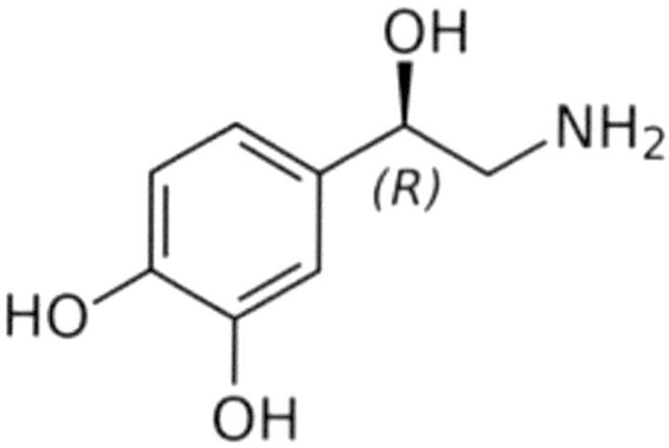

Fig. (109). Noradrenaline (Norepinephrine).

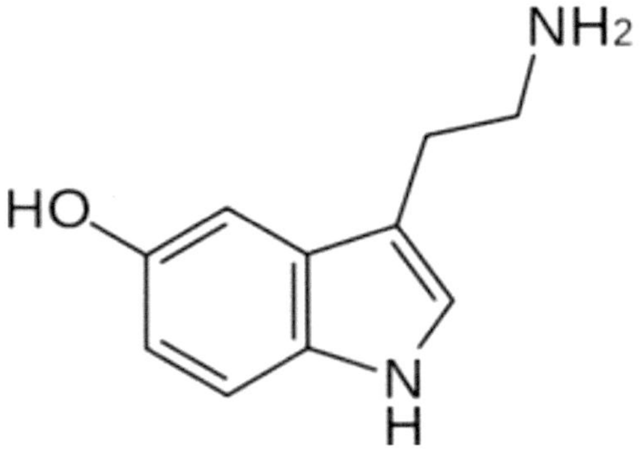

Fig. (110). Serotonin, a neurotransmitter.

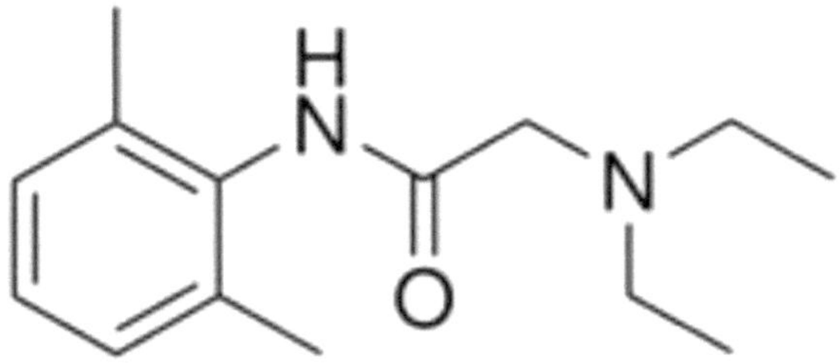

Fig. (111). Lidocaine, a local anesthetic.

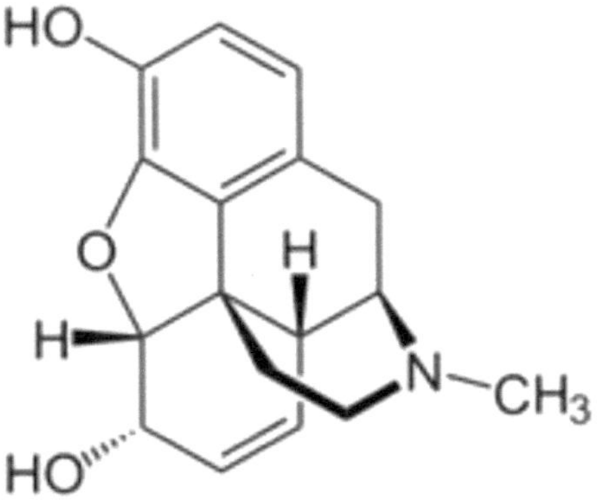

Fig. (112). Morphine, an anesthetic.

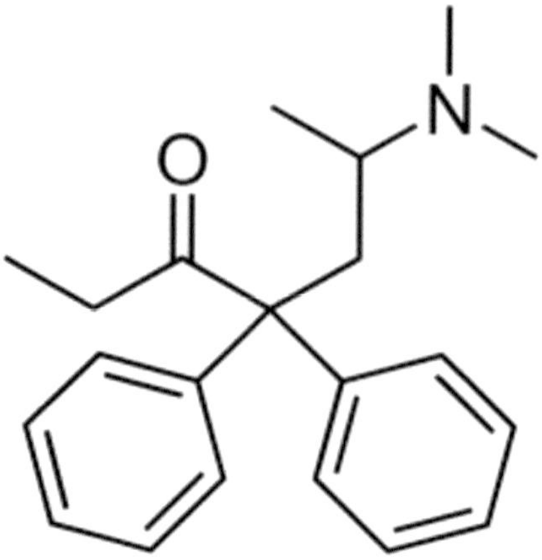

Fig. (113). Methadone, a synthetic opioid.

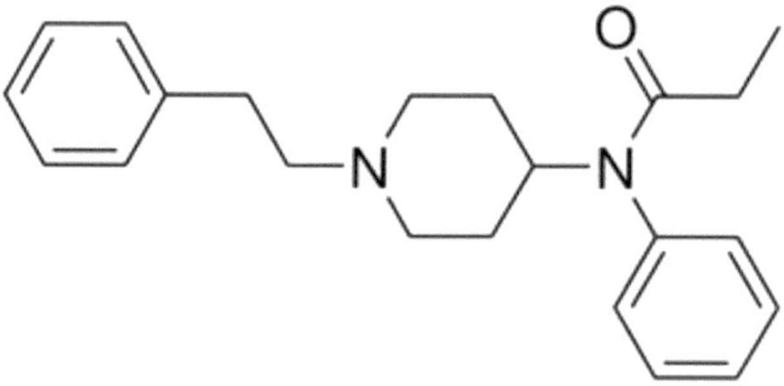

Fig. (114). Fentanyl (Duragesic®) also binds to opioid receptors.

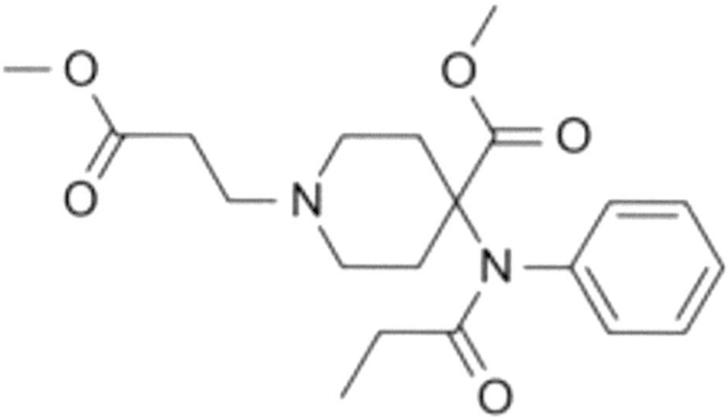

Fig. (115). Remifentanil, is the shortest acting opioid (5-10 min) available for human use.

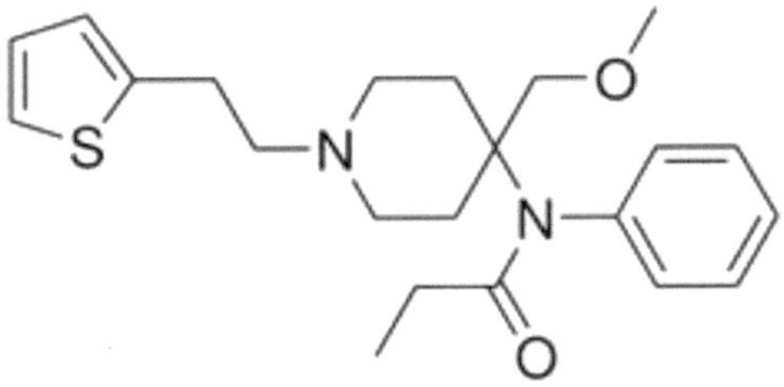

Fig. (116). Sufentanil is 1000 times as potent as morphine and is used in heart surgery.

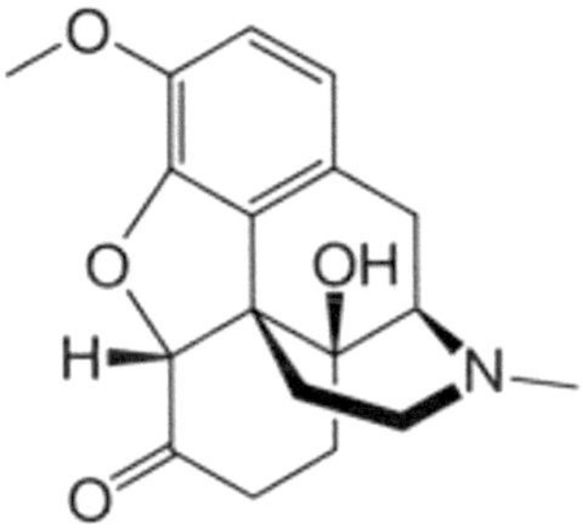

Fig. (117). Oxycodone is a semi-synthetic opioid.

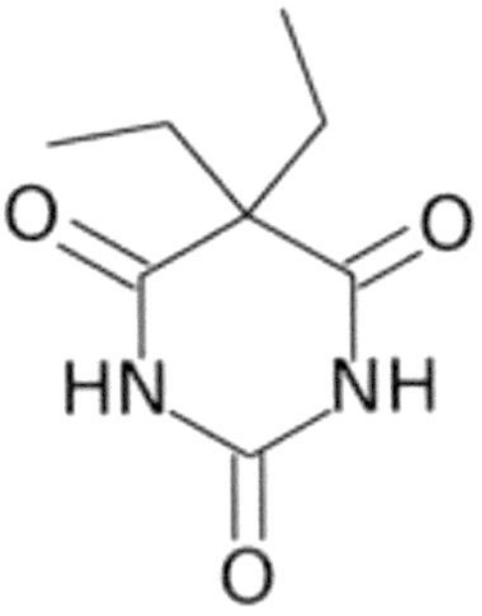

Fig. (118). Barbital, a barbituate It was marketed in 1904 by Bayer as a sleeping aid called Veronal.

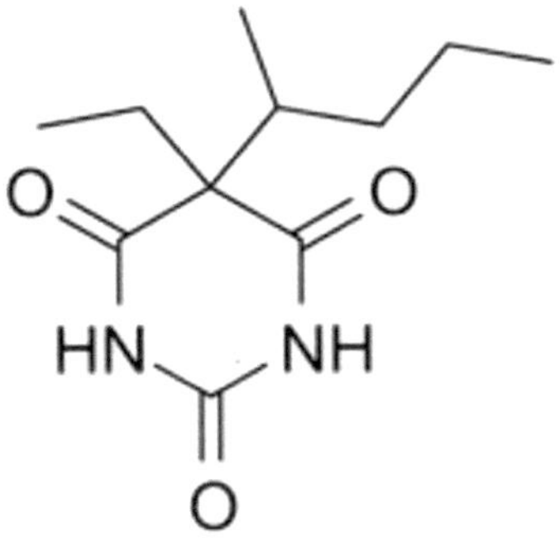

Fig. (119). Pentobarbital, sold under the name Nembutal. Later, an oxygen was replaced by a sulfur to make thiopental, a very short-acting barbiturate.

Fig. (120). Thiopental. An oxygen was replaced by a sulfur to make thiopental, a very short-acting barbiturate. However, propofol has replaced it, because patient's recover from it much better.

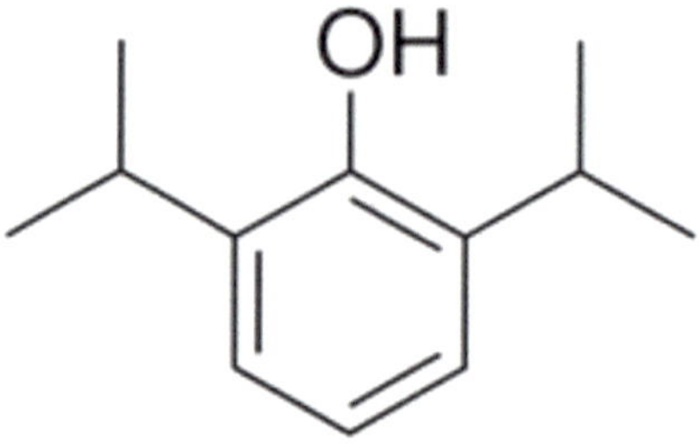

Fig. (121). Propofol, an allo-network drug and anesthetic.

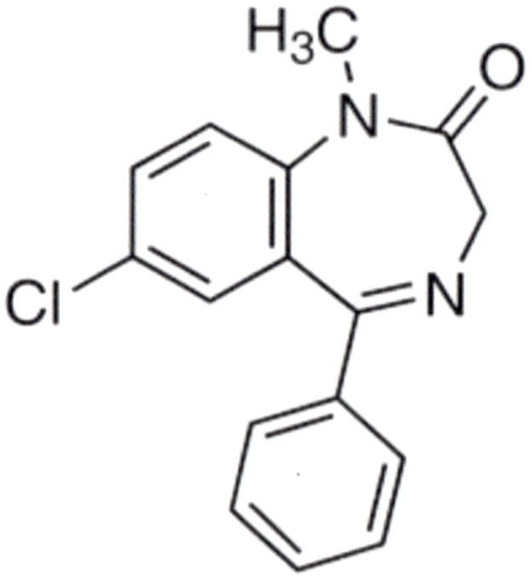

Fig. (122). Diazepam (Valium®), an anxiolytic.

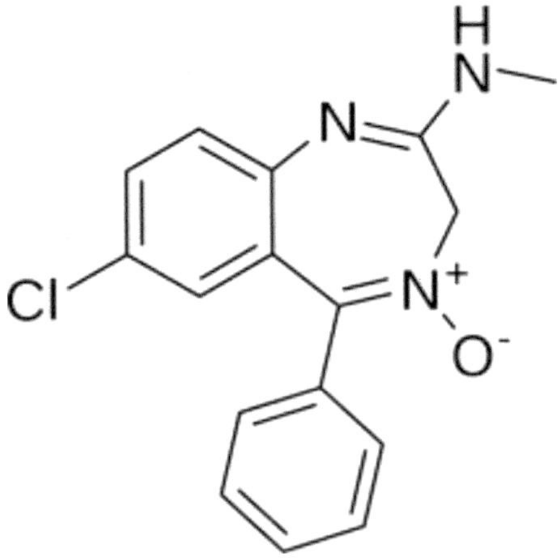

Fig. (123). Chlordiazepoxide, Librium®, a sedative.

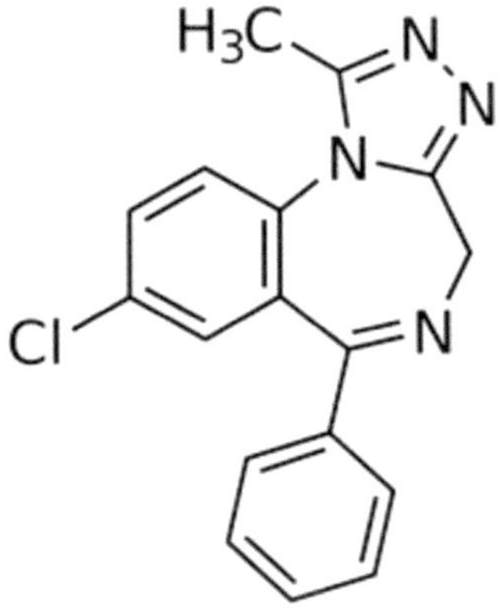

Fig. (124). Alprazolam, Xanax®, an anxiolytic.

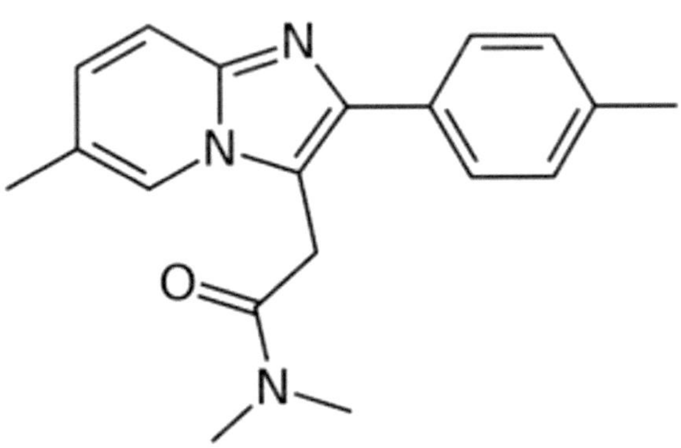

Fig. (125). Zolpidem (Ambien®) binds to one of the subunits (α subunit) of the $GABA_A$ receptor in the brain, inducing sleep.

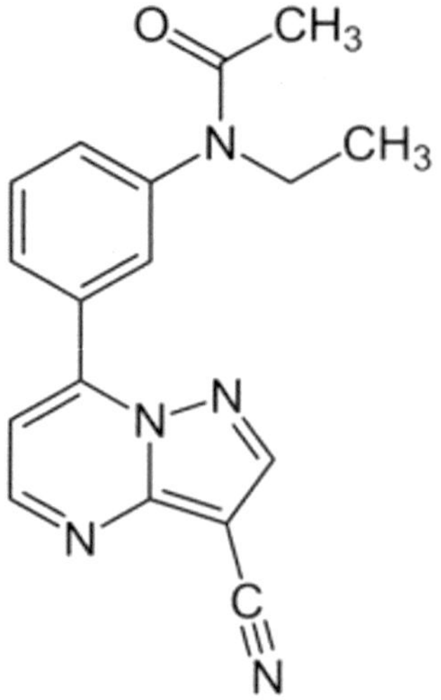

Fig. (126). Zaleplon, a sedative and hypnotic.

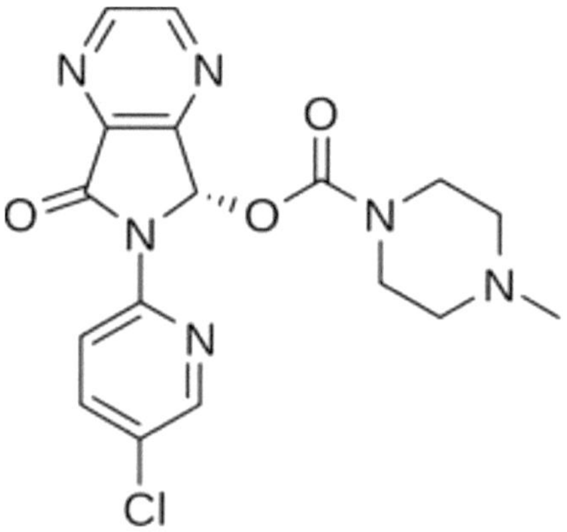

Fig. (127). Eszopiclone (Lunesta®), a hypnotic and sleeping aid.

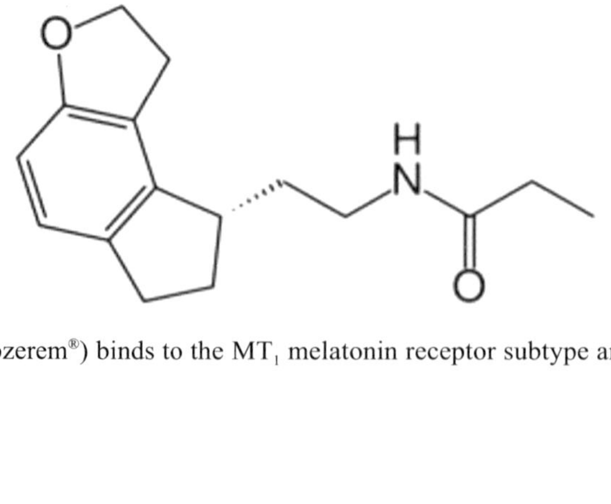

Fig. (128). Ramelteon (Rozerem®) binds to the MT_1 melatonin receptor subtype and helps people fall asleep and stay asleep.

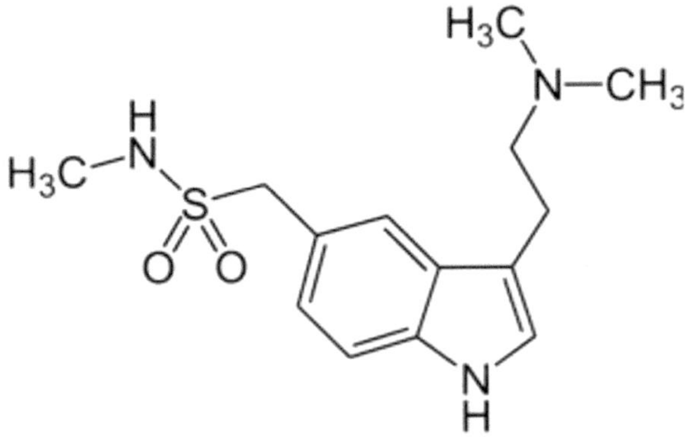

Fig. (129). Sumatriptan (Imitrex®). Selective agonists of the 5-HT_{1D} and 5-HT_{1B} receptors alleviate migraine headaches.

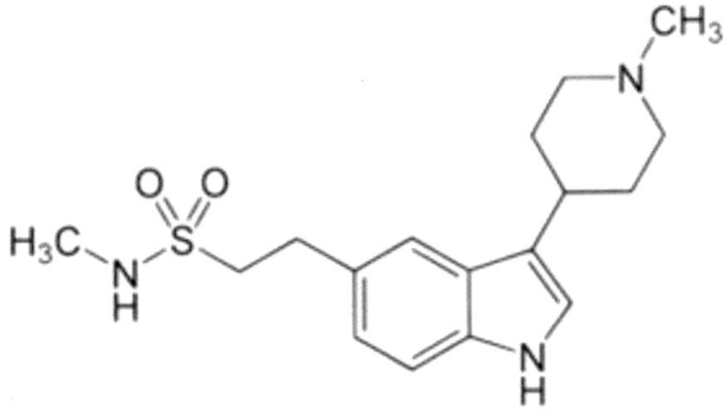

Fig. (130). Naratriptan, for migraine headaches.

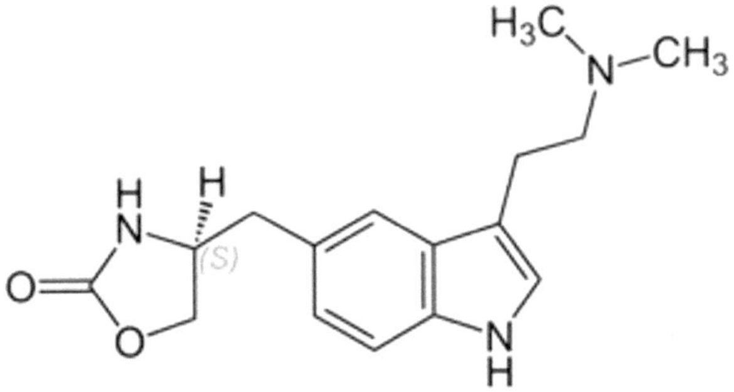

Fig. (131). Zolmitriptan, for migraine headaches.

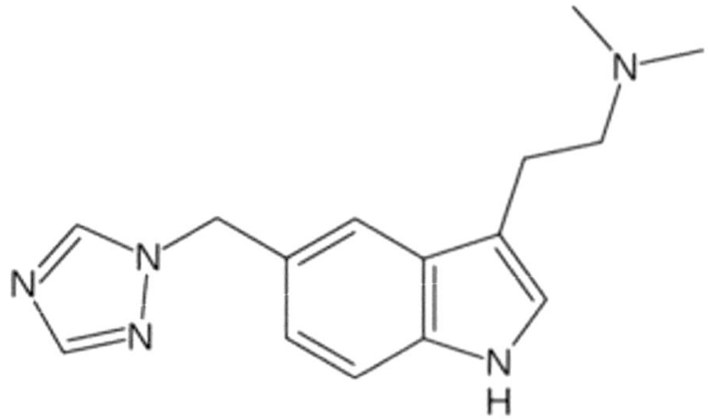

Fig. (132). Rizatriptan, for migraine headaches.

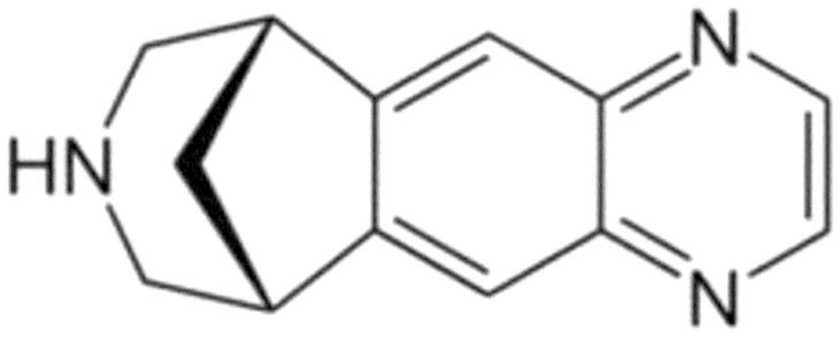

Fig. (133). Varenicline (Chantix®) to help people stop smoking. It binds to the α4β2 subtype of the nAChR.

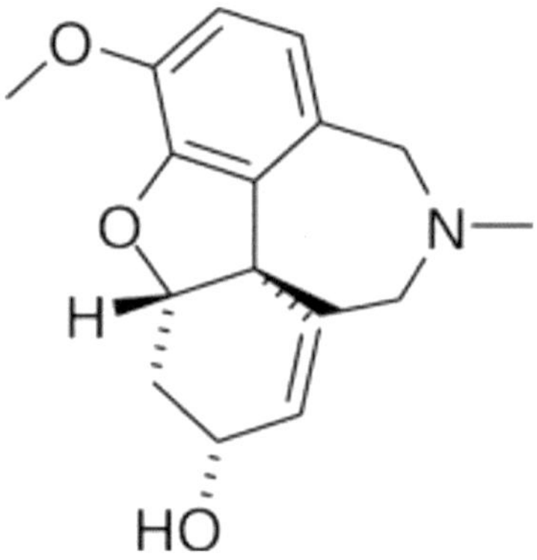

Fig. (134). Galantamine inhibits acetyl choline esterase (AchE) and is used to treat Alzheimer's disease.

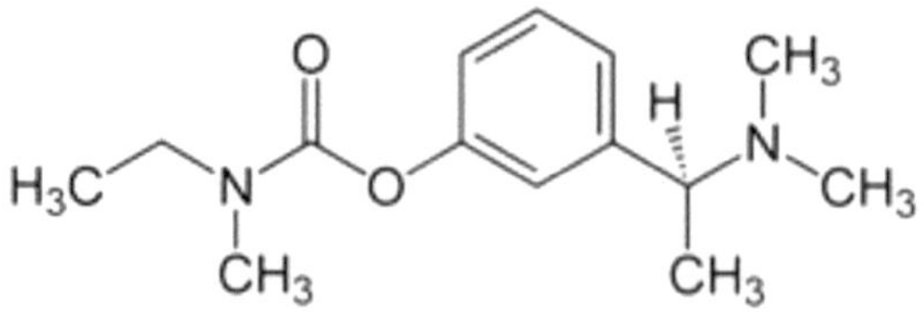

Fig. (135). Rivastigmine, for Alzheimer's disease and Parkinson's disease.

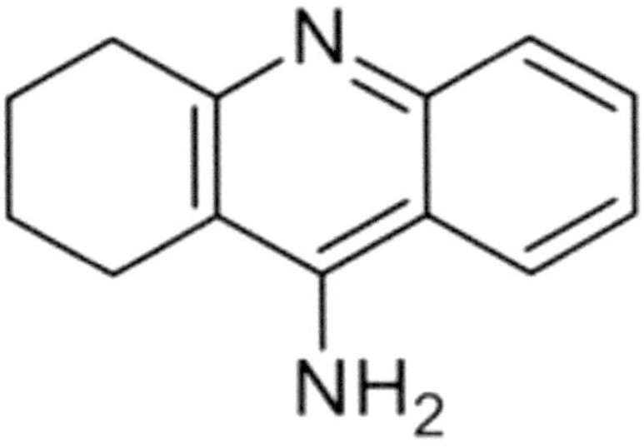

Fig. (136). Tacrine, for Alzheimer's disease.

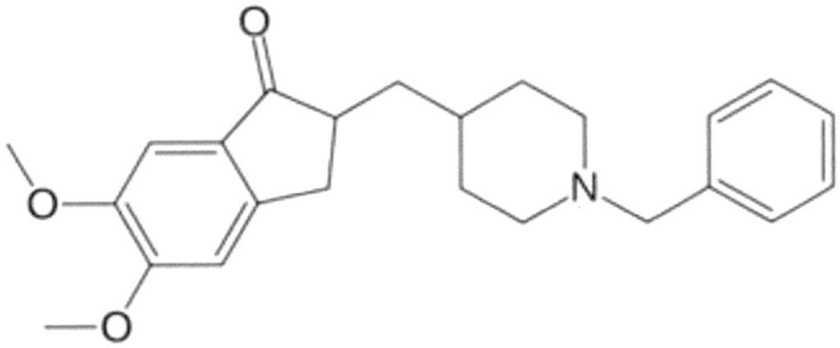

Fig. (137). Donepezil (Aricept®) is a reversible anti-AChE for Alzheimer's Disease.

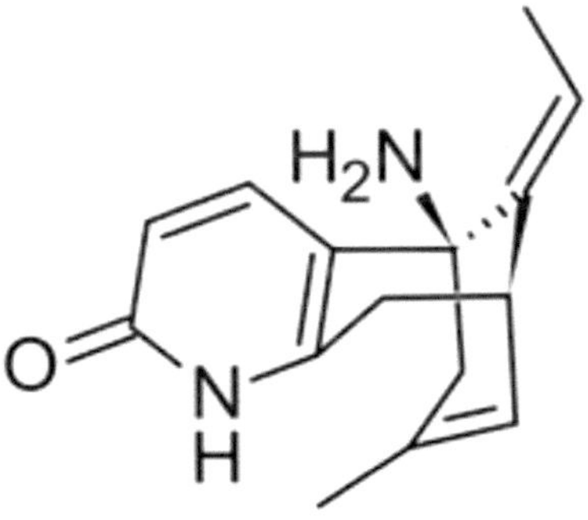

Fig. (138). Huperzine is a natural product and potent reversible inhibitor of AChW.

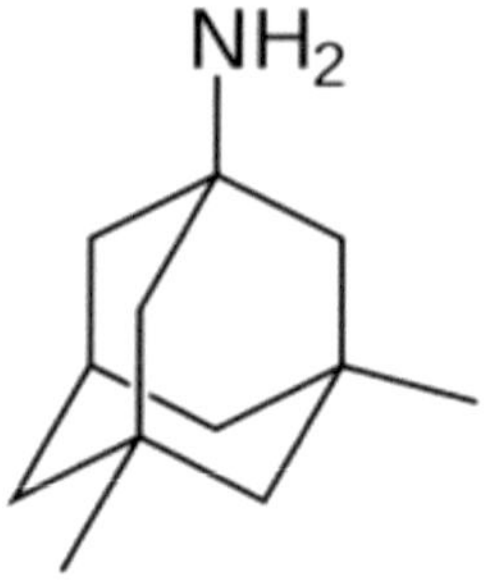

Fig. (139). Memantine prevents over-stimulation of the NMDA subtype of the glutamate receptor. It is used to treat Alzheimer's disease.

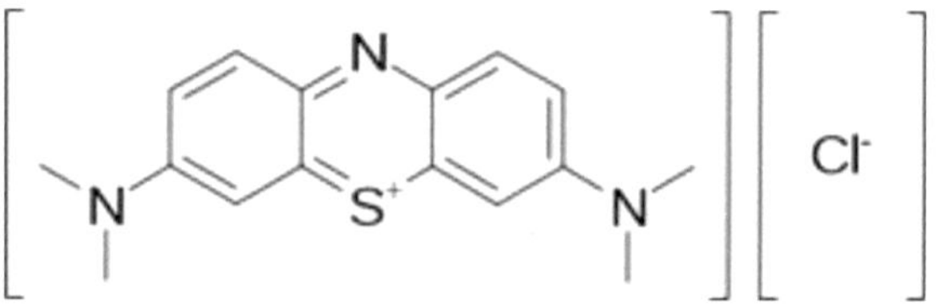

Fig. (140). Rember, or MTC (methylthioninium chloride), inhibits tau protein aggregation.

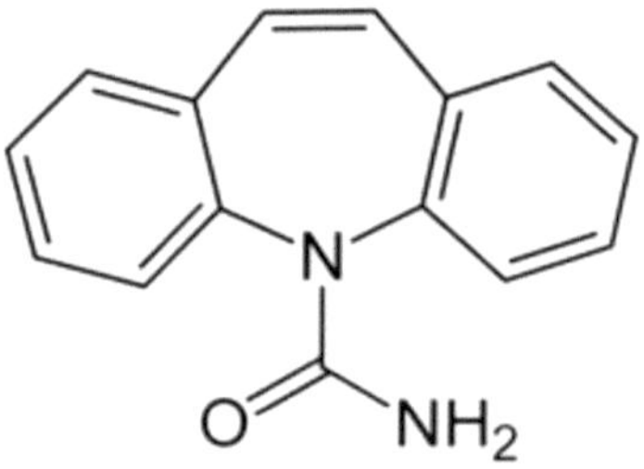

Fig. (141). Carbamazepine carbamazepine (Tegretol) was introduced in the USA in 1968. It also acts on the Na^+ channels.

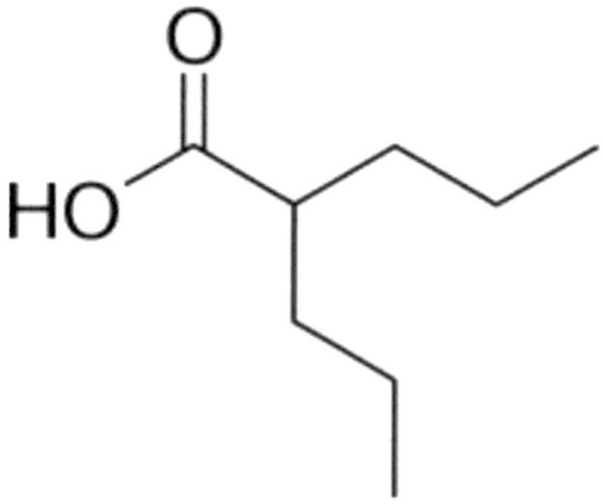

Fig. (142). Valproic acid and its sodium salt (Depakene) is an anticonvulsant that prolongs the recovery of Na^+ channels, mediates voltage-gated Ca^{2+} channels and causes an increase in the amount of GABA in the brain.

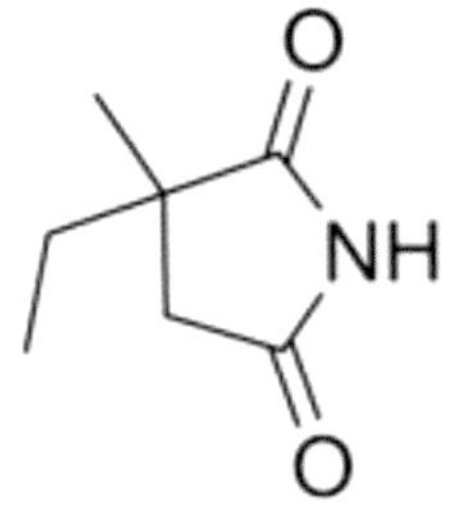

Fig. (143). Ethosuximide (Zarontin®) is an antiepileptic drug.

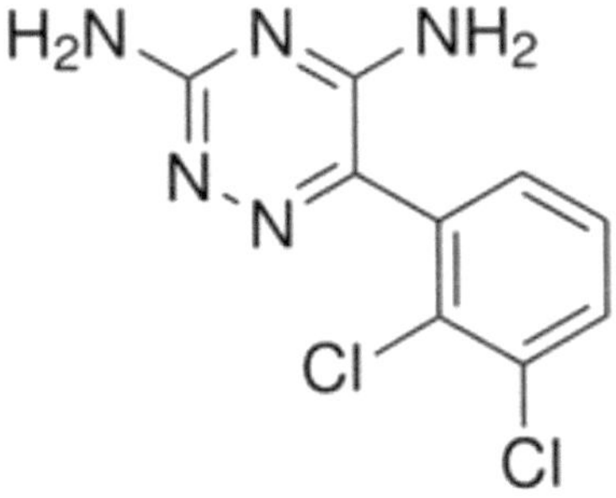

Fig. (144). Lamotrigine (Lamictal®) is an antiepileptic drug.

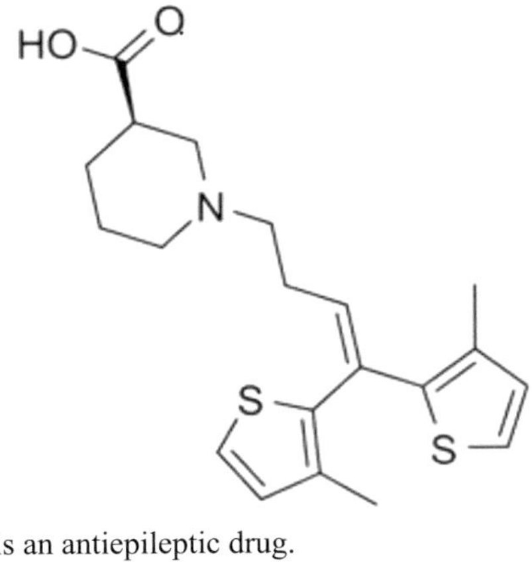

Fig. (145). Tiagabine (Gabitril®) is an antiepileptic drug.

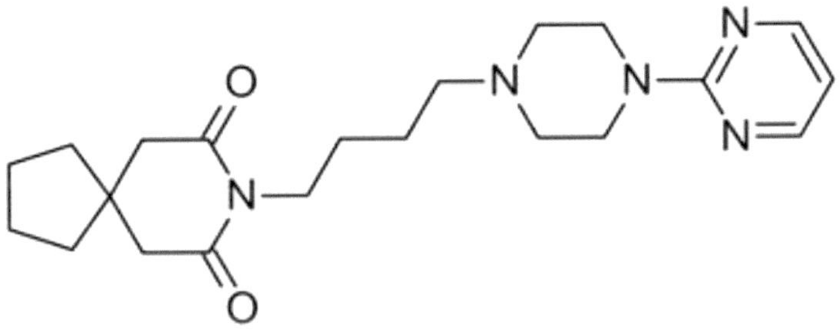

Fig. (146). Buspirone (BuSpar®, or BMS) is used to treat generalized anxiety disorder (GAD).

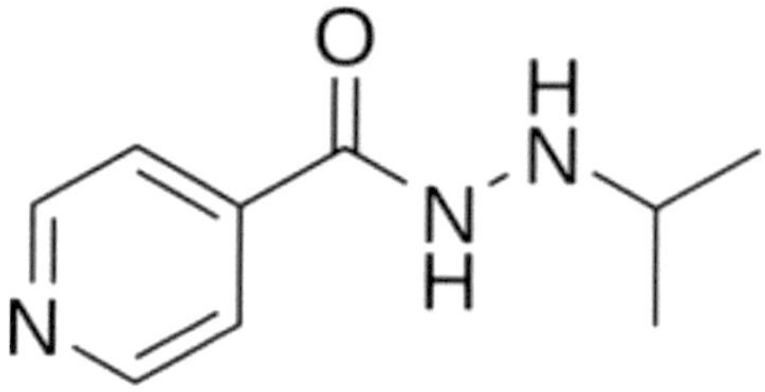

Fig. (147). Iproniazid (Marsilid®) was the first entirely synthetic antidepressant. It is a MAOI.

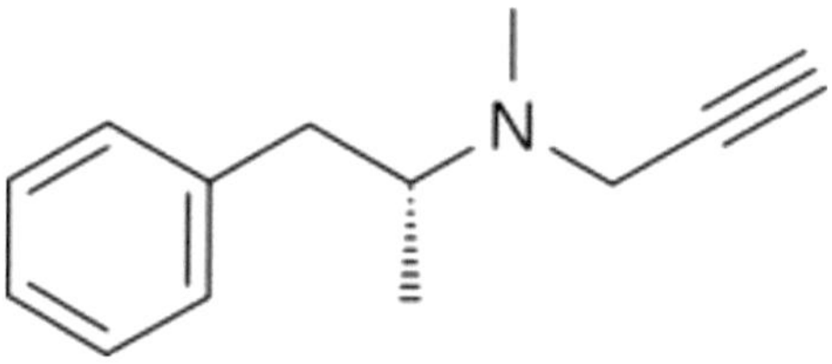

Fig. (148). In 2006, another, Selegiline (Emsam®) is a MAOI for treating major depression.

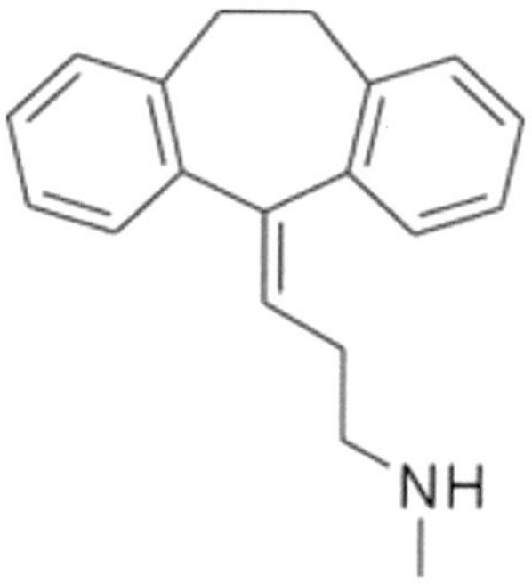

Fig. (149). Notriptyline (Pamelor®) selectively inhibits norepinephrine uptake.

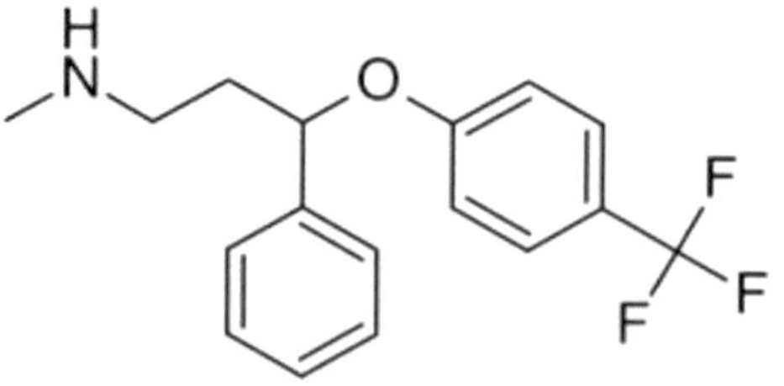

Fig. (150). Fluoxetine (Prozac®) is a second-generation SSRI antidepressant.

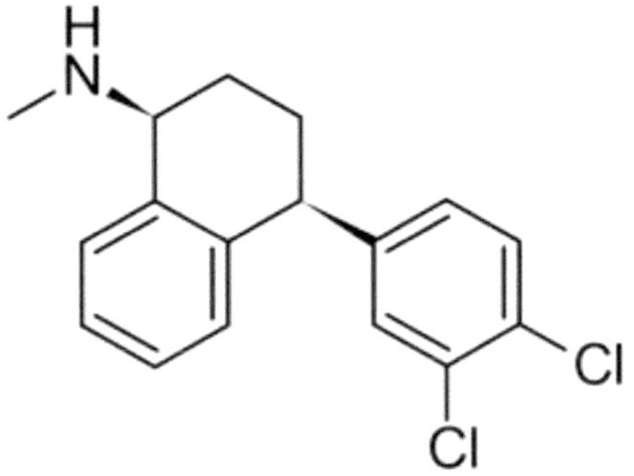

Fig. (151). Sertraline (Zoloft®) is a second-generation SSRI antidepressant.

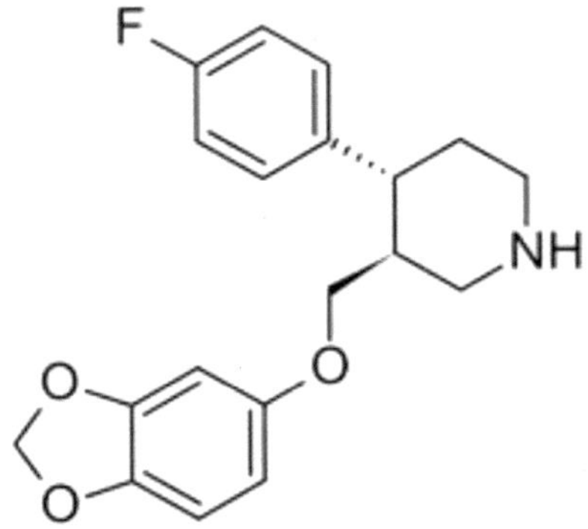

Fig. (152). Paroxetine (Paxil®) is a second-generation SSRI antidepressant. It was also approved as the drug named Brisdelle® for treating vasomotor symptoms of menopause.

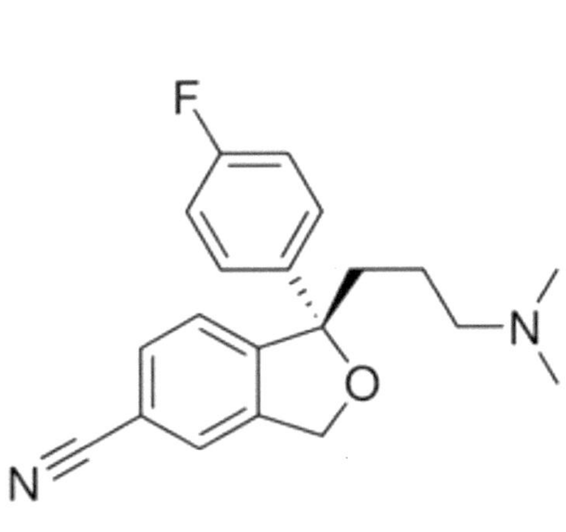

Fig. (153). Escitalopram (Lexapro®) is a second-generation SSRI antidepressant.

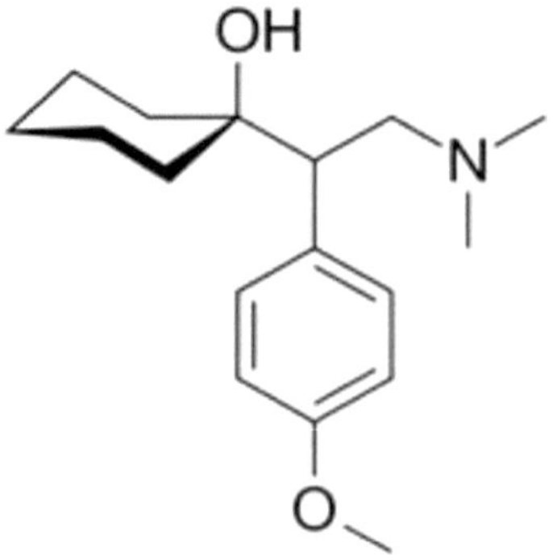

Fig. (154). Venlafaxine (Effexor®) is an SNRI antidepressant.

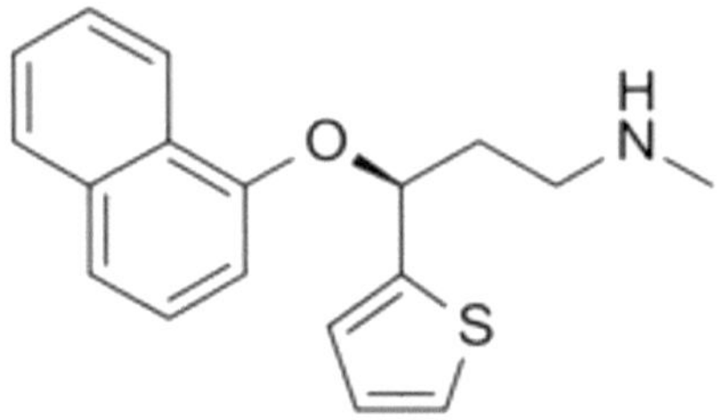

Fig. (155). Duloxetine (Cymbalta®) is an SNRI antidepressant.

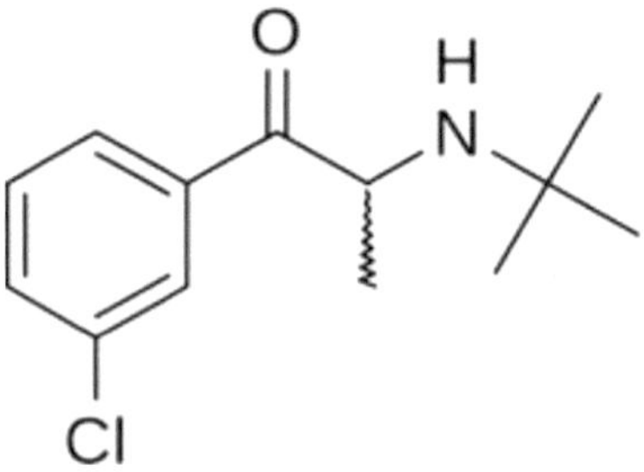

Fig. (156). Bupropion is a widely used inhibitor of reuptake of dopamine and norepinephrine. It is used as both an antidepressant and an aid to stop smoking.

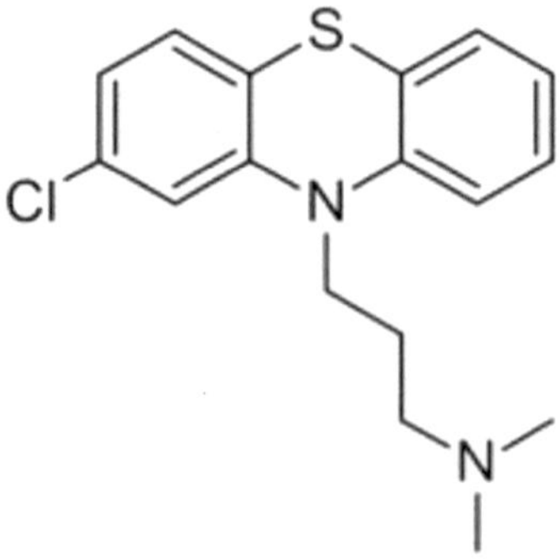

Fig. (157). Chlorpromazine is an antipsychotic.

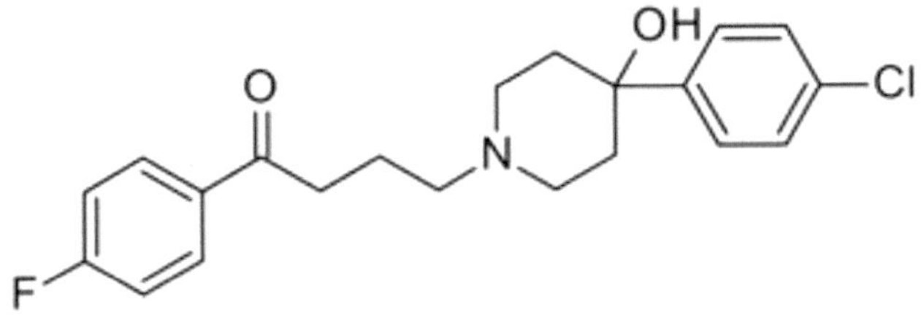

Fig. (158). Haloperidol is an antipsychotic.

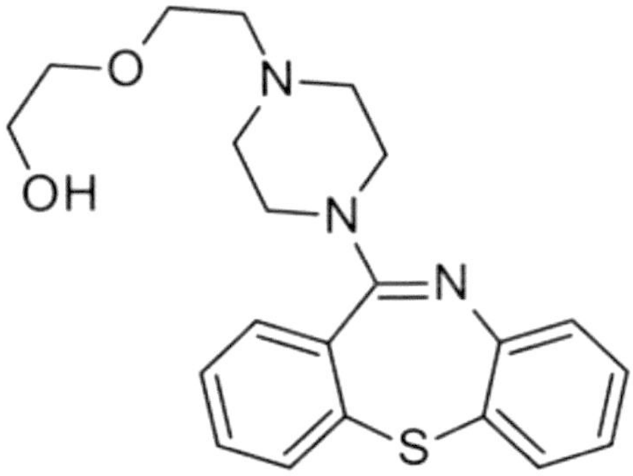

Fig. (159). Quetiapine (Seroquel®) is an atypical antipsychotic.

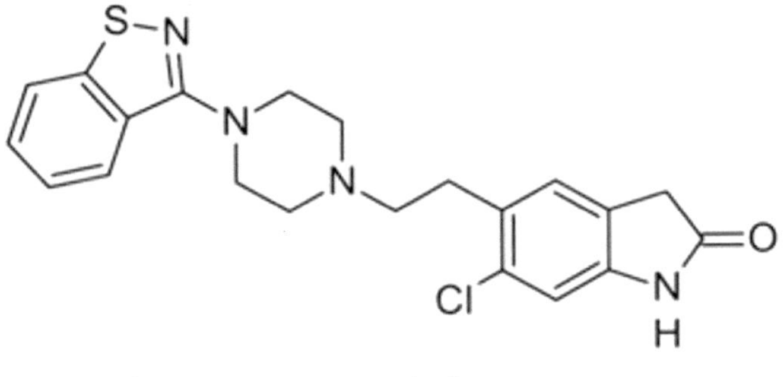

Fig. (160). Ziprasidone (Geodon®) is an atypical antipsychotic.

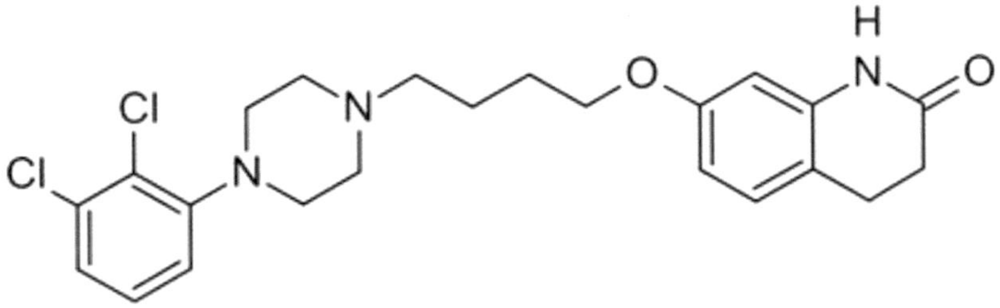

Fig. (161). Aripiprazole (Abilify®) is an atypical antipsychotic.

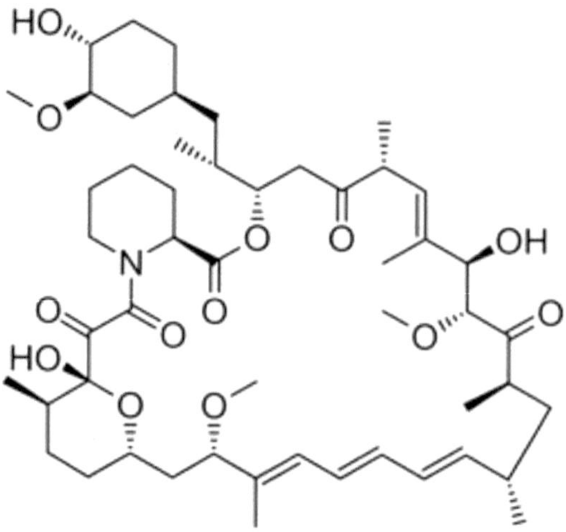

Fig. (162). Rapamycin (Sirolimus) an allosteric inhibitor of mTOR is an immunosuppressant.

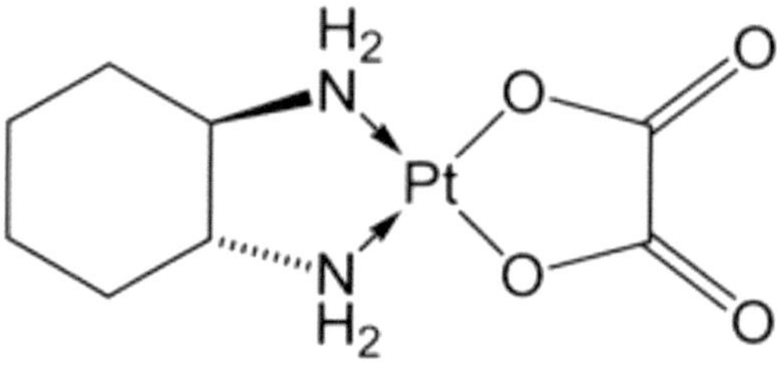

Fig. (163). Oxaliplatin is used to treat colorectal and pancreatic cancer.

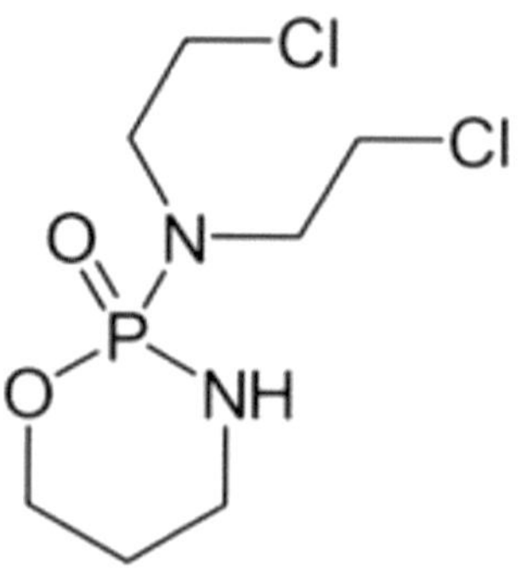

Fig. (164). Cyclophosphamide, a nucleoside analogue for lupus erythematosus.

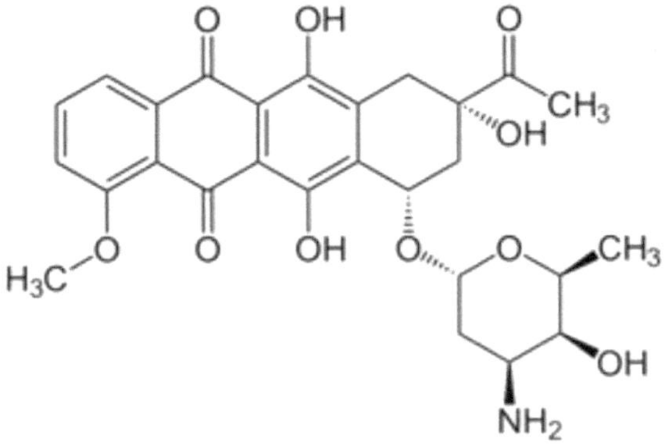

Fig. (165). Daunorubicin, anticancer drug.

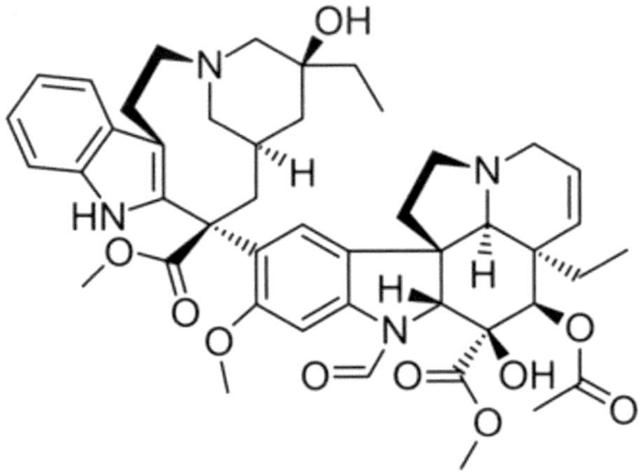

Fig. (166). Vincristine, a microtubule inhibitor, is used to treat non-Hodgkin's lymphoma.

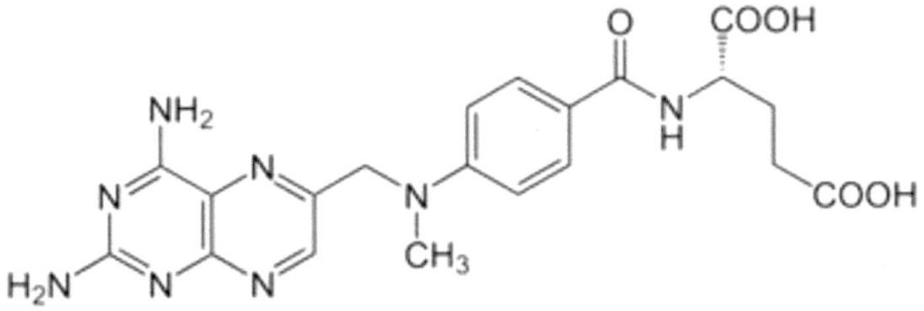

Fig. (167). Methotrexate is a second-generation folate antagonist and immunosuppressant. It is used to treat several autoimmune disorders, including rheumatoid arthritis.

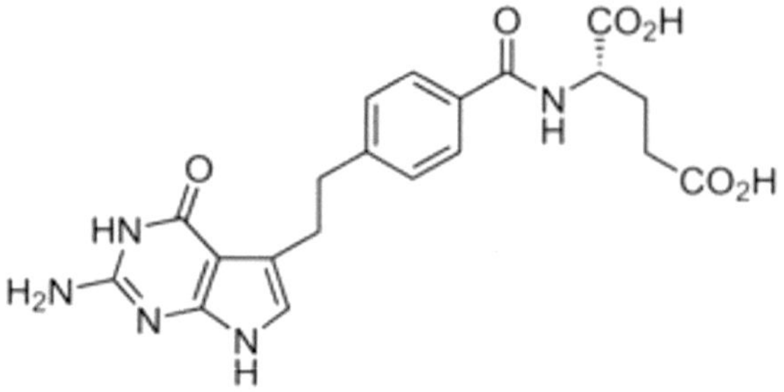

Fig. (168). Pemetrexed (Alimta®) is an anticancer drug.

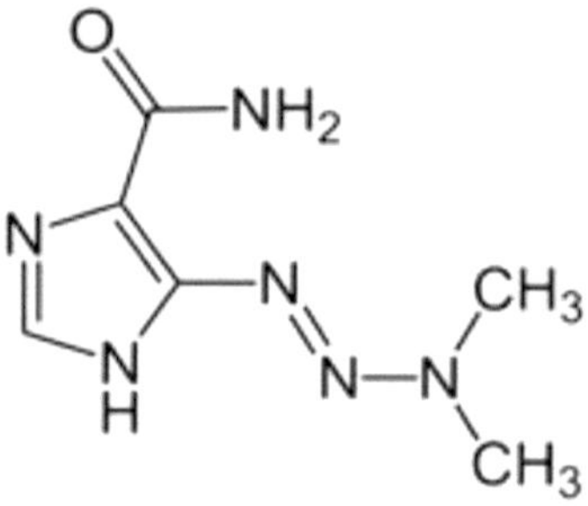

Fig. (169). Dacarbazine is an anticancer drug.

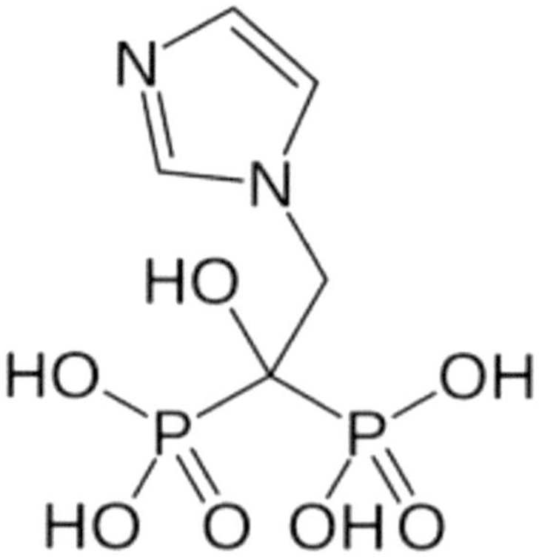

Fig. (170). Zoledronate is a nitrogen bisphosphonates that is prescribed for osteoporosis.

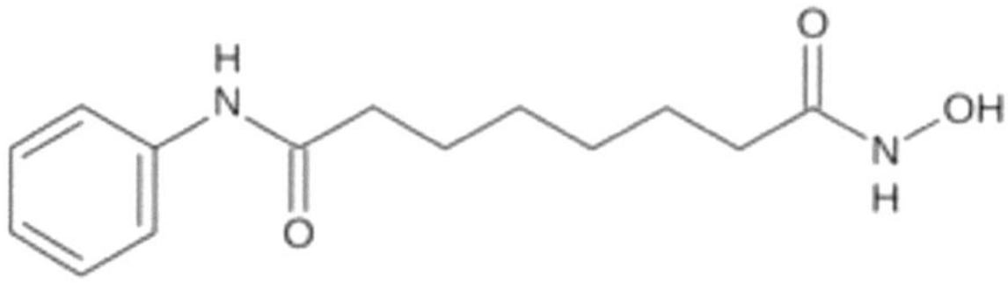

Fig. (171). Vorinostat (Zolinza®) is a histone deacetylase inhibitor.

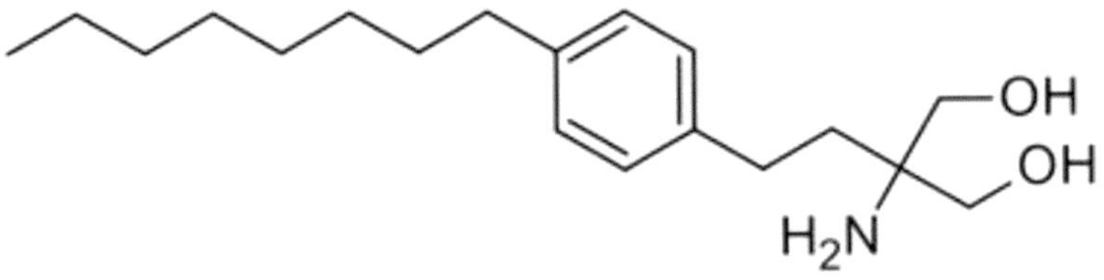

Fig. (172). Fingolimod (Gilenya®) modulates the sphingosine1-phosphate receptor and is approved for treating MS.

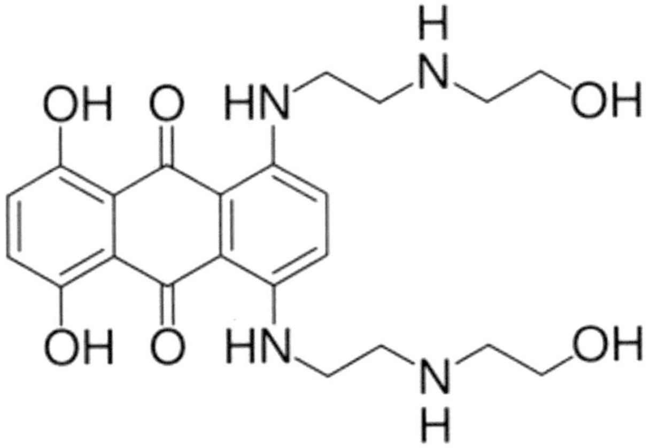

Fig. (173). Mitoxantrone (Novantrone®) is an immunosuppressant that is also used in cancer chemotherapy. It suppresses the activity of T cells, B cells, and macrophages that are thought to lead the attack on the myelin sheath. It is approved for treating MS.

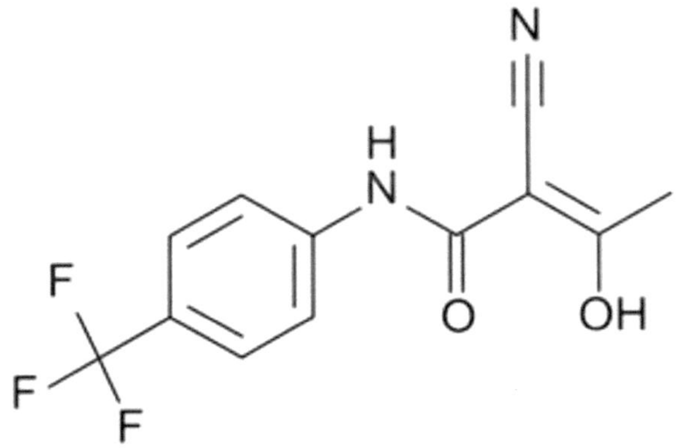

Fig. (174). Teriflunomide reversibly inhibits the enzyme dihydroorotate dehydrogenase, which catalyzes a key step in the biosynthesis of pyrimidines needed for DNA synthesis. So, it reduces the activation, proliferation and function of T- and B-cells.

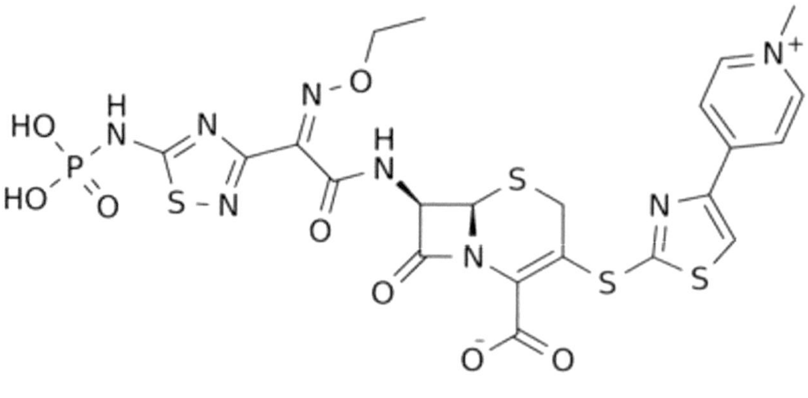

Fig. (175). Ceftaroline fosamil (Teflaro®) is an antibiotic that is used to treat MRSA.

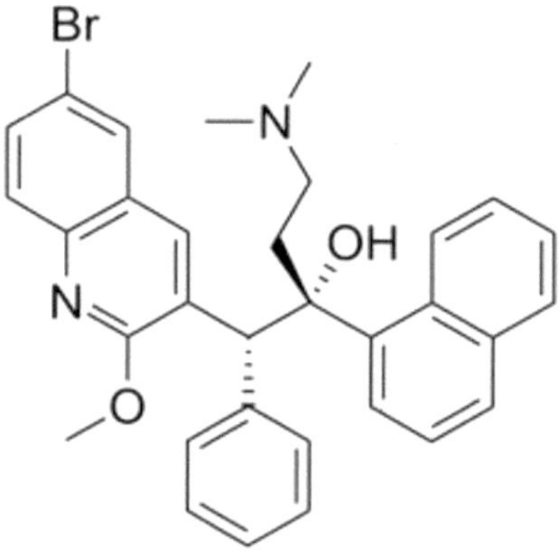

Fig. (176). Bedaquiline is an antibiotic that is used for treating multi-drug resistant tuberculosis.

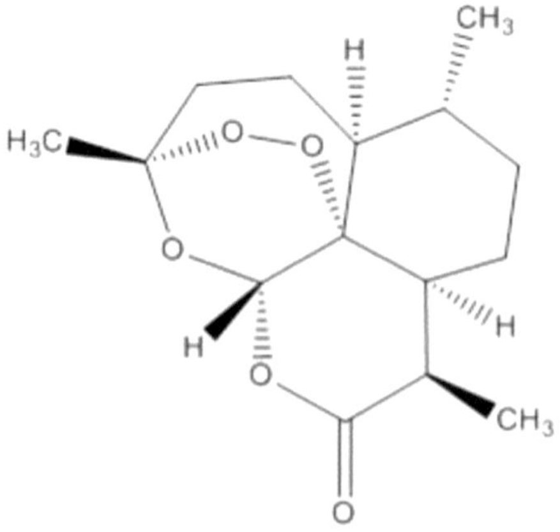

Fig. (177). Artemisinin antimalarial.

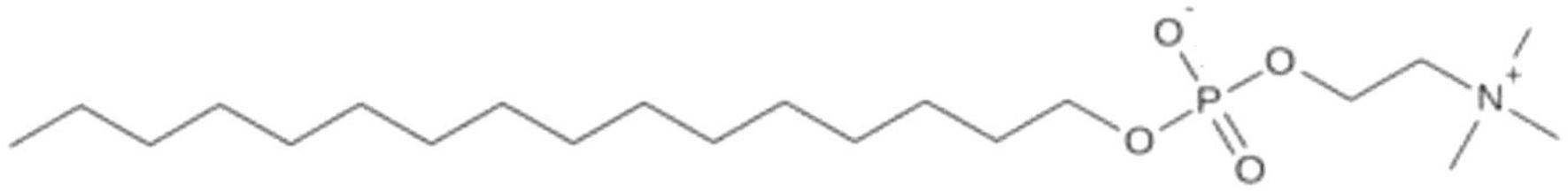

Fig. (178). Miltefosine, (Impavido®) is effective against several stages in the life cycle of *Leishmania* parasites.

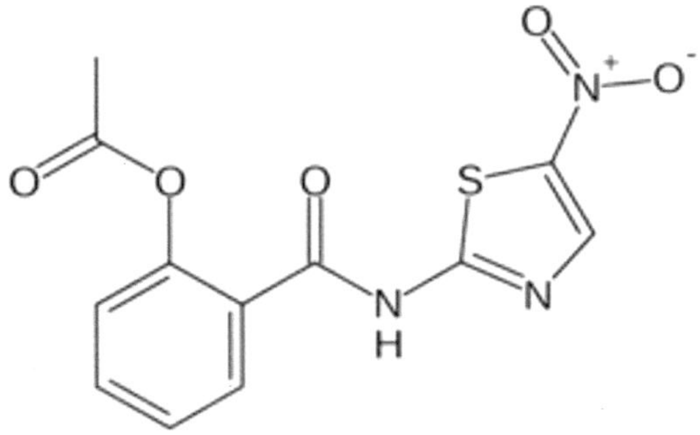

Fig. (179). Nitazoxanide (Alinia®) treats the protozoan parasites *Cryptosporidium parvum*, which causes diarrhea, and *Giardia intestinalis*, which causes intestinal infections.

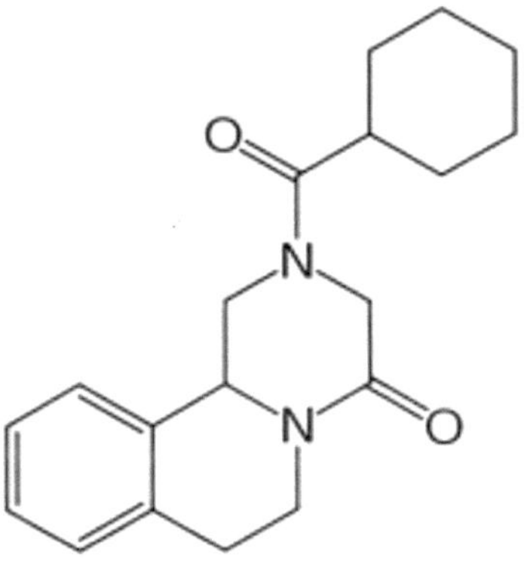

Fig. (180). Praziquantel (Biltricide®) is used to treat schistosomiasis.

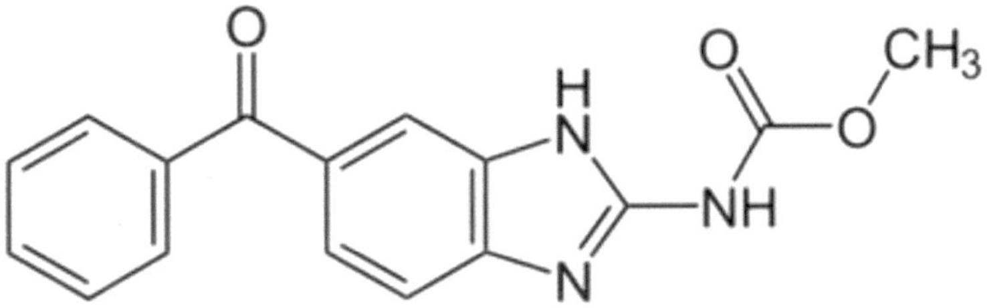

Fig. (181). Mebendazole (Vermox®) is used to treat ascariasis by blocking the uptake of glucose in the worm that causesit.

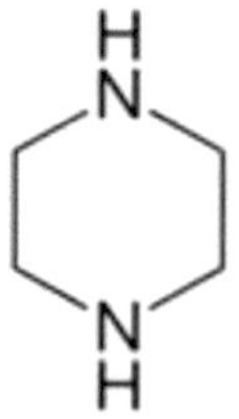

Fig. (182). Piperazine is another anti-helminthic agent. It is an acetylcholine antagonist.

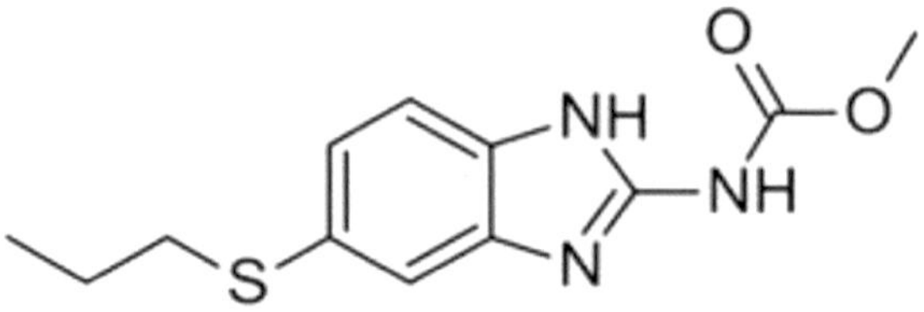

Fig. (183). Albendazole is a broad-spectrum antihelminthic agent that decreases ATP production in the worm. Elephantitis is treated with albendazole, in combination with either.

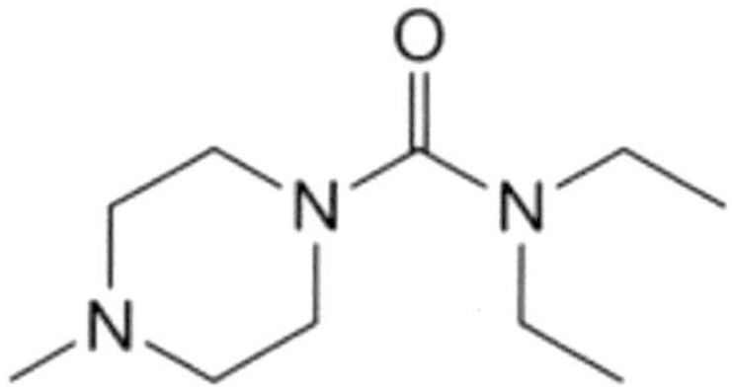

Fig. (184). Diethylcarbamazine is an antihelminthic drug.

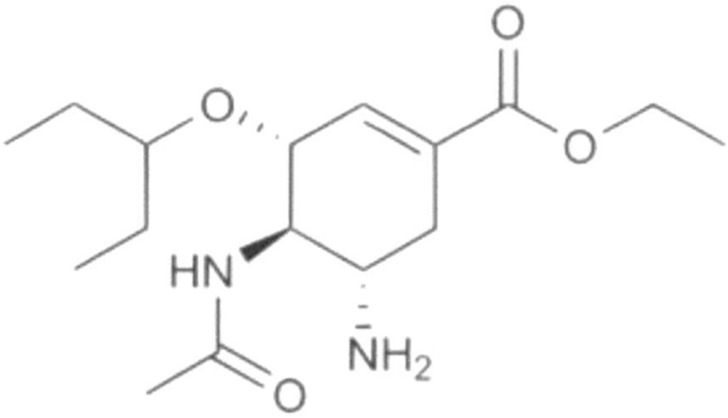

Fig. (185). Oseltamivir (Tamiflu®) is a neuraminidase inhibitor that is effective against the swine flu.

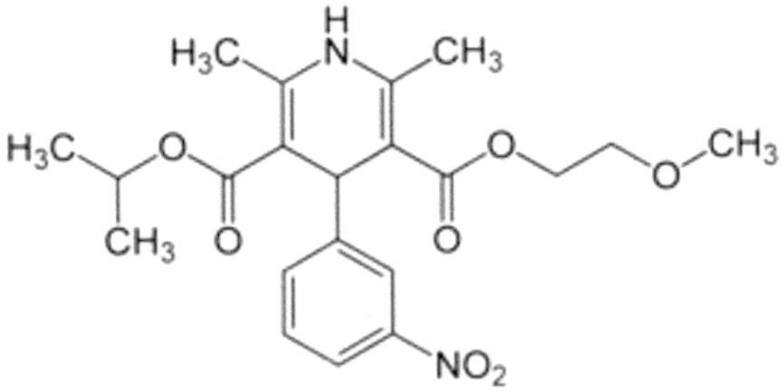

Fig. (186). Nimodipine (Nymalize®) is a Ca^{2+} channel blocker that is used to treat subarachnoid hemorrhage.

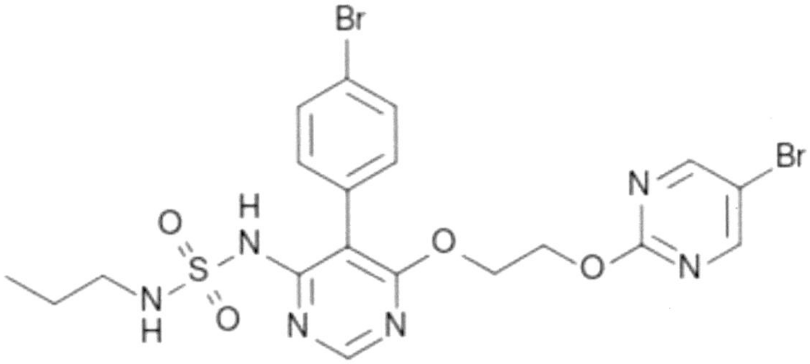

Fig. (187). Macitentan (Opsumit®) is used to treat pulmonary arterial hypertension. It its metabolite are antagonists of the endothelin receptors ET_A and ET_B.

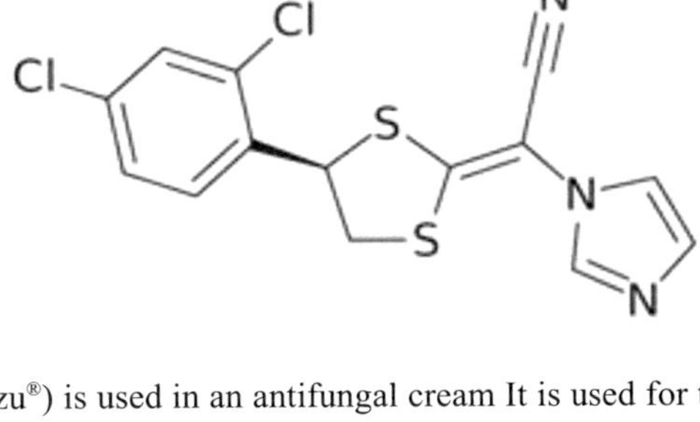

Fig. (188). Luliconazole (Luzu®) is used in an antifungal cream It is used for topical treatment of interdigital tinea pedis, tinea cruris, and tinea corporis caused by the organisms *Trichophyton rubrum* and *Epidermophyton floccosum*, in adults 18 years of age and older.

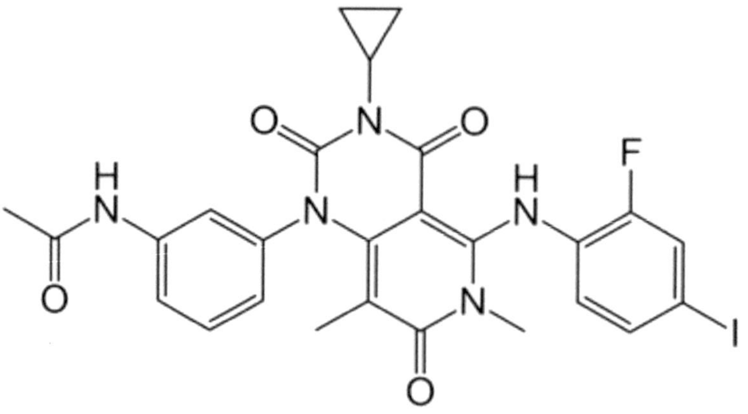

Fig. (189). Trametinib (Mekinist®) was approved for treating unresectable or metastatic melanoma with BRAF V600E or V600K mutations. It inhibits mitogen-activated protein kinase kinases 1 and 2 (MEK-1 and MEK-2).

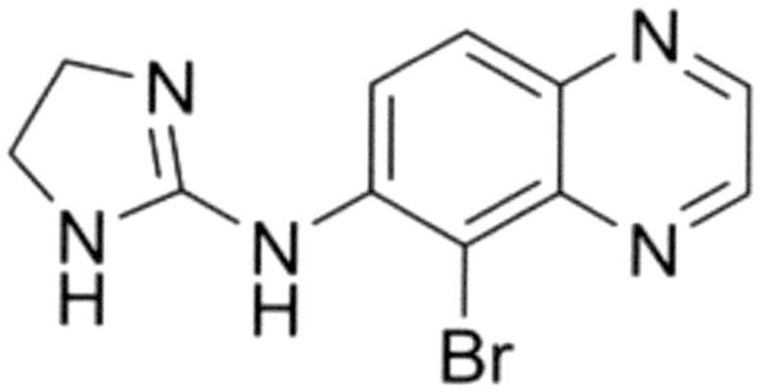

Fig. (190). Brimonidine (Mirvaso®) was approved for treating the facial erythema of rosacea. It is an agonist of the α2-adrenergic receptor with potent vasoconstrictive activity.

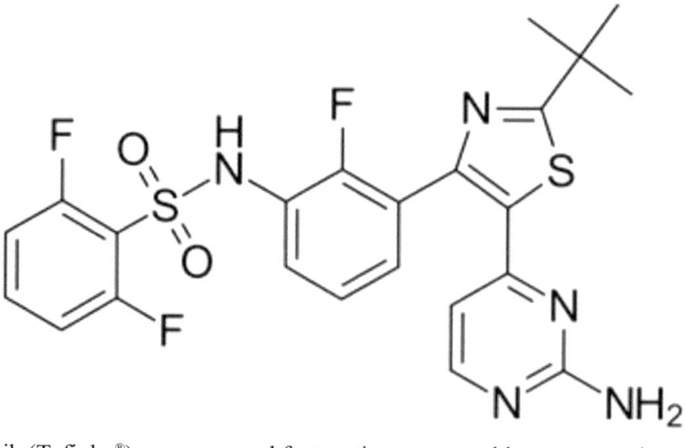

Fig. (191). Dabrafenib (Tafinlar®) was approved for treating unresectable or metastatic melanoma with BRAF V600E mutation.

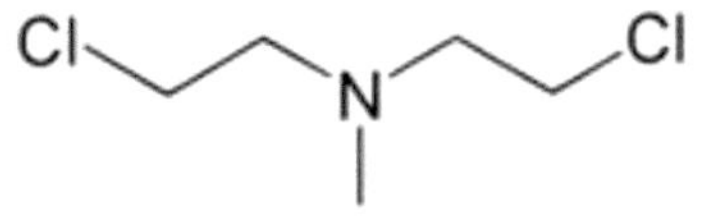

Fig. (192). Mechlorethamine (Valchlor®) is used in a gel to treat Stage IA/IB mycosis fungoides-type cutaneous T-cell lymphoma. It is an alkylating agent that inhibits the growth of rapidly proliferating cells.

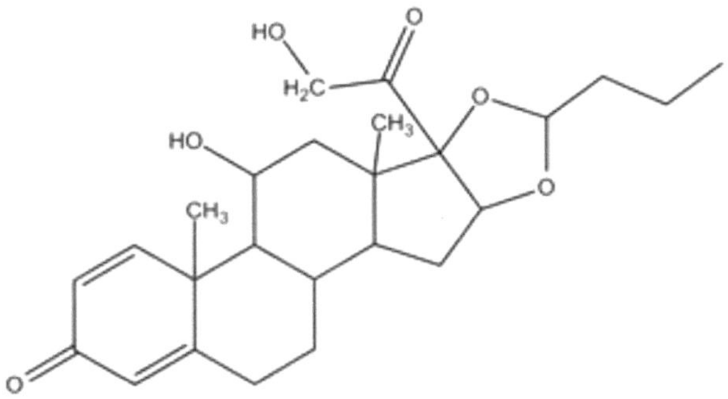

Fig. (193). VLX-600 is a first in its class anticancer drug that was approved by the FDA to enter phase 1 and 2 trials. It kills populations of sleeping cells in regions of solid tumors that are metabolically compromised due to poor vascularization.

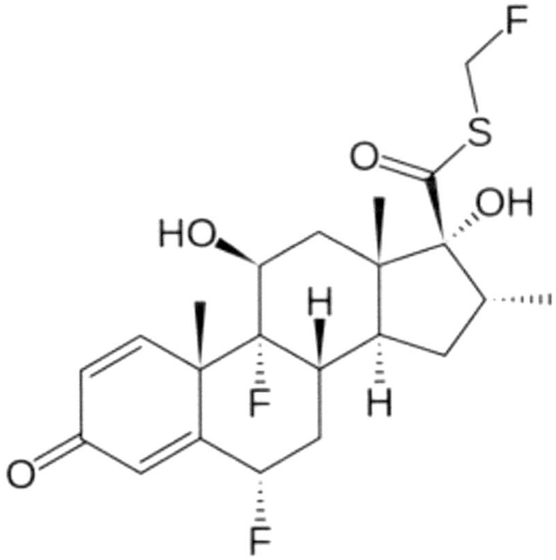

Fig. (194). Fluticasone furoate is an inhaled corticosteroid (ICS) that is used with vilanterol to treat COPD.

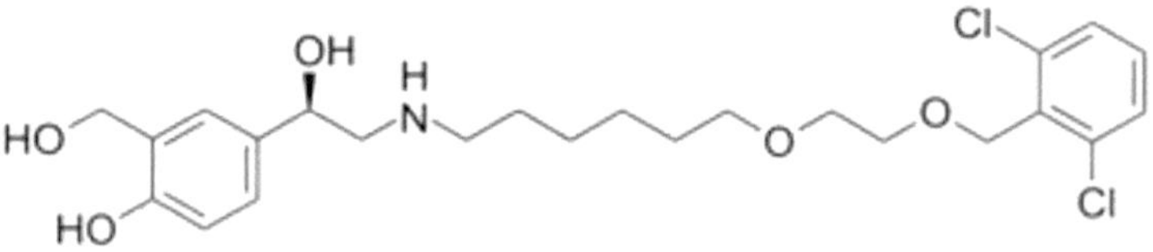

Fig. (195). Vilanterol is a long-acting β2-adrenergic receptor agonist (LABA) that is used with fluticasone to treat COPD. The combination is called Breo Ellipta®.

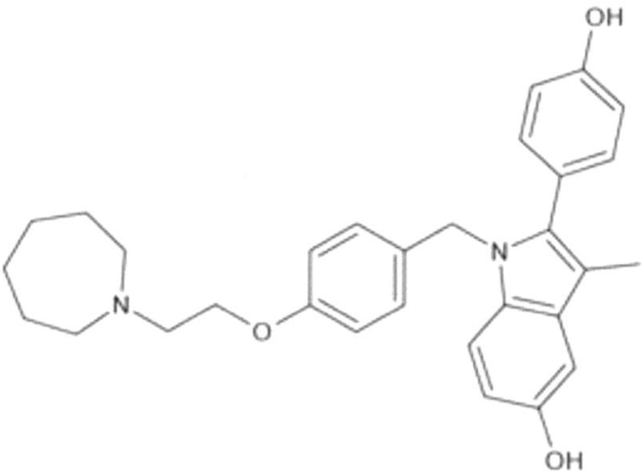

Fig. (196). Bazedoxifene is a selective estrogen receptor modulator (SERM) that is used with conjugated estrogens in the drug Duavee ® which is used to treat the vasomotor symptoms of menopause.

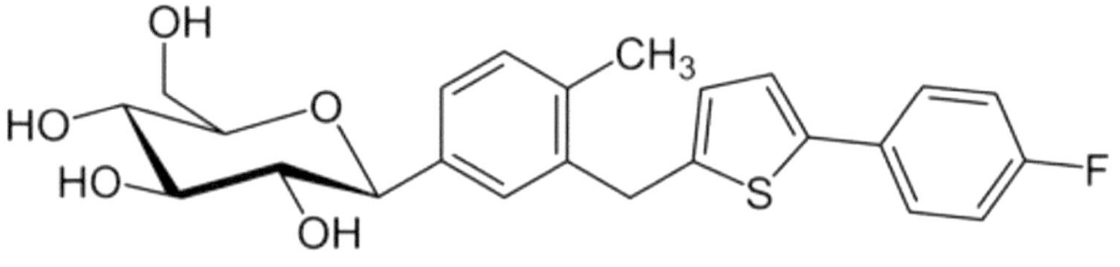

Fig. (197). Canagliflozin (Invokana®) is an inhibitor of sodium-glucose co-transporter 2 that was approved for treating type-2 diabetes.

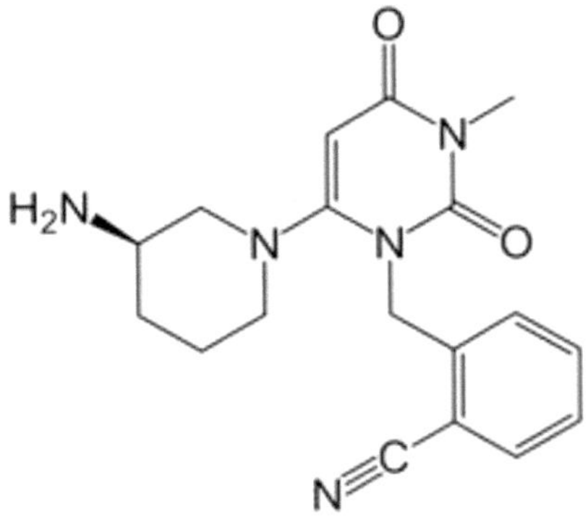

Fig. (198). Alogliptin (Nesina®) is a dipeptidyl peptidase IV (DPP IV) inhibitor that is used to treat type-2 diabetes.

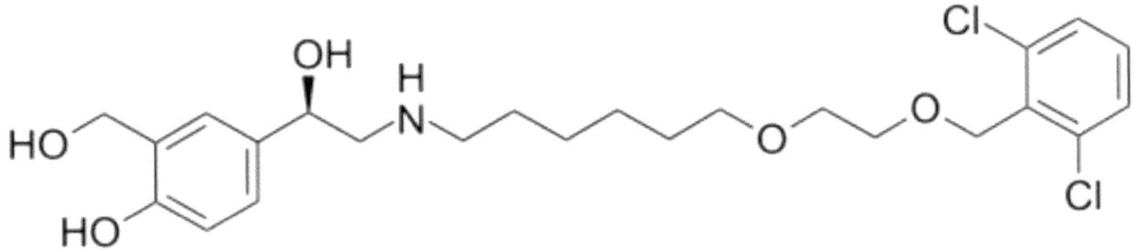

Fig. (199). Ospemifene (Osphena®) binds to estrogen receptors and is used to treat dyspareunia and vulvar and vaginal atrophy due to menopause.

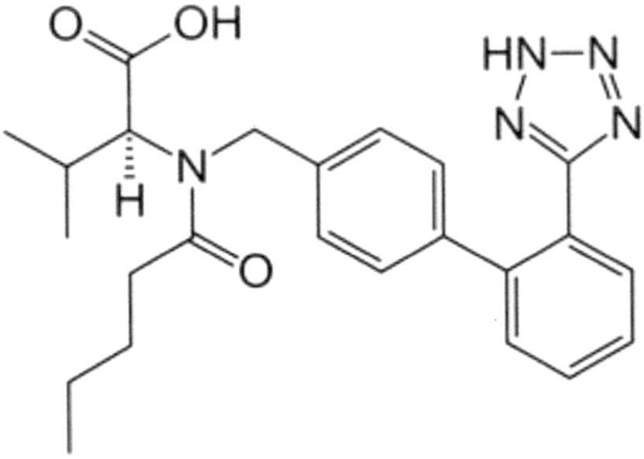

Fig. (200). Valsartan (Dovan®) for high blood pressure and congestive heart failure.

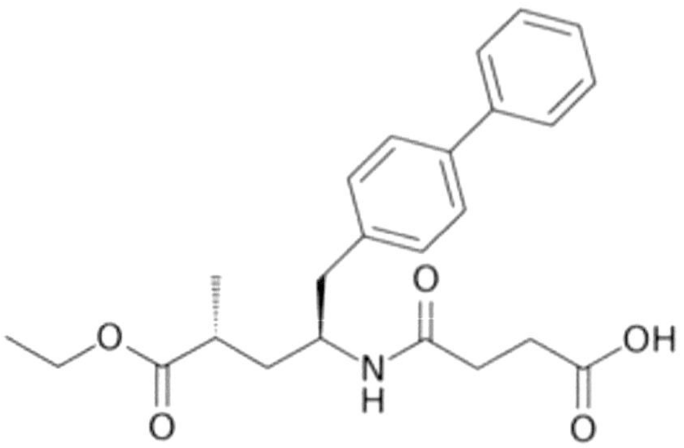

Fig. (201). Sacubirtil.

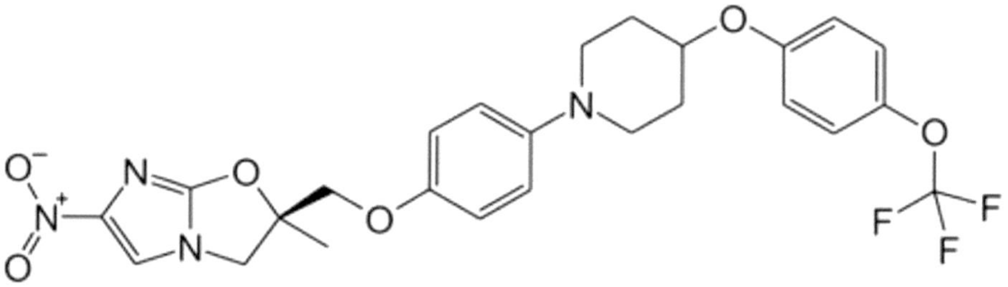

Fig. (202). Delamanid (Deltyba®), for for pulmonary multi-drug-resistant tuberculosis (MDR-TB).

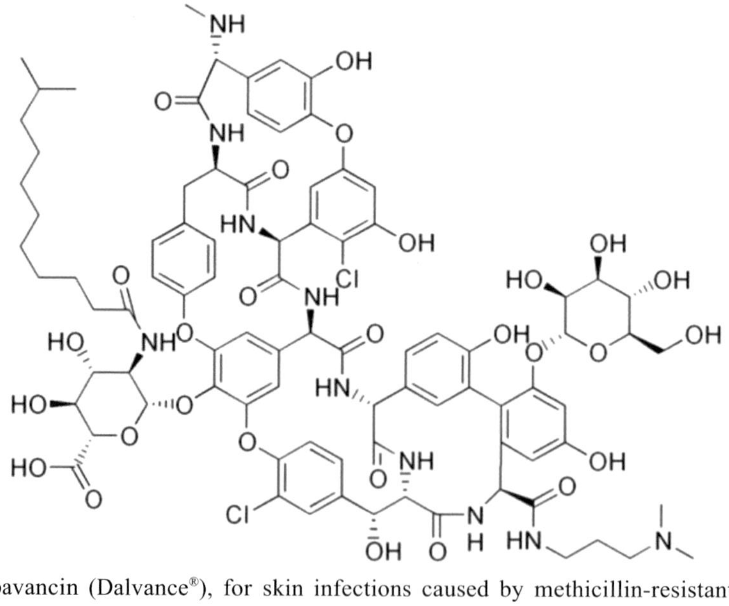

Fig. (203). Dalbavancin (Dalvance®), for skin infections caused by methicillin-resistant *Staphylococcus aureus* (MRSA).

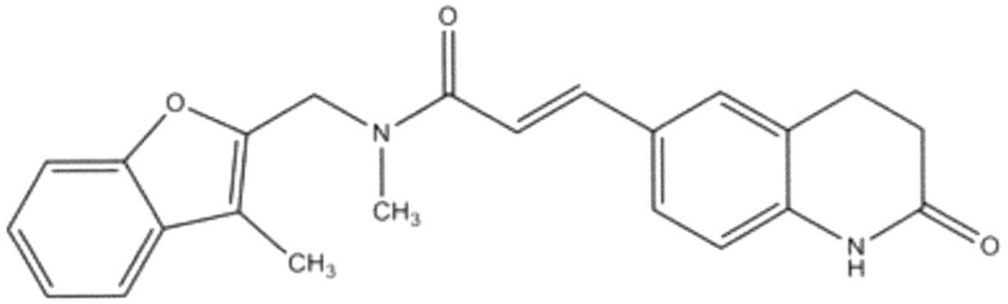

Fig. (204). Debio 1450 (previously known as AFN-1720) is a highly potent oral/IV anti-infective agent that is specifically active against all Staphylococcus species, including all known resistant strains such as methicillin-resistant *S. aureus* (MRSA) and vancomycin-intermediate *S. aureus* (VISA).

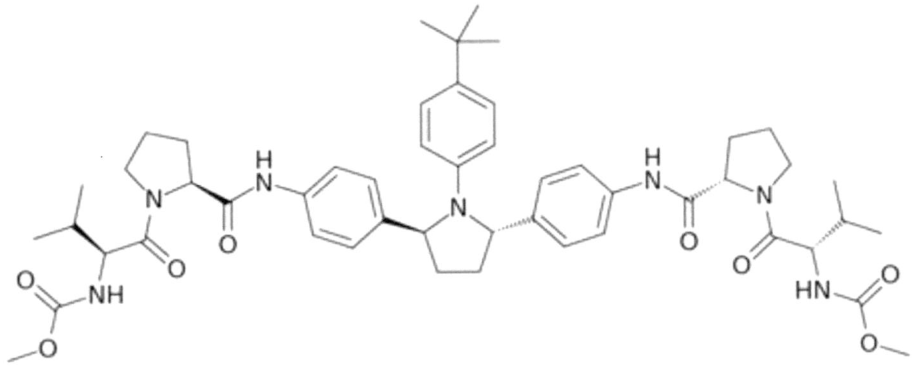

Fig. (205). Ombitasvir (ABT-450).

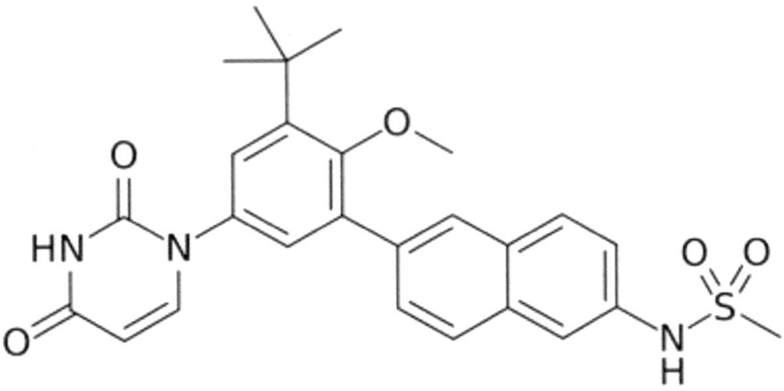

Fig. (206). Dasabuvir.

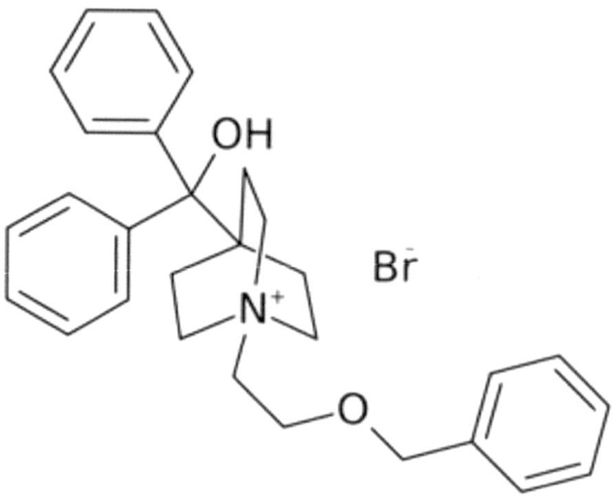

Fig. (207). Umeclidinium bromide (Incruse Ellipta®) is an anticholinergic for treating chronic obstructive pulmonary disease (COPD).

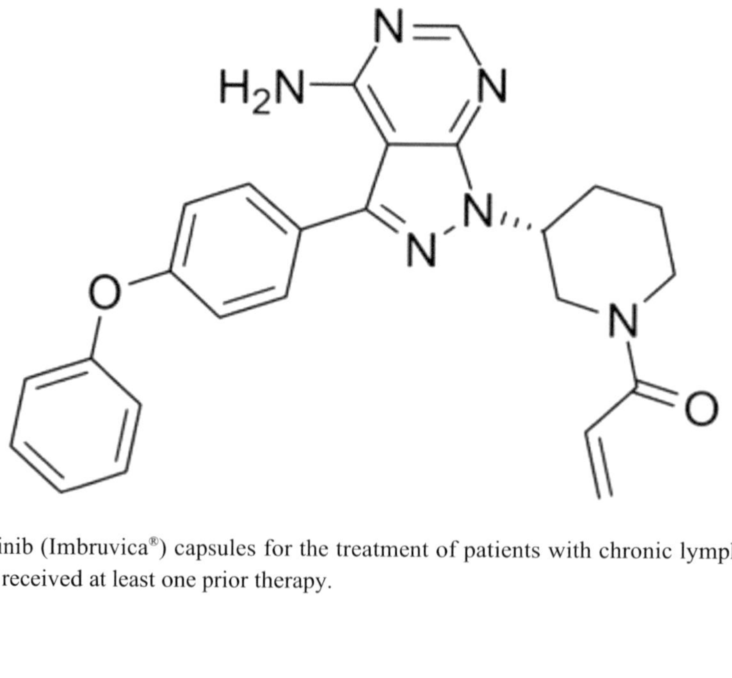

Fig. (208). Ibrutinib (Imbruvica®) capsules for the treatment of patients with chronic lymphocytic leukemia (CLL) who have received at least one prior therapy.

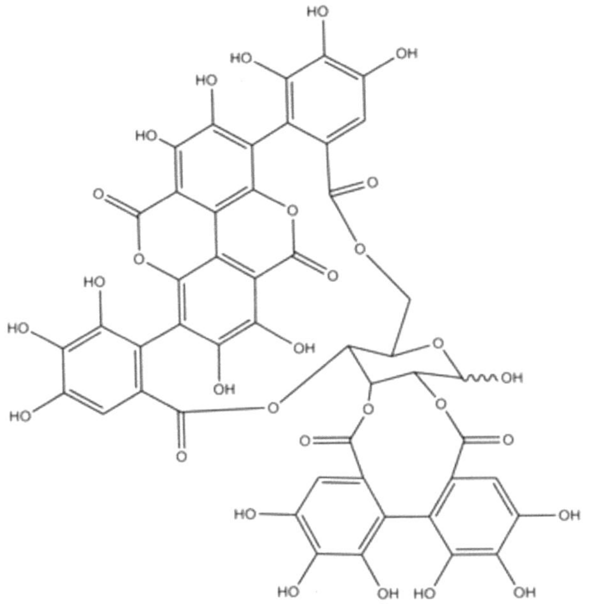

Fig. (209). Punicalagin from pomegranates for neuroinflammation and neurodegenerative diseases.

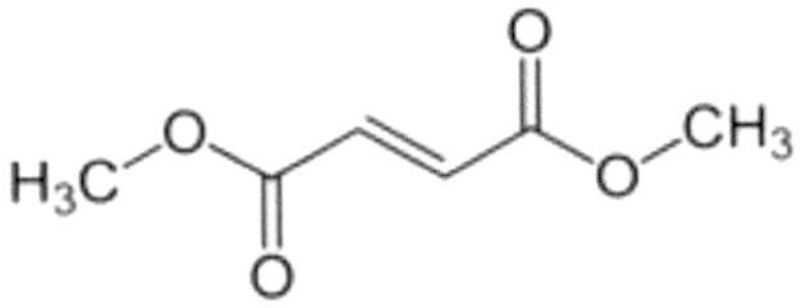

Fig. (210). Tecfidera (dimethyl fumarate) has been approved by the European Commission (EC) as a first-line oral treatment for people with relapsing-remitting multiple sclerosis (RRMS), the most common form of multiple sclerosis (MS).

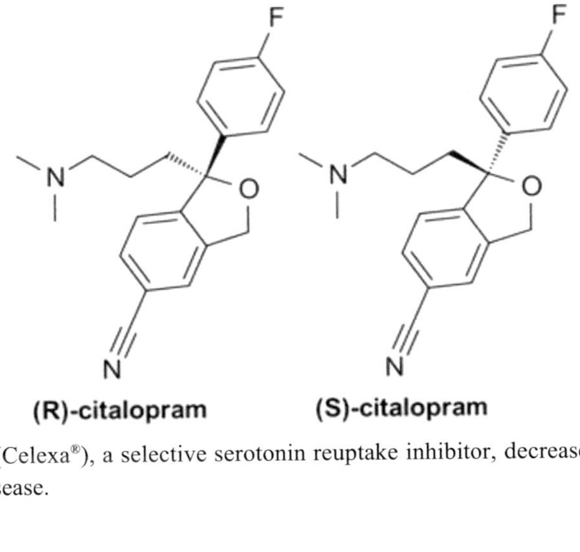

Fig. (211). Citalopram (Celexa®), a selective serotonin reuptake inhibitor, decreased Aβ in brain interstitial fluid for Alzheimer's disease.

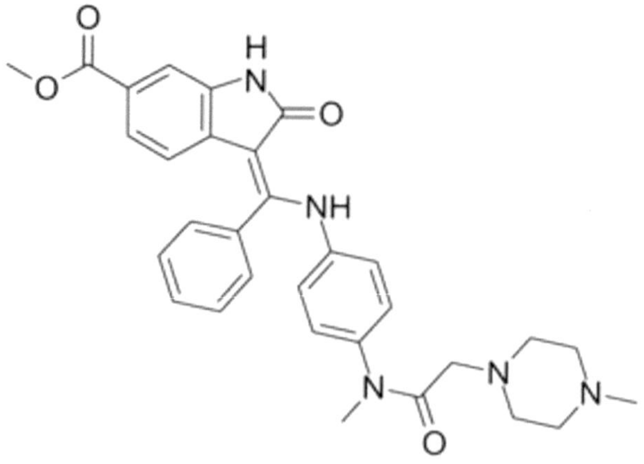

Fig. (212). Nintedanib (Ofev®) for idiopathic pulmonary fibrosis.

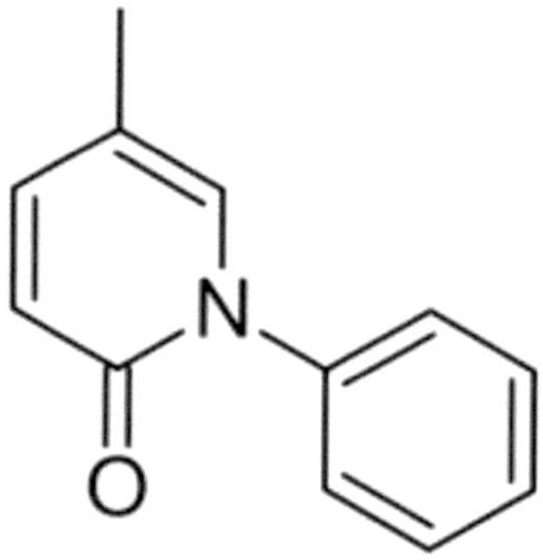

Fig. (213). Pirfenidone for idiopathic pulmonary fibrosis.

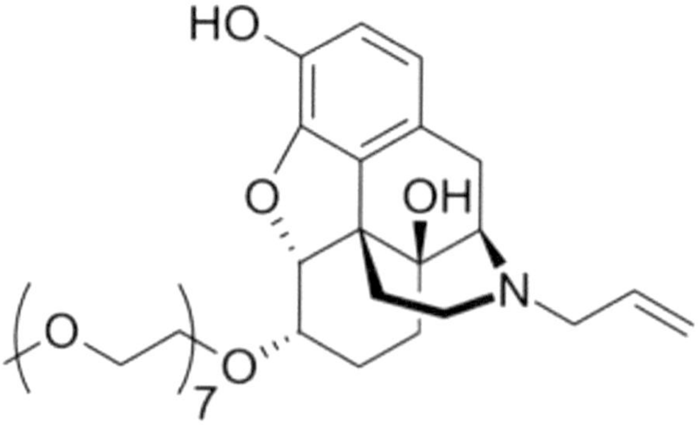

Fig. (214). Naloxegol for opioid-induced constipation in patients with noncancer pain.

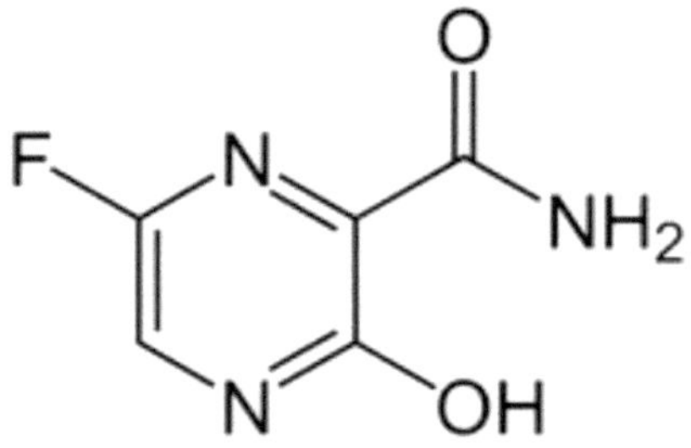

Fig. (215). Favipiravir for influenza and Ebola viruses could have a new target: norovirus, often known as the winter vomiting virus.

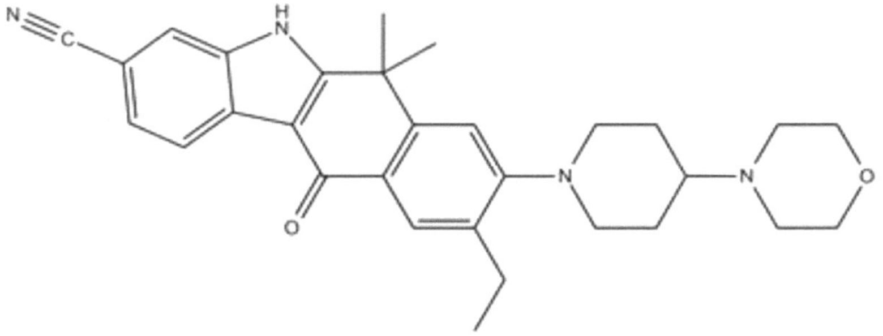

Fig. (216). Alectinib for the treatment of people living with non-small cell lung cancer (NSCLC).

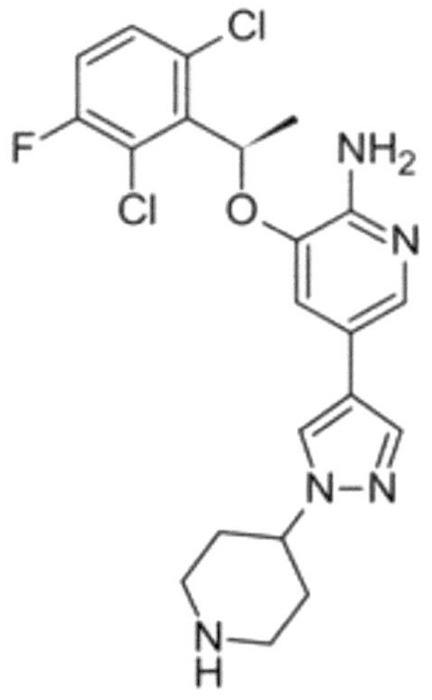

Fig. (217). Crizotinib (Xalkori®) for treating cancer.

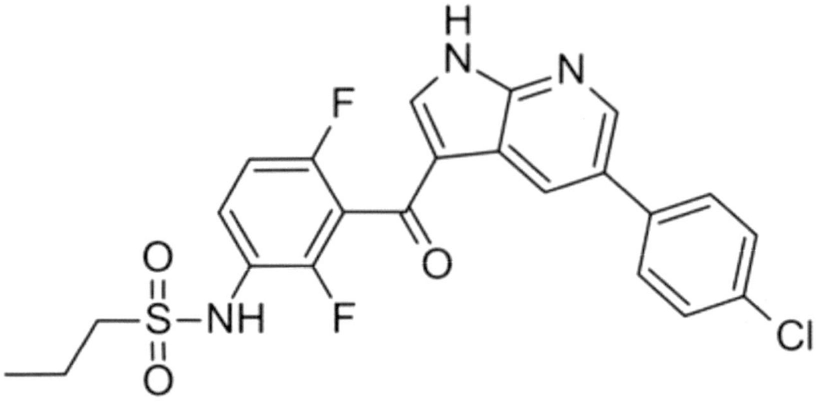

Fig. (218). Vemurafenib (Zelboraf®) for treating cancer.

SUBJECT INDEX

A

Adaptive 9, 78, 82, 108, 121, 147, 148, 221, 330, 331, 338, 360, 370, 376, 383, 386, 430
Alzheimer's disease 35, 288
AMA 435, 458, 483, 484
American Medical Association 435, 458, 462, 483, 484
Arachidonic acid 74, 76, 282, 294, 351, 538, 541
Arthritis 69, 72, 83, 97, 130, 134, 290, 355, 367, 373, 380, 439, 445, 459, 514, 533, 645
Autoimmune diseases ii, 10, 78, 235, 290, 300, 323, 325, 342, 344, 350, 352, 354, 362, 377, 378, 387, 414, 544
Autonomic nervous system 198, 199, 209, 226, 256, 264, 265, 268, 275, 311, 313
Autopoiesis ii, 219, 257, 327, 505, 552, 559, 566, 570, 571, 573, 575, 589, 591
Avastin 3, 39, 145, 188, 362

B

Bacteria 39, 48, 70, 75, 83, 151, 235, 253, 260, 278, 338, 340, 348, 364, 377, 378, 380, 386, 393, 395, 413, 415, 417, 419, 429, 440, 457, 478, 483, 494, 497, 499, 500, 515, 538, 552, 561, 570, 587-589
Biomarkers 3, 4, 7, 9, 13, 14, 17, 29, 49, 57, 67, 82, 138, 150, 161, 174, 194, 218, 253, 371, 504, 509, 532

C

Calcium 75, 126, 127, 131, 187, 198, 203, 222, 223, 227, 318, 453, 455, 460, 468, 483, 488, 512, 538, 540, 604
cAMP 191, 203, 205, 208, 228, 538, 545, 546, 548
Campath 3, 39
Cancer i, ii, 3, 5, 7, 25, 26, 45, 46, 48, 49, 51, 54, 72, 78, 97, 114, 131, 135, 211, 212, 216, 225, 236, 237, 240, 243, 245, 248, 253, 254, 280, 281, 283, 287, 301, 317, 326, 327, 335, 344, 346, 350, 352, 365, 366, 372, 373, 377, 379, 381, 384, 385, 398, 405, 408, 422, 434, 435, 439, 442, 465, 471, 473, 475, 478, 482, 484, 500, 530, 532, 535, 537, 562, 569, 591, 617, 644, 647, 661, 662
Ceramide 538, 541
cGMP 217, 229, 278, 314, 508, 528, 538, 541, 545, 546
Chicken pox 412, 417, 431

D

Diabetes 33, 35, 53, 67, 69, 72, 76, 83, 86, 88, 92, 93, 96, 98, 100, 101, 113, 114, 116, 118, 119, 130, 131, 203, 205, 216, 225, 230, 235, 253, 294, 366, 370, 436, 445, 446, 473, 507, 511, 514, 520, 522, 533, 536, 540, 544, 599, 654, 655
Diacyl glycerol 102, 133, 160, 538, 541
Diphtheria 425, 426, 431, 432
DSHEA 435, 469, 470, 478, 479, 484

E

Ebola virus 35, 494, 529, 536, 537
Eicosanoids 71, 74, 75, 112, 116, 538, 541
Emergent properties 494, 552, 554, 556-559
Erbitux 3, 39, 145, 188, 356, 362

F

FDA ii, 3, 5, 6, 8, 9, 27, 28, 42, 43, 51, 52, 58, 61, 62, 97, 98, 103, 118, 119, 129, 132, 145, 149, 162, 163, 167, 168, 173, 186, 187, 190, 196, 203, 239, 243, 246, 248, 252, 253, 281, 285, 286, 295, 298, 303, 304, 312, 314, 320, 321, 332, 337, 339, 357, 360, 362, 367, 372, 377, 384, 386, 389, 401, 402, 405, 409, 414, 415, 421, 425, 427, 432, 433, 435, 443, 456, 473, 475, 478, 479, 484, 492, 508, 509, 513, 526, 528, 534, 537, 653
Flu virus 370, 397, 412, 423, 424, 431, 495
Folic acid 292, 293, 435, 456, 457, 459, 460, 482
Fungi 323, 325, 326, 332, 395, 399, 405, 408, 409, 426, 570, 576

G

Gaia 552, 554, 570, 574, 575, 589, 592
Gemeprost 226, 234, 241, 249, 251, 626
Gene chips 3
Genetic modification 451, 494, 497, 527, 528
Global warming 370, 441, 450, 494, 496, 497, 501, 526, 530
GPCR 77, 84, 97, 132, 186, 202, 208, 228, 237, 252, 302, 388, 538, 543, 550

H

HAART 323, 374, 375, 381, 382, 387, 496
Heart disease ii, 11, 17, 69, 72, 73, 76, 78, 80, 83, 86, 88, 92, 94, 96, 103, 120, 122, 125, 130, 131, 172, 244, 287, 335, 428, 439, 444, 447, 459, 466, 467, 473, 474, 482, 489
Hepatitis A 412, 413, 417, 418, 425, 431, 432
Hepatitis B 81, 138, 151, 188, 375, 382, 406, 431, 432
Herceptin 3, 5, 37, 39, 145, 161, 162, 183, 188
HIV 12, 16, 58, 184, 323, 336, 338, 358, 359, 386, 387, 392, 393, 398, 414, 415, 418, 426, 428, 429, 431, 495, 496, 503, 511, 516, 530, 534
HPV 42, 60, 184, 339, 412, 414, 415, 417, 418, 422, 426, 427, 431, 434
Hubs 538, 552, 576, 577, 579, 580, 582, 583, 587, 590
Human papilloma virus 42, 60, 138, 151, 188, 339, 412, 415, 418, 426, 431

I

Immune system i, 42, 70, 71, 75, 78, 79, 82, 93, 103, 108, 111, 116, 121, 139, 147, 148, 152, 167, 205, 218, 219, 234, 235, 238, 257, 266, 271, 273, 274, 290, 292, 300, 310, 339, 342, 349, 370, 371, 373, 374, 377, 380, 386, 396, 406, 417, 426, 430, 457, 458, 460, 461, 489, 540, 558, 617
Infectious diseases i, ii, 341, 389, 395, 396, 399, 408, 410, 587
Inflammation i, ii, 48, 92, 96, 98, 102, 112, 113, 119, 129, 148, 216, 275, 289, 294, 300, 324, 325, 361, 371, 378, 390, 394, 416, 426, 439, 444, 452, 459, 471, 476, 513
Innate 75, 77, 79, 81, 82, 84, 86, 87, 108, 147, 221, 235, 330, 331, 338, 340, 342, 354, 357, 358, 360, 364, 365, 370, 386, 389, 396, 430, 558
IP3 133

L

Levonorgestrel 226, 234, 249, 251, 623
Lysophosphatidic acid 538, 541

M

Measles 58, 398, 412, 417, 418, 424, 431, 432, 434

Meningococcus 412, 417, 429, 431
Metabolic syndrome i, ii, 88, 91, 92, 94, 98, 100, 106, 109, 112, 116, 117, 124, 129, 131, 135, 231
Metabolome 3, 5, 29, 46, 57, 64, 65
Mifepristone 226, 234, 241, 249, 251
mTOR 104, 138, 145, 161, 189, 190, 193, 194, 337, 345, 346, 388, 519, 608, 616, 619, 644
Mumps 58, 398, 412, 417, 418, 424, 426, 431, 432
Mycobacterium 395, 402, 404, 429
Myo-inositol 435

N

Nanotechnology 428, 494, 527, 530
National Institutes of Health 6, 16, 58, 63, 315, 435, 462, 483, 489, 524, 529
Neuron 256, 257, 259, 261, 262, 265, 266, 273, 277, 279, 285, 289, 297, 306, 311, 317, 372, 539
Next generation sequencing 3, 9, 26, 63
NIH 6, 13, 15, 16, 58, 63, 248, 300, 315, 399, 410, 433, 435, 478, 484, 489, 492, 524, 529, 556
Nitric oxide 71, 73, 78, 81, 85, 86, 101, 102, 121, 133, 202, 244, 277, 314, 316, 467, 491, 538, 541, 545, 546, 548, 549
Nodes 71, 139, 145, 146, 150, 177, 257, 323, 326, 328, 330, 340, 342, 347, 348, 350, 354, 358, 361, 364, 430, 538, 552, 555, 563, 587, 589, 590

O

Orthomyxoviridae 412, 421, 423, 431

P

Parkinson's disease 49, 61, 67, 256, 312, 314, 322, 346, 474, 489, 491, 494, 508, 520, 527, 535, 544, 627, 636
Pertussis 412, 413, 425, 431, 432
PI3K 17, 95, 96, 99, 104, 132, 138, 154, 176, 189, 193, 345, 346, 360, 388, 481, 482, 485, 493, 545, 546, 616
Pneumococcus 412, 431
Polio 331, 369, 376, 377, 383, 398, 412, 413, 417, 419, 420, 431, 432, 452, 497
Progestin 226, 234, 249, 251, 252, 254, 514, 624, 625
Prostaglandin E 75, 226, 234, 242, 249, 251, 252, 623

R

RAS 41, 81, 99, 105, 107, 133, 138, 154, 160, 189, 211, 229, 482, 512
Rituxan 3, 39, 373, 380
ROS 81, 91, 101, 102, 108, 119, 122, 123, 132, 151, 158, 190, 294, 300, 307, 315, 317, 340, 375, 382, 406, 410, 444, 447, 485
Rotavirus 350, 398, 412, 413, 417, 418, 420, 421, 431-433
Rubella 58, 412, 417, 418, 424, 431, 432

S

S. aureus 395, 400, 401, 409, 657
Saturated fats 69, 76, 78, 80, 83, 98, 353, 439
Scale-free network 552, 577, 579, 582, 588, 590
Soy 226, 234, 247, 250, 252, 437, 439, 444, 445, 460, 469, 477, 478, 492
Stroke 69, 71, 72, 74, 78, 81, 88, 92, 98, 102, 103, 117, 119, 120, 125, 126, 136, 211, 212, 243, 245, 272, 279, 292, 299, 309, 320, 371, 439, 442, 459, 466, 475, 476
Systems biology ii, 4, 6, 8, 35, 50, 56, 66, 67, 73, 86, 319, 376, 383, 392, 428, 430, 467, 489, 552, 555, 588

T

Tetanus 403, 425, 431, 432
Tissue plasminogen activator 88, 120,

124, 130, 132
tPA 88, 120, 132
Trans fats 69
tuberculosis 395, 410, 414, 471, 515, 534, 648, 656
Tyrosine kinase 18, 37, 138, 167, 175, 179, 188, 190, 341, 518, 522, 607, 609, 610

V

Vaccination 10, 280, 331, 365, 379, 384, 385, 412, 415, 421, 530
Varicella 412, 417, 418, 425, 431, 434
Vectibix 3, 39, 145, 188
VEGF 38, 39, 60, 82, 96, 111, 128, 129, 132, 134, 138, 145, 146, 158, 162, 163, 189, 190, 310, 519, 529
Viruses 39, 41, 81, 151, 154, 188, 342, 346, 349, 351, 352, 370, 375, 377, 379, 383, 385, 386, 393, 406, 408, 428, 431, 433, 452, 496, 497, 530, 568, 570, 588, 661

Made in the USA
Las Vegas, NV
03 November 2026

57424494R00404